The Acoustic Bubble

The Acoustic Bubble

T. G. Leighton

Institute of Sound and Vibration Research
The University, Southampton, UK
formerly at : *Cavendish Laboratory* and
Magdalene College, University of Cambridge, UK

ACADEMIC PRESS

San Diego London Boston
New York Sydney Tokyo Toronto

This book is printed on acid-free paper.

First printing paperback edition 1997

Academic Press, Inc.
525 B Street, Suite 1900, San Diego, California 92101-4495, USA
http://www.apnet.com

Academic Press Limited
24 – 28 Oval Road, London NW1 7DX, UK
http://www.hbuk.co.uk/ap/

ISBN 0-12-441920-8 (hbk)
0-12-441921-6 (pbk)

A catalogue record for this book is available from the British Library

Typeset by Technical Typesetters, Ashford, Kent, UK

97 98 99 00 01 02 EB 9 8 7 6 5 4 3 2 1

Transferred to digital printing 2006
Printed and bound by CPI Antony Rowe, Eastbourne

To my parents

Contents

Preface

The interaction of acoustic fields with bubbles in liquids is of interest to a wide range of people, from those involved with the fragmentation of kidney stones and the sensitisation of explosives, to researchers who record the underwater sound of rain-drop impact on lakes. The indications are that the range is widening, and though there are several excellent publications which expound aspects of the field (for example, the biomedical, chemical or erosive implications) and others which extensively review the literature, there is a place for a somewhat gentler approach to supplement these. The object of this book is neither to provide an exacting grounding of any particular aspect of underwater acoustics, nor to summarise and review historic and current literature on acoustic cavitation. It is instead written with the intention of giving the reader a 'physical feel' for the acoustic interactions of acoustic fields with bubbles. With the multidisciplinary nature of the field there will be some who require analogy, rather than formulation, to appreciate a given phenomenon. Though such mental pictures are often not rigorous, if the alternative is simply to trust the end result of a mathematical analysis and hope that in applying it one does so in a manner appropriate to the prevailing environment, then the degree of adaptability that follows from having some appreciation of the physical processes involved can only be good for the science, and the spirit. A physical understanding is also valuable to the mathematically adept who are perhaps new to a field: it takes time to find out where one stands in a new discipline, time in which one assimilates the origins of the formulations so readily employed by others, and to appreciate how the assumptions which form the basis of the mathematical models relate to the physical world. This book therefore attempts to engender a basic level of understanding through imagery. However, whilst some readers will find such descriptions sufficient, the field itself requires that a thorough mathematical description also be available, and this is provided. Because of the size of the field and the rate at which new formulations are developed, selection has been made to include those formulations which may best illustrate the physical mechanisms involved, and which the reader can in many instances follow back to source. As a consequence of this approach, individual readers will encounter sections that are too rudimentary or too involved, depending on their background, but it is hoped that regardless of that background there will be much that is comprehensible and interesting. The range of phenomena discussed is extensive, but the basic interaction of sound with bubbles is the same. If the object is to develop a physical appreciation of this interaction, there is much to be said for examining manifestations of it that are perhaps outside one's immediate experience: for example, the bubble-mediated phenomena of acoustic shielding, bulk property and parametric effects encountered in the use of clinical ultrasound are in fact current issues on a larger scale in ocean acoustics. Even so, space limitations preclude the inclusion of little more than an introduction to several topics, such as hydrodynamic, electrohydraulic, optical and biological cavitation.

The book is divided into five chapters. The first incorporates a basic introduction to acoustics, along with a description of some of the more esoteric phenomena that can be seen when high-frequency high-intensity underwater sound is employed. The second chapter discusses the nucleation of cavitation, and basic fluid dynamics. The third chapter draws together the acoustics and the bubble dynamics to discuss the free oscillation of a bubble, and the acoustic emissions from such activity. Examples are drawn from bubble entrainment through injection, rainfall, and wave action, and the chapter ends with a discussion of oceanic bubble populations. In addition to the natural emissions from these bubbles, acoustic probes are often applied to study such populations, and the behaviour of a bubble when an externally-applied acoustic field drives it into oscillation is the topic of Chapter 4. When such a bubble is forced into a stable oscillation, a variety of phenomena (such as radiation force interaction, rectified diffusion, surface wave activity, microstreaming, subharmonic emission and chaotic oscillation) may occur, and these are discussed. Energetic transient collapse may also occur. Chapter 4 closes with a summary of the theory of a single bubble, and examples of the population phenomena that can arise in practical sound fields. Following the discussion of these behaviours in Chapter 4, Chapter 5 outlines a variety of effects associated with acoustically-induced bubble activity. Bubble detection, sonoluminescence, sonochemistry and pulse enhancement are included. The chapter closes with a discussion of cavitation erosion, and cavitation bioeffects, relating particularly to the clinical use of ultrasound. The emphasis throughout the chapter is on understanding the mechanisms through which the effects may arise and, where alternative mechanisms have been proposed, to assess them in the light of the available experimental data.

In order to convey these ideas I have drawn upon my own research, much of which was undertaken at the Cavendish Laboratory, Cambridge, and my thanks goes to those people there who have assisted me over the years in my work, but above all for their friendship. I would especially like to acknowledge my gratitude to John Field, for his friendship and support over the years. I count myself very lucky in that, in addition to John Field, I have the privilege to thank the following people for their invaluable friendships and criticisms of sections of this book: Phil Nelson, Steve Thorpe, Joe Hammond, Mike Buckingham, Charlie Church, Christy Holland, Roy Williams, Stan Barnett, Hugh Pumphrey, Ron Roy, Andy Phelps, Chris Beton, and Knud Lunde. I could not have been more fortunate in meeting such people. For their advice and assistance during my research I would also like to thank Alan Walton, Mike Pickworth, Phillip Dendy, Larry Crum, Martin Lesser, Karel Vokurka, Nick Safford, Colin Seward, Jim Pickles, Kelvin Fagan, Dave Johnson, Ray Flaxman, Arthur Stripe, Bob Marrah, Roger Beadle, Rik Balsod, Alan Peck and Dick Smith. I hope they share with me fond and perhaps amusing memories of my requests for sometimes unorthodox assistance. I am very grateful to Sue Hellon for the patient assistance she has given me with the references, and to Chris Rice, Maureen Strickland, Paul White, Chris Morfey, Steve Elliott, Frank Fahy, Mike Fisher, Peter Davies, Mike Russell, Anne Barrett, Dean Thomas and Terry Vass for their encouragement since my arrival at the Institute of Sound and Vibration Research, Southampton. For their help since the move I also thank Shiela and Vic Fisher, and John and Kath Morrant. And Siân Lloyd Jones, who read the proofs.

I would very much like to thank the Master of Magdalene and Lady Calcutt, and the Fellows of Magdalene College, who extended immeasurable friendship, and trust, and who from the first gave such enthusiastic support for a young man's interest in acoustic bubbles.

And for their support Annie, Nick and Linda, Justine and John, Anne, Peter, Deniol, Hilda, Patty, and John and Christine.

T.G. LEIGHTON
June 1992

Symbols and Abbreviations

Symbols

$\vec{a}$	acceleration of a particular piece of fluid
$\vec{a}_i$, $\vec{a}_s$	unit vectors along the incident and scattered directions respectively
$a(t)_{nm}$	time-dependent amplitude of a purturbation which is the spherical harmonic of order $\boldsymbol{n}$, $\boldsymbol{m}$, superimposed on a spherically symmetric *pulsating* sphere
a_n'	reduced forms of $a(t)_{nm}$
a_1, a_2	coefficients associated with derivation of the mechanical index
A	amplitude of the particular integral describing steady-state displacement response of a general oscillator (may be complex)
$\hat{A}(t)$	amplitude function associated with velocity potential of raindrop crater
A	plane area perpendicular to a given axis
A_ε	the ratio $A_\varepsilon = R_{max}/R_0$ used to characterise bubble oscillation
A_n	amplitude of spherical harmonic of order $\boldsymbol{n}$ superimposed on a *stationary* spherical shape ($\boldsymbol{n}$ = 1, 2, 3 ...)
b	resistive dissipation constant
b_{tot}	total resistive dissipation constant
b_{VP}	resistive dissipation constant in the volume–pressure frame
b_{RF}^{rad}	radiation dissipative constant in the radius–force frame, equal to $Re\{Z_{RF}^{rad}\}$
b_{RP}^{rad}	radiation dissipative constant in the radius–pressure frame
b_{VF}^{rad}	radiation dissipative constant in the volume–force frame
b_{VP}^{rad}	radiation dissipative constant in the volume–pressure frame
b_{VP}^{th}	thermal dissipation constant in the volume–pressure frame
b_{VP}^{vis}	viscous dissipation constant in the volume–pressure frame
$\boldsymbol{b}$	amplitude acoustic attenuation constant
$\boldsymbol{B}$	bulk modulus
$\boldsymbol{B_s}$	adiabatic bulk modulus
$\boldsymbol{B_T}$	isothermal bulk modulus
$\boldsymbol{B_c}$	bulk modulus of the bubbly liquid
$\boldsymbol{B_w}$	bulk modulus of the bubble-free liquid
$\boldsymbol{B_{bub}}$	a component of bulk modulus incorporating the pressure-mediated volume change which is due to the entire bubble population
B/A	second-order nonlinearity ratio of a liquid
c	speed of acoustic waves

c_c	speed of sound in a bubbly medium
c_d	speed of sound in a solid wall
c_g	speed of sound in the gas contained within the bubble
c_H	speed of wavelets in Huygens construction
c_o	speed of sound at infinitesimal amplitudes
c_L	speed of sound in the liquid at bubble wall
c_w	speed of sound in the bubble-free liquid
c_s	shock speed
c_ω	a function of the acoustic frequency with the dimensions of speed
C	concentration of gas dissolved within a liquid
C_∞	initial uniform gas dissolved within a liquid at time $t = 0$
C_R	dissolved gas concentration in liquid at bubble wall
C_{R_o}	dissolved gas concentration in liquid at bubble wall when bubble has equilibrium radius ($r = R_o$)
C_p	specific heat capacity of gas within bubble at constant pressure
$C_{p,m}$	molar heat capacity of gas within bubble at constant pressure
C_v	specific heat capacity of gas within bubble at constant volume
C_{cap}	electrical capacitance
d_{rad}	radiation damping constant
d_{th}	thermal damping constant
d_{vis}	viscous damping constant
d_{tot}	total damping constant: $d_{tot} = d_{rad} + d_{th} + d_{vis}$
D	diffusion coefficient for dissolved gas within a liquid
D_{air}	thermal diffusivity of air
D_{dip}	dipole strength defined specifically for bubble sources (see text)
$D_{dip,i}$	initial dipole strength defined specifically for bubble sources (see text)
D_g	thermal diffusivity of a gas
D_l	thermal diffusivity of a liquid
$\mathrm{D}/\mathrm{D}t$	$\equiv \partial/\partial t + \vec{v}.\vec{\nabla}$, the so-called 'material derivative', a differential which takes account of changes both in time and position
e	exponential constant, ≈ 2.718
f	ratio of bubble gas density to liquid density at equilibrium
f_{Dop}	fractional Doppler shift
f_1, f_2	functions that are solutions to the wave equation
f	the complex scattering function of the bubble
F	force
F_a	force applied to a 'spring'
F_s	force exerted by a 'spring'
F_A	amplitude of driving force (may be complex)
F_{B1}	Primary Bjerknes force
F_D	drag force on a rigid sphere in potential flow
F_o	amplitude of general driving force of an oscillator
$\Sigma\vec{F}_{ext}$	the vector sum of the external forces, both volume and surface, per unit mass on an infinitesimal element of liquid
$\bar{F}$	time-averaged radiation force

$\bar{F}_{trav}$	time-averaged radiation force on a bubble in a travelling-wave field
$\bar{F}_{B1}$	time-averaged radiation force on a bubble in a standing-wave field
$\bar{F}_{part}$	time-averaged radiation force exerted by a pulsating bubble on a particle
$\bar{F}_{rad2}$	time average of F_{rad2}
F_{rad2}	instantaneous value of radiation force on bubble 2 due to field radiated by bubble 1
$\{\mathcal{F}r\}$	Froude number
$F(\xi)$	a time-dependent function representing the difference between the concentration of dissolved gas at the bubble wall, and the concentration far from the bubble
g	dimensionless multiplicative factor which accounts for effect of surface tension on bubble stiffness
$\vec{g}$	gravitational acceleration
G_{th}	a dimensionless constant associated with thermal damping
G_1, G_2, G_3, G_4, G_5	constants associated with thermal damping such that:
G_1	a parameter proportional to the square of the ratio of the thickness of the layer where conduction will cause significant temperature changes, to the acoustic wavelength in the gas
G_2	a parameter proportional to the square of the ratio of the bubble radius to the thermal penetration depth
h	distance from a plane surface (e.g. transducer faceplate, liquid surface)
h_{ent}	entrainment depth
h_{max}	maximum depth to which dirty bubbles are carried in ocean
h_o	depth at which equilibrium occurs for gas flux from oceanic bubble
h_{obs}	depth of observation below the sea surface
h_R	a substitution parameter, $h_R = (r^3 - R^3)/3$
h_s	saturation depth for gas flux from oceanic bubble
H_s	significant wave height
H	liquid enthalpy
H_D	Henry's constant
$\boldsymbol{H}$	time-dependent locus of a plane gas/liquid interface
i	$i^2 = -1$
I	acoustic intensity
I_{ref}	reference acoustic intensity
I_{index}	an index associated with the likelihood of cavitation
I_{MI}	mechanical index, given by $\sqrt{I_{index}}$
j	mass flux of a fluid (mass crossing unit area in unit time)
J	flux of gas dissolved in liquid
J_n	Bessel function of order n
k	wavenumber $= 2\pi/\lambda$
$\vec{k}$	wavevector ($\lvert\vec{k}\rvert = k$)
$\vec{k}_1$	wavevector from source 1 of a pair
$\vec{k}_2$	wavevector from source 2 of a pair

$\vec{k}$	wavevector from an image
k_c^{comp}	complex wavenumber of sound in a medium containing a uniform distribution of bubbles
k_c	wavenumber of a coherent acoustic field within a sparse distribution of scattering bubbles
k_w	wavenumber in bubble-free seawater
k_B	Boltzmann's constant
k	spring constant
k_R	polytropic stiffness of a bubble for radial displacement
k_{VP}	stiffness of a bubble for volume–pressure frame (effects of surface tension not included)
k_{VP}^{ad}	adiabatic limit of k_{VP} (effects of surface tension not included)
k_{VP}^{iso}	isothermal limit of k_{VP} (effects of surface tension not included)
k_{VP}	stiffness of a bubble in volume–pressure frame
K	a function in the expansion of the radial oscillation
K_g	thermal conductivity of gas within bubble
K_l	thermal conductivity of the liquid
l	length
l_D	width of thermal boundary layer
l_o	length of an unloaded spring
L	a length
L_c	average penetration depth of a bubble cloud
L_d	diameter of a liquid drop
L_{dis}	discontinuity length
L_h	diameter of active area of a hydrophone
L_g	diameter of a molecule
L_{ms}	thickness of acoustic boundary layer over which microstreaming acts
L_s, L_x, L_y	dimensions of an acoustic source
L_w	distance between the point of bubble formation and the wall
L_{wave}	peak-to-trough average wave height
L_ω	a function of the acoustic frequency with the dimensions of length
L_{ind}	electrical inductance
m	inertia of a general oscillator
m_b	mass of gas contained within a bubble
m_g	mass of a gas molecule
m_i	all inertial components of a pulsating bubble excluding the radiation mass
m_{RF}^{rad}	radiation mass in the radius–force frame
m_{RP}^{rad}	radiation mass in the radius–pressure frame
m_{VF}^{rad}	radiation mass in the volume–force frame
m_{VP}^{rad}	radiation mass in the volume–pressure frame
m_v	mass of a vapour molecule
m_λ	mass contained in one acoustic wavelength
M	the acoustic Mach number
m	an integer
n	an integer

n_b	number of bubbles per cubic metre in a population of identical bubbles
$n_b^{gr}(z,R_o)dR_o$	number of bubbles per unit volume at depth z having radii between R_o and $R_o + dR_o$
n_b^{μ}	number of bubbles per cubic metre per micrometre increment in radius
n_g	number density of gas molecules within the bubble
n_s	number density of liquid molecules near a liquid surface (specifically, near the bubble wall)
n_o	equilibrium, constant number density of vapour molecules during isothermal and isobaric collapse conditions for a vapour cavity
N_{Av}	Avogadro's number $\approx 6.023 \times 10^{23}$
N_b	total number of bubbles
N_v	total number of vapour molecules that must condense out to obtain isothermal and isobaric conditions within a collapsing vapour cavity
N_m	total number of moles (especially contained within a bubble)
$\boldsymbol{p}$	momentum of acoustic wave
$\boldsymbol{p}_\lambda$	momentum associated with one acoustic wavelength
p	pressure
p_{atm}	atmospheric pressure at level of sea surface
p_A	the amplitude of the oscillation of the pressure of the *permanent gas phase* within the bubble
p_c	a constant with dimensions of pressure
p_g	instantaneous gas pressure within a pulsating bubble
$p_{g,e}$	pressure of gas phase within the bubble at equilibrium
$p_{g,m}$	initial gas pressure inside cavity before Rayleigh-like collapse under hydrostatic pressure, when $R = R_m$ and $\dot{R} = 0$
$p_{g,max}$	maximum pressure reached by a gas during collapse
p_i	instantaneous total pressure within a pulsating bubble
$p_{i,e}$	pressure within the bubble at equilibrium
p_{impact}	pressure radiated by initial impact of raindrop
p_{in}	instantaneous pressure inside a bubble undergoing shape oscillations
p_{out}	instantaneous pressure outside a bubble undergoing shape oscillations
p_L	liquid pressure just outside a bubble or cavity, at the wall
p_{L_o}	external pressure outside the cavity when it begins to collapse
p_n	the liquid pressure which brings about the onset of nucleation from a conical crevice
p_{N_2}	partial pressure of nitrogen in the water far from the bubble
p_{in}	pressure at wall inside bubble
$p_{rad,abs}$	radiation pressure for absorption
$p_{rad,refl}$	radiation pressure for reflection
p_v	vapour pressure within the bubble
$p_{w\text{-}h}$	water-hammer pressure
p_ε	small-amplitude pressure fluctuation inside bubble
p_σ	pressure due to surface tension (the Laplace pressure)
p_∞	pressure in the liquid far from the bubble
p_o	hydrostatic liquid pressure outside the bubble
p'	the pressure existing at some boundary within the liquid
p_r, p_θ, p_φ	the three principal stresses in the spherical coordinate directions

p_1	fluid pressure at inlet of a flow tunnel
p_2	fluid pressure at constriction within a flow tunnel
p_3	gas pressure above the liquid in the vertically vibrating cell
P	acoustic pressure
P_1, P_2	acoustic pressure amplitudes of two insonating fields
P_{2nd}	acoustic pressure amplitude of 2nd harmonic component in a nonlinearly propagating wave
$\check{P}$	acoustic pressure from an image source
P_A	acoustic pressure amplitude
$P_{A,L}$	amplitude of time-varying component of p_L
P_b	acoustic pressure radiated by a pulsating bubble
P_{b1}	acoustic pressure radiated by bubble 1
P_B	Blake threshold pressure for bubble nucleation
P_d	threshold acoustic pressure for bubble growth by rectified diffusion
P_{A2}	threshold acoustic pressure to generate the subharmonic at half the driving frequency
P_g	acoustic pressure in the gas contained within the bubble
P_{opt}	smallest peak negative pressure required to cause prompt transient cavitation in bubble of radius R_{opt}
P_I	acoustic pressure amplitude of incident plane wave
P_R	acoustic pressure amplitude of reflected plane wave
P_t	threshold acoustic pressure amplitude for transient collapse
P_T	acoustic pressure amplitude of transmitted plane wave
P_{ant}	acoustic pressure at pressure antinode in a complete or partial standing-wave field
P_{nod}	acoustic pressure at pressure node in a complete or partial standing-wave field
P_{neg}	maximum negative pressure encountered in a given sound field
P_{ref}	reference acoustic pressure
P_n	acoustic pressure field associated with *n*th shape oscillation mode
ΔP_{wall}	time-averaged pressure difference across the bubble wall
$\hat{P}_A$	the acoustic pressure amplitude of the incident plane wave which, upon reflection from a free or rigid boundary, sets up the standing-wave field
P_n	Lengendre polynomial
P_n^m	associated Legendre polynomials
q	a wavenumber
q_v	probability that, on colliding with a liquid surface, a given vapour molecule will stick and so condense out
$dq_H, \Delta q_H$	heat added to a unit volume of gas within bubble
Q	the quality factor
Q_s	complex source strength
$\vec{r}$	the general position coordinate vector
$\vec{r}'$	vector describing displacement from sound source to point of observation
$\vec{r}_1$	the position coordinate relative to source 1 of a pair
$\vec{r}_2$	the position coordinate relative to source 2 of a pair

r	the radial coordinate in the spherical frame, with origin at centre of a bubble (if present)
$\check{r}$	distance from an image source to point of observation
r_b	the separation of the centres of two bubbles
r_c	radius of flat-ended cylinder model of a liquid jet
r_{cur}	the local radius of curvature of the front of the original liquid surface of impacting liquid-drop or jet
R	radius of curvature of liquid/gas interface (e.g. radius of a spherical bubble)
R_B	Blake threshold bubble nucleation radius (quasi-static approximation for lower threshold radius for transient collapse)
R_c	radius of circle of meniscus of gas pocket in a cone when contact angle is θ_r
R_{c1}	radius of circle of meniscus of gas pocket in a cone when contact angle is θ_a
R_{crit}	critical radius for stability of a bubble with respect to changes in hydrostatic pressure
R_d	threshold bubble radius for growth by rectified diffusion
R_g	the gas constant, equals 8.31441 ± 0.00026 J K^{-1} mol^{-1}
R_I	the inertial radius
R_m	the initial radius of a Rayleigh-like cavity (the radius of the cavity when the wall velocity is zero, prior to collapse)
R_{max}	the maximum radius reached by a bubble during expansion, prior to the collapse phase
R_{min}	the minimum radius attained on collapse before the bubble rebounds
R^n_{max}	the maximum radius attained by a pulsating bubble after the *n*th rebound
R_n	the radius of curvature of meniscus of gas pocket in a cone at the onset of nucleation
$R_{\boldsymbol{n}}$	bubble wall radial function associated with ***n***th shape oscillation mode
R_{opt}	optimum equilibrium bubble radius for prompt transient cavitation (the one requiring minimum acoustic pressure to undergo transient growth and collapse)
R_r	the radius of a bubble that would be in pulsation resonance with the incident sound field
R_t	upper threshold equilibrium bubble radius for transient collapse
R_o	the radius of a rigid sphere or equilibrium radius of a spherical bubble
R_ε	displacement of bubble radius from the equilibrium, such that $R(t) = R_o + R_\varepsilon(t)$
$R_{\varepsilon o}$	radial displacement amplitude of wall of spherical bubble
R_{o1}	the equilibrium radius of spherical bubble 1
$R_{\varepsilon 1}$	displacement of radius of spherical bubble 1 from equilibrium
$R_{\varepsilon oa}$	wall pulsation ampliude of spherical bubble at pressure antinode in a standing-wave field
R_o^{dhtr}	equilibrium bubble radius of a daughter bubble following the fragmentation of the parent
R_1	radius of concave meniscus of gas pocket in a cone
$\hat{R}_1$, $\hat{R}_2$	the two principal local radii of curvature of a liquid surface
$\hat{R}$	function describing position of regions of bubble wall with respect to bubble centre

R	pressure amplitude reflection coefficient
R_ε	displacement amplitude reflection coefficient
$\mathcal{R}e$	the Reynolds number
s	substitution parameter
s_0, s_1, s_3 ...	coefficients
s_{N_2}	percent oversaturation of nitrogen in the water
S	entropy
$d\vec{S}$	a vector normal to an elemental surface of area $\lvert d\vec{S} \rvert$
t	time
t_{col}	time for which pressure at the centre of an impacting jet remains at its initial value (duration of compressible behaviour)
t_{cur}	the total time over which the liquid behaves compressibly when a curved liquid surface impacts a rigid target
t_{grow}	net time for bubble growth in prompt response to negative sinusoidal pressure pulse
t_{max}	maximum lifetime of oceanic bubble
t_{Ray}	collapse time of a Rayleigh cavity
t_{RD}	time taken for a bubble to double its radius through growth by rectified diffusion
t_1	time taken, after start of negative-halfcycle of a sinusoidal pulse, for the magnitude of the negative acoustic pressure to exceed P_B
t_2	time until the liquid ceases to be in tension after start of negative-halfcycle of a sinusoidal pulse
Δt_I	delay in bubble growth caused by inertial effects
Δt_η	delay in bubble growth caused by viscosity
Δt_σ	delay in bubble growth caused by surface tension
T	absolute temperature (measured in kelvin)
T_m	initial gas temperature inside bubble before collapse under static pressure
T_{max}	maximum gas temperature inside bubble during collapse
T_o	equilibrium absolute temperature within bubble
T_∞	absolute temperature within liquid reservoir
ΔT^{air}_{water}	temperature of the air minus that of the water at surface
T	pressure amplitude transmission coefficient
T_ε	displacement amplitude transmission coefficient
u_b	translational speed of a bubble or spherical body in a fluid
u_d	impact speed of a liquid drop
$\langle u_v \rangle$	average speed of vapour molecule
u_w	local water speed around oceanic bubble
u	dimensionless function associated with pressure radiated by the initial impact of a raindrop
$dU, \Delta U$	increase in internal energy of a unit volume of gas within bubble
U_C	a parameter representing the difference in the mass of dissolved gas (rather than the concentration) from the initial conditions
U_o	amplitude of oscillation of velocity in a liquid about a pulsating sphere
$U_0^{R_0}$	amplitude of oscillation of wall velocity of a pulsating sphere

$U_{o,i}$	amplitude of oscillation of velocity in a liquid about a pulsating sphere at time $t = 0$
$U^{R_o}_{o,i}$	amplitude of oscillation of wall velocity of a pulsating sphere at $t = 0$
$\vec{v}$	fluid particle velocity
$\dot{\vec{v}}$	rate of change of fluid velocity with time at a fixed position: $\dot{\vec{v}} = \left.\frac{\partial \vec{v}}{\partial t}\right\vert_{\vec{r}}$
$\vec{v}_0$	uniform uniaxial flow velocity of a fluid
$-\vec{v}_0$	uniform uniaxial flow velocity of a rigid sphere in a previously stationary fluid
v_1	fluid speed at inlet of a flow tunnel
v_2	fluid speed at constriction within a flow tunnel
v_W	windspeed over laboratory breaking waves
$\Delta\vec{v}_r$	change in fluid particle velocity with change in spatial coordinate
$\Delta\vec{v}_t$	change in fluid particle velocity with change in temporal coordinate
v_r	radial component of fluid velocity
v_θ	tangential component of fluid velocity
V	volume
V'	elemental volume within gas bubble
dV_1	a volume element which is fixed within the fluid
dV_2	a volume element which is moving with the fluid
V_a	apparent additional volume of a body in fluid, so that ρV_a is the apparent added mass
V_b	volume of a body in fluid
V_{min}	minimum volume of a pulsating bubble
V_p	volume of a solid particle
V_o	equilibrium volume of a pulsating bubble
V_ε	volume displacement, so that bubble volume equals $V_o + V_\varepsilon$
$V_{\varepsilon o}$	amplitude of volume displacement
V_1	the volume of bubble 1
V_2	the volume of bubble 2
V_{o1}	the equilibrium volume of bubble 1
$V_{\varepsilon o1}$	volume amplitude of pulsation of bubble 1
V_{o2}	the equilibrium volume of bubble 2
$V_{\varepsilon o2}$	volume amplitude of pulsation of bubble 2
w	a real constant associated with the stability of the surface of a transient cavity
w_{10}	wind speed at 10 m elevation
W_n	work done on a bubble in its compression from initial maximum radius at the start of the *n*th cycle to its next minimum radius
ΔW	work done on a unit volume of gas within bubble
ΔW_c	mechanical work done on or by a gas bubble in a complete cycle
$\dot{W}$	power (also $\dot{W}_1$ etc.)
$\langle \dot{W} \rangle$	time-average power
$\{\mathcal{W}e\}$	Weber number
x	a distance coordinate in the cartesian frame
x_0	a fixed position given by this value of the x-coordinate

x_i — the general cartesian distance coordinate (i = 1, 2, 3)
X — a real amplitude associated with displacement of the general oscillator (also X_1 etc.)

y — a distance coordinate in the cartesian frame
y_i — the locus of a perturbed gas/liquid interface
Y — a real number associated with amplitude of a propagating waveform (also Y_1 etc.)
Y_{mod} — Young's modulus
$Y_n^m(\theta,\varphi)$ — spherical harmonic
¥(t) — the general response of a bubble to an acoustic field

z — a distance coordinate in the cartesian frame
z_ε — the complex displacement
Z — specific acoustic impedance
Z_a — acoustic impedance of air
Z_{RF}^{rad} — radiation impedance for diverging spherical waves
$Z_{RP}^{s.s.}$ — specific acoustic impedance for diverging spherical waves
Z_w — acoustic impedance of water
Z_1 — acoustic impedance of medium 1
Z_2 — acoustic impedance of medium 2

$\hbar$ — the general driving force of an oscillator

α — a coefficient in the expansion of the radial response
α_d — a constant relating to crevice nucleation
α_{th} — the dimensionless multiplicative factor which corrects for the effects of heat flow in the stiffness and resonance frequency of a bubble

β — a resistive constant leading to damping, equal to $b/2m$
β_{VP}^{rad} — a resistive constant leading to thermal damping of bubble in the volume–pressure frame
β_{tot} — total resistive constant leading to damping
β_n — resistive constant associated with nth shape oscillation mode

ς — a constant associated with over-damping
ς_a — the ratio of the largest axis to the smallest in an oblate spheroid

ζ — a substitution variable

γ — ratio of specific heat of a gas at constant pressure to that at constant volume

¢ — a parameter incorporating both the material and convective nonlinearities such that phase speed equals c_0 + ¢ v
¢$_s$ — a parameter associated with the shock speed in a liquid

δ — damping constant , $\delta = Q^{-1}$
δ_{rad} — radiation damping constant for pulsation at resonance frequency

δ_{th}	thermal damping constant for pulsation at resonance frequency
δ_{vis}	viscous damping constant for pulsation at resonance frequency
δ_{tot}	total damping constant for pulsation at resonance frequency: $\delta_{tot} = \delta_{rad} + \delta_{th} + \delta_{vis}$
δ_{rad}^{dip}	dipole radiation damping constant
$\partial a/\partial b$	'partial differential of *a* with respect to *b*'
da/db	'exact differential of *a* with respect to *b*'
Δ	'a finite increment of ... ' (such that $\Delta a \equiv$ 'a finite increment of *a*')
$\vec{\nabla}$	the vector differential
Δ_{log}	logarithmic increment
Δ_{log}^{nonlog}	nonlinear logarithmic damping decrement
$\varepsilon_{o,n}$	the displacement amplitude of the *n*th oscillation of damped system
ε_{oa}	the radial displacement amplitude for a pulsating bubble situated at the pressure antinode of a standing-wave acoustic field
ε	displacement in an oscillatory system
ε_I	displacement of incident wave
ε_R	displacement of reflected wave
ε_T	displacement of transmitted wave
ε_o	general displacement amplitude in an oscillatory system
$\ddot{\varepsilon}_b$	the acceleration of a body (e.g. bubble 1 in a two-bubble system), due to the accelerations of the surrounding fluid medium
$\ddot{\varepsilon}_f$	fluid acceleration at a body (e.g. fluid acceleration at bubble 1, due to oscillations of a second bubble, bubble 2)
$\varepsilon_r', \varepsilon_\theta', \varepsilon_\varphi'$	the three principal rates of strain in the spherical coordinate system
$\ni_{RF}^{rad}$	the radiation reactance, equal to $\mathrm{Im}\{Z_{RF}^{rad}\}$
π	pi, ≈ 3.14159
Π	a function describing frequency response of a forced oscillator
ϑ	a phase factor
ϑ_b	phase difference between the pulsations of two bubbles
ϑ_1	phase of oscillation of bubble 1 relative to driving sound field
θ	a general angle (in the spherical frame referring specifically to the angular coordinate, the measure from the *z*-axis)
θ_a	the advancing contact angle
θ_c	the half-angle of a conical crevice
θ_r	the receding contact angle
θ_I	angle of incidence
θ_R	angle of reflection
θ_T	angle of transmission
θ_{cone}^{perp}	angle of the free fluid surface to the *z*-axis in a conical crater, induced by liquid-drop impact, at the critical configuration where the pressure gradient and particle velocity are parallel
θ_1, θ_2	angles

Θ	change in temperature of bubble gas from equilibrium absolute temperature T_o
φ	an angular coordinate in the spherical frame
ϕ	energy
ϕ_K	kinetic energy of an oscillator system (specifically of the fluid around a freely pulsating bubble)
ϕ_B	potential energy per unit mass of fluid (incorporating body forces)
ϕ_P	potential energy of an oscillator (specifically of the gas in a freely pulsating bubble)
$\phi_{K,max}$	maximum value of ϕ_K
$\phi_{P,max}$	maximum value of ϕ_P
$\phi_{K,\lambda}$	kinetic energy associated with one wavelength
ϕ_T	total energy of an oscillatory system
$\phi_{T,n}$	total system energy of *n*th oscillation of damped oscillator
ϕ_L	latent heat of evaporation per molecule of liquid
Φ	the velocity potential (where $\vec{v} = \vec{\nabla}\Phi$)
Φ_{cone}	the velocity potential of a conical crater, induced by liquid drop impact
Φ_g	the velocity potential of the gas within an oscillating cavity which is perturbed from the spherical
Φ_r	a sub-function of Φ depending only on *r* where $\Phi = \Phi_r\,\Phi_\theta\,\Phi_\varphi$ is a solution of Laplace's equation in spherical polars
Φ_s	a velocity potential of spherical character, the component in a flow resulting from the presence of a rigid sphere
Φ_θ	a sub-function of Φ depending only on θ where $\Phi = \Phi_r\,\Phi_\theta\,\Phi_\varphi$ is a solution of Laplace's equation in spherical polars
Φ_φ	a sub-function of Φ depending only on φ where $\Phi = \Phi_r\,\Phi_\theta\,\Phi_\varphi$ is a solution of Laplace's equation in spherical polars
Φ_n	velocity potential associated with *n*th shape oscillation mode
Φ_o	the velocity potential defining the state of uniform, unperturbed flow (where $\vec{v}_o = \vec{\nabla}\Phi_o$)
Φ_1	a velocity potential (specifically used to mean the velocity potential of the liquid outside an oscillating cavity which is perturbed from the spherical)
Φ_2	a velocity potential
Φ_{in}	velocity potential inside a bubble undergoing shape oscillations
Φ_{out}	velocity potential outside a bubble undergoing shape oscillations
κ	polytropic index
σ	surface tension of a liquid
σ_c	the cavitation number
$\vec{\Omega}$	the vorticity, $\vec{\Omega} = \vec{\nabla} \wedge \vec{v}$
Ω_b^{ext}	the extinction cross-section of a bubble
Ω_b^{scat}	the scattering cross-section of a bubble
$\Omega_b^{b\text{-}s}$	the backscatter cross-section of a bubble
Ω_b^{th}	acoustic cross-section associated with thermal dissipation in a bubble

Ω_b^{vis}	acoustic cross-section associated with viscous dissipation in a bubble
Ω_b^{abs}	absorption cross-section of the bubble
Ω_c^{ext}	the extinction cross section for sound propagating through a bubble cloud
Ω_{cV}^{b-s}	the backscattering cross-section per unit volume (i.e. the volume backscattering coefficient)
$\wp_1, \wp_2$	dimensionless parameters associated with regular entrainment
$\Gamma_1, \Gamma_2, \Lambda_1, \Lambda_2, \forall_1, \forall_2$	parameters associated with thermal damping
λ	wavelength
ν	linear frequency
ν_p	insonation pump frequency
ν_o	linear resonance frequency for pulsating bubble
ν_{rep}	pulse repetition frequency
η	shear viscosity
η_k	kinematic viscosity $= \eta/\rho$
η_B	the bulk (or volume) viscosity
η_{th}	an additional thermal contribution to the effective viscosity
ρ	density (generally of a fluid, specifically of the fluid surrounding a bubble)
ρ_o	equilibrium fluid density
ρ_w	density of a given bubble-free liquid
ρ_d	density of solid wall
ρ_p	density of a solid particle
ρ_1	density of gas in bubble 1
ρ_{1e}	density of gas in bubble 1 at equilibrium
ρ_g	density of the permanent gas within the bubble
ξ	substitution parameter, equal to $\int_0^t R(t')^4 \mathrm{d}t'$
ξ_b	the value of ξ for one bubble period τ_b
τ	time constant of decay
τ_ν	acoustic period $= 1/\nu$
τ_b	characteristic periodicity in bubble pulsation
τ_p	the pulse length or on-time for pulsed ultrasound
τ_s	the off-time of pulsed ultrasound
ω	circular frequency $= 2\pi\nu$
ω_b	oscillatory frequency of a damped system: $\omega_b = \sqrt{(\omega_o^2 - \beta^2)}$
ω_c	a difference frequency
ω_i	imaging insonation frequency
ω_p	pump insonation frequency
$\omega_1, \omega_2, \omega_3$	acoustic frequencies of different insonating fields

ω_0	resonance circular frequency of an oscillator (specifically that of the breathing mode of a bubble)
ω_n	frequency of the shape oscillation described by a spherical harmonic perturbation of order n on a sphere
ϖ_n	frequency of the mode n oscillation of a bubble cloud
Ξ	a complex amplitude associated with displacement of the general oscillator (also Ξ_1 etc.)
Ψ	a complex amplitude associated with propagation of a general wave (also Ψ_1 etc.)
ψ	an amplitude
χ	the phase of $Z_{RP}^{s.s.}$
χ_0	value of χ at $r = R_0$
Σ	'the sum of', e.g. ΣP = 'the sum of the acoustic pressures'

Abbreviations

fps	frames per second
IF	inertial function (component of radial acceleration)
{IL}	intensity level
Im{ }	the imaginary component of a complex expression
PA	pulse average
PF	pressure function (component of radial acceleration)
PRF	pulse repetition frequency
Re{ }	the real component of a complex expression
RF	relates to an equation of motion where radius displacements are driven by an acoustic force
RP	relates to an equation of motion where radius displacements are driven by an acoustic pressure
SATA	spatial average, pulse average
{SPL}	sound pressure level
SPPA	spatial peak, pulse average
SPTA	spatial peak, temporal average
SPTP	spatial peak, temporal peak
SWR	standing wave ratio
TA	temporal average
VF	relates to an equation of motion where volume displacements are driven by an acoustic force
{VF}	void fraction
VP	relates to an equation of motion where volume displacements are driven by an acoustic pressure

1
The Sound Field

1.1 The Acoustic Wave

Introduction: Acoustic Regimes

Whether it be the bang of an explosion or the steady sine-wave tone of a whistle, an underwater sonar pulse or the output of a clinical foetal scanner, 'sound' is a waveform consisting of density variations in an elastic medium, propagating away from a source. Perhaps the most immediate property of sound is its frequency, or pitch, which is usually measured in cycles per second, or hertz (Hz). Not only is our perception acutely sensitive to differences in pitch, but we tend also to categorise sound into various regimes depending on the frequency. The restriction of a particular type of sound to a particular frequency regime arises through the limitations of the devices we use to generate and detect sound. To generate sound one must do mechanical work on the medium (in speaking, for example, by setting the air into motion through vibration of the vocal cords). Similarly in the detection of sound, the wave does mechanical work on the detector (in hearing through acoustic pressure signals causing the eardrum to vibrate). Devices such as the vocal cord and the eardrum, which convert energy from one form to another, are called transducers. Thus sound represents the transmission of mechanical energy and relies on both the elastic and inertial properties of the medium through which it travels. Since it is a relatively low-energy phenomenon,[1] detection of sound relies very often on utilising a mechanical resonance. Therefore, like any such resonator, a sound transducer will be limited in the range of frequencies to which it has a good response.[2]

Figure 1.1 is a logarithmic representation of the sensitive frequency regimes of some transducer systems, spanning nine decades in frequency. Human hearing is said to operate in the range 20–20 000 Hz (20 kHz), though we lose sensitivity to the higher frequencies as we age. From this range, the terms *sonic, ultrasonic* and *infrasonic* are defined. Sonic describes sounds in the nominal frequency range of the human ear, ultrasonic refers to sounds of frequencies greater than 20 kHz, and infrasonic relates to those at less than 20 Hz.

Whilst the salamander can hear only sounds in the range 50–220 Hz, the common house spider can hear from 20 Hz to 45 kHz. The loudspeaker in a television has an upper limit of around 10 kHz, half that of the human ear. Echo-location is used by many mammals. Bats

[1]The human ear can hear a 1 kHz tone at intensities as low as 10^{-12} W/m^2. The intensity of daylight at ground level is around 1000 W/m^2, which is 10^{15} times greater.
[2]See Chapter 4, section 4.1.

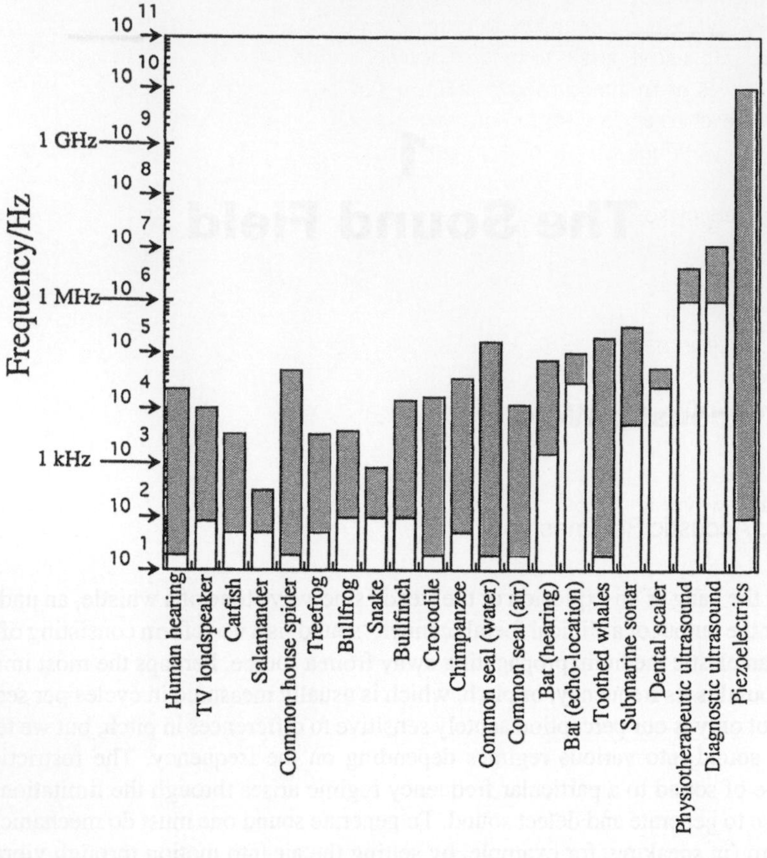

Figure 1.1 Shaded regions indicate the frequency regimes in which various transducer systems operate efficiently (data from reference [1]). The prefix system common to all SI units applies in the text, so that 1 THz = 10^{12} Hz, 1 GHz = 10^9 Hz, 1 MHz = 10^6 Hz, 1 kHz = 10^3 Hz, 1 mHz = 10^{-3} Hz, 1 μHz = 10^{-6} Hz, 1 nHz = 10^{-9} Hz, 1 pHz = 10^{-12} Hz and 1 fHz = 10^{-15} Hz.

employ frequencies in the range 30–80 kHz, whilst toothed whales use frequencies up to 200 kHz [1]. The reason for the difference is that the speed of sound in water is three times that in air, and so to obtain the same degree of spatial resolution, higher frequencies must be employed in water.[3]

Piezoceramic materials are commonly used to generate ultrasonic waves, transforming electrical energy into mechanical. A given piezoceramic crystal is limited by its resonance properties in the frequencies to which it can respond. However, by using a range of crystals, frequencies from the sonic up to 10 GHz can be generated.[4] In fact, air at atmospheric pressure and temperature ceases to transmit sound at around 2 GHz. At such high frequencies the theoretical wavelength is very small (1.7×10^{-7} m). Gas molecules communicate information about temperature, pressure etc. through collisions one with another, and the average distance they travel without colliding is called the *mean free path*. In the atmosphere at low altitudes this

---

[3] See section 1.1.1(a).
[4] 10 GHz = 10^{10} Hz – see the caption to Figure 1.1.

distance is about 7×10^{-8} m, which is comparable to one-half wavelength of 2-GHz sound. Thus at frequencies greater than this in the low-level atmosphere, the particulate nature of the air becomes noticeable, and it ceases to behave as a continuum.

Infrasonic waves of frequency below 1 Hz are generated by tidal motion, by earthquakes, and by explosive charges for use in seismology. Wind currents over mountain ranges can generate infrasound, which may be detected by birds for navigation purposes. Infrasound down to 1 Hz might indeed be detectable by humans: though the ear cannot strictly hear the sound, intense low-frequency sound can affect the inner ear and lead to dizziness. This may possibly occur when a car is driven with the window open [2].

1.1.1 The Longitudinal Wave

The term *wave* usually calls to mind ripples on a water surface, or on a rope in response to the end being 'flicked'. These waves are called *transverse*, because the displacement of the particles in the medium through which the wave travels is perpendicular to the direction of motion of the wave. Sound waves are in contrast *longitudinal*, in that the particles are displaced parallel to the direction of motion of the wave. It is important to note that, in both cases, the particles themselves are merely displaced locally, or oscillate: it is the *wave* that travels from source to detector, not the particles. Therefore if one sings a loud, steady note at a lighted candle from a distance of a few centimetres (careful of that moustache!) the flame barely flickers, since it is local vibrations, of less[5] than 1 μm, which are transmitted: there is no net flow of air, which would correspond to an extinguishing 'blow'.

(a) Description of the Wave

An example of a type of longitudinal wave is shown in Figure 1.2. The system shown is a series of bobs of equal mass, connected in a line by massless springs. The model therefore comprises the two necessary elements of any medium through which a sound wave will pass: inertia and elasticity. Only a section of the infinite line of bobs is shown.

Figure 1.2(a) shows the bobs equally spaced in the equilibrium position. In Figure 1.2(b) a longitudinal wave is passing through the medium, and the bobs are shown frozen at an instant in time (as if photographed). Arrows between parts a and b show how each bob has been displaced. If we now want to consider a wave travelling in a continuous medium, where the spacing of individual particles/bobs is insignificant in comparison with the wavelength, we can think of the bobs shown in Figures 1.2(a) and 1.2(b) as being samples, so that we have drawn only every millionth bob. If we wish to investigate waves in a continuum, we imagine there to be an infinite number of bobs between each one shown in the figure. Thus in Figure 1.2(c) we can represent the concentration of particles as continuous changes in density: the darker the regions, the greater the density. This now represents a longitudinal plane travelling wave, plane because the density variations occur in one direction only, the direction of propagation. Figure 1.2(d) shows the displacement (the solid line) as a function of the equilibrium position for the continuum, a displacement to the right being positive by convention. The pressure is similarly plotted in Figure 1.2(e). Regions of high pressure (*compressions*) in Figure 1.2(e) correspond to points of high population density in Figure 1.2(c). Similarly, low pressure regions (*rarefactions*) occur at points with a low number density of particles. The displacement and pressure

[5]See section 1.1.3 for the calculation.

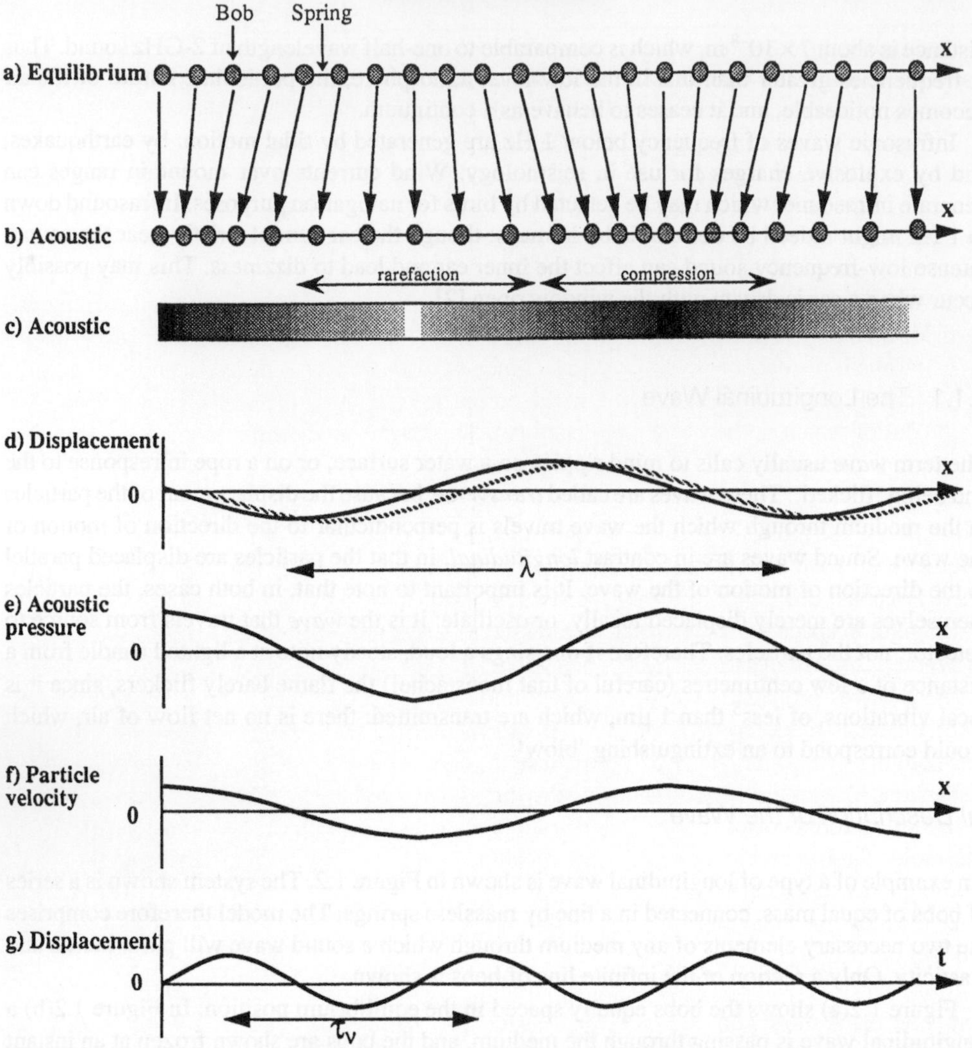

Figure 1.2 Model system for propagation of a one-dimensional travelling wave, using an infinite line of identical bobs attached by identical springs. (a) The bobs at equilibrium. (b) The bobs displaced by the passage of a travelling wave, frozen at some instant in time. Arrows from (a) show the displacement. (c) The density of a continuum through which the same wave passes: darker shadings represent higher densities. (d) Solid line: the displacement in the continuum, plotted as a function of the equilibrium position of the matter. Displacements to the right are taken to be positive. Dotted line: the displacement plotted a small time later. (e) The acoustic pressure plotted as a function of the equilibrium position. (f) The particle velocity plotted as a function of the equilibrium position. (g) The displacement as a function of time.

plots are sinusoidal and in quadrature.[6] This schematic demonstrates an important point in acoustics, that one must take care to specify whether one is referring to pressure or displacement: in the figure, positions of zero displacement correspond to maximum or minimum pressure. If unqualified, common terms such as amplitude, node, or antinode[7] could apply to either displacement or pressure.

---

[6] That is, if one is a sine wave, the other is a cosine wave.
[7] See section 1.1.6.

The figure illustrates the concept of wavelength, that is the distance between two points on a sinusoidal wave showing the same disturbance and doing the same thing (i.e. the disturbance is increasing in both, or decreasing, or stationary). The wavelength λ is shown on the figure.

This wave is travelling to the right, and a small time later the displacement curve has moved to the dotted curve in Figure 1.2(d). The difference in the ordinate between the solid and dotted curves in Figure 1.2(d) is proportional to the velocity. Thus we see that the particles in the compressed regions are moving forwards, and those in rarefaction moving backwards. Particle speed is greatest at the regions of zero displacement, and reduces to zero at the points of maximum displacement in either direction. The particle velocity is shown in Figure 1.2(f). All particles in this wave undergo an identical oscillation, though with different phases (i.e. starting times). Thus if displacement is plotted as a function of *time* instead of distance, it will also be sinusoidal (Figure 1.2(g)). The time for one complete oscillation is called the *period* (τ_ν), the reciprocal of which is the linear frequency (ν). This corresponds to the number of complete oscillations performed by a given particle per second.

Earlier the *phase* of a point in the wave was likened to a measure of the 'starting time' of its oscillation. Strictly speaking, the phase measures the proportion of a complete cycle of the wave, whether that wave is described as a function of space (Figure 1.2(d)) or time (Figure 1.2(e)). There is a phase difference of 2π between two points on a wave that are separated by a complete wavelength in space or, if the wave is expressed as a function of time, by a complete period.

Figure 1.2 illustrates a one-dimensional model of a longitudinal wave of infinite length. True acoustic waves propagating through gases, liquids and solids cause displacement of atoms. Since matter is made up of molecules, no real medium is continuous, but in most cases the wavelength of the sound is so large as to make the discrete nature of the medium irrelevant, and we can treat the sound as though it propagates through a continuum. The 'springs' which provide the restoring force correspond to the interatomic forces in solids. In an ideal gas there are no intermolecular forces, the motion of the molecules being governed by statistical thermodynamics though, as stated earlier, usually the frequency is so low and the wavelength so large as to make the medium behave as a continuum, employing the measurable bulk properties of density and compressibility.

Most acoustic waves are three-dimensional, and would spread out spherically from a point source. A more complicated source, such as a plane oscillator, can be thought of as a summation of point sources, and the subsequent propagation of the sound wave can be found by superimposing the resulting waves. Huygens (1629–1695) went further and stated a principle by which wave motion may be approximated: as a medium is traversed by waves, each point on a wavefront can be treated as a point source of secondary wavelets, and so the position of the wavefront a small time later can be found as the envelope of these wavelets. This treatment is adequate for all manner of simple waves, and a nice illustration can be seen on the beach, where transverse water waves tend to come into the shore parallel to it regardless of their initial direction. This is because the more shallow the water, the slower the wave travels. Imagine therefore a wave coming in at an angle θ_1 to the shore. Individual points on that wavefront act as point sources, from which wavelets propagate out. Figure 1.3 shows the positions of the wavelets some small time Δt later: they form slightly distorted semicircles of radius $c_H\Delta t$ about their source, where c_H is the speed of the wavelets. The nearer the source is to the shore, the smaller the value of c_H, and so the smaller the radius. The new position of the wavefront is given by the envelope of these wavelets, which is now at some smaller angle θ_2. This process will continue as the water depth decreases, so turning the wavefronts parallel to the shore.

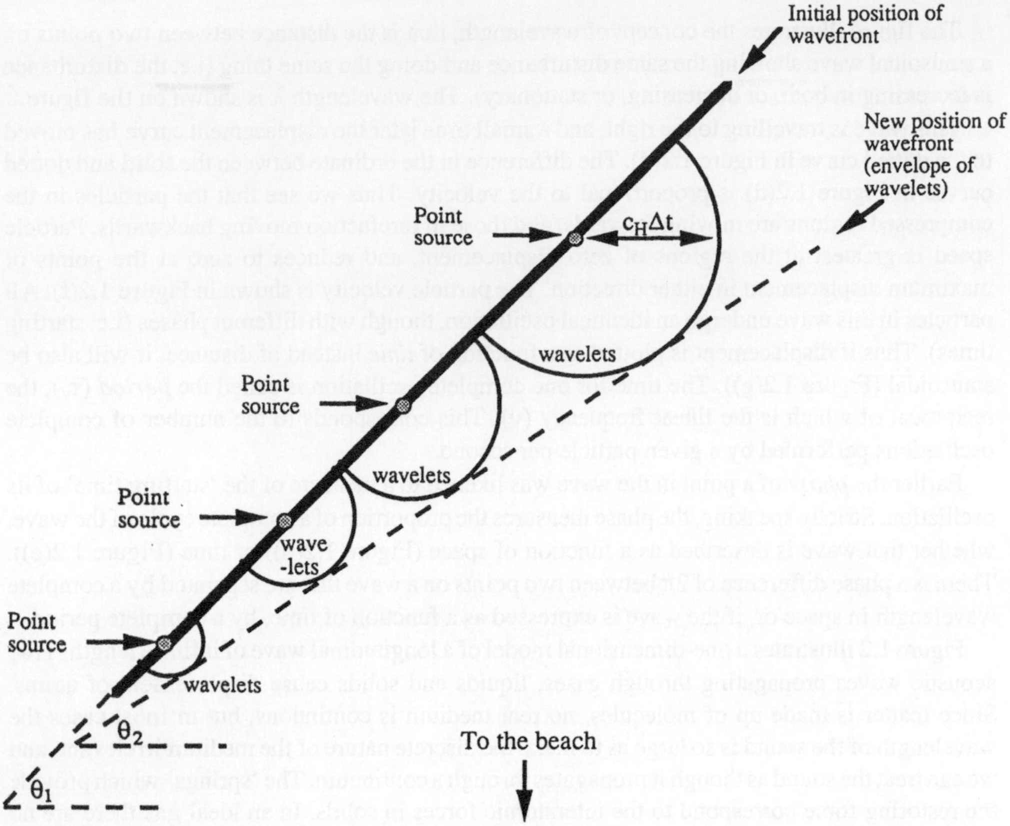

Figure 1.3 Use of the Huygen's construction to illustrate how, as a result of the decrease in the speed of surface waves on water with decreasing depth of the water, the waves tend to arrive travelling in a direction perpendicular to the beach.

Sinusoidal waves of infinite length propagating in one dimension, such as those described in this section, we will term simple waves. For these there is a simple relation between the wave speed (c), the frequency (ν) and the wavelength (λ):

$$c = \nu\lambda = \omega/k \tag{1.1}$$

where $\omega = 2\pi\nu$ is the circular or angular frequency of the wave, and $k = 2\pi/\lambda$ is the wavenumber.

Strictly, c is the *phase speed* of the wave, the speed of a sinusoidal wave of infinite length. There is only a single frequency associated with such simple waves, the frequency of the sinusoid. If all simple waves travel at the same phase speed, regardless of their frequency, the medium is called *non-dispersive*. A pulse or wavepacket, which is not infinite in length, can be thought of as being made up by the summation of many simple waves, each of infinite length but different frequency, which cancel each other out to give no net displacement in the regions of space beyond the confines of the wavepacket.[8] If the medium is *dispersive*,

[8] Any waveform that is not a sinusoid of infinite extent will contain more than one frequency, and can be built up by summing proportions of infinite sinusoids of various frequencies. Amongst the techniques which determine what frequency components make up a given waveform is Fourier analysis [3–5].

in that the phase speed varies with frequency, the component simple waves that make up a wavepacket will each individually travel at their own phase speed. The faster simple waves will propagate ahead of the slower ones, and the cancellation process that serves to confine the wavepacket to a small region of space will not be exact at the edges of the wavepacket. Thus the wavepacket will tend to spread out as it propagates (Figure 1.4). The centre of the wave packet will travel at the *group velocity*, given by $\partial\omega/\partial k$ evaluated at the main frequency and wavelength (or k) of the wavepacket, as shown in Figure 1.4.

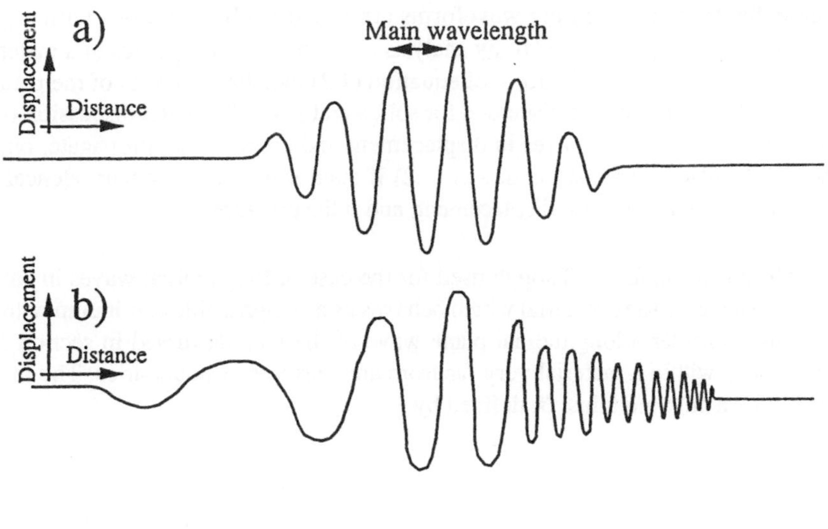

Figure 1.4 Schematic illustration of the propagation of a pulse (plotted for displacement against distance) in a dispersive medium, for the example where the phase speed increases with increasing frequency/decreasing wavelength. (a) The initial shape of the pulse. (b) The pulse some time later: having propagated over a distance in the medium, the pulse has spread out, the lower frequencies tending to lag behind. The main frequency/wavelength of the pulse, at which the group velocity $\partial\omega/\partial k$ is evaluated, is indicated.

In this book, c will now refer specifically to the phase speed of acoustic waves. The speed of sound in air at 20° C and under 1 atmosphere pressure is 344 m s^{-1}, and so equation (1.1) indicates that the wavelength at 80 kHz is 4.3 mm. In echo-location, the limit of spatial resolution is roughly given by the size of the wavelength used, so bats can echo-locate objects down to a size of a few millimetres, corresponding to their diet of insects. Interestingly, those bats which eat fruit, rather than insects, tend not to use echo-location but instead find their larger, stationary food by sight. The speed of sound in water is roughly 1480 m s^{-1}, and so at 80 kHz the wavelength is nearly 2 cm. To achieve a similar degree of spatial resolution as bats, toothed whales must exploit higher frequencies: at 200 kHz, the maximum frequency toothed whales exploit, the wavelength in water is around 7 mm.

An introduction to some basic features of acoustic waves is given in the remainder of section 1.1, to provide a basis for the discussion of the interaction of acoustic waves with bubbles. Treatments which detail this material can be found in the references [3, 6–12].

(b) Dynamics of the Longitudinal Wave

The equation

$$\frac{\partial^2 \varepsilon}{\partial t^2} = \ddot{\varepsilon} = c^2 \frac{\partial^2 \varepsilon}{\partial x^2} \tag{1.2}$$

is known as the one-dimensional linear wave equation,[9] where $\ddot{\varepsilon}$ represents the second derivative of ε with respect to time. Any parameter ε, a function of position x and time t, which satisfies equation (1.2) will propagate as a wave at a speed c in the $+x$ or $-x$ direction. Mathematically, these propagating waveforms are described by functions with arguments of $(ct - x)$ or $(ct + x)$ respectively.[10] If, by analysis of the physical properties of a material, one can generate equations having the form of equation (1.2) then linear waves of the parameter ε will propagate. This will now be discussed for solids and gases. From the discussion of Figure 1.2, which demonstrates how waves in displacement and pressure can propagate, one would expect both parameters to satisfy equation (1.2) if such waves are to occur. Henceforth the symbol ε will represent a general displacement, and p the pressure.

(i) Sound Waves in Solids. Though used for the case of longitudinal waves in solids, this proof can be extended to any material which behaves as a compressible continuum with a given bulk modulus. Consider a longitudinal plane wave of the type discussed in section 1.1.1(a), propagating deep within a material very far from any surfaces or boundaries. The material is characterised by a bulk modulus, $\boldsymbol{B}$, defined by

$$\boldsymbol{B} = -V\frac{\mathrm{d}p}{\mathrm{d}V} \tag{1.3}$$

where V is volume, and $\mathrm{d}p$ the change in pressure from the equilibrium value in response to a volume change $\mathrm{d}V$. Consider a prismatic volume element of cross-sectional area A and aligned with the x-direction, terminated at the coordinates x_0 and $x_0 + \Delta x$, which is in equilibrium (Figure 1.5(a)). If a sound wave passes through the material in the $+x$ direction, then points in the element undergo displacement, the ends moving to coordinates $x_0 + \varepsilon$ and $x_0 + \Delta x + \varepsilon + \Delta\varepsilon$ respectively (Figure 1.5(b)). The element, initially of volume $\{(x_0 + \Delta x) - (x_0)\}A$, changes to a volume $\{(x_0 + \Delta x + \varepsilon + \Delta\varepsilon) - (x_0 + \varepsilon)\}A$, a difference of $\Delta V = A\Delta\varepsilon$. If the differential Δ becomes very small, application of equation (1.3) to the element gives

$$\boldsymbol{B} = -A\Delta x \frac{\Delta p}{A\Delta\varepsilon} \tag{1.4}$$

The change in pressure from equilibrium, which equals the acoustic pressure, is the ratio of force-to-area impressed upon a cross-section of the prism (i.e. in the yz plane) at a given x-coordinate. Therefore the force F upon the plane at position x_0 is

$$F(x_0) = A\mathrm{d}p = -\boldsymbol{B}A\frac{\partial \varepsilon}{\partial x}\bigg|_{x = x_0} \tag{1.5}$$

---

[9] See Chapter 2, section 2.2.3.

[10] The argument $(x - ct)$ can equivalently be used instead of $(ct - x)$ to represent a wave travelling in the $+x$ direction.

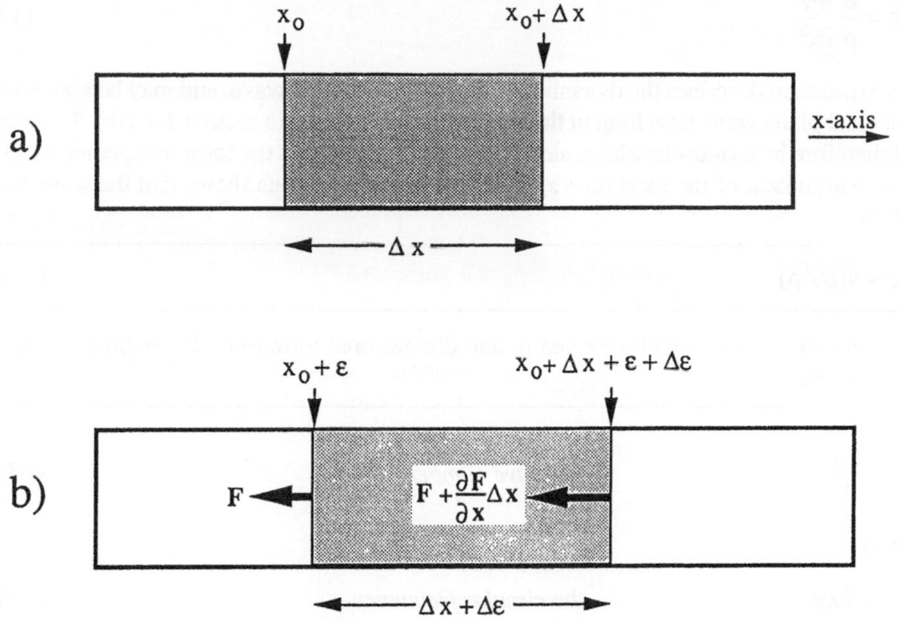

Figure 1.5 A bulk material shown (a) at equilibrium, is (b) distorted by the passage of an acoustic wave.

the notation signifying evaluation of the differential at the position $x = x_0$. First-order Taylor expansion gives the force upon the plane at $x = x_0 + \Delta x$ to be

$$F(x_0 + \Delta x) = F(x_0) + \Delta x \left.\frac{\partial F}{\partial x}\right|_{x=x_0} \tag{1.6}$$

From Newton's Second Law of Motion, the net force on the volume element will be the product of its mass and acceleration, i.e.

$$F(x_0) - F(x_0 + \Delta x) = -\Delta x \left.\frac{\partial F}{\partial x}\right|_{x=x_0}$$

$$= \rho A \Delta x \left.\frac{\partial^2 \varepsilon}{\partial t^2}\right|_{x=x_0} \tag{1.7}$$

where the mass of the element is the product of the density (ρ) with the volume ($A\Delta x$). Substituting for F from equation (1.5) and employing the usual notation for the time derivative (i.e. $\ddot{\varepsilon} = (\partial^2\varepsilon/\partial t^2)$) gives

$$\rho A \Delta x\, \ddot{\varepsilon} = BA \frac{\partial^2 \varepsilon}{\partial x^2} \Delta x \tag{1.8}$$

that is

$$\ddot{\varepsilon} = \frac{B}{\rho}\frac{\partial^2 \varepsilon}{\partial x^2} \tag{1.9}$$

This expression describes the dynamics of longitudinal bulk[11] wave, and may be applied to the sinusoidal plane wave travelling in the $+x$ direction discussed in section 1.1.1(a). The solution will therefore be a sinusoid which, since equation (1.9) follows the form of equation (1.2), will have an argument of the form $(ct - x)$. Comparing the equations shows that the wave speed c will be

$$c = \sqrt{(B/\rho)} \tag{1.10}$$

The argument is conveniently written in non-dimensional form $(\omega t - kx) = 2\pi(ct - x)/\lambda$ using the parameters

$$k = \frac{2\pi}{\lambda} \quad \text{the } \textit{wave number} \tag{1.11}$$

and

$$\omega = 2\pi\nu \quad \text{the } \textit{circular frequency} \tag{1.12}$$

The solutions are therefore waves of displacement having the form $\cos(\omega t - kx)$.

(ii) Sound Waves in a Gas. As sound travels through a gas, regions will be compressed and others undergo rarefaction. The gas in such regions will tend to be heated and cooled respectively.[12] As such volume and pressure changes take place in a gas, heat will tend to flow between a region of the gas and its surroundings if there is a temperature difference between them, flowing from hotter to cooler regions. In an acoustic field, heat will therefore tend to flow from compressions to rarefactions. The propagation of acoustic waves in a gas is now considered in the two extreme cases, where heat flow is either prohibited, or it is unhindered.

If volume and pressure changes take place in a fixed mass of gas without heat being able to transfer between that gas and its surroundings, then the process is called *adiabatic*. Such processes are characterised by the equation

$$pV^{\gamma} = \text{constant} \quad \text{(adiabatic conditions)} \tag{1.13}$$

where γ is the ratio of the heat capacity of the gas at constant pressure to that at constant volume. Differentiation of this equation, and substitution into equation (1.3), gives the *adiabatic bulk modulus* B_s

[11] If, instead of being within the bulk of an infinite medium, the waves are in a bar (with free surfaces parallel to the direction of propagation) the wavespeed is characterised by Young's Modulus, Y_{mod}. This is the ratio of the stress (force per unit area) to the strain (the ratio of the extension to the original length) at some particular point in the material. Therefore, if the element in Figure 1.5(a) were within a rod, so that the upper and lower boundaries were free, then the force $F(x = x_0)$ would equal $AY_{mod}(\Delta\varepsilon/\Delta x)$. Substitution of this force into equation (1.6) would give a wave speed of $c = \sqrt{(Y_{mod}/\rho)}$.

[12] Such a phenomenon will be familiar to anyone who has felt the gas in a bicycle pump become hot as the tyre is pumped up, or who has felt the cold stream of air that issues from an opened valve as gas previously compressed in a car tyre expands out into the atmosphere.

$$B_s = \gamma p \tag{1.14}$$

In contrast, if heat transfer between the gas and its surroundings is unhindered, and if those surroundings are assumed to be an infinite reservoir, then compression or expansion of the gas will occur at constant temperature. Such a process is called *isothermal*. A perfect gas (in which there are no intermolecular forces, and the molecules are points of no finite size) is characterised by the equation

$$pV = N_m R_g T \qquad \text{(perfect gas)} \tag{1.15}$$

where N_m is the total number of moles of gas present in volume V, R_g the gas constant, and T the absolute temperature (measured in kelvin).[13] For a fixed mass of gas at constant temperature, this equation reduces to

$$pV = \text{constant} \qquad \text{(isothermal conditions)} \tag{1.16}$$

Differentiation of this, and substitution into equation (1.3), gives the *isothermal bulk modulus* B_T

$$B_T = p \tag{1.17}$$

Provided the frequency is low enough, the gas can be treated as a continuum and the analysis given above for solids will apply also to the gas. Substitution of equations (1.14) and (1.17) into equation (1.10) gives the speed of sound under adiabatic conditions to be $\sqrt{\gamma p/\rho}$ and under isothermal conditions to be $\sqrt{p/\rho}$. Experimental measurements show that the speed of audio sound waves under room conditions agrees with $\sqrt{\gamma p/\rho}$, showing that the wave process is adiabatic: there is very little conduction of heat between adjacent regions of compression and rarefaction. However, as one employs higher frequencies, the wavelength becomes less. As the distance between compressions and rarefactions decreases, heat conduction between them increases. The propagation of sound waves in air under standard conditions of pressure and temperature tends to the isothermal at frequencies of around 7×10^8 Hz, and the speed of sound then approaches $\sqrt{p/\rho}$.

There is an important point to be shown by this. Rarely is a process completely adiabatic (no heat flow) or isothermal (unhindered heat flow). In practice there is often limited heat flow, and it is sometimes useful to summarise all the above formulation by the definition of a *polytropic index*,[14] κ, where

$$pV^{\kappa} = \text{constant} \tag{1.18}$$

---

[13]Combination of $PV^{\gamma} = \text{constant}$ (equation (1.13)) with $PVT^{-1} = \text{constant}$ (equation (1.15)) for a fixed amount of gas yields

$$TV^{\gamma-1} = \text{constant} \tag{1.f1}$$

and

$$p^{1-\gamma}T^{\gamma} = \text{constant} \tag{1.f2}$$

[14]The polytropic index represents a useful notation, rather than a fundamental quantity. Indeed, during a single process the value of κ may vary with the heat flow, in which case the act of assigning it a constant value is an approximation, the validity of which depends on the circumstances. Within the limits of this polytropic approximation we may, in analogy with equations (1.f1) and (1.f2), write

$$TV^{\kappa-1} = \text{constant} \tag{1.f3}$$

and

$$p^{1-\kappa}T^{\kappa} = \text{constant} \tag{1.f4}$$

The polytropic index can vary in value from γ (corresponding to the adiabatic case) to unity (the isothermal case), and for processes of limited heat flow takes some intermediate value. In the adiabatic and isothermal limits, equation (1.18) reduces to equations (1.13) and (1.16) respectively. The speed of sound in a gas can therefore be written

$$c=\sqrt{\frac{\kappa p}{\rho}} \tag{1.19}$$

(iii) Sound Waves in a Liquid. The propagation of sound through a liquid is discussed fully in Chapter 2, section 2.2. In order to reduce the complicated formulations obtained from physical arguments into a form matching the linear wave equation (equation (1.2)) it is necessary to assume that changes in liquid density are negligible when compared with the equilibrium value of the density. Therefore when this condition is satisfied, simple linear waves of the type described above will propagate.

This approximation is, however, unusual, in that it is in essence stating that the liquid is assumed to be incompressible. Obviously the existence of sound in a medium requires density fluctuation (see Figure 1.2), and so at first sight a treatment of sound in a medium which is assumed incompressible seems self-contradictory. In actual fact, the meaning is that waves *do* propagate in liquids, but they are not the simple waves described by the linearised wave equation. However, if the density fluctuations are small compared with the fluid density, then the waves approximate to simple linearised waves, with an appropriate wavespeed c.

(c) The Complex Representation of Harmonic Oscillations

The preceding sections have discussed a model longitudinal wave where displacement and pressure vary sinusoidally in space and time, described mathematically as harmonic solutions to the one-dimensional linear wave equation. At some fixed position, the temporal variation of a particle's displacement may be described by $\varepsilon = \varepsilon_0 \cos\omega t$, which describes a harmonic oscillation.

The simplest representation of this motion is found by taking a circle of radius ε_0 and resolving a radius vector onto one chosen axis. The radius vector rotates steadily, so that the angle θ between it and the fixed axis increases linearly with time (Figure 1.6(a)). For example, we may set $\theta = \omega t$. This angle is a measure of the phase of the oscillation. A second axis, perpendicular to the first, delineates the origin at its intersection with the first axis, and it is here that the base of the vector is fixed. In the simplest case, the two perpendicular axes are chosen to be the cartesian axes, and the x-coordinate represents the amplitude of the simple harmonic motion, to give the result $\varepsilon = \varepsilon_0 \cos\omega t$. If at time $t = 0$ the displacement is not equal to ε_0, then a constant phase angle ϑ can be added to give $\varepsilon = \varepsilon_0 \cos(\omega t + \vartheta)$. This phase angle simply represents the 'head start' for the rotating vector (Figure 1.6(b)).

Any two perpendicular axes could be used to resolve the rotating radius vector, and so define the simple harmonic oscillation. It is often more convenient to use the axes of the Argand diagram, rather than the cartesian axes. The space is therefore the complex plane, with the real and imaginary axes defining a point $z_\varepsilon = Y_a + iY_b$, where Y_a and Y_b are real constants, and where i is taken[15] to be the positive square root of -1. The radius vector may therefore be given by

$$z_\varepsilon = Ye^{i(\omega t+\vartheta)} = Y\cos(\omega t + \vartheta) + iY\sin(\omega t + \vartheta) \tag{1.20}$$

[15] Different conventions may apply in other texts.

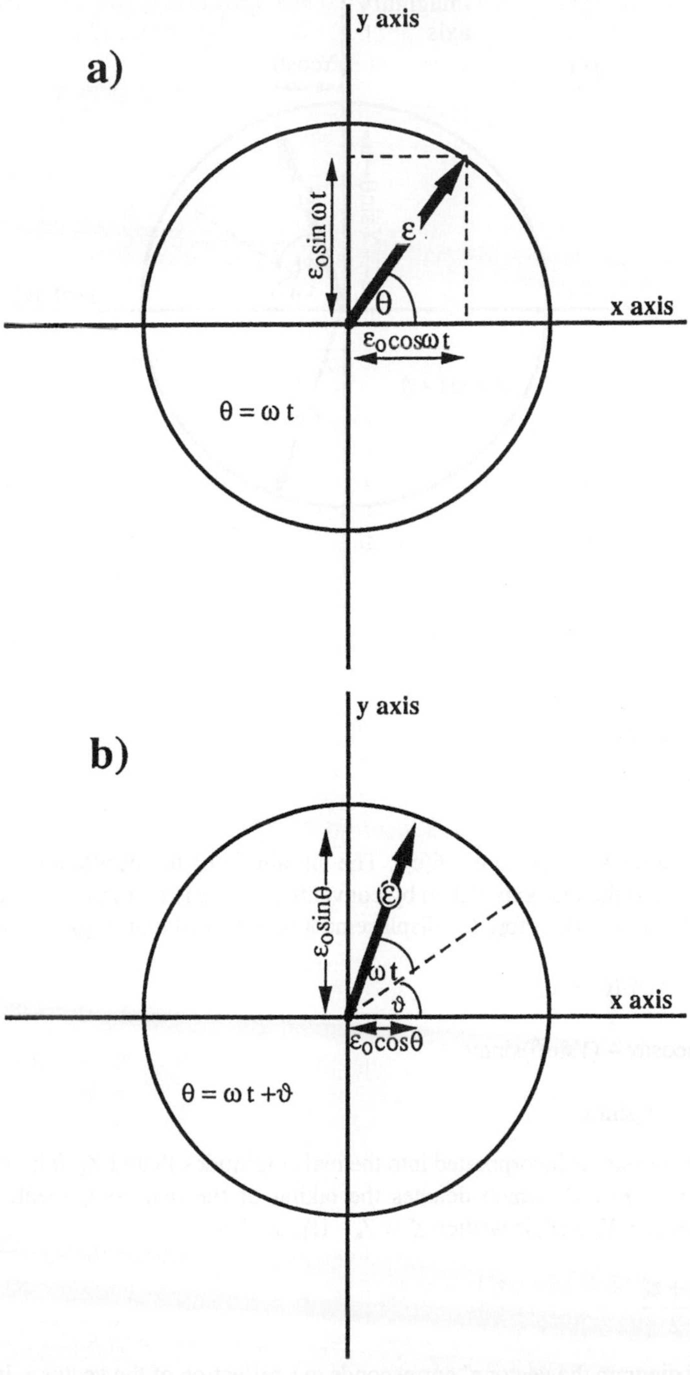

Figure 1.6 Representation of simple harmonic motion by the resolution of the displacement of a circling radius onto a fixed axis, in (a) using the cartesian frame. In (b) the same frame is used, and a phase factor ϑ is incorporated. In (c) on p. 14 the same oscillation is represented in the complex plane. The complex conjugate of z, labelled z^*, is also shown.

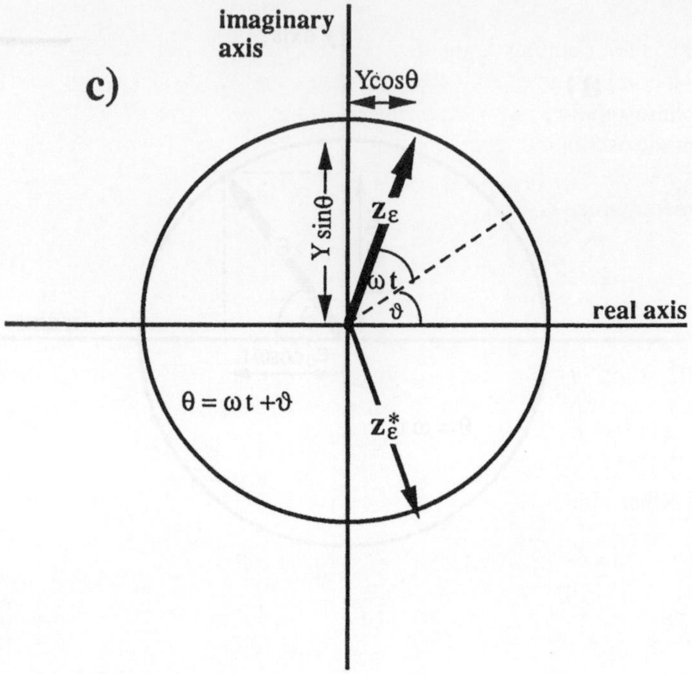

Figure 1.6 *continued*

where Y is a real constant (Figure 1.6(c)). The amplitude of the oscillation is therefore the resolution of z_ε onto the real axis (taken by convention; there is no reason why the imaginary axis could not be used). Therefore the displacement in the oscillation is given by

$$\varepsilon = \text{Re}\{z_\varepsilon\} = Y\cos(\omega t + \vartheta)$$

$$= (Y\cos\vartheta)\cos\omega t - (Y\sin\vartheta)\sin\omega t$$

$$= Y_1\cos\omega t - Y_2\sin\omega t \qquad (1.21)$$

Now the phase constant is incorporated into the real amplitudes Y_1 and Y_2. It is conventional to omit the notation 'Re{ }', which denotes the taking of the real component. The *complex conjugate*, z_ε^*, of $z_\varepsilon = Y_a + iY_b$ is written $z_\varepsilon^* = Y_a - iY_b$, so that

$$\text{Re}\{z_\varepsilon\} = \frac{z_\varepsilon + z_\varepsilon^*}{2} \qquad (1.22)$$

On the Argand diagram the vector z_ε^* corresponds to a reflection of the vector z_ε in the real axis (Figure 1.6(c)): it has the same amplitude, but the phase has been multiplied by -1.

The usefulness of this complex representation of the oscillation is that equation (1.20) can be rearranged as

$$z_\varepsilon = Ye^{i\vartheta}e^{i\omega t} = \Psi e^{i\omega t} \tag{1.23}$$

where $\Psi = Ye^{i\vartheta}$ is the complex amplitude. The phase angle ϑ is therefore hidden within the notation using a complex amplitude. This can be exploited when a driving force is applied to examine the phase relation between force and response.[16] The use of complex notation to represent harmonic oscillations is further discussed in Chapter 3, section 3.1.2(a). The complex representation can be employed only when mathematically linear operations are performed upon the parameter which is represented.[17]

(d) The Complex Representation of Harmonic Waves

As outlined in section 1.1.1(b), solutions of the one-dimensional linear wave equation have arguments of the form $(ct - x)$ or $(\omega t - kx)$ for waves travelling in the $+x$ direction, and $(ct + x)$ or $(\omega t + kx)$ for waves travelling in the $-x$ direction. The complex formulation for the harmonic oscillator employed in the previous section can therefore be extended to describe one-dimensional harmonic waves,[18] expressing the dependence on *both* space and time and allowing for propagation in either direction:

$$z_\varepsilon = \Psi_1 e^{i(\omega t + kx)} + \Psi_2 e^{i(\omega t - kx)} \tag{1.24}$$

The complex amplitudes Ψ_1 and Ψ_2 can incorporate phase constants. If there were no waves propagating in the $-x$ direction, Ψ_1 would equal zero. If one were to examine the form of the wave at some fixed moment in time, say when t equals t_0, then the complex formulation gives the variation in x to be

$$z_\varepsilon = (\Psi_1 e^{i\omega t_0})e^{ikx} + (\Psi_2 e^{i\omega t_0})e^{-ikx}$$

$$= \Psi_3 e^{ikx} + \Psi_4 e^{-ikx} \tag{1.25}$$

where $\Psi_3 = \Psi_1 e^{i\omega t_0}$ and $\Psi_4 = \Psi_2 e^{i\omega t_0}$ express the complex amplitude contributions of the waves travelling in $-x$ and $+x$ direction respectively at the instant $t = t_0$. The real component of z_ε in equation (1.25) would therefore give the displacement of all particles at some fixed instant in time (i.e. it is like a 'snapshot' photograph). Having therefore used equation (1.24) to derive the variation of displacement in space at some fixed time, we will now examine the time-dependence of displacement at some fixed point in space, just as described in section 1.1.1(c). This would correspond to following the motion of one specific bob in Figure 1.2. At a fixed position $x = x_0$, equation (1.24) reduces to

$$z_\varepsilon = (\Psi_1 e^{ikx_0} + \Psi_2 e^{-ikx_0})e^{i\omega t} = \Psi e^{i\omega t} \tag{1.26}$$

---

[16] See Chapter 3, section 3.2.1.

[17] Consider a parameter u_1 that is a function of the coordinate x. Say that an operation G transforms u_1 into w_1, so that $G(u_1) = w_1$. Similarly the operator might act on a second parameter, u_2, such that $G(u_2) = w_2$. If G is a linear operator, then $G(u_1 + u_2) \propto w_1 + w_2$. Thus examples of linear operations on e^{ikx} are: multiplication by a real constant ($G(u_1) = su_1$); differentiation by x ($G(u_1) = du_1/dx$); double-differentiation by x ($G(u_1) = d^2u_1/dx^2$); and displacement in x ($G(u_1(x)) = u_1(x + s)$), where s is a real constant. However, an important operation which is *not* linear is squaring ($G(u_1) = u_1^2$) and related functions: thus in equations (1.33) and (1.63), the real part must be taken prior to the multiplication process.

[18] Equivalent real sine and cosine descriptions of the wave may of course still be employed as an alternative to the complex notation.

in agreement with equation (1.23). Therefore expressing the waveform in the complex notation of equation (1.24) allows description of one dimensional harmonic waves travelling in either direction. It is the usual notation adopted in acoustics, facilitating simple formulation of the type shown in section 1.1.5(a), where the use of $(\omega t - kx)$ and $(\omega t + kx)$ to describe the waveforms ensures that the calculation readily reduces to the simple form of equation (1.46) when the boundary condition is specified at a fixed position ($x = 0$ in section 1.1.5).

1.1.2 Acoustic Impedance

The concept of impedance in physics is a common one, and refers to the ratio of a general driving force to the velocity response. One might therefore arbitrarily express the impedance of reading this book to be the desire to learn about acoustic cavitation (if one could quantify such a thing!), divided by the rate at which pages are read (note that the 'velocity response' must contain a temporal component). A more practical case of impedance is found in Ohm's Law, where the ratio of the driving force (voltage) to the velocity response (current) is the impedance (called the resistance for simple devices). In acoustics, the driving force is the acoustic pressure amplitude (P), and the velocity response is the velocity of particles in the medium. The impedance is therefore

$$Z = \text{Acoustic pressure/Particle velocity} = P/\dot{\varepsilon} \tag{1.27}$$

for the model employed in section 1.1.1(a), where $\dot{\varepsilon} = \partial\varepsilon/\partial t$ is the first derivative of ε with respect to time, which is the particle velocity. This is strictly the *specific acoustic impedance*.[19] Substituting from equation (1.5) using the fact that the acoustic pressure P equals the deviation from equilibrium pressure Δp, and assuming harmonic motion for the displacement (i.e. $\varepsilon = \varepsilon_0 e^{i(\omega t - kx)}$), gives

$$P = \boldsymbol{B}ik\varepsilon \qquad \text{and} \qquad \dot{\varepsilon} = i\omega\varepsilon \tag{1.28}$$

Equation (1.27) then becomes

$$Z = \boldsymbol{B}k/\omega \tag{1.29}$$

Substitution for $\boldsymbol{B}$ from equation (1.10) gives

$$\boxed{Z = \rho c} \tag{1.30}$$

The specific acoustic impedance of a material equals the product of the density with the speed of sound in that material. The acoustic impedance of several materials is given in Table 1.1.[20]

Substitution of $\varepsilon = \varepsilon_0 e^{i(\omega t - kx)}$ into equation (1.28), noting from section 1.1.1(c) that $i = e^{i\pi/2}$, shows that the pressure varies as

$$P = i\boldsymbol{B}k\varepsilon_0 e^{i(\omega t - kx)} = \boldsymbol{B}k\varepsilon_0 e^{i(\omega t - kx + \pi/2)} \tag{1.31}$$

or equivalently

$$P = -i\omega\rho c\varepsilon_0 e^{i(\omega t - kx)} = \omega\rho c\varepsilon_0 e^{i(\omega t - kx - \pi/2)} = \omega\rho c\varepsilon e^{-i\pi/2} \tag{1.32}$$

---

[19] See Chapter 3, section 3.2.1(c)(iii).
[20] Data from Wells [13], Kaye and Laby [14] and Lide [15].

Table 1.1 Specific acoustic impedance and attenuation of some materials.

	Density (kg/m^3)	Speed of sound (m/s)	Specific acoustic impedance (kg/m^2 s)	Amplitude attenuation coefficient (neper/m) at 1 MHz
Air at STP	1.2	330	400	138
Water	1000	1480	1.5×10^6	0.0253
Castor oil	950	1500	1.4×10^6	10.9
PMMA	1190	2680	3.2×10^6	23.0
Aluminium	2700	6400	1.7×10^7	0.0207
Brass	8500	4490	3.8×10^7	0.230

Data from Wells [13]. Data on other materials, and the dependence on frequency and temperature, can be found in Kaye and Laby [14] and Lide [15].

The pressure wave therefore also propagates harmonically i.e. as $P = P_A e^{i(\omega t-kx)}$. The pressure amplitude P_A is $\omega\rho c$ times the amplitude of the displacement wave, and the waves are in quadrature: the displacement leads the pressure by a phase angle of $\pi/2 = 90°$ for plane waves travelling in the $+x$ direction. The factor $e^{-i\pi/2}$ in equation (1.32) agrees with the waves derived in Figure 1.2.

1.1.3 Acoustic Intensity

As little as 10^{-18} J of energy is enough to excite the human ear, equivalent to the work done in lifting a mass of only 10^{-13} grams to a height of 1 mm against gravity. The ear can functionally detect sounds across fourteen magnitudes of energy: when the work done on the ear by the sound reaches around 10^{-4} J, irreversible damage occurs [1].

Our perception of the strength of a sound wave, its loudness, is based upon the acoustic *intensity*, which is the rate at which energy in the wave crosses a unit area perpendicular to the direction of propagation. As shown in Figure 1.2, the fundamental properties of an acoustic wave are the variation in pressure and density, and the underlying basis of the intensity measurements is the acoustic pressure amplitude of the wave, P_A (Figure 1.2(e)). In normal speech, this corresponds to around 0.1 Pa, or a millionth of an atmosphere. By substituting this value into equation (1.32) it is clear that this is equivalent to a displacement amplitude of less than 1 μm. Thus the candle flame discussed in section 1.1.1 is not extinguished.

Since the displacement in a one-dimensional acoustic wave can be expressed as $\varepsilon = \varepsilon_0 e^{i(\omega t-kx)}$, then the velocity of a layer at position x is $\dot{\varepsilon} = i\omega\varepsilon_0 e^{i(\omega t-kx)}$. Note that $\dot{\varepsilon} = \omega\varepsilon e^{i\pi/2}$, demonstrating that velocity and displacement are in quadrature as demonstrated previously in Figure 1.2. Knowledge of the velocity can be used to derive the energy in the wave, which will now be undertaken for plane waves in the absence of attenuation.[21]

Consider a plane element one wavelength long in the x-direction, cross-sectional area A, of mean density ρ, which starts at $x = L$. The kinetic energy of this element is

$$\phi_{K,\lambda} = \tfrac{1}{2}\,(\text{mass}).|\text{velocity}|^2 \tag{1.33}$$

Therefore

[21] The effect of attenuation on intensity is discussed in section 1.1.7.

$$\phi_{K,\lambda} = \frac{1}{2}A\rho \int_L^{L+\lambda} (\mathrm{Re}\{\dot\varepsilon\})^2\,dx = \frac{1}{2}A\rho \int_L^{L+\lambda} \omega^2\varepsilon_0^2\cos^2(\omega t - kx)\,dx$$

$$= \frac{1}{2}A\rho \int_L^{L+\lambda} \omega^2\varepsilon_0^2\,\frac{1}{2}\,(1+\cos 2(\omega t - kx))\,dx \qquad (1.34)$$

Since the integral of the cosine term over a wavelength gives zero, then

$$\phi_{K,\lambda} = \frac{1}{4}A\rho\omega^2\varepsilon_0^2\lambda = \frac{1}{4}m_\lambda\omega^2\varepsilon_0^2 \qquad (1.35)$$

where $m_\lambda = A\rho\lambda$ is the mass of this element, which has volume $A\lambda$. Therefore

$$\text{Mean kinetic energy per unit volume} = \frac{1}{4}\rho|\dot\varepsilon|^2_{max} \qquad (1.36)$$

The model of an acoustic wave given earlier in this chapter was of bobs connected by springs (section 1.1.1(a)). The energy balance in such a harmonic oscillating system is discussed in Chapter 3, section 3.1.2(b), where it is shown that the total energy is twice the mean kinetic energy. Therefore the total energy density in the plane progressive acoustic wave is

$$\text{Total energy density} = \frac{1}{2}\rho|\dot\varepsilon|^2_{max} \qquad (1.37)$$

If the energy is flowing in the $+x$ direction at the wavespeed c, then in time Δt the amount crossing a segment of the xy-plane of area A will be the total energy density multiplied by $Ac\Delta t$ (Figure 1.7). Therefore the intensity of the plane wave I (the energy crossing a unit area in unit time) will be

$$I = (\text{Total energy density}) \times c$$

$$= \frac{1}{2}\rho c|\dot\varepsilon|^2_{max} = \frac{1}{2}Z(P_A/Z)^2 \qquad (1.38)$$

using $\{P_A/|\dot\varepsilon|_{max}\} = Z$, the specific acoustic impedance. Thus acoustic intensity I for a plane wave can be expressed as

$$\boxed{I = \frac{P_A^2}{2Z}} \qquad (1.39)$$

As will be shown in Chapter 3, section 3.2.1(c)(iii), equation (1.39) holds for spherical as well as for plane waves.

Two similar waves can be compared by the ratio of their powers, and this is the basis of the decibel scale. If two signals have powers $\dot W_1$ and $\dot W_2$ respectively, then their relative level in bels is the quantity $[\log_{10}(\dot W_1/\dot W_2)]$, and in decibels is $[10\log_{10}(\dot W_1/\dot W_2)]$. Therefore if a signal of magnitude $\dot W_1$ is detected, its strength can be expressed in decibels by comparing it with the power of some reference signal. Since the area over which the measured sound is collected is constant in such cases (for example, in human hearing), the sound is often quantified as the *intensity level* {IL}

$$\{\text{IL}\} = 10\log_{10}\left(\frac{I}{I_{\text{ref}}}\right) \tag{1.40}$$

The ratio is that of the sound intensity I to some reference intensity I_{ref}. Since the intensity is proportional to the square of the pressure amplitude, this measure is equivalent to $20\log_{10}(P_1/P_2)$ where P_1 and P_2 are the acoustic pressures of the two signals. For convention, the *sound pressure level* {SPL}, in decibels, is taken as the ratio of the acoustic pressure to a reference P_{ref}

$$\{\text{SPL}\} = 20\log_{10}\left(\frac{P}{P_{\text{ref}}}\right) \tag{1.41}$$

Obviously the acoustic pressure and the reference must be measured in the same way (e.g. amplitude, or r.m.s.[22] pressure). If the reference intensity and the reference pressure represent the same physical wave, then IL is equivalent to SPL.

In air, the reference standard is taken as $I_{\text{ref}} = 10^{-12}$ W m^{-2}, which is approximately the threshold intensity for normal human hearing at 1 kHz. This threshold corresponds to an acoustic pressure amplitude of 28.9 μPa (equation (1.39)) for plane and spherical travelling waves. The SPL is usually taken from the ratio of the r.m.s. acoustic pressure to a reference pressure of 20 μPa (which is the nearest integer μPa corresponding to the I_{ref} intensity, the r.m.s. acoustic pressure of a sinusoidal wave having amplitude $P_A = 28.9$ μPa being 20.4 μPa). Because of this rounding, SPL is almost, but not exactly, equal to the IL for plane and spherical waves. In underwater acoustics, reference pressures of 20 μPa, 1 μbar and 1 μPa, equivalent to intensities of 2.70×10^{-16}, 6.76×10^{-9} and 6.76×10^{-19} W m^{-2} respectively, are used. The latter is now the more common [16]. However, use of SPL and IL should be accompanied by the quoted reference pressure.

If the waveform is more complicated than the simple plane and spherical travelling waves, for example, in standing wave fields where equation (1.39) does not hold, the measurements of SPL and IL can disagree. In most situations when the interaction of such a sound field with a bubble is considered, the size of the bubble is significantly less than the lengthscale over which significant pressure changes occur, and the response time of the bubble is less than or comparable with the acoustic period. The behaviour of an individual bubble is therefore determined by the instantaneous value of the local pressure field, and it is more appropriate to refer to the acoustic pressure measurement; hydrophones[23] give an instantaneous voltage representation of the local field.

The advantage of this scale is that it is logarithmic, and so can more readily express the vast range in intensities (approximately fourteen orders) to which our hearing can respond. In addition, the human sensory perception of loudness is logarithmic, in that we judge one sound to be so many *times* louder than another [17].

1.1.4 Radiation Pressure

In the preceding section the energy associated with a wave was calculated. The wave transmits energy, and the absorption of that energy will generate a force upon the absorber.

---

[22] r.m.s. means 'root mean square', which is calculated by squaring the acoustic pressure over some interval, finding the mean of this, then taking the square root of that mean so the result has dimensions of pressure. For a sinusoidal wave, the r.m.s. pressure is $1/\sqrt{2}$ times the acoustic pressure amplitude.

23See section 1.2.2(a)(i).

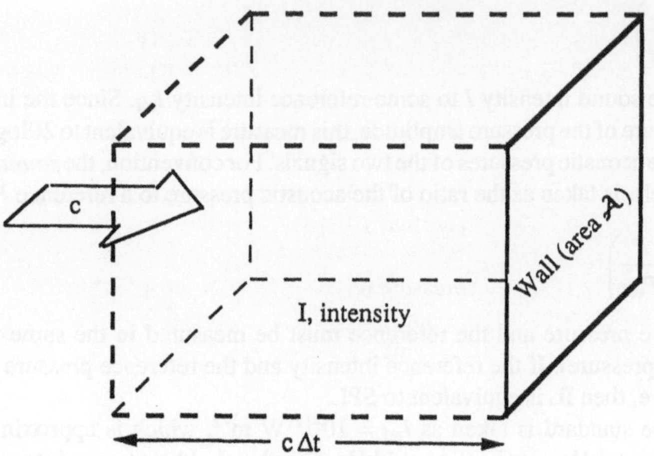

Figure 1.7 The energy in a plane travelling wave approaches a plane of area *A*, which is perpendicular to the direction of motion of the wave, at speed *c*.

Consider again the plane wave, travelling in the $+x$ direction, approaching a wall in the yz-plane of area A (Figure 1.7). The wave energy is completely absorbed by the wall. If the wave has intensity I, then the energy absorbed by the wall in a time Δt is $IA\Delta t$. The wall must have applied a force F_r in the $-x$ direction to stop the wave motion, which in time Δt acted over a distance $c\Delta t$. Therefore the work done by the wall on the wave is $F_r c\Delta t$. Equating this to the energy absorbed, we obtain $F_r = (IA/c)$. From Newton's Third Law of Motion, this must be equal and opposite to the force exerted by the wave on the wall. Therefore upon absorption the wave exerts a *radiation pressure* in the direction of its motion, of magnitude

$$p_{\text{rad,abs}} = \frac{I}{c}, \quad \text{for normal incidence of plane waves.} \tag{1.42}$$

The force F_r exerted by the wall can also be thought of as acting upon the wave to absorb its momentum. In time Δt the wall absorbs a length $L = c\Delta t$ of the wave, exerting an impulse $F_r\Delta t = IAL/c^2$ upon the wave, causing a change in momentum of Δp. Since after absorption the momentum of wave is zero, then the momentum associated with one wavelength of the wave is

$$p_\lambda = \frac{IA\lambda}{c^2} \tag{1.43}$$

If the wave is reflected, instead of being absorbed, this momentum must be not simply absorbed but reversed. The wall must exert twice as much force upon the wave, and so the radiation pressure felt by the reflector is

$$p_{\text{rad,refl}} = \frac{2I}{c} \tag{1.44}$$

for total reflection of normally incident waves back along the line of incidence.

If a wave is partially reflected and partially absorbed, the radiation pressure is intermediate between the two values given by equations (1.42) and (1.44). In one practical case, where the radiation pressure is used to measure the acoustic intensity (see section 1.2.2(a)(ii)), the sound may be incident at 45° to the reflector. The wall exerts the force F_r in the direction of incidence (to absorb the wave momentum in that direction), and exerts another force of equal magnitude perpendicular to the incidence, to generate the momentum for the reflected wave (Figure 1.8). The total radiation force is therefore $\sqrt{2}F_r$ normal to the reflector, giving an appropriate radiation pressure of $\sqrt{2}(I/c)$.

Beissner [18] calculates the radiation pressure resulting from geometries other than the plane wave, where diffraction effects must be considered. The simplest of these results is for a baffled[24] circular plane piston of radius L_s, which gives

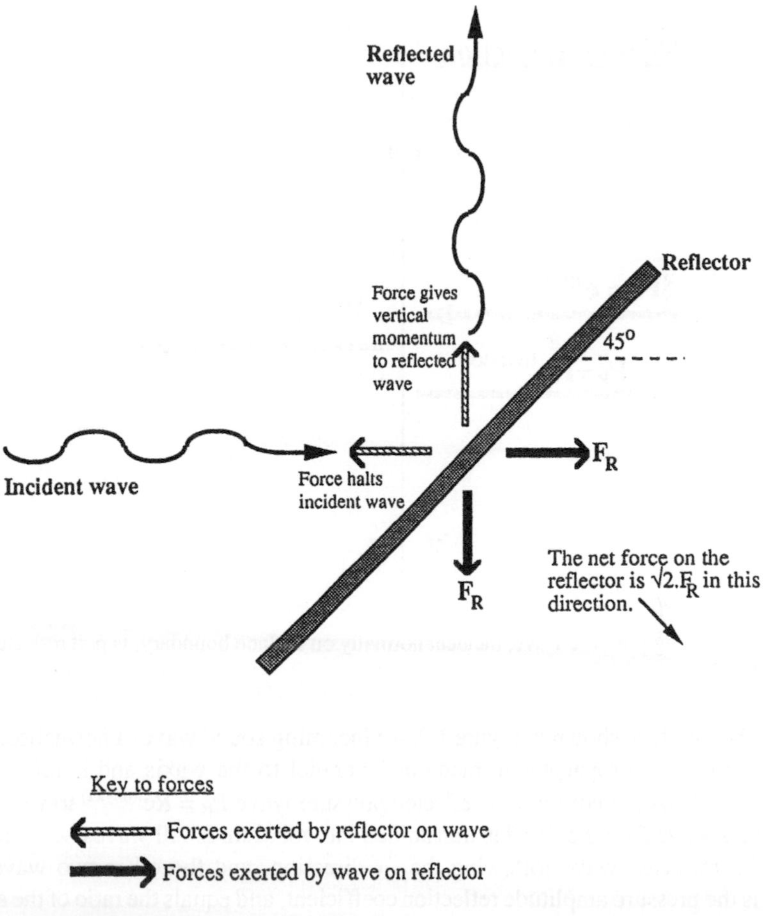

Figure 1.8 A travelling wave is reflected through 90° by a perfect reflector which is angled at 45° to the original direction of motion of the wave.

[24] A baffled source is one very close to an infinite rigid boundary, which thus emits into a half-space (see section 1.2.1(a), and Chapter 3, sections 3.3.2(a) and 3.3.2(b)).

$$p_{\mathrm{rad,abs}} = \frac{I}{c}\left\{\frac{1 - J_0^2(kL_s) - J_1^2(kL_s)}{1 - J_1(2kL_s)/kL_s}\right\} \tag{1.45}$$

where J_n is the Bessel function of order n [19].

1.1.5 Reflection

(a) Reflection at Normal Incidence

As with any waveform, sound waves can be reflected at interfaces between two differing media, giving rise at audio frequencies to the familiar echo effect. In acoustics, the criterion which distinguishes the difference between media is the acoustic impedance, as this section will show.

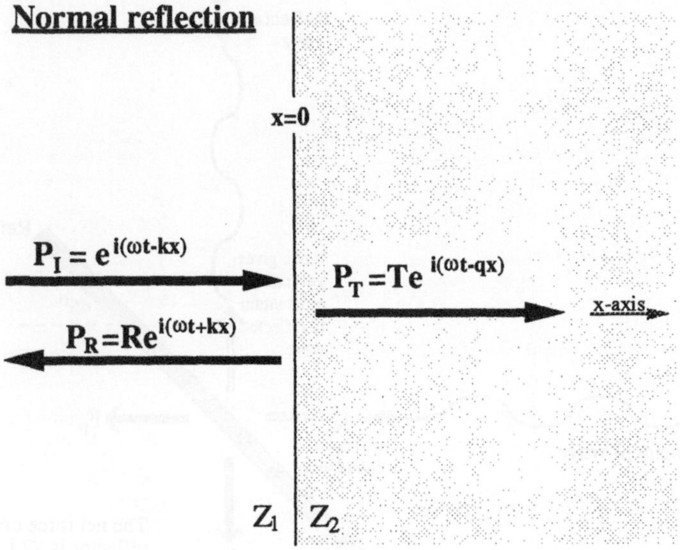

Figure 1.9 A travelling pressure wave, incident normally on a plane boundary, is part reflected and part transmitted.

Consider the interface shown in Figure 1.9. An incoming sound wave of normalised pressure amplitude $P_{\mathrm{I}} = e^{i(\omega t - kx)}$ propagates in medium 1 parallel to the x-axis and is reflected at the boundary (at $x = 0$) with medium 2. A reflected pressure wave $P_{\mathrm{R}} = \boldsymbol{R}e^{i(\omega t + kx)}$ travels back into medium 1 and a wave $P_{\mathrm{T}} = \boldsymbol{T}e^{i(\omega t - qx)}$ is transmitted into medium 2. All waves here travel along the x-axis, the reflected wave following the $-x$ direction, and the other two waves the $+x$ direction. $\boldsymbol{R}$ is the pressure amplitude reflection coefficient, and equals the ratio of the amplitude of the reflected to the incident pressure wave. $\boldsymbol{T}$ is the corresponding transmission coefficient, and is numerically equal to the ratio of the amplitude of the transmitted to the incident pressure wave. The specific acoustic impedances of the two media are Z_1 and Z_2. Since there can be no discontinuity in pressure at the massless interface ($x = 0$), then

$$P_{\mathrm{I}} + P_{\mathrm{R}} = P_{\mathrm{T}} \qquad \Rightarrow$$

$$1 + \boldsymbol{R} = \boldsymbol{T} \qquad \text{at } x = 0 \tag{1.46}$$

The velocities must match at the boundary ($x = 0$) at all times as the media stay in contact. Thus from equation (1.27)

$$\left(\frac{1}{Z_1}\right)(1 - \boldsymbol{R}) = \left(\frac{1}{Z_2}\right)\boldsymbol{T} \tag{1.47}$$

the negative sign before the reflection coefficient appearing because the reflected wave is travelling in the $-x$ direction. Combining equations (1.46) and (1.47) gives

$$\boldsymbol{T} = \frac{2Z_2}{(Z_1 + Z_2)} \tag{1.48}$$

and

$$\boldsymbol{R} = \frac{(Z_2 - Z_1)}{(Z_1 + Z_2)} \tag{1.49}$$

for the pressure amplitude transmission and reflection coefficients for normal incidence.

If $Z_1 = Z_2$, then the wave will be completely transmitted ($\boldsymbol{R} = 0$, $\boldsymbol{T} = 1$). There will be no reflected component, and the two media are said to be 'impedance matched'. However, consider an air–water interface. The impedance of air (Z_a) is 4×10^2 kg m^{-2} s^{-1}, whilst that of water (Z_w) is 1.5×10^6 kg m^{-2} s^{-1}. Therefore a wave in air, impinging upon water, will have a pressure amplitude reflection coefficient of 0.999. Alternatively an acoustic wave, travelling in water, will have a pressure amplitude reflection coefficient of –0.999, at an air–water interface. Therefore in both cases the pressure wave is almost entirely reflected, though in the second case it is also inverted. Because the acoustic impedances of air and water differ by a factor of nearly 4000, any mechanism designed to couple to one medium is unlikely to couple to the other. This might perhaps be why the common seal appears to employ two different hearing mechanisms, one for aerial hearing and one for aquatic, with maximum audible frequencies of 12 kHz and 160 kHz respectively [1]. These frequency ranges coincide approximately with the acoustic emissions of their predators in these environments.

When $Z_1 \gg Z_2$, then $\boldsymbol{R}$ tends to -1, and the interface is termed a *pressure release* or *free* boundary. When $Z_1 \ll Z_2$, then $\boldsymbol{R}$ tends to 1, and the interface is termed a *fixed* or *rigid* boundary.

The power reflection coefficient, giving the proportion of the incident energy reflected, is $\boldsymbol{R}^2$, so that the proportion of the energy transmitted is $1 - \boldsymbol{R}^2$ (which, it should be noted, is not equal to $\boldsymbol{T}^2$).

As in all acoustics, one must take care whether *pressure* or *displacement* amplitude reflection coefficients are being discussed. By following through a similar analysis for waves of displacement, it can readily be shown that the reflection coefficient for displacement is a factor of -1 times that for pressure. To illustrate this, the displacement reflection and transmission coefficients will now be derived for the condition of oblique incidence.

(b) Oblique Reflection

The incident (normalised), the reflected and the transmitted waves of displacement are $\varepsilon_I = e^{i(\omega t - kx\cos\theta_I + ky\sin\theta_I)}$, $\varepsilon_R = \boldsymbol{R}_\varepsilon e^{i(\omega t + kx\cos\theta_R + ky\sin\theta_R)}$ and $\varepsilon_T = \boldsymbol{T}_\varepsilon e^{i(\omega t - qx\cos\theta_T + qy\sin\theta_T)}$ respectively

(Figure 1.10). They are angled to the normal of the interface at the appropriate angles θ_I, θ_R and θ_T. The ratio of the amplitude of the reflected to the incident displacement wave is $\boldsymbol{R}_\varepsilon$, the displacement amplitude reflection coefficient. $\boldsymbol{T}_\varepsilon$ is the corresponding transmission coefficient, and is numerically equal to the ratio of the amplitude of the transmitted to the incident displacement wave. For continuity of the normal displacement at the interface for all times:

$$\varepsilon_I\cos\theta_I + \varepsilon_R\cos\theta_R = \varepsilon_T\cos\theta_T \qquad \Rightarrow$$

$$e^{i(ky\sin\theta_I)}\cos\theta_I + \boldsymbol{R}_\varepsilon\cos\theta_R e^{i(ky\sin\theta_R)} = \boldsymbol{T}_\varepsilon\cos\theta_T e^{i(qy\sin\theta_T)} \tag{1.50}$$

Since equation (1.50) must be true for all values of y, then the exponents in each of the terms must equate and we therefore deduce that:

$\sin\theta_I = \sin\theta_R$ — the *law of reflection*, and

$\dfrac{\sin\theta_I}{c_1} = \dfrac{\sin\theta_T}{c_2}$ — a statement of *Snell's Law*

where $c_1 = \omega/k$ and $c_2 = \omega/q$ are the speeds of sound in medium 1 and 2 respectively. Since the exponents in equation (1.50) are equal, the equation reduces to

$$\cos\theta_I + \boldsymbol{R}_\varepsilon\cos\theta_R = \boldsymbol{T}_\varepsilon\cos\theta_T, \quad \text{at the interface } (x = 0). \tag{1.51}$$

Using equation (1.27) to obtain the equalisation of pressure on both sides of the interface (and again taking into account the direction of the reflected wave), gives

$$Z_1(\varepsilon_I - \varepsilon_R) = Z_2\varepsilon_T \Rightarrow Z_1(1 - \boldsymbol{R}_\varepsilon) = Z_2\boldsymbol{T}_\varepsilon \tag{1.52}$$

where Z_1 and Z_2 are the acoustic impedances in media 1 and 2 respectively. The law of reflection implies that the angle of incidence will equal the angle of reflection ($\theta_I = \theta_R$), and this must be incorporated when equations (1.51) and (1.52) are combined to give

$$\boldsymbol{T}_\varepsilon = \frac{2Z_1\cos\theta_I}{(Z_1\cos\theta_T + Z_2\cos\theta_I)} \tag{1.53}$$

and

$$\boldsymbol{R}_\varepsilon = \frac{(Z_1\cos\theta_T - Z_2\cos\theta_I)}{(Z_1\cos\theta_T + Z_2\cos\theta_I)} \tag{1.54}$$

the transmission and reflection displacement amplitude coefficients.

At normal incidence (i.e. $\theta_I = \theta_R = \theta_T = 0$), comparison of (1.48) with (1.53), and of (1.49) with (1.54), shows that

$$\boldsymbol{R} = -\boldsymbol{R}_\varepsilon \tag{1.55}$$

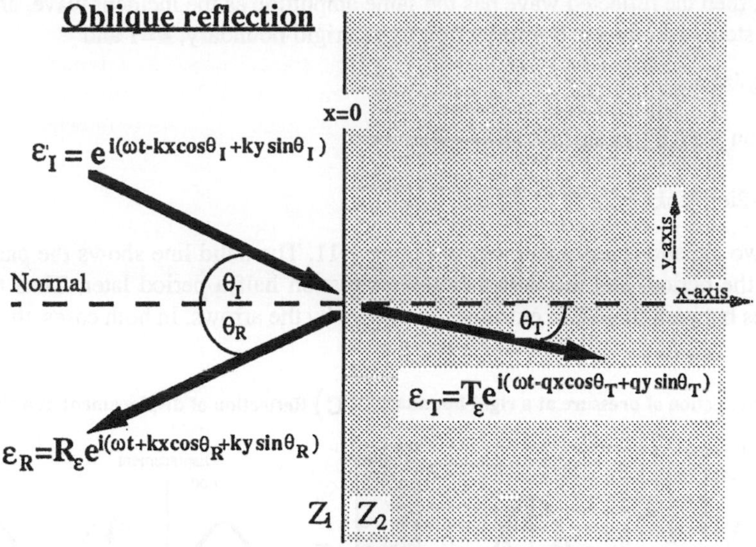

Figure 1.10 A wave of displacement is incident obliquely on a plane boundary and is part reflected and part transmitted.

In summary, normal reflection at a plane rigid boundary causes a reversal of the particle velocity and of the wave velocity, and of the particle displacement. The displacement at the boundary is always zero. The pressure amplitude is a maximum at the boundary.

Reflection from a free interface causes a reversal in normal wave velocity, and in pressure (i.e. a compression is reflected as a rarefaction). The particle velocity and displacements are unchanged, and the pressure at the interface is always zero.

1.1.6 Standing Waves

The waves considered so far are of a familiar type, in that they transmit energy from one position to another, and are therefore called 'travelling' or 'progressive' waves. However, if two identical travelling waves, travelling in opposite directions, are superimposed there is clearly no net flow of energy in any direction. Such a field is then called 'standing wave'.

The preceding section demonstrates how acoustic waves can be partially or wholly reflected at the interface between two media of differing acoustic impedance. The reflected wave can interfere with the incident wave, such that the pressure amplitude (for normalised incident wave) varies as

$$P = P_I + P_R = e^{i(\omega t - kx)} + \boldsymbol{R}e^{i(\omega t + kx)} \tag{1.56}$$

Rearrangement gives

$$2P = (1+\boldsymbol{R})\,(e^{i(\omega t - kx)} + e^{i(\omega t + kx)}) + (1-\boldsymbol{R})\,(e^{i(\omega t - kx)} - e^{i(\omega t + kx)}) \tag{1.57}$$

$$2P = e^{i\omega t}\,(1+\boldsymbol{R})\,2\cos kx + (1-\boldsymbol{R})(-2i\,\sin kx)e^{i\omega t} \tag{1.58}$$

If $|\boldsymbol{R}|=1$, then the reflected wave has the same amplitude as the incident wave, and a standing-wave system is produced. For reflection from a rigid boundary, $\boldsymbol{R}=1$ and

$$P = 2e^{i\omega t}\cos kx \tag{1.59}$$

Reflection from a free surface ($\boldsymbol{R} = -1$) gives

$$P = -2ie^{i\omega t}\sin kx = 2e^{i(\omega t - \pi/2)}\sin kx \tag{1.60}$$

These two situations are illustrated in Figure 1.11. The solid line shows the parameter at the start of the period, and the dotted line the situation half a period later. Thus the parameter oscillates between these two extremes, as shown by the arrows. In both cases, there are a series

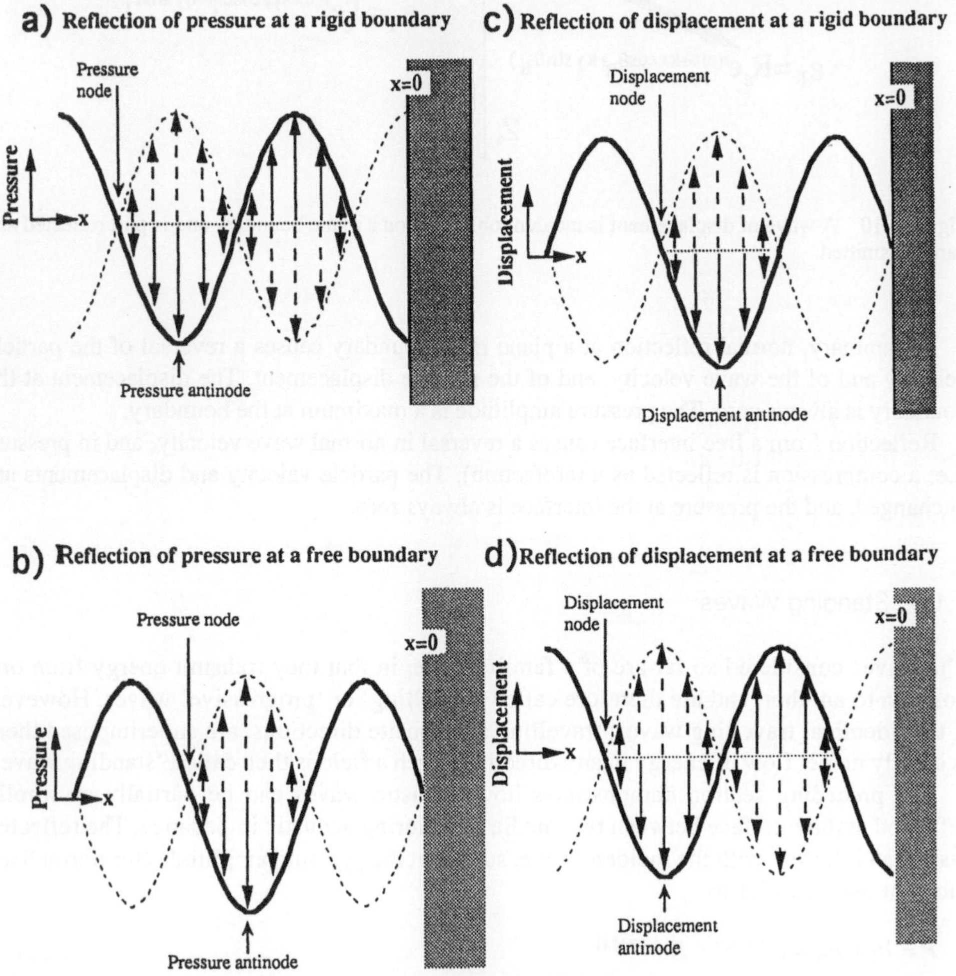

Figure 1.11 A standing-wave system is formed by: (a) reflection of a pressure wave from a rigid boundary; (b) reflection of a pressure wave from a free boundary; (c) reflection of a displacement wave from a rigid boundary; (d) reflection of a displacement wave from a free boundary.

of nodes where the parameter does not vary in time, and a series of antinodes where the temporal variation is a maximum. The spacing between a node and an adjacent antinode is $\lambda/4$. There is a pressure antinode at the rigid boundary (Figure 1.11(a)), and a pressure node at the free surface (Figure 1.11(b)). The plots for displacement amplitude are shown for rigid and free surfaces in Figures 1.11(c) and 1.11(d) respectively. Note that for a given reflector, displacement nodes occur at pressure antinodes, and vice versa.

If $\boldsymbol{R} \neq \pm 1$, then partial reflection occurs. Phase changes may occur upon reflection, so that the reflection coefficient may be written $\boldsymbol{R} = |\boldsymbol{R}|e^{i\vartheta}$. The pressure in the medium is then given by

$$P = P_I + P_R = e^{i(\omega t - kx)} + |\boldsymbol{R}|e^{i(\omega t + kx + \vartheta)} = e^{i(\omega t + \vartheta/2)}\left(e^{-i(kx + \vartheta/2)} + |\boldsymbol{R}|e^{i(kx + \vartheta/2)}\right) \tag{1.61}$$

so that the square of the pressure amplitude, which indicates the intensity, will take the form

$$|P|^2 = (1 + |\boldsymbol{R}|)^2\cos^2(kx + \vartheta/2) + (1 - |\boldsymbol{R}|)^2\sin^2(kx + \vartheta/2) \tag{1.62}$$

Equation (1.62) shows that there are pressure antinodes of amplitude $(1 - |\boldsymbol{R}|)$ when $(kx + \vartheta/2) = \boldsymbol{n}\pi$, and pressure nodes of amplitude $(1 - |\boldsymbol{R}|)$ when $(kx + \vartheta/2) = (\boldsymbol{n} + 1/2)\pi$, where $\boldsymbol{n}$ is a positive or negative integer. The magnitude of the phase change that occurs upon reflection will determine the position of the nodes and antinodes, so that if the incident wave is travelling in $x < 0$ towards the boundary at $x = 0$, the position of the first pressure node will be at x_0 where $2kx_0 = -(\pi + \vartheta)$. If the magnitude of the reflection coefficient $|\boldsymbol{R}|$ is less than unity, part of the wave is transmitted, and part reflected to combine with the incident wave. The resulting field is similar to the standing-wave field, with the placement of the nodes and antinodes determined by the phase of $\boldsymbol{R}$ as outlined above, but the amplitude at the nodes does not fall to zero. The field might be thought of as being part standing-wave, and part progressive wave, the latter component being responsible for the finite amplitudes at the nodes. The percentage of the field which is standing-wave can be calculated:

$$\frac{(P_{ant} - P_{nod})}{(P_{ant} + P_{nod})} \times 100\% \tag{1.63}$$

where P_{ant} is the peak-to-peak pressure variation at the pressure antinode, and P_{nod} the variation at the node. Close to the reflecting interface, this is numerically equal to the reflectivity. Further from the boundary, attenuation effects may reduce the reflected component.

An alternative description of the field is the standing wave ratio (SWR), which is equal to

$$\mathrm{SWR} = \frac{P_{ant}}{P_{nod}} \tag{1.64}$$

which equals $(1 + \boldsymbol{R})/(1 - \boldsymbol{R})$ in the absence of attenuation. The SWR varies between 1 (for perfect absorption at the interface) and ∞ (for perfect reflection).

The SWR will also vary throughout the medium if the medium itself absorbs the acoustic energy. This phenomenon is discussed in the next section.

1.1.7 Attenuation

As a sound wave propagates through a medium it can lose energy to the medium, a process known as attenuation.[25] A progressive pressure wave, travelling in the x-direction, might therefore be described as

$$P = P_0 e^{i(\omega t - qx)} e^{-bx} \qquad (1.65)$$

If intensity is proportional to the square of the pressure, for example in plane waves, then the intensity will decay as e^{-2bx}. Values of b, the amplitude attenuation constant, are given for various materials in Table 1.1. Having dimensions of $[\text{length}]^{-1}$, it is usually measured in nepers per centimetre (Np cm^{-1}). The phenomenon may also be quantified by an attenuation coefficient, measured in decibels per centimetre (dB cm^{-1}), which is numerically equal to $20b(\log_{10}e)$. It should be noted that the penetration of sound in water is much greater than that in air.

In seawater the acoustic amplitude attenuation coefficient takes values of $b \approx 6 \times 10^{-5}$ Np/m at 10 kHz, and $b \approx 0.03$ Np/m at 1 MHz; and it is interesting to compare the decay of these acoustic waves with the decay of radio waves in seawater. The amplitude of the electric field of electromagnetic waves decays[26] approximately as $e^{-0.0016 \times \sqrt{\omega}}$. Consider, for example, the distance over which the wave can travel before it is attenuated to such an extent that its energy is only 1% of the initial value. For a 10 kHz sound wave, that distance is 38 km; whilst for a 10 kHz radio wave, it is only 5.8 m. The corresponding distance for a 1 MHz acoustic wave is 77 m. The intensity of BBC Radio 3 FM in the UK, which has a frequency of 91 MHz, will decay exponentially such that the energy is 1% of its initial value after only 6 cm.

Attenuation in a standing-wave system would mean that, further from the reflector, the incident wave would be larger and the reflected wave smaller, since the latter has travelled a greater distance through the attenuating medium. Thus the proportion of standing-wave in the field will decrease with increasing distance from the reflector.

There are several mechanisms through which the energy can be lost from the wave. The first is the transformation of the mechanical energy into heat as the wave does work against the viscous forces which oppose internal motion in the medium. Secondly, as described earlier, in section 1.1.1(b)(ii), heat flow can occur between the compressions and rarefactions in the sound wave, causing an increase in entropy and therefore a dissipation in energy. This dissipation will not occur only if conditions are purely adiabatic or purely isothermal. Dissipation can also arise as heat is radiated from compressions to rarefactions, and through intermolecular energy exchange [21]. If the medium is non-homogeneous, several other mechanisms can also occur, such as energy loss from the beam due to scattering of the sound, and to frictional heating as a result of relative motions between species in the medium. In addition, relaxation and hysteresis processes may be involved [22]. With increasing frequency, both the timescales and the pressure-rarefaction distances become smaller. From the above description of the dissipative mechanisms it is therefore not surprising that attenuation is strongly frequency-dependent. Higher frequencies tend to be attenuated to a greater extent than lower ones, in air and water varying approximately with the square of the frequency for up to 1 MHz. Attenuation in the MHz range is discussed in the literature [23, 24].

---

[25] The attenuation of an acoustic wave as it propagates through a medium should not be confused with the damping of an oscillator, as discussed in Chapter 3, section 3.1.3

[26] If the electrical conductivity of a medium is high, the amplitude of the electrical field decays as $\exp(-\sqrt{\mu_0 \omega \sigma_0 / 2}\,.x)$ where $\mu_0 = 4\pi \times 10^{-7}$ is the permeability of free space, and σ_0 is the electrical conductivity (which for seawater takes the value of 4 $\text{ohm}^{-1}\text{metre}^{-1}$) [20].

In general, if an oscillatory process is lossy, there will be a time lag or phase difference between the driving force and the velocity response. This manifests itself as a complex acoustic impedance, as is clear from equation (1.27). Similarly, equation (1.29) implies that a complex impedance yields a complex wavenumber, that is $k = q + ib$. If this is the case, then the wave takes the form $e^{ikx} = e^{iqx}e^{-bx}$, which is the reason for the structure of equation (1.65).

The work done is the product of force and displacement, in general $\int F(x)dx$. If there is a phase difference between the force and the response, the product of their real parts gives the work. The rate of doing work, the *power*, is therefore given by

$$\text{Power} = \frac{1}{\tau_v}\int_0^{\tau_v} \mathrm{Re}\{F\}\mathrm{Re}\{\dot{\varepsilon}\}dt \tag{1.66}$$

Division of the power by the area gives the intensity. Therefore since the pressure is $P = F/A$, assuming $\mathrm{Re}\{P\}$ and $\mathrm{Re}\{\dot{\varepsilon}\}$ vary sinusoidally, the intensity of this attenuated progressive plane wave will be

$$I = \frac{1}{\tau_v}\int_0^{\tau_v} \mathrm{Re}\{P\}\mathrm{Re}\{\dot{\varepsilon}\}dt$$

$$= \frac{1}{\tau_v}\int_0^{\tau_v} \tfrac{1}{2}(P+P^*)\tfrac{1}{2}(\dot{\varepsilon}+\dot{\varepsilon}^*)\,dt$$

$$= \pm\tfrac{1}{4}\frac{1}{\tau_v}\int_0^{\tau_v} (Z\dot{\varepsilon} + \{Z\dot{\varepsilon}\}^*)(\dot{\varepsilon}+\dot{\varepsilon}^*)\,dt$$

$$= \pm\tfrac{1}{4}(Z+Z^*)\dot{\varepsilon}\dot{\varepsilon}^*$$

$$= \pm\tfrac{1}{2}\dot{\varepsilon}\dot{\varepsilon}^*\mathrm{Re}\{Z\} \tag{1.67}$$

where P^*, $\dot{\varepsilon}^*$ and Z^* are the complex conjugates of P, $\dot{\varepsilon}$ and Z respectively, such that $\mathrm{Re}\{P\} = (P + P^*)/2$ etc., as discussed in section 1.1.1(c).

As discussed earlier in section 1.1.3, the mean energy density will be twice the mean kinetic energy density, and since the intensity equals the mean energy density multiplied by the velocity (section 1.1.3), then

$$I = \pm 2.\tfrac{1}{8}\rho c\frac{1}{\tau_v}\int_0^{\tau_v} (\dot{\varepsilon}+\dot{\varepsilon}^*)(\dot{\varepsilon}+\dot{\varepsilon}^*)\,dt = \pm\tfrac{1}{2}\rho c\dot{\varepsilon}\dot{\varepsilon}^* \tag{1.68}$$

Comparing equations (1.67) and (1.68) reveals

$$\mathrm{Re}\{Z\} = \rho c \tag{1.69}$$

in agreement with equation (1.30), which assumed conditions of no attenuation.

1.2 Practical Ultrasonic Fields

What follows is an introduction to some aspects of real acoustic fields, in order to prepare those unfamiliar with the subject for the discussion of acoustic cavitation. Readers requiring more detailed discussions of sound fields and apparatus are encouraged to consult specialised texts (for example, Wells [23] and Hill [25] for biomedical, and Urick [26] for sonar, aspects of the field).

1.2.1 Characteristics

(a) Near- and Far-field

As stated in section 1.1.1(a), any sound source can be subdivided into a collection of point sources, each of which can be considered to radiate spherical acoustic waves independently. By summing the wavelets generated from these point sources, the resulting acoustic field can be calculated. In general, at a point M very far from any extended source, the difference in the distances travelled by acoustic wavelets from two points S_1 and S_2 on that source (that is, S_2N_1) can be considered negligible compared with the wavelength (Figure 1.12). The phase difference $2\pi S_2N_1/\lambda$ becomes vanishingly small as S_2M_1 tends to infinity, since the angle θ then becomes vanishingly small. Therefore the wavelets add at the point M_1 with the phase relation that their respective point sources exhibited when they were emitted. However, if the point of interest is near to the source (M_2), an additional phase difference has set in as a result of the path difference S_2N_2, the extra distance travelled by the wavelets from source S_2. In summing the wavelets from all the point sources, such phase differences must be included in the calculation. Since the value of the path difference is dependent on the exact location of point M_2, then the sound field close to any extended sound source will tend to be complicated, and simpler further out. These regions are called the near-field and the far-field respectively. In practice, the extent of each region depends not just on the physical dimensions of the sound source, but also on the magnitude of the wavelength of sound employed. This is because the phase difference that results from a path difference is proportional to the ratio of that path difference to the wavelength, $2\pi S_2N_1/\lambda$ in the preceding example. The smaller the wavelength, the greater the distance over which the near-field region extends. As shown in Figure 1.12(b), if one were taking measurements on the axis of a transducer the division between near and far-field might be expected to occur at a distance L_s^2/λ from the face of a plane transducer. If, for example, the source S_2 were at the centre of a cylindrical disc transducer, and S_1 were a point on the rim, and all points on the disc emit in phase, then the calculation shown in Figure 1.12(b) gives the maximum phase difference that would be found on axis. The field on axis at points closer to the transducer than L_s^2/λ would experience phase differences of greater than one wavelength between centre and rim, and a small change in position can bring about a large change in the phase difference, leading to large changes in acoustic pressure amplitude. The boundary between near- and far-field for such a transducer occurs at a distance L_s^2/λ from the transducer.

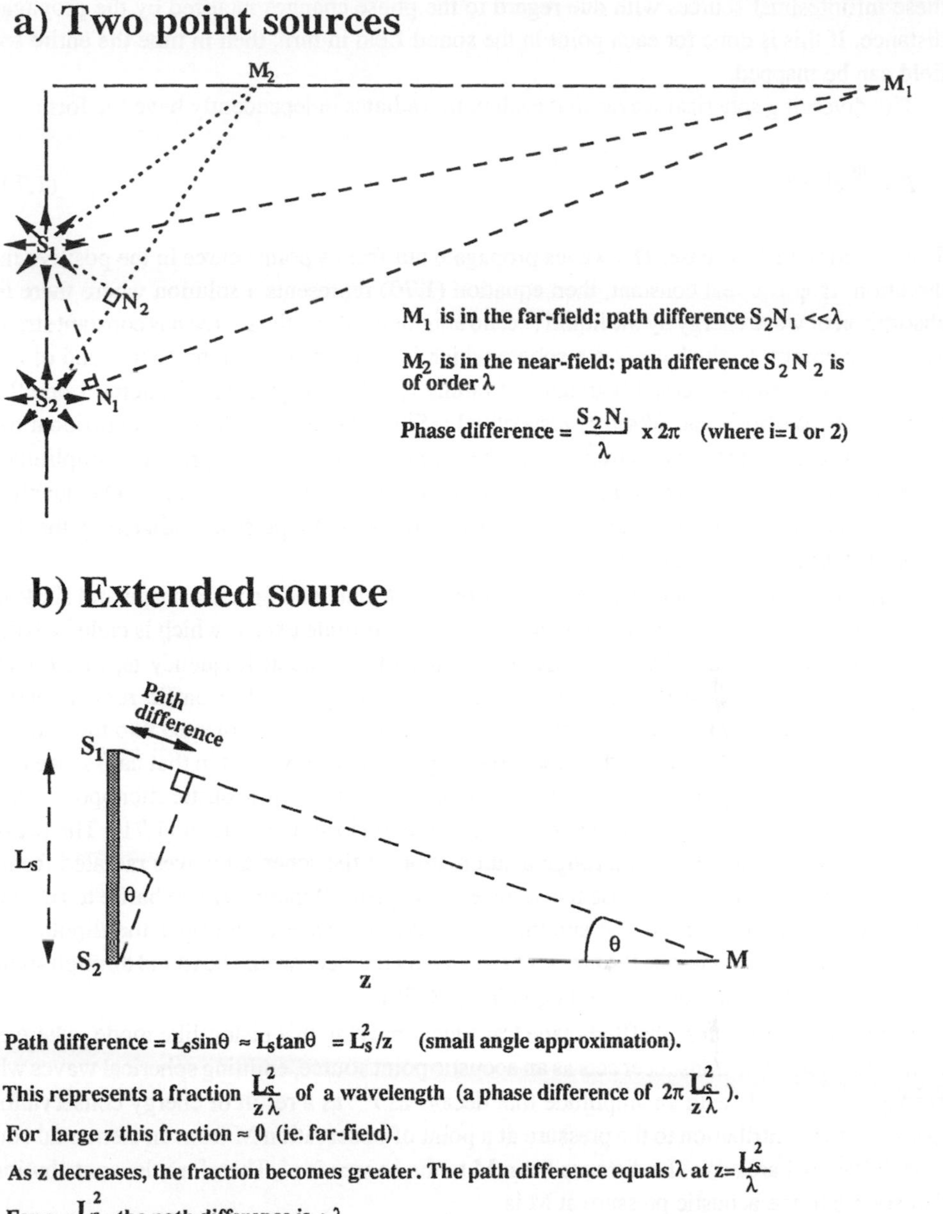

Path difference = $L_s \sin\theta \approx L_s \tan\theta = L_s^2/z$ (small angle approximation).

This represents a fraction $\frac{L_s^2}{z\lambda}$ of a wavelength (a phase difference of $2\pi \frac{L_s^2}{z\lambda}$).

For large z this fraction ≈ 0 (ie. far-field).

As z decreases, the fraction becomes greater. The path difference equals λ at $z = \frac{L_s^2}{\lambda}$

For $z < \frac{L_s^2}{\lambda}$, the path difference is $>\lambda$.

Figure 1.12 Phase difference resulting from path difference between point of observation and sources of sound. (a) Two point sources. (b) An extended source, where the maximum phase difference encountered at M is between the two extreme point sources S_1 and S_2.

It is not difficult to extend the ideas illustrated in Figure 1.12 to construct the pressure field radiated by some extended source, of finite size. The latter is considered to be made up of an infinite number of point sources, spread out over its surface. The sources radiate independently. The pressure at a given point in the sound field is found by adding the pressure from each of these infinitesimal sources with due regard to the phase changes incurred by the propagation distance. If this is done for each point in the sound field in turn, then in time the entire sound field can be mapped.

The diverging spherical waves that each point radiates independently have the form

$$P = \frac{\Psi}{r} e^{i(\omega t - kr)} \tag{1.70}$$

This equation makes sense. The waves propagate out from a point source in the positive radial direction. If ψ is a real constant, then equation (1.70) represents a solution where there is no absorption of wave energy by the liquid (i.e. no attenuation) and the source has constant strength. The same energy ϕ which crosses a sphere which is equicentric with the source and of radius r_1, crosses a similar spherical boundary of radius r_2. The energy flux at spheres 1 and 2 will therefore be $\phi/(4\pi r_1^2)$ and $\phi/(4\pi r_2^2)$ respectively. Since the energy flux is proportional to the square of the acoustic pressure,[27] then the ratio of the acoustic pressure amplitudes is $P_A(r_1)/P_A(r_2) = r_2/r_1$. Thus the acoustic pressure amplitude will decay as r^{-1}. The quantity ψ, which has units of [Pa.m], is numerically equal to the acoustic pressure radiated by the source a unit distance from that source.

Consider a source which is a two-dimensional flat plate of a specific shape and finite size, mounted on or very close to a rigid plane boundary of infinite extent, which is called a baffle[28] [27]. The plate oscillates harmonically in a piston-like mode at frequency ω, in a direction normal to the baffle, all points on the plate moving in phase. The baffle reflects acoustic emissions radiated by the plate such that all the acoustic energy is projected into the half-space in front of the plate. Because of this, the acoustic pressure at any point in that half-space can be found by considering the surface of the plate to be made up of acoustic monopole sources, radiating spherical diverging waves in the manner described by equation (1.71). The pressure in the half-space can be found through a summation of the spherical waves radiated from the monopole sources which compose the surface of the plate. If there were no baffle to reflect the radiation into the half-space, the summation would also have to consider the dipole source component to the surface, and the pressure radiated by the plate would be found through solution of the Kirchhoff–Helmholtz integral equation [28, 29].

Figure 1.13 illustrates a baffled plane transducer moving in a piston-like mode, where each elemental area of the transducer acts as an acoustic point source, emitting spherical waves which propagate outwards with an amplitude that decays as r^{-1} as a result of energy conservation.[29] Consider the contribution to the pressure at a point of observation M from an elemental source of infinitesimal area dS. The distance from M to the source is r'. Therefore the contribution of that source to the acoustic pressure at M is

$$dP \propto \frac{e^{i(\omega t - kr')}}{r'} \, dS \tag{1.71}$$

[27] This has been shown for plane waves in section 1.1.6. It will be shown to be true also for these spherical diverging waves in Chapter 3, section 3.2.1(c)(ii).

[28] See Chapter 3, sections 3.3.2(a) and 3.3.2(b).

[29] This is true for point sources only. A source of finite size, such as a bubble, might be considered as a collection of point sources, so that the system of spheres over which one would measure energy conservation are not concentric. Therefore, the radiation from sources of finite size approximates to the r^{-1} law of decay only in the far-field, when the separation of the multiple point sources appears negligible to the observer.

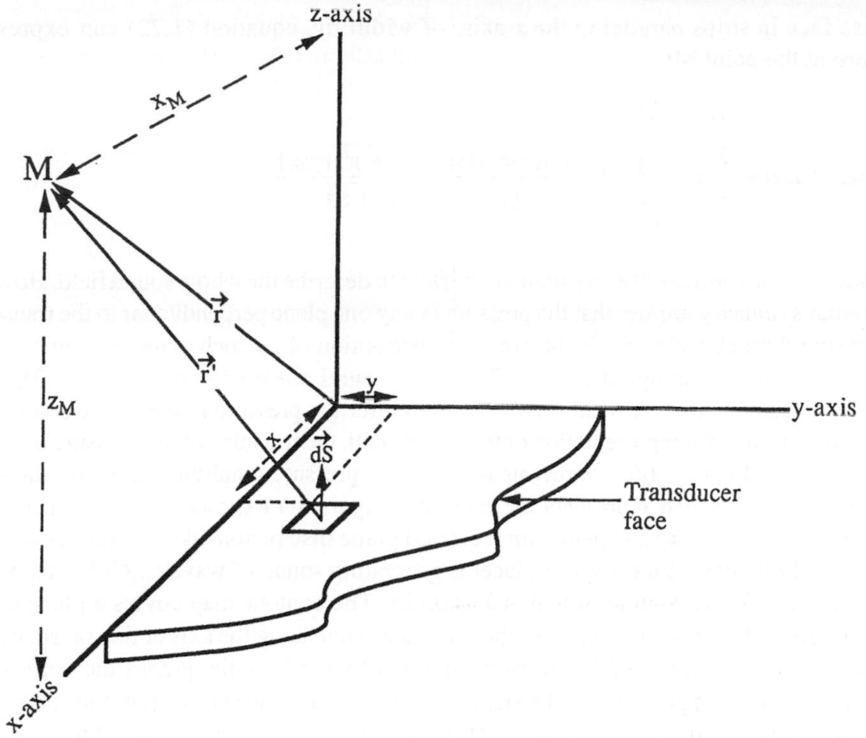

Figure 1.13 Geometry for the calculation of the pressure field at a point M in the x–z plane, as radiated from a baffled plane piston-like source, the front of which lies at equilibrium in the x–y plane.

the constant of proportionality reflecting the ratio of the source strength ψ to the area of the transducer. Thus the total amplitude at M is simply the sum of all these point sources

$$P \propto \oint \frac{e^{i(\omega t - kr')}}{r'} dS \tag{1.72}$$

where the integral is taken over the whole surface of the transducer. To solve this we simply have to incorporate into the integral that geometry which is appropriate to the transducer in question. As shown in Figure 1.13, the transducer lies in the x–y plane, centred on the origin, the z-axis extending out into the fluid. We will find the pressure at a point of observation M in the positive x–z plane, at position vector $(x_M, 0, z_M)$ relative to the origin. The sound is emitted from infinitesimal regions of the transducer at position vector $(x, y, 0)$. These regions have area $dx.dy$, the source strength obviously being proportional to the area since the infinitesimal monopole sources are considered to be distributed continuously and evenly.

The vector $\vec{r}'$ is therefore simply the difference between the position vectors of source and observation point. Thus $\vec{r}' = (x_M - x, -y, z_M)$, the magnitude being $\sqrt{(x_M - x)^2 + y^2 + z_M^2}$. These can be substituted into equation (1.72) and the integration performed over the face of the transducer.

Consider firstly a disc transducer of radius L_s. For this there is the additional constraint $x^2 + y^2 \leqslant L_s$, confining the source points to the disc surface. If the integration is done over the disc face in strips parallel to the x-axis, of width dy, equation (1.72) can express the pressure at the point M:

$$P(x_M, 0, z_M) \propto \int_{-L_s}^{L_s} \mathrm{d}y \int_{-\sqrt{L_s^2-y^2}}^{+\sqrt{L_s^2-y^2}} \mathrm{d}x \frac{\exp\{-ik\sqrt{(x_M-x)^2+y^2+z_M^2}\}}{\sqrt{(x_M-x)^2+y^2+z_M^2}} \tag{1.73}$$

This integral must be done at every point in the fluid to describe the whole sound field. However, the circular symmetry implies that the pressure in any one plane perpendicular to the transducer and passing through the origin is the pressure distribution of all such planes perpendicular to the disc and passing through the origin. Thus the integral rendered in equation (1.73), which gives the pressure in the x–z plane, can be used to infer the pressure in the whole sound field. To obtain a meaningful representation of the sound field, the modulus of the pressure $|P|$, which as is clear from Figure 1.6(c) represents the acoustic pressure amplitude, must be shown; or alternatively $|P|^2$, which represents the intensity. Figure 1.14 shows a contour map of the pressure amplitude in the x–z plane, for a baffled plane disc piston-like transducer of radius $L_s = 1$ unit. In Figure 1.14(a) the transducer is generating sound of wavelength $\lambda = 0.250$, and in Figure 1.14(b) it is emitting sound at $\lambda = 0.125$. The contour map covers a plane section extending from the origin (the centre of the transducer) out along the axis of the transducer (the z-axis) a distance 8 units, and in the x-direction out to $x = 2$. In the graphs the vertical axis represents the acoustic pressure amplitude, so that the height of the peaks represents the acoustic pressure amplitude experienced at the point in the x–z plane vertically below that peak. Plots (i) and (ii) show the map from two different viewpoints. Since the calculation is done over a finite number of grid elements, the map is not a perfect representation of the acoustic pressure amplitude: for example, in Figures 1.14(a)(ii) and 1.14(b)(ii) the dips on axis in the near-field should go to zero, but they do not quite do so in the plot since the nearest grid point at which the calculation was undertaken did not coincide with the coordinates where the function actually equals zero. Nevertheless some trends are immediately apparent. Close to the transducer the near-field region is complicated, as we would expect, with large variations in acoustic pressure amplitude accompanying small changes of the point of observation. As one moves outwards along the z-axis, the near-field region ends at some broad maximum. Beyond this, in the far-field, the variation of acoustic pressure amplitude with position is more gentle. The transition from near- to far-field can be seen to occur at $z = L_s^2 / \lambda$. The smaller the wavelength and the higher the acoustic frequency, the further into the medium the near-field region extends. Off-axis in the far field, as the angular deviation from the z-axis increases, the sound pressure amplitude undergoes fairly regular oscillation. The power appears to be channelled in preferred directions. If the acoustic pressure amplitude were plotted as a function of angle the graph would have the appearance of 'lobes' along these directions. As the wavelength is reduced and the frequency increased, the width of these lobes decreases and their number increases.[30] Also, as the wavelength decreases, the width of the central beam tends to decrease. This illustrates why it is easier to produce a tight beam of high-frequency sound than it is with lower frequencies.

The same formulation can be adapted to find the pressure resulting from a rectangular transducer of sides $2L_x$ and $2L_y$. The transducer lies in the x–y plane, and the point of observation

[30]In Figure 3.11 in Chapter 3, such lobes can be seen developing for a different type of extended sound source, which instead of being a plane piston is made up simply of two adjacent point sources. The principle is the same in both cases, and a comparison is worthwhile.

in the x–z plane, as before, so that once more $\vec{r}' = (x_M - x, -y, z_M)$. Again performing the integration over elemental strips on the transducer face, the pressure at M is given by

$$P(x_M, 0, z_M) \propto \int_{-L_x}^{L_x} dx \int_{-L_y}^{L_y} dy \frac{\exp\{-ik\sqrt{(x_M - x)^2 + y^2 + z_M^2}\}}{\sqrt{(x_M - x)^2 + y^2 + z_M^2}} \tag{1.74}$$

Figure 1.15 shows contour maps of the acoustic pressure amplitude in the x–z plane for a transducer of height $2L_y = 4$ units and width $2L_x = 2$ units, for $\lambda = 0.250$ units in 1.15(a), and

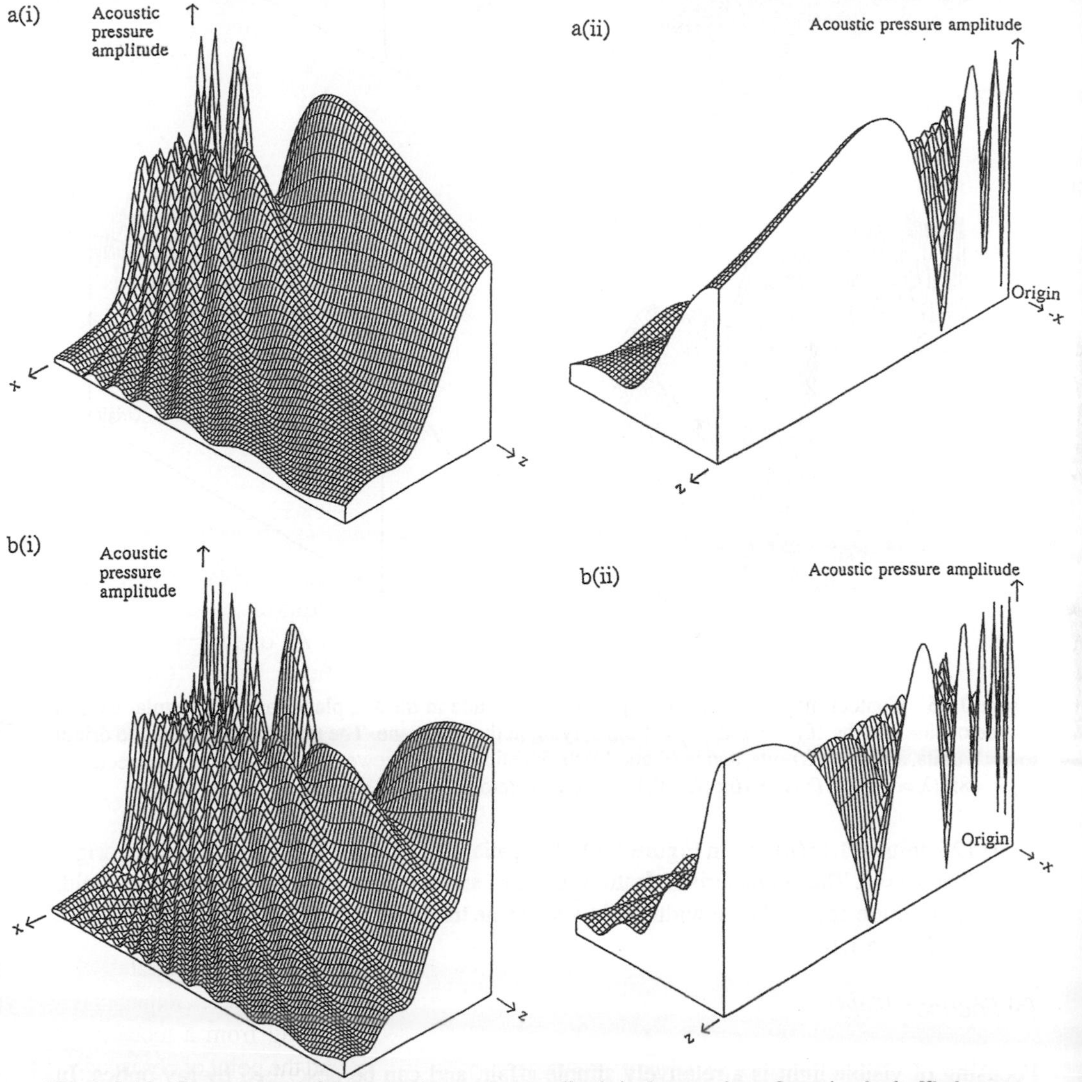

Figure 1.14 Contour map of the acoustic pressure amplitude in the x–z plane, for a circular baffled source of radius 1 unit lying in the x–y plane. The map extends from the origin to $x = 2$ units, and to $z = 8$ units, and in (i) and (ii) the identical plot is viewed from two different directions. (a) $k = 8\pi$ ($\lambda = 0.250$). (b) $k = 16\pi$ ($\lambda = 0.125$). (CWH Beton and TG Leighton.)

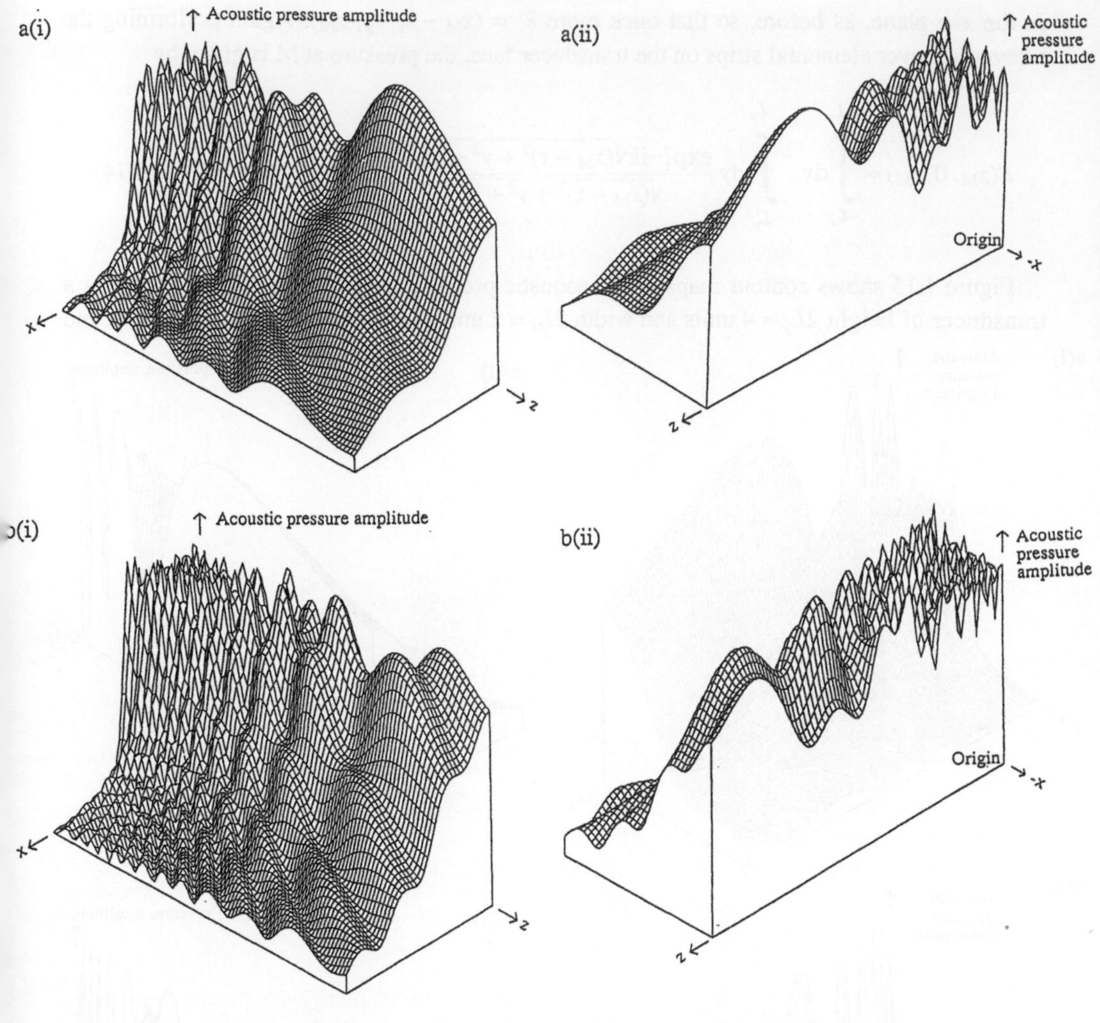

Figure 1.15 Contour map of the acoustic pressure amplitude in the x–z plane for a rectangular baffled source of side lengths $2L_x = 2$ and $2L_y = 4$ units, lying in the x–y plane. The map extends from the origin to $x = 2$ units, and to $z = 8$ units, and in (i) and (ii) the identical plot is viewed from two different directions. (a) $k = 8\pi$ ($\lambda = 0.250$). (b) $k = 16\pi$ ($\lambda = 0.125$). (CWH Beton and TG Leighton.)

$\lambda = 0.125$ units in 1.15(b). As in Figure 1.14, the spatial extension of the map is from the origin to $x = 2$ and $z = 8$. The characteristic features of near- and far-field, the extent of the near-field, the lobe structure and the beam width follow the same trends as the wavelength varies as were shown in Figure 1.14.

(b) Focused Fields

Focusing of visible light is a relatively simple affair, and can be described by ray optics. In acoustics, however, the wavelength is much larger, and so even the production of a beam of ultrasound is not simple. In all cases of focused ultrasound, the size of the focus is dependent on the ratio of the size of the transducer to the acoustic wavelength, the focus being smaller the larger the ratio. Ultrasound may be focused using curved transducers, acoustic mirrors or acoustic lenses. Where the size of the focus becomes comparable with the size of the instrument used to measure that focus, care must be taken to avoid spatial averaging (see section 1.2.2(a)(i)).

(c) Pulsed Ultrasonic Fields

Many applications of ultrasound use pulsed, rather than continuous-wave, ultrasound. The main pulsing characteristics can be illustrated using the idealised representation of a pulsed ultrasonic signal shown in Figure 1.16.

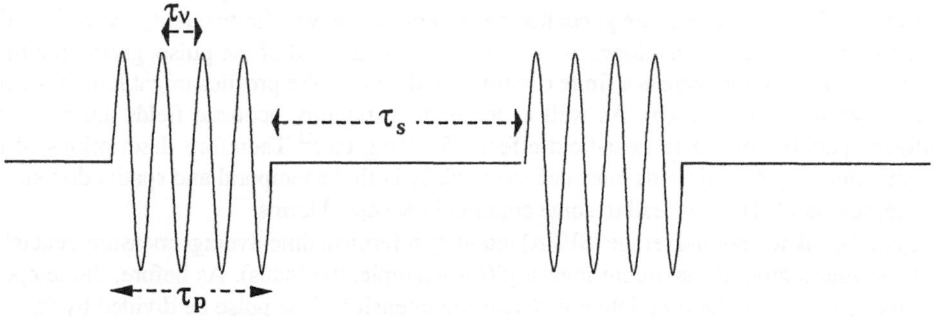

Figure 1.16 Idealised acoustic pressure pulses.

The basic ultrasonic wave has a fundamental acoustic frequency of ν, with period $\tau_\nu = 1/\nu$ as shown in Figure 1.16. The pulse persists for a time τ_p, the pulse length or on-time. The off-time, τ_s, separates the pulses, and together τ_p and τ_s define the *pulse repetition frequency* $\nu_{rep} = 1/(\tau_p + \tau_s)$. The *duty cycle* is the ratio τ_p:τ_s.

To discuss the acoustic intensity of this wave, care must be taken. For example, we might discuss the pulse average (PA) intensity, that is the acoustic intensity in one pulse averaged over the pulse length τ_p. This would be found by integrating the acoustic intensity over the pulse, then dividing by τ_p. Thus if the pulse is assumed to start at time $t = 0$, then the pulse average intensity of the idealised pulses in Figure 1.16 is

$$I_{PA} = \frac{1}{\tau_p} \int_{t=0}^{\tau_p} I(t)\mathrm{d}t \qquad (1.75)$$

Alternatively, we might discuss the *time average* (TA) intensity. Here the integral is taken from the start of one pulse to the start of the next, or for a time interval of duration $(\tau_p + \tau_s)$. Thus for the idealised pulses shown in Figure 1.16, where the moments when the pulse start and end are clearly defined, the time average intensity is

$$I_{TA} = \frac{1}{\tau_p + \tau_s} \int_{t=0}^{\tau_p+\tau_s} I(t)\mathrm{d}t \qquad (1.76)$$

However, for the pulses in Figure 1.16 it is clear that $I(t)$ is finite only for the interval $t = 0$ to $t = \tau_p$. Thus the integral in equation (1.76) is zero for $t = \tau_p$ to $t = \tau_p + \tau_s$ so that the equation reduces to

$$I_{TA} = \frac{1}{\tau_p + \tau_s} \int_{t=0}^{\tau_p} I(t)\mathrm{d}t = \left(\frac{\tau_p}{\tau_p + \tau_s} \right) I_{PA} \tag{1.77}$$

The TA intensity will be less than the PA since the average is now taken effectively over the off-time as well as the on-time. However, it must be stressed that these simple relations expressed in equations (1.75) and (1.76) hold true only for idealised pulses that start and end sharply. Real acoustic pulses are generated by systems that have a finite quality factor,[31] so that with each acoustic pulse envelope, ringing will occur at the end of the pulse, giving it a finite decay time. In addition there is a finite rise time, so the real pulse profiles might look more like that shown in Figure 1.25(b). As well as temporal variations, acoustic fields are also non-uniform spatially, owing to near-field effects, focusing etc.[32] Therefore descriptions of the acoustic intensity of real fields must define carefully both the temporal and spatial domains of the measurement. This has lead to some commonly accepted terms.

Spatial-peak temporal-average (SPTA) intensity refers to a time-average measurement taken at the spatial region of maximum intensity (for example, the focus). As before, the temporal average requires that the time-integrated acoustic intensity of the pulse be divided by $(\tau_p + \tau_s)$ or, more accurately, multiplied by the pulse repetition frequency, since τ_p and τ_s might themselves be difficult to define individually as a result of the ringing and rise-time effects. Definitions exist to enable pulses to be interpreted in a standard manner as regards pulse length etc. [30].

The spatial-peak temporal-peak (SPTP) intensity is based on the maximum intensity encountered in the field. For example, it would be calculated from the measurement of the maximum acoustic pressure during the pulse, measured at the region where the sound field is the greatest (for example, the maximum acoustic pressure for the pulse profile encountered at the focus of the transducer).

Two other common values, the spatial-average temporal-average (SATA) and the spatial-peak pulse-average (SPPA) intensities suffer from the fact that their values depend on definitions of the cross-sectional area of the beam and of the pulse length respectively, and these vary between diverse regulatory bodies. These disparities can give rise to variations as great as 40% in measurements of SATA and SPPA [31].

1.2.2 Transducers: the Generation and Detection of Underwater Sound

In principle, the generation of sound at any frequency is simply a case of producing a mechanical oscillation at the correct frequency, and of coupling that oscillation to the desired medium so that displacements are generated in that medium, leading to an acoustic wave. The poorer the coupling, the less efficient the energy transfer, and the greater the power required of the initial displacement. Consideration must be given as to where the excess energy will go if the coupling is inefficient: a physiotherapeutic piezoelectric transducer, designed to transmit sound into flesh

---

[31] See Chapter 3, section 3.4.1.
[32] See section 1.2.1.

through the use of a film of coupling gel at the interface, will become hot if driven in air, and may be damaged.

The piezoelectric nature of quartz was discovered by J. and P. Curie in the 1880s [32, 33]. There are electric charges fixed within the structure of the crystal lattice. If pressure is applied to a piezoelectric crystal, giving rise to a displacement of crystal planes, equal and opposite electric charges appear on opposing faces of that crystal, resulting in an electrical potential between them. Artificially grown crystals, such as lithium sulphate, are also piezoelectric. Other artificial piezoelectric materials are the ferroelectrics, where the effect results from the alignment of charge domains. Ceramic ferroelectrics, such as lead zirconate titanates, are commonly used for the generation of ultrasound.

Piezoelectrics can be used to generate ultrasound by exploiting the fact that the production of electric charges through mechanical stress is reversible. Thus if an electric voltage is applied to selected faces of a piezoelectric material, that material will alter shape. Faces will be displaced, and so the electrical energy will be converted into mechanical energy. An oscillating electrical signal will cause an appropriate piezoelectric material to undergo geometrical changes at that frequency. If adequately coupled to a medium, the displacements of the piezoelectric will cause acoustic waves of that frequency to propagate in the medium. The piezoelectric may have a natural frequency which for a given material is determined by its geometry. The largest amplitude sound waves are generated when the crystal is driven near to resonance, or to a harmonic.

A similar way of generating acoustic waves through the transformation of electrical energy into mechanical wave energy is through magnetostriction. The magnetostrictive effect was first observed by Joule in 1847 [34]. A bar of ferromagnetic material undergoes a length change when subjected to a magnetic field. An oscillating magnetic field can therefore be used to generate an oscillating displacement, and so an acoustic wave. In general, magnetostriction is efficient only at frequencies of below about 30 kHz, owing to losses in the core, which generate heat.

As stated, these transducers, which convert electrical to mechanical energy, are in general reversible. Thus, for example, electrical signals can be applied to piezoelectric materials in order to generate sound, whilst a similar piezoelectric material could be subjected to acoustic waves and the resulting electrical output used to monitor the pressure changes. In biomedical acoustics, the term *transducer* is often restricted to devices which generate sound, whilst detector devices are specifically named (e.g. hydrophones, force balances etc.). This convention does not apply to other branches of underwater acoustics, the term *projector* being employed to specify sound sources in sonar technology. It should be noted that a single transducer is in some circumstances employed as both acoustic source and detector.

(a) Measurement of Underwater Sound

Presented here is a brief outline of equipment, and some comments on the practical aspects of measurement. Fuller discussions of equipment can be found in the references cited earlier [23, 26].

(i) The Hydrophone. The hydrophone takes a real-time[33] measurement of the acoustic pressure against time, for some small region of the sound field (generally much smaller than

[33] That is, the hydrophone continuously monitors the acoustic pressure, any variations in that input showing in the output almost instantaneously.

the acoustic wavelength). Such outputs can be seen in Figures 1.24 and 1.25 for continuous-wave and pulsed ultrasound respectively. Ideally, the pressure data is linearly transformed into a time-varying voltage; if the conversion factor (voltage/pressure) is known, then the hydrophone is calibrated.[34] The voltage output from the hydrophone can then be analysed by other equipment to give information on acoustic pressures, frequency components, pulse shapes etc. Hydrophones can be used to take readings at several positions in the sound field to determine the spatial variation of the field. For accurate readings the hydrophone must be significantly smaller than any spatial variations in the field, caused for example by focusing of the ultrasonic beam, or near-field variations. In particular, the hydrophone should be smaller than the acoustic wavelength. Hydrophones are readily available with sensitive elements of diameter of 0.5 mm, and there are some with diameter 0.1 mm. Care must be taken when measuring high-frequency fields, as the wavelength of ultrasound in water is about 1.5, 0.15 and 0.05 mm at 1 MHz, 10 MHz and 30 MHz respectively. This is to minimise the disturbance to the measured field caused by the presence of the hydrophone itself, and also to ensure that *spatial averaging* does not occur. The ideal hydrophone would be infinitesimally small, as its output is interpreted as a point measurement. Hydrophones do, however, have a finite size, and if one were, for example, to try to measure the maximum pressure at an acoustic focus smaller than the hydrophone, then the hydrophone would actually sense not just the maximum pressure but simultaneously, over its sensitive area, experience the lesser pressures surrounding the focus (Figure 1.17). The signal is an average of the sound pressure experienced over the area of the active element. The output of the hydrophone would therefore register some intermediate pressure. The instrument could only measure the true maximum pressure at the focus if that pressure were sustained over the whole sensitive area of the hydrophone.

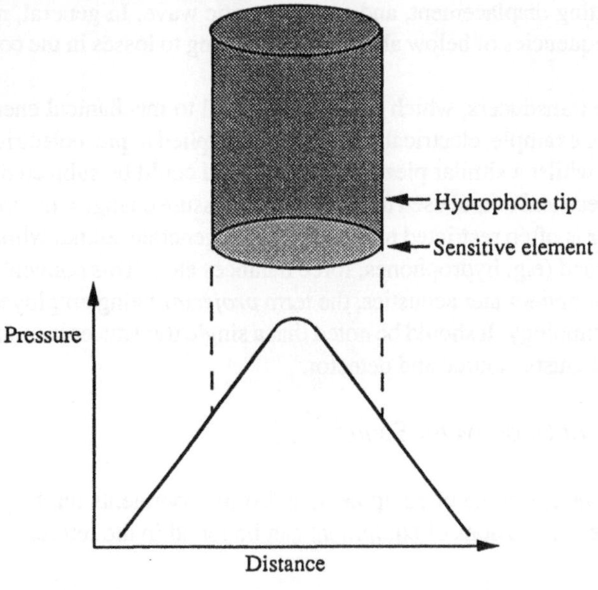

Figure 1.17 Schematic illustration of the source of spatial averaging during measurement with a hydrophone.

[34]Hydrophones are calibrated under a specific set of conditions, e.g. insonation frequency, range of pressure levels where a linear response holds, and in a medium of a certain acoustic impedance. If the hydrophone is employed outside this range of parameters, the calibration may not hold true.

Standard guidelines exist for the maximum size a hydrophone can be before spatial averaging becomes significant [30]. Depending upon the recommending body, the criteria are

$$L_h < \frac{\lambda h}{L_t} \qquad \text{(AIUM}^{35}\text{/NEMA}^{36}\text{)} \qquad (1.78)$$

or

$$L_h < \frac{\lambda h}{2L_t} \qquad \text{(IEC}^{37}\text{) if } h > L_t \qquad (1.79)$$

with

$$L_h < \frac{\lambda}{2} \qquad \text{(IEC) if } h < L_t \qquad (1.80)$$

where L_h is the diameter of the active element of the hydrophone, h is the distance from the insonating transducer to the hydrophone (focal length for AIUM/NEMA) and L_t is the dimension of the source (e.g. diameter of a disc transducer, length of a long edge for a rectangular transducer etc.). Harris [30] recommends the more conservative IEC guideline, and points out that under it spatial averaging will be a factor in many types of measurement of diagnostic ultrasound system.

In addition, hydrophones will not respond uniformly to all frequencies. Quoted calibrations will always have an associated frequency range over which they apply. A typical hydrophone for audio work may be roughly uniform between 2 Hz and 200 kHz. A hydrophone for clinical ultrasonics may typically have an approximately uniform response in the range 1–10 MHz. One should be aware that, given a particular signal, this bandwidth limitation will be relevant not just for the fundamental frequency of a signal, but for all frequency components. Thus, for example, a measured pulse shape may not accurately represent the true temporal profile owing to lower sensitivity to high-frequency components. Given sufficient information about the spatial and temporal response of a given hydrophone, it may be possible to correct for these limitations by transforming the hydrophone output: however this is often not a simple process.

The response of a hydrophone placed within a sound field is usually directional, in that it has different sensitivities to sound arriving from different directions. Calibrations should contain directionality information, and users should be careful to employ a calibration relevant to the geometry of use. The smaller the size of the hydrophone in relation to the wavelength, the less directional the instrument. As a rough guide, the response is generally adequately omnidirectional if the element dimension is less than $\sim\lambda/10$ [35]. Therefore the lower the frequency, the less directional the response, and there are commercially available audio frequency hydrophones that approach uniformity in directional sensitivity. The sensitive element in hydrophones in this audio frequency range are usually piezoceramics, impedance matched to water through a specialised rubber surround (Figure 1.18(a)).

To study the MHz regime there are two major types of probe, the membrane and the needle hydrophone. The membrane hydrophone (Figure 1.18(b)) is described by Preston *et al.* [36]. The prototype was reported in 1977 by De Reggi *et al.* [37], and further devices have been produced [38]. The sensitive element is made of polyvinylidene fluoride (PVDF), a strongly

[35] American Institute of Ultrasound in Medicine.
[36] National Electrical Manufacturers Association (USA).
[37] International Electrotechnical Commission.

a

b

c

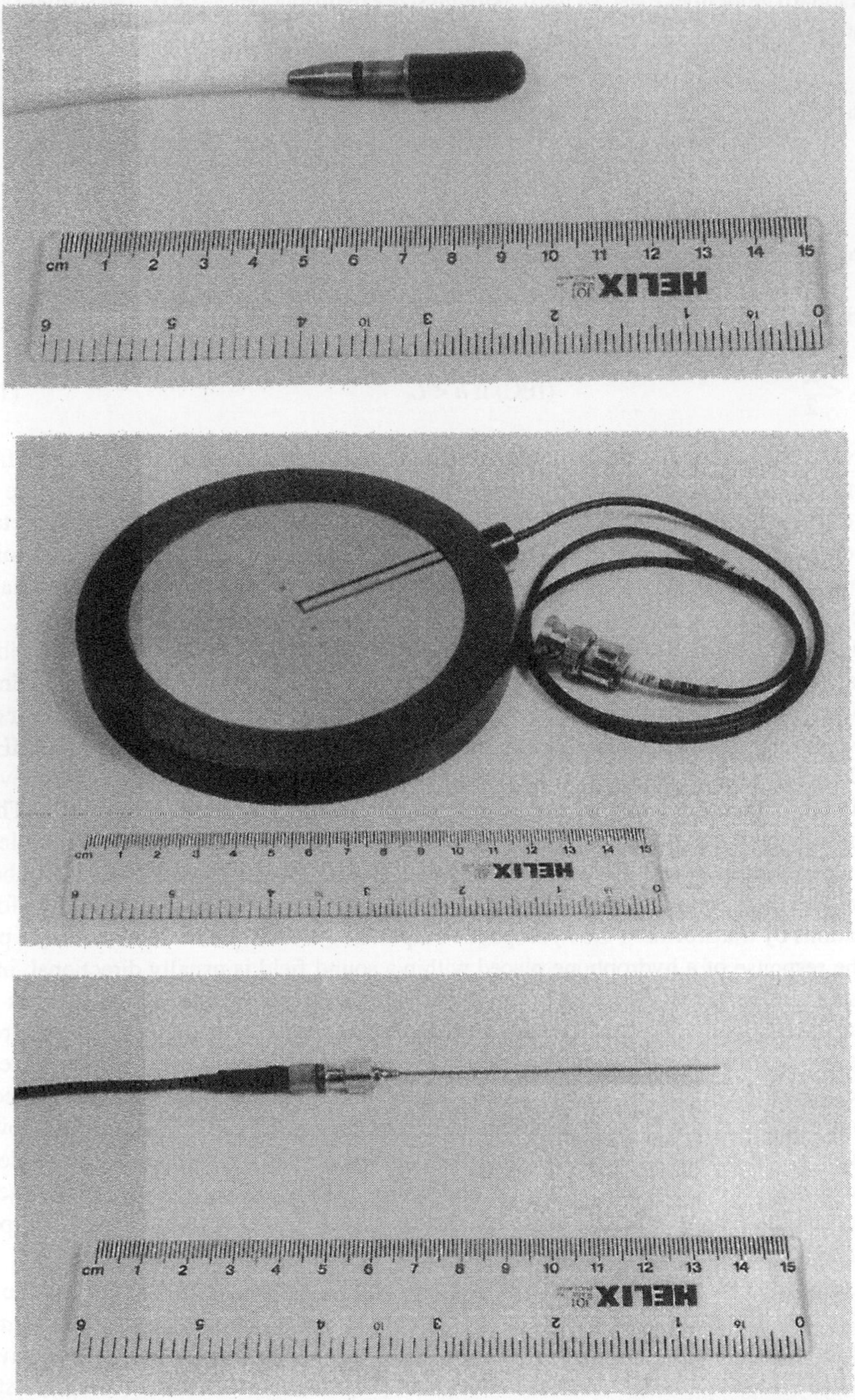

Figure 1.18 Three types of hydrophone. (a) Bruel and Kjaer type 8103 miniature hydrophone, suitable for measurements from roughly 50 Hz to 200 kHz. (b) PVDF bilaminar shielded membrane hydrophone (GEC–Marconi Research Centre, Hydrophone Type Y-34- 3598), calibrated from 1 MHz to 30 MHz. (c) Needle hydrophone (Dapco NP10-3) (for measurement in the approximate range of 1 MHz to 10 MHz).

piezoelectric polymer [39, 40]. The membrane hydrophone [41] is made from thin sheets (~5–100 μm thick) of piezoelectric foil, which unlike ceramic has an impedance well-matched with that of water (the polymer impedance is typically 4×10^6 kg m^{-2} s^{-1} [36]). A thin film of PVDF is stretched in a circular frame of about 5 cm radius. Conducting leads are evaporated onto both sides of the membrane. The central sensitive area, poled at an elevated temperature to generate piezoelectric sensitivity, is the region of overlap of these leads. In the bilaminar models, two films are stretched within the frame.

The pressure sensor on the needle probe (Figure 1.18(c)) is also commonly made of PVDF [42]. Platte [43] describes a PVDF-coated metallic 'needle', the tip of the needle being the polarised sensitive element. Alternatively, a piezoceramic can be used [44]. Needle hydrophones must be carefully designed to avoid internal oscillations, radial resonances, reflections within the casing and reverberation. Electrical shielding must also be effective. The pickup of electrical noise is often less of a problem than the confusion that can arise from the fact that, because the ultrasonic signal is generated by application of a high-voltage electrical signal to the transducer, electrical pickup by the hydrophone can strongly resemble the expected acoustic signal. A first-order check for this is to repeat the experiment in air, rather than water, when the acoustic component of the signal is eliminated. Needle hydrophones have the advantage of being admissible to confined spaces (even, for example, *in vivo* insertion through a biopsy needle), but prolonged immersion of a needle hydrophone into a liquid can lead to the absorption of liquid by the device, which can change its properties. Membrane hydrophones are often preferred as reference units because of their stability.

This book concerns the way in which gas bubbles in a liquid can interact with sound in that liquid. Because of the strength of that interaction, the measurement of that sound will be impaired by the presence of bubbles. Small bubbles may attach themselves to the hydrophones, distorting the acoustic signal they record. At higher ultrasonic intensities, the action of bubbles can be more dramatic. High number densities can scatter acoustic waves, and significantly alter the acoustic bulk properties of the medium.[38] As will be seen in later chapters, the sound can drive bubbles into oscillation. Such bubbles will emit pressure waves, in some cases very strongly. Hydrophones can be damaged by violent cavitation. Care must be taken, therefore, in both the use of hydrophones and the interpretation of their measurements if bubbles are present.

Hydrophones should be constructed to a stated polarity, usually giving an increased positive voltage in response to an increase in pressure. This is not always the case however, and the hydrophone can be checked by application of a static pressure, or by observation of nonlinear propagation, where the cycle that is shocked corresponds to compression, and if the hydrophone is polarised in the above manner the pulse as displayed should resemble that shown in Figure 1.26. If it appears inverted, the polarity is reversed somewhere in the detection system.

Hydrophone measurements give time-series pressure data, ideally at a point. Information such as the temporal peak pressure and the pulsing characteristics (e.g. duty cycle) are available, but unless a slow and laborious examination of a large portion of the sound field is made, it is difficult to obtain measurements of the total power output of a transducer. Harris [30] demonstrates how I_{SPPA} and other intensities can be calculated from hydrophone measurements for fixed beam and scanning diagnostic systems. The hydrophone can be used to find spatial peak values with relative ease, but for spatial average measurements the user must make assumptions about beam shape, pulse length etc., for which standardised interpretations (e.g. from AIUM/NEMA) can be employed.

---

[38] See Chapter 4, section 4.1.2(e).

(ii) The Force Balance. Quick, simple and direct measurements of the total acoustic power output from a transducer can be made using a force balance (also called a radiation pressure balance). The principle is to measure the radiation force (as described in section 1.1.4) through absorption or reflection. One simple design [45] reflects a vertical ultrasonic beam through 90°, the reflector being mounted on a simple beam balance, being counterweighted across a fulcrum (Figure 1.19). The radiation force causes a moment that is compensated for by adjustment of the rider, giving a measure of radiation force. The reflector is constructed from two metal sheets, sealed together, with dry paper sandwiched inside to ensure an impedance mismatch with the water. Other balances have been described by Newell [46], Kossoff [47] and Hill [48], amongst others. Wells [49] reviews these and other techniques based upon radiation pressure. Commercial portable electronic balances are available.

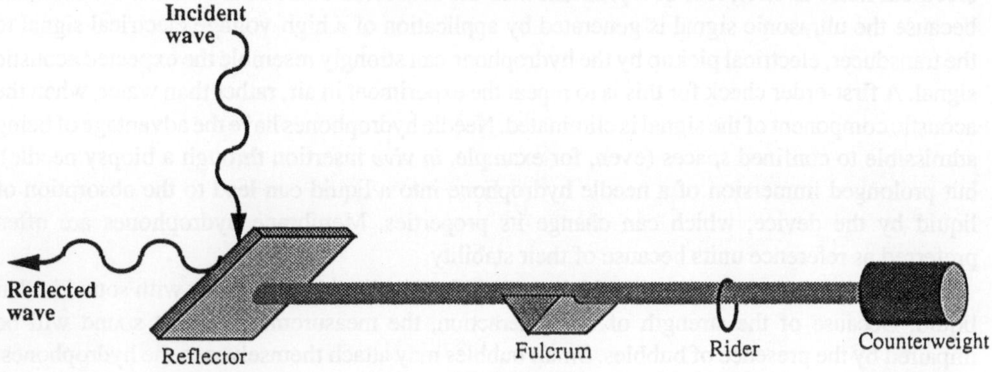

Figure 1.19 Illustration of a basic force balance (after Wells *et al.* [45]).

The radiation balance gives no information on pulse shapes, acoustic frequency, duty cycle etc. Therefore to characterise an ultrasonic field it is necessary, as recommended by Hill [48], to employ both force balance (to measure the total energy flux through a defined cross-section) and hydrophone techniques (to make 'point' measurements in the beam with temporal resolution).

Most force balances are equipped with an acoustic window, placed between the transducer and the sensor. In the absence of this window, the measured pressure increases owing to acoustic streaming.[39] Sjöberg *et al.* [50] developed a device to measure ultrasonic power through streaming, using it to levitate a plastic bead against gravity in the flow.

(iii) Other Devices. There are many other techniques, both quantitative and qualitative, which have been used to investigate underwater acoustic beams. A noteworthy device, which is becoming increasingly popular in the measurement of lithotripsy fields, is the elastic sphere radiometer [51, 52]. The acoustic radiation force on the sphere can be used to obtain spatial average and almost-point measurements, depending on the relative dimension of sphere and acoustic beam. A radiometer designed specifically to measure low output powers from therapy equipment is described by Shotton [53].

[39] See section 1.2.3(b).

Other techniques employ thermometric methods, which exploit the fact that absorption of ultrasound generates heat. This can be used to measure acoustic power, if the temperature rise can be separated from that resulting from the heating of the inefficient transducer [45]. Fry and Fry [54] developed a thermocouple probe, embedded in a known absorber, and Szilard [55] produced a thermister probe, again embedded in absorber. The changes in density as sound passes through a transparent medium lead to changes in the refractive index. These can be visualised through Schlieren photography. Other methods, including the monitoring of sonochemical effects, are reviewed by Wells [49]. All require technology and expertise. In 1984 Breyer and Devčić [56] commented that, on the whole, physiotherapists are unable to check their devices routinely beyond a simple qualitative observation of bubble activity or ripples on the water surface when the device is activated, expertise and equipment for routine quantitative checking being unavailable. They propose a semiquantitative device to monitor the output of physiotherapeutic devices using a sheath liquid crystal thermometer.

(b) The Generation of Ultrasound: Simple Focused Transducer

As with detection, the generation of ultrasound is a subject about which interested readers should consult specialist books (e.g. [23, 25, 26]). However, in this section the construction of two systems for generating underwater oscillating pressure fields is described: a simple focused acoustic transducer operating at around 10 kHz, and a vibrating device to produce a low-frequency (100 Hz) simulated sound field. The cavitation which can be seen to occur within these systems is used to illustrate several points in later chapters.

(i) A Focusing Piezoceramic Cylinder. It is a relatively simple matter to construct a piezoelectric transducer. Piezoelectric materials are readily available from commercial sources in a variety of shapes (disc, cylinder, flat plate etc.) and with a consequent wide range of natural resonances and focusing characteristics. Crystals generating either compressional or shear waves are available. However, shear waves do not propagate significantly in liquids.

Two faces of the transducer are coated with a conductive material, and application of an oscillating electric field to this will generate displacement. The electrical impedance of the crystal must be considered, and adjusted to match the output impedance of the driver.

The Piezoceramic Cylinder. Consider the cylindrical piezoceramic shown in Figure 1.20. This can readily be used to generate a simple focused underwater acoustic field. Two piezo cylinders, each about 3 cm high, can be glued to a cylindrical glass section which acts as a window for observation and illumination. The stiffer the glue, the higher the mechanical Q of the system, and the sharper the mechanical resonance.[40] The action of the sound can erode the glue bonds over a period of time. The test liquid is placed in a cylindrical cell to be insonated. The cylinder has a height of 9.3 cm, an inner diameter of 6.9 cm and an outer one of 8.3 cm. The base is a glass plate, mounted on rubber feet to improve the resonance.

The inner surfaces of the cylinder represent one electrode, the outer surfaces the other. Electrical contact to the piezoceramic is made through copper strips soldered on directly. It is very important to insulate the cylinder, contacts and leads adequately to prevent electrical shock. The outer ceramic surfaces are connected in parallel, as are the inner. This inner surface, and the liquid it contains, are earthed, for two reasons: firstly, to reduce any pickup by the hydrophones, and secondly to ensure standardised conditions (electric fields may, for example,

[40] See Chapter 3, section 3.1.

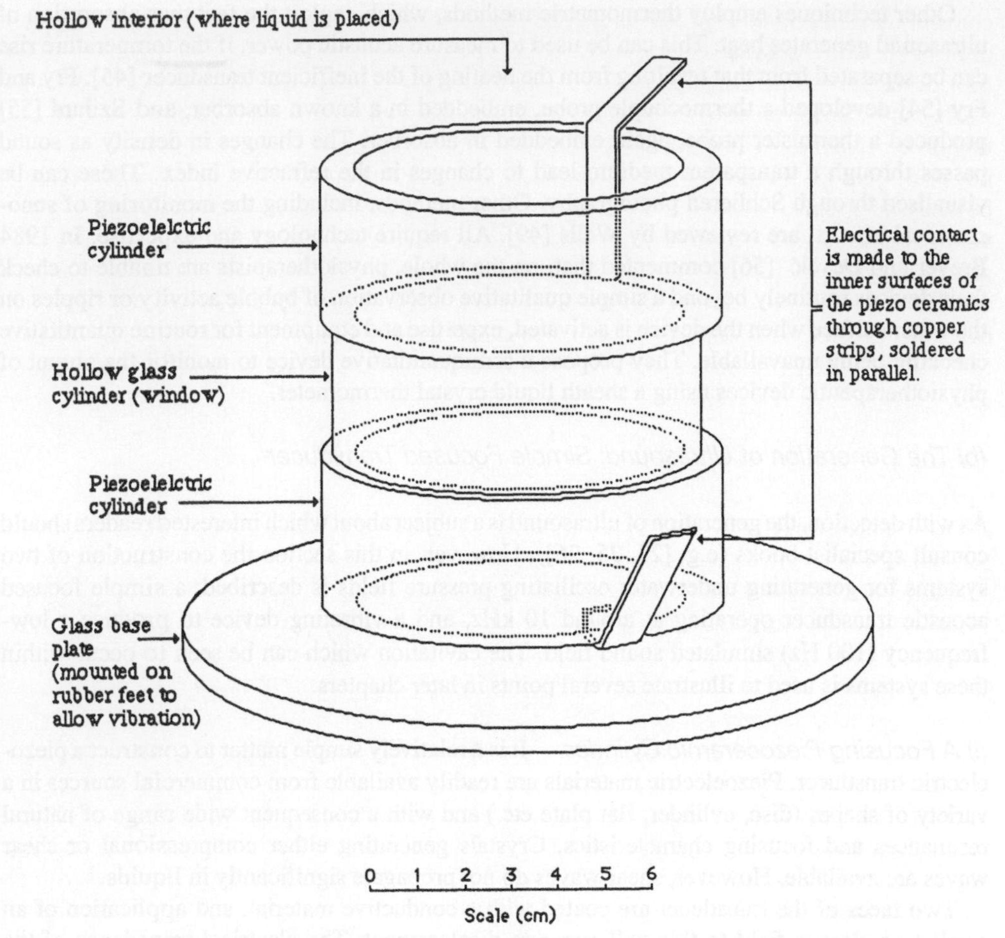

Figure 1.20 Transducer system for producing cylindrically focused sound in liquids (necessary safety precautions not illustrated for clarity – after Leighton *et al.* [57]).

affect cavitation [58, 59, 60]). This cylinder has acoustic normal modes governed by the Bessel function, and as a result the desired driving frequency ω can be chosen to obtain an axial focus.

The Electrical Signal. The cylinder described above produces sound when an appropriate electrical signal is applied across the piezoceramic. The piezoceramic cylinder is mainly capacitative in nature, of magnitude 0.025 μF. Standard amplifiers operate most efficiently into a real impedance of up to about 75 Ω (depending on the amplifier). Therefore to dissipate high power in the cylinder it is simplest to incorporate the piezo into an $L_{\text{ind}}C_{\text{cap}}$ resonance system, the value of the inductance L_{ind} being set roughly equal to $1/(\omega^2 C_{\text{cap}})$ where C_{cap} is the capacitance of the cylinder. The cylinder can be placed in series with a tuneable solenoid to allow the operating frequency to be varied. An appropriate inductor can be fashioned by winding a coil (whose dimensions can be estimated by simple calculation) onto a hollow paxolin rod. The self-inductance of this solenoid is varied by the partial insertion of a ferrite rod. The $L_{\text{ind}}C_{\text{cap}}$ system in resonance will therefore represent a real load, and so be amenable for driving by a conventional amplifier. The response of the transducer to a low-amplitude signal as the length of inserted ferrite is varied can be used to finely tune the inductor. Once tuned, the system can be used at high power. Caution must be exercised, however. When at high power, the magnetic fields can be powerful enough to draw the ferrite into the inductor, de-tuning the system and therefore potentially damaging. As a consequence, the ferrite must be locked in place when high power is used. To obtain high fields, high voltages are necessary and suitable precautions must be taken, particularly in the presence of water (which can, for example, be earthed, the oscillating voltage being applied to the insulated external surface). This system will readily generate sound pressure amplitudes in excess of 1 atmosphere, and so is suitable for studying acoustic cavitation. Whether filled with water or not, the leakage of sound to the air is considerable, and care must be taken to wear hearing protection sufficient to meet the needs, particularly if audio frequencies are employed. High-intensity sound at ultrasonic frequencies may still cause damage. All such devices should be professionally checked for electrical and acoustical safety.

(ii) The Simulated Sound Field. In this section an oscillating pressure field is produced in a liquid by vibrating it vertically. The oscillating pressure in the water mimics an acoustic field at the vibration frequency, and any pressure sensor in the liquid will respond to it as such. In particular, as shown in later chapters, a bubble in the liquid will react as though in a driving sound field. A simple picture of the process is to imagine the bubble in the vessel: when the vessel is still, the bubble 'feels' the pressure of the hydrostatic head of liquid above it. If the vessel is vibrated vertically up and down, the bubble experiences alternating high and low pressures (just as a man in an elevator, carrying a pile of books, 'feels' the books become heavier as the elevator accelerates up, and lighter as the lift drops downwards).

Consider the vertical vibration of a tube of liquid at a set frequency ω (in practice, 100 Hz) and with an amplitude of oscillation ε_0. The vertical displacement of the cell from equilibrium is therefore $\varepsilon = \varepsilon_0 e^{i\omega t}$, and the acceleration $\ddot{\varepsilon} = -\varepsilon_0 \omega^2 e^{i\omega t}$. Consider a vertical prismatic column within the liquid, with a cross-sectional area of A and a length h, which extends down from the liquid surface (Figure 1.21). Then if ρ is the liquid density, g the acceleration due to gravity, p_3 the pressure above the liquid and p the pressure at a distance h below the liquid surface, then from Newton's Second Law

$$(p - p_3)A - \rho h g A = \rho h A \ddot{\varepsilon} \tag{1.81}$$

Rearrangement of this gives

$$p = (p_3 + \rho h g) - \rho h \omega^2 \varepsilon_0 e^{i\omega t} \tag{1.82}$$

A small bubble situated at a distance h below the surface of a vertically-vibrating fluid therefore experiences the pressure p given by equation (1.82). The two terms which add to make p correspond to an oscillating pressure of amplitude ($\rho h \omega^2 \varepsilon_0$) and frequency ω, superimposed on a static pressure of ($p_3 + \rho h g$). Generally, the larger the static pressure at a bubble, the smaller the amplitude of the radial oscillations. Thus to obtain high-amplitude, and therefore nonlinear, bubble pulsations in the simulated acoustic field, the quantity ($p_3 + \rho h g$) must be minimised. This is done experimentally by reducing the pressure head p_3 above the liquid. The apparatus is shown in Figure 1.22. The test liquid is placed to a level of about 3/4. The cell, made of PMMA and aluminium, has internal dimensions of 90 mm long and 27 mm diameter, and contains windows for illumination and viewing. Piping from the top plate of the cell leads to the pumping train. The gas head above the liquid is pumped down to about 1 torr ($\approx$100 Pa). Extended springs attached to the top of the cell pull up against its weight, so that at rest the cell

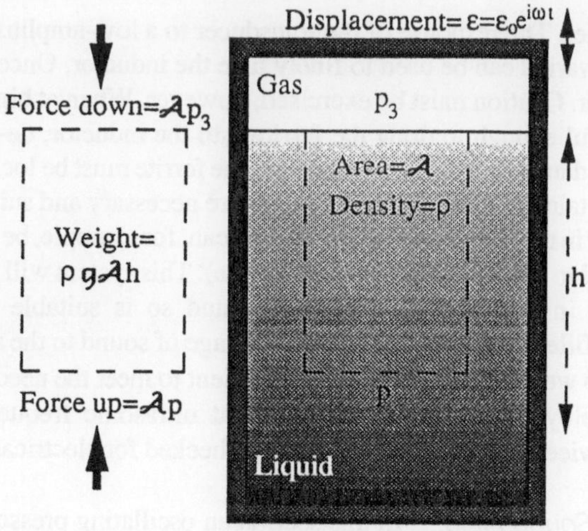

Figure 1.21 The cell in which a pressure field made up of static and oscillating components is produced at depth in a vertically vibrated liquid. (After Leighton *et al.* [61]. Reprinted by permission from *European Journal of Physics*, vol. 11, pp. 352–358; Copyright © 1990 IOP Publishing Ltd.)

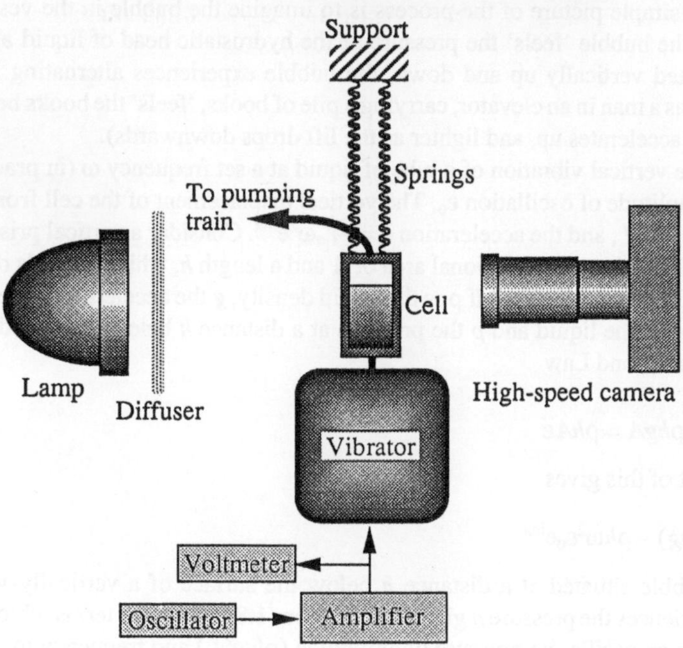

Figure 1.22 Apparatus to vibrate the cell vertically and to photograph the bubble oscillations. (After Leighton *et al.* [61]. Reprinted by permission from *European Journal of Physics*, vol. 11, pp. 352–358; Copyright © 1990 IOP Publishing Ltd.)

lies at the equilibrium position of the vibrator. Without these springs, the vibrator bottoms out at high amplitude, and the motion ceases to be sinusoidal. The sinusoidal nature of the motion can be checked by illuminating the cell with a stroboscope tuned to the vibration frequency, and then plotting the cell displacement against the phase delay of the stroboscope. The vibrator (Goodmans Industries type 390) is driven by an electrical sinusoid at around 100 Hz from a Brookdeal Signal Source (type 471), which is amplified by a Quad 510 power amplifier.

The advantage of this system is that it can produce high-amplitude bubble oscillations at low frequency, so that many of the interesting nonlinear effects described in later chapters can be observed directly. To generate these effects using a piezoelectric driver system under atmosphere pressure, such as was described earlier in section 1.2.2(b)(i), would produce hazard to hearing.

1.2.3 Nonlinear Effects in Underwater Ultrasound Beams

As sources of higher power were developed, and as our ability to analyse and process acoustic signals improved, it became clear that the description of simple linear acoustics, as outlined so far in this chapter, is inadequate to explain many phenomena associated with the propagation sound of finite amplitude. Such nonlinear effects are reviewed by Bjørnø [62] and Hamilton [63]. Though less amenable to simple explanation, these nonlinear phenomena have over the past few decades proved a valuable area of research. Returns have included the parametric array, which is a form of nonlinear sonar; the enhanced resolution of acoustic microscopes [64]; and the application of the nonlinear properties of tissue for diagnostic [65] and therapeutic [66] purposes. Though such uses have only appeared in the last few decades, it has been the subject of consideration since the eighteenth century [67]. Probably the first discussions were published in 1759 by Euler [68], who considered the nonlinear wave equation for finite amplitude sound.

(a) The Propagation of Finite Amplitude Waveforms

The linear acoustics derived so far in this chapter are based on the one-dimensional linear wave equation (equation (1.2)) and, as discussed in section 1.1.1(b), any parameter which satisfies that equation will propagate as a linear wave, a single-frequency sine wave of that parameter propagating in a medium, without change, at a fixed speed c. In Chapter 2, section 2.2 the dynamics of a sound wave are formulated for specific assumed conditions, and it is found that density and pressure do not in fact satisfy this linear wave equation, except in the limit of very small amplitudes. The dynamics of a one-dimensional sound wave in an inviscid fluid with no body force effect can be expressed in Euler's equation (equation (2.61))

$$\frac{\partial P}{\partial x}+v\rho\frac{\partial v}{\partial x}+\rho\dot{v}=0 \tag{1.83}$$

and by combining the equation of continuity (equation (2.51)) with the wavespeed equation (2.89), where the fact that $c^2=\partial p/\partial\rho$ implies that $\dot{\rho}c^2=\dot{p}$, to give

$$\frac{\partial v}{\partial x}+\frac{v}{\rho}\frac{\partial\rho}{\partial x}+\frac{1}{\rho c^2}\dot{p}=0 \tag{1.84}$$

Here $v=\dot{\varepsilon}$ is the particle velocity and $\dot{v}=\ddot{\varepsilon}$ the particle acceleration [69]. If the nonlinear terms $v\rho(\partial v/\partial x)$ and $(v/\rho)(\partial\rho/\partial x)$ are negligible, these equations reduce to

$$\frac{\partial P}{\partial x}+\rho_0\dot{v}=0 \tag{1.85}$$

and

$$\frac{\partial v}{\partial x}+\frac{1}{\rho_0 c^2}\dot{v}=0 \qquad (1.86)$$

where ρ_0 is the equilibrium fluid density. Equations (1.84) and (1.85) combine to give

$$\frac{\partial^2 P}{\partial x^2}-\frac{1}{c^2}\frac{\partial^2 P}{\partial t^2}=0 \qquad (1.87)$$

which describes a pressure wave propagating at speed c, as discussed in section 1.1.1. However, where the nonlinear terms are not negligible, the wave deviates from the classical longitudinal case. This has two important effects when the propagation of a sinusoidal acoustic wave is considered. There is firstly a convection effect. In simple terms, if v/c is not negligible, then the parts of the wave tend to propagate as $c + v$. The particle velocity varies throughout the wave as shown in Figure 1.2, and so the greater the local acoustic pressure, the greater the velocity of migration of that section of the wave. Therefore regions of compression (where, from Figure 1.2, the particle velocity is in the same direction as the wave velocity) tend to migrate faster than the regions of rarefaction (where v is opposite to c). The pressure peak travels with the greatest speed, the trough with the least. The second effect arises because, when a fluid is compressed, its bulk modulus and stiffness increase. This causes an increase in sound speed (equation (1.10)), and this effect too will cause the pressure peaks to travel at greater speed than the troughs, and tend to try to catch up and encroach upon them. A continuous wave that is initially sinusoidal will therefore distort in the way shown in Figure 1.23. The sketches illustrate the shape of three cycles of the wave, plotting the pressure as a function of distance in the manner used for Figure 1.2(e). The waveform is initially sinusoidal (Figure 1.23(a)). As it propagates through the medium, each compressive region gains upon the preceding rarefactive half-cycle, an accumulated steepness of the waveform developing between the two (Figure 1.23(b)). After propagating a certain distance (the ***discontinuity length***) the waveform becomes saw-tooth (Figure 1.23(c)). A shock wave develops. With increasing time, any further compressional advance leads to dissipation and results in a reduction in the amplitude of the shock (Figure 1.23(d)). In 'old age', the energy transfer is not sufficient to maintain the shock, and the wave approaches a low-amplitude sinusoidal form (Figure 1.23(e)) [69]. This behaviour is termed the nonlinear propagation of finite-amplitude waves.

If hydrophone measurements are made of such continuous-wave fields then a positively-polarised hydrophone (that is one for which the output response to an increase in pressure is a positive voltage) will produce signals of the form shown in Figure 1.24. The measured acoustic pressure from a continuous-wave 1-MHz source in water is plotted at 1.0, 2.0 and 2.5 m from the transducer faceplate, with the corresponding frequency spectra shown. A pure sine wave would have energy invested only in a single frequency. From the figure, distortion can be seen to be associated with the presence of other frequencies (*harmonics*), which are here at multiples of the basic frequency of the sinusoid (the *fundamental*).[41] The

[41] We know that two singers or instruments which are producing a note of the same pitch, nevertheless sound different. The voices of the singers or instruments have different qualities, or *timbres*. Thus when an orchestra tunes up, a flute playing 'A' at 440 Hz has a different sound from the oboe, which is playing a note of the same pitch. If the spectra of the waveforms were analysed, both instruments would be found to be generating the same *fundamental*, at 440 Hz, but would be generating different amounts of the harmonics. It is the differences in these harmonics which we perceive to be differences in timbre. Neither instrument produces a perfectly sinusoidal waveform: if such a waveform is generated, our perceptions translate the source to be a 'whistle'. On a matter of nomenclature, the 'first harmonic' is in fact the fundamental: the harmonic which is twice the frequency of the fundamental is the 'second harmonic', that at three times the 'third harmonic', etc. These frequencies are discussed further in Chapter 4, section 4.4.7.

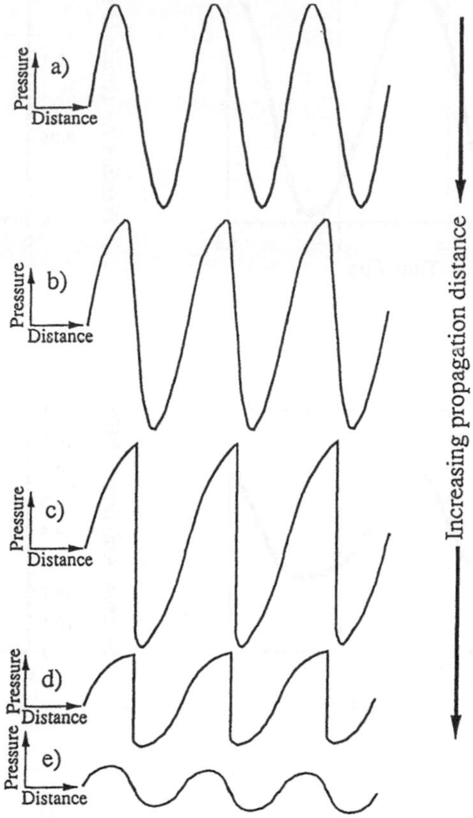

Figure 1.23 Schematic illustration of the evolution of a nonlinearly-distorting waveform, the acoustic pressure being plotted against distance, as would be encountered at progressively increasing distances from the source. The initially sinusoidal waveform (a) shocks in (c) and enters old-age in (e).

onset of nonlinear distortion, and its decay, can be seen as the observation distance increases. The shape of the waveform, as seen on the oscilloscope, is inverted from the waveforms shown in Figure 1.23 since in that figure the pressure is plotted against distance, whereas in Figure 1.24 it is plotted against time: the pressures at the front of the waveforms shown in Figure 1.23 arrive first at the hydrophone, and those at the tail in the spatial frame arrive at the later times in Figure 1.24. As stated above, this distortion process may be thought of as the channelling of energy from lower to higher harmonic frequencies. A plane wave of amplitude P_1 which has propagated a distance h in an aqueous medium will have a second-harmonic component P_{2nd} of magnitude [70]

$$\left[\frac{P_{2nd}}{\text{bar}}\right] = \frac{1}{1910}\frac{h}{\lambda}\left[\frac{P_1}{\text{bar}}\right]^2 \tag{1.88}$$

The fact that higher frequencies are more strongly attenuated than lower ones contributes to the severe attenuation that sets in after the discontinuity length has been exceeded, and which

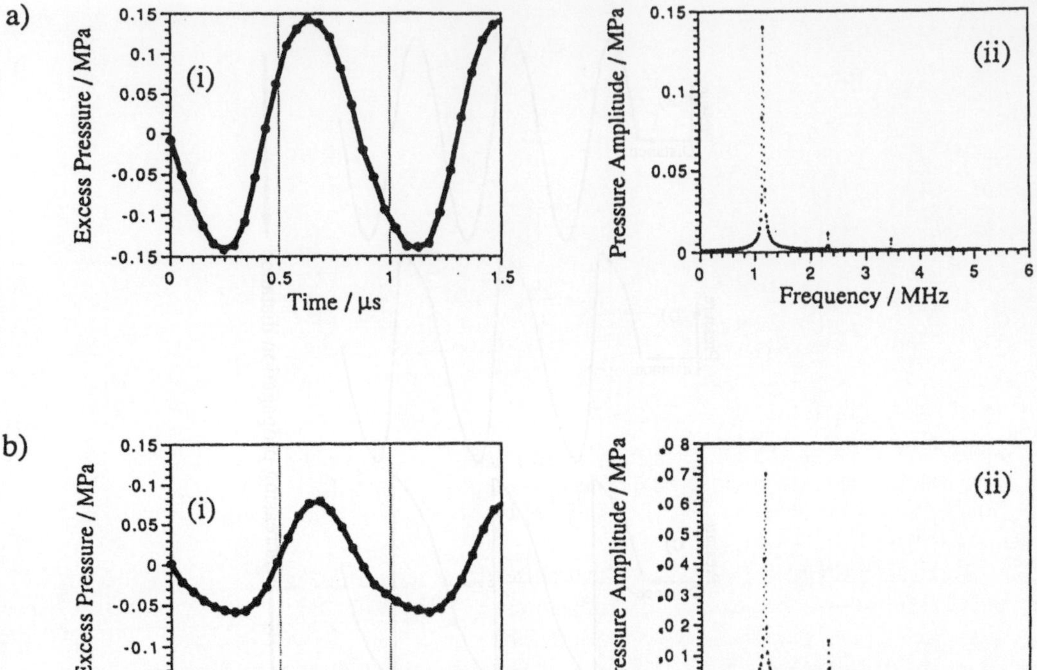

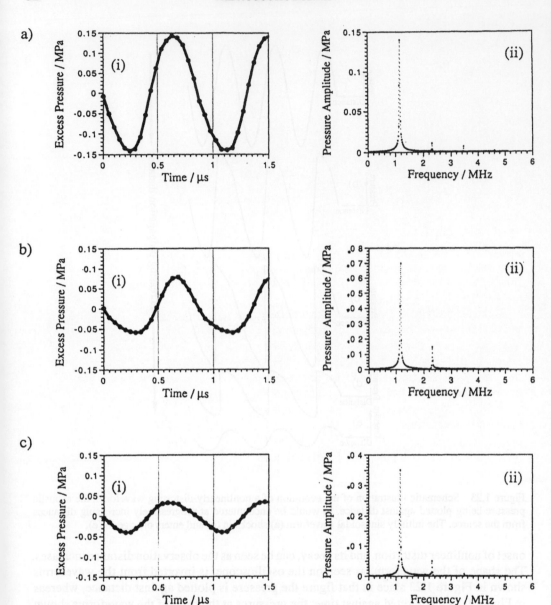

Figure 1.24 Data for the (i) waveforms and (ii) the corresponding spectra recorded at (a) 1.0 m, (b) 2.0 m and (c) 2.5 m from the faceplate of a Therasonic 1030 ultrasound transducer, operating in water at 1 MHz. The apparent sharpness of the distortion may have been reduced by the limited bandwidth of the Dapco NP10-3 needle hydrophone (TG Leighton and A Hardwick).

causes appearance of the 'old age' waveform. An additional feature arises from the action of dissipation and diffraction, which cause phase shifts in the various frequency components of the wave. As a result, the distorted waveform is not symmetrical about the zero-line: in general the trough, which corresponds to negative values of the acoustic pressure, becomes rounded whilst the positive peak is augmented.

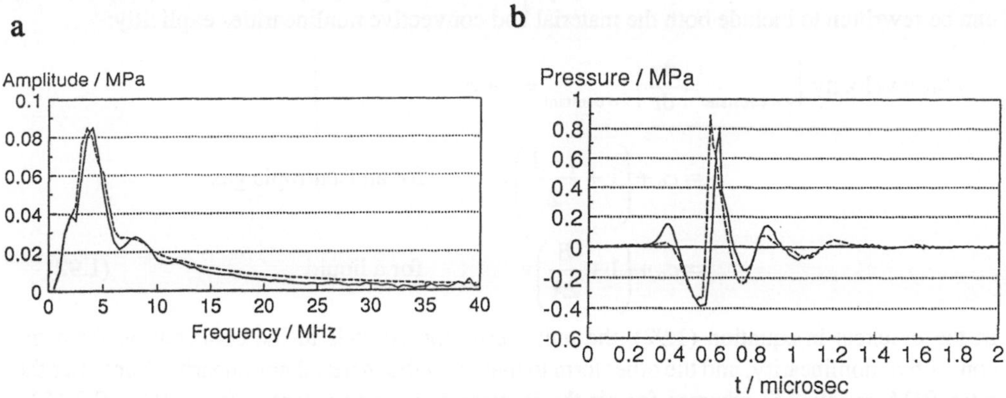

Figure 1.25 The (a) spectral magnitudes and (b) waveform on axis at a distance of 100 mm from diagnostic transducer faceplate: — experiment; - - - theory. (After Baker [71]. Reprinted by permission from *Physics in Medicine and Biology*, vol. 36, pp. 1457–1464; Copyright © 1991 IOP Publishing Ltd.)

This analysis also explains the typical pulse shape shown in Figure 1.25(b), where when plotted against time the edge of the positive pressure pulse at smaller times (i.e. on the left) has a steeper gradient than that at later times (on the right). Note the very significant asymmetry about the zero-axis evident in the distortion of such high-amplitude pulses, generated through the interaction of diffraction and nonlinear distortion [72, 73]. Such traces as these can therefore be used to test whether an unknown hydrophone has positive or negative polarity.

Nonlinear propagation may effect beam shape, and consideration must be given when applying a beam measurement taken in one medium to another [74]. Nonlinear propagation occurs not just in water, but also in tissue, Duck *et al.* [75] having observed it both *in vitro* (ox-liver) and *in vivo* (human calf). Such nonlinear distortion has been shown to have important effects at biomedical frequencies and intensities [76–78].

As outlined above, there are two contributions to the change in phase speed of the wave: the change in stiffness (or bulk modulus) and the convection. These are called the *material nonlinearity* and the *convective nonlinearity* respectively, and are formulated as follows. The phase velocity of a point in the wave having particle velocity v is

$$\text{Phase velocity}\bigg|_{v=\text{constant}} = \frac{dx}{dt}\bigg|_{v=\text{constant}} = v + c \tag{1.89}$$

where a convection component v is added to c, the local speed of sound. This local speed has in turn been affected by the change in bulk modulus, and is related to c_0, the sound speed for waves of infinitesimal amplitude, by

$$c = c_0 + \frac{\gamma - 1}{2} v \tag{1.90}$$

Though equation (1.89) applies to a perfectly isentropic gas, it may also apply to a liquid where the ratio of specific heats, γ, for the liquid is given by

$$\gamma = 1 + \frac{B}{A} \tag{1.91}$$

where B/A is the second-order nonlinearity ratio of the liquid [79]. Therefore equation (1.89) can be rewritten to include both the material and convective nonlinearities explicitly:

$$\text{Phase velocity}\bigg|_{v=\text{constant}} = \frac{dx}{dt}\bigg|_{v=\text{constant}} = v + c$$

$$= c_0 + \left(1 + \frac{\gamma - 1}{2}\right) v \qquad \text{for an isentropic gas}$$

$$= c_0 + \left(1 + \frac{B}{2A}\right) v \qquad \text{for a liquid} \qquad (1.92)$$

In the brackets in equation (1.92), the unity term corresponds to the contribution from the convective nonlinearity, and the other term to that from the material nonlinearity. For water the ratio $B/2A$ equals 2.5, whereas for air the isentropic gas equivalent is $(\gamma - 1)/2 = 0.2$ [63]. Therefore in water the dominant cause of the waveform distortion is nonlinearity of the equation of state (the material nonlinearity), whereas in air the distortion arises mainly through convection.

The sum of the material and convective nonlinearities is given by the parameter $\not{c}$, where from equation (1.90)

$$\text{Phase velocity}\bigg|_{v=\text{constant}} = \frac{dx}{dt}\bigg|_{v=\text{constant}} = c_0 + \not{c}\, v$$

$$\text{where } \not{c} = \frac{\gamma + 1}{2} \qquad \text{for an isentropic gas}$$

and

$$\text{where } \not{c} = \left(1 + \frac{B}{2A}\right) \qquad \text{for a liquid.} \qquad (1.93)$$

For air $\not{c} = 1.2$, and for water $\not{c} = 3.5$. From this, one might expect the distortion to be more obvious in water than in air; however the attenuation of high frequencies must be taken into account. The discontinuity length, which is taken to be the distance propagated when an infinite slope first appears in the waveform, is given by

$$L_{dis} = \frac{1}{\not{c} M k} \qquad (1.94)$$

where $k = \omega / c_0$ and M is the peak acoustic Mach number of the source, the ratio of the amplitude of the particle velocity at the source to c_0 [63]. Thus the shock forms closer to the source as the amplitude increases, because the speed differential between the peaks and the troughs increases. Similarly L_{dis} decreases with increasing frequency, since this signifies decreasing wavelength, and so the peaks have to travel a shorter distance to catch up with the troughs. If the point of observation is at a distance greater than L_{dis}, then as either amplitude or frequency is increased the discontinuity length decreases, the point where the shock forms moving closer to the source. This means that increasing amounts of energy can be dissipated at the shocks beyond L_{dis} before the wave arrives at the point of observation. If the energy supplied to the wave at the source is continually increased, there comes a point at which the increased dissipation between the discontinuity point and the point of observation outweighs the increase in energy supplied to the wave at the source. Beyond this critical source power, the wave ceases to be dependent upon the source amplitude: any additional energy supplied to the wave by the source is lost at the shock front, and *acoustic saturation* is said to have occurred [63].

There are two main techniques for determining the nonlinearity parameter B/A [80]: the thermodynamic method [81], and the finite-amplitude method, which uses measurement of the second harmonic generated as a propagating sine wave is distorted [82]. One should bear in mind that if the nonlinear propagation is viewed as the pumping of acoustic energy into higher frequencies in the wave [83], then measurement of the wave will be limited by the frequency response of the detector. For example, hydrophones become less sensitive above a certain frequency.[42] Therefore the higher frequencies, which are manifest in the sharper or discontinuous features of the wave, may go undetected. The shape of the wave, as measured, may differ from the real pulse shape. More importantly, as energy is transferred into the higher frequencies, it may become 'invisible' to the detection system. Bandwidth and impulse response limitations in the detection system can therefore lead to errors in the measurement of the peak positive acoustic pressure [84]. Follet [85] used optical Schlieren techniques to study the pulse shapes. Models of nonlinear propagation exist to aid in interpreting such measurements (see, for example, Bacon [86, 87] and Lucas and Muir [88]). The models agree well with experiment for transducers which act as perfect pistons in both continuous-wave [89] and pulsed [90] modes. Comparison with the output of clinical transducers has also been made in continuous-wave mode by Bacon and Baker [91], who found that the experimental output of a 3.35-MHz focused transducer was intermediate between piston and Gaussian[43] models for the source. Figure 1.25 shows the experimentally measured pulse 100 mm from a 3.35-MHz clinical transducer operating in water, compared with theory as modelled by Baker [71].

As energy is transferred to the harmonics these, being of higher frequency, suffer greater absorption, and so can enhance heating of the medium [92–94]. The absorption of acoustic energy by the medium during nonlinear propagation depends on the geometry of the sound field, the propagation distance and the nonlinearity of the medium, as well as the intensity [95]. Bacon and Carstensen [96] considered the effect that this would have on the heating of tissues subjected to diagnostic pulsed ultrasound, their results suggesting an enhancement of up to threefold in the heating of tissue-mimicking gel after propagation through water, which was used to mimic the scanning of a foetus through a full bladder. With a 3.6-MHz focused beam, this enhancement lead to a maximum temperature rise of 2° C. The result from this worst-case model, where there is little attenuation along the path to the focus, can be compared with the AIUM guideline [97] which postulates that in an afebrile patient (that is, one not having a fever) a temperature rise of 1° C resulting from exposure to diagnostic ultrasound is 'safe'. The exposure was for 3 minutes with I_{SPTA}=1 W cm^{-2}, an intensity exceeded in one survey by half the clinical pulsed-Doppler units tested [98]. Nonlinear enhancement of heating in tissues can be greater in focused fields than in plane-wave ones [99], and in the bladder-model of Bacon and Carstensen [96] the maximum heating was found to occur in the pre-focal, rather than the focal, region. Lewin *et al.* [66] model nonlinear propagation with regard to tumour therapy through hyperthermia. Ultrasonically induced hyperthermia is further discussed in Chapter 5, section 5.4.2.

[42]See section 1.2.2(a)(i).

[43]Unlike a piston source, where the displacement is uniform across the plane face of the transducer, the displacement across the face of a Gaussian plane transducer varies so that it is greatest in the centre, and decreases towards the edges. In the model, the form of this decrease follows a precise mathematical function, called the Gaussian: if a plot were made of the displacement amplitude as one took measurements across one diameter of the source, the graph would have a characteristic 'bell-shaped' profile.

(b) Acoustic Streaming

As the acoustic wave travels through the medium, it may be absorbed, as discussed in section 1.1.7. However, the momentum absorbed from the acoustic field manifests itself as a flow of the liquid in the direction of the sound field, termed *acoustic streaming*. Since it is more usual to consider energy rather than momentum in the context of acoustic waves, the process is conventionally thought of as the setting up of an energy gradient in the direction of propagation when energy is absorbed from the beam during its passage through an attenuating liquid. A gradient in energy corresponds to a force, and when this acts upon the liquid the streaming flow results. The force per unit volume, F/V, equals the gradient in pressure, $\vec{\nabla} p$, which causes the liquid to accelerate in the direction of propagation:

$$\vec{\nabla} p = \frac{F}{V} = \frac{2Ib}{c} \tag{1.95}$$

[100, 101]. From equation (1.95) it is clear that if both intensity and attenuation can vary spatially throughout a sound beam in a uniform medium, then so will the streaming forces and flows. An increase in either parameter will increase the streaming. The intensity variations in near- or focused fields have already been discussed. As will be evident from the preceding section, the pumping of energy into higher frequencies, which are more strongly absorbed than the fundamental, means that attenuation may vary spatially in an acoustic field of finite amplitude [102].

Finite-amplitude effects will also affect streaming through the formation of shocks. Starritt *et al.* [103] observed the enhancement of streaming in high-amplitude diagnostic pulsed ultrasonic fields which have formed shocks in water.

Streaming speeds of up to around 10 cm/s can be demonstrated from clinical ultrasonic equipment. Figure 1.26 [102] shows a plan view of dye, carried along by the streaming flow, in three clinical underwater ultrasound beams. The visualisation technique is described by Merzkirch [104], utilising the electrolysis of water containing dissolved thymol blue indicator. Through the insertion of acoustically transparent 'clingfilm' windows in the diagnostic *B*-scan field, Starritt *et al.* [102] demonstrated that in addition to local energy absorption, a more significant contribution to unimpeded flow in the far-field region beyond the focus comes from a narrow jet of flowing liquid which is generated near the focus and flows onwards from there with much the same speed. They therefore concluded that the acoustic stream is powered significantly by a near-focus 'source pump', which results from the enhanced absorption of the high-frequency components of the distorted finite-amplitude pulses there.

In materials, such as tissue, which are not free to flow, stresses may still be set up by these processes, and consideration must be given to the response of the medium [102].

There is a second type of streaming associated not with the spatial attenuation of a wave in free space, but which instead occurs near small obstacles placed within a sound field, or near small sound sources, or vibrating membranes or wires. It arises instead from the frictional forces between a boundary and a medium carrying vibrations of circular frequency ω, and unlike the streaming described earlier, this time-independent circulation occurs only in a small region of the fluid, being generally confined to an acoustic microstreaming boundary layer of thickness

$$L_{ms} = \sqrt{2\eta / \rho\omega} \tag{1.96}$$

where η and ρ are the shear viscosity and density respectively of the liquid [105]. Because of the restricted scale of the circulation (often but not exclusively microscopic), it is commonly

(a)

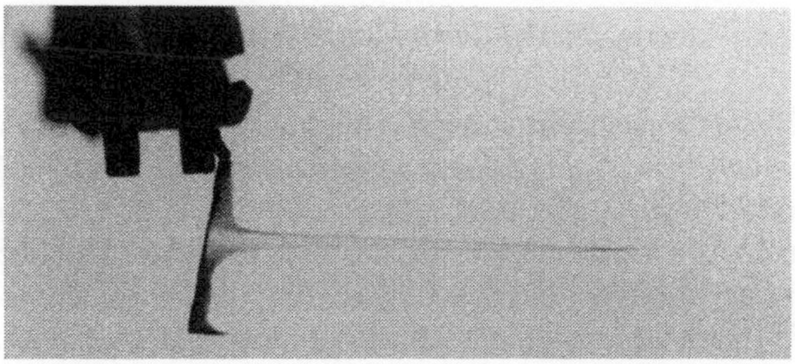

(b)

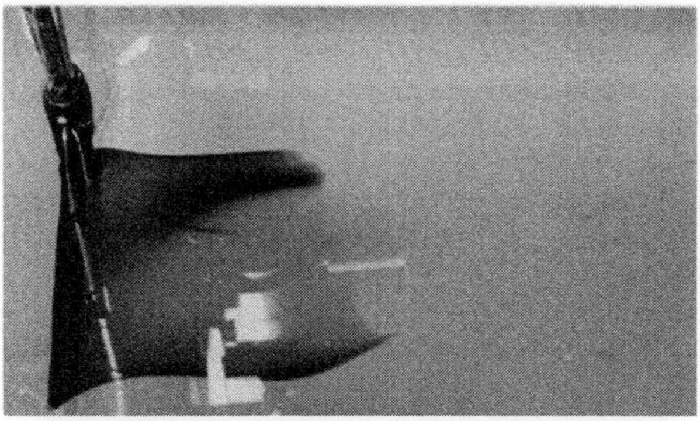

(c)

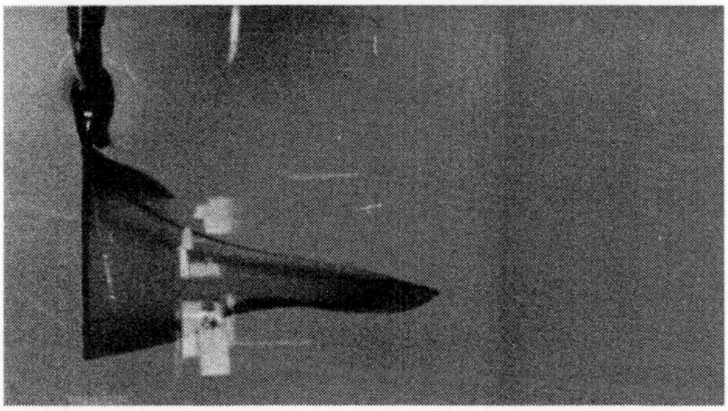

Figure 1.26 Acoustic streaming patterns in water, visualised using thymol blue indicator. The sound fields were generated by (a) a diagnostic *B*-scan transducer, (b) a scanned diagnostic array and (c) a physiotherapy transducer. (After Starritt *et al.* [102]. Reprinted by permission from *Physics in Medicine and Biology*, vol. 36, pp. 1465–1474; Copyright © IOP Publishing Ltd.)

termed *microstreaming*.[44] Microstreaming can bring about a number of important effects. The shear forces set up within the liquid may disrupt DNA (deoxyribonucleic acid) [106], disaggregate bacteria [107], disrupt human erythrocytes and platelets *in vitro* and *in vivo* [108, 109], and cause other bioeffects [110]. Microstreaming can specifically occur as a result of the oscillations

[44]For the remainder of this book, the term 'streaming' will be reserved for the bulk fluid flow described at the beginning of this section, and the time-independent circulation which occurs locally around a source or obstacle will be called 'microstreaming' regardless of whether or not the boundary layer is microscopic. These fluid flows should not be confused with 'streamers' and 'microstreamers', terms introduced in Chapter 4 to describe lines of bubbles driven into translatory motion by an intense acoustic field.

of an acoustically driven bubble in a sound field. This can also lead to bioeffects [111], and bring about other important effects.[45]

(c) Self-interaction, Parametric and Stimulated Scattering Effects in Liquids

Bunkin *et al.* [112] cite the nonlinear propagation discussed in section 1.2.3(a) as being indirectly responsible for the relative scarcity of nonlinear acoustic phenomena, compared with optical ones, as sound is non-dispersive[46] in many pure media at frequencies for which absorption over the distance of a wavelength is small. As a result, high-pressure fields are converted into shock waves as energy is pumped into the higher harmonics, as outlined earlier. Since absorption tends to be greater at higher frequencies, the high-frequency oscillations are strongly absorbed, so that strong sound waves are rapidly dissipated in the liquid. This effect can be reduced by engineering dispersion into the medium, for example by propagating the sound through a waveguide, or by introducing gas bubbles [113].[47]

Nevertheless there are several other nonlinear effects associated with the propagation of ultrasound, such as self-interaction and parametric phenomenon, stimulated scattering and phase conjugation [112, 114]. Examples of some of these are given below.

One self-interaction effect is illustrated by the self-focusing of acoustic beams [115–119]. Thermal self-focusing occurs because, by absorption of acoustic energy, the medium through which the sound beams passes becomes warmer. In most liquids this causes the sound-speed to fall (or equivalently causes the acoustic refractive index to increase), so that the beam is focused in towards the axis as a result of total internal reflection at its perimeter. In water, however, the sound speed passes through a maximum at 74° C (Figure 1.27), so that thermal self-focusing occurs only at temperatures in excess

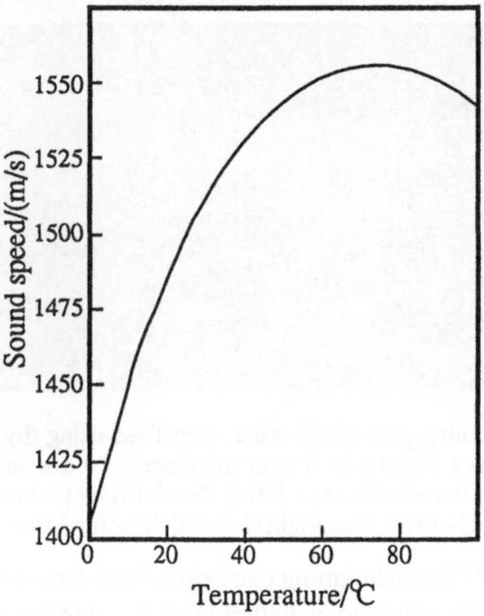

Figure 1.27 The temperature variation of the speed of sound in pure water, for linear propagation under one atmosphere (after Wells [23]), using data of del Grosso and Mader [156].

[45] These will be discussed in Chapter 4, sections 4.4.3(c)(iii) and 4.4.4, and Chapter 5, section 5.4.2.
[46] This is to say, for given conditions of temperature, pressure etc., the phase speed of sound does not vary with frequency.
[47] See Chapter 4, section 4.4.1(e).

of this: at lower temperatures, the inverse phenomenon of self-defocusing occurs [120]. Acoustic streaming imparts a defocusing effect by increasing the sound speed near the beam axis.

There are mechanisms other than thermal which can bring about self-focusing, including some resulting from the presence of bubbles [121]. For example, bubbles can cause a reduction in sound speed on the axis of a beam.[48] The presence of bubbles can also bring about self-transparency, where the absorption decreases with increasing intensity [122], though this can arise in bubble-free media, such as glycerin, owing to the temperature-dependence of the absorption coefficient [112, 118, 119]. Self-focusing can occur not just in the bulk of a liquid, but also at an interface between two media [123, 124], where, for example, distortion of the surface of a water sample by the beam can lead to a focusing effect simply as a result of local angling of the reflecting surface (Figure 1.28(a)). In the extreme case, when the reflected beams are directed upwards, it can lead to *acoustic self-concentration* and the formation of a fountain (Figure 1.28(b)).

Sound may be suppressed by sound [12, 125, 126], modulated, or scattered [127–130]. The parametric array was conceived in 1960 by Westervelt [131, 132], and following examples of applications suggested by Berktay [133–135] in the mid-1960s is now perhaps the most familiar application of nonlinear acoustics [136, 137]. If two collimated acoustic beams of finite amplitude and high frequencies, ω_1 and ω_2, are transmitted in the same direction, nonlinear interaction between them generates sound at $\omega_c = \omega_1 - \omega_2$. Thus a parametric antennae, driven at two frequencies, provides a highly directional extended source of low-frequency sound: the source itself can be relatively small, and the sidelobe features (which are apparent in Figures 1.14 and 1.15) greatly reduced. The low-frequency ω_c can be detected in the medium, along with such combinations as $\omega_1 \pm \omega_c$. However, these other frequencies, being higher, are more strongly attenuated and over long propagation distances the lowest frequency, ω_c, dominates. The possibility of increasing the efficiency of these systems by the addition of bubbles has been investigated [138–140].

Parametric receivers exhibit the same high directionality in the detection of sound [131, 132]. It can be formed from two transducers, one being the source of a collimated acoustic beam of high intensity, and the other a hydrophone positioned in this beam. The high-amplitude sound in the beam interacts nonlinearly with any acoustic signal that crosses its path, and the resulting modulation is detected by the hydrophone. The amplitudes of the sum- and difference-signals so detected are greatest when the parametric receiver is aligned with the incident signal [63].

Electroacoustic echo can arise if two high-frequency pulses, separated by an interval t_1, are introduced into a piezoelectric material; a third pulse, the echo, can be detected a time t_1 after the second pulse [141–143]. The possibility of observing such echoes on bubbly media has been examined [144–147].

The presence of gas bubbles can introduce dispersion and increase the degree of nonlinearity of the medium, and can bring about the parametric amplification of sound. If a medium contains a monodisperse[49] population of bubbles with a resonance frequency half that of the driving frequency, that resonance can be detected in the sound spectrum. Increasing the size distribution of bubbles weakens the dispersion, increases beam absorption and decreases the parametric amplification.

Stimulated acoustic scattering can occur through the formation of temperature inhomogeneities in high-viscosity liquids which act as a grating, leading to Rayleigh-like scattering

[48] Acoustic radiation forces cause small bubbles to aggregate on the axis of a beam, where the acoustic pressure amplitude is greatest (Chapter 4, section 4.4.1). The high concentration of bubbles reduces the sound speed there (Chapter 4, section 4.1.2(e)).

[49] That is, all bubbles are of the same size.

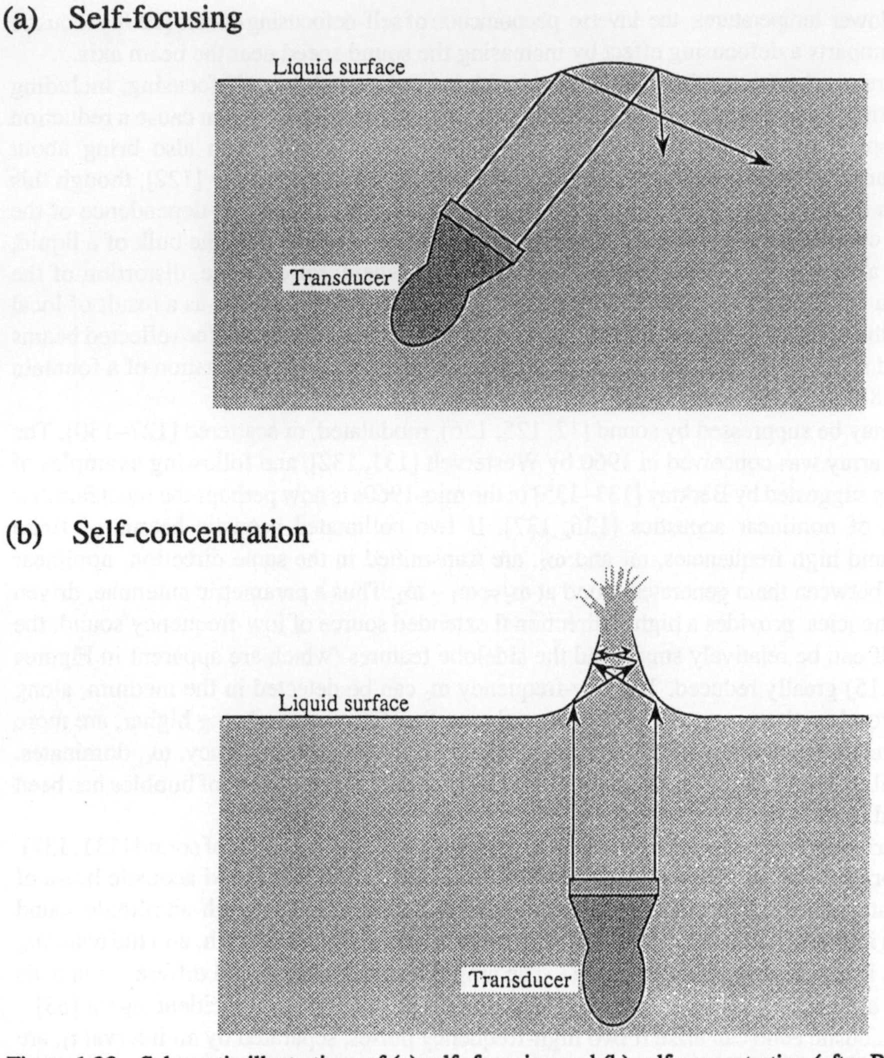

Figure 1.28 Schematic illustrations of (a) self- focusing and (b) self-concentration (after Bunkin *et al.* [112]).

(Stimulated Temperature Acoustic Scattering, STAS). Similar effects include the Stimulated Acoustic Scattering by Acoustic Streaming (SASAS) [148], and through the resonance effects of air bubbles (Stimulated Acoustic Raman Scattering, SARS) [149, 150].

Conclusion

A sound wave is generated, propagates through a medium and can then be detected. As the sound traverses the medium it can experience reflection or attenuation, and also more esoteric effects such as self-focusing and the generation of streaming in liquids.

Ultrasound is employed widely both in nature and technology. The use of echo-location by a variety of animals has been discussed, and analogies exist in underwater sonar and in medical diagnostic scanning. The Doppler change in frequency that occurs when ultrasound is reflected from a moving object can be used, for example, to measure the blood flow in the heart of a foetus. Whilst the latter processes ideally obtain information about the target without affecting it, ultrasound can be used in medicine to change the target, i.e. for therapy. Ultrasound is commonly employed in physiotherapy, and lithotripsy[50] can be used to break kidney stones. High intensity ultrasound can cause cell disruption [151], an effect which is today in routine use in laboratories [152], and cause a wide range of effects in industry, for example by degassing the melt and refining grain structure in welding joints [153–155].

The activation of bubbles is involved in several of the latter effects, and it is to explain the interaction of sound with bubbles that is the purpose of this book. This chapter gave a brief introduction to the acoustic wave. In the next, the basics of the bubble will be outlined. From Chapter 3 onwards, the interaction between the sound and the bubble will be studied.

References

[1] The New Encyclopaedia Britannica, Macropaedia, Vol. 17. Encyclopaedia Britannica, William Benton, Helen Hemmingway Benton (Publishers), 15th edn, 1974, pp 19–51
[2] The New Encyclopaedia Britannica, Macropaedia, Vol. 17. Encyclopaedia Britannica, William Benton, Helen Hemmingway Benton (Publishers), 15th edn, 1974, p 843
[3] Morse PM and Ingard KU. Linear acoustic theory. In: Handbuch der Physik, Vol. II, Part 1 (S Flügge, ed.). Springer, Berlin, 1961
[4] Nelson PA and Elliott SJ. Active Control of Sound. Academic Press, New York, 1992, chapter 2
[5] Arfken G. Mathematical Methods for Physicists. Academic Press, New York, 2nd edn, 1970, chapter 14
[6] Skudrzyk E. The Foundations of Acoustics. Springer, New York, 1971
[7] Pierce AD. Acoustics: An Introduction to its Physical Properties and Applications. McGraw-Hill, New York, 1981
[8] Kinsler LE, Frey AR, Coppens AB and Sanders JV. Fundamentals of Acoustics. Wiley, New York, 3rd edn, 1982
[9] Dowling AP and Ffowcs-Williams JE. Sound and Sources of Sound. Ellis Horwood, Chichester, 1983
[10] Morse PM and Ingard KU. Theoretical Acoustics. McGraw-Hill, New York, 1968
[11] Hall DE. Basic Acoustics. Wiley, New York, 1987
[12] Nelson PA and Elliott SJ. Active Control of Sound. Academic Press, New York, 1992
[13] Wells PNT. Biomedical Ultrasonics. Academic Press, London, 1977, pp 14 and 136
[14] Kaye GWC and Laby TH. Tables of Physical and Chemical Constants. Longmans, Green, 13th edn, 1968 impression, pp 60–73
[15] Lide DR. CRC Handbook of Chemistry and Physics. CRC Press, Boca Raton, Florida, 72nd edn, 1991
[16] Kinsler LE, Frey AR, Coppens AB and Sanders JV. Fundamentals of Acoustics. Wiley, New York, 3rd edn, 1982, p 116
[17] Kinsler LE, Frey AR, Coppens AB and Sanders JV. Fundamentals of Acoustics. Wiley, New York, 3rd edn, 1982, pp 270–272
[18] Beissner K. Radiation force calculations. Acustica 1987; 62: 255–263
[19] Beissner K. Acoustic radiation pressure in the near field. J Sound Vib 1984; 93: 537–544
[20] Grant IS and Phillips WR. Electromagnetism. Wiley, New York, 1975, pp 379–383
[21] Wood A. Acoustics. Blackie, London, 1940. References to 1960 re-issue, p 124
[22] Wells PNT. Biomedical Ultrasonics. Academic Press, London, 1977, p 22
[23] Wells PNT. Biomedical Ultrasonics. Academic Press, London, 1977

[50] See Chapter 5, sections 5.1.2 and 5.4.2(b)(iv).

[24] Bamber JC. Attenuation and absorption. In: Physical Principles of Medical Ultrasonics (Hill CR, ed.). Ellis Horwood, Chichester (for Wiley, New York), 1986, chapter 4
[25] Hill CR (ed.). Physical Principles of Medical Ultrasonics. Ellis Horwood, Chichester (for Wiley, New York), 1986
[26] Urick RJ. Principles of Underwater Sound. McGraw-Hill, New York, 3rd edn, 1983
[27] Kinsler LE, Frey AR, Coppens AB and Sanders JV. Fundamentals of Acoustics. Wiley, New York, 3rd edn, 1982, pp 167–182
[28] Pierce AD. Acoustics: An Introduction to its Physical Properties and Applications. McGraw-Hill, New York, 1981, chapter 4
[29] Nelson PA and Elliott SJ. Active Control of Sound. Academic Press, New York, 1992, pp 281–284
[30] Harris GR. A discussion of procedures for ultrasonic intensity and power calculations from miniature hydrophone measurements. Ultrasound Med Biol 1985; 11: 803–817
[31] Livett AJ and Preston RC. A comparison of the AIUM/NEMA, IEC and FDA (1980) definitions of various acoustic output parameters for ultrasonic transducers. Ultrasound Med Biol 1985; 11: 793–802
[32] Curie J and Curie P. Comptes Rendus 1880; 91: 294
[33] Curie J and Curie P. Comptes Rendus 1881; 93: 1137
[34] Joule JP. On the effects of magnetism on the dimensions of iron and steel bars. Phil Mag 1847; 30: 76–87 and 226–241
[35] Wells PNT. Biomedical Ultrasonics. Academic Press, London, 1977, p 82
[36] Preston RC, Bacon DR, Livett AJ and Rajendran K. PVDF membrane hydrophone performance properties and their relevance to the measurement of the acoustic output of medical ultrasonic equipment. J Phys E 1983; 16: 786–795
[37] De Reggi AS, Edelman S, Roth SC, Warner H and Wymn J. Piezoelectric polymer receiving arrays for ultrasonic applications. J Acoust Soc Am 1977; 61(Suppl 1): S17
[38] Shotton KC, Bacon DR and Quilliam RF. A PVDF membrane hydrophone for operation in the range 0.5 MHz to 15 MHz. Ulrasonics 1980; 18: 123–126
[39] Kawai H. The piezoelectricity of polyvinylidene fluoride. Japan J Appl Phys 1969; 8: 975–976
[40] Murayama N, Nakamura K, Obara H and Segawa M. The strong piezoelectricity in polyvinylidine fluoride (PVDF). Ultrasonics 1976; 14: 15–23
[41] Bacon DR. Characteristics of a PVDF membrane hydrophone for use in the range 1–100 MHz. IEEE Trans Sonics Ultrasonics 1982; SU-29: 18–25
[42] Lewin PA. Miniature piezoelectric polymer ultrasonic hydrophone probes. Ultrasonics 1981; 19: 213–216
[43] Platte M. A polyvinylidene fluoride needle hydrophone for ultrasonic applications. Ultrasonics 1985; 23: 113–118
[44] Filmore PR and Chivers RC. Measurements on batch produced miniature ceramic ultrasonic hydrophones. Ultrasonics 1986; 24: 216–229
[45] Wells PNT, Bullen MA, Follet DH, Freundlich HF and James JA. The dosimetry of small ultrasonic beams. Ultrasonics 1963; 1: 106–110
[46] Newell JA. A radiation pressure balance for the absolute measurement of very low ultrasonic power outputs. Phys Med Biol 1963; 8: 215–221
[47] Kossoff G. Calibration of ultrasonic therapeutic equipment. Acustica 1962; 12: 84–90
[48] Hill CR. Calibration of ultrasonic beams for biomedical applications. Phys Med Biol 1970; 15: 241–248
[49] Wells PNT. Biomedical Ultrasonics. Academic Press, London, 1977, pp 88–94
[50] Sjöberg A, Stahle J, Johnson S and Sahl R. The treatment of Meniere's disease by ultrasonic radiation. Acta oto-lar 1963; Suppl 178
[51] Fry WJ and Dunn F. Ultrasound: analysis and experimental methods in biological research. In: Physical Techniques in Biological Research, Vol. IV (Nastuk WL, ed.). Academic Press, New York, 1962, chapter 6, pp 261–394
[52] Dunn F, Averbuch AJ and O'Brien WD Jr. A primary method for the determination of ultrasonic intensity with an elastic sphere radiometer. Acustica 1977; 38: 58–61
[53] Shotton KC. A tethered float radiometer for measuring the output power from ultrasonic therapy equipment. Ultrasound Med Biol 1980; 6: pp 131–133

[54] Fry WJ and Fry RB. Determination of absolute sound levels and acoustic absorption coefficients by thermocouple probes – experiment. J Acoust Soc Am 1954; 26: 294–310
[55] Szilard J. A new device for monitoring ultrasound dosage. Abstr 8th Int Congr Acoust 1974; 352
[56] Breyer B and Devčić B. A simple device for checking the acoustic beam of ultrasonic therapy equipment. Ultrasonics 1984; 22: 285–286
[57] Leighton TG, Walton AJ and Field JE. The high-speed photography of transient excitation. Ultrasonics 1989; 27: 370-373
[58] Apfel RE. Acoustic cavitation inception. Ultrasonics 1984; 22: 167–173
[59] Chang LS and Berg JC. Electroconvective enhancement of mass or heat exchange between a drop or bubble and surroundings in the presence of an interfacial tension gradient. AIChE 1985; 31: 149–151
[60] Chang LS and Berg JC. The effect of interfacial tension gradients on the flow structure of single drops or bubbles translating in an electric field. AIChE 1985; 31: 551–557
[61] Leighton TG, Wilkinson M, Walton AJ and Field JE. The forced oscillations of bubbles in a simulated acoustic field. Eur J Phys, 1990; 11: 352–358
[62] Bjørnø L. Nonlinear acoustics: ancient foundations – modern objectives – exciting applications. Ultrasonics Symposium Proceedings 1983, pp 338–345
[63] Hamilton MF. Fundamentals and applications of nonlinear acoustics. In: Nonlinear Wave Propagation in Mechanics – AMD, Vol. 77 (Wright TW, ed.). [Book Number H00346] American Society of Mechanical Engineers, 1986
[64] Rugar D. Resolution beyond the diffraction limit in the acoustic microscope: a nonlinear effect. J Appl Phys 1984; 56: 1338–1346
[65] Ichida N, Sato T, Miwa H and Murakami K. Real-time nonlinear parameter tomography using impulsive pumping waves. IEEE Trans Sonics Ultrasonics 1984; SU-31: 635–641
[66] Lewin PA, Schafer ME and Haran ME. Nonlinear propagation models in ultrasound hyperthermia. Proc Inst Acoustics 1986; 8: 85–93
[67] Blackstock DT. History of nonlinear acoustics and a survey of Burgers' and related equations. In: Nonlinear Acoustics, Proceedings of the Symposium held at Applied Research Laboratories, The University of Texas at Austin, 1969 (Muir TG, ed.) [AD 719 936], pp 1–27
[68] Euler L. 1759 Mem Acad Sci Berlin 1766; 15: 185–209
[69] Pfeiler M, Matura E, Iffländer H and Seyler G. Lithotripsy of renal and biliary calculi: physics, technology and medical–technical application. Electromedica 1989; 57: 52–63
[70] Nyborg WL. Physical principles of ultrasound. In: Ultrasound: its Application in Medicine and Biology (Fry FJ, ed.). Elsevier, New York, 1978
[71] Baker AC. Prediction of non-linear propagation in water due to diagnostic medical ultrasound equipment. Phys Med Biol 1991; 36: 1457–1464
[72] Sutin AM. Influence of nonlinear effects on the properties of acoustic focusing systems. Sov Phys Acoust 1978; 24: 334–339
[73] Duck FA and Starritt HC. Acoustic shock generation by ultrasonic imaging equipment. Brit J Radiol 1984; 57: 231–240
[74] Reilly CR and Parker KJ. Finite-amplitude effects on ultrasound beam patterns in attenuating media. J Acoust Soc Am 1989; 86(6): 2339
[75] Duck FA, Starritt HC and Hawkins AJ. Observations of finite-amplitude distortion in tissue. Proc Inst Acoustics 1986; 8: 71–77
[76] Carstensen EL, Law WK, McKay ND and Muir TG. Demonstration of nonlinear acoustical effects at biomedical frequencies and intensities. Ultrasound Med Biol 1980; 6: 359–368
[77] Muir TG. Nonlinear effects in acoustic imaging. In: Proc 9th Int Symp Acoustical Imaging (Wang K, ed.). Plenum, New York, 1980
[78] Carstensen EL, Beecroft SA, Law WK and Barbee DB. Finite amplitude effects on the thresholds for lesion production in tissues by unfocused ultrasound. J Acoust Soc Am 1981; 70: 302–309
[79] Bjørnø L. Nonlinear acoustics. In: Acoustics and Vibration Progress, Vol 2. Chapman and Hall, London, 1976
[80] Sarvazyan AP, Chalikiann TV and Dunn F. Acoustic nonlinearity parameter *B/A* of aqueous solutions of some amino acids and proteins. J Acoust Soc Am 1990; 88: 1555–1561
[81] Beyer RT. Parameter of nonlinearity in fluids. J Acoust Soc Am 1960; 32: 719–721

[82] Law WK, Frizzel LA and Dunn F. Determination of the nonlinearity parameter of biological media. Ultrasound Med Biol 1985; 11: 307–318
[83] Humphrey VF, Burgess M and Sampson N. Harmonic generation due to non-linear propagation in a focused ultrasonic field. Proc Inst Acoustics 1986; 8: 47–54
[84] Bacon DR. Finite amplitude distortion of the pulsed fields used in diagnostic ultrasound. Ultrasound Med Biol 1984; 10: 189–195
[85] Follett DH. Light diffraction as evidence of finite amplitude distortion. Proc Inst Acoustics 1986; 8: 55–62
[86] Bacon DR. Finite amplitude distortion of the pulsed fields used in diagnostic ultrasound. Ultrasound Med Biol 1984; 10: 189–195
[87] Bacon DR. Nonlinear ultrasonic fields: theory and experiment. Proc Inst Acoustics 1986; 8: 39–46
[88] Lucas BG and Muir TG. Field of a finite-amplitude focusing source. J Acoust Soc Am 1983; 74: 1522–1528
[89] Bacon AC, Anastasiadis K and Humphrey VF. The nonlinear pressure field of a plane circular piston: theory and experiment. J Acoust Soc Am 1988; 84: 1483–1487
[90] Baker AC and Humphrey VF. Nonlinear Propagation in Pulsed Ultrasonic Fields in Frontiers of Nonlinear Acoustics, 12th ISNA (Austin, Texas, 1990) (Hamilton MF and Blackstock DT, eds.). Elsevier, New York, pp 185–190
[91] Bacon DR and Baker AC. Comparison of two theoretical models for predicting non-linear propagation in medical ultrasound fields. Phys Med Biol 1989; 34: 1633–1643
[92] Carstensen EL, McKay ND, Dalecki D and Muir TG. Absorption of finite amplitude ultrasound in tissues. Acoustica 1982; 51: 116–123
[93] Swindell W. A theoretical study of non-linear effects with focused ultrasound in tissues: an 'acoustic Bragg peak'. Ultrasound Med Biol 1985; 11: 121–130
[94] Hynynen K. Demonstration of enhanced temperature elevation due to nonlinear propagation of focussed ultrasound in dog's thigh *in vivo*. Ultrasound Med Biol 1987; 13: 85–91
[95] Dalecki D, Carstensen EL, Parker KJ and Bacon DR. Absorption of finite amplitude focused ultrasound. J Acoust Soc Am 1991; 89: 2435–2447
[96] Bacon DR and Carstensen EL. Increased heating by diagnostic ultrasound due to nonlinear propagation. J Acoust Soc Am 1990; 88: 26–34
[97] American Institute of Ultrasound in Medicine (AIUM), Bioeffects Committee. Bioeffects considerations for the safety of diagnostic ultrasound. J Ultrasound Med 1988; 7(Suppl 9): S1–S38
[98] Duck FA. Output data from European studies. Ultrasound Med Biol 1989; 15(Suppl 1): 61–64
[99] Carstensen EL and Muir TG. The role of non-linear acoustics in biomedical ultrasound. In: Tissue Characterisation with Ultrasound (Greenleaf JG, ed.). CRC, Boca Raton, Florida, 1986, pp 57–79
[100] Nyborg W. Acoustic streaming. Physical Acoustics, Volume 11 (Mason WP, ed.). Academic Press, New York, 1975
[101] Beyer RT. Nonlinear Acoustics. Department of the Navy, Sea Systems Command, Washington DC, 1974, pp 23–268
[102] Starritt HC, Duck FA and Humphrey VF. Forces acting in the direction of propagation in pulsed ultrasound fields. Phys Med Biol 1991; 36: 1465–1474
[103] Starritt HC, Duck FA and Humphrey VF. An experimental investigation of streaming in pulsed diagnostic ultrasound beams. Ultrasound Med Biol 1989; 15: 363–373
[104] Merzkirch W. Flow Visualisation. Academic Press, London, 1987
[105] Nyborg WL. Acoustic streaming near a boundary. J Acoust Soc Am 1958; 30: 329–339
[106] Williams AR. DNA degradation by acoustic microstreaming. J Acoust Soc Am 1974; 55: S17A
[107] William AR and Slade JS. Ultrasonic dispersal of aggregates of *Sarcina lutea*. Ultrasonics 1971; 9: 85–87
[108] Williams AR, Hugh DE and Nyborg WL. Hemolysis near a transversely oscillating wire. Science 1974; 169: 871–873
[109] Williams AR. Intravascular mural thrombi produced by acoustic microstreaming. Ultrasound Med Biol 1977; 3: 191–203

[110] Rooney JA. Shear as a mechanism for sonically-induced biological effects. J Acoust Soc Am 1972; 52: 1718–1724
[111] Clarke PR and Hill CR. Physical and chemical aspects of ultrasonic disruption of cells. J Acoust Soc Am 1970; 50: 649–653
[112] Bunkin FV, Kravstov YuA and Lyakhov GA. Acoustic analogues of nonlinear-optics phenomena. Sov Phys Usp 1986; 29: 607–619
[113] Kotel'nikov IA and Stupakov GV. Nonlinear effects in the propagation of sound in a liquid with gas bubbles. Sov Phys JETP 1983; 57(3): 555
[114] Jackson DR and Dowling DR. Phase conjugation in underwater acoustics. J Acoust Soc Am 1991; 89: 171–181
[115] Askar'yan GA. Self-focusing and focusing of ultrasound and hypersound. JETP Lett 1966; 4: 99–101 (Pis'ma Zh Eksp Teor Fiz 1966; 4: 144)
[116] YuV Gulyaev. Nonlinear theory of ultrasonic amplification in semiconductors. Sov Phys Solid State 1970; 12: 328–337 (Fiz Tverd Tela (Leningrad) 1970; 12: 415)
[117] Proclov VV and Mirgorodsky YI. Acta Politechn Scand Ser Appl Phys 1979; 128: 8
[118] Assman VA, Bunkin FV, Vernik AB, Lyakhov GA and Shiplov KF. Observation of a thermal self-effect of a sound beam in a liquid. JETP Lett 1985; 41: 182–184 (Pis'ma Zh Eksp Teor Fiz 1985; 41: 148)
[119] Assman VA, Bunkin FV, Vernik AB, Lyakhov GA and Shiplov KF. Self-action of a sound beam in a high-viscosity liquid. Sov Phys Acoust 1986; 32: 86 (Akust Zh 1986; 32: 138)
[120] Zabolotskaya EA and Khokhlov PV. Thermal self-action of sound waves. Sov Phys Acoust 1976; 22, 15–17 (Akust Zh 1976; 22: 28)
[121] Ciuti P, Iernetti G and Sagoo MS. Optical visualisation of nonlinear acoustic propagation in cavitating liquids. Ultrasonics 1980; 18: 111–114
[122] Kobelev YuA, Ostrovskii LA and Sutin AM. Self-illumination effect for acoustic waves in a liquid with gas bubbles. JETP Lett 1979; 30: 395–398 (Pis'ma Zh Eksp Teor Fiz 1979; 30: 423)
[123] Bunkin FV, Vlasov DV and Kravtsov YuA. Sov Tech Phys Lett 1982; 7: 138 (Pis'ma Zh Tekh Fiz 1982; 7: 325)
[124] Andreeva NA, Bunkin FV, Vlasov DV, Karshiev KN, Kravtsov YuA and Shurygin EA. Preprint 163, Fiz Inst Akad Nauk SSSR, M.1983
[125] Schaffer ME. The suppression of sound with sound. Tech Rep ARL-TR-75-64, Applied Research Laboratories, The University of Texas at Austin, 1984 [ADA 023 128]
[126] Moffett MB, Konrad WL and Carlton LF. Experimental demonstration of the absorption of sound by sound in water. J Acoust Soc Am 1978; 63: 1048–1051
[127] Gorelik AG and Zverev VA. On the problem of neutral interaction between soundwaves. Sov Phys Acoust 1955; 1: 353–358 (Akust Zh 1955; 1: 339)
[128] Zverev VA and Kalachev VI. Measurement of the scattering of sound by sound in the superposition of parallel beams. Sov Phys Acoust 1968 ; 14: 173–178 (Akust Zh 1968; 14: 214)
[129] Zverev VA and Kalachev VI. Sound radiation from the region of interaction of two sound beams. Sov Phys Acoust 1969; 15: 322–327 (Akust Zh 1969; 15: 369)
[130] Zverev VA and Kalachev VI. Modulation of sound by sound in the intersection of sound waves. Sov Phys Acoust 1970; 16: 204–208 (Akust Zh 1970; 16: 245)
[131] Westervelt PJ. Parametric end-fire array. J Acoust Soc Am 1960; 32: 934–935
[132] Westervelt PJ. Parametric acoustic array. J Acoust Soc Am 1963; 35: 535–537
[133] Berktay HO. Possible exploitation of non-linear acoustics in underwater transmitting applications. J Sound Vib 1965; 2: 435–461
[134] Berktay HO. Parametric amplification by the use of acoustic non-linearities and some possible applications. J Sound Vib 1965; 2: 462–470
[135] Berktay HO. Proposals for underwater transmitting applications of non-linear acoustics. J Sound Vib 1967; 6: 244–254
[136] Muir TG. Nonlinear acoustics: a new dimension in underwater acoustics. In: Science, Technology and the Modern Navy. Office of Naval Research, Arlington, Virginia, 1976
[137] Bjørnø L. Parametric acoustic arrays. In: Aspects of Signal Processing. Reidel, Dordrecht, The Netherlands, 1977, p 33
[138] Kozyaev EA and Naugol'nykh KA. Parametric sound radiation in a two-phase medium. Sov Phys Acoust 1980; 26: 48–51 (Akust Zh 1980; 26: 91)
[139] Kustov LM, Nazarov VE, Ostrovskii LA *et al.* Acoust Lett 1982, 6(2): 15

[140] Lerner AM and Sutin AM. Influence of gas bubbles on the field of a parametric sound radiator. Sov Phys Acoust 1983; 29: 388–392 (Akust Zh 1983; 29: 657)

[141] Fossheim K and Holt KM. Physical Acoustics: Principles and Methods, Vol. 16 (Mason WP, ed.). Academic Press, New York, 1982, p 217

[142] Kajimura K. Physical Acoustics: Principles and Methods, Vol. 16 (Mason WP, ed.). Academic Press, New York, 1982, p 295

[143] Melcher RL and Shiren NS. Physical Acoustics: Principles and Methods, Vol. 16 (Mason WP, ed.). Academic Press, New York, 1982, p 341

[144] Lopatnikov SL. Acoustic phase echo in a liquid with gas-bubbles. Sov Tech Phys Lett 1980; 6: 270 (Pis'ma Zh Tekh Fiz 1980; 6: 623–626)

[145] Nemstov BE and Eidman BYA. Spatial echo effect in a liquid containing gas bubbles. Sov Phys Acoust 1982; 28: 396–397 (Akust Zh 1982; 26: 669)

[146] Manykin EA, Ozhovan MN and Poluektov RR. Echo-like coherent phenomena in ensembles of bubbles in liquids. Sov Phys Tech Phys 1983; 28: 466–469 (Zh Tekh Fiz 1983; 53: 738)

[147] Kotel'nikov IA. Echo effect in gas-bubble liquid. Izv Vtssh Uehebn Zaved Radiofiz 1983; 26: 1227–1234 [Sov Radiophys]

[148] Bunkin FV, Volyak KI, Lyakhov GA and Romanovskii MYu. Stimulated scattering of sound by an acoustic flow of a viscous liquid. JETP Lett 1982; 36: 469–471 (Pis'ma Zh Eksp Teor Fiz 1982; 36: 389)

[149] Bunkin FV, Vlasov DV, Zabolotskaya EA and Kravtsov YuA. Active acoustic spectroscopy of bubbles. Sov Phys Acoust 1983; 29: 99–100 (Akust Zh 1983; 29: 169)

[150] Zabolotskaya EA. Interaction of gas bubbles in a sound field. Sov Phys Acoust 1984; 30: 365–368 (Akust Zh 1984; 30: 650)

[151] Harvey E and Loomis A. High speed photomicrography on living cells subjected to supersonic vibrations. J Gen Physiol 1932; 15: 147

[152] Alliger H. Ultrasonic disruption. American Laboratory, October, 1975

[153] Hiedemann, EA. Metallurgical effects of ultrasonic waves. J Acoust Soc Am 1954; 26: 831

[154] Davies GJ and Garland JG. Review 196: Solidification structures and properties of fusion welds. Int Metall Rev 1975; 20: 83

[155] Swallowe GM, Field JE, Rees CS and Duckworth A. A photographic study of the effect of ultrasound on solidification. Acta Metall 1989; 37(3): 961–967

[156] del Grosso VA and Mader CW. Speed of sound in pure water. J Acoust Soc Am 1972; 52: 1442–1446

2
Cavitation Inception and Fluid Dynamics

2.1 The Bubble

2.1.1 Surface Tension Pressure

Consider the bubble of radius R shown in Figure 2.1 . There is an internal pressure p_i within the bubble as a result of the pressure of gas (p_g) and the pressure of liquid vapour (p_v), so that

$$p_i = p_g + p_v \tag{2.1}$$

The pressure within a bubble at rest is greater than the pressure in the liquid immediately outside the bubble as a result of surface tension forces. If the pressure in the liquid outside the bubble, at the bubble wall, is p_L, that within the bubble is

$$p_i = p_L + p_\sigma \tag{2.2}$$

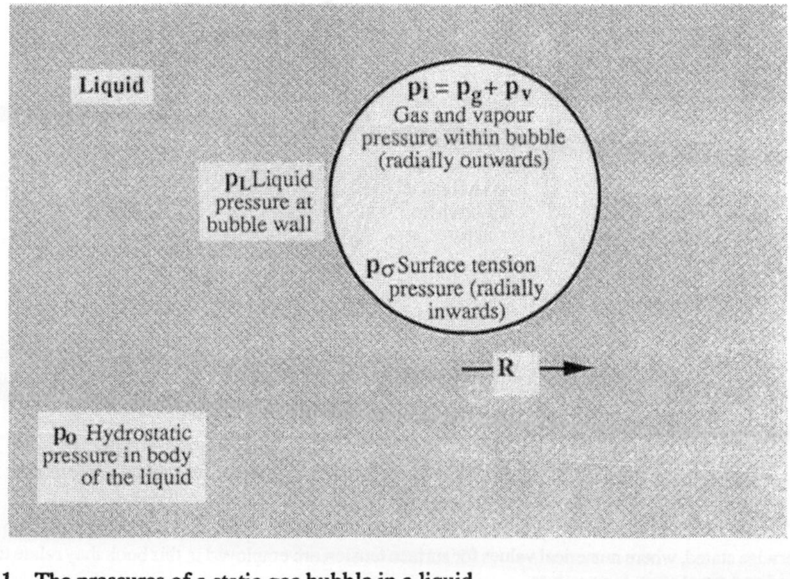

Figure 2.1 The pressures of a static gas bubble in a liquid.

If an imaginary cut is made which divides the bubble in half, the excess pressure p_σ would tend to push the two halves of the bubble away from one another. The force which balances the excess pressure and so keeps the static bubble intact is the surface tension.

The energy associated with a surface in a liquid is given by the product of the surface tension with the surface area. Surface tension σ is numerically equal to the force per unit length acting perpendicular to one side of a straight line in a liquid surface.[1] The force along the imaginary cut is therefore $2\pi R\sigma$.

The force which this balances is $\pi R^2 p_\sigma$, that is the excess pressure multiplied by the effective area seen in the direction of the 'push', πR^2 (Figure 2.2). This can be shown more rigorously by integrating the force on annuli such as the one shown in Figure 2.3(a). The radius of the annulus is $2\pi R\sin\theta$ (Figure 2.3(b)). Only the x-component of the force is non-zero, so that the force on such an annulus is

$$dF_x = (2\pi R\sin\theta)(p_\sigma\cos\theta)(Rd\theta) \tag{2.3}$$

Integrating this from $\theta = 0$ to $\theta = \pi/2$ gives the required net force of $\pi R_o^2 p_\sigma$. Equating this to the surface tension force gives

$$p_\sigma = \frac{2\sigma}{R} \tag{2.4}$$

the excess pressure inside a bubble which results from surface tension, also known as the *Laplace pressure*.

It is interesting to compare the surface tension pressure of the gas bubble in a liquid with that of the soap bubble. This contains air, and exists in a medium of air. The bubble now contains two surfaces, an inner and an outer, so that the length of the 'cut' is now $4\pi R_o$. As a result, the excess pressure resulting from surface tension in the soap bubble is $4\sigma/R_o$.

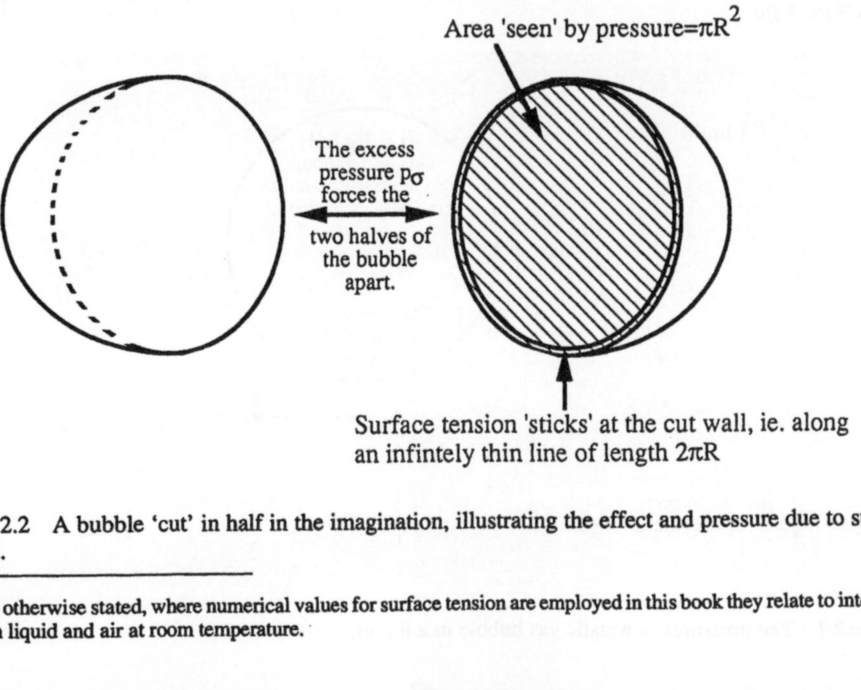
Area 'seen' by pressure=πR²
The excess pressure pσ forces the
two halves of the bubble apart.
Surface tension 'sticks' at the cut wall, ie. along an infintely thin line of length 2πR

Figure 2.2 A bubble 'cut' in half in the imagination, illustrating the effect and pressure due to surface tension.

[1]Unless otherwise stated, where numerical values for surface tension are employed in this book they relate to interfaces between liquid and air at room temperature.

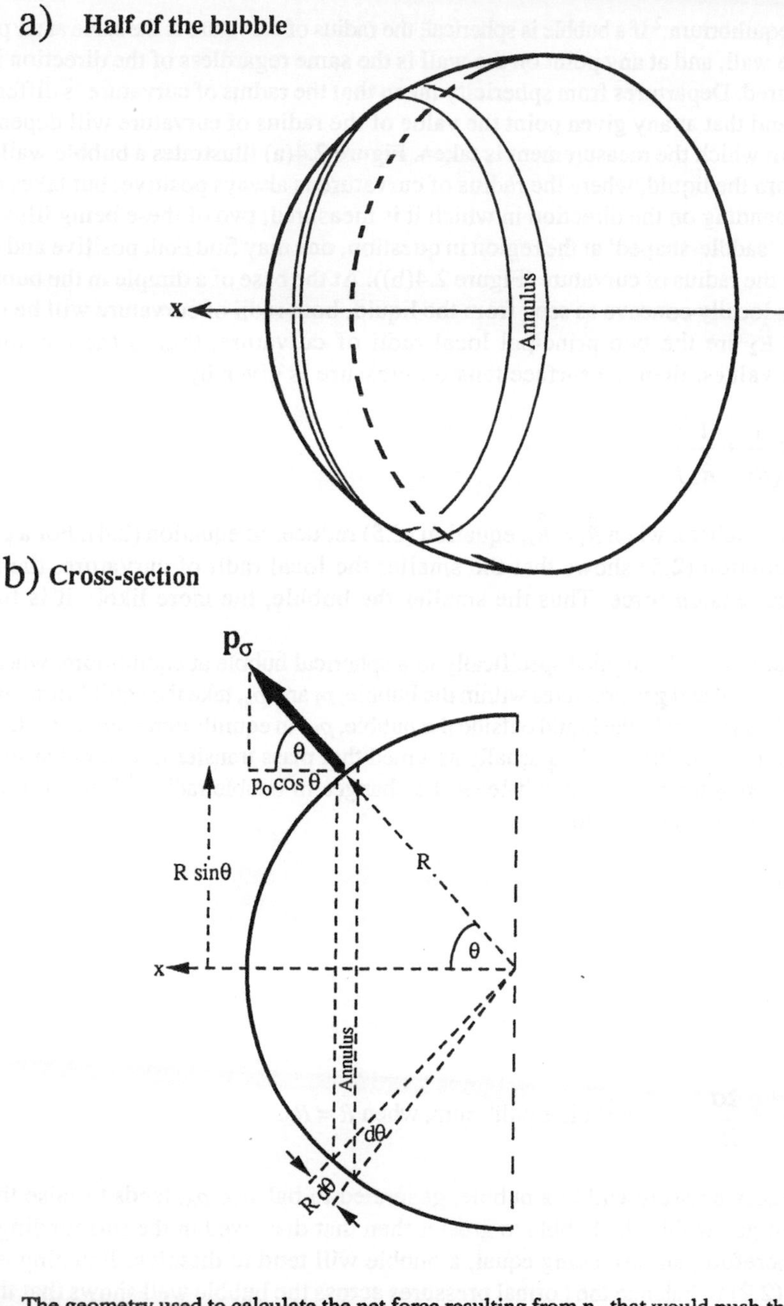

Figure 2.3 The geometry used to calculate the net force resulting from p_σ that would push the halves of the hypothetically cut bubble apart. (a) The annulus. (b) The geometry.

If the surface of a gas bubble in a liquid has non-uniform curvature (i.e. the bubble is non-spherical), there is a local surface tension pressure associated with each part of the surface. In the absence of other forces (like gravity/buoyancy, adhesion to walls etc.) the imbalance of these pressures will tend to return the bubble to a spherical shape. Overshoot resulting from inertia of the liquid will cause oscillation of the bubble shape about the spherical equilibrium.[2] If a bubble is spherical, the radius of curvature is the same at all points on the bubble wall, and at any point on the wall is the same regardless of the direction in which it is measured. Departures from sphericity mean that the radius of curvature is different over the wall, and that at any given point the value of the radius of curvature will depend on the direction in which the measurement is taken. Figure 2.4(a) illustrates a bubble wall, convex as seen from the liquid, where the radius of curvature is always positive, but takes different values depending on the direction in which it is measured, two of these being illustrated. If the wall is 'saddle-shaped' at the region in question, one may find both positive and negative values for the radius of curvature (Figure 2.4(b)). At the base of a dimple in the bubble wall, where it is locally concave as seen from the liquid, both radii of curvature will be negative. If $\hat{R}_1$ and $\hat{R}_2$ are the two principal local radii of curvature, that is the maximum and minimum values, then the surface tension pressure is given by

$$p_\sigma = \sigma\left(\frac{1}{\hat{R}_1} + \frac{1}{\hat{R}_2}\right) \tag{2.5}$$

Clearly for a sphere, when $\hat{R}_1 = \hat{R}_2$, equation (2.5) reduces to equation (2.4). For a perturbed sphere, equation (2.5) shows that the smaller the local radii of curvature, the stronger the surface tension force. Thus the smaller the bubble, the more likely it is to tend to sphericity.

The analysis can be applied specifically to a spherical bubble at equilibrium, when $R = R_0$. Here the internal and gas pressures within the bubble, p_i and p_g, take the equilibrium values, $p_{i,e}$ and $p_{g,e}$. The pressure in the liquid outside the bubble, p_L, in equilibrium is equal to the pressure remote from the bubble, p_∞. It is usually assumed that mass transfer is so rapid as to keep the the vapour pressure p_v constant whatever the changes in bubble radius.[3] Therefore equations (2.1), (2.2) and (2.4) reduce to

$$p_{i,e} = p_{g,e} + p_v \tag{2.6}$$

$$p_\sigma = \frac{2\sigma}{R_0} \tag{2.7}$$

and

$$p_{i,e} = p_\infty + \frac{2\sigma}{R_0} \quad \text{at static equilibrium, when } R = R_0. \tag{2.8}$$

The excess pressure within a bubble, generated to balance p_σ, tends to raise the partial pressure of gas within the bubble to greater than that dissolved in the surrounding body of liquid. Therefore, all else being equal, a bubble will tend to dissolve. Equating equations (2.1) and (2.2) to balance the normal pressures across the bubble wall shows that the partial pressure of gas within a bubble will exceed $p_L - p_v$. Henry's Law[4] implies that for equilibrium conditions, the concentration of gas in solution will be proportional to the partial pressure of the adjacent gas phase at a given temperature [1]. As a result, therefore, of the excess gas partial pressure within the bubble, the concentration of dissolved gas at

---

[2] See section 2.2.4(c) and Chapter 3, section 3.6.
[3] See section 2.1.3(b).
[4] See Chapter 4, section 4.4.3(b)(ii).

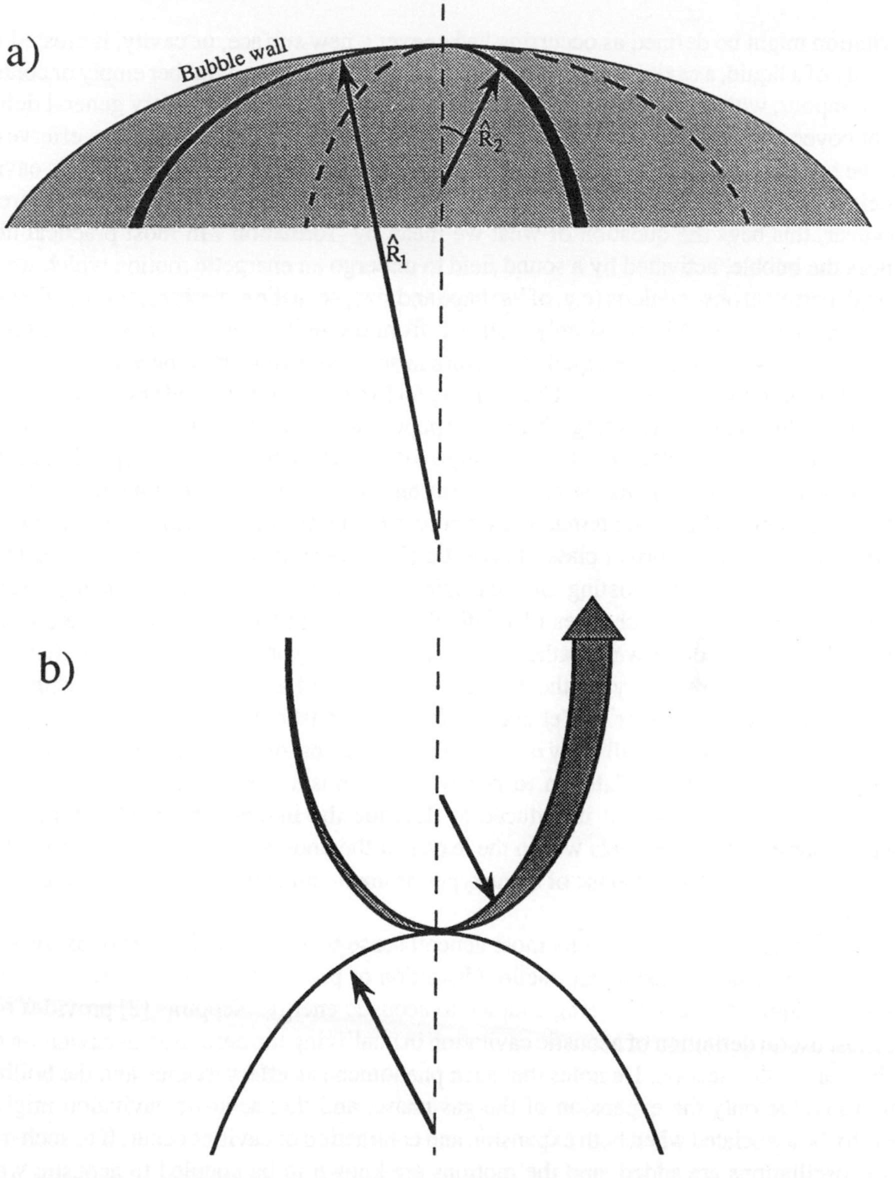

Figure 2.4 The two principal radii of curvature on aspherical surfaces. (a) If the bubble wall is convex as seen from the liquid, both radii of curvature are positive. (b) A positive and a negative radius of curvature at a saddle point.

the bubble wall exceeds the concentration far from the bubble, and as a result of this concentration gradient the bubble will dissolve. The bubble size is reduced as a result, p_σ increases, and so the excess gas partial pressure becomes greater (from equation (2.4)). Thus a gas bubble in a pure liquid will tend to dissolve away.[5]

[5]See section 2.1.2(b), and Chapter 4, section 4.4.3(b).

2.1.2 Cavitation Inception

Cavitation might be defined as occurring "whenever a new surface, or cavity, is created within the body of a liquid, a cavity being defined as any bounded volume, whether empty or containing gas or vapour, with at least part of the boundary being liquid" [2]. This very general definition might cover such wide-ranging phenomena as underwater explosions, and also effervescence and the boiling of a liquid. To open discussion, Apfel [3] initially refers to acoustic cavitation as being the formation of a vapour cavity or bubble in response to an acoustic pressure field. However, this begs the question of what we mean by 'formation'. In most practical circumstances the bubble, activated by a sound field to undergo an energetic motion which we detect through pertinent observations (e.g. of its shape and size, acoustic emissions, sonoluminescence, erosive properties etc.) is not simply 'formed' from the bulk liquid, but is seeded from some pre-existing gas pocket in the liquid. Therefore in some situations this 'new surface' is created by expansion of existing surfaces. Therefore Apfel [3] generalises his definition to "encompass any observable activity involving a bubble or population of bubbles stimulated into motion by an acoustic field." Whilst the above example of the growth of macroscopic bubbles from microscopic nuclei in response to relatively strong pressure fluctuations illustrates an acoustic bubble interaction where an external sound field can evoke a dramatic response (volume change) from the bubble, an important class of acoustic phenomena involving bubbles arises from the passive emission of pre-existing macroscopic bubbles which are mechanically excited to undergo shape or volume changes of relatively small amplitude. Such processes, where the bubble excites the sound wave rather than vice versa, might be classed as *passive acoustic cavitation*. They represent by far the most common form of acoustic interaction with bubbles, and are discussed in Chapter 3. Yet another form of acoustic cavitation is the class *gas body activation* [4] in which a small body of gas, stabilised against dissolution and buoyancy through a supportive structure, oscillates in response to an acoustic field. Such bodies are often not spherical. The class was first introduced to describe the interaction of ultrasound with the intercellular gas-filled channels within the leaves of the underwater plant *Elodea* [5], and may provide a model for the response of some types of mammalian intracellular bubbles to clinical ultrasound fields.[6]

Acoustic cavitation might in its most general sense therefore be thought of as the creation of new surfaces or expansion/contraction/distortion of pre-existing ones in a liquid within the body of a liquid, the process being coupled to acoustic energy. Neppiras [2] provides perhaps the most useful definition of acoustic cavitation in qualifying the definition of cavitation quoted at the start of this section. He notes that such phenomena as effervescence and the boiling of a liquid involve only the expansion of the gas phase, and that acoustic cavitation might most usefully be associated when both expansion and contraction of cavities occur. If to such motions shape oscillations are added, and the motions are known to be coupled to acoustic waves in some way, then acoustic cavitation is adequately described.[7] Neppiras [2] notes that the term cavitation is often loosely applied to the effects, as well as the formation and motion, of cavities.

If one starts with a pure, homogeneous liquid, the process of *initiating* acoustic cavitation might in its simplest sense therefore be thought of as 'tearing' the liquid apart, the cavity so formed filling with liquid vapour and/or any gases that are dissolved within the liquid. One would therefore expect acoustic cavitation inception to occur only at acoustic pressure amplitudes

---

[6]See Chapter 5, sections 5.4.2(b) and 5.4.2(c).

[7]Definitions of acoustic cavitation are still the subject of controversy, and the subject will be further discussed in Chapter 4, section 4.4.8.

which generate tension within the liquid equal to or greater than the tensile strength of the liquid. However, in interpreting this phase, it is often not the properties of the *pure* liquid *per se* which are important. The relevant tensile strength relates to the complete liquid system, and may depend critically on inhomogeneities and contaminants in the liquid, and sometimes even on the container walls.

(a) Tensile Strength of a Liquid

The tensile strength is that tension which a liquid can support without breaking (cavitating). However, in most practical situations the liquid involved is a complex system, containing many impurities which can act as weaknesses, and it is these which limit the tension that the liquid can support.[8] As regards acoustic phenomena, to a first approximation a liquid is put into tension when the acoustic pressure amplitude exceeds the static ambient pressure, such that the pressure in the liquid becomes negative (Figure 2.5). Initial studies were, however, done on the application of static stresses.

The history of tensile strength measurement for fluid is expounded in detail by Trevena [6]. In the sixteenth century, Euler predicted that a flowing liquid might be subject to tensile stresses

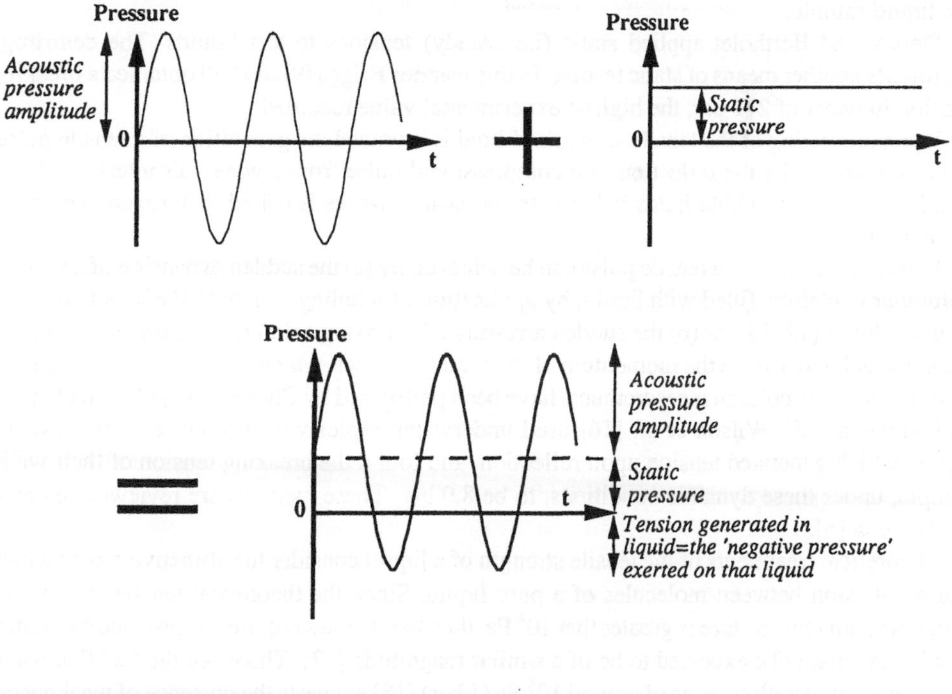

Pressure: 1 atmosphere = 1.013 bar = 1.013×10^5 Pa.

Figure 2.5 When an acoustic wave passes through a liquid, the pressure in a liquid is the sum of static and oscillating terms. If the acoustic pressure amplitude is greater than the static pressure, then the pressure in the liquid will be negative for part of each acoustic cycle, when the liquid will be in tension.

[8] 'The strength of a chain is governed by its weakest link'.

if the flow velocity were high enough.[9] In 1846, Francois Donny performed experiments on stationary liquids under tension [7]. A column of sulphuric acid, filling one arm of a closed U-tube, supported its own weight after the gas pressure in the other arm had been reduced with a vacuum pump as a result of *cohesion* of the liquid molecules to one another, and *adhesion* between the liquid and the glass. Donny found that if the liquid were free of dissolved gas it could sustain greater tension (up to about 1/4 bar[10]) than if it were not. In the latter case the appearance within the liquid of many bubbles, which expanded as the pressure was reduced, eased the tension in the liquid.

In 1850, Bertholet [8] heated a closed glass tube which was almost filled with liquid, the remainder of the volume being gas. On heating, the liquid expanded more than the glass, forcing the gas into the liquid, so that the latter filled the vessel. On cooling, the liquid adhered to the glass: since the liquid was thus restrained from contraction, tension was generated within it. The tension increased as the liquid cooled, until cavitation occurred. Thus Bertholet measured the tensile strength of the water sample to be around 50 bar, though since cavitation initiated at the walls of the tube, rather than in the body of the liquid, it was the forces of adhesion between glass and liquid that were overcome, not the cohesion between the liquid molecules.

This demonstrates an important point: that it is not the properties of the liquid *per se* that determine the maximum tension a liquid can sustain, but often the other bodies present within the liquid sample.

Donny and Bertholet applied static (i.e. steady) tensions to test liquids. The centrifuge represents another means of static testing. In this manner Briggs [9] in 1950 obtained a breaking tension in water of 277 bar, the highest experimental value recorded.

Dynamic testing of the tensile strength of liquids involved the generation of a tensile pulse, either directly or by the reflection of a compressional pulse from a water–air interface.[11] The rapid formation of bubble fields behind strong tension waves is called 'heterogeneous shock cavitation' [10, 11].

Direct application of a tensile pulse can be achieved by (a) the sudden expansion of a vertical columnar container, filled with liquid, by application of a falling weight to the lower end-wall of the column [12, 13], or (b) the sudden arresting of a moving column of liquid: adhesion and cohesion act to dissipate the momentum of the liquid particles, which sets up tension within the liquid.[12] Moving column measurements have been performed by Chesterman [14] and Overton and Trevena [15]. Wilson *et al.* [16] used underwater explosives to produce compressional pulses which generated tension upon reflection, and found the breaking tension of their water sample, under these dynamic conditions, to be 8.0 bar. These methods are reviewed in detail by Trevena [6].

Theoretical treatments of the tensile strength of a liquid consider the attractive forces which cause cohesion between molecules of a pure liquid. Since the theoretical tensile strength of water at room temperature is greater that 10^8 Pa, then the stresses required to produce cavitation would reasonably be expected to be of a similar magnitude [17]. Therefore the fact that water can be cavitated with stresses of around 10^5 Pa (1 bar) [18] suggests the presence of weaknesses or cavitation nuclei within the liquid. Overton *et al.* [19] have shown a marked dependence of

---

[9] The physics of this process will be discussed in section 2.2.2.

[10] The SI unit of pressure is the pascal (Pa), equivalent to 1 newton per square metre. A common historical unit is the bar, equal to 10^5 Pa. One atmosphere equals a head of 760 mm of mercury, equal to 101 325 Pa = 1.013 25 bar. However, a standard atmosphere is sometimes taken by meteorologists to mean 1 bar = 10^5 Pa.

[11] See Chapter 1, section 1.1.5.

[12] In much the same way as a gymnast's shoulders are subjected to tension when she catches herself on a bar during a fall.

liquid tensile strength on its cavitation history, owing to the effect on the nuclei population. These nuclei are discussed within the next section.

(b) Cavitation Nuclei

The nature of the pre-existing seed nuclei within a liquid from which bubbles can grow is still the subject of some controversy. Free-floating bubbles containing a mixture of both water vapour and gases that are dissolved in the liquid are unsuitable candidates for two reasons. Firstly, buoyancy would remove them from the liquid. Flynn [20] showed that water which has been left standing for a few hours contains only bubbles of radius less than 5×10^{-6} m. Secondly, as already discussed, the excess internal pressure which counteracts contractual forces due to tension in the bubble wall would cause the bubble to dissolve away.[13] Epstein and Plesset [24] demonstrated theoretically that a bubble of a few micrometers radius will dissolve in water within a few seconds.[14] Even when the liquid is supersaturated with dissolved gas, bubbles smaller than some critical size will dissolve [24, 25].

When discussing the nucleation of bubbles in liquids we are generally discussing nuclei of micron-order size. Clearly, from the above timescales, dissolution is the more immediate threat to a tiny bubble nucleus, buoyancy being of secondary importance, a consideration that is reduced in the presence of motions and currents within the liquid. Thus, although freshly drawn tap water may contain free-floating bubble nuclei, such entities cannot be considered viable general seeds for most liquids.

Some workers have attempted to measure the gas nuclei population through the absorption of sound [26–28]. Il'in *et al.* [29] did likewise, in addition using hydrodynamically generated pressure reductions to measure both the concentration and strength of cavitation nuclei. Apfel [30], however, points out that these studies determine neither the mechanism of stabilisation nor the dependence of cavitation strength on nuclei size.

Messino *et al.* [31] were the first to conduct a statistical study of ultrasonic cavitation nucleation in 1963. Carpenedo *et al.* [32] applied Messino's statistical methods to tensile strength measurements, cavitating water acoustically and then producing nuclei-size distribution spectra by statistical means. Finkelstein and Tamir [33] reduced the pressure above supersaturated solutions of gases in liquids until bubbles appeared. These results they related to a mechanism based on clusters of gas molecules acting as nucleation centres.

In the somewhat different system of a cavitation tunnel, Hentschel *et al.* [34] were able to generate population histograms for bubble nuclei in a given size range, using digital analysis of pulsed off-axis holographic records. Holography[15] allowed the entire population of bubbles in the diameter range 20–250 μm to be sampled simultaneously throughout a given volume (up to 220 mm^3), and sized to ±5 μm, provided the particle concentration in the cavitation tunnel was not too great. A special degassing and nuclei-seeding system allowed counts to be taken under a range of conditions, for flow with a model of a propeller running at frequencies of 20

---

[13] In an interesting and controversial theoretical discussion, it has been suggested that even in the absence of external stabilising forces a bubble might not dissolve completely. Wentzell [21] hypothesised that, if the van der Waals equation of state is valid for the bubble, microbubbles would dissolve to a very small size, but then become stabilised when the van der Waals forces are in stable equilibrium with the surface tension. In water at 20°C, bubbles of CO_2 and Xe are stabilised by this theory when they reach radii of around 0.023 and 0.028 μm respectively. Lubkin [22] argues instead that at this critical radius, the bubble is not stabilised but instead experiences liquefaction, with fast shrinkage, followed by slow but complete dissolution. Wentzell [23] counters that liquefaction is prohibited at temperatures exceeding some critical value.

[14] See Chapter 4, section 4.4.3(b)(ii).

[15] See Chapter 5, section 5.1.

and 30 Hz. Though differing in several major ways regarding the type of bubble under consideration, this study demonstrates the potential for holography in this field.

Clearly, if one is to test the tensile strength of a pure liquid *per se*, it is important to remove alien bodies from the liquid. Tap water contains between 50 000 to 100 000 of these solid *motes* per cubic centimetre [3, 35]. The majority of these can be removed by very careful filtering, enabling the resulting water to withstand stresses of 1.6×10^7 Pa for up to a minute, and as high as 2.1×10^7 Pa for a few seconds [36]. One would expect that if such motes were the sole source of stabilised nuclei, such filtration would enable water to withstand tensions equivalent to its theoretical tensile strength. However, the result of Greenspan and Tschiegg [36] is still only a fifth of the theoretical tensile strength, the discrepancy being due to nuclei produced by radiation [35].

Experiments using radioactive sources have previously shown the importance of radiation-induced cavitation inception. West and Howlett [37] observed no memory effect in the liquid, whilst Sette and Wanderlingh [38], Lieberman [39], Finch [40] and Barger [41] did see such an effect. An appropriate mechanism using a 'thermal spike' model was suggested by Seitz [42] in 1958. In it, ionising radiation generates a positive ion, the motion of which excites neighbouring atoms, producing a thermal spike and nucleating a vapour bubble. If the liquid is in tension, this bubble will grow. The production of free-floating microbubbles by radiation represents cavitation inception in the absence of stabilised nuclei. This can also occur through tribonucleation [43] as a result of frictional processes when solid surfaces are rubbed together, and when tension is generated at the walls of the container through impact [44].

A suitably shielded unfiltered liquid sample, left to stand for a long period prior to testing, would be free of nucleation by radiation or by unstabilised gas or vapour cavities as a result of dissolution and buoyancy. The fact that the cavitation threshold of such a sample has been observed to be constant in time [41, 45, 46] suggests stabilised nuclei. Similarly once denucleated, a liquid remains so for extended periods [47, 48], indicating the lack of any mechanism to generate fresh nuclei rapidly. This implies that the original nuclei must have been present for a significant time prior to denucleation, and were therefore stabilised. Of the candidate models for the putative stabilised nucleus, two have proved overwhelmingly successful. These are referred to as the variably permeable skin model, and the crevice model.[16]

(i) Variably Permeable Skin Model. Most water samples contain organic impurities such as fatty acids. These might accumulate on the wall of a bubble. As the bubble dissolves, the area of the bubble wall becomes smaller, and at some point the organic impurities will form a complete 'skin' surrounding the bubble. Further decrease in bubble size is inhibited by this skin: a decrease in bubble wall area would necessitate a decrease in the configurational entropy of the skin. However, if the liquid were subjected to tension, the bubble is free to grow by an influx

[16]Which model is predominantly relevant depends on the details of the circumstances, primarily on the nature of the contaminant in the liquid (whether surface-active chemicals or hydrophobic solid particles) and on the type of bubble behaviour which is relevant to the situation in question. Two extreme examples can be found in later sections of this book. In the consideration of oceanic bubbles (Chapter 3, section 3.8), one can envisage a situation where the quantity of surface-active agent is very considerable, generated, for example, by biological or industrial processes, and where the acoustic interactions consist of freely floating bubbles with relatively low-amplitude sound (for example, in the bubble-mediated waveguiding of ambient noise). In such situations, the acoustic pressures would be unlikely to produce the tensions required to generate a freely floating bubble from a crevice. For small bubbles, the energetic motion of the seawater will be more significant than buoyant forces. As a result, one would expect the significance of the crevice model to be reduced. In contrast, in Chapter 5, section 5.1.2 one finds examples of the purposeful and controlled seeding of otherwise very clean liquids with several forms of solid and stabilised gaseous nucleus. The acoustic pressure amplitudes that are employed are in this case very high, and the liquid will go into tension.

of gas previously dissolved in the liquid. Liquids such as alcohol, in which fatty acids would dissolve and so be ineffective in nuclei stabilisation, would accumulate a skin of protein.

In this manner a bubble nucleus would be stabilised against dissolution, but buoyancy might still remove it from the liquid. Small bubbles, however, would probably not be affected much by buoyancy: as the size decreases, liquid currents and Brownian motion become increasingly important.

The model was first proposed in 1954 by Fox and Herzfeld [49] who envisaged a rigid skin of organic compounds. Herzfeld subsequently withdrew the model as being inconsistent with the measurements of Strasberg [46]. In addition, Barger [41] measured no difference between the thresholds for distilled and tap water, though from the original model one would have expected a noticeable difference in the thresholds of two liquids that differed so much in chemical purity.

In 1970, Sirotyuk [50] modified the model to incorporate a skin of specially orientated surface-active polar molecules. The restriction of a rigid skin was removed. Akulichev [51] proposed a model where ionic charges present in the liquid migrate to the surface and bring about stabilisation through Coulombic repulsion.

The most successful of these 'skin' models was put forward by Yount [52–54], who proposed an elastic organic surface-active skin which, though initially permeable to gas, became impermeable as the concentration of organic molecules increased on the contracting bubble wall, thus stabilising the bubble against dissolution. This variably permeable skin allows the model to give results compatible with observations of gas diffusion [55], rectified diffusion,[17] and Strasberg's [46] observations on the effects of hydrostatic pressure and degassing [30]. The surface-active molecules are stored in a reservoir located just outside the skin. Molecules can be exchanged between the skin and the reservoir as required when the volume of the bubble, and therefore the area of the bubble wall, changes. Thus, unlike ordinary gas bubbles, Yount's nuclei can be subjected to changes in pressure and stabilised at several different pressures, provided that the threshold for bubble formation is not exceeded. This fact can be used to distinguish such stabilised nuclei from ordinary gas bubbles [54]. Yount's model is in agreement with the observations of Johnson and Cooke [56], who noted that some bubbles, when injected into seawater, after a certain time ceased to dissolve any more (others dissolved completely). These nuclei had radii of the order of μm. Although changes to the nuclei distribution occurred over the next 20 hours or so, these bubbles nevertheless appeared to be stabilised.

Nuclei counts have been performed by Yount and Strauss [48] and Yount and Yeung [57]. Yount *et al.* [54] undertook an investigation of microscopic bubble nuclei, using both light and electron microscopy. Free-floating nuclei, which could be stabilised at several pressures and so were not ordinary gas bubbles, were found in both distilled water and Knox and agarose gelatin,[18] ranging in radius from 10^{-9} to 10^{-6} m. The number density decreases exponentially with increasing radius. That these nuclei are gas-filled is evidenced by their change in size in response to pressure fluctuations in the liquid. Clusters and binaries are observed: this and the absence of motes suggests stabilisation through surface-active agents. From observed bilayer septa dividing joined nuclei, the thickness of the surfactant films is estimated to be $(20 \pm 7) \times 10^{-10}$ m. The occurrence of binary and trinary clusters in regions darkened by osmium-tetroxide staining is suggestive of surfactant reservoirs, since osmium is often employed to enhance the contrast of surface-active materials in electron microscopy.

[17] See Chapter 4, section 4.4.3.

[18] Indeed, Yount's original model [52] was proposed to explain observations on supersaturations in gelatin [48], relevant to decompression sickness studies in humans.

(ii) Crevices in Motes. Bubbles rise by buoyancy because the mass of fluid displaced by the bubble is less than the mass of gas within the bubble. If, however, the bubble were adhering to a small solid particle (called a mote) of greater density than the liquid, then the bubble might not rise.[19] If the motes have diameter less than about 10 μm, Brownian rather than gravitational effects dominate their motion, and the motes do not settle out [20, 58]. Therefore a bubble might be stabilised with regard to buoyancy by attachment to a mote, or also by adhering to the wall of the container. Most people will be familiar with the occurrence of visible bubbles in their overnight glass of water or gin, stabilised against buoyancy by attachment to the wall of the glass. Such bubbles are generally larger than the microscopic nuclei of relevance in this chapter, but nevertheless warrant discussion for illustrative purposes. The solubility of air in water decreases with increasing temperature. These bubbles form when gas, which readily dissolves into the water or gin in the cold night, comes out of solution in the warmth of the morning.[20] However, though liquid samples may possess motes to stabilise small bubbles against buoyancy, they do not in general benefit from temperature fluctuations to stabilise the bubbles against dissolution. It is here, however, that the real importance of the mote model for nucleus stabilising becomes evident.

In section 2.1.1 it was shown how there is a pressure p_σ generated by surface tension on a curved surface which acts in towards the centre of the radius of curvature. To counteract this, the pressure within the bubble must be greater by p_σ than the pressure in the liquid. This gas pressure imbalance means that gas will dissolve out of the bubble into the liquid. However, if the curvature of the bubble were the other way (i.e. concave as seen from the liquid, which we will take to indicate $R < 0$), the surface tension pressure would act into the liquid, reducing the gas pressure within the bubble, so that dissolved gas would tend to leave the liquid and enter the bubble.

Figure 2.6 illustrates a gas pocket in a crevice. The radius of curvature is shown, the arrow signifying the direction of the surface tension pressure. If the meniscus is concave as seen from the liquid, the radius of curvature is negative and so, therefore, from equation (2.4), is p_σ, which thus acts to reduce the gas pressure (Figure 2.6(a)). The angle of contact θ, of the meniscus against the solid material of the mote is shown. In Figure 2.6(b), the meniscus and crevice geometries are such that the meniscus is flat. This corresponds to an infinite radius of curvature, and so from equation (2.4) the surface tension pressure is zero. The pressure within the gas pocket will equal the pressure in the liquid. Figure 2.6(c) illustrates a circumstance when the radius of curvature is positive, and the meniscus convex as seen from the liquid. The surface tension pressure will act to increase the pressure within the gas pocket above that in the liquid.

It is therefore not difficult to see how a crevice in a mote might stabilise a gas pocket against dissolution, even in an undersaturated liquid, by enabling the meniscus to take a concave form as seen from the liquid. In turn, when circumstances are favourable, this crevice can generate free-floating bubbles which pass into the body of the liquid. Anyone who has watched bubbles rising in beer will not be surprised that discrete sites can act thus as nucleation points.[21]

The model was first put forward in 1944 by Harvey *et al.* [47]. Studies of gas pockets contained within small-angled crevices in imperfectly wetted hydrophobic solids followed. The

---

[19] In the unlikely event of the average mass of the bubble–mote system equalling the mass of fluid displaced, there would be no buoyant force on the bubble. It would neither rise nor sink, but stay suspended, much like the system of a lighter-than-air balloon attached to a man in a basket. One might think that in practice there is likely to be a net mass imbalance, which would cause a slow rise or fall, however, the translatory motion of particles of such small sizes as these will be dominated by the random walk introduced through Brownian motion.

[20] See Chapter 5, section 5.2.2(b)(iii).

[21] The gas flux here resulting from oversaturation of dissolved carbon dioxide in the liquid.

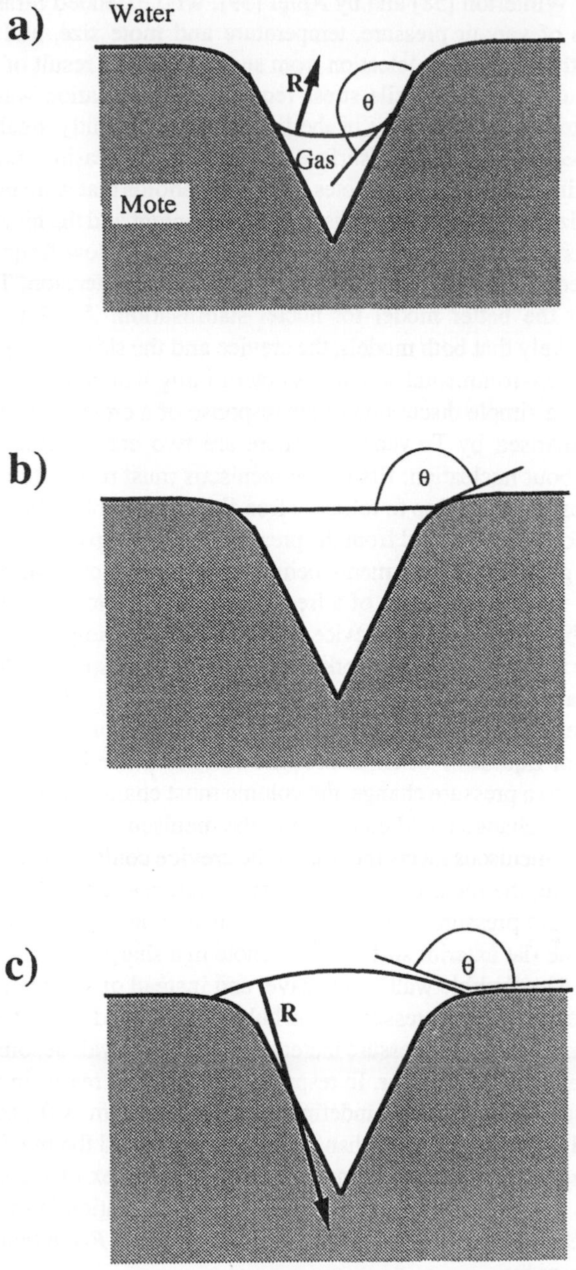

Figure 2.6 Schematic of a gas-pocket within a crevice in a mote, illustrating the radii of curvatures R that may be found: (a) R is negative; (b) R is infinite; (c) R is positive.

hydrophobicity arises either through the nature of the particle itself, or through hydrophobic organic contaminants which are adsorbed onto the mote surface. Strasberg [46] attempted to predict the effect of imperfectly wetted conical crevices on the cavitation threshold. Strasberg considered the concentration of dissolved gas, and the degree of pre-pressurisation. Further work was done by Winterton [58] and by Apfel [59], who extended Strasberg's arguments to include the effects of vapour pressure, temperature and mote size, and obtained qualitative predictions of the threshold for nucleation from such seeds as a result of changes in the static pressure. Apfel found that the tensile stress required for nucleation was independent of the dissolved gas content and the history of the liquid for sufficiently small and readily wetted motes, but increased with increasing surface tension and decreasing mote size. At the other extreme, large and imperfectly wetted motes caused thresholds that were independent of surface tension and mote size, but for which the dissolved gas content and the history of the liquid were critical. The results were then generalised to cavitation under low-frequency acoustic fields. Crum [60] extended the model to include the effect of surface tension. Though some authors have considered it the better model for nuclei stabilisation, Atchley and Prosperetti [61] conclude that it is likely that both models, the crevice and the skin of varying permeability, are valid, the particular environmental conditions determining which predominates.

The following is a simple discussion of the response of a crevice gas-pocket to quasi-static pressures, as summarised by Trevena [6]. There are two conceivable criteria that must be satisfied to bring about nucleation: firstly, the meniscus must recede to the top of the crevice, and secondly it must then develop to release a free-floating bubble. The threshold condition in Trevena's exposition is determined from the pressure reduction required to cause the meniscus to recede, simple geometrical arguments being given to suggest that this criterion is more stringent than that for the generation of a free-floating bubble to nucleate the liquid once the meniscus has reached the top of the crevice. However, this assumption may be too simplistic, and the discussion will end with the work of Atchley and Prosperetti [61], who consider the criterion for generation of a free-floating bubble.

Trevena's model relates the pressure changes that occur within the fluid, as happens during the tensile testing of liquids or when an acoustic wave is passed, to the behaviour of the gas pocket. In response to a pressure change, the volume must change. There are two ways through which such a volume change could come about: the meniscus shape could change, or the line of contact where the meniscus meets the wall of the crevice could move. Energy considerations show that, if possible, the meniscus will move in preference. Consider a crevice in a pocket, under a positive liquid pressure, where the crevice is a cone of half-angle θ_c, having straight walls which meet the flat exterior surface of the mote in a sharp corner. The angle of contact is θ_l (Figure 2.7(a)). The 'bubble wall' is concave, and instead of surface tension increasing the pressure within the bubble, the pressure in the bubble is reduced. Thus the bubble is stabilised against dissolution. If the liquid pressure increases, the bubble wall becomes more concave, the radius of curvature becoming greater. In response to further increases in the positive pressure, the angle of contact cannot increase indefinitely: the maximum is θ_a, the *advancing contact angle.* If the liquid pressure is greater than that required to bend the meniscus to θ_a, the line of contact between meniscus and crevice moves towards the apex of the cone. The gas pocket shrinks, the liquid advancing down into the crevice. This situation is shown in Figure 2.7(b), The meniscus touches the wall of the cone in a circle of radius R_{c1}, which as can be seen from Figure 2.7(b) is given by

$$R_{c1} = R_1 \sin(\theta_a - \theta_c - \pi/2) \tag{2.9}$$

where R_1 is the radius of curvature of the meniscus at this moment. In considering the tensile strength and nucleation of a liquid, we are clearly dealing with the case of *reduced* liquid pressures, not the increased ones discussed above. If the liquid pressure is reduced, then the bubble changes volume by initially changing the shape of the meniscus, as before. If the liquid pressure is reduced slightly, the wall will simply become less concave. However, if the pressure reduction is considerable, the bubble wall becomes convex (i.e. the radius of curvature of the bubble wall becomes positive). The minimum angle is θ_r, the *receding contact angle*. At this liquid pressure or less, the liquid recedes up the cavity.

In the most basic analysis, the whole question of nucleating liquids from these microscopic bubble nuclei is one of overcoming the surface tension forces. When the bubble wall is convex, the net surface tension pressures are inwards, and the surface tension is acting to contain the gas within the bubble. Thus, it has been argued that if the liquid pressure is reduced such that the gas–liquid pressure imbalance can overcome the surface tension when the latter is acting

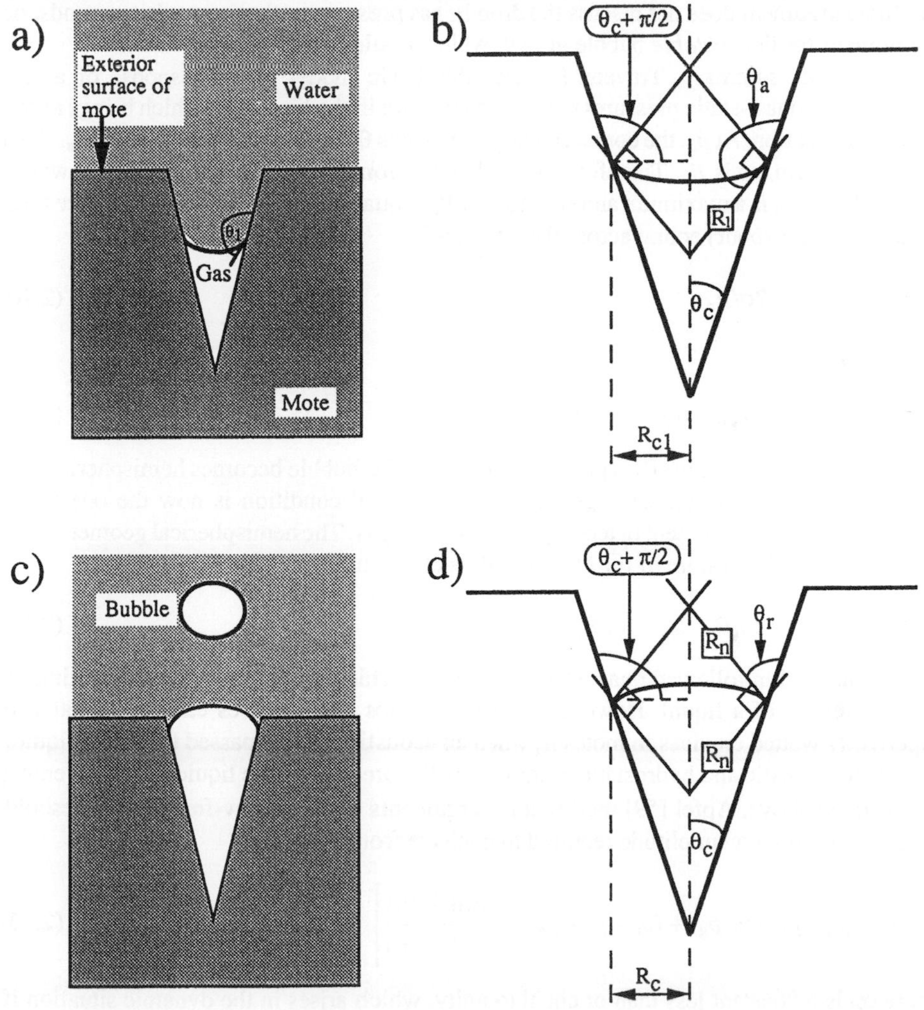

Figure 2.7 Geometry of a gas-pocket within a crevice. (a) Originally, the bubble wall is static and concave, with negative radius of curvature and angle of contact θ_1. (b) Under an increased pressure, the bubble wall advances into the crevice. The radius of curvature of the wall is R_1 (negative), and the meniscus touches the crevice wall in a circle of radius R_{c1}. (c) Under a reduced pressure, the bubble wall is convex, and recedes up the crevice. The figure illustrates the situation at the top of the crevice, where a free-floating bubble may be nucleated. (d) Geometry of the receding wall.

most potently to contain the gas, then in the simplest analysis the liquid will become nucleated. As can be seen from section 2.1.1, the surface tension pressure is greatest when the radius of curvature of the bubble wall is least. This occurs in our particular crevice model just at the moment when the meniscus attains an angle of contact of θ_r. Before this stage, it can still bow outwards further. After it, the circle of contact of the meniscus with the crevice begins to move outwards, increasing the base area of the meniscal dome and so decreasing its radius of curvature. Thus if the gas–liquid imbalance can take the bubble beyond that critical stage when the angle of contact first equals θ_r, then surface tension will not be able to contain the gas and the liquid will be nucleated: On reaching the top of the crevice, a bubble is pinched off the top of the gas pocket (Figure 2.7(c)). The liquid has been nucleated. It should be noted that this simplistic argument does not discuss the drop in gas pressure as the gas pocket expands, or the requirement for the *unstable* bubble growth which would lead to the nucleation.

In a simple summary, Trevena [6] uses this basic model of the response of a crevice gas-pocket to quasi-static pressures to find the pressure in the liquid, p_n, which brings about the onset of nucleation. At p_n, the contact angle just reaches θ_r, and the bubble radius is R_n. Because θ_r is a minimum, so is R_n. Therefore the surface tension pressure, acting inwards towards the apex of the cone, is a maximum and equal to $2\sigma/R_n$. Equating the net pressures (surface tension, liquid, gas and vapour) acting across the interface

$$p_n = p_g + p_v - 2\sigma/R_n \tag{2.10}$$

where

$$R_c = R_n \sin(\theta_c + \pi/2 - \theta_r) \qquad \text{(Figure 2.7(d))} \tag{2.11}$$

If the half-angle of the cone θ_c is greater than θ_r, then the bubble becomes hemispherical before reaching the receding contact angle. The hemispherical condition is now the one at which surface tension is most potent in tending to contain the gas. The hemispherical geometry implies that the radius of curvature of the bubble wall is at that moment equal to R_c. Thus

$$p_n = p_g + p_v - 2\sigma/R_c \qquad \text{for } \theta_c > \theta_r \tag{2.12}$$

This discussion followed quasi-static arguments, which are relevant to static testing of the tensile strength of a liquid. However, it also indicates how bubbles can be nucleated from imperfectly wetted crevices in motes if, when an acoustic wave is passed through a liquid, the sum of the acoustic and hydrostatic terms causes the pressure in the liquid to fall far enough.

As stated above, Apfel [59] modified his arguments to give a low-frequency threshold for the acoustic pressure amplitude required to nucleate from a crevice:

$$P_A = (p_0 - p_v - \alpha_d p_g) + (p_0 - p_v - p_g)\left|\frac{\cos(\theta_r - \theta_c)}{\cos(\theta_a - \theta_c)}\right| \tag{2.13}$$

where α_d is a constant less than or equal to unity, which arises in the dynamic situation if, as the cavity expands outwards, diffusion does not occur rapidly enough to maintain the gas partial pressure at equilibrium. Obviously, in the static case $\alpha_d = 1$.

Application of the above formulation for crevice nucleation was limited by the difficulty in determining the values of θ_a and θ_r. Crum [60] showed that the results of Bargeman and Van Voorst Vader [62] could be incorporated into Apfel's result, giving θ_a and θ_r in terms of the surface tension and properties of nonpolar solids, enabling comparison with experiment [63].

Atchley and Prosperetti [61] point out that the above discussions of the crevice model have an incomplete criterion for nucleation. As the meniscus recedes, not only does the Laplace pressure fall, but so does the partial pressure of the gas within the cavity. If the latter decreases more rapidly than p_σ, the interface will recede only slowly and will cease to recede as soon as the liquid pressure stops falling, as will occur at some point if the pressure changes are acoustic in origin. The growth will, however, be unstable if the partial pressure of the gas decreases less rapidly than the Laplace pressure, and the gas pocket will grow very rapidly.[22] The rate at which the gas pressure falls is determined by the rate of diffusion of dissolved gas out of the liquid and into the gas pocket. If it is rapid, the earlier treatments are valid, but if it is not then the model should incorporate the mandate for mechanically unstable bubble growth. Atchley and Prosperetti [61] therefore define two thresholds for nucleation. The first is derived from the requirement that both the equilibrium solution should be unstable with respect to perturbations, which generate rapid growth, and also that the angle of contact should equal θ_r. If the pocket becomes mechanically unstable before $\theta = \theta_r$, then their threshold coincides with that of previous workers. However, in those cases where the pocket is stable when $\theta = \theta_r$, the theory of Atchley and Prosperetti [61] predicts that a greater pressure drop is required to cause nucleation.

Atchley and Prosperetti [61] model the behaviour of the gas-pocket when the meniscus reaches the top of the crevice. The instability condition at this point may in certain circumstances require the pressure reduction to cause further growth to exceed that required for the initial gas-pocket growth by unstable motion of the contact line. As a result, a second threshold is defined for growth out of the crevice, and this second threshold may be more stringent than the first. In the general case, it is the threshold which requires the greater reduction in pressure which will determine whether nucleation occurs. In the light of the range of behaviours they predict for the crevice model, Atchley and Prosperetti [61] suggest that current understanding of the model is only superficial.

The importance of unstable growth to the nucleation process was first introduced by Blake [25] in 1949, in a discussion of the response of a freely floating spherical bubble to a quasi-static reduction in the liquid pressure. Since discussions of the tensile strength of a liquid, and its nucleation from seeds, have the natural end point of placing a free-floating spherical bubble in the body of the liquid, such a response will now be considered. Though, in the general case, the free-floating bubble can respond to the pressures in the liquid that are static, changing quasi-statically, or acoustically driven, in keeping with the rest of this chapter the discussions will be confined to the response to static and quasi-static pressures in the liquid.

Two situations are discussed. The first considers a bubble containing gas and vapour, in equilibrium with the liquid, which then responds to a slow change in the liquid pressure. The condition of the threshold for unstable growth is examined. The second case is the famous 'Rayleigh cavity', which is simply an empty void collapsing under the ambient liquid pressure. The effect of introducing a gas content within such a collapsing cavity is then investigated.

2.1.3 Response of a Bubble to a Static Pressure

Pressure fluctuations in the liquid, brought about by flow, or by the passage of an acoustic wave through the liquid etc., may cause the generation of free-floating bubbles from stabilised seed nuclei. We have seen that these seeds can be either microbubbles stabilised by a skin of

[22] Section 2.1.3(a) discusses a similar form of unstable growth in relation to a spherical free-floating bubble.

impurities, or gas-pockets trapped within crevices in motes or the container walls. The free-floating bubbles can now respond in some manner to the pressure in the liquid. The two simple cases to be considered may be thought of as the response of a cavity to a steady reduced pressure, and to a steady over-pressure.

(a) Blake Threshold Pressure

Once cavitation has occurred, processes may happen so rapidly that the effects of buoyancy and dissolution can be ignored. Blake [25] and Neppiras and Noltingk [64] made this assumption when they considered the dependence of the pressure required to maintain a bubble in static equilibrium, on the radius of that bubble.

Consider a bubble in static equilibrium. The existence of the surface tension pressure sets up a pressure difference between the gas within the bubble and the external liquid pressure, with the result that the bubble slowly dissolves.

If the liquid is initially in a static equilibrium, the pressure throughout that liquid, including the pressure at very large distances from the bubble (p_∞) and that immediately outside the bubble (p_L), equals the hydrostatic pressure p_0. If the liquid has surface tension σ, then from equation (2.8) the internal pressure of a bubble (radius R_0) within that liquid is

$$p_{i,e} = p_0 + \frac{2\sigma}{R_0} \quad \text{(equilibrium conditions)} \tag{2.14}$$

where, as discussed in section 2.1.1, the equations are applied at the equilibrium position when $R = R_0$, $p_i = p_{i,e}$ and $p_g = p_{g,e}$. If the vapour pressure is always p_v, then from equations (2.6) and (2.14) the pressure of the gas phase within the bubble is

$$p_{g,e} = p_0 + \frac{2\sigma}{R_0} - p_v \quad \text{at equilibrium.} \tag{2.15}$$

Let us say that the pressure in the liquid changes from p_0 in a quasi-static[23] manner such that outside the bubble it has value p_L. Since the change is quasi-static, the pressure throughout the liquid must be spatially uniform and so also equal to p_L. In response to the change in liquid pressure, the bubble radius will change from R_0 to R, and the gas pressure within the bubble will be

$$p_g = p_{g,e}\left(\frac{R_0}{R}\right)^{3\kappa} = \left(p_0 + \frac{2\sigma}{R_0} - p_v\right)\left(\frac{R_0}{R}\right)^{3\kappa} \tag{2.16}$$

assuming the gas obeys a polytropic law (equation (1.18)). Substitution into equation (2.1) gives the pressure in the bubble to be

[23]Every sample of matter contain molecules. Depending on the characteristics of the sample (solid, liquid, gas etc.) the particles will be spatially distributed in a certain way. Impressed forces and thermal motions will impart to each molecule a certain velocity. If a sample of matter were undergoing a quasi-static change, then if an observer was for a single instant able to know the positions and velocities of all the particles within that sample (for example, by taking a special 'snap-shot' that recorded not just molecular positions but also velocities), the observer would not be able to tell from that observation whether or not the sample was indeed changing. In practice, quasi-static changes are those performed slowly enough for dynamic effects to be negligible.

$$p_i = \left(p_0 + \frac{2\sigma}{R_0} - p_v\right)\left(\frac{R_0}{R}\right)^{3\kappa} + p_v \tag{2.17}$$

and thus from equations (2.2) and (2.4) the pressure in the liquid immediately beyond the bubble wall is

$$p_L = \left(p_0 + \frac{2\sigma}{R_0} - p_v\right)\left(\frac{R_0}{R}\right)^{3\kappa} + p_v - 2\sigma/R \tag{2.18}$$

As stated, since we are considering the response of a bubble to a quasi-static pressure change, then the pressure in the liquid remote from the bubble must equal the liquid pressure at the bubble wall, i.e. $p_\infty = p_L$.

Equation (2.18) determines the position of the new equilibrium after the liquid pressure has changed uniformly and quasi-statically from p_0 to p_L. If we were discussing an increase in pressure, then the liquid pressure would still be positive but of greater value than before, and the bubble would have contracted. This position of equilibrium is stable, since the positive pressure will assist the surface tension pressure[24] in 'confining' the gas.

If we had been discussing a small decrease in pressure that still kept $p_L > 0$, the bubble would have expanded but would still be stable: a slight perturbation from this new equilibrium size would generate forces which would return the bubble back to the new equilibrium. In fact, even if the liquid pressure were reduced to zero, the bubble would still be confined by the surface tension pressure. However, a large negative pressure in the liquid will work against this confining effect, and the equilibrium need not be stable.

If the liquid pressure becomes negative ($p_L < 0$), it can begin to counteract the confining effect of p_σ. If $p_L < 0$, then since $p_i = p_L + p_\sigma = p_g + p_v$, after the bubble is larger than some critical radius the pressure balance across the bubble wall cannot be maintained, and the bubble will grow explosively. The bubble is unstable, bubble growth engendering further bubble growth. This situation cannot, of course, continue unabated. After the start of such instability, the bubble behaviour cannot be predicted by the above equilibrium theory. The situation is no longer quasi-static, and we must have recourse to bubble dynamics to describe the growth. Such bubble dynamics are discussed in Chapter 4. In reality, it is impossible to maintain a negative liquid pressure around the explosively expanding cavity for more than a fraction of a second. Thus, in practice, the cavity would expand to some maximum size and then collapse.

To describe such behaviour fully, the inertial and viscous effects of the liquid have to be included. The Blake model ignores these, accounting only for surface tension effects in the quasi-static case. It can therefore never describe the explosive growth. However, it is valid for predicting the onset of explosive growth in situations that are quasi-static and where surface tension dominates inertial and viscous effects. This occurs in the limit of very small bubbles. In these, $p_\sigma = 2\sigma/R_0$ tends to very large values, dominating the other effects. In addition, such bubbles have high resonance frequencies (see equation (3.36)) so that most pressure changes will be over timescales that incorporate frequencies much less than the bubble resonance. As a result, with respect to the bubbles own timescale (determined by its resonance), such pressure changes will be 'felt' by the bubble to be quasi-static.

---

[24] The wall of this spherical bubble is always convex, as seen from the liquid, so that the Laplace pressure is radially inwards.

Understanding, therefore, that the Blake treatment applies only to very small bubbles where p_σ dominates, it is possible to find the threshold value of p_L such that a negative pressure of greater magnitude than this threshold will cause a suitable bubble to undergo explosive growth.

Physically it is clear that these small cavities must be greater than some critical size, since as $R_0 \rightarrow 0$ the surface tension pressure, which tends to provide confinement and stability against growth, increases without limit. This critical radius, R_{crit}, which a nucleus must exceed if it is to become unstable with respect to further expansion, is found by differentiating equation (2.18) with respect to R:

$$\frac{\partial p_L}{\partial R} = -3\kappa\left(p_0 + \frac{2\sigma}{R_0} - p_v\right) R_0^{3\kappa} R^{-(3\kappa+1)} + \frac{2\sigma}{R^2} \tag{2.19}$$

The quantity $\partial p_L / \partial R$ equals zero at the critical radius R_{crit} such that for $R > R_{crit}$ the bubble is unstable with respect to small changes in radius, but stable if $R < R_{crit}$. So

$$(R_{crit})^{3\kappa-1} = \frac{3\kappa}{2\sigma}\left(p_0 + \frac{2\sigma}{R_0} - p_v\right) R_0^{3\kappa} \tag{2.20}$$

If conditions are isothermal ($\kappa = 1$), the critical radius is given by

$$R_{crit} = \sqrt{\frac{3R_0^3}{2\sigma}\left(p_0 + \frac{2\sigma}{R_0} - p_v\right)} \tag{2.21}$$

Similarly, there is a minimum in p_L (with respect to R) from the locus of points defined by equation (2.18). This threshold is found by substituting R_{crit} from equation (2.21) into equation (2.18). If p_L is less than this minimum (i.e. more negative than it), there will be rapid uncontrolled bubble growth. It is conventional to define this critical value of the liquid pressure using P_B, the difference between the critical value and the original hydrostatic pressure in the liquid, such that $p_L = p_0 - P_B$. The critical liquid pressure is therefore

$$p_L = p_0 - P_B = p_v - \frac{4\sigma}{3R_{crit}} = p_v - \frac{4\sigma}{3}\sqrt{\frac{2\sigma}{3\left(p_0 + \frac{2\sigma}{R_0} - p_v\right)}} \tag{2.22}$$

applicable when surface tension dominates, making inertial and viscous effects negligible (satisfied as $R_0 \rightarrow 0$), and in the quasi-static regime (for example, when ω_0 is very much greater than ω, the frequency of the driving force, again applicable as $R_0 \rightarrow 0$). Since to bring about this growth p_L must be negative, the vapour pressure is clearly less than the $4\sigma/(3R_{crit})$ term in equation (2.22). The term P_B is known as the *Blake threshold pressure*. If surface tension dominates (i.e. $2\sigma/R_0 \gg p_0$, as expected for the regime where the Blake threshold is valid), and in addition vapour pressure is neglected, then equation (2.22) reduces [65] to

$$P_B \approx p_0 + 0.77\frac{\sigma}{R_0} \tag{2.23}$$

(b) The Rayleigh Collapse

Whilst considering the sound of water boiling in a kettle, Lord Rayleigh was introduced to the problem of cavitation behind screw propellers. As a result in 1917 he published the pioneering paper on cavitation collapse [66]. For the thirty years following this, all the theoretical work on cavitation was concerned with hydrodynamically generated cavities, culminating in the work of Plesset [67]. It was only then that work began on acoustically generated cavities with the work of Blake [25], outlined above.

(i) Collapse of an Empty Cavity. Rayleigh [66] considered the collapse of an empty cavity[25] which remains spherical at all times, located in an incompressible liquid. The empty cavity, at rest, is envisaged to be "as if a spherical portion of the fluid is suddenly annihilated," to quote Besant [68]. The cavity at this time, when the wall velocity is zero, is assumed to have a radius R_m. Since the cavity contains no gas, the liquid pressure p_L just outside the cavity is zero (if surface tension is assumed to be negligible). Thus the work done by the hydrostatic pressure p_∞ from that time until the cavity has contracted to a radius R, which is given by $4\pi p_\infty(R_m^3 - R^3)/3$, will simply equal the kinetic energy of the liquid, which is found by integrating the energy over spherical shells of liquid, of thickness Δr, mass $4\pi r^2\rho\Delta r$, and speed $\dot{r}$:

$$\frac{4\pi}{3}p_\infty(R_m^3 - R^3) = \frac{1}{2}\int_R^\infty \dot{r}^2\rho 4\pi r^2 \mathrm{d}r \tag{2.24}$$

where ρ is the liquid density. If the liquid is taken to be incompressible, then at a given instant the rate of mass of liquid flowing through any spherical surface (radius r) equicentric with the bubble must be a constant. In time Δt, a mass of liquid $4\pi r^2\rho\dot{r}\Delta t$ flows across a surface at some general radius r outside the bubble. Equating this to the flow at the bubble wall gives

$$\dot{r}/\dot{R} = R^2/r^2 \tag{2.25}$$

By substituting equation (2.25) in (2.24), integration yields the kinetic energy to be $2\pi\rho\dot{R}^2R^3$, giving:

$$\dot{R}^2 = \frac{2p_\infty}{3\rho}\left(\frac{R_m^3}{R^3} - 1\right) \tag{2.26}$$

To find the wall velocity, $\dot{R}$, the negative root of equation (2.26) is taken since, because the wall motion is inwards, $\dot{R}$ must be negative. Integration of equation (2.26) with respect to time gives the collapse time t_{Ray} of the cavity:

---

[25] Some authors use 'cavity' to describe a void within a liquid that does not contain a permanent gas phase, that is, it is either empty or contains only vapour ('convention 1'). They reserve the term 'bubble' for a space the contents of which include a permanent gas phase. Flynn [20] generally uses 'cavity' to describe the model, and 'bubble' to describe the physical object that may be represented by the model cavity. In either case (as Flynn admits) the authors generally do not adhere to their rules for terms like transient and stable cavitation. Therefore though in this book 'convention 1' is generally followed, the phraseology employed is modified at times in keeping with the author's experience of current trends. For example, a small nucleus containing a gas phase that expands to many times its original size, so that the pressure of this gas phase is then very small, is sometimes termed a cavity. Of course, the timescales involved might allow mass-diffusion to generate a significant gas pressure within what would then be a bubble (see Figure 5.14). Inviolate definitions are not the rule in cavitation!

$$t_{\text{Ray}} = \int_{R_m}^{R=0} \frac{dR}{\dot{R}} \approx 0.915 R_m \sqrt{\frac{\rho}{p_\infty}} \tag{2.27}$$

The assumption that the liquid is incompressible means that the derivation becomes invalid once the velocity of the cavity wall approaches the speed of sound in the liquid.[26] From equation (2.26) it can be seen that this will always occur at some point during the collapse of an empty spherical cavity. Therefore Rayleigh suggested the presence of some permanent gas within the cavity.

(ii) Collapse of a Gas Cavity. **Rayleigh [66] proposed that at an initial radius R_m, when $\dot{R} = 0$, the gas cavity would collapse, and then rebound, and from thence oscillate between a maximum and a minimum value, $R = R_{max}$ and $R = R_{min}$ respectively, the wall speed being zero at the two extremes. In the absence of dissipation, $R_{max} = R_m$. At the start of the collapse, when $R = R_m$ and $\dot{R} = 0$, the gas within the bubble has pressure $p_{g,m}$ and temperature T_m. If there is no heat flow across the bubble wall, the gas pressure p_g follows an adiabatic law:**

$$p_g = p_{g,m} \left(\frac{R_m}{R} \right)^{3\gamma} \tag{2.28}$$

Since there is now a gas phase within the bubble, the decrease in potential energy, $4\pi p_\infty (R_m^3 - R^3)/3$, now equals the sum of the kinetic energy of the liquid, $2\pi\rho\dot{R}^2R^3$, *plus* the work done in compressing the gas as the radius changes from R_m to R. The energy balance might therefore be written

$$-\int_{R_m}^{R} p_\infty 4\pi R^2 dR = 2\pi R^3 \dot{R}^2 \rho - \int_{R_m}^{R} p_L 4\pi R^2 dR \tag{2.29}$$

The final term, which is the work done in adiabatically compressing the gas as the radius changes from R_m to R is, explicitly

$$-\int_{R_m}^{R} p_L dV = \frac{1}{\gamma - 1} \frac{4\pi R_m^3}{3} p_{g,m} \left\{ \left(\frac{R_m}{R} \right)^{3(\gamma-1)} - 1 \right\} \tag{2.30}$$

The energy balance is therefore [69]

$$\int_{R_m}^{R} p_\infty 4\pi R^2 dR = 2\pi R^3 \dot{R}^2 \rho + \frac{1}{\gamma - 1} \frac{4\pi R_m^3}{3} p_{g,m} \left\{ \left(\frac{R_m}{R} \right)^{3(\gamma-1)} - 1 \right\} \tag{2.31}$$

---

[26] The assumption of incompressibility is equivalent to stating that the sound speed in the liquid is infinite. This should be clear from Figure 1.2; if the spacing between the bobs on a line is fixed, then motion of one end of the line results in simultaneous identical motion of the remote end. If the bubble wall velocity approaches the speed of sound in the liquid, then clearly the finite value of the sound speed will have a critical influence on the dynamics, and the assumption of incompressibility is not valid.

From this energy balance, one may in the limits of negligible vapour pressure and surface tension, obtain the energy equation for the collapse:

$$\frac{3\rho\dot{R}^2}{2} = p_\infty\left\{\left(\frac{R_m}{R}\right)^3 - 1\right\} - p_{g,m}\frac{1}{1-\gamma}\left\{\left(\frac{R_m}{R}\right)^3 - \left(\frac{R_m}{R}\right)^{3\gamma}\right\} \quad (2.32)$$

if the external pressure p_∞ is taken as constant [69, 70]. Recognising that the polytropic exponent would change throughout the cavitation event, Noltingk and Neppiras [70] reasoned that since a process would be more nearly adiabatic as the rate of change of volume increases, they would assume the collapse phase as described by equation (2.32) to be adiabatic, justifying the use of γ for κ in their formulation.

Thus the bubble starts the collapse with a zero wall velocity and a maximum radius R_m, the gas within the bubble having pressure $p_{g,m}$ and temperature T_m. The wall velocity will next be zero at the minimum radius R_{min}, when the pressure and temperature of the gas within the bubble are a maximum, $p_{g,max}$ and T_{max} respectively. The positions of maximum and minimum radius are found by setting $\dot{R} = 0$ in equation (2.32), so that

$$p_\infty(\gamma - 1)\left\{\left(\frac{R_m}{R}\right)^3 - 1\right\} = p_{g,m}\left\{\left(\frac{R_m}{R}\right)^{3\gamma} - \left(\frac{R_m}{R}\right)^3\right\} \qquad (R = R_{max}, R_{min}) \quad (2.33)$$

One solution to equation (2.33) describes the position of R_{max} which, as expected from energy considerations, in the absence of dissipation equals R_m, the initial radius. The other solution occurs at $R = R_{min} \ll R_m$, which allows equation (2.33) to be simplified to

$$p_{g,m}\left(\frac{R_m}{R_{min}}\right)^{3(\gamma-1)} = p_\infty(\gamma - 1) \quad (2.34)$$

Applying equation (1.13) when the bubble is at the minimum and initial maximum radii gives the gas pressure $p_{g,max}$ corresponding to this instant of $R = R_{min}$, when the bubble attains the minimum volume:

$$p_{g,max} = p_{g,m}\left(\frac{R_m}{R_{min}}\right)^{3\gamma} \approx p_{g,m}\left(\frac{p_\infty(\gamma-1)}{p_{g,m}}\right)^{\frac{\gamma}{\gamma-1}} \quad (2.35)$$

The maximum temperature attained in the collapse, T_{max}, is obtained by application of equation[27] (1.f1) ($TV^{\gamma-1} = \text{constant}$) to the conditions of initial and minimum radius. Maintaining the assumption of an adiabatic collapse, this gives

$$\frac{T_{max}}{T_m} = \left(\frac{R_m}{R_{min}}\right)^{3(\gamma-1)} \quad (2.36)$$

and so by substituting equation (2.34) into equation (2.36) the temperature T_{max} is obtained in terms of the gas pressure at the start of the collapse, $p_{g,m}$:

$$T_{max} \approx \left(\frac{p_\infty(\gamma-1)}{p_{g,m}}\right) T_m \quad (2.37)$$

[27] See footnote to Chapter 1, section 1.1.1(b)(ii).

For practical application of the above formulation, it is important to consider how the bubble reached the conditions for the start of the collapse. When a bubble collapses violently in a sound field, the collapse phase is often termed 'Rayleigh-like' because of the dominance of inertial forces.[28] The slow growth phase prior to such a collapse is often assumed to be isothermal, and indeed Noltingk and Neppiras [70] came to this conclusion by the same reasoning that lead them to believe the collapse to be adiabatic.

Let us therefore consider the start of this phenomenon to be not the condition when the bubble is at maximum size, but when it is at equilibrium in the liquid, with radius R_o and $\dot{R} = 0$, and internal gas pressure and temperature $p_{g,o}$ and T_o (which is also the ambient temperature of the liquid). In response to an acoustic tension being passed through the liquid, the bubble grows to some maximum size, when $\dot{R}$ is again equal to zero. This is clearly prior to a Rayleigh-like collapse, so this maximum radius is equivalent to R_m in the above formulation. We shall now, however, call it R_{max}, since the situation being discussed has developed so that it has now become less like the collapse of an empty cavity and more like the response of an oscillator to a driving force. Reaching and recognising such a distinction represents an important landmark in this discussion of the acoustic bubble.

If the growth phase is isothermal, then at $R = R_{max}$ the gas temperature is still T_o. Application of equation (1.16) to the situation at equilibrium and when the bubble has grown to a maximum size R_{max} gives

$$p_{g,o} R_o^3 = p_{g,m} R_{max}^3 \tag{2.38}$$

Substitution in equation (2.37), and assuming that surface tension is negligible so that $p_{g,o} = p_\infty$, gives

$$T_{max} = T_o(\gamma - 1)\left(\frac{R_{max}}{R_o}\right)^3 \tag{2.39}$$

as derived by Apfel [71] for the response of a bubble to an acoustic wave where surface tension, viscous effects and vapour pressure are assumed to be negligible, where the bubble grows isothermally and then undergoes a Rayleigh-like adiabatic collapse, and where the pressure outside the cavity is ambient during the collapse (which is to say that the acoustic component is zero on collapse, being finite only during the growth).

Therefore it is clear that the above analysis provides an adequate basis on which to discuss the important phenomenon of violent collapse in acoustic fields. Certain basic features of violent acoustic cavitation can be predicted from the results so far. The presence of a permanent gas component will tend to 'cushion' the collapse, so that each individual collapse becomes more violent as the gas content is reduced. However, as the content is reduced, there will be fewer appropriate bubble nuclei. Thus reducing the gas content liquid reduces the number of collapses, but would tend to make each individual collapse more violent [20]. One would therefore expect an optimum gas content for which activity in the bubble field as a whole is a maximum. If the violent collapse is preceded by a slow growth phase then during that growth, when the pressure within the bubble is low, dissolved gas will come out of the liquid and enter the bubble. This will reduce the pressure differential across the bubble wall and, appealing to equations (2.31) and (2.33), increase $p_{g,m}$, the gas pressure when the bubble radius is a maximum at the start of the collapse. Thus the violence of the collapse will be reduced, an increase in $p_{g,m}$ engendering a reduction in $p_{g,max}$ and T_{max}. Therefore dissolving a gas such as carbon dioxide, which will

---

[28] See Chapter 4, sections 4.3.1(b), 4.3.1(c) and 4.3.2.

readily come out of solution during the growth phase, will again cause a reduction in the violence of individual collapses. To complete this section on Rayleigh-like collapses, the final component of the bubble interior, the liquid vapour, is now discussed.

(iii) Collapse of a Vapour Cavity. In the derivation of the Blake threshold pressure and the accompanying consideration of the bubble in equilibrium, the processes of evaporation and condensation were assumed to occur so rapidly as to keep the vapour pressure constant at all times. As an illustration of the condensation process, it is interesting to investigate through simple estimates the extreme limit of this idea: whether, during a Rayleigh-like collapse, condensation of vapour can take place rapidly enough to ensure conditions of constant pressure and temperature within the bubble.

The discussion will begin with the vapour cavity at some maximum size, R_m. This cavity may have formed by isothermal growth from a nucleus, for example, or though a hydrodynamically generated or acoustic pressure drop. The formulation will estimate the conditions necessary for the collapse to be isothermal and isobaric, and discuss whether they can be met.

If the vapour is a perfect gas then

$$p_v = n_o k_B T \tag{2.40}$$

where k_B is Boltzmann's constant. If the collapse were to be isothermal (at temperature T) and maintain constant pressure p_v, then the number density of vapour molecules n_o within the bubble must remain constant. Thus the total number of vapour molecules that must condense into the liquid as the bubble radius reduces from R_m to R_{min}, the minimum radius, would be

$$N_v = n_o.\frac{4\pi}{3}(R_m^3 - R_{min}^3) \tag{2.41}$$

If N_v cannot condense out during the time the collapse takes, then the process cannot be isothermal at constant pressure.

When a vapour molecule strikes a surface of its liquid, there is a finite chance that it will 'stick'. If this happens, the molecule has condensed into the liquid phase. The following calculation enables the probability q_v that a vapour molecule will condense if it strikes a surface, to be determined. It can readily be shown [72] that if ϕ_L is the latent heat of evaporation per molecule of the liquid, T is the temperature of both the vapour and the liquid reservoir, and k_B is Boltzmann's constant, then the saturated vapour pressure p_v is given by

$$p_v = \frac{\pi n_s k_B T}{q_v} exp\left(\frac{-\phi_L}{k_B T}\right) \tag{2.42}$$

where n_s is the number density of molecules near the liquid surface, within the liquid.

Therefore by plotting $\ln(p_v/\mathrm{Pa})$ against $(T/\mathrm{K})^{-1}$ (data from Walton [72]) the intercept will give the value of $\ln(\pi n_s k_B T/q_v)$. This can then be used to determine q_v. This plot is made in Figure 2.8.

The intercept value is 24.8. From this, at 275 K $q_v = 6.73 \times 10^{-3}$, and therefore less than 0.7% of vapour molecules colliding with the liquid surface at 275 K can be expected to stick, and so condense. For comparison, at 470 K the value of q_v is 9.9×10^{-3}.

The rate at which molecules condense out will be given by the sticking probability, q_v, multiplied by the rate at which vapour molecules collide with the liquid surface. This rate of

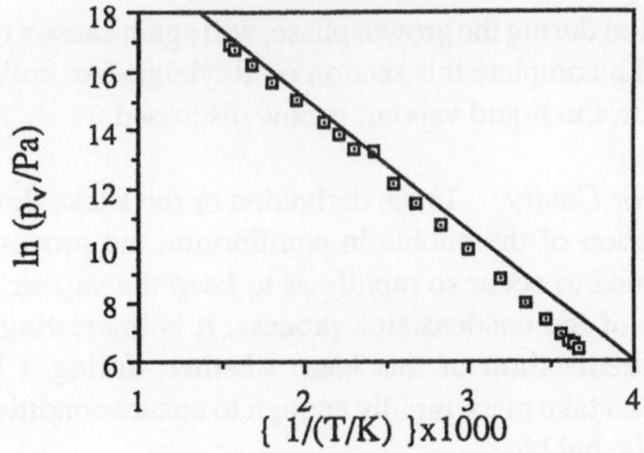

Figure 2.8 The vapour pressure of water as a function of the reciprocal of the temperature (data from Walton [72]).

collision will be the product of the bubble wall area with the vapour molecule flux, $n_o < u_v >/4$ (where $< u_v > = \sqrt{8k_BT/\pi m_v}$ is the average vapour molecule velocity and m_v is the mass of a vapour molecule). Thus the rate of condensation will be

$$q_v 4\pi R^2 . \frac{n_o < u_v >}{4} \tag{2.43}$$

Substituting for $<u_v>$ gives the number of vapour molecules condensing in time Δt to be

$$q_v n_o \pi R^2 = \sqrt{\frac{8k_BT}{\pi m_v}}\ \Delta t = q_v n_o R^2 \sqrt{\frac{8\pi k_BT}{m_v}}\ \frac{dR}{\dot{R}} \tag{2.44}$$

If the collapse is assumed to be Rayleigh-like, then equation (2.26) can be substituted into equation (2.44) to give the number condensing in time dt as

$$-q_v n_o R^2 \sqrt{\frac{8\pi k_BT}{m_v}} \sqrt{\frac{3\rho}{2p_\infty}} \sqrt{\frac{R^3}{R_m^3 - R^3}}\ dR \tag{2.45}$$

(the negative root being chosen since $\dot{R} < 0$). Therefore the total number condensing out during the collapse is

$$\int_{R_{min}}^{R_m} q_v n_o \sqrt{\frac{12\pi k_BT\rho}{p_\infty m_v}} \left\{1 - \left(\frac{R}{R_m}\right)^3\right\}^{-1/2} \left(\frac{R}{R_m}\right)^{3/2} R^2 dR \tag{2.46}$$

Using the substitution $R^3 = R_m^2 \sin^2\zeta$ gives, on integration, the total number of molecules condensing out to be

$$q_v n_o \pi \sqrt{\frac{12\pi k_BT\rho}{p_\infty m_v}}\ \frac{R_m^3}{6} \tag{2.47}$$

If the isothermal isobaric process is to be possible, then this quantity must be greater than or equal to that given by equation (2.41), i.e.:

$$q_v n_0 \pi \sqrt{\frac{12\pi k_B T \rho}{p_\infty m_v}} \frac{R_m^3}{6} \geqslant n_0 \cdot \frac{4\pi}{3}(R_m^3 - R_{min}^3) \tag{2.48}$$

Substituting appropriate values into this gives $0.7 \geqslant 4$, a condition which is clearly not satisfied. Thus this estimate indicates that Rayleigh-like collapse of water vapour cavities cannot be maintained at constant pressure and temperature.

The above derivation is meant simply as a brief illustration of some of the factors that must be considered in the dynamics of vapour cavities. Plesset and Prosperetti [73] describe the dynamics of vapour cavities, and include thermal considerations, such as the latent heat of evaporation. The latter is required for conversion of liquid to vapour, so that during this process heat will be lost from a shell of liquid of width equal to the thermal diffusion length surrounding the bubble. Similarly, condensation will release thermal energy. However, the vapour pressure within the bubble is dependent on the temperature of this liquid shell. During bubble growth, Plesset and Prosperetti [73] calculate that in water at 15°C the temperature drop is of the order of 0.2°C, which causes a 1% decrease in vapour pressure, and so has negligible effect on the bubble dynamics. However, if the liquid temperature is 100°C the temperature fall in the shell is around 13°C, causing a 50% reduction in the vapour pressure. Where thermal, rather than inertial, effects dominate, Plesset and Prosperetti [73] employ the term 'vapour bubble', as opposed to 'cavitation bubble'. Though the modelling of the growth phase of vapour bubbles is simplified by the fact that the lengthscale of the temperature gradients in the liquid is much smaller than the bubble radius, this is not the case during their collapse.

The first half of this chapter deals with the response of the bubble to static pressure. The acoustic character of a bubble arises in its interaction with oscillating pressure fields, both in its response to them and its generation of them. The propagation of these oscillating acoustic pressure fields through fluids is the subject of the next section.

2.2 Fluid Dynamics

In Chapter 1, the physical properties of solids and gases were related to the linearised wave equation to show how sound could propagate through those media. The following section examines the fluid dynamics of the acoustic process, to illustrate how sound can propagate through a fluid.

2.2.1 The Equation of Continuity

Consider a volume element dV_1 which is *fixed* within the fluid (Figure 2.9), relative to world. That it is to say, if the fluid within the element flows off, then it leaves the element behind. Such elements should be small enough to ensure that thermodynamic properties are spatially uniform across them, but large enough for the fluid to behave as a continuum. The rate of change of

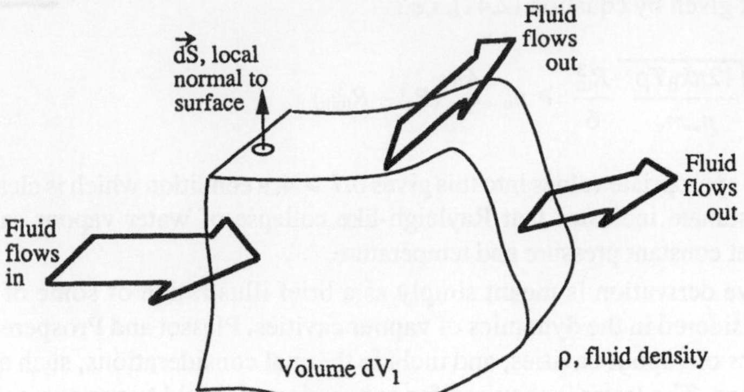

Figure 2.9 The fluid flux through a volume element dV_1 which is stationary.

mass within the volume is a result of the mass flux of fluid crossing the surface of the volume element, or of local changes in density. Therefore

$$-\int_V \dot{\rho}\, dV = \oint_S \rho\vec{v}.d\vec{S} \tag{2.49}$$

where $\vec{v}$ is the fluid particle velocity, and where the integrals are taken over the entire volume and surface of the element respectively. Use of the vector quantity $\rho\vec{v}$ in Gauss's theorem [74] gives

$$\oint_S (\rho\vec{v}).d\vec{S} = \int_V \vec{\nabla}.(\rho\vec{v})\, dV \tag{2.50}$$

which enables equation (2.49) to be written in the form

$$\dot{\rho} + \vec{\nabla}.(\rho\vec{v}) = 0 \qquad \textit{the equation of continuity.} \tag{2.51}$$

Equation (2.51) simply states that mass is neither created nor destroyed within dV_1. In other words, there are no sources or sinks within the volume element. The term $(\rho\vec{v})$ is often known as $\vec{j}$, the mass flux. The differential operator $\vec{\nabla}$ is simple to understand in the familiar cartesian coordinate system, with axes labelled x, y and z. In this system, the axial elements of distance are Δx, Δy and Δz, and these correspond to the lengths of the sides of a volume element at position (x_0, y_0, z_0) (Figure 2.10). The vector differential operator therefore takes the form

$$\vec{\nabla} = \left(\frac{\partial}{\partial x}, \frac{\partial}{\partial y}, \frac{\partial}{\partial z}\right) \tag{2.52}$$

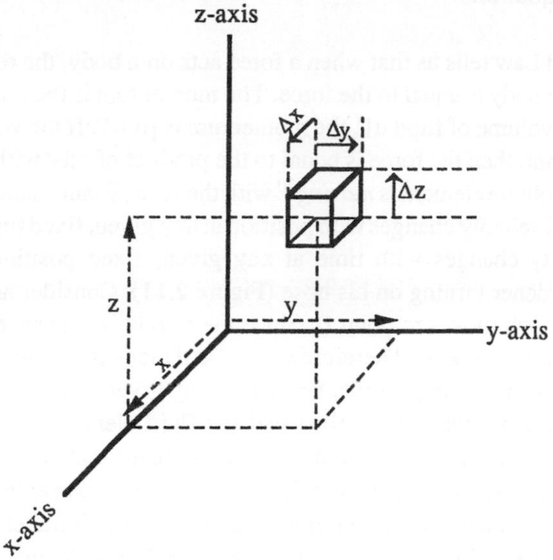

Figure 2.10 The elemental volume in the cartesian system.

Grad operates upon some scalar[29] Φ to give a vector

$$\mathbf{grad}\ \Phi = \vec{\nabla}\Phi = \left(\frac{\partial \Phi}{\partial x}, \frac{\partial \Phi}{\partial y}, \frac{\partial \Phi}{\partial z}\right) \tag{2.53}$$

whilst the operator **div** acts upon a vector, for example $\vec{v} = (v_x, v_y, v_z)$, to give

$$\mathbf{div}\ \vec{v} = \vec{\nabla}.\vec{v} = \frac{\partial v_x}{\partial x} + \frac{\partial v_y}{\partial y} + \frac{\partial v_z}{\partial z} \tag{2.54}$$

which is a scalar. Thus **grad** operates on a scalar to give a vector which shows by magnitude and direction the spatial variation of that scalar at the position in question. **Div** operates on a vector to give a scalar value corresponding to how that vector varies in space at the region where it is evaluated. The operator **del**-squared, acting on some scalar Φ, is equal to **div.grad** Φ, and corresponds to the scalar

$$\nabla^2\Phi = \vec{\nabla}.(\vec{\nabla}\Phi) = \left(\frac{\partial^2 \Phi}{\partial x^2} + \frac{\partial^2 \Phi}{\partial y^2} + \frac{\partial^2 \Phi}{\partial z^2}\right) \tag{2.55}$$

In the above examples, space was interpreted through the cartesian system of three orthogonal rectilinear axes. The $\vec{\nabla}$ operator is not restricted to this one coordinate system, as it is capable of representing physical phenomena that exist regardless of the frame employed. In section 2.2.4(b), $\vec{\nabla}$ will be interpreted through the spherical coordinate system.

---

[29] A *scalar* has only magnitude (e.g. 'the speed of the car is 50 km/h'). A *vector* has both magnitude and direction (e.g. 'the velocity of the car is 50 km/h due east'). A vector can conveniently be represented on a diagram by an arrow, the length of the arrow indicating the magnitude of the quantity.

2.2.2 Eulers Equation

Newton's Second Law tells us that when a force acts on a body, the resulting rate of change of momentum of the body is equal to the force. The momentum is the product of the mass and the velocity, so for a volume of fluid dV, the momentum is $\rho\vec{v}dV$. If the volume and the density are considered constant, then the force is equal to the product of ρdV with the rate of change of $\vec{v}$. However, if the volume element is *moving*[30] with the flow, $\vec{v}$ can change for two reasons: firstly because the liquid velocity changes with position at any given, fixed time; and secondly because the liquid velocity changes with time at any given, fixed position. This can be seen by considering a gardener turning on his hose (Figure 2.11). Consider an element of fluid that is at point A, and which moves to point B. Since the nozzle is narrower at B than at A, the fluid is flowing faster at B than at A. Therefore as our fluid element moves from A to B, it will speed up as a result of its change in position, because at any given time the fluid speed at B has to be greater than at A. However, all the time that the fluid element is moving from A to B, the gardener is turning on the tap more and more. This means that at any fixed point, A or B, the fluid speed at a later time will be greater than it was at any time earlier. Therefore there will be two contributing factors to the acceleration of the element as it travels from A to B: one is that at any given time the fluid speed at B must be greater than at A, and the second is that as time goes by the fluid speed at any given point in the pipe will be increasing.

This can be formulated for the general case by considering a volume element in a simple fluid, where viscous[31] and thermal conduction effects are negligible, of volume dV_2, which is *moving with the fluid*. The net instantaneous force due to unbalanced fluid pressure on such an element is simply $-dV.\vec{\nabla}p(\vec{r},t)$, where p is the pressure in the liquid. If $\vec{g}$ is the vector of gravitational acceleration, then application of Newton's Second Law yields the vector equation

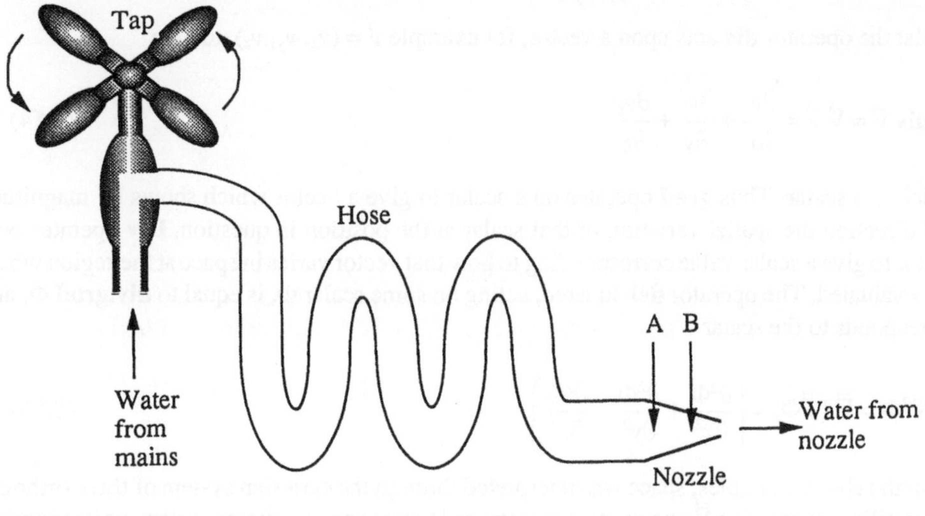

Figure 2.11 Changes in fluid velocity in time at a fixed position, and the differences in simultaneous fluid speeds at different positions, as illustrated through the flow of water through a hose nozzle as the tap is turned on.

30 In contrast to the fixed element dV_1 earlier, in section 2.2.1.
31 The effect of viscosity is considered in section 2.3.2.

$$(\rho\vec{g} - \vec{\nabla}p(\vec{r}, t)) \,.\, dV_2 = \rho dV_2 \,.\, \vec{a} \tag{2.56}$$

where $\vec{a}$ is the acceleration of a *particular piece of fluid*, and the vector $\vec{g}$ is in the downwards direction. To retain generality and incorporate any other body forces, let us say that ϕ_B is the potential energy per unit mass of fluid. For example, if gravity is the only body force then $\phi_B = gh$, where h is the height (upwards being increasingly positive), and $g = |\vec{g}|$. The force per unit volume resulting from this potential energy is given by $-\rho\vec{\nabla}\phi_B$. Therefore, in general, Newton's Second Law gives

$$-\vec{\nabla}p - \rho\vec{\nabla}\phi_B = \rho\vec{a} \tag{2.57}$$

As outlined above, since the particle velocity is a function of both time and space, the acceleration is a sum of two terms. The first is a result of the change in fluid velocity in a small time Δt at a fixed point in space, which is given by $\Delta\vec{v}_t$. The second term comes from the fact that the volume element under consideration is moving with the fluid. If the position changes by $\Delta\vec{r}$, then there is a corresponding change in the fluid particle velocity of $\Delta\vec{v}_{\vec{r}}$. Therefore the total change in particle velocity is given by the partial differential equation

$$\Delta\vec{v} = \Delta\vec{v}_t + \Delta\vec{v}_{\vec{r}} = \left.\frac{\partial\vec{v}}{\partial t}\right|_{\vec{r}} \Delta t + \left.\frac{\partial\vec{v}}{\partial\vec{r}}\right|_t \Delta\vec{r} \tag{2.58}$$

the notation signifying that the two partial differentials are taken at constant position ($\vec{r}$) and constant time (t) respectively. The exact time derivatives can be approached by relating the infinitesimal changes in equation (2.58) to the accompanying change in time

$$\frac{\Delta\vec{v}}{\Delta t} = \dot{\vec{v}} + \frac{\Delta\vec{r}}{\Delta t}\left.\frac{\partial\vec{v}}{\partial\vec{r}}\right|_t \tag{2.59}$$

where $\dot{\vec{v}} = (\partial\vec{v}/\partial t)|_{\vec{r}}$. For infinitesimal changes, the acceleration of a particular piece of fluid can be more conveniently written

$$\vec{a} = \dot{\vec{v}} + (\vec{v}.\vec{\nabla})\vec{v} \tag{2.60}$$

and represents the so-called 'material derivative' of the fluid particle velocity.[32] As a result, equation (2.57) implies that, under inviscid conditions, the pressure in the fluid can be expressed in terms of flow and body forces thus:

$$-\vec{\nabla}p(\vec{r},t) = \rho(\dot{\vec{v}} + (\vec{v}.\vec{\nabla})\vec{v} + \vec{\nabla}\phi_B) \qquad \textit{Euler's equation} \tag{2.61}$$

(a) Reduced Forms of Euler's Equation

Euler's equation assumes inviscid conditions. It can be reduced to several other useful forms appropriate to given specific sets of additional conditions relevant to irrotational, steady and incompressible flows. To do this it is first necessary to consider vorticity and circulation in the liquid.

---

[32]See section 2.3.2(a)(i).

Vorticity and Circulation. Applying the vector identity [75]

$$(\vec{v}.\vec{\nabla})\vec{v} = (\vec{\nabla} \wedge \vec{v}) \wedge \vec{v} + \vec{\nabla}(\vec{v}.\vec{v})/2 \tag{2.62}$$

to equation (2.61) [76] allows Euler's equation to be recast in terms of the vector

$$\vec{\Omega} = \vec{\nabla} \wedge \vec{v} \tag{2.63}$$

which is known as the *vorticity*:

$$-\vec{\nabla}p(\vec{r},t) = \rho(\dot{\vec{v}} + \vec{\Omega} \wedge \vec{v} + \tfrac{1}{2}\vec{\nabla}v^2 + \vec{\nabla}\phi_B) \tag{2.64}$$

The *circulation* around any closed loop in the fluid is given by the line integral of the fluid velocity along the complete loop:

$$\oint \vec{v}.\mathrm{d}\vec{l} \tag{2.65}$$

where $\mathrm{d}\vec{l}$ is an elemental vector length along the loop, and $\vec{v}.\mathrm{d}\vec{l}$ is the product of the magnitude of $\mathrm{d}\vec{l}$ with the component of the fluid velocity which is in the direction of $\mathrm{d}\vec{l}$. Using Stoke's theorem [77]

$$\oint \vec{v}.\mathrm{d}\vec{l} = \int (\vec{\nabla} \wedge \vec{v}).\mathrm{d}\vec{S} = \int \vec{\Omega}.\mathrm{d}\vec{S} \tag{2.66}$$

where $\mathrm{d}\vec{S}$ is an infinitesimal area, the perimeter of which is the loop over which the circulation is calculated. Equation (2.63) shows that the circulation per unit area for an infinitesimal loop equals the vorticity, and so the vorticity is the circulation around a unit area perpendicular to the vector $\vec{\Omega}$ [76], and in general equals twice the angular rotation of a fluid element [78].

If the vorticity is zero everywhere, the flow is *irrotational*, and

$$\vec{\Omega} = \vec{\nabla} \wedge \vec{v} = 0 \tag{2.67}$$

Since the curl of the gradient of any function is zero, when the flow is irrotational it is possible to define a scalar function Φ, the *velocity potential*, such that

$$\vec{v} = \vec{\nabla}\Phi \tag{2.68}$$

(i) Inviscid, Irrotational Flow. The advective term $(\vec{v}.\vec{\nabla})\vec{v}$ in Euler's equation can be simplified by the use of a vector identity [75]

$$2(\vec{v}.\vec{\nabla})\vec{v} = \vec{\nabla}(\vec{v}.\vec{v}) - 2\vec{v} \wedge (\vec{\nabla} \wedge \vec{v}) \tag{2.69}$$

Now $\vec{v}.\vec{v} = v^2$, and if the fluid is irrotational then $(\vec{\nabla} \wedge \vec{v}) = 0$. Substitution of this vector identity, and equation (2.68), into Euler's equation (equation (2.61)) gives

$$-\vec{\nabla}p = \rho\vec{\nabla}\,(\dot{\Phi} + \tfrac{1}{2}(v^2) + \phi_B) \tag{2.70}$$

Spatial integration of equation (2.70) gives

$$p + \rho\dot{\Phi} + \tfrac{1}{2}\rho(v^2) + \phi_B\rho = p_c \tag{2.71}$$

where the term p_c is a constant of integration with the dimensions of pressure. This is an absolute constant, arising from the integration process.

(ii) Inviscid Steady Flow. If the flow is steady, that is the fluid velocity at any fixed point does not change in time, then $\dot{\vec{v}} = 0$. The lines of motion therefore coincide with the particle paths, and these lines of motion are constant in time. Since particle paths (e.g. smoke streams) will therefore follow, and can be used to visualise, the lines of motion, these lines are called ***streamlines***. If the dot product of both sides of equation (2.64) is taken, then since $\vec{v}.\vec{\Omega} \wedge \vec{v} = 0$

$$\vec{v}.\vec{\nabla}\left(p + \frac{v^2}{2} + \phi_B\right) = 0 \tag{2.72}$$

This equation states that the quantity in brackets is constant for small displacements in the direction of fluid velocity. However, in steady flow any displacement must be along a streamline. Therefore, taking the example where gravity is the only body force, equation (2.68) tells us that

$$\frac{p}{\rho} + gh + \frac{v^2}{2} = \text{constant} \tag{2.73}$$

for all points along a streamline (*Bernoulli's equation*).

Equation (2.73) shows how a rapid flow could cause a drop in pressure. Such hydrodynamic pressure changes are not acoustic, they do not require the medium to be elastic and they do not propagate to a distance.[33] Unlike the constant p_c in equation (2.71), the 'constant' in Bernoulli's equation is constant along any particular streamline, but may vary from one streamline to another. Since the derivation does not assume irrotational flow, this formulation is applicable without assuming the existence of a velocity potential.

Bernoulli's equation can be used to define a number which characterises the probability that cavitation will occur as a result of hydrodynamically generated pressure drops. This is the ***cavitation number***, σ_c.

Consider two points, labelled 0 and 1, on a streamline in an incompressible flow where $\dot{\vec{v}} = 0$. From equation (2.69) (Bernoulli's equation) if the change in gravitational potential is small (i.e. both points of consideration are at roughly the same height)

$$\frac{p_1}{\rho} + \frac{v_1^2}{2} = \frac{p_2}{\rho} + \frac{v_2^2}{2} \tag{2.74}$$

If the point 1 corresponds to the inlet of a water tunnel, and the point 2 to a point of constriction, then since the flow at the constriction must be greater than at the inlet ($v_2 > v_1$), the pressure must be less ($p_2 < p_1$). The cavitation number is defined as

[33] 'Pressure mines' are triggered by the non-acoustic hydrodynamic pressure fluctuations generated close to a ship, which contain frequencies generally below 1 Hz. This is in contrast to 'acoustic mines', which are triggered by the actual acoustic emissions from a ship [79].

$$\sigma_c = \frac{p_2 - p_v}{\frac{1}{2}\rho v_2^2} \qquad \text{for point 2} \tag{2.75}$$

The bubbles are generated by a pressure drop. The numerator in σ_c represents the static pressure which opposes any reduction in pressure and the generation of bubbles. The denominator represents the dynamic term, which causes the pressure drop (see equation (2.73)) and so promotes cavitation inception. The cavitation number can therefore be thought of as the ratio of terms inhibiting cavitation to those promoting it [80]. Therefore the lower the value of σ_c, the more likely cavitation is to occur as a result of the flow.

(iii) Inviscid Incompressible Flow. If density variations within the liquid are negligible compared with the density of the liquid in equilibrium (ρ_o); if particle velocity perturbations are small such that $|(\vec{v}.\vec{\nabla})\vec{v}| \ll |\dot{\vec{v}}|$, and if in addition the only component of the liquid pressure with a finite spatial variation is the acoustic component, and the effect of body forces is negligible, such that $\vec{\nabla}p = \vec{\nabla}P$ (where P is the acoustic pressure) – then Euler's equation reduces to

$$\vec{\nabla}P = -\rho_o\dot{\vec{v}} \qquad \textit{the linear inviscid force equation.} \tag{2.76}$$

Taking the divergence of this equation gives

$$\nabla^2 P = -\rho_o \vec{\nabla}.\dot{\vec{v}} \tag{2.77}$$

which will be used later in this chapter (section 2.2.3) to derive the linear wave equation for sound in liquid.

(iv) Inviscid, Incompressible, Irrotational Flow. Equation (2.76) was derived for flow in an incompressible fluid. If in addition the flow is irrotational, then a velocity potential can be defined. Since the curl of the gradient of any function is zero then in those conditions

$$\vec{\nabla} \wedge (\vec{\nabla}P) = -\rho_o \vec{\nabla} \wedge \dot{\vec{v}} = 0 \tag{2.78}$$

Therefore $\vec{\nabla} \wedge \vec{v} = 0$ in incompressible, irrotational conditions. Kelvin's theorem states that inviscid motions in a conservative[34] field which are initially irrotational (i.e. $\vec{\Omega} = \vec{\nabla} \wedge \vec{v} = 0$) remain so [81]. Substitution of equation (2.68) into (2.76) gives

$$\vec{\nabla}(\rho_o\dot{\Phi} + P) = 0 \tag{2.79}$$

The expression in parentheses is therefore isotropic in that it has zero gradient, and if there is no acoustic excitation can be set to zero. Thus

$$P = -\rho_o\dot{\Phi} \tag{2.80}$$

Potential flow occurs when the vorticity is zero. If the liquid is incompressible, then $\rho = \rho_o$, a constant in both space and time, so that equation (2.51) implies

[34] A conservative field is one in which the work done by the field on a body in taking it between two points is independent of the path taken between those two points – an example of such a field is the gravitational one.

$$\vec{\nabla}.\vec{v} = 0 \tag{2.81}$$

Substitution from equation (2.68) gives

$$\nabla^2 \Phi = 0 \qquad \text{which is } \textit{Laplace's equation.} \tag{2.82}$$

Most of the remainder of this chapter will be taken up with the discussion of these two cases. Firstly, the linear inviscid force equation will be used to obtain the equation describing the propagation of low-amplitude acoustic waves through an inviscid incompressible liquid. Secondly, velocity potential solutions to Laplace's equation will be used to describe flow conditions in inviscid, incompressible, irrotational fluids.

2.2.3 The General Wave Equation

By combining the linear inviscid force equation and the equation of continuity, the pressure fluctuations in a fluid can be shown to propagate as a wave, provided that terms relating to the spatial variation in density are negligible. In this way the equation of continuity (2.51) becomes

$$\frac{\dot{\rho}}{\rho} + \vec{\nabla}.\vec{v} = 0 \tag{2.83}$$

which, on differentiation with respect to time, becomes

$$\frac{\ddot{\rho}}{\rho} - \left(\frac{\dot{\rho}}{\rho}\right)^2 + \vec{\nabla}.\dot{\vec{v}} = 0 \tag{2.84}$$

If the quadratic term in equation (2.84) is deemed negligible, then eliminating $\vec{\nabla}.\dot{\vec{v}}$ between equations (2.77) and (2.84) gives

$$\frac{\ddot{\rho}}{\rho} + \frac{\nabla^2 P}{\rho_0} = 0 \tag{2.85}$$

If $(\rho/\rho_0) \approx 1$, then equation (2.85) can be combined with the definition of the adiabatic bulk modulus B_s

$$B_s = -V \left.\frac{\partial p}{\partial V}\right|_S = \rho \left.\frac{\partial p}{\partial \rho}\right|_S \tag{2.86}$$

since $V \propto \rho^{-1}$ for a fixed mass of fluid, where the subscript S indicates the adiabatic condition. The result is

$$\nabla^2 P = \frac{1}{c^2}\ddot{P} \qquad \text{the } \textit{linearised wave equation} \text{ in pressure} \tag{2.87}$$

where the wave speed is given by

$$c = \sqrt{\frac{B_S}{\rho_o}} \qquad \text{(if } (\rho/\rho_o) \approx 1) \tag{2.88}$$

or, substituting for B_S from equation (2.86)

$$c = \sqrt{\left.\frac{\partial p}{\partial \rho}\right|_S} \tag{2.89}$$

As discussed in Chapter 1, section 1.1.1(b), any parameter which satisfies the linearised wave equation will propagate as a waveform without loss through a medium. In order for the acoustic pressure to satisfy the linearised wave equation, the above derivation requires the assumptions that fluid particle velocity fluctuations are very much less than the speed of sound in the liquid, and also that density fluctuations are negligible in comparison with the equilibrium density. Obviously, density is not constant during the propagation of an acoustic wave. Particle displacements cause local variations in density which, as shown in Figure 1.2, are periodic for a sinusoidal wave. Therefore the meaning of this derivation, which appears to give the equation for waves in an incompressible fluid, is that the smaller the amplitude of the acoustic wave, the more nearly it approximates to the simple case discussed here. Finite-amplitude sound waves will propagate in a fluid, but their physics is more complicated.[35]

In the limit of very high frequency, the acoustic wavelength becomes comparable with the mean free path of the fluid molecules,[36] and significant thermal conduction can take place between adjacent regions of compression and rarefaction in a longitudinal wave. In this limit the isothermal bulk modulus[37]

$$B_T = -V \left.\frac{\partial p}{\partial V}\right|_T = \rho \left.\frac{\partial p}{\partial \rho}\right|_T \tag{2.90}$$

(since $V \propto \rho^{-1}$ for a fixed mass of fluid) is the appropriate one to use, the subscript indicating constant temperature. The wave speed is therefore

---

[35] See Chapter 1, section 1.2.3(a).

[36] See Chapter 1, section 1.1.1(b)(i)

[37] The ratio of the adiabatic to the isothermal bulk modulus is simply γ, the ratio of the specific heats at constant pressure, C_p, and volume, C_V. By definition

$$C_p/C_V = T\left.\frac{\partial S}{\partial T}\right|_p \Big/ T\left.\frac{\partial S}{\partial T}\right|_V$$

By the Reciprocity Theorem, therefore

$$C_p/C_V = \left(\left.\frac{\partial S}{\partial p}\right|_T \cdot \left.\frac{\partial p}{\partial T}\right|_S\right) \Big/ \left(\left.\frac{\partial S}{\partial V}\right|_T \cdot \left.\frac{\partial V}{\partial T}\right|_S\right)$$

Cancelling gives

$$C_p/C_V = \left(-V\left.\frac{\partial p}{\partial V}\right|_S\right) \Big/ \left(-V\left.\frac{\partial p}{\partial V}\right|_T\right) = B_S/B_T$$

the ratio of the adiabatic and isothermal bulk moduli.

$$c = \sqrt{\frac{B_T}{\rho_o}} \quad \text{(if } (\rho/\rho_o) \approx 1) \tag{2.91}$$

or, substituting for B_T from equation (2.90)

$$c = \sqrt{\left.\frac{\partial p}{\partial \rho}\right|_T} \quad \text{for isothermal propagation in a fluid.} \tag{2.92}$$

For an incompressible fluid, the sound speed tends to ∞, in which case the linearised wave equation reduces to Laplace's equation, applicable to inviscid, incompressible, irrotational flow.

2.2.4 Laplace's Equation

Because of the restrictive conditions imposed by its derivation, Laplace's equation $\nabla^2\Phi = 0$ (equation (2.82)) is valid only for inviscid, incompressible, irrotational flow in conservative fields. The advantage of having a potential by which the flow can be defined, and for which Laplace's equation holds, are twofold. Firstly, much of the formulation from electrostatics can be applied to fluids; secondly, the Uniqueness Theorem is applicable.

(a) The Uniqueness Theorem

This theorem states that if a potential function Φ_1 is a solution of Laplace's equation which satisfies the boundary conditions, then it is the only solution.

The Uniqueness Theorem is most readily proved by trying to show the existence of a second solution, Φ_2. At some bounding surface to the volume in question (which may be infinite), the velocities $\vec{v}_1 = \vec{\nabla}\Phi_1$ and $\vec{v}_2 = \vec{\nabla}\Phi_2$ have the same vector component normal to the surface. Now consider the potential function $\Phi = \Phi_1 - \Phi_2$, which is also (by substitution) a solution of Laplace's equation. Substitution of the vector quantity $\Phi\vec{\nabla}\Phi$ into Gauss's theorem (see equation (2.50)), with application of the simple differential vector identity

$$\vec{\nabla}.(\Phi\vec{\nabla}\Phi) = \Phi\vec{\nabla}.(\vec{\nabla}\Phi) + (\vec{\nabla}\Phi)^2 \tag{2.93}$$

gives

$$\oint_S \Phi\vec{\nabla}\Phi \, d\vec{S} = \int_V \Phi\vec{\nabla}.(\vec{\nabla}\Phi)\, dV + \int_V (\vec{\nabla}\Phi)^2\, dV \tag{2.94}$$

This is a form of Green's Theorem [74]. The term on the left equals zero as a result of the stated conditions on $\vec{v}_1 = \vec{\nabla}\Phi_1$ and $\vec{v}_2 = \vec{\nabla}\Phi_2$ at the boundary. Similarly, the first term on the right equals zero, since $\vec{\nabla}.(\vec{\nabla}\Phi) = \nabla^2\Phi = 0$. Thus, equation (2.94) reduces to

$$\int_V (\vec{\nabla}\Phi)^2\, dV = 0 \tag{2.95}$$

Since $(\vec{\nabla}\Phi)^2$ must be greater than zero everywhere, then to give a zero-valued integral $\vec{\nabla}\Phi$ must be equal to zero at every point within the bounded space in question. Thus $\vec{v}_1 = \vec{\nabla}\Phi_1 = \vec{\nabla}\Phi_2$

which implies that $\vec{v}_1 = \vec{v}_2$. Also, $\vec{\nabla}\Phi$ equalling zero implies that throughout the volume the difference between Φ_1 and Φ_2 must be constant. It is almost always the case that some boundary condition will specify the value of the solution that satisfies it, so that at that point $\Phi_1 = \Phi_2$. Since this boundary condition stipulates that the difference between them is zero at one point, it must be zero everywhere. Thus Φ_1 and Φ_2 are identical. If you have found one solution, any other solution must be identical.

This means that, instead of solving Laplace's equation, it is possible to guess the likely answer, and then test it through substitution with the boundary conditions. If the solution is thus found to be an appropriate solution to Laplace's equation in a particular set of circumstances, it is the only solution, and you need look no further. This approach is particularly useful as, with a little experience, the likely solutions can readily be guessed from the family of functions which satisfy Laplace's equation.

One such family, of particular interest to bubble considerations, are the spherical harmonics. However, before discussing solutions to Laplace's equation in the spherical frame, it is necessary to formulate the operator $\vec{\nabla}$ in that frame.

(b) Spherical Differentials

The cartesian form of the operator has already been discussed. With bubbles it is often convenient to employ a spherical coordinate system, rather than a cartesian one, to utilise any symmetry in flow arising from the bubble shape. The coordinates r, θ and φ relate to the cartesian system as $z = r\cos\theta$, $x = r\sin\theta\cos\varphi$ and $y = r\sin\theta\sin\varphi$ (Figure 2.12). The locus of points r = constant represents a sphere, centred at the origin; θ = constant represents a cone, vertex at the origin, the axis of symmetry being the z-axis; φ = constant represents vertical half-planes, terminating on the z-axis. If both r and φ are constant, the locus of points resembles lines of longitude on a sphere; the locus for θ and r constant resemble lines of latitude. The elemental distances, and the lengths of the sides of the volume element, are now Δr, $r\Delta\theta$ and $r\sin\theta\Delta\varphi$ (Figure 2.12). As a result, in this spherical system the operator **grad** becomes

$$\vec{\nabla} = \left(\frac{\partial}{\partial r}, \ \frac{1}{r}\frac{\partial}{\partial \theta}, \ \frac{1}{r\sin\theta}\frac{\partial}{\partial \varphi} \right) \tag{2.96}$$

and the operator **del-squared**, acting on a potential Φ, has the form [82]

$$\nabla^2\Phi = \frac{1}{r^2}\frac{\partial}{\partial r}\left(r^2 \frac{\partial \Phi}{\partial r} \right) + \frac{1}{r^2\sin\theta}\frac{\partial}{\partial \theta}\left(\sin\theta \frac{\partial \Phi}{\partial \theta} \right) + \frac{1}{r^2\sin^2\theta}\frac{\partial^2 \Phi}{\partial \varphi^2} \tag{2.97}$$

Similarly, functions can also be expressed in terms of the spherical, rather than the cartesian, coordinate system. One particular group of functions, which form a solution to Laplace's equation in the spherical coordinate system, is the spherical harmonics.

(c) Spherical Harmonics

Laplace's equation (equation (2.82)) can be expressed in terms of the spherical coordinates r, θ and φ. If the components in r, θ and φ of the function which is a solution to this are independent, such a solution Φ can be expressed as the product of three component functions: Φ_r (which is a function of r only, and independent of θ and φ), Φ_θ (depends on θ only) and Φ_φ (a function

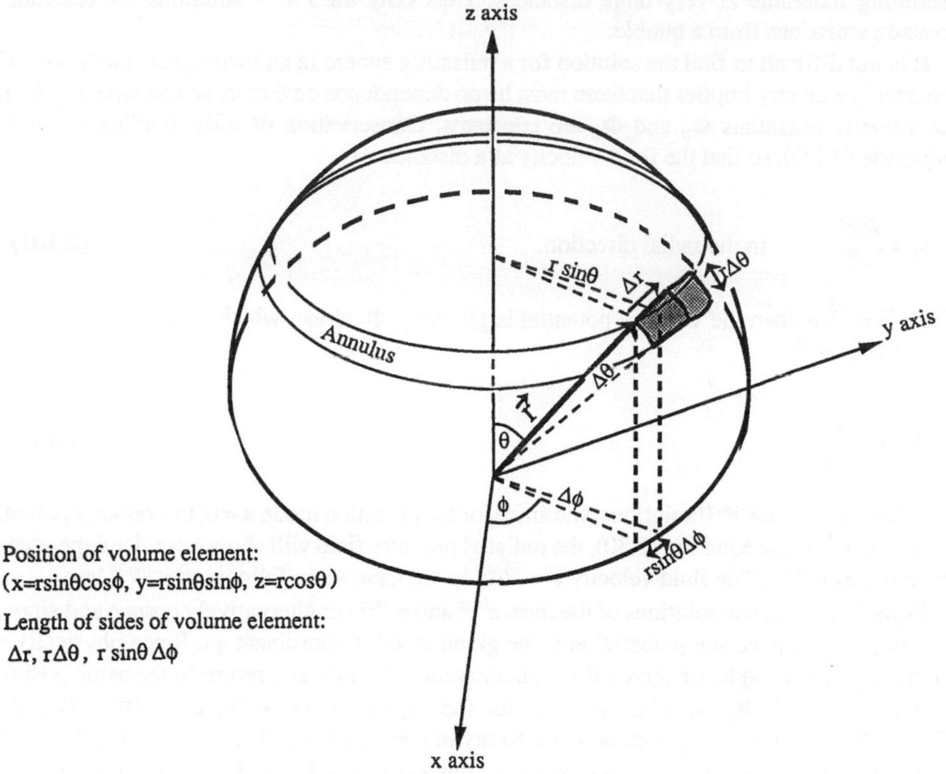

Figure 2.12 The elemental volume in the spherical coordinate system.

of φ only). Substitution of $\Phi = \Phi_r\Phi_\theta\Phi_\varphi$ into Laplace's equation, using the spherical form of del-squared given in equation (2.97), yields three separate equations:

$$\frac{1}{r^2}\frac{\partial}{\partial r}\left(r^2\frac{\partial \Phi_r}{\partial r}\right) - \frac{n(n+1)}{r^2}\Phi_r = 0 \tag{2.98}$$

$$\frac{\partial^2 \Phi_\varphi}{\partial \varphi^2} + m^2\Phi_\varphi = 0 \tag{2.99}$$

$$\frac{1}{\sin\theta}\frac{\partial}{\partial \theta}\left(\sin\theta\frac{\partial \Phi_\theta}{\partial \theta}\right) + \left(n(n+1) - \frac{m^2}{\sin^2\theta}\right)\Phi_\theta = 0 \tag{2.100}$$

where m and n are integers, taken as greater than or equal to zero.

It was noted in section 2.2.3 that, in the incompressible limit, the linearised wave equation reduces to Laplace's equation. Equation (2.98) has solutions of the form r^n or $r^{-(n+1)}$. Although the r^n modes are possible solutions to Laplace's equation, if the bubble is the source of the acoustic energy it is obviously unphysical actually to imply any solution where the potential, and therefore the velocity and the energy, increase with increasing distance from the bubble, becoming indefinite at very large distances. Thus only the $r^{-(n+1)}$ solutions are relevant to acoustic emissions from a bubble.

It is not difficult to find the solution for a pulsating sphere in an incompressible liquid. The spherical symmetry implies that there must be no dependence on θ or φ, so that $v_\theta = v_\varphi = 0$, and the velocity potentials Φ_θ and Φ_φ are constants. Conservation of mass implies $\dot{r}r^2 = \dot{R}R^2$ (equation (2.25)), so that the fluid velocity at a distance r is

$$v_r = \frac{\dot{R}R^2}{r^2} \quad \text{in the radial direction.} \tag{2.101}$$

Since $\vec{v} = \vec{\nabla}\Phi$, then the velocity potential is given by Φ_r alone, which from equations (2.26) and (2.101) is[38]

$$\Phi_r = -\frac{\dot{R}R^2}{r} \tag{2.102}$$

As this is one of the $r^{-(n+1)}$ solutions, and since for the pulsation mode $n = 0$, the velocity potential decays as r^{-1}. From equation (2.80), the radiated pressure field will show a similar dependence, decaying as $r^{-(n+1)}$. The fluid velocity $\vec{v} = \vec{\nabla}\Phi$, as we have seen, follows an $r^{-(n+2)}$ law.

Equation (2.99) has solutions of the form $e^{im\varphi}$ and $e^{-im\varphi}$ (or alternatively $\cos m\varphi$ and $\sin m\varphi$). We may thus express the potential at some given angular coordinate φ_1. Since physically we wish our potential to be unaltered if we circumvent the z-axis and return to the same position, then the potential Φ_φ must be the same for the angles φ_1, $\varphi_1 + 2\pi$, $\varphi_1 + 4\pi$ ... etc. Thus $e^{im\varphi} = e^{im(\varphi+2\pi)} = e^{im(\varphi+4\pi)}$ ... etc., which is to say that m must be either zero or integer.

The solutions to equation (2.100) are called the Associated Legendre polynomials, P_n^m, with argument $\cos\theta$, where $P_n^m = \sin^m\theta\, P_n$ and

$$P_0(\cos\theta) = 1 \tag{2.103}$$

$$P_1(\cos\theta) = \cos\theta \tag{2.104}$$

$$P_2(\cos\theta) = (3\cos^2\theta - 1)/2 \tag{2.105}$$

$$P_3(\cos\theta) = (5\cos^3\theta - 3\cos\theta)/2 \tag{2.106}$$

are the first few solutions for P_n, the Legendre polynomials. As with m, n must be zero or integer. The θ and φ dependence can be suitably normalised and combined to give a function proportional to the product $\Phi_\theta\Phi_\varphi$, which is called the *spherical harmonic*

$$Y_n^m(\theta,\varphi) = (-1)^m \sqrt{\frac{1}{4\pi}} \sqrt{\frac{(n-m)!}{(n+m)!}} \sqrt{2n+1}\, P_n^m(\cos\theta) e^{im\varphi} \tag{2.107}$$

The first few spherical harmonics are therefore

$$Y_0^0(\theta,\varphi) = \sqrt{\frac{1}{4\pi}} \tag{2.108}$$

[38]See also Chapter 4, section 4.4.2(b)(i).

$$Y_1^0\ (\theta,\varphi) = \sqrt{\frac{3}{4\pi}}\ \cos\theta \tag{2.109}$$

$$Y_1^1\ (\vartheta,\varphi) = -\sqrt{\frac{3}{8\pi}}\ \sin\theta e^{i\varphi} \tag{2.110}$$

$$Y_2^0\ (\theta,\varphi) = \sqrt{\frac{5}{4\pi}}\left(\frac{3}{2}\cos^2\theta - \frac{1}{2}\right) \tag{2.111}$$

Spherical harmonics with $m = 0$ have no φ-dependence. Thus if there exists a coordinate (θ_0,φ_0) for which the value of the spherical harmonic is zero, it will be zero for all values of φ where $\theta = \theta_0$. As a result, a plot of this harmonic will contain a nodal line circumventing the z-axis, at $\theta = \theta_0$. There will be a similar nodal line at each value of θ where the value of the spherical harmonic is zero, and these nodal lines will divide the plot of the spherical harmonic into zones of alternating sign (+ then –). These spherical harmonics with $m = 0$ are called *zonal harmonics* [83]. If, as is often the case in bubble physics, there is an axis of symmetry (taken as the z-axis), then the zonal harmonics are the relevant ones to describe flow conditions.

A brief discussion will show how the zonal spherical harmonics can be used to describe the bubble. There are two main uses. Firstly, as is clear from the preceding discussion, the spherical harmonics can be used to describe the velocity potential of the fluid about the oscillating bubble, and thence the fluid velocity and the emitted pressure waves using, for example, equations (2.70) and (2.80) respectively. The application of the different orders of n is clear from the fact that, if $n = 0$, there is no dependence of the potential on θ and ϕ: we are dealing with a spherically symmetric system, and so the bubble oscillation is a pulsation. As n increases, the angular dependence of the fluid motion becomes more complex.

The second use of the spherical harmonics is to describe the bubble shape in terms of perturbations from the spherical. The spherical harmonics represent a complete orthogonal set, so can in principle be combined to describe a given shape. They are not the only set, but in this system give particularly concise formulation since they also represent the normal modes.

For example, in a simple situation the bubble wall might be described by

$$R = R_0 + \Sigma A_n Y_n^m(\theta,\varphi)(\theta,\varphi) \tag{2.112}$$

where A_n is the time-dependent amplitude coefficient. The A_0 coefficient would control the proportion of the motion that was spherically symmetric, for example a pulsation. Deviations from sphericity would be described by the higher modes. This is simplest to see by considering the case when only a single mode is active, that is $A_n = 0$ for all n except a single one.

Consider a spherical bubble. The changes that a bubble has undergone which have been discussed so far in this book have preserved the spherical shape. This can be described by perturbing the bubble with the zero-order spherical harmonic, Y_0^0.

Figure 2.13(a) shows the Y_0^0 spherical harmonic, plotted in 3-D. The positive z-axis goes vertically upwards. The three-dimensional plots are interpreted by realising that the magnitude of the function $Y_n^m(\theta,\varphi)$ for a particular angular combination (θ,φ) is given by the distance of the surface out from the origin in a particular direction (specified by θ and φ). Since from equation (2.108) the Y_0^0 function does not vary in θ or φ, the plot is spherical. Therefore if an initially spherical bubble is perturbed by the zero-order spherical harmonic, the bubble remains spherical, but the radius changes. A two-dimensional half-space plot of a sphere of radius 1 is

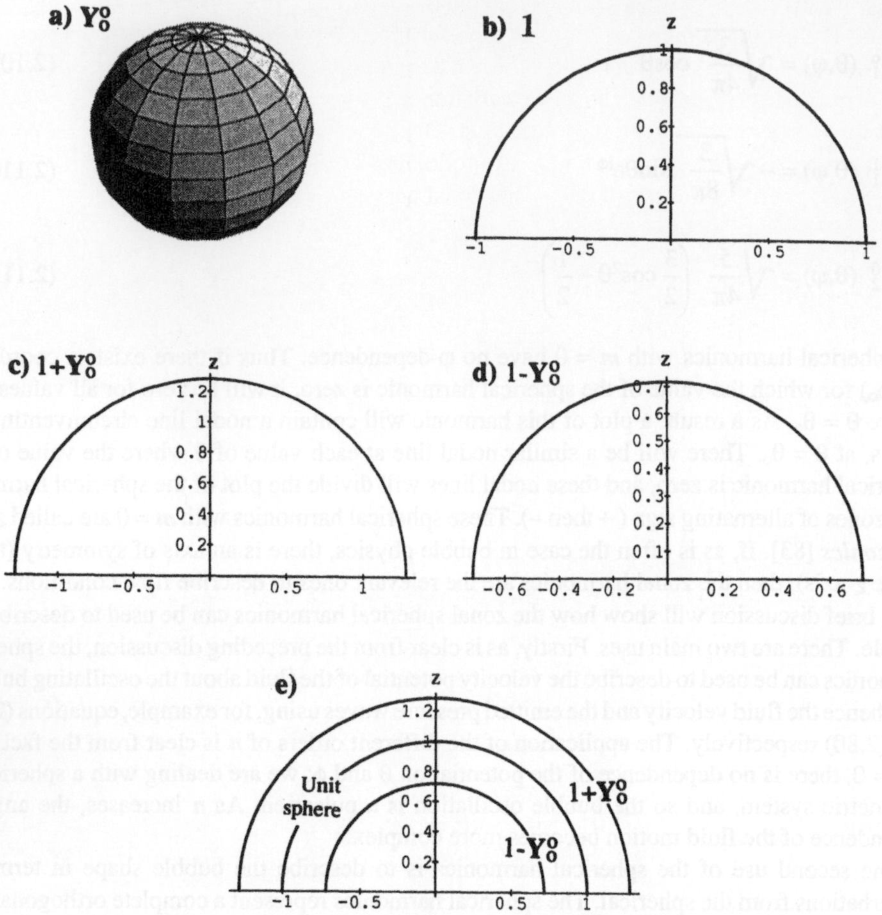

Figure 2.13 Y_0^0, the zero-order axisymmetric spherical harmonic ($n = m = 0$). (a) The function Y_0^0 in 3-D. The distance of the surface from the origin represents the value of the function in that particular direction (i.e. for that particular combination of θ and φ). (b) The function Y_0^0, plotted in a 2-D half-space (i.e. $z \geqslant 0$ only shown). The full function in 2-D would be a circle, centred at the origin., Parts (c), (d) and (e) are shown in a similar half-space. (c) The spherical harmonic added onto a sphere of radius 1. (d) The spherical harmonic subtracted from the unit sphere. (e) Parts (c) and (d) shown together with the unit sphere.

shown in Figure 2.13(b), and in 2.13(c) the Y_0^0 function has been *added* onto this unit sphere, which has remained spherical but expanded. If this harmonic is added in an oscillatory fashion, so that the radius varies as, for example, $1 + Y_0^0\cos\omega t$, this will describe a pulsation, and for pulsations of greater amplitude, greater amounts of Y_0^0 are added or subtracted from the initial sphere. If Figure 2.13(c) were to represent the extreme of the expansion of a bubble pulsating as $1 + Y_0^0\cos\omega t$, i.e. $1 + Y_0^0$, then Figure 2.13(d) illustrates the other extreme, the contraction $1 - Y_0^0$. To summarise this pulsation, Figure 2.13(e) shows the equilibrium position of a sphere of radius 1, superimposed upon which are the maximum and minimum positions the pulsating sphere $1 + Y_0^0\cos\omega t$. To obtain different amplitudes of oscillation, different proportions of the harmonic would be added. For obvious reasons, this $n = 0$ perturbation upon a sphere is often called the 'breathing mode'.

The higher harmonics can be used to describe more complicated bubble shapes. Figure 2.14(a) shows a 3-D representation of the $n = 2$ harmonic, with a 2-D half-space representation of Y_2^0 included for clarity. It shows the variation of the spherical harmonic with θ (which varies between 0 and π) for some fixed value of φ. The exact value of φ does not, of course, matter for these axially symmetric plots. In the following figures, the three-dimensional plots are obtained by rotating the two-dimensional zonal harmonic plots about the z-axis.

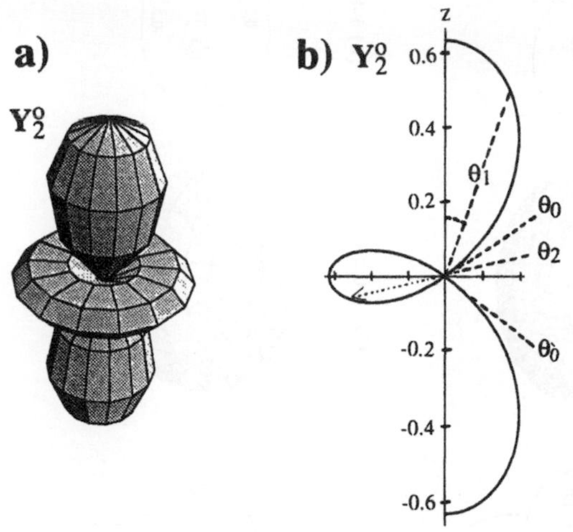

Figure 2.14 Y_2^0 the second-order axisymmetric spherical harmonic ($n = 2$, $m = 0$). (a) The function shown in 3-D. (b) The function shown in a 2-D cross-section in the spherical coordinate system, plotted for $0 \leqslant \theta \leqslant \pi$ There is no variation in φ and so the 3-D plot of the harmonic shown in (a) can be reconstructed by a complete rotation of the plot in (b) about the z-axis. The value of the function at any given θ is given by the radial distance from the origin to the plot. Examples illustrated show that $Y_2^0(\theta_1,\varphi) = 0.52$, $Y_2^0(\theta_0,\varphi) = 0$, $Y_2^0(\theta_2,\varphi) = -0.27$, $Y_2^0(\theta_0',\varphi) = 0$.

Consider an initially spherical bubble onto which is *added* the second-order spherical harmonic, Y_2^0 . The variation in θ, as plotted in Figure 2.14(b), can be used to assess the perturbed shape of the bubble. As can be seen, at $\theta = 0$ the spherical harmonic Y_2^0 has magnitude 0.63 (found from the distance from the origin to the curve at that angle). This is positive, so that on addition the unit sphere will be extended outwards in that direction. As θ increases, this extension decreases as the value of Y_2^0 which is added onto the spherical bubble decreases. For example, at the angle $\theta = \theta_1$ (≈20°), the radial distance from the origin to the curve is reduced to 0.52. This is still positive, so on addition the unit sphere will be extended in that angle. At the angle θ_0 (≈55°) the value of Y_2^0 is zero, that is, the bubble wall at this angle is not perturbed from the position it had when the bubble was spherical, and will have a nodal line circumscribing the z-axis at this angle. As the angle increases to greater than θ_0, the value of Y_2^0 becomes negative. The bubble wall at these angles has not moved outwards, but *inwards*. For example, at $\theta = \theta_2$ (≈75°), $Y_2^0 = -0.26$, so a bubble that was initially spherical with radius equal to 1 will be perturbed so that the distance from the origin to the bubble wall is now $1 - 0.26 = 0.74$. There will be a local contraction for all angles between the two nodal lines, the second one occurring at $\theta = \theta_0'$ (≈130°). The

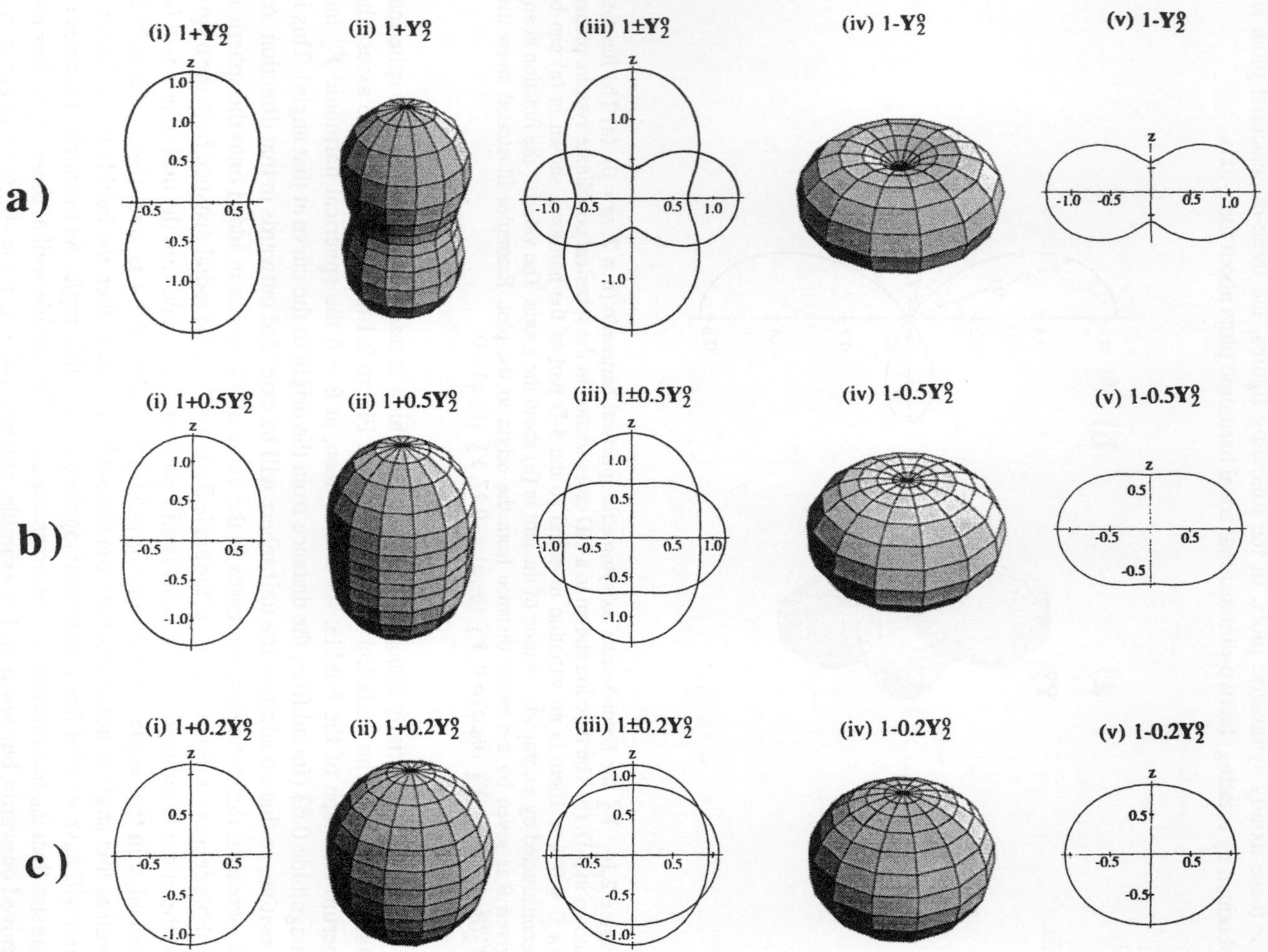
a)
b)
c)
(i) 1+Y$_2^0$
(ii) 1+Y$_2^0$
(iii) 1±Y$_2^0$
(iv) 1-Y$_2^0$
(v) 1-Y$_2^0$
(i) 1+0.5Y$_2^0$
(ii) 1+0.5Y$_2^0$
(iii) 1±0.5Y$_2^0$
(iv) 1-0.5Y$_2^0$
(v) 1-0.5Y$_2^0$
(i) 1+0.2Y$_2^0$
(ii) 1+0.2Y$_2^0$
(iii) 1±0.2Y$_2^0$
(iv) 1-0.2Y$_2^0$
(v) 1-0.2Y$_2^0$
z

function $1 + Y_2^0$ is plotted in Figures 2.15(a)(i) and 2.15(a)(ii). To obtain the shape of the bubble in 2.15(a)(ii), the curve in 2.15(a)(i) should be rotated about the z-axis.

Now if the initially spherical bubble is perturbed in an oscillatory manner by the $n = 2$ harmonic, the bubble radius varying as $1 + Y_2^0 \cos\omega t$, then the shape will oscillate between $1 + Y_2^0$ and $1 - Y_2^0$. The former, as already explained, is shown in Figures 2.15(a)(i) and 2.15(a)(ii). The latter is shown in Figures 2.15(a)(iv) and 2.15(a)(v). The two extremes of the oscillation are superimposed in Figure 2.15(a)(iii). Since the function $\cos\omega t$ equals zero at $\omega t = \pi, 3\pi/2$, twice in an oscillation of the type $1 + A_n Y_n^m \cos\omega t$, the bubble will pass through the equilibrium unperturbed spherical position twice every cycle.

The magnitude of the perturbation can of course vary, depending on the energy of the excitation. Figure 2.15(b) shows the extremes of the $0.5 \times Y_2^0$ perturbation (i.e. $A_n = 0.5$) and Figure 2.15(c) those of the $0.2 \times Y_2^0$ perturbation, where $A_n = 0.2$. It is clear that oscillatory perturbations of the Y_2^0 type refer to the famous 'needle-to-pancake' mode of bubble oscillation. Though Figures 2.15(a), 2.15(b) and 2.15(c) can be thought of as representing the extremes of the similar motions with ever decreasing amplitudes, by looking in turn at the 3-D representation, Figures 2.15(a)(ii), 2.15(b)(ii), 2.15(c)(ii), 2.15(c)(iii), 2.15(b)(iii) and 2.15(a)(iii), the reader can gain an impression of one half-cycle of the $1 + Y_2^0 \cos\omega t$ oscillation, where the bubble would pass through the equilibrium spherical shape between 2.15(c)(ii) and 2.15(c)(iii). Similarly, a bubble with an initial distortion of the form $1 + 0.5 Y_2^0 \cos\omega t$ and zero wall velocity would pass through, in sequence, Figures 2.15(b)(ii), 2.15(c)(ii), 2.15(c)(iii), 2.15(b)(iii), and then return.

For completeness, perturbations of a sphere of the form $1 \pm Y_3^0$ are shown in Figure 2.16.

Readers may have observed that the Y_1^0 was passed over in the above discussion as it differs from the others, in that, unlike Y_2^0 and higher modes, it cannot describe a shape oscillation about some fixed centre. Just as both r^n and $r^{-(n+1)}$ are radial potential solutions for Laplace's equation, yet only the $r^{-(n+1)}$ functions are physically relevant to the bubble, in just such a manner the $n = 1$ mode, whilst being a mathematically acceptable spherical harmonic, has no place in the physical description of shape oscillations of bubbles: the coefficient A_1 is always zero. What is physically acceptable is a subset of what is mathematically possible.

The use of spherical harmonics to describe oscillatory shape changes in a bubble will be discussed in Chapter 3, where the frequency of these oscillations is found (equation (3.236)). Substitution of $n = 1$ into equation (3.236) gives the frequency for this mode to be zero, showing that it cannot be employed in the description of oscillatory shape changes in a bubble. The zero frequency suggests that it represents some 'DC' effect. That effect is clear when one considers that, although $1 + A_1 Y_1^0$ describes no oscillatory perturbation of the bubble shape, the function Y_1^0 on its own may still be used as a velocity potential to describe a fluid flow. Clearly, a 'DC' flow is an unchanging, unidirectional steady motion.

Figure 2.15 Addition of (a) 100%, (b) 50% and (c) 20% of the second-order axisymmetric spherical harmonic Y_2^0 onto the unit sphere. (i), (iii) and (v) are 2-D plots in spherical coordinates, for $0 \leqslant \theta \leqslant 2\pi$. The relevant 3-D plots can be found by rotating these 2-D ones around the z-axis, and these are shown in (ii) and (iv). Plots (i) and (ii) show the harmonic added to the unit sphere, and (iv) and (v) show it subtracted from the unit sphere. Plot (iii) shows (i) and (v) superimposed.

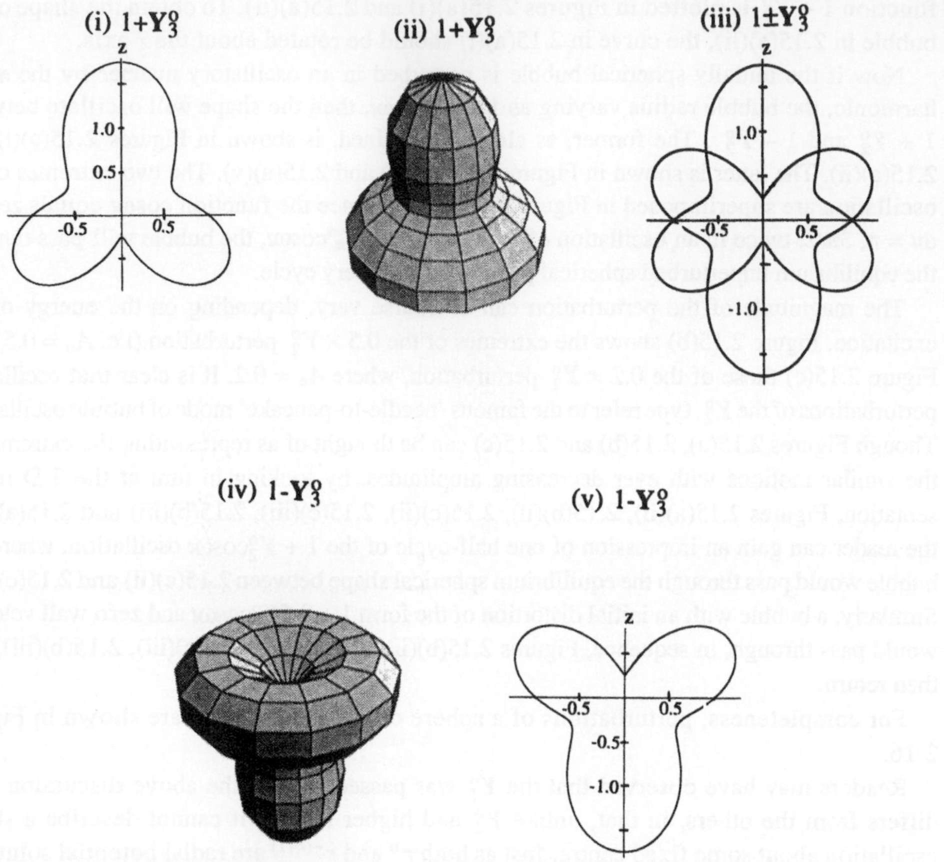

Figure 2.16 A similar construction as in 2.15, but now for addition and subtraction of 100% of the third-order axisymmetric spherical harmonic Y_3^0 with the unit sphere.

Thus Y_1^0 could be employed in bubble physics to describe the component of flow around a bubble that is uniform[39] or, looking at the situation from the point of view of the fluid, the unidirectional motion of a bubble through a stationary fluid.

Bubble translations are a topic in their own right, and motions, dealing as they do more in fluid dynamics than acoustics, cannot be dealt with in any detail within this book, which concentrates on the interactions between acoustic fields and bubbles. Readers interested in the translations of bubbles should consult the referenced works, for example reference [84]. However, a simple analysis of rigid spheres in potential flow, followed by a brief discussion of more complicated flows, will serve to illustrate to the reader the complexities involved in a full treatment of the bubble as an inertial body under the influence of an impressed unidirectional force.

---

[39] In the presence of a bubble, a flow cannot be unidirectional everywhere, owing to the disturbing presence of the sphere itself. However, Y_1^0 can describe that component of the flow which is uniform, onto which other flow descriptions can be added until the complete fluid motion is described. It may be thought of as being that uniform flow which existed prior to the presence of the bubble.

2.3 The Translating Bubble

2.3.1 Fluid Dynamics of a Rigid Sphere

At first sight, bubbles appear not to fall into the category of rigid spheres, as they are deformable and often not spherical. These two properties, which tend to be more pronounced the larger the bubble, make the dynamics complicated, and a background knowledge of the fluid dynamics of a rigid sphere provides a useful introduction to the behaviour of a gas-pocket in a liquid. In fact, a real bubble is rarely so simple an entity as a gas-pocket, and in addition to the gas and the bulk liquid, its surface is likely to have complex properties of its own. Most water samples (domestic, industrial and in the natural world) will contain surface active materials, and bubbles rapidly absorb these onto their walls [85, 86]. As a result, the wall can sustain a stress. Once the material covers the bubble, its dynamics do in fact tend to those of a rigid particle because of the immobilisation of the interface and the resulting viscous boundary layer on the bubble. The internal circulation is damped out by the surface-active contaminant[40] (an effect which is more important in the dynamics of liquid droplets). Bubbles whose dynamics are significantly affected by surface-active materials in the wall are termed 'dirty bubbles'. Those possessing no surface-active material are called 'clean bubbles'. Beginning with the simplest situation therefore, the motion of a rigid sphere in inviscid potential flow will now be discussed.

(a) Potential Flow Around a Rigid Sphere

Consider a rigid sphere of radius R_0. It is stationary in a fluid which flows so that, if the sphere were not present, all particles in the fluid would flow with a uniform velocity $\vec{v}_0$ aligned with the z-axis. The centre of the sphere is at the origin. The flow is irrotational, so that a velocity potential Φ can be defined, and the fluid is incompressible so that $\nabla^2\Phi = 0$ (Laplace's equation). Let Φ_0 be the potential defining the state of uniform flow that would exist in the absence of the sphere. Since $\vec{v}_0 = \vec{\nabla}\Phi_0$, and the only non-zero component of $\vec{v}_0$ is in the z-direction, then

$$v_0 = \frac{\partial \Phi_0}{\partial z} \Rightarrow \Phi_0 = v_0 z \tag{2.113}$$

the constant of integration being set to zero. Since $z = r\cos\theta$ (Figure 2.12), then

$$\Phi_0 = v_0 r \cos\theta \tag{2.114}$$

In the presence of the sphere, the potential is Φ. This can be thought of as being the sum of the potential due to the original flow (Φ_0) and a potential due to presence of the sphere (Φ_s). However, the potential flow due to a sphere in a flow satisfying Laplace's equation is given by a sum of the solutions discussed in section 2.2.4(c). Each solution term in the summation is a product of the form $\Phi = \Phi_r \Phi_\theta \Phi_\varphi$ so that, assuming from the symmetry no dependence on φ, one would guess[41] the solution to be

$$\Phi_s = \frac{s_1}{r} + \frac{s_2}{r^2}\cos\theta + \frac{s_3}{2r^3}(3\cos^2\theta - 1) + \ldots \tag{2.115}$$

[40] See section 2.3.2(b)(ii).
[41] The result of the Uniqueness Theorem lending confidence to this guessing process.

where s_1, s_2, s_3 etc. are coefficients yet to be determined. Since the sphere is rigid, the fluid flow normal to the surface of the sphere (i.e. in the radial direction) must be zero at the sphere's surface, that is

$$\left.\frac{\partial \Phi}{\partial r}\right|_{r=R_o} = 0 \tag{2.116}$$

the notation signifying that the partial differentiation is evaluated at the position $r = R_o$ only, that is, on the surface of the sphere. However, $\Phi = \Phi_o + \Phi_s$, so that equation (2.116) implies

$$\left.\frac{\partial \Phi_o}{\partial r}\right|_{r=R_o} = -\left.\frac{\partial \Phi_s}{\partial r}\right|_{r=R_o} \tag{2.117}$$

Substitution of equations (2.114) and (2.115) into equation (2.117), with the required condition of $r = R_o$, gives

$$v_o\cos\theta = \frac{s_1}{R_o^2} + \frac{2s_2}{R_o^3}\cos\theta + \frac{3s_3}{2R_o^4}(3\cos^2\theta - 1) + \ldots \tag{2.118}$$

This equation must be satisfied at all points on the sphere's surface, that is at $r = R_o$ and for all values of θ. This second condition means that since the left-hand side of the equation contains terms in $\cos\theta$ only, then only terms in $\cos\theta$ on the right-hand side can be nonzero. Therefore the coefficients s_1, s_3 ... are zero, only s_2 being finite. This is hardly surprising from the discussion in the previous section (2.2.4(c)): there is no pulsation component to the motion, so s_1 must be zero; similarly there are no higher modes denoting shape oscillations. There is only a translational motion, which we can now see is described by a potential employing the $n = 1$ harmonic. The coefficients s_1, s_3 etc. having been determined, solution of equation (2.118) gives $2s_2 = v_oR_o^3$, so that by adding equations (2.114) and (2.115) with this knowledge we obtain

$$\Phi = \Phi_o + \Phi_s = \left(r + \frac{R_o^3}{2r^2}\right)v_o\cos\theta \tag{2.119}$$

the velocity potential for flow around a sphere which, in the absence of the sphere, would be uniform. It should be noticed that far from the sphere, when r becomes very large, the potential tends to that of uniform flow in that $\Phi \rightarrow \Phi_o$, as given by equation (2.114). This is as one would expect.

Knowing the potential (equation (2.119)), it is possible to find the fluid velocity from $\vec{v} = \vec{\nabla}\Phi$ (equation (2.70)). Therefore the radial velocity is

$$v_r = \frac{\partial \Phi}{\partial r} = \left(1 - \frac{R_o^3}{r^3}\right)v_o\cos\theta \tag{2.120}$$

Similarly, the angular component in θ can be found, the elemental length for differentiation in this direction being $r\Delta\theta$ (Figure 2.12). This component of the velocity is

$$v_\theta = \frac{1}{r}\frac{\partial \Phi}{\partial \theta} = -\left(1 + \frac{R_o^3}{2r^3}\right)v_o\sin\theta \tag{2.121}$$

There is no dependence of the potential on the angle φ, so that there are no nonzero components of the velocity in this direction as one would expect from symmetry. If the fluid flow is steady,

$\vec{v}_o = 0$ and so $\dot{\Phi} = 0$. However, because there is a flow velocity, differences in fluid pressure may still arise over the surface (equation (2.71)). Since this inviscid flow is steady, Bernoulli's equation is applicable. The fluid pressure field depends only on the square of the magnitude of the fluid velocity (equation (2.71)), that is, on $|\vec{v}|^2 = v_r^2 + v_\theta^2$. From equations (2.120) and (2.121) it is clear that this will vary not in terms of $\cos\theta$ and $\sin\theta$, but as the squares of these terms. This leads to a mirror-symmetry in the pressure before and behind the sphere, so that if $\vec{v}_o$ is constant in time there is no net force on the body. This result is quite contrary to experience, as one expects a finite drag when liquid flows uniformly past a stationary sphere. This is d'Alembert's paradox. However, it should be remembered that this formulation was based on Euler's equation, which assumes inviscid flow. The paradox is resolved when the fluid is considered to have finite viscosity, which results in the physically expected drag force.

(b) A Rigid Sphere Moving Through a Liquid

Consider a sphere moving at constant velocity $-\vec{v}_o$ along the z-axis, in the direction of $-z$ (i.e. by convention to the left). In the absence of the sphere, the fluid would be stationary. We have in fact already met this problem, though in a slightly different form, in the previous section. In the former case the sphere was stationary in a fluid which would, in the absence of the sphere, move with uniform velocity along the z-axis in the direction of $+z$, that is, to the right. If our coordinate axis and origin were now to move with that uniform velocity along the z-axis, we would have the current situation of a sphere moving along $-z$ in a fluid that would otherwise be stationary. Simply moving our point of observation this way translates the problem of a stationary sphere in a previously uniform flow, to one of a uniformly moving sphere in a previously stationary flow. The potential for the system of a moving sphere in an otherwise stationary fluid will simply be that derived in the previous section (2.3.1(a)), minus the potential corresponding to the motion of the coordinate axes, i.e. $v_o r\cos\theta$. The potential of this system is therefore

$$\Phi = \frac{R_o^3}{2r^2} v_o \cos\theta \qquad (2.122)$$

Therefore the potential of an initially stationary fluid containing a sphere which moves in the direction $+z$ (i.e. to the right) is -1 times the potential of when the sphere moves to the left, i.e.

$$\Phi = -\frac{R_o^3}{2r^2} v_o \cos\theta \qquad (2.123)$$

This gives the radial and tangential components of the velocity respectively to be

$$v_r = \frac{\partial \Phi}{\partial r} = \frac{R_o^3}{r^3} v_o \cos\theta \qquad (2.124)$$

and

$$v_\theta = \frac{1}{r}\frac{\partial \Phi}{\partial \theta} = \frac{R_o^3}{2r^3} v_o \sin\theta \qquad (2.125)$$

Knowing Φ and v, the pressure resulting from the fluid flow in these conditions can therefore be found from equation (2.71) to be

$$p = p_c - \rho\dot{\Phi} - \tfrac{1}{2}\rho\langle v^2\rangle - \phi_B\rho \tag{2.126}$$

Applying equation (2.126) to flow around the rigid sphere, it can be shown that, despite the fact that the liquid may be frictionless, a drag force can arise. The term in ρv^2 is symmetrical as one goes around the sphere, since from equations (2.124) and (2.125),

$$v^2 = v_r^2 + v_\theta{}^2 = \left(\frac{R_o}{r}\right)^6 v_o^2\left(\cos^2\theta + \frac{\sin^2\theta}{4}\right) = \left(\frac{R_o}{r}\right)^6 v_o^2\left(\frac{1}{4} + \frac{3\cos^2\theta}{4}\right) \tag{2.127}$$

Equation (2.127) shows that ρv^2 can be thought of as either the sum of two terms in squared quadrature, or as a constant plus a $\cos^2\theta$ term. Either way, it will be symmetrical as θ varies, so that any pressures generated by this term at one point on the sphere's surface will be balanced by corresponding pressure elsewhere on the sphere. Similarly, the terms $p_c - \rho\phi_B$ will be constant over the sphere surface, giving rise to no net directional force, variations in gravitational terms being deemed negligible. Therefore if the sphere moves at constant velocity through the liquid, $\dot{\vec{v}}_o = 0$ and so $\dot{\Phi} = 0$, and from equation (2.126) there is no net force on the sphere. In fact, if $\dot{\Phi} = 0$, equation (2.126) reduces to Bernoulli's equation, which was used to show that there is no force on a stationary sphere in steady inviscid flow. However, if the sphere accelerates or decelerates, $\dot{\vec{v}}_o \neq 0$ and the $\dot{\Phi}$ term in equation (2.126) may give rise to pressures on the sphere. The net drag force on the sphere due to $-\rho\dot{\Phi}$ is therefore

$$F_D = \int_0^\pi (2\pi R_o \sin\theta).(\tfrac{1}{2}\rho\dot{v}_o R_o\cos\theta).(\cos\theta).(R_o d\theta) \tag{2.128}$$

$$F_D = \frac{2\pi}{3}\rho R_o^3 \dot{v}_o \tag{2.129}$$

Even if the flow is inviscid, a finite drag force[42] arises if $\dot{\vec{v}}_o \neq 0$. In other words, if a sphere is travelling through a fluid that would be stationary but for the presence of the sphere, the inertia of the sphere is apparently increased by half the mass of the fluid displaced. This 'apparent mass' arises because, in order to move, the sphere must set the fluid into motion, displacing it away from the space the sphere will occupy, and moving fluid into the space it has vacated.

The magnitude of the apparent mass is dependent on the shape of the body. Whereas for a sphere it corresponds to half the fluid displaced, for a prolate ellipsoid of revolution it is less. If the axes are in the ratio of 1:2, the apparent mass is only a fifth of the fluid displaced.

(c) The Mass of a Bubble

The concept of mass is not a simple one. The aspect that first comes to mind represents a *gravitational* mass of a body: this is taken to be proportional to the gravitational force of attraction experienced between the body in question and some standard mass. In most circumstances this is a negligibly small 'mass' to consider in the discussion of the gas within a bubble. However, it is always important to consider the whole picture, and a bubble is not simply a

---

[42] Note that in the previous example of a stationary sphere in a previously spatially and temporally uniform flow, a drag force would have arisen if $\dot{\vec{v}}$ had not been equal to zero.

spherical space containing gas and vapour: an integral part of the system we call a bubble is the surrounding liquid. If we consider the effect of gravity on the bubble as a system, then the effect of gravity on the surrounding fluid is important, leading to the phenomenon of buoyancy. Indeed the nature of the fluid surrounding that gas determines many dynamic properties of the bubble. In consideration of the translational motion of a bubble, we may from Newton's Second Law define an inertial mass which represents the ratio of some unidirectional net force impressed upon a body to the resulting linear acceleration. The simple treatment of a rigid sphere in an inviscid fluid, given in the previous section (2.3.1(b)), shows that the sphere has an added inertial mass equal to half the mass of fluid displaced. It arises because, for the sphere to translate within the fluid, elements of that fluid must be set into motion and move around the sphere. If we are considering the translation of a ball through air, then density of the fluid surrounding the ball is so much less than that of the ball itself that the effect of the added mass is negligible. However, when a spherical gas body translates through a liquid, since the liquid is much denser than the gas within the bubble, this added mass is far larger than the mass of the gas/vapour content of the bubble, and so dominates. It should be remembered that the mass of the translating bubble is seldom so simple as the added mass derived above from potential flow around rigid spheres: the translation of real bubbles is discussed in section 2.3.2(b).

The above masses correspond to the translational response of a bubble to a net impressed unidirectional force: the relevant motion is a displacement of the centre of mass of the bubble. However, most of the acoustic effects of bubbles involve a different kind of motion: in the simplest case, the bubble changes volume, expanding or contracting, so that the centre of mass of the bubble is stationary and it is the fluid surrounding the bubble which is displaced. If the bubble is spherical, this liquid motion is in a radial direction, outwards or inwards, with respect to the centre of the bubble. Therefore if an oscillator is emitting acoustic pressure waves into the surrounding fluid, there will be an additional mass associated with the inertia of the oscillating fluid elements, called a 'radiation mass'. It will be seen in Chapter 3 that the radiation mass associated with a pulsating gas bubble in a liquid dominates the mass of other elements, which have negligible inertia in comparison. Thus it is this radiation mass which characterises the inertia of the pulsating bubble, and so determines the resonance.

2.3.2 Flow in Viscous Liquids

(a) Rigid Spheres

(i) High Reynolds Numbers (Rigid Spheres). In many cases, flow past rigid spheres differs from the potential flow discussed above. Even so, the treatment given earlier may in some instances approximate to the true flow. Though real liquids are not inviscid, under certain circumstances viscous effects may be negligible throughout large regions of the flow field. Such conditions occur when the Reynolds number, $\mathcal{Re}$, which can be interpreted as expressing the ratio of the inertial to the viscous forces, is large. The Reynolds[43] number is given by

[43] When flow occurs relative to a body in a viscous fluid, it is possible to apply a scaling procedure using the non-dimensional quantity $\mathcal{Re}$. In incompressible fluids it is the Reynolds number which characterises the flow, so that if a scale model of some engineering design is to be built for a system with this type of flow, the experimenter has to choose the ratio of $\rho v/\eta$ used in the model and the full-size rig to be the reciprocal of the ratio of the length scaling factor employed. If the liquid is compressible, then both the Reynolds number and the Mach number (that is, the ratio of the particle velocity to the sound speed, i.e. v/c) must be the same in both the model and full-scale.

$$\mathcal{R}e = \frac{\rho}{\eta} v_0 L \tag{2.130}$$

where L and v_0 are characteristic lengths and speeds pertinent to the body, and η and ρ are the shear viscosity and density of the fluid. At high Reynolds numbers, extensive portions of the flow field can be treated as inviscid. The effect of viscosity is only considered in thin layers around boundaries in the flow. Thus when the translation of a bubble is characterised by a large Reynolds number, the viscous effects may be confined to a *boundary layer*, close to the bubble wall: the majority of the liquid in which the bubble exists may be treated as inviscid.

The reason why approximations of inviscid flow correlate well with experience at high Reynolds numbers can be seen from the Navier–Stokes equation of motion. The Navier–Stokes equation describes the motion of an infinitesimal volume element within a viscous compressible liquid where the viscosity is constant throughout the liquid[44] [87]

$$\rho \frac{D\vec{v}}{Dt} = \rho \left\{ \frac{\partial \vec{v}}{\partial t} + (\vec{v}.\vec{\nabla})\vec{v} \right\} = \rho \Sigma \vec{F}_{ext} - \vec{\nabla} p + \left\{ \frac{4\eta}{3} + \eta_B \right\} \vec{\nabla}(\vec{\nabla}.\vec{v}) - \eta \vec{\nabla} \wedge \vec{\nabla} \wedge \vec{v} \tag{2.131}$$

D/Dt is the so-called 'material derivative', a differential which, as described in section 2.2.2, takes account of changes in time at a fixed point through $\partial/\partial t$, and changes following the fluid motion by the convective term $\vec{v}.\vec{\nabla}$. The term $\Sigma \vec{F}_{ext}$ represents the vector sum of the external forces, both volume and surface, per unit mass on an infinitesimal element of liquid. The terms η and η_B are the *shear viscosity coefficient* and the *bulk* (or *volume*) *viscosity*, respectively. Both have units of [Pa.s]. The shear viscosity is responsible for the transfer of momentum through molecular collision between elements of a liquid which are in contact and are moving at different velocities. The dissipation of energy through mechanisms involving turbulence, streaming and vorticity is accounted for by its use in the $\eta \vec{\nabla} \wedge \vec{\nabla} \wedge \vec{v}$ term [87]. Though the shear viscosity will describe the ratio of the stresses generated in different fluids under the same circumstances, if one wishes to understand how viscosity will modify an existing fluid motion one must incorporate the relative inertias of the different liquids. This is done through the *kinematic viscosity*, η_k, which for a fluid of density ρ is equal to η/ρ [89]. The bulk viscosity is manifested in the loss of mechanical energy when a fluid is compressed or dilated. If the energy transfer in the fluid can only take place through processes associated with the translational kinetic energy of molecules, then the volume viscosity of that fluid is zero. Making this assumption (the *Stokes assumption*) and employing the differential vector identity [75]

$$\vec{\nabla} \wedge \vec{\nabla} \wedge \vec{v} = \vec{\nabla}(\vec{\nabla}.\vec{v}) - \nabla^2 \vec{v} \tag{2.132}$$

gives

$$\rho \frac{D\vec{v}}{Dt} = \rho \left\{ \frac{\partial \vec{v}}{\partial t} + (\vec{v}.\vec{\nabla})\vec{v} \right\} = \rho \Sigma \vec{F}_{ext} - \vec{\nabla} p + \frac{\eta}{3} \vec{\nabla}(\vec{\nabla}.\vec{v}) + \eta \nabla^2 \vec{v} \tag{2.133}$$

If the liquid is assumed to be incompressible, then both $\dot{\rho}$ and $\vec{\nabla}\rho$ equal zero, so by continuity $\vec{\nabla}.\vec{v} = 0$ (equation (2.2)). If, in addition, gravity is taken to be the only external force acting, then equation (2.133) reduces to

[44] See Batchelor [88] for a discussion of the case where the viscosity is not spatially uniform.

$$\rho \frac{D\vec{v}}{Dt} = \rho\left(\frac{\partial}{\partial t} + \vec{v}.\vec{\nabla}\right)\vec{v} = \rho\vec{g} - \vec{\nabla}p + \eta\nabla^2\vec{v} \tag{2.134}$$

Clearly, if $\mathcal{R}e = \rho v_0 L/\eta$ is very large, the final term $\eta\nabla^2\vec{v}$ can be expected to be insignificant in comparison with the others, and the Navier–Stokes equation reduces to Euler's equation (equation (2.61)), which was used as the basis of the above inviscid discussion of flow past rigid spheres. In this sense the inviscid approximation is a good one at high Reynolds numbers, where viscous effects are confined to a boundary layer: however, at sufficiently high Reynolds numbers, the main stream of fluid, which behaves inviscidly, ceases to follow the wall of the body, and instead breaks away. Rather than remaining close to the surface, the boundary layer (where viscous effects occur) separates from the body, and forms a downstream *wake* [90]. When this *separation* occurs, inviscid theory does not predict the drag to any accuracy. The flow in the boundary layer itself may be laminar or, as the flow undergoes *transition*, become turbulent.

(ii) Low Reynolds Numbers (Rigid Spheres). At the other extreme, as $\mathcal{R}e \to 0$, the viscous effects dominate, whilst the fluid inertia and the convective term are negligible. In this regime of 'creeping flow', the Navier–Stokes equation reduces to

$$\vec{\nabla}(p - p_0 - \rho g h) = \eta\nabla^2\vec{v} \tag{2.135}$$

At such low Reynolds numbers, there is no wake at the rear of the particle. In fact, both (2.135) and the other key equation, that of continuity, are reversible, in that both are still satisfied if $\vec{v}$ is replaced by $-\vec{v}$. Thus if a particle, like a sphere, has fore- and aft-symmetry, flow fields in the regime of creeping flow will be expected to show a like symmetry.

Stokes [91] examined this regime for the case when the flow is steady, that is $\dot{\vec{v}} = 0$, and found that the drag force on a rigid sphere of radius R_0, travelling at speed v_0, is $6\pi\eta R_0 v_0$.

(iii) Intermediate Reynolds Numbers (Rigid Spheres). Though the flow at very high and very low Reynolds numbers can be approached through inviscid and creeping approximations respectively, the flow at intermediate Reynolds numbers is not so simple. As $\mathcal{R}e$ increases, asymmetry sets in (the vorticity becomes asymmetrical at $\mathcal{R}e \sim 1$, and the streamlines at $\mathcal{R}e \sim 10$ [83]). For $1 < \mathcal{R}e < 20$, the flow is unseparated. Flow separation, as indicated by a change of sign of vorticity, occurs at around $\mathcal{R}e \approx 20$, at a stagnation point behind the sphere. As the Reynolds number increases, this recirculating region increases in size, the wake lengthening and the vorticity being convected downstream. The steady wake exists for $20 < \mathcal{R}e < 130$. As $\mathcal{R}e$ increases further, one of the vortices behind the sphere becomes so long that it breaks off and travels downstream with the fluid. The fluid curls in behind the sphere and forms a new vortex. Thus for $130 < \mathcal{R}e < 400$, the wake has becomes unstable and, diffusion and convection of vorticity no longer being able to keep up with the rate of generation of vorticity, vortices are shed. The value of $\mathcal{R}e$ when vortex shedding commences is known as the 'lower critical Reynolds number', though the onset is often not well-defined. The tip of the wake oscillates at low frequency, the amplitude of oscillation increasing with Reynolds number. Associated with pulsations of the fluid circulating in the wake, large vortices begin to form and travel downstream at $\mathcal{R}e \approx 270$. For $400 < \mathcal{R}e < 3.5 \times 10^5$, vortices are regularly shed from alternate sides of a plane which slowly precesses about the axis. Strong periodicity can be observed in the wake. For $\mathcal{R}e > 3.5 \times 10^5$, a turbulent boundary layer can form, with an associated drop in drag occurring between $\mathcal{R}e = 2 \times 10^5$ and $\mathcal{R}e = 4 \times 10^5$. At higher Reynolds numbers, the drag

coefficient tends to increase towards a constant value. Introductory and more involved discussions are given respectively by Sabersky *et al.* [92] and Clift *et al.* [93].

(b) Flow Past Real Bubbles

In the natural world, clean bubbles of radii less than ~100 μm become 'dirty' in a few tens of seconds. Whilst $\sigma = 7.2 \times 10^{-2}$ N/m for clean bubbles in water, the walls of dirty bubbles in the ocean may have surface tension reduced to typically 3.6×10^{-2} N/m [94]. On larger bubbles, the surface-active material accumulates in a cap, resulting in local changes in stress. Larger bubbles may behave like 'clean' bubbles [94]. The motion of real bubbles differs from that of rigid spheres in that the bubbles may deform, and that the liquid–gas interface is mobile.

(i) Deformation. A bubble can deform from the spherical. It may, for example, adopt a nonspherical equilibrium shape (ellipsoidal, skirted, spherical cap etc. [95]). It can also change shape as time goes by, for example in response to liquid flow variations, or it may undergo dynamic shape oscillations[45] in an otherwise quiescent liquid as a result of an initial distortion from the equilibrium bubble shape. In practice, bubbles will deform and break up at high Reynolds numbers.

Air bubbles in water having an equivalent radius (i.e. that radius the bubble would assume if it were a sphere of the same volume) between roughly 0.5 and 7 mm tend to assume ellipsoidal shapes. Large bubbles, with an equivalent radius larger than about 9 mm in water, tend to undergo severe distortions, and the shape can usually be approximated by a segment of a sphere (e.g. 'spherical-cap'). Thin annular films, called 'skirts', may form from a dimple in the rear of the bubble, which affect the wake. Clearly the formation of such a distortion of the bubble will depend on competition between the viscous forces, which act to distort the bubble, and the surface tension, which makes the bubble tend to the spherical. Thus whether skirts form depends not only on the Reynolds number, but also on the Weber number, $\mathcal{W}e$, which reflects the surface tension. The Weber number is given by $\mathcal{W}e = v_0^2 L\rho/\sigma$, the length scale L being taken to be twice the equivalent radius of the drop. The ratio of the Weber to the Reynolds number is called the capillary, or skirt, number of a bubble. Figure 2.17 shows the shape assumed by some bubbles injected underwater. The upper bubble has formed a skirt, whilst the shape of the lower contains a significant proportion of the $n = 2$ spherical harmonic, and resembles Figure 2.15(a)(ii). Smaller bubble fragments can be seen in the picture.

Clift *et al.* present the terminal velocities of spherical, ellipsoidal and some spherical-cap air bubbles in water [96]. Increased oblateness tends to promote the formation of an attached wake and the onset of wake shedding [97].

Steady-state solutions for the distortion of a bubble in a uniaxial, inviscid straining flow have been found by Miksis [98] and Ryskin and Leal [99]. The latter found that for $\mathcal{R}e \geqslant 10$, steady converged solutions could not be obtained if the Weber number exceeded a critical value. Kang and Leal [100] found that in this regime of high Weber numbers, the bubble is continuously elongated, and will continue to be so even at Weber numbers smaller than the critical one if the initial deformation were large enough. Oscillatory shape changes could be found at low Weber numbers in the inviscid limit, the oscillatory frequency decreasing as $\mathcal{W}e$ increased, approaching zero at the critical value. Kang and Leal [101] provide a formulation to describe small-amplitude shape perturbations of a gas bubble in a uniaxial, inviscid straining flow, investigating the

[45] See section 2.2.4(c), and Chapter 3, section 3.6.

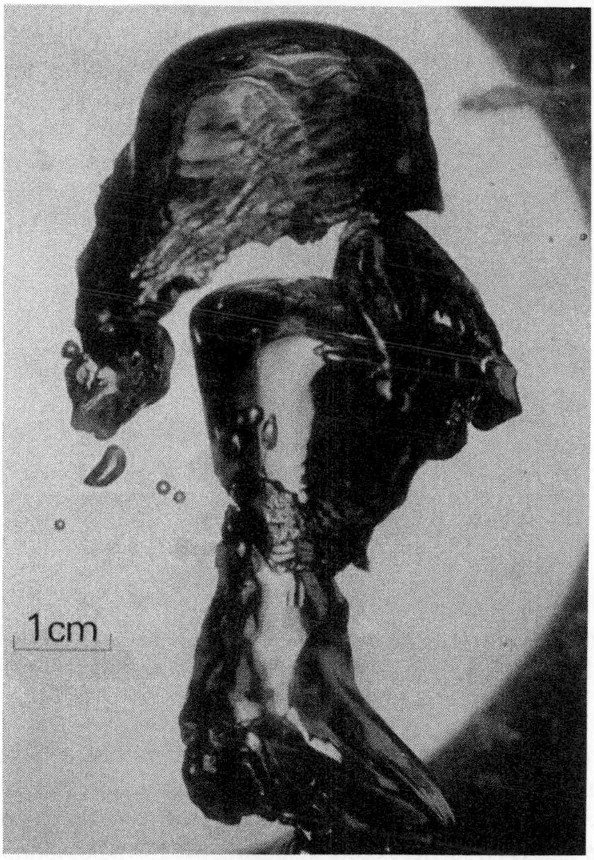

Figure 2.17 Photograph of two large bubbles injected from a nozzle. The upper large bubble shows a skirt: the lower one has a shape containing components of $1 + Y_2^0$, as illustrated in Figure 2.15(a)(ii).

frequency of the oscillatory motions of a nearly-spherical bubble in detail. They too find that it decreases with increasing Weber number.

(ii) Mobility at the Liquid–Gas Interface. The wall of a clean bubble is mobile: even if the bubble as a whole is not moving, the various fluid elements within the wall can flow relative to one another. This mobility means that the constraint that applies to rigid spheres, that the fluid velocity tangential to the surface must be zero at the sphere's surface, is not applicable to the clean bubble. Motion at the interface drives internal circulation of the gas contents within the bubble. If the bubble is dirty, surface contaminants immobilise the interface, affecting the flow field on either side of it: internal circulation within the bubble is reduced or eliminated, and the effect on the liquid flow field outside the bubble causes the change in dynamics already discussed, so that, for example, small dirty bubbles behave as rigid spheres. If an initially clean bubble moves through a medium containing surface-active agents, these substances will tend to accumulate at the liquid–gas interface. Consequently a translating bubble will tend to collect surfactant from the liquid through which it is moving. These substances are swept to the rear of the bubble, and the concentration gradient so formed will set up tangential stresses in the bubble wall as the surface tension varies from the front to the rear of the bubble. These stresses will make the interface rigid.

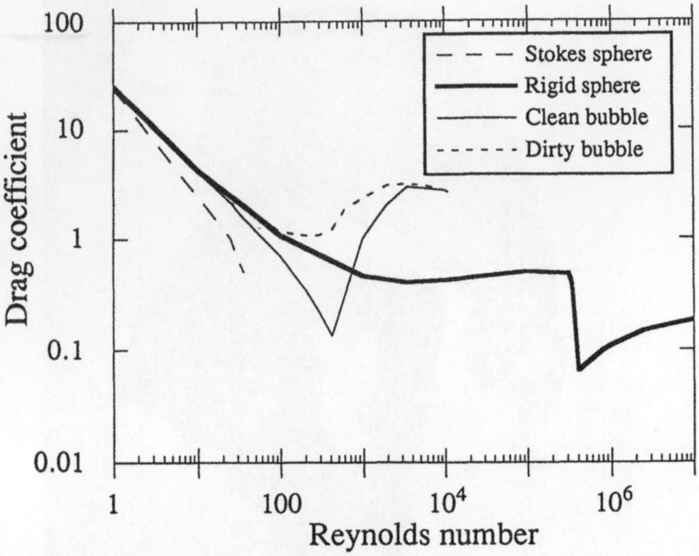

Figure 2.18 The variation in drag coefficient with Reynolds number for: a clean bubble (———); a dirty bubble (- - - -); a rigid sphere (———). Stoke's law for a sphere is given for comparison (– – –). (After Clift et al. [84].)

As mentioned above, increasing oblateness in a bubble tends to promote the formation of an attached wake and the onset of wake shedding. Similarly, the presence of surface-active contaminants encourages the formation of an attached wake and subsequent points where the wake begins to be shed [97], so these occur at much lower Reynolds numbers for dirty bubbles than for clean ones.

(iii) Effect of Deformation and Interface Mobility on Drag. The drag is characterised by the drag coefficient, C_D, which equals the ratio of the drag force to the product $(\rho L v_0^2/2)$, where L is a characteristic dimension of the body and v_0 the relative speed of the body and the fluid. The drag coefficient is a function of the Reynolds number, and estimates of that function can be found in the literature [102, 103].

Figure 2.18 shows the variation of the drag coefficient, C_D, with the Reynolds number, $\mathcal{Re}$, for (a) rigid spheres, (b) air bubbles in pure water and (c) air bubbles in contaminated water. The Stoke's Law drag is shown for comparison (d). Interface mobility tends to reduce the drag by reducing the size of the wake, delaying the onset of vortex shedding, and delaying boundary layer separation. In real bubbles, any deformation from the spherical shape tends to increase the drag. At larger bubble sizes the gradients are less pronounced, and the effect of contaminants less important, as can be seen in Figure 2.18.

(iv) Secondary Motions. As well as the primary translation, Clift *et al.* [97] describe two classes of secondary motion, which may occur simultaneously. The first, 'rigid body' movement, incorporates the sideways rocking motion, or tendency to follow a zig-zag or spiral path, commonly observed in rising bubbles. The wakes of zig-zagging and spiralling air bubbles in water are visualised in Figures 2.19 and 2.20 respectively by allowing the bubbles to first rise

(i) (ii)

(a)

(b)

(c)

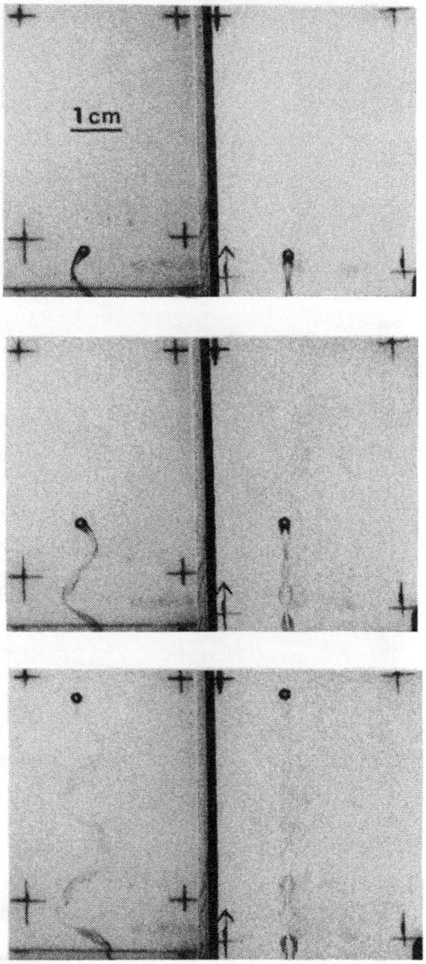

Figure 2.19 A bubble rising in water, having first risen through a 1% (weight) salt solution coloured by food dye (beneath the frame). Salt solution can be seen to be dragged along behind the bubble. Three frames (a), (b) and (c) are shown, taken at 800 frames per second (f.p.s.). The bubble is viewed in two perpendicular vertical mirrors, so that each frame is divided into two halves, (i) and (ii), which give simultaneous views of the bubble from perpendicular directions. In this way it can be seen that the bubble rises in a zig-zag manner in this figure. (Photograph: K. Lunde, private communication.)

through a 1% (by weight) salt solution coloured by food dye. Consecutive frames labelled (a), (b) and (c), taken from an 800 frames per second (f.p.s.) high-speed film, show the bubbles as they emerge into clear water (free of salt and dye), trailing their wakes. The rising bubble was photographed in two perpendicular planes simultaneously, one view (labelled (ii)) being direct, and the other (labelled (i)) through a mirror set at 45° to the line of sight. This enables the zig-zagging bubble (Figure 2.19) to be distinguished from the spiralling one (Figure 2.20). The wake of the zig-zagging bubble in Figure 2.19 consists of regularly shed vortex loops, a process akin to vortex shedding behind bluff bodies. The wake of the spiralling bubble in Figure 2.20

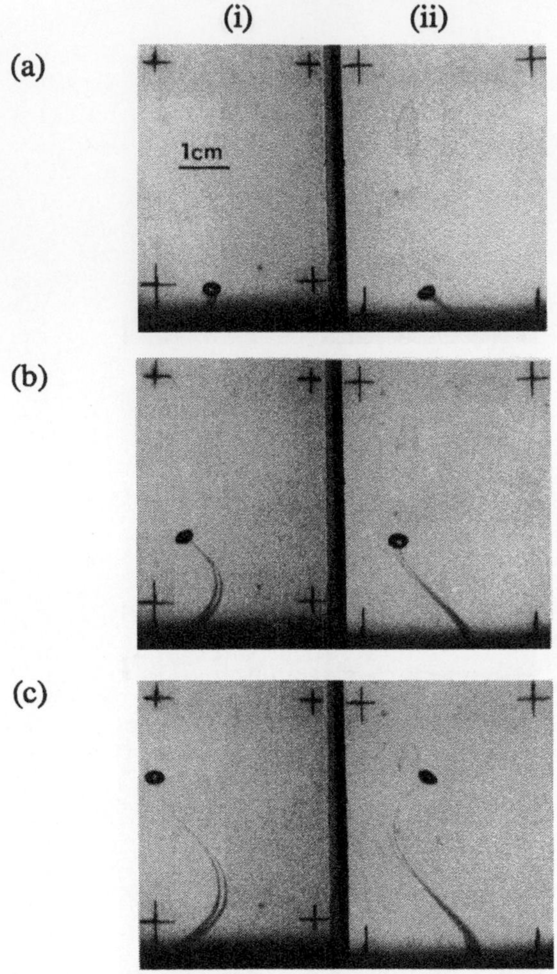

Figure 2.20 The path of a rising bubble is imaged using the same technique as for Figure 2.19. The bubble rises in a spiral path in this figure. (Photograph: K. Lunde, private communication.)

consists of two stable, attached vortices, similar to the trailing vortices found behind aeroplane wings.

The second class of secondary motion, shape oscillations, have already been introduced earlier in this chapter (section 2.2.4(c)). These motions are probably associated, in translating bubbles, with wake shedding, though there are other theories. Oscillations have been observed to set in at the same time as vortex shedding from the wake [104–106], but though both are periodic the two phenomena may differ in frequency.

Rectilinear motion in air bubbles rising in water has been found to be unstable, the bubbles taking zig-zag paths which become spiral for larger bubbles. Details of the release mechanism of the bubbles may influence the secondary motions.

Secondary motions increase drag, and heat and mass transfer.

This section has examined the fluid dynamics of translating rigid spheres, and indicated how a real bubble can differ from such spheres. In acoustics, the translation of a bubble is of secondary importance to its oscillation, which will be examined in the remainder of this book.

References

[1] Steiner LE. Introduction to Chemical Thermodynamics. McGraw-Hill, New York, 1941, pp 277–286

[2] Neppiras EA, Acoustic cavitation: an introduction. Ultrasonics 1984; 22: 25–28

[3] Apfel RE. Acoustic cavitation inception. Ultrasonics 1984; 22: 167–173

[4] Miller DL. Gas body activation. Ultrasonics 1984; 22: 261–269

[5] Miller DL. The effects of ultrasonic activation of gas bodies in Elodea leaves during continuous and pulsed irradiation at 1 MHz. Ultrasound Med Biol 1977; 3: 221–240

[6] Trevena DH. Cavitation and Tension in Liquids. Hilger, Bristol 1987

[7] Donny FML. Ann Chim Phys 1846; 16(3): 167–190

[8] Bertholet M. Ann Chim Phys 1850; 30(3): 232–237

[9] Briggs LJ. Limiting negative pressure of water. J Appl Phys 1950; 21: 721–722

[10] Rein M and Meier GEA. On the dynamics of heterogeneous shock cavitation. Acustica 1990; 71

[11] Rein M and Meier GEA. On the influence of different parameters on heterogeneous shock cavitation. J Acoust Soc Am 1990; 88(4): 1921

[12] Lackmé C. Proc Colloquium, Ispra 1977. Commission of the European Communities, Luxembourg, 1978

[13] Favreau C. Transmission d'une impulsion de tension dans une colonne d'eau. Rev Phys Appl 1984; 19: 951–961 (in French)

[14] Chesterman WD. The dynamics of small transient cavities. Proc Phys Soc 1952; B65: 846–858

[15] Overton GDN and Trevena DH. Cavitation phenomena and the occurrence of pressure-tension cycles under dynamic stressing. J Phys D 1981; 14: 241–250

[16] Wilson DA, Hoyt JW and McKune JW. Measurement of tensile strength of liquids by an explosion technique. Nature 1975; 253: 723–725

[17] Kwak H-Y and Panton R L. Tensile strength of simple liquids predicted by a model of molecular interactions. J Phys D 1985; 18: 647

[18] Trevena DH. Cavitation and the generation of tension in liquids. J Phys D 1984; 17: 2139–2164

[19] Overton GDN, Williams PR and Trevena DH. The influence of cavitation history and entrained gas on liquid tensile strength. J Phys D 1984; 17: 979–987

[20] Flynn HG. Physics of acoustic cavitation in liquids. In: Physical Acoustics, Vol. 1, Part B (Mason WP, ed.). Academic Press, New York, 1964, pp 57–172

[21] Wentzell RA. Van der Waals stabilisation of bubbles. Phys Rev Lett 1986; 56: 732–733

[22] Lubkin E. Comment on 'Van der Waals stabilisation of bubbles'. Phys Rev Lett 1986; 56: 2653

[23] Wentzell RA. Wentzell responds. Phys Rev Lett 1986; 56: 2654

[24] Epstein PS and Plesset MS. On the stability of gas bubbles in liquid–gas solutions. J Chem Phys 1950; 18: 1505–1509

[25] Blake FG. Technical Memo 12, Acoustics Research Laboratory, Harvard University, Cambridge, Massachusetts, USA, September 1949

[26] Iyengar KS and Richardson EG. Measurements on the air-nuclei in natural water which give rise to cavitation. Brit J Appl Phys 1958; 9: 154

[27] Gavrilov LR. On the size distribution of gas bubbles in water. Sov Phys Acoust 1969; 15: 22–24

[28] Barabanova GYa, Il'in VP, Levkovskii YL and Chalov AV. Relationship between the strength and size of cavitation nuclei. Sov Phys Acoust 1981: 27: 25–28

[29] Il'in VP, Levkovskii YL and Chalov AV. Influence of the concentration and distribution of cavitation nuclei on the inception and noise of bubble cavitation. Sov Phys Acoust 1981: 27: 220–222

[30] Apfel RE. Acoustic cavitation inception. Ultrasonics 1984: 22; 167–173

[31] Messino D, Sette D and Wanderlingh F. Statistical approach to ultrasonic cavitation. J Acoust Soc Am 1963; 35: 1575–1583

[32] Carpenedo L, Ciuti P and Iernetti G. Cavitation nuclei distribution in liquids by tensile strength measurements. Acoust Lett 1983; 7: 51–55

[33] Finkelstein Y and Tamir A. Formation of gas bubbles in supersaturated solutions of gases in water. AIChE J 1985; 31: 1409

[34] Hentschel W, Zarschizsky H and Lauterborn W. Recording and automatical analysis of pulsed off-axis holograms for determination of cavitation nuclei size spectra. Opt Comm 1985; 53: 69–73
[35] Apfel, RE. Acoustic cavitation. In: Methods in Experimental Physics, Vol. 19 (Edmonds PD, ed.). Academic Press, New York, 1981, pp 355–413
[36] Greenspan M and Tschiegg CE. Radiation-induced acoustic cavitation; apparatus and some results. J Res Natl Bur Stand Sect C 1967; 71: 299–311
[37] West C and Howlett R. Some experiments on ultrasonic cavitation using a pulsed neutron source. Brit J Appl Phys Ser 2 1968; 1: 247–254
[38] Sette D and Wanderlingh F. Nucleation by cosmic rays in ultrasonic cavitation. Phys Rev 1962; 125: 409–417
[39] Lieberman D. Radiation-induced cavitation. Phys Fluids 1959; 2: 466–468
[40] Finch RD. Influence of radiation on the cavitation threshold of degassed water. J Acoust Soc Am 1964; 36: 2287–2292
[41] Barger J. Acoust Res Lab Harvard Uni Tech Memo 57, April 1964
[42] Seitz F. On the theory of the bubble chamber. Phys Fluids 1958; 1: 2–13
[43] Hayward ATJ. Tribonucleation of bubbles. Brit J Appl Phys 1967; 18: 641–644
[44] Hayward ATJ. The role of stabilised gas nuclei in hydrodynamic cavitation inception. J Phys D 1970; 3: 574–579
[45] Galloway WJ. An experimental study of acoustically induced cavitation in liquids. J Acoust Soc Am 1954; 26: 849–857
[46] Strasberg M. Onset of ultrasonic cavitation in tap water. J Acoust Soc Am 1959; 31: 163–176
[47] Harvey EN, Barnes DK, McElroy WD, Whiteley AH, Pease DC and Cooper KW. Bubble formation in animals. J Cell Comp Physiol 1944; 24: 1–22
[48] Yount DE and Strauss RH. Bubble formation in gelatine: a model for decompression sickness. J Appl Phys 1976; 47: 5081–5088
[49] Fox FE and Herzfeld KF. Gas bubbles with organic skin as cavitation nuclei. J Acoust Soc Am 1954; 26: 984–989
[50] Sirotyuk MG. Stabilisation of gas bubbles in water. Sov Phys Acoust 1970; 16: 237–240
[51] Akulichev VA. Hydration of ions and the cavitation resistance of water. Sov Phys Acoust 1966; 12: 144–149
[52] Yount DE. Skins of varying permeability: a stabilisation mechanism for gas cavitation nuclei. J Acoust Soc Am 1979; 65: 1429–39
[53] Yount DE. On the evolution, generation, and regeneration of gas cavitation nuclei. J Acoust Soc Am 1982; 71: 1473–81
[54] Yount DE, Gillary EW and Hoffman DC. A microscopic investigation of bubble formation nuclei. J Acoust Soc Am 1984; 76: 1511–1521
[55] Lieberman L. Air bubbles in water. J Appl Phys 1957; 28: 205–217
[56] Johnson BD and Cooke RC. Generation of stabilised microbubbles in sea water. Science 1981; 213: 209–211
[57] Yount DE and Yeung CM. Bubble formation in supersaturated gelatine: A further investigation of gas cavitation nuclei. J Acoust Soc Am 1981; 69: 702–15
[58] Winterton RHS. Nucleation of boiling and cavitation. J Phys D 1977; 10: 2041–56
[59] Apfel RE. The role of impurities in cavitation-threshold determination. J Acoust Soc Am 1970; 48: 1179–1186
[60] Crum LA. Tensile strength of water. Nature 1979; 278: 148–149
[61] Atchley AA and Prosperetti A. The crevice model of bubble nucleation. J Acoust Soc Am 1989; 86: 1065–1084
[62] Bargeman D and Van Voorst Vader F. Effect of surfactants on contact angles at nonpolar solids. J Colloid Sci 1973; 42: 467–472
[63] Crum LA. Acoustic cavitation thresholds in water. In: Cavitation and Inhomogeneities in Underwater Acoustics (Lauterborn W, ed.). Springer, New York, 1980
[64] Neppiras EA and Noltingk BE. Cavitation produced by ultrasonics: theoretical conditions for the onset of cavitation. Proc Phys Soc 1951; B64: 1032–1038
[65] Walton AJ, Reynolds GT. Sonoluminescence. Adv Phys 1984; 33: 595–660
[66] Rayleigh Lord. On the pressure developed in a liquid during the collapse of a spherical cavity. Phil Mag 1917; 34: 94–98

[67] Plesset MS. The dynamics of cavitation bubbles. J Appl Mech 1949; 16: 277–282
[68] Besant. Hydrostatics and Hydrodynamics. Cambridge University Press, London, 1859, §158
[69] Vokurka K. On Rayleigh's model of a freely oscillating bubble. I. Basic relations. Czech J Phys 1985; B35: 28–40
[70] Noltingk BE and Neppiras EA. Cavitation produced by ultrasonics. Proc Phys Soc 1950; B63: 674–685
[71] Apfel RE. Possibility of microcavitation from diagnostic ultrasound. IEEE Trans Ultrasonics Ferroelectrics Freq Control 1986; UFFC-32(2): 139–142
[72] Walton AJ. Three phases of matter. Clarendon, Oxford, 2nd edn, 1983
[73] Plesset MS and Prosperetti A. Bubble dynamics and cavitation. Ann Rev Fluid Mech 1977; 9: 145–185
[74] Arfken G. Mathematical Methods for Physicists. Academic Press, New York, 2nd edn, 1970, pp 48–49
[75] Spiegel MR. Theory and Problems of Vector Analysis and an Introduction to Tensor Analysis. McGraw-Hill, New York, SI Metric edition, 1974, p 58
[76] Feynman RP, Leighton RB and Sands M. The Feynman Lectures on Physics, Vol. 2. Addison-Wesley, Reading, Massachusetts, 1981, pp 40–44
[77] Arfken G. Mathematical Methods for Physicists. Academic Press, New York, 2nd edn, 1970, p 56
[78] Clift R, Grace JR and Weber ME. Bubbles, Drops and Particles. Academic Press, New York, 1978, p 6
[79] Urick RJ. Principles of Underwater Sound. McGraw-Hill, New York, 3rd edn, 1983, pp 7–8
[80] Lichtarowicz A and Pearce ID. Proc Conf Cavitation 1976. Institution of Mechanical Engineers, London, pp 129–144
[81] Lamb H. Hydrodynamics. Cambridge University Press, London, 5th edn, 1924, p 34
[82] Stephenson G. Mathematical Methods for Science Students. Longman, London, 2nd edn, 1982, p 362
[83] Morse PM and Feshbach H. Methods in Theoretical Physics. McGraw-Hill, New York, 1953, p 1264
[84] Clift R, Grace JR and Weber ME. Bubbles, Drops and Particles. Academic Press, New York, 1978
[85] Detwiler A and Blanchard DC. Ageing and bursting bubbles in trace-contaminated water. Chem Eng Sci 1978; 33: 9–13
[86] Detwiler A. Surface-active contamination on air-bubbles in water. In: Surface Contamination; Genesis, Detection and Control, Vol. 2 (Mittal KL ed.). Plenum, New York, 1979, pp 993–1007.
[87] Kinsler LE, Frey AR, Coppens AB and Sanders JV. Fundamentals of Acoustics. Wiley, New York, 3rd edn, 1982. p 146
[88] Batchelor GK. An Introduction to Fluid Dynamics. Cambridge University Press, London, 1967, p 147
[89] Lamb H. Hydrodynamics, Cambridge University Press, London, 5th edn, 1924, p 545
[90] Sabersky RH, Acosta AJ and Hauptmann EG. Fluid Flow. Macmillan, London, 3rd edn, p 186
[91] Stokes GG. On the motion of pendulums, section IV. In: Mathematical and Physical Papers III. Cambridge University Press, London, pp 55–75 (previously appeared in Trans Camb Phil Soc 1851; 9)
[92] Sabersky RH, Acosta AJ and Hauptmann EG. Fluid Flow. Macmillan, London, 3rd edn, pp 297–300
[93] Clift R, Grace JR and Weber ME. Bubbles, Drops and Particles. Academic Press, New York, 1978, pp 99–100
[94] Thorpe S. On the clouds of bubbles formed by breaking wind-waves in deep water, and their role in air–sea gas transfer. Phil Trans Roy Soc 1982; A304: 155–210
[95] Clift R, Grace JR and Weber ME. Bubbles, Drops and Particles. Academic Press, New York, 1978, p 27
[96] Clift R, Grace JR and Weber ME. Bubbles, Drops and Particles. Academic Press, New York, 1978, pp 171–172
[97] Clift R, Grace JR and Weber ME. Bubbles, Drops and Particles. Academic Press, New York, 1978, pp 184–185
[98] Miksis MJ. A bubble in an axially symmetric shear flow. Phys Fluids 1981; 24: 1229–1231

[99] Ryskin G and Leal LG. Numerical solution of free-boundary problems in fluid mechanics. Part 3. Bubble deformation in an axisymmetric straining flow. J Fluid Mech 1984; 148: 37–43
[100] Kang IS and Leal LG. Numerical solution of unsteady free-boundary problems at finite Reynolds number – I. Finite-difference scheme and its application to the deformation of a bubble in an uniaxial straining flow. Phys Fluids 1987; 30: 1929–1940
[101] Kang IS and Leal LG. Small-amplitude perturbations of shape for a nearly spherical bubble in an inviscid straining flow (steady shapes and oscillatory motion). J Fluid Mech 1988; 187: 231–266
[102] Levich VG. Physiochemical Hydrodynamics. Prentice-Hall, Englewood Cliffs, New Jersey, 1962
[103] Moore DW. The boundary layer on a spherical gas bubble. J Fluid Mech 1963; 16: 161–176
[104] Edge RM and Grant CD. Terminal velocity and frequency of oscillation of drops in pure systems. Chem Eng Sci 1971; 26: 1001–1012
[105] Schroeder RR and Kintner RC. Oscillations of drops falling in a liquid field. AIChE J 1965; 11: 5–8
[106] Winnikow S and Chao BT. Droplet motion in purified systems. Phys Fluids 1966; 9: 50–61

3
The Freely-oscillating Bubble

Introduction

To consider a freely-oscillating bubble as undriven is perhaps misleading. In order to oscillate, an oscillator must at some time receive some exciting force. What characterises the free oscillators under consideration in this chapter is that they oscillate not as a result of some periodic driving force, but instead respond to some initial impulse. At all other times the forcing term is zero, so that prior to the impulse the oscillator is stationary, and afterwards it responds in a characteristic manner that depends on system properties such as the resonance frequency and the damping. The ideal impulse is of infinitesimally short duration, so that whilst the system undergoes subsequent oscillation it effectively experiences no driving force. It is this ideal case which has lead to the concept of free oscillation.

A bubble in a liquid will contain either liquid, vapour or gas, or a mixture. If the bubble is mechanically excited, for example, by a pressure impulse, it will oscillate. Though, as discussed in Chapter 2, there are many modes of oscillation, we will consider initially only the zero-order mode: the bubble pulsates, its volume oscillating about the mean equilibrium value in approximately simple harmonic motion. A simple analogy for this is to consider a bob of mass m attached to a spring. The height of the bob relates to the position of the bubble wall, so that the extension ε from the equilibrium position corresponds to the difference in the bubble radius R_ε between its present and equilibrium values (given by R and R_0 respectively). The spring, which provides the restoring force, models the gas within the bubble. Upon compression and rarefaction, the gas causes pressure differentials which act to restore the bubble to the equilibrium volume.

Care must be taken with this analogy, however, when considering the inertia, because in the case of the bob the body is more dense than its surrounding medium (air), whilst in the case of the bubble the body is less dense than the surrounding medium. Therefore in the spring–bob system, inertial and momentum effects are dominated by the mass of the bob, the inertial effects associated with the motion of the surrounding fluid (air) being negligible. In contrast with the bubble, it is the inertia associated with the motion of the surrounding fluid (water) which dominates the inertia, the mass of the gas within the bubble being negligible.

3.1 The Unforced Oscillator

3.1.1 The Equilibrium Position

When considering a bob on a spring, the action of gravity on the bob does not enter the mathematics. This can be seen by considering the unloaded, massless spring, of length l_0 (Figure 3.1(a)), which extends to a length l when the mass m is added (Figure 3.1(b)). Equating forces at the equilibrium position

$$mg = k(l - l_0) \tag{3.1}$$

where k is the spring constant or *stiffness* of the oscillator. From this, it can be seen that if a force F_a is slowly applied to a spring, the spring will stretch so that the bob undergoes a displacement ε to some new equilibrium position. The product of the displacement ε and the spring constant k equals the tension generated in the spring, which is equal and opposite to the applied force, and so restores an equilibrium. Therefore

$$F_a = k\varepsilon \tag{3.2}$$

for a new equilibrium. Since it is an equilibrium, the extended spring must be exerting a balancing force F_s equal and opposite to the applied force F_a, such that

$$F_s = -k\varepsilon \tag{3.3}$$

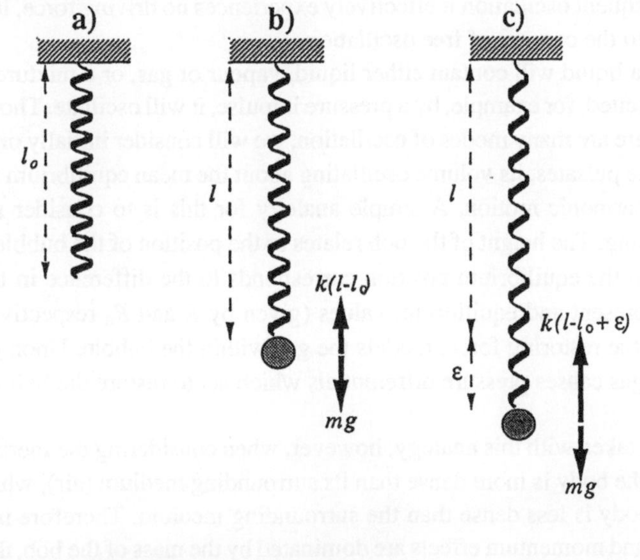

Figure 3.1 The forces on a spring. (a) The unloaded spring has length l_0 at equilibrium. (b) When loaded by a mass m, the equilibrium length of the spring is l. (c) If the loaded spring is extended by ε from equilibrium, the net force exerted by the spring on the bob is $k\varepsilon$ upwards.

If the spring is now extended from the equilibrium position, and then released, oscillation occurs. Applying Newton's Second Law at some displacement ε (Figure 3.1(c)), and taking downwards as the positive direction, the summation of forces gives

$$\Sigma F = mg - k(l + \varepsilon - l_0) = m\ddot{\varepsilon} \tag{3.4}$$

Substitution of equation (3.1) into (3.4) gives

$$k\varepsilon = -m\ddot{\varepsilon} \tag{3.5}$$

The gravity terms are cancelled by the initial extension of the spring, and the motion of the oscillator can be considered in terms of the displacements from equilibrium only.[1] In the same way, the bubble compensates for a small change in the static pressure simply with a small change in the equilibrium volume.

3.1.2 The Undamped Oscillator

(a) The Resonance Frequency

Equation (3.5) is the equation of motion of an undamped simple harmonic oscillator, for which a solution is

$$\varepsilon = e_0 e^{i\omega_0 t} \tag{3.6}$$

where substitution into equation (3.5) gives

$$\boxed{\omega_0^2 = \frac{k}{m}} \tag{3.7}$$

The circular frequency ω_0 represents the resonance (strictly, the natural) frequency of the oscillator, and the only constraint on ω_0 is that it must satisfy equation (3.7). Therefore two solutions are mathematically possible, which oscillate as $e^{i\omega_0 t}$ and $e^{-i\omega_0 t}$ respectively. Having no reason to discard either, the general solution to equation (3.5) is therefore

$$\varepsilon = \Xi_1 e^{i\omega_0 t} + \Xi_2 e^{-i\omega_0 t} \tag{3.8}$$

where Ξ_1 and Ξ_2 may be complex. On the Argand diagram these two solutions represent vectors rotating in the anticlockwise ($e^{i\omega_0 t}$) and clockwise ($e^{-i\omega_0 t}$) directions. These will combine to give a harmonic oscillation along the real axis if Ξ_1 is the complex conjugate of Ξ_2 (see Figure 1.6(c)), since then $\Xi_2 e^{-i\omega_0 t}$ is the complex conjugate of $\Xi_1 e^{i\omega_0 t}$: thus, on summation in equation (3.8), the imaginary terms cancel and the result is a real displacement [1]. When considering the Fourier analysis of a given waveform, it is convenient to express both $\Xi_1 e^{i\omega_0 t}$ and $\Xi_1^* e^{-i\omega_0 t}$ terms explicitly [2]. However, for the description of the time-dependence of a harmonic oscillation the solution can be expressed in a simpler form. If the amplitude of the first solution is written in terms of its magnitude X (a real number) and phase $-\vartheta$ such that $\Xi_1 = Xe^{-i\vartheta}$, then since complex

---

[1] If an oscillator does not respond to a change in a relevant force by a change in equilibrium position, then that force will feature in the equation of motion of the small-amplitude harmonic oscillations. Thus whilst the frequency of oscillation of a bob on a spring is independent of the gravitational acceleration g, the resonance frequency of a simple pendulum of length L is given by $\sqrt{(L/g)}$.

conjugates have the same magnitude but opposite phase, the amplitude of the second solution is $\Xi_2 = Xe^{i\vartheta}$. Therefore if the solution is a harmonic oscillation, equation (3.8) may be re-expressed as

$$\begin{aligned}\varepsilon &= Xe^{i(\omega t-\vartheta)} + Xe^{-i(\omega t-\vartheta)} \\ &= 2X\cos(\omega_o t - \vartheta) \\ &= \mathrm{Re}\{2\Xi_1 e^{i\omega_o t}\} \\ &= \mathrm{Re}\{\Xi_3 e^{i\omega_o t}\} \qquad (3.9)\end{aligned}$$

Similarly, this is also equivalent to $\mathrm{Re}\{2Xe^{i(\omega t-\vartheta)}\}$, $\mathrm{Re}\{\Xi_3 e^{-i\omega_o t}\}$ and $\mathrm{Re}\{2Xe^{-i(\omega t-\vartheta)}\}$. The time dependence of a harmonic oscillator can therefore be expressed as the real part of a solution that varies as $e^{i\omega_o t}$ and which either has a complex amplitude, or a real amplitude with an explicit phase constant in the argument of the exponential. As an alternative, the $e^{-i\omega_o t}$ may be used, though both solutions cannot be used together unless their amplitudes are complex conjugates. In this book, the $e^{i\omega_o t}$ will conventionally be employed.

(b) The Energy of the Oscillator

The potential energy ϕ_P of the spring displaced from equilibrium by a distance ε is the integral product of the force and the displacement

$$\phi_P = \int_0^{\varepsilon} F\,d\varepsilon' = \int_0^{\varepsilon} k\varepsilon'\,d\varepsilon' = \frac{k\varepsilon^2}{2} \qquad (3.10)$$

The kinetic energy of the bob is given by

$$\phi_K = \frac{m\dot{\varepsilon}^2}{2} \qquad (3.11)$$

The kinetic energy is zero when the bob is stationary, which occurs at maximum displacement, when $\varepsilon = \pm\varepsilon_0$. At this point all the energy is potential, taking the maximum value

$$\phi_{P,max} = \frac{k\varepsilon_0^2}{2} \qquad (3.12)$$

Similarly the potential energy is zero at the equilibrium position ($\varepsilon = 0$). At this position all the energy is kinetic, taking a maximum value $\phi_{K,max}$. Since from equation (3.6) the velocity is $\dot{\varepsilon} = i\omega\varepsilon_0 e^{i\omega t}$ it takes a maximum value of $\varepsilon_0\omega$, giving

$$\phi_{K,max} = \frac{m\omega^2\varepsilon_0^2}{2} \qquad (3.13)$$

In the undamped oscillator, no energy is lost from the system: it merely transforms back and forth from kinetic to potential. Thus

$$\phi_{K,max} = \phi_{P,max} = \phi_T \qquad (3.14)$$

where ϕ_T is the total energy of the system. Substitution of equations (3.12) and (3.13) into equation (3.14) gives, once again, the resonance frequency (equation (3.5)). Since displacement varies sinusoidally, the time average $\langle\varepsilon^2\rangle \equiv \langle\varepsilon_0^2\sin^2\omega t\rangle = \varepsilon_0^2/2$. Thus the mean kinetic energy and the mean potential energy both equal $\phi_T/2$.

In real systems, energy is usually lost from oscillators through various dissipative mechanisms. From the above formulation it is clear that as the energy in the system decreases, the amplitude of oscillation decays. This phenomenon is studied in the following section.

3.1.3 The Damped Oscillator

In the spring-and-bob analogy for a pulsating bubble, damping may be thought of as being due to a viscous medium within which the bob moves.[2] The damping force is proportional to the velocity of the bob, and in the direction opposite to the velocity. Taking the downwards direction as positive, application of Newton's Second Law to this system (Figure 3.2) gives

$$\sum F = -k\varepsilon - b\dot{\varepsilon} = m\ddot{\varepsilon} \quad (3.15)$$

where b is a real positive constant. Thus if the velocity is downwards ($\dot{\varepsilon} > 0$), the drag force will be upwards ($-b\dot{\varepsilon} < 0$), and vice versa. Rearrangement gives the equation of motion of a damped harmonic oscillator:

$$m\ddot{\varepsilon} + b\dot{\varepsilon} + k\varepsilon = 0 \quad (3.16)$$

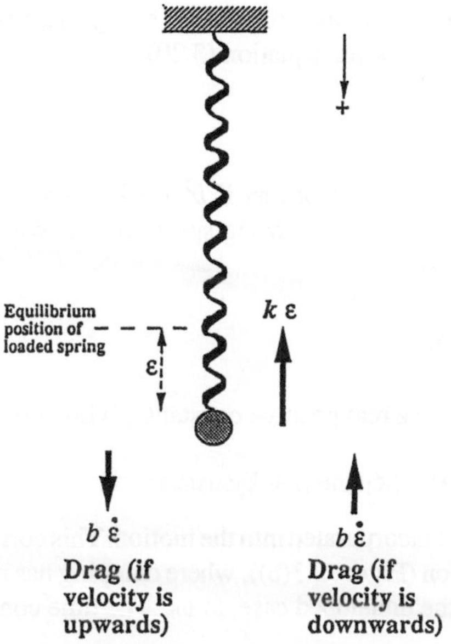

Figure 3.2 The forces on a bob in the spring–bob system oscillating in a viscous medium.

[2]Bobs moving through air, water and honey respectively would experience increasing amounts of damping.

Substitution of a trial solution $\varepsilon = Ae^{st}$ into equation (3.16) gives

$$s = \frac{-b \pm \sqrt{b^2 - 4km}}{2m} \tag{3.17}$$

Three distinct types of behaviour can be expected, depending on the relative sizes of b (the magnitude of the damping), k and m (the spring constant and the mass of the bob), and therefore the resonance frequency ($\omega_o = \sqrt{k/m}$).

Overdamped Oscillation. This occurs if $b^2 > 4km$, that is when the damping is very strong (for example, if a light bob on a weak spring moves within a viscous fluid). Equation (3.17) reduces to give

$$s = -\frac{b}{2m} \pm \varsigma \tag{3.18}$$

where

$$\varsigma = \frac{\sqrt{b^2 - 4km}}{2m} \tag{3.19}$$

is a real positive constant. This gives a solution for the motion of

$$\varepsilon = (\Xi_1 e^{\varsigma t} + \Xi_2 e^{-\varsigma t})\, e^{-\beta t} \tag{3.20}$$

using the substitution $\beta = b/(2m)$, where both solutions for s are incorporated into the motion. β has the dimensions [time]$^{-1}$. The solution embodied in equation (3.20) corresponds to a linear combination of two exponential decays, one decay being more rapid than the other (Figure 3.3(a)). This can clearly be seen when equation (3.20) is recast as

$$\varepsilon = \Xi_1 e^{-(\beta-\varsigma)t} + \Xi_2 e^{-(\beta+\varsigma)t} \tag{3.21}$$

Lightly Damped Oscillation. This occurs if $b^2 < 4km$, that is when the damping is very weak, for example, if a heavy bob on a strong spring moves within a fluid of low viscosity. Equation (3.17) reduces to give

$$s = -\frac{b}{2m} \pm i\omega_b = -\beta \pm i\omega_b \tag{3.22}$$

where $\omega_b = \sqrt{(4km - b^2)}/2m$ is a real positive constant, giving a solution of

$$\varepsilon = (\Xi_1 e^{i\omega_b t} + \Xi_2 e^{-i\omega_b t})\, e^{-\beta t} = (X_1 \sin\omega_b t + X_2 \cos\omega_b t)\, e^{-\beta t} \tag{3.23}$$

where both solutions for s are incorporated into the motion. This corresponds to an exponentially decaying harmonic oscillation (Figure 3.3(b)), where damping has reduced the frequency of the free oscillation from ω_o in the undamped case, to ω_b. The time constant of the decay is β^{-1}.

Critical Damping. This occurs if $b^2 = 4km$. Equation (3.17) reduces to give two identical solutions for s:

$$s = -\frac{b}{2m} = -\beta \quad \text{for both roots.} \tag{3.24}$$

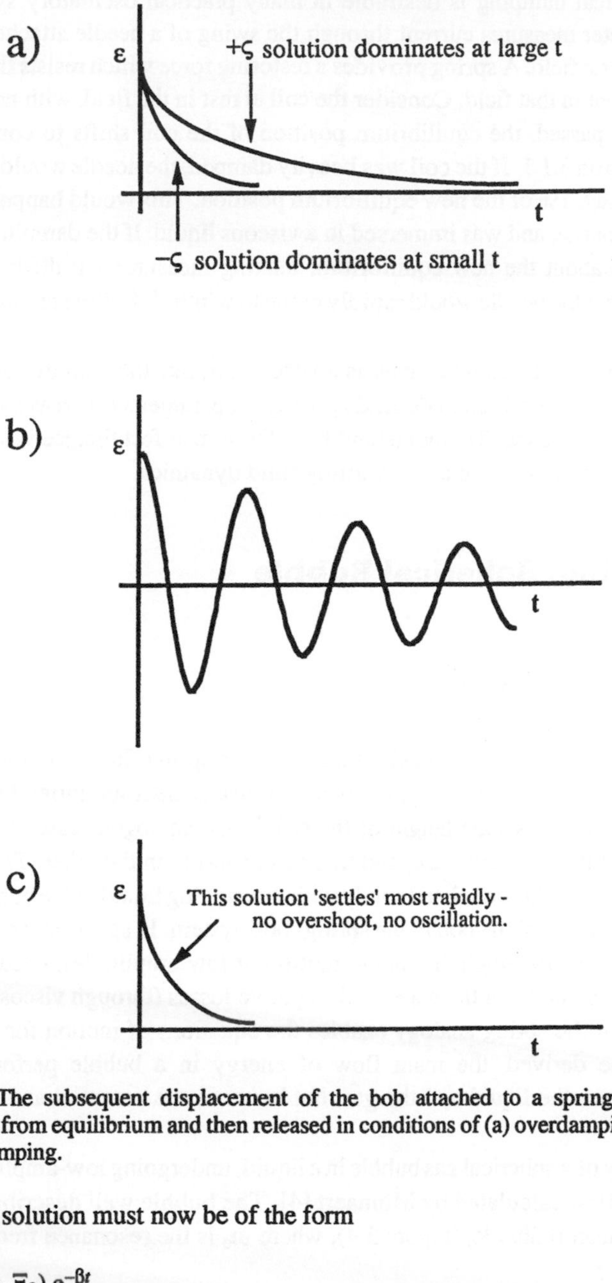

Figure 3.3 The subsequent displacement of the bob attached to a spring if it is given an initial displacement from equilibrium and then released in conditions of (a) overdamping, (b) light damping and (c) critical damping.

The general solution must now be of the form

$$\varepsilon = (\Xi_1 t + \Xi_2)\, e^{-\beta t} \tag{3.25}$$

since if neither amplitude was a function of time, they could add to give an equivalent amplitude ($\Xi_1 + \Xi_2 = \Xi_3$, say) and one solution would be lost. This motion leads to rapid decay of the displacement (Figure 3.3(c)). Though theoretically none of the three solutions leads to a stationary oscillator in the equilibrium position (i.e. $\ddot{\varepsilon} = \dot{\varepsilon} = 0$), in critical damping the oscillator most rapidly decays to within any given small displacement from equilibrium position without overshoot.

The condition of critical damping is desirable in many practical oscillatory systems. For example, the galvanometer measures current through the swing of a needle attached to a coil, which resides in a magnetic field. A spring provides a restoring force which resists the magnetic force acting upon a current in that field. Consider the coil at rest in the field, with no current. If a steady current is then passed, the equilibrium position of the coil shifts to compensate, a process described in section 3.1.1. If the coil was heavily damped, the needle would take a long time to come to within, say, 1% of the new equilibrium position. This would happen if the coil had a small moment of inertia, and was immersed in a viscous liquid. If the damping was light, the coil would oscillate about the new equilibrium, making measurement difficult. Critical damping would ensure that the needle would rapidly come to within 1% of the required reading, and would not oscillate.

The unforced gas bubble in a liquid behaves as an oscillator, like the ones discussed above. It is clear that the nature of the oscillation depends greatly on parameters such as the resonance frequency, amount of damping etc. To understand how these manifest themselves in the case of the bubble, it is necessary to analyse the governing fluid dynamics.

3.2 The Pulsating, Spherical Bubble

Introduction: the Minnaert Frequency

A gas bubble in a liquid is an oscillator, much like a bob on a spring. In the simplest case of a bubble which remains spherical at all times, the bubble pulsates after an initial displacement, the bubble wall oscillating as does the height of the bob in the analogous case. The restoring force is the elasticity of the gas in this case, and that of the spring in the other. The inertia, as already outlined, is assumed to be associated solely with the moving liquid in the bubble system, and to reside solely in the mass of the bob in the spring–bob system. In spring and bubble cases, the oscillation approximates to simple harmonic motion at low amplitudes, occurring at the natural frequency. It is assumed that there are no dissipative losses (through viscosity, thermal conduction etc.). Though the spring analogy enables the equations of motion for the linearly oscillating bubble to be derived, the main flow of energy in a bubble performing such oscillations is not between the liquid and the gas, but between the potential and the internal energies [3].

The natural frequency of a spherical gas bubble in a liquid, undergoing low-amplitude simple harmonic motion, was first calculated by Minnaert [4]. The bubble wall describes a motion $R_\varepsilon = -R_{\varepsilon 0} e^{i\omega_0 t}$ about a mean radius R_0 (Figure 3.4), where ω_0 is the resonance frequency. The bubble radius is therefore

$$R = R_0 + R_\varepsilon(t) = R_0 - R_{\varepsilon 0} e^{i\omega_0 t} \tag{3.26}$$

the negative sign being used since a quasi-static increase in pressure causes a decrease in bubble radius. The kinetic energy of the liquid is found by integrating over shells of liquid from the bubble wall to infinity. A given shell at radius r has thickness dr, and mass $4\pi r^2 \rho dr$. The kinetic energy of the liquid is therefore

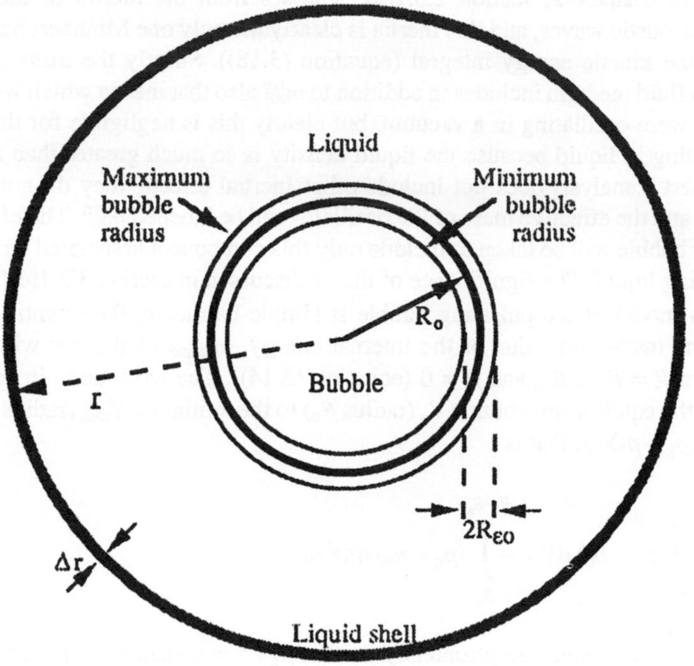

Figure 3.4 A bubble of radius R_0, the wall of which is undergoing small-amplitude linear oscillations of amplitude $R_{\varepsilon 0}$, is surrounded by spherical shells of liquid. One of these, having radius r and width Δr, is shown.

$$\phi_K = \frac{1}{2}\int_R^\infty (4\pi r^2 \rho \, dr)\dot{r}^2 \tag{3.27}$$

The mass of liquid flowing in time dt through any spherical surface around the bubble is $4\pi r^2 \dot{r}\rho dt$. If the liquid is assumed to be incompressible, then by conservation of mass this general flow can be equated to the flow at the bubble surface to give $\dot{r}/\dot{R} = R^2/r^2$ (equation (2.25)). Substitution of this into equation (3.27) gives

$$\phi_K = 2\pi R^3 \rho \dot{R}^2 \tag{3.28}$$

which takes a maximum value at the equilibrium position, when $R = R_0$ and $\dot{R} = i\omega_0 R_{\varepsilon 0} e^{i\omega_0 t}$, implying that $|\dot{R}|^2 = (\omega_0 R_{\varepsilon 0})^2$. Thus the maximum value of the kinetic energy is

$$\phi_{K,max} = \frac{1}{2} m_{RF}^{rad} (R_{\varepsilon 0}\omega_0)^2 \tag{3.29}$$

where

$$m_{RF}^{rad} = 4\pi R_0^3 \rho \tag{3.30}$$

is termed the *radiation mass* of the bubble in the radius–force frame.[3] This mass is the effective inertia of the liquid component of the oscillating system which the pulsating bubble represents. As outlined in Chapter 2, section 2.3.1(c), it arises from the inertia of the liquid that is transmitting acoustic waves, and this inertia is clearly the only one Minnaert has considered in formulating the kinetic energy integral (equation (3.18)). Strictly the mass of an oscillator vibrating in a fluid medium includes in addition to m_{RF}^{rad} also that inertia which would be present if the source were oscillating in a vacuum, but clearly this is negligible for the case of a gas bubble pulsating in liquid because the liquid density is so much greater than that of the gas. Since Minnaert's analysis does not include other inertial effects, they do not appear in this formulation, and the effective mass of the oscillator can be labelled m_{RF}^{rad}. The effective mass of the pulsating bubble will be taken to include only this component associated with the inertia of the surrounding liquid. The significance of this is discussed in section 3.2.1(c)(iii).

Since this model of the pulsating bubble is simple harmonic, the quantity $\phi_{K,max}$ can be equated to the maximum value of the internal energy, $\phi_{P,max}$, of the gas within the bubble, attained when $R = R_0 \pm R_{\varepsilon o}$ and $\dot{R} = 0$ (equation (3.14)). The work done in compressing the bubble from the equilibrium volume V_0 (radius R_0) to the minimum V_{min} (radius $R_0 - R_{\varepsilon o}$) is the integral of $-(p_g - p_0)\mathrm{d}V$, that is

$$\phi_{P,max} = -\int_{V_0}^{V_{min}} (p_g - p_0)\,\mathrm{d}V = -\int_{R_0}^{R_0 - R_{\varepsilon o}} (p_g - p_0)\,4\pi r^2 \mathrm{d}r \tag{3.31}$$

Assuming the gas behaves polytropically so that $p_g V^{\kappa}$ = constant, then since $R_\varepsilon = R - R_0$, equating the pressure and volume conditions at equilibrium to those when the bubble attains minimum volume gives

$$p_g(R_0 + R_\varepsilon)^{3\kappa} = p_0 R_0^{3\kappa} \tag{3.32}$$

or

$$\frac{p_g}{p_0} = \left(1 + \frac{R_\varepsilon}{R_0}\right)^{-3\kappa} \tag{3.33}$$

Binomial expansion of equation (3.33) gives

$$p_0 - p_g = \frac{3\kappa R_\varepsilon p_0}{R_0} \tag{3.34}$$

Substitution of this into equation (3.31), with the use to first order of $R_\varepsilon = R - R_0$ coordinates, gives

$$\phi_{P,max} = \int_0^{R_{\varepsilon o}} \frac{3\kappa p_0 R_\varepsilon}{R_0}\,4\pi R_0^2 \mathrm{d}R_\varepsilon = 6\pi\kappa p_0 R_0 R_{\varepsilon o}^2 \tag{3.35}$$

Equating $\phi_{K,max}$ (equation (3.29)) to $\phi_{P,max}$ (equation (3.35)) gives

[3]The relevance of the *frame* is discussed in section 3.2.1(a).

$$\omega_o \approx \frac{1}{R_o}\sqrt{\frac{3\kappa p_o}{\rho}} \tag{3.36}$$

or

$$\nu_o \approx \frac{1}{2\pi R_o}\sqrt{\frac{3\kappa p_o}{\rho}} \tag{3.37}$$

where $\nu_o = \omega_o/(2\pi)$ is the linear resonance frequency for bubble pulsations. The effect of heat conduction from the bubble is only incorporated through the use of the unknown κ, and surface tension is assumed to be negligible. These approximate equations were first derived in the adiabatic form by Minnaert [4], who similarly assumed negligible surface tension. The assumption of negligible heat flow (adiabatic conditions) has since been shown[4] to be a good approximation for freely-oscillating bubbles that are commonly encountered (including millimetre-size air bubbles in water), and for this reason it is convenient to define the *Minnaert resonance frequency*:

$$\omega_M = \frac{1}{R_o}\sqrt{\frac{3\gamma p_o}{\rho}} \tag{3.38}$$

Thus $\omega_M \approx \omega_o$ for bubbles where surface tension and heat conduction are negligible, and where the kinetic energy resides solely in the motion of the water. For air bubbles in water under one atmosphere, equation (3.38) reduces to a memorable form:

$$\nu_o R_o \approx 3 \text{ m.s}^{-1} \tag{3.39}$$

Minnaert went on to examine how such bubble oscillations might give rise to natural acoustic emissions, in his own words "the murmur of the brook, the roar of the cataract, or the humming of the sea." Previous speculations on the source of these sounds had been made by Sir William Bragg [5], a report based on his 1919 Royal Institution children's Christmas lectures. Bragg had suggested that the sounds emitted by running water might be produced from cavities generated by liquid drop impact upon a water surface.[5] Worthington [6, 7] had photographed such cavity formation, and Bragg cites the previously unpublished work of Sir Richard Paget who modelled these cavities from plasticine. By blowing across openings in the models, Paget had produced sounds which resembled those heard when objects were dropped into water. It was Minnaert, however, who treated the bubbles as simple harmonic pulsators and derived the associated acoustic emissions.

Minnaert's calculation has shown that the bubble can, at small amplitudes, approximate to the classical simple harmonic oscillator which was represented at the start of this chapter by a bob on a massless spring. We will now proceed to characterise the bubble in terms of the familiar spring parameters, k and m. This allows a physical and quantitative understanding of how the fundamental properties of stiffness and inertia arise in the bubble. As in Minnaert's derivation, the effects of surface tension and of dissipation (through viscosity, thermal conduction etc.) are deemed to be negligible.

[4]See section 3.4.2.
[5]An idea which has since proved fruitful in its own right (see section 3.7.2).

3.2.1 The Pulsating Bubble as a Simple Oscillator

(a) Frames for the Equation of Motion

Consideration of the bubble as a simple spring–bob oscillator can lead to an equation of motion for the bubble. However, the exact values of the terms stiffness, mass and the dissipative constant appropriate to the bubble depend on the way the system is defined. In particular, there are two alternative ways to define each of a particular pair of system parameters, leading in combination to four commonly used reference frames.

Firstly, the driving term can be defined as either a force or an acoustic pressure. Secondly, the displacement can either be in terms of the bubble volume or of the radius displacement. Referring to the displacement system defined in equation (3.26) and assuming a small-amplitude linear response, the bubble radius is given by $R = R_o + R_\varepsilon(t) = R_o - R_{\varepsilon o}e^{i\omega_v t}$, and by analogy the volume by $V = V_o + V_\varepsilon(t) = V_o - V_{\varepsilon o}e^{i\omega_v t}$, the negative sign in both cases indicating that an increase in the acoustic pressure causes bubble contraction. Four equations of motion are therefore possible:

The 'radius–pressure frame' equation:

$$m_{RP}^{rad}\ddot{R}_\varepsilon + b_{RP}\dot{R}_\varepsilon + k_{RP}R_\varepsilon = P_A e^{i\omega t} \tag{3.40}$$

The 'volume–pressure frame' equation:

$$m_{VP}^{rad}\ddot{V}_\varepsilon + b_{VP}\dot{V}_\varepsilon + k_{VP}V_\varepsilon = P_A e^{i\omega t} \tag{3.41}$$

The 'radius–force frame' equation:

$$m_{RF}^{rad}\ddot{R}_\varepsilon + b_{RF}\dot{R}_\varepsilon + k_{RF}R_\varepsilon = F_A e^{i\omega t} \tag{3.42}$$

The 'volume–force frame' equation:

$$m_{VF}^{rad}\ddot{V}_\varepsilon + b_{VF}\dot{V}_\varepsilon + k_{VF}V_\varepsilon = F_A e^{i\omega t} \tag{3.43}$$

Simply from dimensional arguments it is clear that in each of the four equations of motion that can be derived from this system, the form of the m, b and k will vary. Though the forms of a given parameter are algebraically different, they represent the same physical phenomenon though in different frames. The frequency, of course, will be invariant. Since if $\lambda \gg R_o$ the acoustic pressure at equilibrium is the ratio of the acoustic force over the bubble surface to the surface area:

$$F_A = P_A.4\pi R_o^2 \tag{3.44}$$

so that by comparing (3.41) and (3.43)

$$\frac{m_{VF}^{rad}}{m_{VP}^{rad}} = \frac{b_{VF}}{b_{VP}} = \frac{k_{VF}}{k_{VP}} = 4\pi R_o^2 \tag{3.45}$$

and

$$\frac{m_{RF}^{rad}}{m_{RP}^{rad}} = \frac{b_{RF}}{b_{RP}} = \frac{k_{RF}}{k_{RP}} = 4\pi R_o^2 \tag{3.46}$$

The R_ε and V_ε terms in equations (3.40) to (3.43) refer, it should be remembered, to *changes* in radius and volume from equilibrium. Since the equilibrium volume V_0 equals $4\pi R_0^3/3$, then

$$V_\varepsilon = dV = 4\pi R_0^2 dR = 4\pi R_0^2 R_\varepsilon \tag{3.47}$$

Therefore by comparing like terms in equations (3.42) and (3.43):

$$\frac{m_{RF}^{rad}}{m_{VF}^{rad}} = \frac{b_{RF}}{b_{VF}} = \frac{k_{RF}}{k_{VF}} = 4\pi R_0^2 \tag{3.48}$$

and

$$\frac{m_{RP}^{rad}}{m_{VP}^{rad}} = \frac{b_{RP}}{b_{VP}} = \frac{k_{RP}}{k_{VP}} = 4\pi R_0^2 \tag{3.49}$$

Since the conversion ratio is the same in both cases, the values of the parameters in the radius–pressure frame equal the values of the corresponding parameters in the volume–force frame:

$$m_{RP}^{rad} = m_{VF}^{rad} \tag{3.50}$$

$$b_{RP} = b_{VF} \tag{3.51}$$

$$k_{RP} = k_{VF} \tag{3.52}$$

(b) The Stiffness of a Pulsating Bubble

The following argument considers the radius response of a bubble to a force, in order to find the stiffness. The formulation therefore takes place in the radius–force frame.

The spring constant, or stiffness, of a spring is the ratio of the force the spring exerts to the extension which produces that force, that is

$$k = -\frac{F_s}{\varepsilon} \tag{3.53}$$

The minus sign arises because the force is restoring, that is, it is in the direction opposed to the extension (equation (3.3)).

The stiffness of a bubble can be found by calculating the same ratio [8]. Suppose a bubble, previously at equilibrium in a liquid of static pressure p_0, is compressed. This corresponds to a change of $-R_\varepsilon$ in radius, from R_0 to $R_0 - R_\varepsilon$. As the volume changes by ∂V, from V_0 to V, the pressure in the bubble therefore changes by ∂p_i. Having at equilibrium the value $p_{i,e} = p_0 + 2\sigma/R_0$ (equation (2.8)) it changes to a new non-equilibrium value of p_i. Assuming a polytropic law, $p_i V^\kappa$ = constant, then

$$\Delta p_i = -\frac{\kappa}{V_0} p_{i,e} \Delta V \tag{3.54}$$

If $R_\varepsilon \ll R_0$, the change in volume ∂V equals $4\pi R_0^2 R_\varepsilon$ (equation (3.47)), so that

$$\frac{\Delta V}{V_0} = \frac{-4\pi R_0^2 R_\varepsilon}{(4/3)\pi R_0^3} = -\frac{3R_\varepsilon}{R_0} \tag{3.55}$$

To a similar approximation, the force exerted upon the bubble to produce this change in radius is the product of the excess pressure with the area over which it acts:

$$F_A = -4\pi R_o^2 \Delta p_i \qquad (3.56)$$

Substitution of equations (3.54) and (3.55) into (3.56) gives the force as being

$$F_A = -12\pi\kappa R_o p_i R_\varepsilon \qquad (3.57)$$

From equation (3.53), this gives the stiffness of a bubble to be

$$k_{RF} = 12\pi\kappa R_o p_{i,e} \qquad (3.58)$$

The value of κ will take the appropriate value to correspond to the amount of heat flow that occurs during the process, as discussed in Chapter 1, section 1.1.1(b)(ii). Equation (3.58) can now be employed in equations (3.46) and (3.52) to give:

$$k_{RP} = k_{VF} = \frac{3\kappa}{R_o} p_{i,e} \qquad (3.59)$$

and

$$k_{VP} = \frac{3\kappa}{4\pi R_o^3} p_{i,e} = \frac{\kappa}{V_o} p_{i,e} \qquad (3.60)$$

where V_o is the equilibrium bubble volume.

If the surface tension pressure p_σ is deemed negligible, then from equation (2.8) $p_{i,e} \approx p_o$, so that

$$k_{RF} \approx 12\pi\kappa R_o p_o \qquad (3.61)$$

$$k_{RP} = k_{VF} \approx \frac{3\kappa p_o}{R_o} \qquad (3.62)$$

and

$$k_{VP} \approx \frac{3\kappa p_o}{4\pi R_o^3} = \frac{\kappa p_o}{V_o} \qquad (3.63)$$

Having found from Minnaert's formulation that the pulsating bubble approximates to a spring–bob system at low amplitude, and knowing now the effective spring constant of the bubble, the question that must now be posed is: 'what is the mass?' To answer that, we must investigate in detail the nature of the spherical acoustic waves which the bubble will emit, and which were detected by Minnaert.

(c) The Inertia of the Pulsating Bubble

Having calculated the stiffness **k** of the bubble, we will proceed to calculate the inertia, *m*. The two of these will then be used to examine the resonance properties of the bubble, since from the spring constant and the inertia of a simple oscillator, the resonance frequency can be found (equation (3.7)).

The determination of the inertia is revealing. As demonstrated in Chapter 2, section 2.3.1(b), the dynamics of any body moving in a medium depend on the properties of both the body and the medium. Whilst the effect of the medium is small if it is much less dense than the body, it can be considerable in the case of the bubble. When a bubble pulsates at radius R, the liquid velocity at radius r is $\dot{R}R^2/r^2$ (equation (2.25)) if the liquid is assumed to be incompressible. Therefore the liquid is still undergoing finite motion at very large distances from the bubble. In fact, the assumption of incompressibility means that the liquid flux through any spherical shell centred on the bubble is a constant.[6] Thus when a bubble pulsates a considerable amount of liquid is set into motion, and this will contribute to the inertia. This is an approximate description, since clearly the liquid is not perfectly incompressible, or else sound would not transmit through it at finite speeds. However, it is possible to quantify the inertia of the liquid around the pulsating bubble through these very acoustic waves. Initially, therefore, we will examine the nature of the spherical waves emitted by a pulsating bubble.

(i) Spherical Pressure Waves. (Based upon Kinsler *et al.* [8]) In Chapter 2, section 2.2.3 it was shown that low-amplitude pressure fluctuations propagate through a fluid in accordance with the linearised general wave equation. A spherical pulsating bubble, the wall motion approximating to simple harmonic, will emit harmonic spherical pressure waves into the liquid. The spherical Laplacian operator is

$$\nabla^2 = \frac{1}{r^2}\frac{\partial}{\partial r}\left(r^2\frac{\partial}{\partial r}\right) + \frac{1}{r^2\sin\theta}\frac{\partial}{\partial\theta}\left(\sin\theta\frac{\partial}{\partial\theta}\right) + \frac{1}{r^2\sin^2\theta}\frac{\partial^2}{\partial\phi^2} \tag{3.64}$$

As a result of the spherical symmetry, the $\partial\theta$ and $\partial\varphi$ terms are zero, and so the wave equation (equation (2.87)) is simplified to

$$\frac{\partial^2 P}{\partial r^2} + \frac{2}{r}\frac{\partial P}{\partial r} = \frac{1}{c^2}\frac{\partial^2 P}{\partial t^2} \tag{3.65}$$

Rewriting gives

$$\frac{\partial^2(rP)}{\partial r^2} = \frac{1}{c^2}\frac{\partial^2(rP)}{\partial t^2} \tag{3.66}$$

The form of this equation is that of the wave equation in one spatial dimension, and the general solution is

$$rP = f_1(ct - r) + f_2(ct + r) \tag{3.67}$$

Comparison with the plane-wave solution discussed in Chapter 1, section 1.1.1(b) is fruitful. For a plane wave in a one-dimensional field, the function f_1 represents a disturbance travelling in the direction of increasing spatial coordinate, whilst f_2 describes a motion in the direction of decreasing spatial coordinate. These correspond to the plane, one-dimensional wave travelling right and left, respectively. In our spherical coordinates, f_1 is associated with waves propagating out from the bubble (in the direction of increasing r), and f_2 with incoming spherical waves which converge at the bubble. In the case of a bubble oscillating in an infinite medium in the absence of an impressed sound field, $f_2 = 0$. Therefore the solution for the spherical pressure waves emitted by a pulsating bubble is

[6] See Chapter 2, section 2.1.3(b)(i).

$$P(r,t) = \frac{1}{r} f_1(ct - r) \tag{3.68}$$

In Chapter 1, section 1.2.1(a) the solution

$$P = \frac{\Psi}{r} e^{i(\omega t - kr)} \tag{3.69}$$

was investigated. The quantity ψ, which has units of [Pa.m], is numerically equal to the acoustic pressure radiated by the source at unit distance from that source. The formulation can be explicitly modified to incorporate inconstant source strength and attenuation. For example, if the source is damped with time constant β^{-1}, and if the attenuation coefficient is b, the pressure of spherical diverging waves is given by

$$P = \frac{\Psi}{r} e^{-\beta(t - r/c)} e^{-br} e^{i(\omega t - kr)} \tag{3.70}$$

where the use of $(t - r/c)$ allows for the propagation time, and $\psi e^{-b[1 \text{ metre}]}$ is the acoustic pressure amplitude 1 metre from the centre of the source at time $t = r/c$.

Having found in equation (3.70) an expression for $P(r,t)$, the liquid particle velocity $\dot{\varepsilon}$ can be found from the *linear inviscid force equation* (equation (2.76)). The oscillatory nature of $\dot{\varepsilon}$ makes it a special case of the generalised flow velocity v described by the linear inviscid force equation, which therefore for this spherical case reduces to

$$\frac{\partial P}{\partial r} = -\rho_0 \ddot{\varepsilon} \tag{3.71}$$

Thus differentiation of equation (3.70) with respect to r, and integration with respect to t, gives the oscillatory liquid particle velocity

$$\dot{\varepsilon}(r,t) = \left(1 - \frac{i}{kr}\right) \frac{P}{\rho_0 c} \tag{3.72}$$

Here one must abandon the ideal point source, since equation (3.72) becomes unphysical as r tends to zero.[7] The pulsating spherical bubble is a source of finite size, the minimum value of r encountered in the liquid occurs at the bubble wall, and $r = 0$ corresponds to the centre of the bubble where there is no liquid. The bubble will emit spherical pressure waves into the surrounding liquid, with an acoustic pressure $P(r,t)$ and fluid particle velocity $\dot{\varepsilon}$. There will therefore be a specific acoustic impedance associated with these spherical waves equal to

$$Z_{RP}^{s.s.} = \frac{P(r,t)}{\dot{\varepsilon}\,(r,t)} \tag{3.73}$$

The notation $Z_{RP}^{s.s.}$ indicates that this is a *specific* acoustic impedance for *spherical* waves. Since the specific acoustic impedance relates the pressure to the radial response, it is relevant to the radius–pressure frame, so the notation RP is redundant, but provided for clarity. Equations (3.71) and (3.72) show that there will be a phase difference between P and $\dot{\varepsilon}$ so that the impedance will be complex. Substitution into equation (3.73) gives

---

[7] The liquid at an infinitesimal point cannot be moving in all directions simultaneously.

$$Z_{\mathrm{RP}}^{\mathrm{s.s.}} = \rho_o c \frac{kr}{kr - \mathrm{i}} = \frac{\rho_o ckr}{1 + (kr)^2}(kr + \mathrm{i}) \tag{3.74}$$

of magnitude

$$|Z_{\mathrm{RP}}^{\mathrm{s.s.}}| = \frac{\rho_o ckr}{\sqrt{1 + (kr)^2}} \tag{3.75}$$

and of phase χ where

$$\tan\chi = \frac{1}{kr} \tag{3.76}$$

(ii) The Intensity of Spherical Waves. (Based upon Kinsler *et al.* [8]) At some fixed radius r, the acoustic pressure is $P = P_A e^{i\omega t}$ where $P_A = \psi e^{-ikr}/r$. The magnitude of the specific acoustic impedance gives the ratio of the acoustic pressure amplitude P_A to the speed amplitude U_o so that

$$P_A = U_o |Z_{\mathrm{RP}}^{\mathrm{s.s.}}| = U_o \rho_o c \cos\chi \tag{3.77}$$

since from equation (3.76) and Figure 3.5

$$\cos\chi = \frac{kr}{\sqrt{1 + (kr)^2}} \tag{3.78}$$

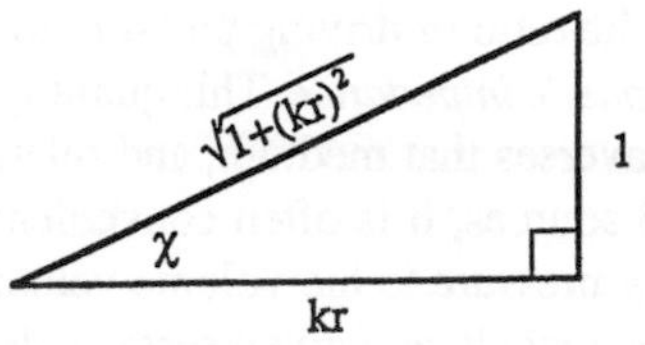

Figure 3.5 The trigonometry associated with the phase of the specific acoustic impedance for radiation of diverging spherical pressure waves. If $kr = kR_o \ll 1$, then $\chi = \chi_o \to \pi/2$. If $r \to \infty$, then $\chi \to 0$.

From the energy conservation arguments discussed in the footnote to section 1.2.1(a) of Chapter 1, this correlates with the failure of the acoustic pressure in equation (3.77) to fall off as r^{-1}. The speed amplitude is easily found. The particle velocity is

$$\dot{\varepsilon} = \frac{P}{Z_{\mathrm{RP}}^{\mathrm{s.s.}}} = \frac{P_A e^{i\omega t} e^{-i\chi}}{\rho_o c \cos\chi} \tag{3.79}$$

the real part of which is

$$\mathrm{Re}\{\dot{\varepsilon}\} = \frac{P_A \cos(\omega t - \chi)}{\rho_o c \cos\chi} \tag{3.80}$$

having amplitude

$$U_o = \frac{P_A}{\rho_o c \cos\chi} \tag{3.81}$$

When $\cos\chi$ is evaluated at $r = R_o$, equation (3.81) gives the speed at the wall of the sphere, $U_o(r = R_o) = U_o^{R_o}$.

The acoustic intensity is the average flow of energy through unit area normal to the direction of propagation of the wave. The work done is the time integral product of the force and the speed, so that division of this by area and time gives the intensity (equation (1.67)). Therefore

$$I = \frac{1}{\tau_v}\int_0^{\tau_v} \text{Re}\{P\}\text{Re}\{\dot{\varepsilon}\}\,dt$$

$$= \frac{1}{\tau_v}\int_0^{\tau_v} (P_A\cos\omega t)\,(U_o\cos(\omega t - \chi))\,dt = \frac{P_A U_o \cos\chi}{2} \quad (3.82)$$

Thus substitution of equation (3.81) into equation (3.82) shows that

$$I = \frac{P_A^2}{2\rho_o c} \quad (3.83)$$

The spherical wave, and the plane wave, both have the same relationship between the acoustic pressure amplitude and the intensity.

Having characterised the pressure waves emitted by the pulsating bubble, it is possible to quantify the impedance, the resistance and inertia, associated with them.

(iii) The Radiation Mass. The ratio of driving pressure to particle velocity, which we have used here, gives the *specific acoustic impedance*. This quantity is a characteristic of the medium and the type of wave which traverses that medium, and relates to the radius–pressure frames. When discussing actual sound sources, it is often convenient to use the *acoustic impedance*, which is the ratio of the driving pressure to the volume velocity[8] of the fluid, and so relates to the volume–pressure frame. As a result, at a given surface, the following holds:

$$\text{Acoustic impedance} = \frac{\text{Specific acoustic impedance}}{\text{Surface area}} \quad (3.84)$$

A third impedance would refer to the ratio of the force to the volume velocity, and so relate to the volume–force frame and be numerically equal to the specific acoustic impedance.

The impedance in the fourth frame is very useful. The *radiation impedance* (Z_{RF}^{rad}) is defined as the ratio of the applied force to the particle speed, and since it relates to the radius–force frame it is numerically equal to the product of the specific acoustic impedance and the area:

$$Z_{RF}^{rad} = (\text{Specific acoustic impedance}) \times (\text{Surface area}) \quad (3.85)$$

It is this impedance which characterises the coupling between the acoustic source and the radiated waves. The oscillation of a source in a vacuum can be characterised by a mechanical impedance through the relationship between the drive and the velocity response. There will be inertial and resistive properties associated with this. If a source is oscillating at frequency ω in

[8] The volume velocity over an area where the fluid flow is uniform, for example, equals the product of the fluid flow speed and the area.

a fluid medium there is an additional impedance, the radiation impedance of the fluid. The resistive and inertial properties associated with this are given by $\mathrm{Re}\{Z_{RF}^{rad}\}$ and $\mathrm{Im}\{Z_{RF}^{rad}\}/\omega$.

Let us say that the gas bubble pulsating in a liquid has an effective inertia $m = m_{RF}^{rad} + m_i$, where $m_{RF}^{rad} = \mathrm{Im}\{Z_{RF}^{rad}\}/\omega$ is the inertia associated with the fluid motion in the radiated pressure waves, and m_i is the inertia from all other contributions (e.g. the gas within the bubble). Evaluating Z_{RF}^{rad} at the bubble surface ($r = R_0$), where

$$Z_{RF}^{rad} = b_{RF}^{rad} + i\ni_{RF}^{rad} \tag{3.86}$$

using equations (3.74) and (3.85) in the limit of $kR_0 \ll 1$ gives

$$b_{RF}^{rad} = 4\pi R_0^2 \rho_0 c (kR_0)^2 \quad \text{is the } \textit{radiation resistance}\text{, and} \quad (kR_0 \ll 1) \tag{3.87}$$

$$\ni_{RF}^{rad} = 4\pi R_0^2 \rho_0 c kR_0 \quad \text{is the } \textit{radiation reactance}. \quad (kR_0 \ll 1) \tag{3.88}$$

If this additional resistance is positive, the power dissipated by the source will be increased above that which is dissipated when the source oscillates in a vacuum, that dissipation arising from the power that is radiated away from the source in the acoustic field.

The reactance $\ni_{RF}^{rad}$ arises because the acoustic waves diverge radially at the source to a considerable extent: energy is stored and released as successive layers of the liquid expand and contract circumferentially, to alter the outward radial displacement. The inertia associated with this process is $m_{RF}^{rad} = \ni_{RF}^{rad}/\omega$ where ω is the bubble pulsation frequency [9]. If $\ni_{RF}^{rad}$ were zero then so too would be m_{RF}^{rad}, and the resonance frequency of the oscillator would be $\sqrt{k_{RF}/m_i}$, where m_i includes all inertial components of the oscillator other than the radiation mass. A positive value for $\ni_{RF}^{rad}$ manifests itself as a mass loading on the system, which reduces the resonance frequency from $\sqrt{k_{RF}/m_i}$ to $\sqrt{k_{RF}/(m_i + m_{RF}^{rad})}$.

The inertial effect of the radiation mass for sound sources in a light medium, such as air, is negligible, and m_i dominates. The resonance in that situation is indeed $\omega_0 \approx \sqrt{k_{RF}/m_i}$. However, in the pulsating bubble, when the radiation mass represents the inertia of the oscillating elements of a dense liquid, and other inertial components arise from the light gas, m_{RF}^{rad} is the only significant inertia. Far from simply lowering the resonance frequency as an additional mass loading, it determines the value of the resonance almost unaided, such that $m_{RF}^{rad} \gg m_i$ and so $\omega_0 \approx \sqrt{k_{RF}/m_{RF}^{rad}}$.

As stated above, ω is the frequency of the bubble pulsation. Though for freely-oscillating bubbles this will equal the bubble resonance, the formulation is valid within its approximations if the bubble is pulsating at some other frequency, for example, if it were being driven off-resonance by a sound field. If the bubble is in resonance, then from equations (1.1) and (3.37) the wavelength of the sound is very much greater than the bubble radius. Such a situation, where $kR_0 \ll 1$, is classed as the low-frequency regime. At such low frequencies the reactive term dominates in Z_{RF}^{rad}. The pressure and particle speed are nearly $\pi/2$ out-of-phase. The radiation mass is equal to three times the mass of the fluid displaced by the sphere:

$$m_{RF}^{rad} = \frac{\ni_{RF}^{rad}}{\omega} = \rho 4\pi R_0^3 \tag{3.89}$$

The complete set of expressions for the radiation mass relevant to the other frames can be found by substitution of equation (3.89) into equations (3.45) to (3.50):

$$m_{\mathrm{RP}}^{\mathrm{rad}} = \rho R_0 \tag{3.90}$$

$$m_{\mathrm{VP}}^{\mathrm{rad}} = \frac{\rho}{4\pi R_0} \tag{3.91}$$

$$m_{\mathrm{RF}}^{\mathrm{rad}} = \rho 4\pi R_0^3 \tag{3.92}$$

and

$$m_{\mathrm{VF}}^{\mathrm{rad}} = \rho R_0 \tag{3.93}$$

The resonance frequency for pulsation can be predicted from simple oscillator theory of section 3.1.2(a), equation (3.7) implying

$$\omega_o = \sqrt{\frac{\kappa_{\mathrm{RF}}}{m_{\mathrm{RF}}^{\mathrm{rad}} + m_i}} \tag{3.94}$$

As the radiation mass dominates ($m_{\mathrm{RF}}^{\mathrm{rad}} \gg m_i$), then the resonance approximately equals

$$\omega_o \approx \sqrt{\frac{\kappa_{\mathrm{RF}}}{m_{\mathrm{RF}}^{\mathrm{rad}}}} \tag{3.95}$$

All inertias other than $m_{\mathrm{RF}}^{\mathrm{rad}}$, such as the inertia associated with the gas within the bubble, having been considered negligible, one would expect the frequency given by the term on the right to equal the Minnaert resonance frequency ω_M. This is because both are derived by considering the radiation mass resulting from the inertia of the liquid to be the only inertia (see the formulation for the kinetic energy in equation (3.27)), and the assumption that surface tension is negligible. This is confirmed by substitution of equations (3.61) and (3.89) into equation (3.95) to yield

$$\omega_0 = \frac{1}{R_0}\sqrt{\frac{3\kappa p_0}{\rho}} \tag{3.96}$$

in agreement with equation (3.28), and equation (3.38) in adiabatic conditions. Thus it will be seen that Minnaert's dynamic assumptions amount to a statement that in considering the effective mass of the bubble as an oscillator, only the inertia associated with the motions of the fluid are of importance.

In response to high-frequency sound, when the bubble would be expected to perform as a stiff system, the radiation impedance is dominated by the real term and in the limit becomes pure resistive, increasing the power dissipated by the acoustic source

$$Z_{\mathrm{RF}}^{\mathrm{rad}} \to 4\pi R_0^2 \rho_0 c \quad \text{as } \omega \gg \omega_0 \tag{3.97}$$

Having characterised both the bubble and the spherical pressure waves, we will examine the relationship and interaction between the two.

3.3 The Radiating Spherical Bubble

3.3.1 The Pressure Radiated by a Pulsator

(a) The Pressure Radiated by a Rigid Spherical Pulsator

Before considering the pulsating bubble, it is worthwhile discussing the pressure radiated by what, for want of better terminology, is termed a rigid pulsator. This is one where, as the volume of the sphere changes during the pulsation, the pressure within the body does not vary: the pressure in the fluid is governed by the interface motion alone. This analysis is based upon Kinsler *et al.* [19]. When the rigid sphere pulsates at low amplitude with circular frequency ω, the wall velocity behaves in simple harmonic fashion with amplitude U_0:

$$R = R_0 + R_\varepsilon(t) = R_0 - R_{\varepsilon 0}e^{i\omega t} \quad \Rightarrow$$

$$\dot{R}(t) = \dot{R}_\varepsilon(t) = -i\omega R_{\varepsilon 0}e^{i\omega t} = U_0 e^{i\omega t} \quad \Rightarrow$$

$$\ddot{R}(t) = \ddot{R}_\varepsilon(t) = \omega^2 R_{\varepsilon 0}e^{i\omega t} = i\omega U_0 e^{i\omega t} \tag{3.98}$$

in analogy to equation (3.26), so that $U_0 = -i\omega R_{\varepsilon 0}$. For continuity, this must equal the radial fluid particle velocity $\dot{\varepsilon}$ at the sphere wall at all times. Therefore the acoustic pressure at the wall can be found from the product of the specific acoustic impedance for spherical diverging waves and the wall velocity (equation (1.27))

$$P(r = R_0, t) = Z_{RP}^{s.s.}.\dot{\varepsilon}(r = R_0, t) \tag{3.99}$$

Substitution from equations (3.75), (3.78) and (3.98) gives

$$P(r = R_0, t) = \rho_0 c U_0 \cos\chi_0 . e^{i(\omega t + \chi_0)} \tag{3.100}$$

where evaluation of equations (3.76) and (3.78) at the sphere wall ($r = R_0$) gives

$$\cot\chi_0 = kR_0 \tag{3.101}$$

and

$$\cos\chi_0 = \frac{kR_0}{\sqrt{1 + (kR_0)^2}} \tag{3.102}$$

However, equation (3.69) will also express this quantity when $r = R_0$

$$P(r = R_0, t) = \frac{\psi}{R_0}e^{i(\omega t - kR_0)} \tag{3.103}$$

Equating this with equation (3.100) gives

$$\psi = \rho_0 c U_0 R_0 \cos\chi_0 . e^{i(kR_0 + \chi_0)} \tag{3.104}$$

which on substitution into equation (3.70) gives the expression for the acoustic pressure at any radius $r \geqslant R_0$

$$P(r,t) = \rho_0 c U_0 \frac{R_0}{r} \cos\chi_0 e^{i(\omega t - k(r-R_0)+\chi_0)} \tag{3.105}$$

the magnitude of which is

$$|P(r,t)| = \frac{\rho_0 c |U_0| R_0^2 k}{r\sqrt{1+(kR_0)^2}} \tag{3.106}$$

The acoustic intensity within the fluid (equation (3.83)) is therefore

$$I = \frac{\rho_0 c |U_0|^2}{2}\left(\frac{R_0}{r}\right)^2 \cos^2\chi_0 = \frac{\rho_0 c |U_0|^2}{2}\left(\frac{R_0}{r}\right)^2 \frac{(kR_0)^2}{(1+(kR_0)^2)} \tag{3.107}$$

In the long-wavelength limit ($kR_0 = \omega_0 R_0/c \ll 1$), $\chi_0 \to \pi/2$ (Figure 3.5, when $r = R_0$), and $e^{i\chi_0} \to i$. Substitution of this into equation (3.105) yields the pressure at radius r in the liquid in this limit:

$$P_A(r,t) \approx \frac{i\rho_0 c k U_0 R_0^2}{r} e^{i(\omega t - kr)} \tag{3.108}$$

Thus as $kR_0 \to 0$, the pressure will be almost $\pi/2$ out-of-phase with the particle velocity. The phase difference can never be exactly $\pi/2$, that is kR_0 can never actually equal zero, since then with pressure and velocity in quadrature, their product will be sinusoidal, giving no net energy flow in the direction of propagation during any complete acoustic cycle.

In the same limit ($kR_0 \ll 1$) the intensity becomes

$$I \approx \frac{\rho_0 c k^2 |U_0|^2}{2}\left(\frac{R_0^2}{r}\right)^2 \tag{3.109}$$

The intensity depends on k^2 (or equivalently ω^2), and R_0^4.

If the wall motion of the sphere is subject to damping,[9] then in general $U_0 = U_{0,i} e^{-\beta t}$, where $U_{0,i}$ is the amplitude of the wall velocity oscillation at $t = 0$. The time of travel of the signal from the wall is strictly equal to $(r - R_0)/c \approx r/c$, this analysis being valid only if the source is of a size such that the time R_0/c is insignificant. Thus modifying equation (3.106) gives the magnitude of the pressure at distance r from the sphere:

$$|P(r,t)| = \frac{\rho_0 c R_0^2 k}{\sqrt{1+(kR_0)^2}} \frac{1}{r} U_{0,i} e^{-\beta(t-r/c)} \tag{3.110}$$

where $\beta = b/2m$, as discussed earlier. The complex nature of the impedance $Z_{RP}^{s.s.}$, as seen from equation (3.74), will result in a complex wavenumber, and therefore leads to attenuation.[10] If $kR_0 \ll 1$ the pressure amplitude is

$$P_A(r,t) \approx \frac{i\rho_0 c k R_0^2}{r} U_{0,i} e^{-\beta(t-r/c)} e^{i(\omega t - kr)} \tag{3.111}$$

the damped analogy to equation (3.108). This is in agreement with equation (3.69).

In the far field, as $r \to \infty$, from equations (3.74) to (3.76) it is clear that the impedance tends to be real ($\tan\chi \to 0$; see Figure 3.5) and equal to $\rho_0 c$, giving a real wavenumber. Equation (3.107)

---
9 See section 3.1.3.
10 See Chapter 1, section 1.1.7.

gives for the acoustic intensity in this limit, as the spherical waves far from a damped source resemble plane waves [8]:

$$I \approx \frac{\rho_0 c k^2}{2}\left(\frac{R_o}{r}\right)^2 \frac{|U_{o,i}|^2 \, e^{-2\beta(t-r/c)}}{1+(kR_o)^2} \tag{3.112}$$

(b) The Pressure Radiated by a Spherical Pulsating Bubble

The previous analysis has been developed for a monopole source, which might be envisaged to be a pulsating sphere which is 'rigid' in that as its volume changes its internal pressure is invariant. In this respect it is like the Rayleigh cavity discussed in Chapter 2, section 2.1.3(b)(i): that cavity is empty, so that the pressure inside is constant whatever the volume. Thus as its volume changes, the decrease in potential energy is balanced by the increase in kinetic energy of the liquid. In the section that followed (Chapter 2, section 2.1.3(b)(ii)), the effect of having a gas phase within the bubble was explained: the decrease in potential energy is due not just to the increase in kinetic energy of the liquid, but also to the work done by the liquid on compressing the gas. Therefore in considering the pressure radiated by a pulsating bubble, one must consider the compression of the gas. This will be done following the analysis of Vokurka, which from the Rayleigh model develops a formulation for the small-amplitude linear oscillations discussed so far in this chapter. Applying equation (2.61) to the spherically-symmetric case, Vokurka [10] relates the pressure in the liquid p to the fluid velocity v:

$$\frac{1}{\rho}\frac{\partial p}{\partial r} = -\frac{\partial v}{\partial t} - v\frac{\partial v}{\partial r} \tag{3.113}$$

Using the incompressibility relation, equation (2.25), Vokurka integrates equation (3.113) to obtain the pressure a distance r from the bubble centre when the instantaneous bubble radius is r:

$$\left(\frac{p}{p_\infty} - 1\right)\frac{p_\infty}{\rho} = \frac{R}{r}(\ddot{R}R + 2\dot{R}^2) - \frac{\dot{R}^2}{2}\left(\frac{R}{r}\right)^4 \tag{3.114}$$

The acoustic pressure radiated by the bubble, P_{b1}, is given by

$$P_{b1}(r,t) = p(r,t) - p_\infty \tag{3.115}$$

and, in the absence of surface tension and vapour pressure, the acoustic pressure in the gas at the bubble wall is

$$P_g(R,t) = p_g(R,t) - p_\infty \tag{3.116}$$

In this limit (where surface tension and vapour pressure are negligible), $p_g(R,t) = p_L(t)$, so that

$$P_g(R,t) = p_L(t) - p_\infty \tag{3.117}$$

It is possible to relate the acoustic pressure in the gas at the bubble wall to the acoustic pressure in the liquid at radius r by obtaining the equation of motion. This can be done by firstly differentiating equation (2.29) with respect to R, noting that

$$\frac{\partial(R^3\dot{R}^2)}{\partial R} = 3R^2\dot{R}^2 + R^3\frac{\partial(\dot{R}^2)}{\partial t}\frac{1}{\dot{R}} = 3R^2\dot{R}^2 + 2R^3\ddot{R} \tag{3.118}$$

to give

$$\frac{p_L - p_\infty}{\rho} = \frac{3\dot{R}^2}{2} + R\ddot{R} \tag{3.119}$$

The pressure in the liquid at the bubble wall, p_L, can be substituted from equation (3.117) into (3.119) to give the acoustic pressure within the gas:

$$\frac{P_g(R,t)}{\rho} = \frac{3\dot{R}^2}{2} + R\ddot{R} \tag{3.120}$$

Substituting (3.120) into (3.114) and (3.115) relates the acoustic pressure in the gas at the bubble wall to the acoustic pressure radiated by the bubble in the liquid at radius r:

$$P_{b1}(r,t) = \frac{R}{r}P_g + \frac{R}{r}\left(\frac{\rho\dot{R}^2}{2}\right) - \left(\frac{R}{r}\right)^4\left(\frac{\rho\dot{R}^2}{2}\right) \tag{3.121}$$

Having derived this equation, Vokurka [10] explains the physical meaning of the terms which contribute to the pressure radiated by the bubble. The first term, $P_g(R/r)$, clearly represents the acoustic component of the pressure in the gas at the bubble wall, communicated directly to the liquid and decaying in a monopole fashion. The second term, $\rho R\dot{R}^2/(2r)$, arises through the dynamics of the bubble wall. Both of these terms decay as r^{-1}, and represent spherical acoustic waves [10]. However, the third term, $\rho R^4\dot{R}^2/(2r^4)$, which also arises through the wall dynamics, decays as r^{-4}, and is significant only at small distances from the bubble. Substitution from equation (2.25) shows how in an incompressible liquid this term equals $\rho v^2/2$, which is the pressure term arising from liquid flow in equation (2.73). Therefore the term $\rho R^4\dot{R}^2/(2r^4)$ is called the Bernoulli pressure or *kinetic wave* [10, 11].

This chapter deals primarily with small-amplitude harmonic pulsations, and Vokurka's analysis can be shown to approximate to this form in the small-amplitude limit. Differentiation of equation (2.32) with respect to time gives another expression for the acceleration of the bubble wall in the compression system. Using R_{max} for R_m in equation (2.32) (differentiated), as described in Chapter 2, section 2.1.3(b)(ii), gives

$$\frac{\rho}{R_{max}^3 p_\infty}\ddot{R} = \frac{p_{g,m}}{p_\infty}\frac{1}{(1-\gamma)}\left\{R^{-4} - \gamma R_{max}^{3(\gamma-1)} R^{-(3\gamma+1)}\right\} \tag{3.122}$$

so that substituting equation (2.32) into equation (3.122) yields

$$\frac{\rho}{p_\infty}R\ddot{R} + \frac{3}{2}\frac{\rho}{p_\infty}\dot{R}^2 = \frac{p_{g,m}}{p_\infty}\left(\frac{R}{R_{max}}\right)^{-3\gamma} - 1 \tag{3.123}$$

for the compression [10], the initial conditions for $R(t)$ being $R(0) = R_{max}$ and $\dot{R}(0) = 0$. A similar analysis yields for the expansion system:

$$\frac{\rho}{p_\infty}R\ddot{R} + \frac{3}{2}\frac{\rho}{p_\infty}\dot{R}^2 = \frac{p_{g,max}}{p_\infty}\left(\frac{R}{R_{min}}\right)^{-3\gamma} - 1 \tag{3.124}$$

the initial conditions for $R(t)$ being $R(0) = R_{min}$ and $\dot{R}(0) = 0$. Vokurka [10] transforms equation (3.123) into a form relating to the equilibrium radius:

$$\frac{\rho}{p_\infty} R\ddot{R} + \frac{3}{2}\frac{\rho}{p_\infty}\dot{R}^2 = \left(\frac{R}{R_0}\right)^{-3\gamma} - 1 \tag{3.125}$$

the initial conditions for $R(t)$ being $R(0) = R_{max}$ and $\dot{R}(0) = 0$.

Vokurka [3, 10] uses the preceding formulation to discuss the pressure radiated by a bubble. As previously shown through discussion of equation (3.121), it follows that on the right-hand side of equation (3.114) the first term represents the acoustic pressure and the second the kinetic wave. At distances far from the bubble, this kinetic term is negligible and the acoustic pressure in the liquid is

$$P_{bl}(t) = p(t) - p_\infty \approx \rho \frac{R}{r}(\ddot{R}R + 2\dot{R}^2) \tag{3.126}$$

Vokurka [10] obtains plots for $P_{bl}(t)$ as a function of r/R_{max} by substituting into equation (3.114) for $\dot{R}$ and $\ddot{R}$ from equations similar to (2.32) and (3.122) respectively. The normalised pressure in the liquid, $p(t)/p_\infty = (P_{bl}(t) + p_\infty)/p_\infty$, is shown in Figure (3.6) for four stages *during a collapse* for which the initial maximum bubble radius is twice the equilibrium radius (i.e. $R_{max}/R_0 = 2$). The pressure is shown as a function of the normalised radial coordinate, r/R_{max}, each plot terminating at the bubble wall. In Figure 3.6(a), when $R = R_{max}$, the pressure decreases towards the bubble wall. As the collapse proceeds, a positive pressure contribution closer to the bubble wall appears, so that the liquid pressure is roughly constant in the liquid, rapidly falling off only close to the bubble wall (Figure 3.6(b)). Though at the closest distances to the bubble wall a rapid decrease in pressure is seen, the effect of the additional positive pressure contribution close to the bubble wall becomes more prominent, creating a maximum. This is demonstrated in Figure 3.6(c), which illustrates the moment when the bubble wall passes through the equilibrium position. The position and amplitude of the maximum can be found by setting the differential of equation (3.114) with respect to r, equal to zero. As the bubble radius decreases, the maximum moves closer to the bubble wall. When the wall velocity is a maximum (and $\ddot{R} = 0$), the pressure maximum enters the bubble wall. Though the formulation predicts a maximum within the bubble, this has no meaning: the pressure simply increases right up to the bubble wall [10]. In Figure 3.6(e), the liquid pressure at the four stages of the collapse are shown, demonstrating how the maximum develops, and how the overall pressure in the liquid increases. In Figure 3.7, the contributory time-varying terms in equation (3.114) are plotted to explain how the pressure in the liquid develops. Plot (i) corresponds to the normalised pressure from the acceleration term, $(R^2\ddot{R}/r)\rho/p_\infty$. Plot (ii) represents the $(2R\dot{R}^2/r)\rho/p_\infty$ term. Together, they sum to give the acoustic term, $(\ddot{R}R + 2\dot{R}^2)(R/r)\rho/p_\infty$, which decays in monopole r^{-1} fashion, and is labelled (ac) on the graph. Finally the non-acoustic component, $-(\dot{R}^2/2)(R/r)^4\rho/p_\infty$, is labelled (k). The normalised liquid pressure plotted in Figure 3.6 is given by (ac) + (k) + 1, or alternatively (i) + (ii) + (k) + 1. The pressure component radiated by the bubble is (ac) + (k) = (i) + (ii) + (k) (see Equation (3.115)). When $R = R_{max}$ (Figures 3.6(a) and 3.7(a)) the wall velocity $\dot{R}$ equals zero. Therefore the only contribution to the pressure comes from the acceleration term. In all the plots, the acceleration term (i) has a consistent form, being negative and falling off as r^{-1}. Similarly, the non-acoustic term remains consistently negative, becoming significant only very close to the wall. In this way it serves to generate the near-bubble pressure fall-off from the maximum. Far from the bubble it is insignificant, so that there the pressure effectively decreases as r^{-1}. The term (ii) is positive, falling off as r^{-1}, and increasing in time. This term

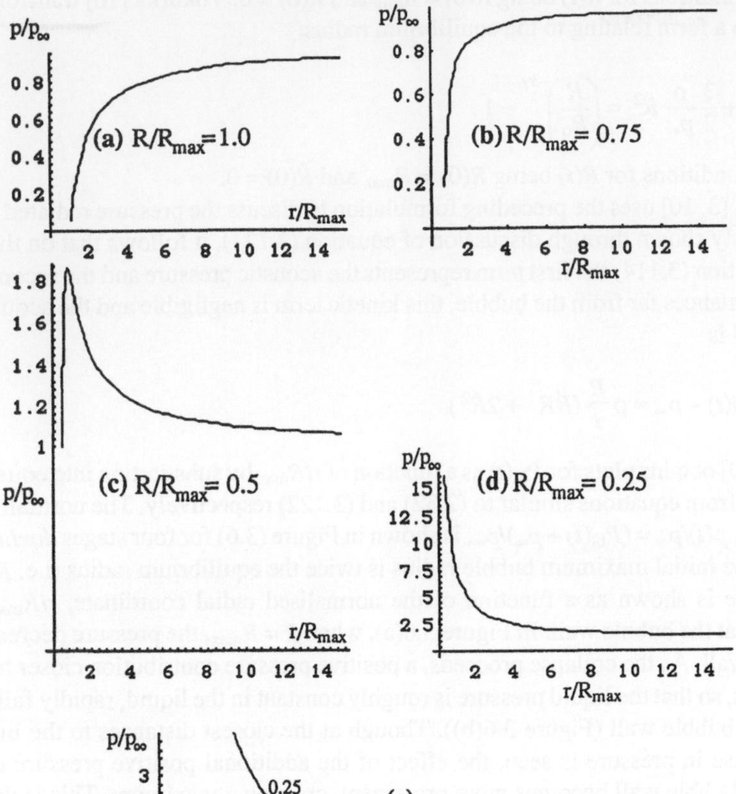

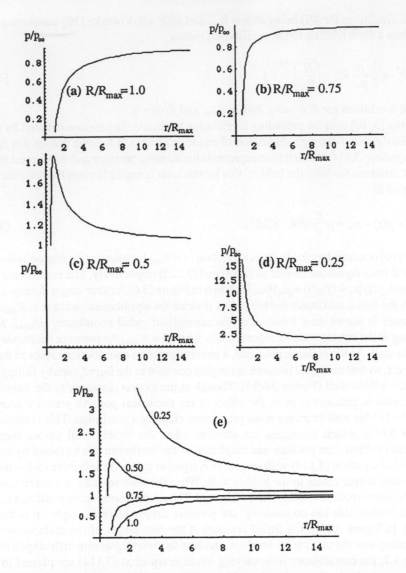

Figure 3.6 The pressure in the liquid, normalised to the static pressure at ∞, at various times in the compression phase of bubble motion, when it collapses from a radius (a) $R = R_{max}$, at which point the wall speed is zero. In this case, the maximum bubble radius is twice the equilibrium radius. The pressure is shown when the bubble radius R equals (b) $3R_{max}/4$, (c) $R_{max}/2$, ($= R_o$), and (d) $R_{max}/4$. In (e), the four plots are shown together.

serves to generate the maximum, and contributes significantly as the collapse proceeds and the wall speed increases. After the bubble has reached a minimum radius, as it expands, the maximum travels back out from the bubble wall into the liquid. The speed at which the maximum travels is related to the wall motion, and not to the speed of sound in the liquid (which is infinite in this model, since we are employing the incompressibility assumption inherent in equation (2.25)) [10]. Vokurka [10] discusses the form of the pressure and velocity fields for various values of the initial expansion ratio, $A_\varepsilon = R_{max}/R_o$. The significance of this parameter is further discussed

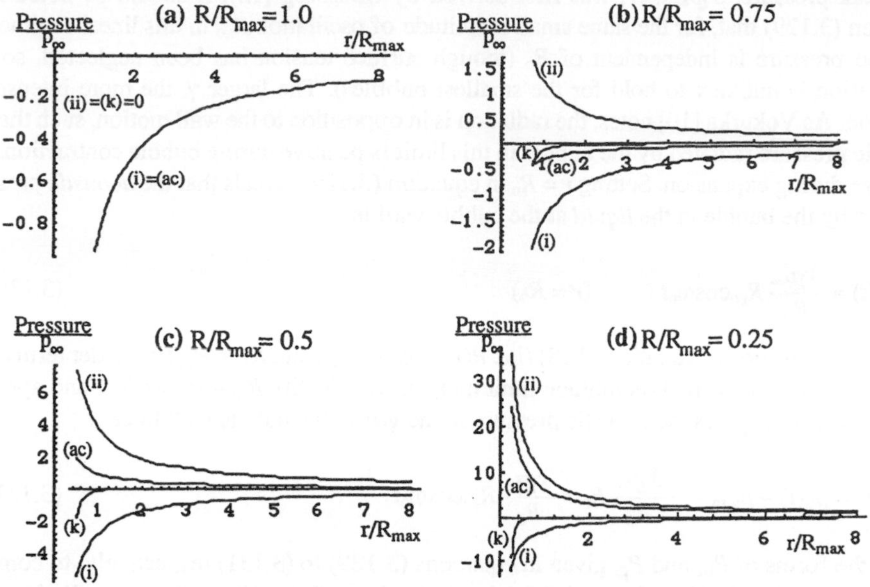

Figure 3.7 The components of the pressure in the liquid for the collapse shown in Figure 3.6, showing the kinetic (k) and acoustic (ac) contributions. The latter is the sum of curves (i) and (ii).

by Vokurka [3]. The empty cavity is characterised by $A_\varepsilon \to \infty$, and small-amplitude linear oscillations by $A_\varepsilon \to 1$. Though the above procedure does clearly illustrate the pressure contributions in nonlinear collapses, most of this chapter deals with such linear oscillations. The equation of motion for small-amplitude oscillation of the bubble wall can be found by recasting equation (3.125) in terms of the displacement R_ε of the wall from equilibrium. Setting $R(t) = R_0 + R_\varepsilon(t)$ in equation (3.125) and assuming small-amplitude pulsations $R_\varepsilon \ll R_0$ allows a binomial expansion. Discarding second-order terms such as $R_\varepsilon \ddot{R}_\varepsilon$ and $\dot{R}_\varepsilon^2$ yields

$$\ddot{R}_\varepsilon + \frac{3\gamma p_\infty}{\rho R_0^2} R_\varepsilon = 0 \tag{3.127}$$

Thus in the small-amplitude limit, the equation of motion is the familiar second-order linear differential equation, which has harmonic solutions of the form $R_\varepsilon = -R_{\varepsilon 0}\cos\omega_0 t$, where ω_0 agrees with ω_M, as given by equation (3.38).

To obtain the pressure radiated by the bubble into the liquid in the limit of small-amplitude pulsations, equation (3.126) can be reduced in the same manner (i.e. $R(t) = R_0 + R_\varepsilon(t)$ and $R_\varepsilon \ll R_0$). Discarding second-order terms yields for the acoustic pressure in the liquid

$$P_{bl}(t) \approx \rho \frac{R_0^2}{r} \ddot{R}_\varepsilon \tag{3.128}$$

Substitution for $\ddot{R}_\varepsilon$ from equation (3.127) gives

$$P_{bl}(t) \approx -\frac{3\gamma p_\infty}{r} R_\varepsilon = \frac{3\gamma p_\infty}{r} R_{\varepsilon 0}\cos\omega_0 t \tag{3.129}$$

The peak pressure, $3\gamma p_\infty R_{\varepsilon 0}/r$, was first derived by Strasberg [12]. It should be noted from equation (3.129) that, for the same small amplitude of oscillation $R_{\varepsilon 0}$ in this linear regime, the radiated pressure is independent of R_0 (though surface tension has been neglected, so this observation is unlikely to hold for the smallest bubbles). The larger γ, the more intense the radiation. As Vokurka [10] notes, the radiation is in opposition to the wall motion, such that the acoustic pressure radiated by the bubble in this limit is positive during bubble contraction, and negative during expansion. Setting $r = R_0$ in equation (3.129) reveals that the *acoustic* pressure radiated by the bubble in the *liquid* at the bubble wall is

$$P_{bl}(t) \approx \frac{3\gamma p_\infty}{R_0} R_{\varepsilon 0} \cos \omega_0 t \qquad (r = R_0) \tag{3.130}$$

Simple linearisation of equation (2.28) (i.e. $R(t) = R_0 + R_\varepsilon(t)$ and $R_\varepsilon \ll R_0$, first order terms only retained), and in this small-amplitude limit making the identity $R_m - R_0 = -R_{\varepsilon 0}$, and approximating $p_{g,m} = p_\infty$, gives the acoustic pressure in the *gas* at the bubble wall to be

$$P_g(t) = p_g(t) - p_{g,m} \approx -\frac{3\gamma p_\infty}{R_0} R_\varepsilon = \frac{3\gamma p_\infty}{R_0} R_{\varepsilon 0} \cos \omega_0 t \tag{3.131}$$

Using the forms of P_{bl} and P_g, given in equations (3.129) to (3.131) respectively, to compare terms in equation (3.121) shows that in the limit of small-amplitude linear oscillations, the pressure radiated by the bubble is determined by the first term in equation (3.121), RP_g/r, the pressure communicated to the liquid from the gas. This is in contrast to the monopole source discussed in section 3.3.1(a), where the pressure within the rigid pulsating sphere was constant, and the pressure in the liquid was the result of interface motion.

Between the two limits of infinity (empty cavity) and unity (small-amplitude linear oscillations) there lies a range for $A_\varepsilon = R_{max}/R_0$ where nonlinear behaviour can be observed. A bubble pulsating at finite amplitude is clearly not a linear oscillator. There is an obvious asymmetry: for example, whilst in expansion the amplitude of wall displacement has no absolute limits, in compression the radial displacement of the wall from equilibrium cannot be greater than the equilibrium radius R_0. Vokurka [13] employs the parameters of the maximum radius reached by a bubble during expansion (R_{max}), and the ratio $A_\varepsilon = R_{max}/R_0$, to characterise the bubble. Vokurka [10] distinguishes approximately linear oscillations, where $A_\varepsilon \leqslant 1.05$, from the nonlinear ($A_\varepsilon > 1.05$), and classifies the bubbles into four broad types depending on the equilibrium radius and the amplitude of the oscillation: microbubbles, macrobubbles, and medium-sized bubbles containing the subclass of scaling bubbles [13]. Figure 3.8 [13] from Vokurka's paper uses R_{max} rather than R_0, since with large amplitudes of oscillation the latter is often easier to measure. The effects of surface tension and viscosity on the dynamics of *microbubbles* cannot be neglected, owing to their small size. Such bubbles are discussed in Chapter 2, section 2.1 regarding seed nuclei for cavitation. The large size of *macrobubbles* results in significant deviations from shape distortions, surface tension forces being less effective at maintaining sphericity and fluid inertia causing overshoot in oscillation. The larger the bubble, the more gravity influences the shape and the dynamics. Large bubbles have been produced by underwater chemical explosions, using explosive charges ranging in size from 1 gram [14] to 1 tonne [11], the latter generating a bubble with a maximum radius of 10 m. After expanding to a maximum size as a result of the explosion, the bubbles then oscillate in the manner described earlier. Bubbles produced by chemical explosives have R_{max} greater than about 0.01–0.3 m, depending on the oscillation amplitude. The maximum radii of bubbles produced by underwater nuclear explosions are of the order of 100 m [15].

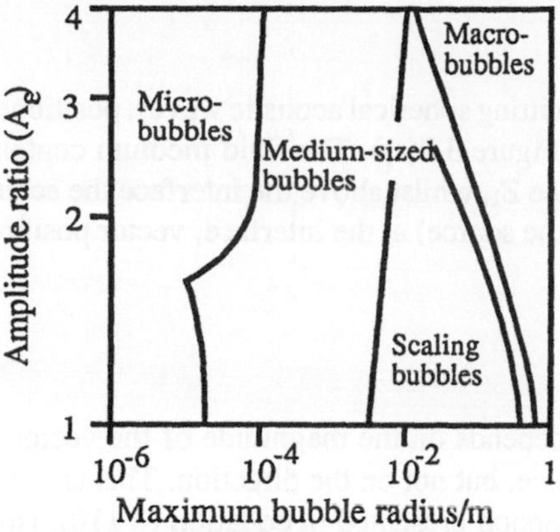

Figure 3.8 The classification of bubbles with respect to the amplitude of oscillation and the maximum size attained (after Vokurka [13]).

The life history of bubbles produced by underwater explosion is as follows. Firstly a shock wave is emitted from the underwater detonation. The gas produced by the explosion forms a bubble called a 'gas globe', which expands to some maximum size and then contracts. It then rebounds, and at that instant emits a positive pressure pulse. The bubble continues to pulsate, each rebound emitting a successively weaker positive pressure pulse as the energy of the bubble is dissipated. It is the nature of this bubble pulsation and rebound that is key to the emission of pressure pulses after the initial shock: no such pulses are detected if the gas globe breaks through the sea surface (and so ceases to be a gas bubble), or if the detonation is contained within a casing that remains intact during the explosion, preventing the formation and expansion of the bubble [16]. As it pulsates, the bubble will migrate upwards, the velocity of motion varying with the instantaneous bubble size.

The illustrative example of acoustic data given in Figure 3.23 comes from bubbles of the *medium-size* class. *Scaling bubbles* represent a subclass, being medium-sized bubbles which are assumed to obey scaling laws [17]. For these bubbles, the radius–time history of the bubble is independent of bubble size, and can simply be scaled. The effects of surface tension, viscosity, gravity and heat conduction are negligible for these bubbles. The effects of heat conduction are minimal in very large and very small bubbles (Figure 3.16), and in scaling bubbles [17]. In medium-sized, non-scaling bubbles, however, the effects of heat conduction during free oscillation cannot be neglected.

Vokurka [3, 10] discusses the components in the pressure field for nonlinear gas bubble oscillations, and for oscillations of vapour bubbles (where if evaporation and condensation occur with infinite speed the pressure within the bubble, in the absence of surface tension, is constant at p_v), and considers the effects of surface tension, viscosity, gravity and heat conduction. Samek [18] formulates the acoustic waves produced by a spherical pulsating bubble in a viscous, compressible unbounded liquid of finite thermal conductivity.

3.3.2 Monopole Close to Boundaries

In many practical situations involving the freely-oscillating bubble, a boundary of some sort is involved. This is because, as will be discussed in section 3.7, in most cases the bubble is excited to sing when it is entrained from some free surface.

(a) Image Sources

Consider a point source emitting spherical acoustic waves, positioned a distance h below a plane pressure-release surface (Figure 3.9(a)). The fluid medium containing the source has a finite specific acoustic impedance Z_1, whilst above the interface the acoustic impedance is zero. The acoustic pressure (due to the source) at the interface, vector position $\vec{r}$ from the source, can be given by

$$P = P_A(|\vec{r}|)\, e^{i(\omega t - \vec{k}.\vec{r})} \tag{3.132}$$

where the amplitude P_A depends on the magnitude of the vector $\vec{r}$, that is upon the distance separating it from the source, but not on the direction. This is to accord with all the factors of damping and energy dispersion presented in equation (3.110). However, the reflections from the pressure-release boundary must give zero pressure at all points on this boundary, and at all times (see Chapter 1, section 1.1.6). The only way to achieve this is by mathematically replacing the boundary with an image source at a point equidistant from the boundary as the original source, on the far side (Figure 3.9(b)). The acoustic pressure from this image source detected at the interface is $\check{P}$, and since the boundary is pressure-release, then $P + \check{P} = 0$. Since this must be true for any point of interest along the line of the original boundary, and since such points are equidistant from both sources, then

$$\check{P} = -P = -P_A(|\vec{r}|)\, e^{i(\omega t - \vec{k}.\vec{r})} = P_A(|\vec{r}|)\, e^{i(\omega t - \vec{k}.\vec{r} + \pi)} \tag{3.133}$$

Let $\vec{\check{r}}$ be the vector position of the point of observation, measured with the image source as the

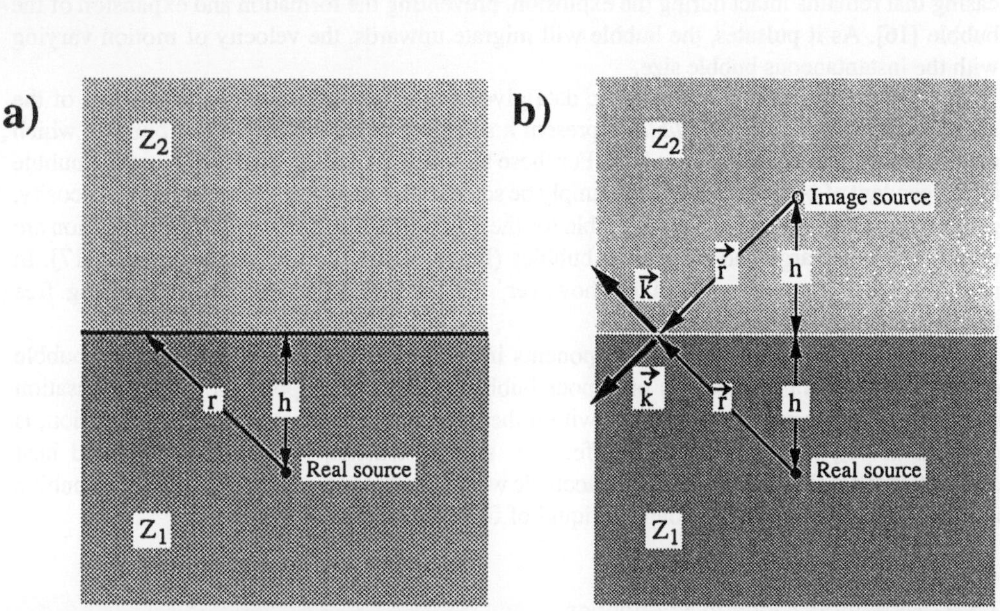

Figure 3.9 (a) A real point source of spherical pressure waves in a medium of specific acoustic impedance Z_1 is a distance h away from the plane interface with a medium of impedance Z_2. The vector $\vec{r}$ extends from the source to a point on the interface. (b) The pressure field at the interface and in medium 1 can be modelled by placing an appropriate image source in medium 2.

origin; clearly $|\vec{r}| = |\vec{\check{r}}|$. If the wavevector of sound from the image source is $\vec{\check{k}}$, then $|\vec{\check{k}}| = |\vec{k}|$, since the sound from both sources has the same wavelength. Therefore since $\vec{r}$ is parallel to $\vec{k}$, and $\vec{\check{r}}$ is parallel to $\vec{\check{k}}$, then $\vec{k}.\vec{r} = \vec{\check{k}}.\vec{\check{r}}$ (Figure 3.9(b)). Thus

$$\check{P} = P_A(|\vec{r}|)\, e^{i(\omega t - \vec{k}.\vec{r} + \pi)} \tag{3.134}$$

The image source is therefore identical to the original, except that it emits sound with a phase difference of π.

If the boundary is rigid (i.e. $Z_2 = \infty$), then the acoustic pressure amplitude at the boundary must be twice that of the sound from the source felt at the boundary (Chapter 1, section 1.1.6). Therefore

$$\check{P} = P = P_A(|\vec{r}|)\, e^{i(\omega t - \vec{k}.\vec{r})} \tag{3.135}$$

that is, the image source now oscillates in phase with the original one. The addition of a suitable image source in this manner models the sound field throughout the medium of the original source, not just on the boundary.

The limiting cases examined here are approached in real terms by sources in water close to air and metal boundaries. In these practical situations, the interfaces are neither perfectly free nor rigid (i.e. $|R| \neq 1$). The appropriate models have image sources which are not identical to the original sources.

If a source is very close to a rigid plane surface, and the boundary extends much further than a wavelength, then the plane is often considered to be infinite. Such a boundary is called a *baffle*.[11] The image and the real source are separated by a distance small compared with the wavelength, and the far-field emission is as if the two occupied the same place. Since they emit in phase, the pressure field in the far-field is therefore double that from the source alone, and the intensity increased fourfold.

(b) Paired Sources and Acoustic Dipoles

(i) Radiation Patterns. If the source is very close to a boundary, relative to the distance to the point of observation, then the field takes an interesting form.[12] The field from a pair of close spherical sources is now discussed. These could be two sources of identical strength and frequency, separated from each other by a distance $2h$, oscillating with a phase difference ϑ. Alternatively the formulation could be describing a real source a small distance h from either a rigid ($\vartheta = 0$) or free ($\vartheta = \pi$) boundary. The two sources are shown in Figure 3.10, oscillating with some phase difference ϑ. Since we are dealing with the far-field, the pressure from each single source approximates to a r^{-1} dependence (i.e. as equation (3.77) reduces to equation (3.111) when $kr \gg 1$).

Setting the coordinate origin at the midpoint between the sources, and defining the z-axis, where $\theta = 0$, to pass through the centre of both sources (Figure 3.10), the net pressure ΣP at the

[11] See Chapter 1, section 1.2.1(a).

[12] In electrostatics, a dipole field would result if a charge were close to a conductor, giving an image charge of the opposite sign. However, in acoustics there is the phase relation of the oscillators to consider: if the boundary is pressure-release, then the image source has the opposite 'sign' to the original (the bubbles oscillate in antiphase), and the field takes dipole form. If the boundary is rigid, then the image has the same 'sign' (the bubbles oscillate in phase) and the field does not have the familiar dipole encountered in electrostatics.

point of observation M, which is at position $\vec{r}_1$ relative to source 1 and at $\vec{r}_2$ relative to source 2, is

$$\Sigma P = \frac{\psi_1}{r_1} e^{i(\omega t - \vec{k}_1 \cdot \vec{r}_1)} + \frac{\psi_2}{r_2} e^{i(\omega t - \vec{k}_2 \cdot \vec{r}_2 + \vartheta)} \tag{3.136}$$

where each source emits as an independent spherical source, as described by equation (3.34). If the sound from both sources has the same wavelength, then $|\vec{k}_1| = |\vec{k}_2|$, which we shall call k. Similarly, if they have the same strength, then $\psi_1 = \psi_2 = \psi$, which is numerically equal to the acoustic pressure amplitude a unit distance away from a single source. As can be seen from Figure 3.10, $r_1 = |\vec{r}_1| = r - h\cos\theta$ and $r_2 = |\vec{r}_2| = r + h\cos\theta$, and $\vec{k}.\vec{r}_1 = k(r - h\cos\theta)$ since $\vec{k}$ is parallel to $\vec{r}_1$ at M. Similarly $\vec{k}_2.\vec{r}_2 = k(r + h\cos\theta)$. Equation (3.136) can therefore be written

$$\Sigma P = \psi e^{i\omega t} \left\{ \frac{e^{-ik(r - h\cos\theta)}}{r - h\cos\theta} + \frac{e^{-ik(r + h\cos\theta - \vartheta)}}{r + h\cos\theta} \right\}$$

$$= \psi e^{i(\omega t - kr)} \left\{ \frac{e^{ikh\cos\theta}(r + h\cos\theta) + e^{i\vartheta} . e^{-ikh\cos\theta}(r - h\cos\theta)}{r^2 - h^2\cos^2\theta} \right\} \tag{3.137}$$

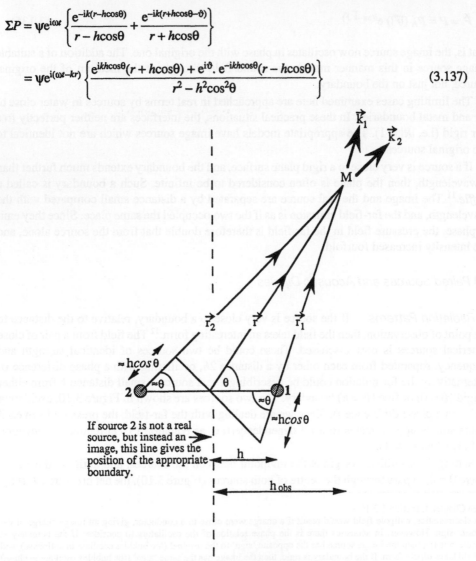

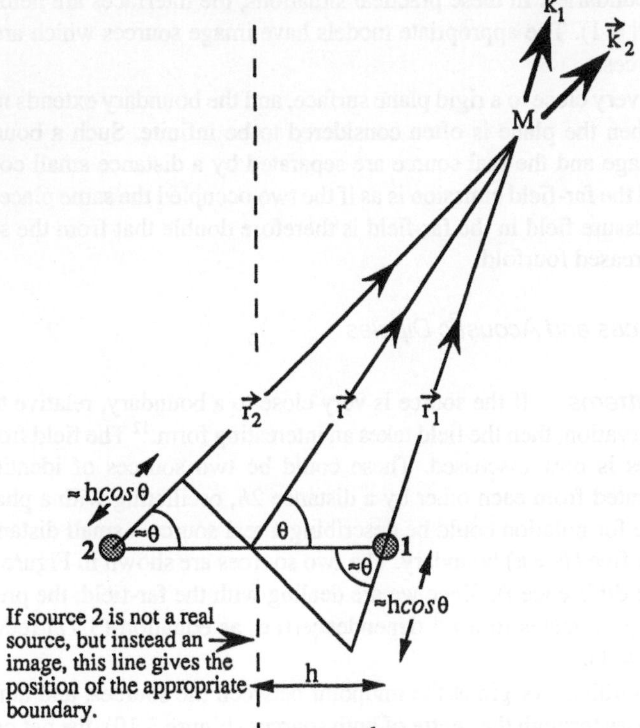

Figure 3.10 The geometry of a dipole source (source 2 may be real or an image).

Assuming $h^2/r^2 \ll 1$ (far-field approximation), then

$$\Sigma P = \frac{\psi}{r^2}\, e^{i(\omega t - kr)} \left\{ h\cos\theta\left(e^{ikh\cos\theta} - e^{i\vartheta}.\, e^{-ikh\cos\theta}\right) + r\left(e^{ikh\cos\theta} + e^{i\vartheta}.\, e^{-ikh\cos\theta}\right)\right\}$$

$$= \frac{\psi}{r^2}\, e^{i(\omega t - kr)}\, e^{i\vartheta/2} \left\{ h\cos\theta\left(e^{i(kh\cos\theta - \vartheta/2)} - e^{-i(kh\cos\theta - \vartheta/2)}\right) + r\left(e^{i(kh\cos\theta - \vartheta/2)} + e^{-i(kh\cos\theta - \vartheta/2)}\right)\right\}$$

$$= \frac{\psi e^{i(\omega t - kr)}\, e^{i\vartheta/2}}{r^2} \left\{ (h\cos\theta)(2i\sin(kh\cos\theta - \vartheta/2)) + r(2\cos(kh\cos\theta - \vartheta/2))\right\} \tag{3.138}$$

If the two monopole sources oscillate in antiphase, then $\vartheta = \pi$, and equation (3.138) gives

$$\Sigma P = \frac{\psi e^{i(\omega t - kr)}\, i}{r^2} \left\{ -(2h\cos\theta).\cos(kh\cos\theta) + 2r\sin(kh\cos\theta)\right\} \tag{3.139}$$

This is the pressure at M radiated from a single monopole a small distance h from a pressure-release surface. It is also the radiation from two identical monopoles, separated by a small distance $2h$ in an infinite liquid, pulsating in antiphase (called an 'acoustic doublet').

If the two monopoles oscillate in phase, then $\vartheta = 0$, and equation (3.138) gives

$$\Sigma P = \frac{\psi e^{i(\omega t - kr)}}{r^2} \left\{ (2ih\cos\theta).\sin(kh\cos\theta) + 2r\cos(kh\cos\theta)\right\} \tag{3.140}$$

This is the pressure at M radiated from a single monopole a small distance h from a rigid surface, or from two identical monopoles, separated by a small distance $2h$ in an infinite liquid, pulsating in phase.

If $h \ll r$, then equations (3.139) and (3.140) can be reduced further. For a single monopole near a pressure-release surface (the image pulsates in antiphase), equation (3.139) gives

$$\Sigma P = \frac{2i\psi\sin(kh\cos\theta)}{r}\, e^{i(\omega t - kr)} \tag{3.141}$$

Since $kh \ll 1$ for the far-field, then equation (3.141) becomes

$$\Sigma P = \frac{2i\psi kh\cos\theta}{r}\, e^{i(\omega t - kr)} \quad \text{(far-field approximation)} \tag{3.142}$$

which is a dipole field.

For a single monopole close to a rigid boundary (the image pulsates in phase), equation (3.140) gives

$$\Sigma P = \frac{2\psi\cos(kh\cos\theta)}{r}\, e^{i(\omega t - kr)} \quad \text{(far-field approximation).} \tag{3.143}$$

These fields are illustrated in Figure 3.11, where the functions $\sin^2(kh\cos\theta)$ (Figure 3.11(a)) and $\cos^2(kh\cos\theta)$ (Figure 3.11(b)) are plotted for varying kh to give the intensity of the radiation $|P|^2$ from equations (3.141) and (3.143) in polar plot. The angle θ is measured from the horizontal axis, which runs through the centre of both sources. In Figure 3.11(a) the sources emit in antiphase, and in Figure 3.11(b) they emit in phase. If the second source is an image, the position of the boundary is given by the vertical axis, and only the solution to the right of the boundary, in the half-space containing the real source, accurately represents the sound field. The three-dimensional pattern can be envisaged by rotating the plot about the horizontal axis.

If $kh \ll 1$, the applicable solutions are seen to be the familiar two-lobe dipole radiation for the free boundary (Figure 3.11(a)(i)), and the isotropic solution for the rigid boundary (Figure 3.11(b)(i)). The rigid boundary solution develops in this limit of $kh\cos\theta \to 0$ such that $\cos(kh\cos\theta \ll 1) \approx 1$, resulting in approximate isotropy. This is as expected, since in such cases the phase difference resulting from the separation of the monopoles is negligible ($kh \approx 0$), the radiation from the sources has propagated almost identical path lengths and thus arrives with the same phase, and so adds again to give a pressure twice that from a single source. In this limit, equation (3.143) reduces to $\Sigma P = 2\psi e^{i(\omega t - kr)}/r$. A source close to such a boundary has been 'baffled'.[13]

As the product kh increases (owing to increased source separation or decreased wavelength) the phase difference between the two sources becomes increasingly significant and lobes develop. Note also that regardless of the value of kh, the radiation at the angles $\theta = \pi/2$ and $\theta = 3\pi/2$ has propagated identical path lengths and thus arrives with the same phase relation with which it departed from the sources. Therefore in Figure 3.11(a) (sources in antiphase) the pressure at $\theta = \pi/2$ and at $\theta = 3\pi/2$ is zero, and in Figure 3.11(b) (sources in phase) the pressure at these angles is twice that from a single source.

The development of the lobes is similar in both cases. The lobe in the horizontal axis (that is, along the line joining the two sources, or perpendicular to the boundary) is distorted inwards as kh increases (Figures 3.11(a)(iii) and 3.11(b)(ii)), becoming dimpled (Figures 3.11(a)(iv) and 3.11(b)(iii)). When the dimple reaches the origin, that is, when zero intensity is radiated along the horizontal axis, the single lobe has become two daughters (Figures 3.11(a)(v) and 3.11(b)(iv)). As kh increases beyond this point, a new lobe forms along the horizontal axis (Figure 3.11(a)(vi)). It grows to some maximum (Figure 3.11(b)(v)), then is distorted inwards (Figures 3.11(a)(viii) and 3.11(b)(vi)), and the cycle repeats until this new lobe has itself formed two daughters. In this way the number of off-axis lobes increases, and the width of each necessarily narrows (Figures 3.11(a)(ix) and 3.11(b)(ix)). The mechanism behind this process is simple to understand (Figure 3.12). An increase in kh will come about either through an increase in the separation of the sources, or a decrease in wavelength. If the sources are pulsating in antiphase, then when the separation equals a whole number of wavelengths their summed pressure will cancel along the axis that passes through their centres (Figure 3.12(a)). If the separation is equals to $(2n+1)\lambda/2$, the pressure will sum in phase along that line and give maximum intensity at $\theta = 0$ and $\theta = \pi$ (Figure 3.12(b)). If the sources are oscillating in phase, cancellation will occur when the separation is $(2n+1)\lambda/2$ (Figure 3.12(c)) and reinforcement when it is $n\lambda$ (Figure 3.12(d)). In between these extremes, the intensity along $\theta = 0$ and $\theta = \pi$ will either be growing to the next maximum, or decreasing towards the next zero. Though at the limit of $kh \to 0$ the two radiation patterns are quite different (compare figures 3.11a(i) and 3.11b(i)), at large kh they resemble one another (Figures 3.11(a)(ix) and 3.11(b)(ix)). The resemblance is, however, only superficial in that the two basic characteristics which distinguish the two radiation patterns are identical in both limits of kh. Firstly because of the phase relation between the sources, at $\theta = \pi/2$ and $3\pi/2$ the intensity must be zero in Figure 3.11(a) and maximum in Figure 3.11(b), regardless of kh. Secondly for $kh > 1$, when the intensity at $\theta = 0$ and π for one plot is a maximum, it must be zero for the other, because of the phase with which

[13]See Chapter 1, section 1.2.1(a).

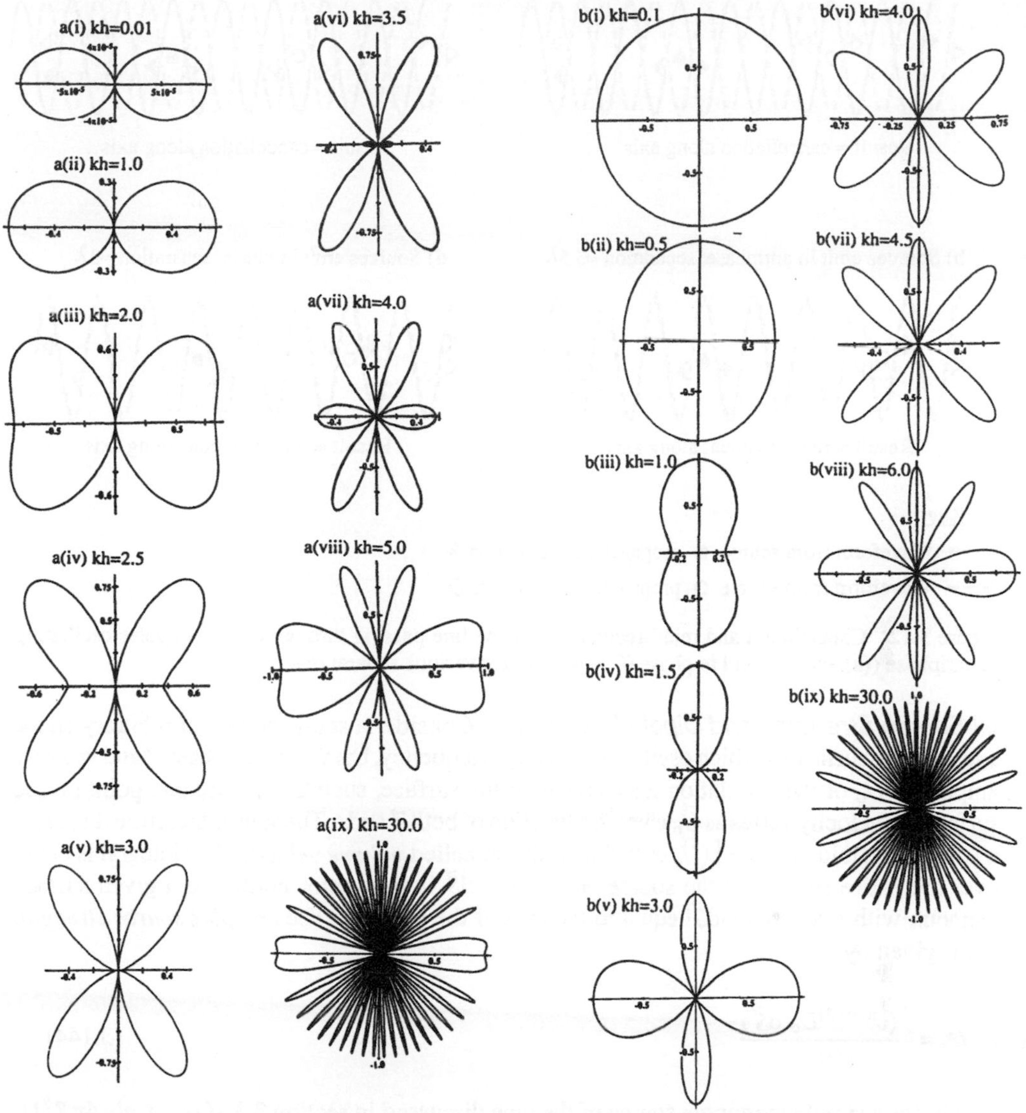

Figure 3.11 The acoustic pressure amplitude radiated from two monopole sources aligned along the horizontal axis, pulsating (a) in antiphase, and (b) in phase, for varying values of the product *kh*.

the waves add in those directions. It should be noticed that, whilst as $kh \rightarrow 0$ the relative intensity for when the sources are in phase approaches ever closer to unity, the intensity from the dipole source becomes ever smaller, going as $(kh)^2$.

Notice from Figure 3.10 that $\cos\theta = h_{obs}/r$, where h_{obs} is the depth below the surface that observations are taken. Therefore in the far-field, equation (3.142) gives a pressure amplitude of magnitude $2\psi kh . h_{obs}/r^2$, so that at a fixed observation depth the dipole radiation falls off as r^{-2} in the far-field, much more quickly than the r^{-1} decay from the monopole source.

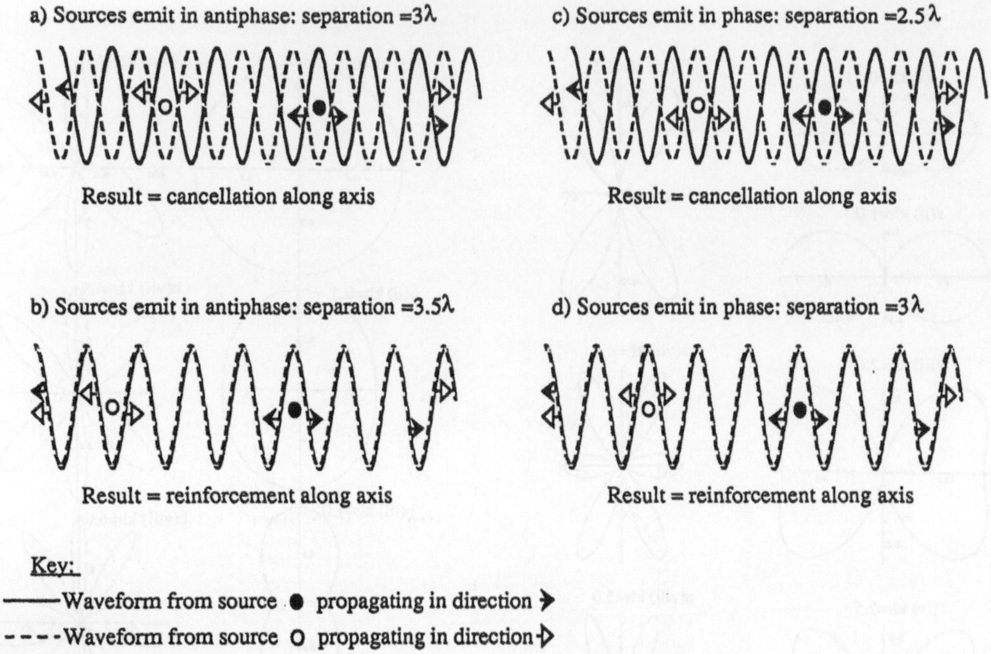
a) Sources emit in antiphase: separation =3λ
Result = cancellation along axis
c) Sources emit in phase: separation =2.5 λ
Result = cancellation along axis
b) Sources emit in antiphase: separation =3.5λ
Result = reinforcement along axis
d) Sources emit in phase: separation =3 λ
Result = reinforcement along axis
Key:
Waveform from source ● propagating in direction
Waveform from source O propagating in direction

Figure 3.12 Cancellation and reinforcement along the line passing through two monopoles oscillating in antiphase ((a) and (b)) and in phase ((c) and (d)), for various separations.

(ii) Source Strengths and Dipole Strengths. Consider a sound source of arbitrary shape and size, the surface of which oscillates at a single frequency, but where the phase ϑ and velocity amplitude $\vec{U}_o$ of the oscillation may vary over the surface, such that at any one point on the surface the velocity varies as $\vec{U}_o\, e^{i(\omega t+\vartheta)}$, a function of both $\vec{r}$ and t. The source therefore displaces a volume of fluid at a rate $\int (\vec{U}_o\, e^{i(\omega t+\vartheta)}).d\vec{S}$, the so-called *volume velocity*. The integral is taken over the entire surface of the source, and where $d\vec{S}$ is the vector normal to a given surface element, with magnitude $|d\vec{S}|$ equal to the area of the element. The *complex source strength*, Q_s, is given by

$$Q_s = \frac{\int (e^{i(\omega t+\vartheta)}\vec{U}_o.d\vec{S})}{e^{i\omega t}} \tag{3.144}$$

which, for a simple monopole source of the type discussed in section 3.3.1(a), equals $4\pi R_o^2 U_o$ [19–21].

The source strength of a dipole is generally taken to equal the product of the strength of each of the individual sources (assumed equal) with the distance between them ($2h$), making the *dipole strength* $Q_s 2h$ [22]. Therefore when measuring the acoustic output from bubbles entrained beneath breaking waves in a laboratory, Medwin and Beaky [23] take the dipole strength to equal $8\pi h R_o^2 U_o^{R_o}$, that is, the product of the separation of the sources ($2h$) with the rate of change of volume at equilibrium ($4\pi R_o^2 U_o^{R_o}$). Defined in this standard way, the dipole strength has dimensions of [length^4/time].

However, in the studies of freely-oscillating bubbles entrained from pressure-release surfaces, the emission has been retroactively related to the source strength by means of a parameter more readily measurable than the bubble wall velocity, i.e. the acoustic pressure amplitude as measured at a known distance from the source. This has, however, lead to the adoption of a nonstandard definition of the dipole strength for these bubbles which departs from the more usual definition, outlined in equation (3.144), which is adopted by the remainder of the acoustics fraternity.

As one moves radially away from the dipole at a given angle θ in the far-field, the acoustic pressure from a dipole falls off as r^{-1}. It is therefore possible to associate a constant with each dipole, given by the product of the radial distance r from the dipole to the constant term in the magnitude of the acoustic pressure radiated from that dipole at distance r. This product is termed the *dipole strength* by many workers in acoustic cavitation, which from equation (3.142) is given in the far-field by

$$D_{dip} = r\Sigma P = 2\psi kh \tag{3.145}$$

The dipole strength so defined has units of [Pa.m]. As it has featured in the literature of acoustic cavitation, it will be this nonstandard terminology which is adopted for the remainder of this book. As defined in equation (3.145), the acoustic dipole strength can, for example, be related to the oscillatory parameters of a pulsating sphere discussed earlier by application of equation (3.81) at the wall. This gives $U_0^{R_0} = (P_A(r = R_0)\sqrt{1+(kR_0)^2}/(\rho ckR_0) \approx (P_A(r = R_0))/(\rho ckR_0)$, since $kR_0 \ll 1$. In equation (3.145), the single-source acoustic pressure amplitude is evaluated at the sphere. Since the single-source acoustic pressure varies as r^{-1}, then $\{P_A(r = 1\text{ m})\} \times [1\text{ metre}] = \{P_A(r = R_0)\} \times R_0$. Since $\{P_A(r = 1\text{ m})\} \times [1\text{ metre}] = \psi$ by definition, then substitution of the above expression for $U_0^{R_0}$ into (3.145) yields

$$D_{dip} = 2kh\{P_A(r = R_0)\} \times R_0 = 2\rho c(kR_0)^2 hU_0^{R_0} \tag{3.146}$$

In this way the magnitude of the pressure from a dipole may be written

$$|P_{dip}(r,\theta,t)| = \frac{D_{dip,i}}{r} e^{-\beta(t-r/c)} \cos\theta . e^{i(\omega t - kr)} \tag{3.147}$$

where $D_{dip,i}$ is the initial value of the dipole strength, which subsequently decays as $e^{-\beta t}$. From equation (3.146), $D_{dip,i} = 2\rho c(kR_0)^2 hU_{0,i}^{R_0}$ where $U_{0,i}^{R_0}$ is the initial wall velocity amplitude of the pulsating sphere (at $t = 0$). Consequently it can be seen that β^{-1} is the decay time of the pulsations of the sphere. Equation (3.147) takes into account the propagation time r/c.

The ratio of the standard acoustic definition of dipole strength to that defined in equation (3.145) for pairs of pulsating spheres is $4\pi/(\rho ck^2)$.

(c) Resonance Effects of Bubble Entrainment from a Free Surface

Bubbles generated by rain or wave action will be close to the parent surface (at depth h). If the surface is approximately planar, and $R_0 \ll h$, a simple antiphase image source will result. Application of image theory shows that the acoustic radiation of one source will interfere and partly cancel the other, so that in the far-field the monopole emissions tend to dipole. The radiation damping of the oscillations will be reduced. Because the inertia in the fluid motion is less, the frequency of pulsation of the bubble will be increased (compare with a reduction of the mass in the simple spring–bob system). A similar reduction in fluid inertia, and increase of frequency, will be found if the liquid surrounding the bubble contains other bubbles.

Strasberg [24] calculated the effect on the resonance frequency. Spherical bubbles pulsating at a small distance h below a free surface have the resonance frequency *increased* from ω_0 to

$$\frac{\omega_o}{\sqrt{1-(R_o/2h)-(R_o/2h)^4}} \tag{3.148}$$

If the bubble centre is at a depth of $4R_o$ beneath the surface, the frequency is raised by around 7%. Strasberg [24] has also shown that a bubble in the shape of an oblate spheroid, with ς_a being the ratio of the largest axis to the smallest, will have a pulsation resonance frequency increased by $\Delta\omega_o$ compared with that of a sphere of the same volume, where

$$\frac{\Delta\omega_o}{\omega_o} = \varsigma_a^{-1/3}\left(\sqrt{\frac{\sqrt{\varsigma_a^2-1}}{\arctan\sqrt{\varsigma_a^2-1}}}\right) - 1 \tag{3.149}$$

A similar formula applies for the prolate spheroid [25].

The presence of the free surface has a greater effect on the damping.[14] Such damping dissipates energy from the bubble pulsation and so, with due regard to the time of travel, P_A in the above equations will vary in time as $e^{-\beta(t-r/c)}$. The damping parameter $\beta = b/2m$ incorporates terms appropriate to dissipation from several sources, for example, viscosity and thermal conduction. In addition, energy will be lost from the bubble through acoustic radiation. This is characterised, for example, in the volume–pressure frame by a contribution b_{VP}^{rad} to the dissipative coefficient b_{VP}. Generally, however, it is simpler to use the dimensionless and frame-invariant form $\delta_{rad} = b_{VP}^{rad}/(m_{VP}^{rad}\omega_o)$. These processes will be discussed in detail in section 3.4.

If the bubble is close to a free surface, the acoustic pressure radiated by the bubble takes dipole form, rather than monopole as seen for a bubble in an infinite medium, and the radiation losses are reduced.

The radiation damping coefficient, which for a monopole source was δ_{rad}, is replaced by a dipole radiation damping constant δ_{rad}^{dip} for a bubble depth h below the liquid surface [26], where

$$\delta_{rad}^{dip} = \tfrac{2}{3}(kh)^2\delta_{rad} \qquad \text{for } kh \ll 1. \tag{3.150}$$

Entrainment of a pulsating bubble close to a free surface, which can lead to the generation of dipole radiation, is not uncommon in the natural world. For example, a bubble entrained by the impact of a raindrop upon a body of water,[15] or when generated near the sea surface by spilling breakers[16] will in fact be close to a pressure-release surface, the surface of the sea. To act as a plane boundary, the sea must be approximately flat [27]. If it is not, then an incoming sound wave incident at θ_1, the angle of incidence measured from the vertical, will experience a reflecting surface which is locally at some angle other than the horizontal. The wave will reflect normally off this local surface, where the angles of incidence and reflection θ_2 relative to the *local* normal are equal, so that the reflected wave will not be at an angle θ_1 to the vertical (Figure 3.13). One way to consider this is to assume that such waves are lost from the summation ΣP, and the formulation can be treated by reducing the strength of the image [28]. The surface roughness can be characterised by the parameter $4L_{wave}\cos\theta_I/\lambda$: if this is less than unity the sea is taken as smooth, and if greater than unity is taken to be rough. L_{wave} is the average height of surface waves, measured peak to trough. As this becomes grazing (that is, $\theta_I \to \pm\pi/2$), surface

---

[14] *Damping*, which refers to the loss of energy from the oscillator, is quite different from *attenuation* which is the loss of energy from the wave as it propagates through the medium.
[15] See section 3.7.2.
[16] See section 3.8.

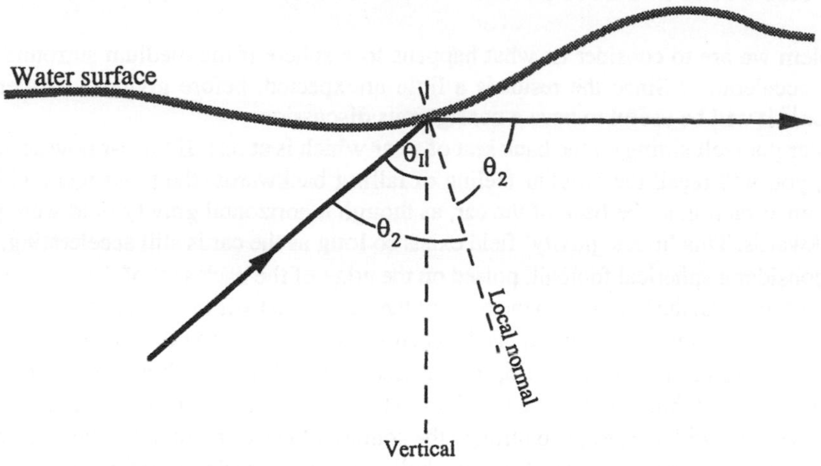

Figure 3.13 Reflection with respect to the local normal to a water surface. If the surface is not flat, this local normal may differ from the vertical.

inhomogeneities and ray bending due to sound speed variations can destroy interference effects [28].

Longuet-Higgins [29] discusses the case of a source situated beneath a plane surface which contains a hemispherical indentation directly above the source, so that there is axial symmetry. This models the geometry for a radiating bubble entrained through drop impact (Figure 3.42). He finds that when the bubble is situated close to the base of the indentation, the dipole field is three times the strength it would be if the bubble were the same distance below a planar surface. Longuet-Higgins obtains the form for the far-field, and for the oscillation frequency which he finds to be in agreement with the results of Oğuz and Prosperetti [30], who model a less simple indentation. The change in frequency due to the presence of the hemispherical indentation is found always to be less than the change in frequency that results from the proximity of a planar surface.

Obviously, the second bubble of the pair discussed in the above analysis need not be an image, but could instead be a real bubble. However, to give fields of the type described by equations (3.142) and (3.143) the two bubbles would have to pulsate at the the same frequency, and would have to be locked together in antiphase or phase respectively. Two real bubbles, oscillating freely, are unlikely to have such closely matched outputs, and so the field will be more complex than that described above. However, if the bubbles are *forced*, they will pulsate at the driving frequency, and have a definite phase relation. This situation is discussed in Chapter 4, section 4.4.1.

We have discussed the image source that arises when a bubble is close to a boundary, and seen the acoustic fields that result. There is a further consideration to make: the real bubble is now experiencing the sound field radiated by the image. This can lead to forces on the bubble which will now be examined.

(d) The Force Between a Bubble and a Plane Boundary

The problem we are to consider is: what happens to a sphere if the medium surrounding the sphere is accelerated? Since the result is a little unexpected, before giving a mathematical description[17] it will be useful to have a nonrigorous discussion.

Consider yourself sitting on the back seat of a car which is at rest. If the car now accelerates forwards, you will recall the familiar feeling of falling backwards: the passengers within the car 'feel' an attraction to the back of the car, as though a horizontal gravity field were pulling them backwards. This 'mock gravity' field exists so long as the car is still accelerating.

Now consider a spherical football, poised on the edge of the back seat of the car. As the car accelerates forwards, the ball rolls to the back of the car: it too feels the mock gravity field. The ball only *appears* to move backwards, as the occupants view it in relation to the motion of the car itself. Relative to the outside world, the ball accelerates forwards, but at a lesser rate than the car and the air within the car. The translational motion of the car is transmitted to the ball through its contact with the seat, and through the contact of the air around it: if both air and seat are accelerating forwards, they will force the ball to do so. Clearly the effect of the air is the lesser one, but for illustrative purposes it is very important, and if the ball were falling in the car, in contact with only the air, it would be the only horizontal force. To see how the air exerts a force upon the ball, we can appeal again to the horizontal mock gravity field. This will tend to make the air towards the back of the car denser and at higher pressure than that at the front: mountaineers forced to carry their own oxygen supplies will be well-acquainted with this continuous decrease in air density and pressure as they move up greater distances in the planetary gravitational field. In the mock gravity field in the car, the air pressure against the rear of the ball will be higher than that against the front, and the pressure imbalance will generate a slight force that will force the ball to move forwards.

But what would happen to a lighter-than-air helium balloon which a child sitting on the back seat is holding by a string? When the car is stationary, the helium balloon will float vertically upwards, trying to float off in the direction *opposite* to the Earth's gravitational field. Thus, when the car accelerates, the balloon will move in the direction opposite to the mock gravity field: it moves forwards. Thus objects which are heavier than air, like the football, lag behind the air in the car as the car accelerates (i.e. they 'fall' in the mock field), whilst lighter-than-air objects accelerate more than the air which surrounds them (they 'rise' in the mock field). An object with the same average density as air (like a dirigible), which would float neither up nor down in a gravitational field when surrounded by air, would accelerate exactly with the car (if you could fit a dirigible in your car), neither leading nor lagging the air in the car. These bodies are illustrated in Figure 3.14, and the formulation is given in equation (4.157).

Consider an air bubble in water. The bubble is less dense than the fluid medium surrounding it, and will rise in the Earth's gravitational field, much as the helium balloon surrounded by the denser air does. Thus if the water is accelerated in a particular direction, the bubble will accelerate ahead of it in the same direction (just as the helium bubble did in the car). The less dense the bubble, the greater the acceleration.

Consider the pulsating bubble near a rigid boundary (Figure 3.15(a)). It is so close to the boundary that the phase change due to propagation distance (i.e. e^{-ikx} when $x = 2h$) is negligible ($2kh \ll 1$). Thus, since the wall displacement of the real bubble (labelled 2) oscillates as $e^{i\omega t}$, so does that of the image bubble (labelled 1). The liquid in the vicinity of the real bubble therefore accelerates as a result of the motion of the image bubble, as the second time differential of the

---

[17]See Chapter 4, section 4.4.1.

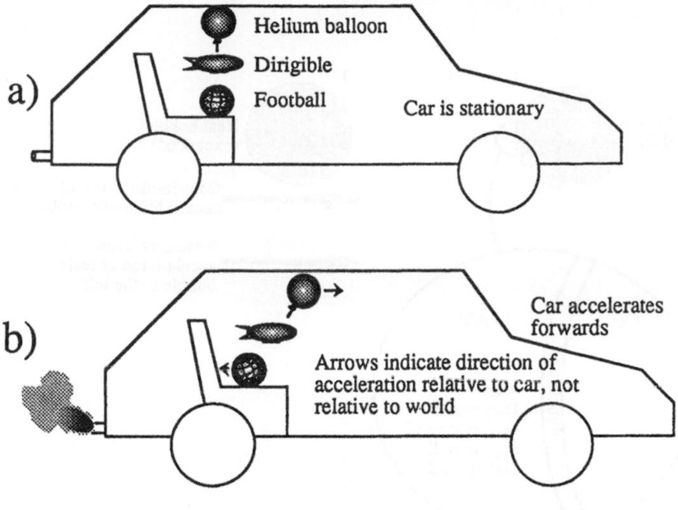

Figure 3.14 (a) A stationary car contains a football and a helium balloon, which are respectively more and less dense than the surrounding air. A dirigible has the same density as the surrounding air. (b) The car accelerates forwards. The bodies in the car all accelerate in the same direction, with respect to the outside world. The balloon accelerates more than the air within the car, the dirigible accelerates at the same rate as the air within the car, and the football accelerates forwards at a rate less than the air. Observers within the car, noting the acceleration of the contents *with respect to the car* (rather than the outside world), therefore see the balloon accelerate forwards, the dirigible remain motionless, and the football accelerate backwards with respect to the car.

displacement, $-\omega^2 e^{i\omega t}$. The liquid acceleration has the opposite sign to the bubble pulsation, equivalent to a phase difference of π. Consider the situation shown in Figure 3.15(a)(i). The image bubble, on the left, is expanded to larger than the equilibrium volume (expansion corresponds to the circled part of the wall being displaced to the right). The circled bubble wall is therefore accelerating to the left (so that if the image bubble is expanding, the rate of expansion is slowing down, and if contracting, the rate of contraction is increasing, in keeping with a pulsation mode). This causes the liquid in the region of the real bubble to accelerate to the left. The real bubble, being lighter than the fluid, will accelerate ahead of the fluid, to the left.

Conversely, when both real and image bubbles are contracted (Figure 3.15(a)(ii)), the liquid near the real bubble will accelerate to the right, and the real bubble will accelerate ahead of it to the right. However, the bubble is less dense when expanded than when contracted, since there is a fixed mass of gas within the bubble. Thus the acceleration to the left when expanded will be greater than that to the right (when the bubble is contracted). The net motion of the bubble will thus be towards the image and the rigid boundary.

If a free boundary is used (Figure 3.15(b)), then the real and image bubbles pulsate in antiphase. When the image is expanded and the liquid near the real bubble is accelerating to the left, the real bubble accelerates to the left as before (Figure 3.15(b)(i)). However, it is now in its contracted form, and so the acceleration is less great that when it accelerates to the right (when the real bubble is expanded, and the image contracted, as shown in Figure 3.15(b)(ii)). Thus the net acceleration of the real bubble will be away from the free boundary and the image bubble.

This simple explanation illustrates why pulsating bubbles tend to be attracted to rigid boundaries but repelled from free ones. During natural oscillations of the type considered in this chapter, the associated translational forces are not great. However, when the bubble is

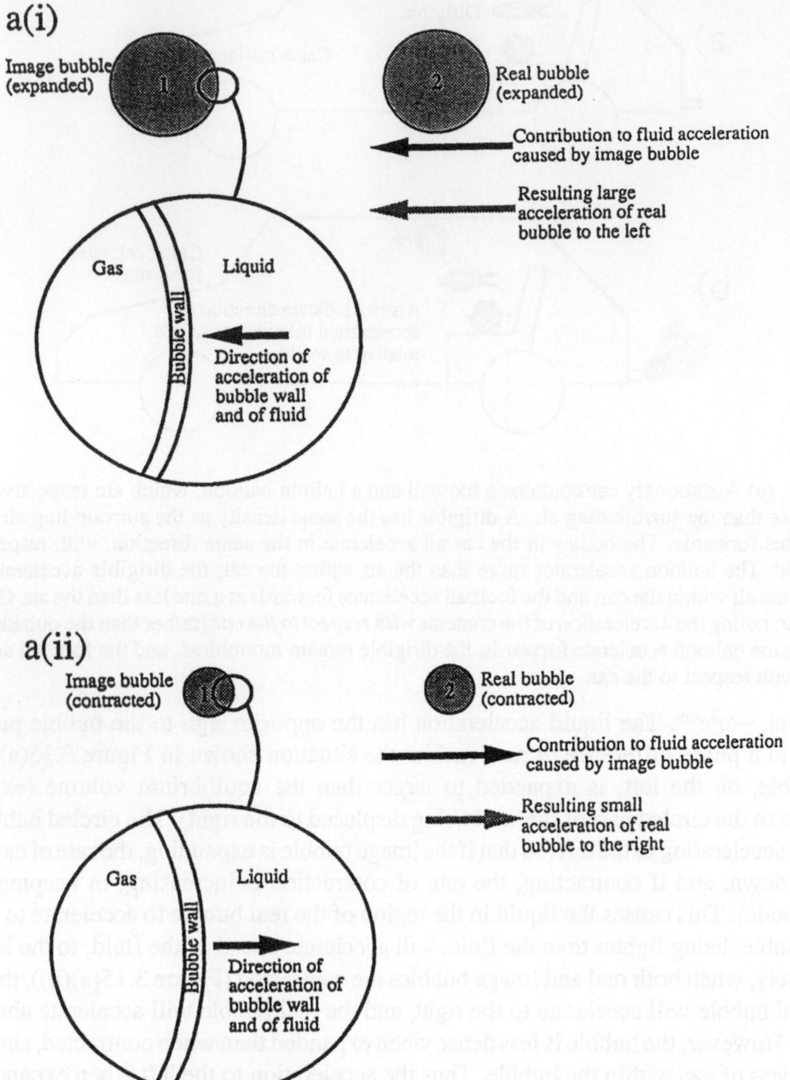

Acceleration to the left, when real bubble is expanded, is larger than the acceleration to the right that occurs when the bubble is contracted. Therefore the net acceleration of a bubble is towards a rigid boundary, or towards a second real bubble if that bubble oscillates in phase.

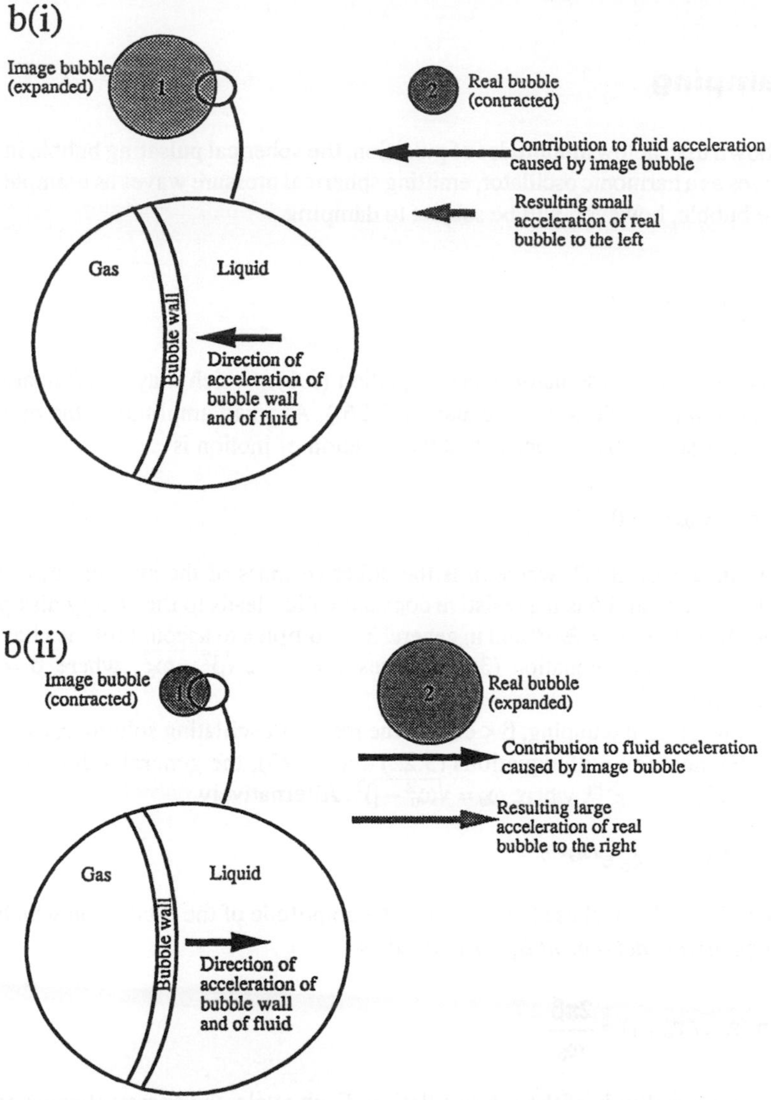

Acceleration to the right, when real bubble is expanded, is larger than the acceleration to the left that occurs when the bubble is contracted. Therefore the net acceleration of a bubble is away a free boundary, or away from a second real bubble if that bubble oscillates in antiphase.

Figure 3.15 The force between a bubble and (a) a rigid, and (b) a free boundary. The component of fluid acceleration at the real bubble which is due to the oscillations of the image bubble is found by examination of the fluid acceleration at the nearest wall of the image bubble. The real bubble accelerates in the direction of the fluid, the acceleration being greater when it is expanded than when it is contracted.

forced, the liquid accelerations are large enough to cause significant bubble translation. This phenomenon will be explored quantitatively in Chapter 4, section 4.4.1.

3.4 Damping

We have shown that, at low amplitudes of pulsation, the spherical pulsating bubble in an infinite liquid behaves as a harmonic oscillator, emitting spherical pressure waves as a simple monopole source. The bubble, however, will be subject to damping.

3.4.1 Basic Physics

The displacement of the oscillator is $\varepsilon(t)$ (equation (3.6)), which may be compared with the displacement of the bubble wall R_ε (equation (3.26)). At small amplitudes, the motion can be approximated as simple harmonic, so that the equation of motion is

$$m\ddot{\varepsilon} + b\dot{\varepsilon} + m\omega_0^2\varepsilon = 0 \tag{3.151}$$

as discussed in section 3.1.3, where m is the effective mass of the system, $\omega_0 = 2\pi\nu_0$ is the resonance frequency, and b is the resistive constant which leads to the energy dissipation. The equation has the solution $\varepsilon = \Xi\,e^{st}$ and in general Ξ is complex to account for the phase constant. Substitution of this into equation (3.151) gives $s = -\beta \pm \sqrt{\beta^2 - \omega_0^2}$, where $\beta = b/2m$ has dimensions of $[\text{time}]^{-1}$.

In conditions of light damping, $\beta < \omega_0$ and the result is oscillating solutions, as expected for the bubble. By analogy with equations (3.22) and (3.23), the general solution is therefore $\varepsilon = (\Xi_1\,e^{i\omega_b t} + \Xi_2\,e^{-i\omega_b t})\,e^{-\beta t}$, where $\omega_b = \sqrt{\omega_0^2 - \beta^2}$. Alternatively

$$\varepsilon = Xe^{-\beta t}\,e^{i(\omega_b t+\vartheta)} = \varepsilon_0\,e^{i(\omega_b t+\vartheta)} \tag{3.152}$$

where X is real and ϑ is a phase factor, and ε_0 the amplitude of the oscillation which decays as $e^{-\beta t}$. The *logarithmic decrement* $\Delta_{\log}$ is defined as

$$\Delta_{\log} = \ln(\varepsilon_{0,n} / \varepsilon_{0,n+1}) = \frac{2\pi\beta}{\omega_b} \tag{3.153}$$

where $\varepsilon_{0,n}$ is the amplitude of the nth oscillation. Each cycle, the energy changes from $\phi_{T,n}$ to $\phi_{T,n+1}$, where the ratio

$$(\phi_{T,n+1} / \phi_{T,n}) = (\varepsilon_{0,n+1} / \varepsilon_{0,n})^2 = e^{-2\Delta_{\log}} \tag{3.154}$$

Therefore the dimensionless Q-factor or *Quality*-factor, which is defined as the number of radians required for the energy to decay by e^{-1}, is

$$Q = \frac{\pi}{\Delta_{\log}} = \frac{m\omega_b}{b} = \frac{\omega_b}{2\beta} \tag{3.155}$$

from equation (3.153). Thus from equation (3.154)

$$\frac{\text{Energy lost per cycle}}{\text{Energy in system at end of cycle}} = \frac{\phi_{T,n} - \phi_{T,n+1}}{\phi_{T,n+1}} = \frac{2\pi}{Q} \tag{3.156}$$

For very light damping, $\beta \ll \omega_0$ so that $\omega_b \approx \omega_0$. Equation (3.155) therefore implies

$$Q \approx \frac{\omega_0}{2\beta} \tag{3.157}$$

where β^{-1} is the time constant of decay for amplitude. As a result, Q may also be defined from the response of the oscillator to being driven at frequencies ω not equal to the resonance. Typical response curves are shown in Figure 4.4. The frequencies on either side of the resonance where the lower is half that at resonance or, equivalently, the amplitude is $1/\sqrt{2}$ of the amplitude at resonance, are conventionally used to describe the width of the resonance. The difference between these two frequencies is called the full-width half-power bandwidth. This gives an alternative- and equivalent[18] expression for the quality factor:

$$Q = \frac{\text{Resonance frequency}}{\text{Half-power bandwidth}} \tag{3.158}$$

For damping at the resonance frequency, as discussed here, the dimensionless *damping constant* δ is defined as

$$\delta = \frac{1}{Q} = \frac{\Delta_{\log}}{\pi} \tag{3.159}$$

3.4.2 Damping of a Spherical Pulsating Bubble

The dimensionless damping constant δ can be used to describe the damping of various oscillatory parameters, such as the oscillation amplitude, or the change in bubble volume. The latter parameter was employed by Devin [31] in his discussion of the damping of bubbles at the resonance frequency. It has the advantage in that, being dimensionless, it is frame independent. It takes the same value regardless of whether it was derived from an equation of motion expressed in terms of pressure, force, radius or volume displacements. For bubbles, the damping is frequency dependent, and so the simple derivation involving Q-factors etc. given in the previous section is not strictly appropriate.

We will now derive the equation of the motion in the volume–pressure frame, in keeping with the standard papers of Devin [31] and of Eller [32]. Although the relevant stiffness and mass parameters, k_{VP} and m_{VP}^{rad} were found in section 3.2.1 by converting from k_{RF} and m_{RF}^{rad}, we will now derive them directly as an introduction to the frame. Consider the bubble discussed in section 3.2. The bubble volume is able to vary about the equilibrium V_0 as $V_1(t) = V_0 + V_\varepsilon(t)$. To find the equation of motion, we will consider the response of the bubble to an excess external pressure, which will later represent the applied acoustic pressure. The arguments will follow

[18] Equations (3.158) and (3.159) can be seen to be equivalent by considering the frequencies that are inherent in a given signal. As discussed earlier, the shorter the signal, the greater, in general, the range of frequencies present in that signal. If the amplitude of a given signal decays with time constant β^{-1}, i.e. it goes as $e^{-\beta t}$, then the power (which is proportional to the square of the amplitude, and so goes as $e^{-2\beta t}$) decays with time constant $1/2\beta$. Therefore the frequency content of the power spectrum will incorporate a spread of frequencies, of width ~2β, centred of course on the fundamental frequency (in this case, the resonance).

those of section 3.1.1, by considering not the oscillation of the system, but rather the shift in equilibrium resulting from the slow application of a small force. The force and resulting displacement will in this case be infinitesimal, allowing the formulation to be expanded to the dynamic situation.

Initially the pressure outside the bubble is p_0. If the surface tension is negligible, then p_0 equals $p_{i,e}$ (equation (2.8)), and the bubble is at equilibrium, with volume $V_i = V_0$. The pressure outside the bubble then slowly changes by an infinitesimal amount $\Delta p = P$, so that the external pressure is now $p_0 + P$. To restore the equilibrium, the bubble responds by a change in volume $\Delta V = V_\varepsilon$, so that in the absence of surface tension the pressure within the bubble increases by $\Delta p_i = P$, to equal that in the liquid. If the gas within the bubble behaves polytropically, then from equation (3.54),

$$\Delta p_i = -\frac{\kappa p_0}{V_0}\,\Delta V \qquad \Rightarrow$$

$$P = -\frac{\kappa p_0}{V_0}\,V_\varepsilon \tag{3.160}$$

This equation is analogous to equation (3.2), derived for the spring–bob system. The magnitude of the ratio of the 'driving force' to the 'displacement' equals the spring constant. Therefore

$$k_{VP} = \frac{\kappa p_0}{V_0} = \frac{3\kappa p_0}{4\pi R_0^3} \tag{3.161}$$

in agreement with equation (3.60). Assuming that only the radiation mass is significant, from equation (3.7) the stiffness in the volume–pressure frame also equals

$$k_{VP} = m_{VP}^{rad}\,\omega_0^2 \tag{3.162}$$

Equating equations (3.161) and (3.162) will give the bubble pulsation resonance within the limits of the approximations, which are specifically that surface tension is negligible and changes are slow. This will agree with equation (3.36), and equation (3.38) in adiabatic circumstances, which is not surprising since these were derived under the same set of assumptions about the inertia, and both assume negligible surface tension.

Having found the mass and spring constant by considering shifts in the equilibrium position, it is possible to write down the linear second-order differential equation by simple analogy with section 3.1.3:

$$m_{VP}^{rad}\ddot{V}_\varepsilon + b_{VP}\dot{V}_\varepsilon + k_{VP}V_\varepsilon = P_A e^{i\omega t} \tag{3.163}$$

For free oscillation, the driving force is zero:

$$m_{VP}^{rad}\ddot{V}_\varepsilon + b_{VP}\dot{V}_\varepsilon + k_{VP}V_\varepsilon = 0 \tag{3.164}$$

Through consideration of the dissipative constant, b_{VP}, the dimensionless damping constant, which is equally applicable to the damping of displacements in any frame, can be obtained. If the bubble is oscillating off-resonance at a frequency ω, in response to a driving force as described by equation (3.163), the dimensionless damping constant is $d = \omega b_{VP}/k_{VP}$. If the bubble is oscillating at the resonance frequency ω_0, for example if the driving frequency is ω_0 in equation (3.163) or during the free oscillation described by equation (3.164), the dimensionless damping constant is given by the value of d at $\omega = \omega_0$, which is given the special symbol $\delta = \omega_0 b_{VP}/k_{VP}$.

Energy losses from the bubble occur through three mechanisms:[19]

(i) Energy is radiated away from the bubble as acoustic waves (*radiation damping*).
(ii) Energy is lost through thermal conduction between the gas and the surrounding liquid (*thermal damping*).
(iii) Work is done against viscous forces at the bubble wall (*viscous damping*).

Dissipative and damping constants, b_{VP}^{rad}, d and δ, can be associated with each of these three processes, giving, for example, d_{rad}, d_{th} and d_{vis} respectively (at resonance, δ_{rad}, δ_{th} and δ_{vis}). The total damping constant of the bubble, d_{tot}, is their sum:

$$d_{tot} = d_{rad} + d_{th} + d_{vis}$$

or, at resonance,

$$\delta_{tot} = \delta_{rad} + \delta_{th} + \delta_{vis} \tag{3.165}$$

Equations (3.163) and (3.164) were derived by Devin by consideration of the Lagrangian of the system, defined as the kinetic energy of the system minus the potential energy, though for an adiabatic rather than a polytropic gas. From the equation of motion the thermal, radiation and viscous damping constants can be found. The derivation that will now be given as an introduction to the damping of resonant bubbles follows that presented by Devin in 1959. Another notable treatment of the free, damped oscillations of spherical bubble is that of Chapman and Plesset [33]. Other models will be explored in Chapter 4 in the discussion of the off-resonance damping of forced bubbles.

(a) Thermal Damping

The derivation of this damping constant is somewhat involved, and will now be outlined before a formal exposition is given.

A pulsating bubble of volume $V_1(t) = V_0 + V_\varepsilon(t)$ is considered. The pressure in the liquid, far from the bubble, is $p_\infty = p_0 + P_A e^{i\omega t}$, the sum of static and acoustic terms which oscillate at some driving frequency ω. The actual pressure at the bubble wall, p_L, differs from this as a result of the inertial reaction of the moving liquid. Initially, the effects of surface tension are neglected.

The flow of heat across the bubble wall is related to the work done on the gas space, and the internal energy of the gas, through the First Law of Thermodynamics (conservation of energy). This equation is then solved to determine the temperature profile within the bubble. Knowledge of the temperature of an element of gas enables the calculation of its expansion, and thus through integration of all gas elements, yields the volume of the bubble as a whole.

Having therefore found an expression for the volume displacement, that expression can be substituted into the equation of motion to solve for the damping and the stiffness. These parameters are formulated not as a function of the equilibrium bubble radius R_0, but of the ratio of R_0 to l_D, the thermal boundary layer thickness, and surface tension is not included. For a complete expression of the solution, surface tension must be incorporated, and this is formulated as a function of R_0. To ally the surface tension term with the previous formulation, the precise relationship between R_0 and l_D must be known. Devin applies an approximate relationship valid for large air bubbles in water at resonance to find expressions for the stiffness, damping constant

[19] Other mechanisms may apply (see, for example, section 3.8.2(a)). However for the damping of a *spherical* bubble, it is usual to consider only these three.

and resonance frequency. These expressions are not, however, independent in their derivation, and for a full solution a method of converging successive approximations must be used. Results from this are presented for comparison with Devin's solutions.

We shall therefore begin by assessing the thermodynamics of the pulsation, which enable the temperature profile within the bubble to be derived, and from that the pulsation amplitude and the damping parameters.

As the volume of the bubble changes, the pressure and temperature of an ideal gas inside will change in accordance with PV^{κ} = constant and $TV^{\kappa-1}$ = constant,[20] where κ is as yet not specified. If there were no heat conduction across the bubble wall, then the process would be adiabatic ($\kappa = \gamma$). If heat conduction between the gas and the infinite fluid reservoir could occur without limit in time or magnitude, then the process would be isothermal ($\kappa = 1$), in which case as much heat would flow into the bubble on expansion as leaves on compression.

For a real gas bubble in a liquid, the gas next to the bubble wall behaves isothermally because of the high thermal conductivity and specific heat of the liquid, whilst towards the centre of the bubble the gas behaves adiabatically. As a result, the polytropic index for the bubble as a whole takes an intermediate value between γ and unity. As the bubble pulsates, there is a hysteresis effect, the driving pressure doing more work in compressing the bubble than the gas in the bubble does in moving the liquid on expansion. There is therefore a net flow of heat from the bubble into the liquid, and this loss of energy represents thermal damping.

Pressure at Bubble Surface. Devin follows the derivation of Pfriem [34] in assuming that the bubble is driven when a liquid of hydrostatic pressure p_0 is subjected to an acoustic pressure $P_A e^{i\omega t}$, so that the pressure far from the bubble is

$$p_\infty = p_0 + P_A e^{i\omega t} \tag{3.166}$$

The pressure in the liquid at the bubble wall, p_L, will be the sum of the static pressure, p_0, and time-varying terms of frequency ω and having an amplitude $P_{A,L}$:

$$p_L = p_0 + P_{A,L} e^{i\omega t} \tag{3.167}$$

The time-varying pressure at the bubble surface, $P_{A,L}$, will be the acoustic pressure minus the inertial reaction of the moving liquid, the inertia of the actual gas within the bubble being negligible:

$$P_{A,L} e^{i\omega t} = P_A e^{i\omega t} + m_{VP}^{rad} \ddot{V}_\varepsilon \tag{3.168}$$

The final term is in fact the acoustic pressure field radiated by the bubble, evaluated at the bubble wall. The following assumptions are made. The bubble responds with small-amplitude pulsations and the wavelength is much larger than the bubble. The liquid has high specific heat capacity and thermal conductivity: it behaves as a reservoir of constant temperature, no matter what heat flows in or out of it, and the temperature of the liquid adjacent to the bubble wall is invariant. Inside the bubble, the gas is assumed to have constant specific heat. The pressure varies only in time, being at a given instant uniform throughout the bubble. Heat transfer occurs only through conduction, there being not enough time for convection. Elimination of convection from the formulation means that the assumption of spherical symmetry is still valid.

The basis for discussion of these processes is the First Law of Thermodynamics, which states that

---

[20] See equations (1.18) and (1.f3), which can be found in footnote 13 of Chapter 1.

$$\Delta U = \Delta q_H + \Delta W \tag{3.169}$$

where ΔU, Δq_H and ΔW are, respectively, the increase in internal energy of, the heat added to, and the work done on, a unit volume of gas within the bubble. Therefore in the limit of infinitesimal changes with respect to time

$$\frac{dU}{dt} = \frac{dq_H}{dt} + \frac{dW}{dt} \tag{3.170}$$

Each of these three quantities will now be evaluated, then combined in equation (3.170) to provide the linear differential equation describing the temperature distribution within the bubble.

Internal Energy. Firstly, if C_v is the specific heat capacity of the gas at constant volume, then a unit volume of gas has a constant-volume heat capacity of $\rho_1 C_v$. Therefore

$$\frac{dU}{dt} = \rho_1 C_v \frac{d\Theta}{dt} \tag{3.171}$$

Θ being the change in temperature of the bubble gas from T_o, the absolute gas temperature at equilibrium, such that $\Theta = T - T_o$.

Heat. From the definition of the thermal conductivity, K_g, of gas within bubble

$$\frac{dq_H}{dt} = K_g \nabla^2 \Theta \tag{3.172}$$

which for our assumed spherical symmetry reduces to

$$\frac{dq_H}{dt} = \frac{K_g}{r}\frac{\partial^2(r\Theta)}{\partial r^2} \tag{3.173}$$

Work. The work done on a gas volume element V' is typically $-p_L \partial V'$, so that the rate of working by the liquid per unit volume of gas is

$$\frac{dW}{dt} = -\frac{p_L}{V'}\frac{\partial V'}{\partial t} \tag{3.174}$$

This quantity can be evaluated in terms of the liquid temperature through the ideal gas equation. If the volume element V' contains N_m moles of gas (total mass $\rho_1 V'$), then from equation (1.15)

$$p_L V' = N_m R_g (T_o + \Theta) \tag{3.175}$$

The ideal gas constant R_g equals the difference between the *molar* heat capacities at constant pressure and at constant volume. Since the N_m moles of gas in volume V' have total mass $\rho_1 V'$, then the ratio of specific to molar heats is $\rho_1 V'/N_m$. Therefore the ideal gas constant is given by

$$R_g = (C_p - C_v)\rho_1 V'/N_m \tag{3.176}$$

Substitution of (3.176) into (3.175) gives

$$p_L V' = (C_p - C_v)\rho_1 V'(T_o + \Theta) \tag{3.177}$$

If the mass of gas within any volume element is constant, then the product $\rho_1 V'$ is time-independent. Differentiation of equation (3.177) with respect to time therefore gives

$$i\omega P_{A,L}\, e^{i\omega t} V' + p_L \frac{\partial V'}{\partial t} = \frac{(C_p - C_v)\rho_1 V'}{r} \frac{\partial (r\Theta)}{\partial t} \qquad (3.178)$$

Substitution of equation (3.178) into equation (3.174) yields

$$\frac{dW}{dt} = -\frac{p_L}{V'}\frac{\partial V'}{\partial t} = -\frac{(C_p - C_v)\rho_1}{r}\frac{\partial (r\Theta)}{\partial t} + i\omega P_{A,L}\, e^{i\omega t} \qquad (3.179)$$

Temperature Profile. Now that we have expressions for the rate of change of internal energy, heat flow and working, we may substitute equations (3.171), (3.173) and (3.179) into equation (3.170) to give the equation that describes the temperature profile within the bubble:

$$\frac{\partial (r\Theta)}{\partial t} = D_g \frac{\partial^2 (r\Theta)}{\partial r^2} + \frac{i\omega r}{\rho_1 C_p} P_{A,L}\, e^{i\omega t} \qquad (3.180)$$

where

$$D_g = \frac{K_g}{\rho_1 C_p} \qquad (3.181)$$

is the thermal diffusivity of the gas within the bubble. Devin presents the following approximate solution to equation (3.180) for the temperature change within the bubble:

$$\Theta \approx P_{A,L}\left(\frac{R_o}{\rho_1 C_p}\right)\left(\frac{r}{R_o} - \frac{\sinh(r\sqrt{i\omega/D_g})}{\sinh(R_o\sqrt{i\omega/D_g})}\right)\left(\frac{1}{r}\, e^{i\omega t}\right) \qquad (r \leqslant R_o) \qquad (3.182)$$

The Volume Displacement. From equation (3.182), the temperature of each shell of gas at radius r within the bubble is known. The conditions within a given shell can be related to the equilibrium conditions through the ideal gas law, and so the volume of the shells can be integrated to give the total bubble volume. The difference between the latter and the equilibrium bubble volume gives the volume displacement V_ε in the form:

$$V_\varepsilon = P_{A,L} \frac{e^{i\omega t}}{\gamma p_o}\left(1 + \frac{3(\gamma - 1)}{i\omega R_o^2/D_g}\left(\sqrt{\frac{i\omega}{D_g}}\, R_o \coth(R_o\sqrt{i\omega/D_g}) - 1\right)\right) \qquad (3.183)$$

Now that the volume displacement is known in terms of the assumed harmonic excitation pressure at the bubble surface, it is possible to use it explicitly in equation (3.183) to find the damping characteristics, and the stiffness of the bubble, which has previously only been expressed in terms of the unknown quantity κ.

Thermal Damping Coefficient. Substitution of equation (3.168) into equation (3.163) yields the equation of motion of the bubble system, rewritten in terms of the gas space alone:

$$b_{VP}^{th}\, \dot{V}_\varepsilon + k_{VP} V_\varepsilon = -P_{A,L}\, e^{i\omega t} \qquad (3.184)$$

where b_{VP}^{th} is the additive thermal component of the dissipative constant in the volume–pressure frame. This equation now relates the properties of the gas to the time-varying component of the liquid pressure at the bubble wall. Equation (3.183) for V_ε can be combined with equation (3.184) to give a formulation for the complex entity $(k_{VP} + i\omega b_{VP}^{th})^{-1}$. This can be separated into real and imaginary parts in the usual manner by multiplication top-and-bottom with the conjugate form $(k_{VP} - i\omega b_{VP}^{th})$, to give

$$\frac{k_{VP} - i\omega b_{VP}^{th}}{(k_{VP})^2 + (\omega b_{VP}^{th})^2} = \frac{V_0}{\gamma p_0}\left[1 + \frac{3(\gamma-1)}{R_0/l_D}\left\{\frac{\sinh(R_0/l_D) - \sin(R_0/l_D)}{\cosh(R_0/l_D) - \cos(R_0/l_D)}\right.\right.$$
$$\left.\left. - \, i\left(\frac{\sinh(R_0/l_D) + \sin(R_0/l_D)}{\cosh(R_0/l_D) - \cos(R_0/l_D)} - \frac{1}{R_0/2l_D}\right)\right\}\right] \quad (3.185)$$

(there is a typographical error in the expression given by Devin). Equating the real terms in equation (3.185) will give the stiffness k_{VP}, whilst equating the imaginary terms will give the dimensionless thermal damping constant $d_{th} = \omega b_{VP}^{th} / k_{VP}$. With the use of the approximation $(\omega b_{VP}^{th} / k_{VP})^2 \ll 1$ (at resonance $d_{th} = \omega_0 b_{VP}^{th} / k_{VP} \sim 0.1$), the imaginary terms give the dimensionless constant for thermal damping:

$$d_{th} = \frac{\omega b_{VP}^{th}}{k_{VP}} \approx \frac{\left(\dfrac{\sinh(R_0/l_D) + \sin(R_0/l_D)}{\cosh(R_0/l_D) - \cos(R_0/l_D)} - \dfrac{1}{R_0/2l_D}\right)}{\left(\dfrac{R_0/l_D}{3(\gamma-1)} + \dfrac{\sinh(R_0/l_D) - \sin(R_0/l_D)}{\cosh(R_0/l_D) - \cos(R_0/l_D)}\right)} \quad (3.186)$$

where $l_D = \sqrt{D_g/2\omega}$ represents the thickness of a thermal boundary layer in the bubble. The parameter R_0/l_D is obviously critical to the value of δ_{th}. For comparatively large values

$$d_{th} \approx \left(1 - \frac{2l_D}{R_0}\right)\left(1 + \frac{R_0}{l_D}\frac{1}{3(\gamma-1)}\right)^{-1} \quad (3.187)$$

accurate to within 1% for $R_0/l_D \geqslant 5$, which in the limit becomes

$$d_{th} \approx \frac{3(\gamma-1)}{R_0/l_D} \quad (3.188)$$

This condition corresponds to large bubbles, where the surface area to volume ratio is small and the bubble radius is large compared with the thermal diffusivity of the gas. Heat flow is restricted, and the situation tends to the adiabatic. Note that as R_0/l_D becomes larger in this limit, δ_{th} becomes small. Therefore in the limit of very large bubbles, the thermal dissipation tends to zero.

At the other end of the bubble size scale ($R_0/l_D \leqslant 2$), equation (3.186) reduces, with accuracy to within 1%, to

$$d_{th} \approx \frac{(\gamma-1)}{\gamma}\frac{R_0^2\omega}{15D_g} = \frac{1}{30}\frac{(\gamma-1)}{\gamma}\left(\frac{R_0}{l_D}\right)^2 \quad (3.189)$$

The condition $R_0/l_D \leqslant 2$ corresponds to small bubbles and isothermal conditions. Note that as R_0/l_D becomes smaller, so in this limit does δ_{th}. In other words, as R_0 tends to zero, so does the thermal dissipation.

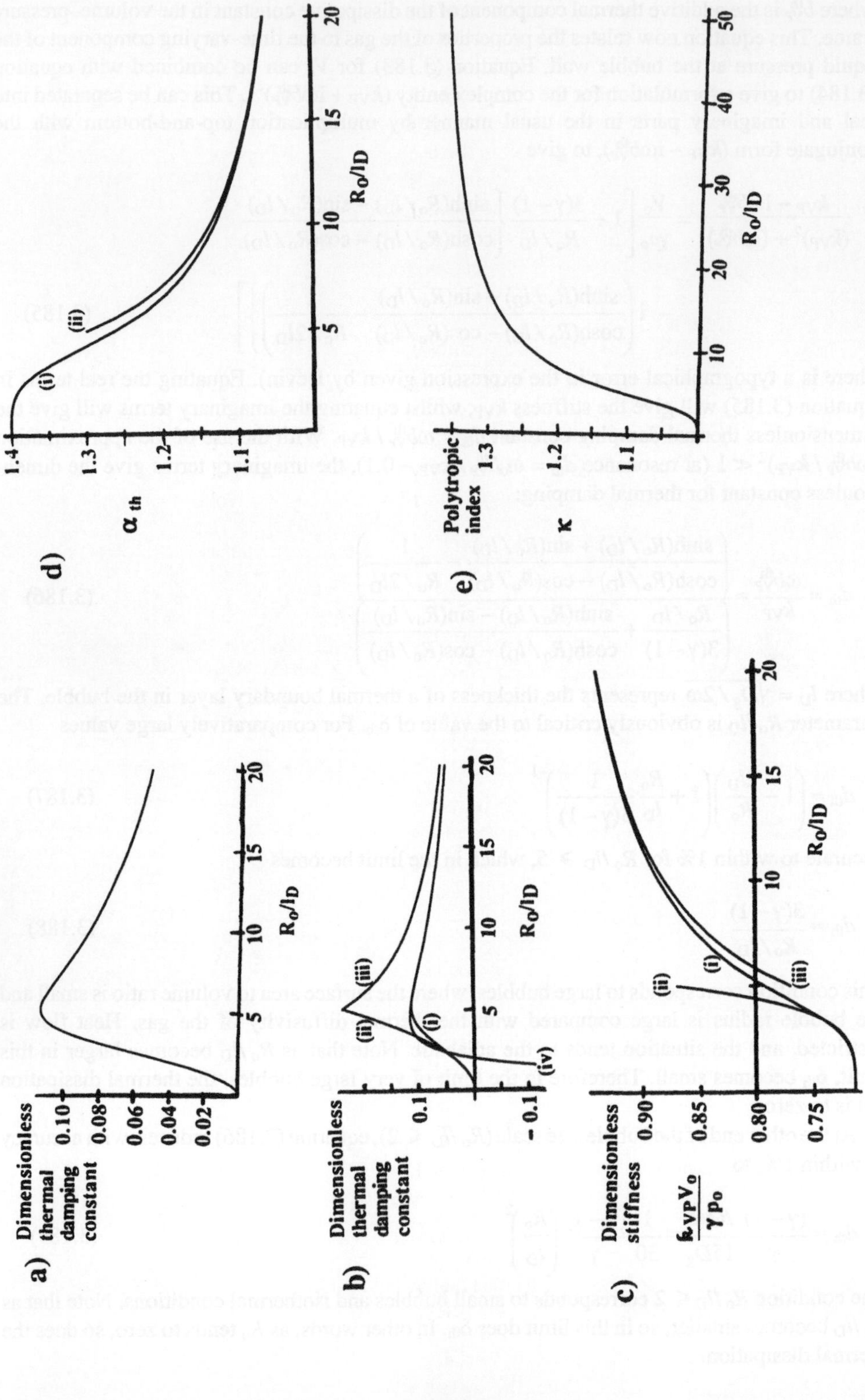

a) Dimensionless thermal damping constant
0.02
0.04
0.06
0.08
0.10
5
10
15
20
Ro/lD
b) Dimensionless thermal damping constant
0.1
− 0.1
(i)
(ii)
(iii)
(iv)
5
10
15
20
Ro/lD
c) Dimensionless stiffness
κVP Vo / γ po
0.75
0.80
0.85
0.90
(i)
(ii)
(iii)
5
10
15
20
Ro/lD
d) α th
1.1
1.2
1.3
1.4
(i)
(ii)
5
10
15
20
Ro/lD
e) Polytropic index κ
1.1
1.2
1.3
10
20
30
40
50
Ro/lD

It is clear therefore that thermal dissipation is small in both very large and very small bubbles, as can be seen in Figure 3.16(a), which plots equation (3.186). In Figure 3.16(b) it is again plotted, together with the approximate forms given by equations (3.169) for $0 \leqslant R_0/l_D \leqslant 5$, and by equations (3.187) and (3.188) for $5 \leqslant R_0/l_D \leqslant 20$. As $R_0/l_D \to 0$ the process is isothermal, and as $R_0/l_D \to \infty$ it is adiabatic, and thermal dissipation is small, the dimensionless damping constant tending to zero. It is clear that in the transition region between the two regimes there is a maximum in thermal dissipation (occurring at around $R_0/l_D \approx 5.5$). A substantial net flow of heat into the liquid does occur, and the gas behaves polytropically.

Stiffness. The stiffness of the bubble can be found by equating the real terms in equation (3.185):

$$\frac{\gamma p_0}{V_0 k_{VP}} = \left\{1 + \left(\frac{\omega b_{VP}^{th}}{k_{VP}}\right)^2\right\}\left[1 + \frac{3(\gamma - 1)}{R_0/l_D}\left\{\frac{\sinh(R_0/l_D) - \sin(R_0/l_D)}{\cosh(R_0/l_D) - \cos(R_0/l_D)}\right\}\right] \tag{3.190}$$

The first bracket equals $(1 + d_{th}^2)$, which tends to unity in the adiabatic and isothermal limits, as R_0/l_D becomes respectively very large or very small. In the adiabatic case of large bubbles ($R_0 \gg l_D$), equation (3.190) gives the *adiabatic stiffness* k_{VP}^{ad} for the volume–pressure frame

$$k_{VP}^{ad}\frac{V_0}{\gamma p_0} \approx \left\{1 + \frac{3(\gamma - 1)}{R_0/l_D}\left(1 + \frac{3(\gamma - 1)}{R_0/l_D}\right)\right\}^{-1} \tag{3.191}$$

In the isothermal limit ($R_0 \ll l_D$), equation (3.191) reduces to give the *isothermal stiffness* k_{VP}^{iso}

$$k_{VP}^{iso}\frac{V_0}{\gamma p_0} \approx \left\{\gamma - \frac{(R_0/l_D)^4}{1890}\left(1 - \frac{2.1(\gamma - 1)^2}{\gamma}\right)\right\}^{-1} \tag{3.192}$$

The terms on the left of equations (3.191) and (3.192) are known as the 'dimensionless stiffnesses'. These limiting forms for air ($\gamma = 1.4$) are plotted, along with the actual value of the dimensionless stiffness ($V_0 k_{VP}/\gamma p_0$) calculated by reciprocating equation (3.190), in Figure 3.16(c). In the limits of small and large bubbles, the dimensionless stiffness approaches γ^{-1} and unity respectively.

(b) The Resonance Frequency

(i) Effect of Heat Conduction. Damping reduces the resonance frequency of a system from ω_0 to ω_b, as expounded in section 3.1.3. In order explicitly to determine the thermal damping

---

Figure 3.16 Graphs of thermal parameters as a function of the ratio of the equilibrium bubble radius to the width of the thermal boundary layer in the gas. (a) Dimensionless thermal damping constant, as given by equation (3.186). In (b), equation (3.186), shown again as plot (b)(i), is compared with approximate forms of the dimensionless thermal damping constant: (b)(ii) shows the plot of equation (3.169) for $0 \leqslant R_0/l_D \leqslant 5$; whilst approximate forms for $5 \leqslant R_0/l_D \leqslant 20$ are given by (b)(iii) equation (3.188) and (b)(iv) equation (3.187). In (c)(i) the dimensionless stiffness is shown (equation (3.190)), with approximate forms (c)(ii) for small bubbles in the isothermal limit (equation (3.192)), and (c)(iii) for large bubbles in the adiabatic limit (equation (3.191)). The parameter α_{th} is plotted in (d)(i) using equation (3.194), with the approximate equation (3.195) shown in (d)(ii). In (e) the polytropic index, as found by combining equations (3.194) and (3.197), is shown, for $\gamma = 1.4$.

constant at resonance, the modified resonance must be calculated. This is done for the bubble through knowledge of the dependency of the stiffness on the heat flow.

In general, of course, the volume stiffness coefficient, from equation (3.190), is

$$k_{VP} = -\frac{\partial P}{\partial V_1} = \frac{\kappa p_0}{V_0} = \frac{\gamma}{\alpha_{th}} \frac{p_0}{V_0} \tag{3.193}$$

where

$$\alpha_{th} = \left\{1 + \left(\frac{\omega b_{VP}^{th}}{k_{VP}}\right)^2\right\}\left[1 + \frac{3(\gamma-1)}{R_0/l_D}\left\{\frac{\sinh(R_0/l_D) - \sin(R_0/l_D)}{\cosh(R_0/l_D) - \cos(R_0/l_D)}\right\}\right] \tag{3.194}$$

is the dimensionless multiplicative constant which accounts for the effect of heat conduction from the bubble. Devin employs the approximate form of this, relevant for $R_0 \gg l_D$, as found from equation (3.191):

$$\alpha_{th} = \left\{1 + \frac{3(\gamma-1)}{R_0/l_D}\left(1 + \frac{3(\gamma-1)}{R_0/l_D}\right)\right\} \tag{3.195}$$

The parameter α_{th} given by equation (3.194) is plotted for air with the approximate form in Figure 3.16(d). It tends to γ in the isothermal limit, and to unity in the adiabatic case.

The resonance frequency is found from the root of the ratio of the stiffness to the effective mass appropriate for the frame, $m_{VP}^{rad} = \rho/(4\pi R_0)$, so that $\omega_0 = \sqrt{k_{VP}/m_{VP}^{rad}}$. Substituting $V_0 = 4\pi R_0^3/3$ into equation (3.193) therefore gives

$$\omega_0 = \frac{1}{R_0}\sqrt{\frac{3\gamma p_0}{\rho \alpha_{th}}} \tag{3.196}$$

It should be remembered that the formulation so far accounts only for heat conduction, and that the effect of surface tension on the resonance frequency has not yet been incorporated. However, equation (3.196) can therefore be directly compared with equation (3.36), which employed the unknown polytropic index to account for heat conduction and also made no allowance for surface tension effects. Such comparison shows that

$$\kappa = \frac{\gamma}{\alpha_{th}} \tag{3.197}$$

The polytropic index for air, as found from equation (3.197), is plotted in Figure 3.16(e) as a function of R_0/l_D. The trend is clear: κ tends to unity in the isothermal limit ($R_0/l_D \to 0$) and γ in the adiabatic one ($R_0/l_D \to \infty$). The effect of thermal conduction has thus been formulated as a function of R_0/l_D. It now remains to incorporate surface tension into equations (3.193) and (3.196), the expressions for k_{VP} and ω_0.

(ii) Effect of Surface Tension. If the effect of surface tension is included in addition to the thermal conduction, then we obtain an expression for the actual stiffness of the bubble in the volume–pressure frame, k_{VP}, defined as

$$k_{VP} = -\frac{\partial p_L}{\partial V} \tag{3.198}$$

So far in this section (3.4.2) it has been assumed that the instantaneous gas pressure within the bubble is equal to the liquid pressure at the bubble wall. The surface tension contributes a pressure, however, so that the pressure within the bubble is given by

$$p_i(t) = p_L(t) + \frac{2\sigma}{R(t)} \tag{3.199}$$

where p_L has static and time-varying terms as expounded in equations (3.167) and (3.168). Equation (3.199), relating as it does to the dynamic situation, is analogous to the quasi-static case derived in Chapter 2, section 2.1.1, equation (2.8). The polytropic equation of state will relate the instantaneous conditions to those at equilibrium which, assuming the vapour pressure term in equation (2.17) is negligible, gives

$$p_i(t) = \left(p_o + \frac{2\sigma}{R_o}\right)\left(\frac{V_o}{V(t)}\right)^{\kappa} \qquad \text{where } \kappa = \gamma/\alpha_{th} \tag{3.200}$$

Elimination of $p_i(t)$ through equations (3.199) and (3.200), followed by substitution into equation (3.198), gives

$$k_{VP} = \frac{\gamma p_o}{V_o}\frac{g}{\alpha_{th}} \tag{3.201}$$

where

$$g = 1 + \frac{2\sigma}{p_o R_o} - \frac{2\sigma}{3\kappa p_o R_o} \tag{3.202}$$

is the dimensionless multiplicative constant which accounts for the effect of surface tension on the bubble stiffness. The full expression for the bubble resonance frequency, taking into account the effects of thermal conduction (through α_{th}) and surface tension (through g), is therefore

$$\omega_o = \frac{1}{R_o}\sqrt{\frac{3\gamma p_o}{\rho}}\sqrt{\frac{g}{\alpha_{th}}} = \omega_M\sqrt{\frac{g}{\alpha_{th}}} \tag{3.203}$$

where, as usual, p_o is the static pressure in the liquid. This value of the resonance frequency departs from Minnaert's adiabatic derivation, where surface tension was neglected, in that $\omega_o/\omega_M = \sqrt{g/\alpha_{th}}$. Substitution of equations (2.14), (3.197) and (3.202) into equation (3.203) gives

$$\omega_o = \frac{1}{R_o}\sqrt{\frac{3\gamma p_o}{\rho}\left(1 + \frac{2\sigma}{p_o R_o}\right) - \frac{2\sigma}{\rho R_o}} \quad \Rightarrow$$

$$\omega_o = \frac{1}{R_o}\sqrt{\frac{3\gamma p_{i,e}}{\rho} - \frac{2\sigma}{\rho R_o}} \tag{3.204}$$

where $p_{i,e}$ is the gas pressure in the bubble at equilibrium. As discussed in section 3.4.2(b)(i), the pulsations of large bubbles tend to be adiabatic. In addition, the surface tension pressure is small in such large bubbles (Chapter 2, section 2.1.1). Therefore Minnaert, who assumed adiabatic conditions and negligible surface tension, obtained good agreement between experiment and his theory (equation (3.38)) for the resonance frequencies of the relatively large bubbles he studied.

Equation (3.202) incorporates R_0 explicitly, and also R_0/l_D implicitly through κ. Therefore to evaluate g and ω_0, one must first relate R_0 to l_D, the thermal boundary layer thickness. The latter equals $\sqrt{D_g/2\omega}$, and so depends on the pulsation frequency and, through D_g, on the nature of the gas (equation (3.181)). Devin chose to relate R_0 to l_D by adopting a particular gas (air) to fix D_g, and a particular frequency, setting ω equal to the pulsation resonance ω_0. Since this uses an approximate ω_0/R_0 to relate R_0 to l_D in order to find ω_0 explicitly as a function of R_0, the formulation is not independent. This problem is discussed in the following section.

(iii) The Thermal Damping Constant at Resonance. At resonance ($\omega = \omega_0$), the critical parameter R_0/l_D is given by

$$\frac{R_0}{l_D} = R_0\sqrt{\frac{2\omega_0}{D_g}} \tag{3.205}$$

Substitution for $\omega_0 = 2\pi\nu_0$ from equation (3.203) gives

$$\frac{R_0}{l_D} = \sqrt{\frac{4G_{th}}{\nu_0}}\sqrt{\frac{g}{\alpha_{th}}} \tag{3.206}$$

where

$$G_{th} = \frac{3\gamma p_0}{4\pi\rho D_g} \tag{3.207}$$

The thermal damping constant at resonance, δ_{th}, can therefore be found by substitution of equation (3.206) into equation (3.186). However, to obtain the value of δ_{th} for a bubble of given radius R_0 from this, one needs to evaluate α_{th} in equation (3.206), for which the value of δ_{th} is required. This lack of independence means that successive approximations must be used if the bubble is to be characterised by R_0 alone, rather than R_0/l_D. This is done to obtain the dimensionless thermal damping constant at resonance in Figure 3.20.

Devin substitutes for R_0/l_D into equation (3.187) to obtain an approximate answer for the damping constant at resonance, accurate when $R_0 \gg l_D$. This is in fact appropriate for resonating air bubble in water under one atmosphere of static pressure, since

$$\frac{R_0}{l_D} = \sqrt{\frac{R_0^2.2\omega_0}{D_{air}}} \approx \sqrt{\frac{2R_0}{D_{air}}2\pi\,(3\text{ Hz . metres})} \tag{3.208}$$

Taking the thermal diffusivity of air, D_{air}, to equal 2.08×10^{-5} m^2 s^{-1}, equation (3.208) reduces to

$$\frac{R_0}{l_D} \approx 1373\sqrt{\frac{R_0}{\text{metres}}} \tag{3.209}$$

so justifying Devin's approximation for bubbles with radii of greater than around 10 μm. However, Devin makes a further approximation in this step of assuming that D_g is a constant, which will later be shown to break down for the smallest bubble sizes.

Devin's approximations can usefully be incorporated to plot the variation in the relevant thermodynamic parameters with equilibrium radius of air bubbles in water at resonance. It should be remembered that since equation (3.209) is used to relate R_o to l_D, and since D_g is assumed constant, the accuracy decreases with decreasing bubble radius. Nevertheless the plots serve to illustrate the trends well. Figure 3.17(a) shows the variation of g with bubble radius, and in Figure 3.17(b) the ratio $\sqrt{g/\alpha_{th}} = (\omega_o/\omega_M)$ is shown. Large bubbles behave adiabatically, and surface tension is negligible. As the equilibrium bubble radius decreases, the resonance frequency decreases slightly below ω_M as a result of thermal conduction. The character of $(1/\sqrt{\alpha_{th}}) \propto \sqrt{\kappa}$, as plotted in Figures 3.16(d) and 3.16(e), dominates. As R_o decreases further to below 10 μm, the effect of surface tension causes the resonance frequency to rise dramatically above ω_M. The factor $\sqrt{g}$, as indicated by Figure 3.17(a), dominates. This effect can be seen in Figure 3.18, where ω_o is compared with plots of equation (3.36) with $\kappa = 1$ (isothermal) and $\kappa = \gamma$ (adiabatic, which then gives ω_M). For large values of R_o ($\gtrsim$ 0.1 mm) the plot of ω_o is indistinguishable from that of ω_M. At $R_o \approx 10$ μm, the value of ω_o is intermediate between the adiabatic and isothermal forms of equation (3.36), and the gas is behaving polytropically. In Figure 3.18(b) the subsequent increase in ω_o at smaller radii is in evidence. The resonance is following a multiplication of the root of Figure 3.17(b) with the ω_M (adiabatic) line in Figure 3.18.

With his large-bubble approximation for R_o/l_D, Devin finds the dimensionless thermal damping constant for the resonant pulsating air bubble to be

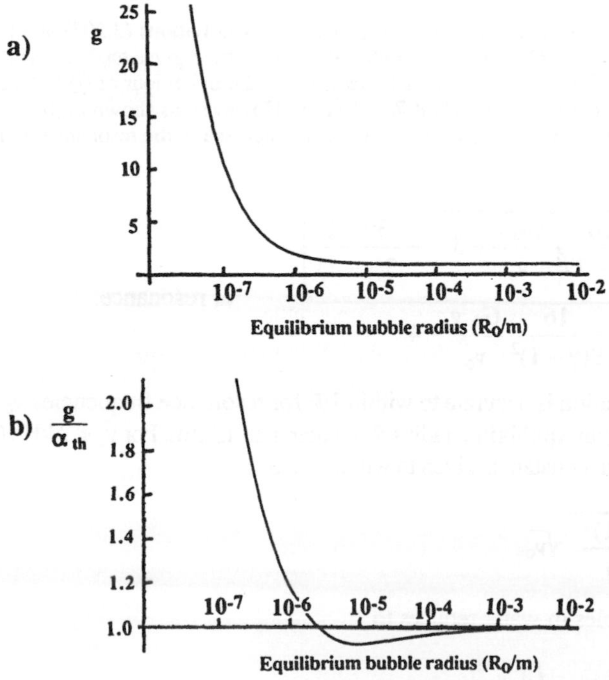

Figure 3.17 (a) The variation of g with bubble radius, indicating the effect of surface tension. (b) The ratio $\sqrt{g/\alpha_{th}} = (\omega_o/\omega_M)$ is shown, indicating the combined effect of surface tension and thermal losses.

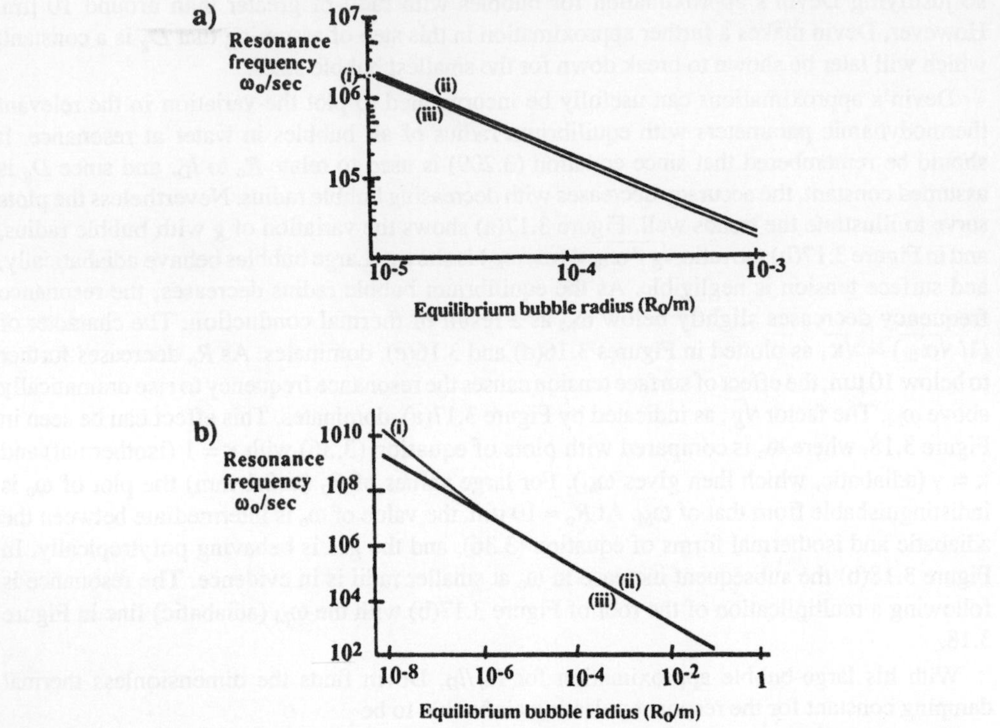

Figure 3.18 (i) The resonance frequency, ω_o, as given by equations (3.203) and (3.204), is compared with plots of equation (3.36) for (ii) $\kappa = \gamma$ (adiabatic, which then gives ω_M), and (iii) $\kappa = 1$ (isothermal). In (a) as the bubble size decreases, thermal effects cause the behaviour of (i) to tend from the adiabatic (for $R_o \sim 0.1$–1 mm) to the isothermal (at $R_o \sim 10$ μm). However, as shown in (b), as the radius becomes smaller, the effect of surface tension causes a dramatic increase in the resonance frequency.

$$\delta_{th} = \frac{2\left(\sqrt{\frac{16}{9(\gamma-1)^2}\frac{G_{th}\,g}{\nu_o} - 3} - \frac{3\gamma-1}{3(\gamma-1)}\right)}{\frac{16}{9(\gamma-1)^2}\frac{G_{th}\,g}{\nu_o} - 4} \quad \text{at resonance.} \tag{3.210}$$

The above formulation is accurate to within 1% for resonance frequencies ν_o less than 240 kHz, that is those having an equilibrium radius R_o greater than 15 μm. For $\nu_o < 7$ kHz (i.e. $R_o > 500$ μm), the resonant thermal constant is given to within 1% by

$$\delta_{th} = \sqrt{\frac{9(\gamma-1)^2}{4G_{th}}}\;\sqrt{\nu_o} \tag{3.211}$$

which for air bubbles in water reduces to

$$\delta_{th} = 4.41 \times 10^{-4}.\;\sqrt{\left[\frac{\nu_o}{\text{Hz}}\right]} \tag{3.212}$$

For $\omega_o \lesssim 40$ kHz, this is accurate to about 10%.

(iv) The Effect of Surface Tension on the Thermal Diffusivity of the Gas. Devin's analysis assumes that the thermal diffusivity of the gas is constant for all bubbles, regardless of the equilibrium size. However, equation (3.181) merits analysis. One mole of gas contains N_{Av} molecules, where N_{Av} is Avogadro's number, approximately equal to 6.023×10^{23}. If the mass of each gas molecule is m_g, then the mass of a mole of gas is $N_{Av}m_g$, and the molar ($C_{p,m}$) and specific (C_p) heat capacities are related through

$$C_{p,m} = N_{Av} m_g C_p \tag{3.213}$$

Substitution into equation (3.161) yields

$$D_g = \frac{K_g N_{Av} m_g}{\rho_1 C_{p,m}} \tag{3.214}$$

The thermal conductivity is given [35] by

$$K_g = \frac{k_B}{\pi L_g^2}\sqrt{\frac{2k_B T}{\pi m_g}} \tag{3.215}$$

where $k_B \approx 1.38 \times 10^{-23}$ J K^{-1} is Boltzmann's constant, and L_g the diameter of the molecule. Therefore since K_g, N_{Av}, m_g and $C_{p,m}$ (the latter being related to the number of degrees of freedom only, being, for example, equal to $9R_g/2$ for a diatomic gas at room temperature [36]) are all independent of the density of the gas, then D_g varies as ρ_1^{-1}.

If therefore the density of gas within bubbles varies between bubbles of different equilibrium radii but otherwise identical conditions, this assumption is not justified. The number density of gas molecules within the bubble, n_g, is found from equation (1.15) by noting that the ratios $N_m/V = R_g/k_B$ both equal Avogadro's number, N_{Av}. Therefore if the gas, at pressure p_g within the bubble, is perfect, then for a bubble at equilibrium in a liquid of temperature T_o

$$p_g = n_g k_B T_o \tag{3.216}$$

Substituting equations (2.6) to (2.8) into equation (3.216) enables the number density n_g of the gas molecules to be related to the external conditions at equilibrium:

$$n_g = \frac{1}{k_B T_o}(p_\infty - p_v + p_\sigma) = \frac{1}{k_B T_o}\left(p_\infty - p_v + \frac{2\sigma}{R_o}\right) \tag{3.217}$$

If the Laplace pressure $p_\sigma = (2\sigma/R_o)$ is negligible, then bubbles under the same static pressure p_∞ will at equilibrium have the same gas density, and Devin's assumption is justified. However, this will not be so for the very smallest bubbles, and the effect on D_g must be incorporated in the analysis.

Figure 3.19 shows the variation of polytropic and resonance frequency with bubble radius where a method of successive approximations, rather than equation (3.189), is used to relate R_o to l_D. The solid line incorporates the variation of D_g with bubble radius. The thermal damping constant is similarly calculated in Figure 3.20.

This exposition started by consideration of the equation of motion of a bubble driven by an acoustic pressure field, and then reduced the resulting damping coefficient to its value at the resonance frequency. A discussion of off-resonance damping is given in Chapter 4, section 4.4.2.

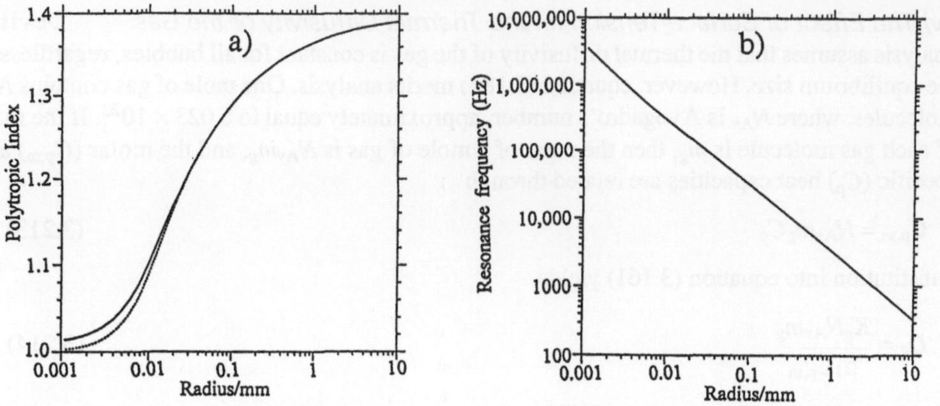

Figure 3.19 The variation of (a) polytropic index and (b) resonance frequency, with bubble radius where a method of successive approximations, rather than equation (3.189), is used to relate R_0 to l_D. While the dashed line follows Devin's calculation, the solid line incorporates the variation of D_g with bubble radius. On this scale the results are almost indistinguishable in (b), except at the smallest radii. (CWH Beton and TG Leighton.)

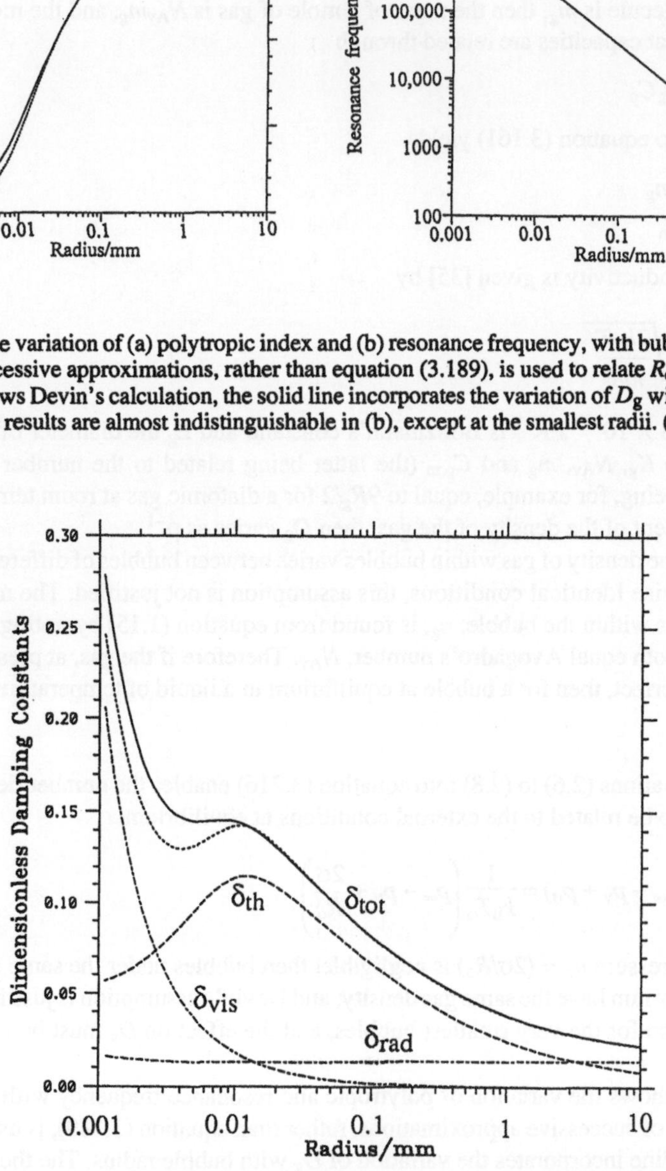

Figure 3.20 The variation of dimensionless damping constants at resonance with bubble radius, where a method of successive approximations, rather than equation (3.189), is used to relate R_0 to l_D. Dashed lines show the constants δ_{rad}, δ_{vis} and δ_{tot} as calculated by Devin's theory. The plot of δ_{th} incorporates the contribution of the variation of D_g with bubble radius, which is also incorporated into the solid-line plot of the total damping constant. (CWH Beton and TG Leighton.)

(c) Radiation Damping

In section 3.2, the spring and acoustic dissipation constants, and the radiation mass, for bubble pulsations in the radius frame were derived from the acoustic impedance for spherical waves. The real part of Z_{RF}^{rad} is the radiation resistance $b_{RF}^{rad} = 4\pi R_0^2 \rho_0 c (kR_0)^2$ (equation (3.87)), the resistive term responsible for the damping that results from the radiation of energy away from the bubble as sound. Using the appropriate system mass m_{RF}^{rad} for this radial system from equation (3.89), the dimensionless radiation damping constant δ_{rad} is therefore

$$\delta_{rad} = \frac{b_{RF}^{rad}}{\omega_0 m_{RF}^{rad}} = \frac{R_0 \omega_0}{c} \tag{3.218}$$

This equation is exact. An expression for the dimensionless thermal damping constant at resonance which incorporates the effects of heat flow and surface tension can be found by substituting for ω_0 from the expression given in equation (3.203), since that formulation for the resonance frequency incorporates those two effects, noting that the inclusion of δ_{rad} changes ω_0 again:

$$\delta_{rad} = \sqrt{\frac{3\gamma p_0}{\rho}} \sqrt{\frac{g}{\alpha_{th}}} \frac{1}{c} \tag{3.219}$$

This is in agreement with the damping constant, derived by Devin through consideration of the velocity potential in a compressible fluid about a pulsating bubble, from which he derives the pressure at the bubble and the equation of motion. This was done in the volume–pressure frame, where the dissipation constant is

$$b_{VP}^{rad} = \frac{\rho \omega^2}{4\pi c} \tag{3.220}$$

giving, at resonance,

$$\delta_{rad} = \frac{b_{VP}^{rad}}{\omega_0 m_{VP}^{rad}} = \frac{\omega_0 R_0}{c} \tag{3.221}$$

which is in agreement with equation (3.219). Note that the mass employed is now m_{VP}^{rad}, not m_{RF}^{rad}, since this calculation is performed in the volume–pressure and not the radius–force frame, and similarly for the resistance. This illustrates the advantages of the dimensionless δ over the case-specific b, in that confusion between, for example, b_{RF}^{rad}, and b_{VP}^{rad} is avoided when only δ_{rad} is quoted.

It should be noted that for large bubbles, when the effects of surface tension and heat conduction are negligible, the quantity $\omega_0 R_0$ is approximately constant (see equation (3.36)). Therefore equation (3.221) tells us that the resonance radiation damping coefficient δ_{rad} is independent of bubble size in this regime. The coefficient δ_{rad} is shown in Figure 3.20.

(d) Viscous Damping

In Chapter 2, section 2.3.2(a)(i) it was shown that application of the Stokes assumption ($\eta_B = 0$) to the Navier–Stokes equation for a fluid of constant viscosity yielded equation (2.133):

$$\rho \frac{D\vec{v}}{Dt} = \rho\left\{\frac{\partial \vec{v}}{\partial t} + (\vec{v}.\vec{\nabla})\,\vec{v}\right\} = \rho\Sigma\vec{F}_{ext} - \vec{\nabla}p + \frac{\eta}{3}\vec{\nabla}(\vec{\nabla}.\vec{v}) + \eta\nabla^2\vec{v} \tag{3.222}$$

If the liquid is assumed to be incompressible, then $\vec{\nabla}.\vec{v} = 0$ (equation (2.51)). If in addition the liquid motion is irrotational (Chapter 2, section 2.2.2(i)) then $\nabla^2\vec{v} = 0$, since in addition to $\vec{\nabla}.\vec{v}$ equalling zero in equation (2.132), so does $\vec{\nabla}\wedge\vec{v}$. Thus for a spherical bubble undergoing (irrotational) radial oscillation in an incompressible viscous liquid, equation (3.222) reduces to

$$\rho \frac{D\vec{v}}{Dt} = \rho\left\{\frac{\partial \vec{v}}{\partial t} + (\vec{v}.\vec{\nabla})\,\vec{v}\right\} = \rho\Sigma\vec{F}_{ext} - \vec{\nabla}p \tag{3.223}$$

Therefore there are no net viscous forces acting in the body of an incompressible viscous liquid around the pulsating spherical bubble (see Chapter 4, section 4.2.1(b)) [31]. Momentum transfer through viscosity will occur, but each liquid volume element receives as much momentum as it loses, so that no net viscous force is manifest within the liquid interior. However, net viscous forces can occur at the liquid surface of the bubble wall, where they result in an excess pressure. Mallock [37] describes how viscous forces cause the distortion of spherical shell volume elements concentric with the bubble.

As the bubble volume increases, such a volume element decreases its radial dimension (thickness) and increases its lateral one (i.e. the area of both inner and outer surfaces of the shell increase). When the bubble contracts, the converse happens. If the liquid is incompressible, these distortions of the element cannot be caused by compression of the liquid comprising the element; rather, they are the result of the viscous stresses. As a result there is a net energy loss on compression. Devin derives the equation of motion for viscous dissipation with the coefficient

$$b_{VP}^{vis} = \frac{\eta}{\pi R_0^3} \tag{3.224}$$

so that the dimensionless viscous damping coefficient is

$$\delta_{vis} = \frac{b_{VP}^{vis}}{\omega_0 m_{VP}^{rad}} = \frac{4\eta}{R_0^2 \rho \omega_0}$$

$$= \frac{4\eta}{3\gamma p_0}\,\omega_0\,\frac{\alpha_{th}}{g} = \frac{4\eta}{\rho R_0}\sqrt{\frac{\rho}{3\gamma p_0}}\sqrt{\frac{\alpha_{th}}{g}} \tag{3.225}$$

This shows that over the range where $\sqrt{\alpha_{th}/g}$ is approximately constant, the viscous damping constant is proportional to R_0^{-1} or equivalently to ω_0, where the latter is given by equation (3.203). As can be seen from Figure 3.17(b), this is true for large bubbles. Figure 3.20 shows the variation of δ_{vis} with bubble radius.

(e) Total Damping Constant

The total damping constant $\delta_{tot} = \delta_{rad} + \delta_{th} + \delta_{vis}$, and its components for an air bubble in water are plotted in Figure 3.20. Successive approximations are used to find ω_0, and the effect of surface tension on D_g is included. As expected, δ_{vis} dominates for small bubbles, and is negligible for large spherical ones. δ_{rad} is approximately constant. For comparison, a plot of δ_{tot} where D_g is assumed to be constant is included. It is clear that the effect, which is due solely to the δ_{th} contribution, is negligible for macroscopic bubbles.

More sophisticated treatments of dissipation, particular as relates to thermal conduction, are discussed in Chapter 4, section 4.4.2. These treatments adhere to the concept of dissipation in a spherical pulsating bubble due to thermal, viscous and acoustic losses.

Experimental measurements of the decay of freely-oscillating bubbles have, however, revealed behaviour beyond that predicted by this theory. Damping may be higher than expected, and in some cases the damping may change during the course of the oscillation. Examples of such experimental observation techniques, which may use the time constant of decay for amplitude, β^{-1}, to find the total damping constant $\delta_{tot} = Q^{-1} \approx 2\beta/\omega_0$ (equation (3.157)). Section 3.5 also shows, however, that such bubbles can exhibit significant deviations from sphericity, and it is to these that Longuet-Higgins [38] attributes the observed departures from predicted behaviour. This is discussed in section 3.8.2(a).

3.5 Experimental Investigations of Bubble Injection

Investigations of the acoustic pressure, and photographs of single injected bubbles, have been made by Minnaert [4], Longuet-Higgins *et al.* [25], Strasberg [39], Fitzpatrick and Strasberg [40], Leighton and Walton [41] and others.

3.5.1 A Simple Experiment for a Pulsating Spherical Bubble

Though care is needed, it is not difficult to devise a simple experiment to investigate the resonance and damping of bubbles injected from a nozzle.[21] The basic apparatus is shown in Figure 3.21.

(a) Techniques

(i) Acoustic Sensors. The principal measuring instrument is a hydrophone (for example, a Bruel and Kjaer type 8103). However, an adequate alternative can be produced for a fraction of the cost by substituting a piezoelectric 'tweeter' loudspeaker suitable for outdoor use (e.g. Radio Spares type 248-325, which has an 85 mm Mylar cone speaker).[22] This makes a suitable hydrophone for this study, though the magnitude of its output is uncalibrated and the inefficient electrical shielding makes it unsuitable for examining electrically generated sound fields. The hydrophone detects the sound from the bubble as it is produced. A Bruel and Kjaer type 8103 hydrophone has a sensitivity of about 30 μV per Pascal: the output can be viewed by a storage oscilloscope, or stored digitally for further analysis. Reverberation in the tank will affect the signal unless the tank is sufficiently large or lined with adequate acoustic absorber.

---

[21] Though this experiment is simple in construction and readily yields data on the acoustic emissions from bubbles, a more detailed investigation would show that the behaviour of a bubble injected from a nozzle can be very complicated. This will be discussed in section 3.5.1(b). Suffice to say for now that this simple experiment requires that the bubbles be released individually and repeatedly from the nozzle.

[22] For a very low-cost sensor, the tweeter from a disused Hi-Fi speaker can be adequately waterproofed by painting the paper cone with aero-modelling dope.

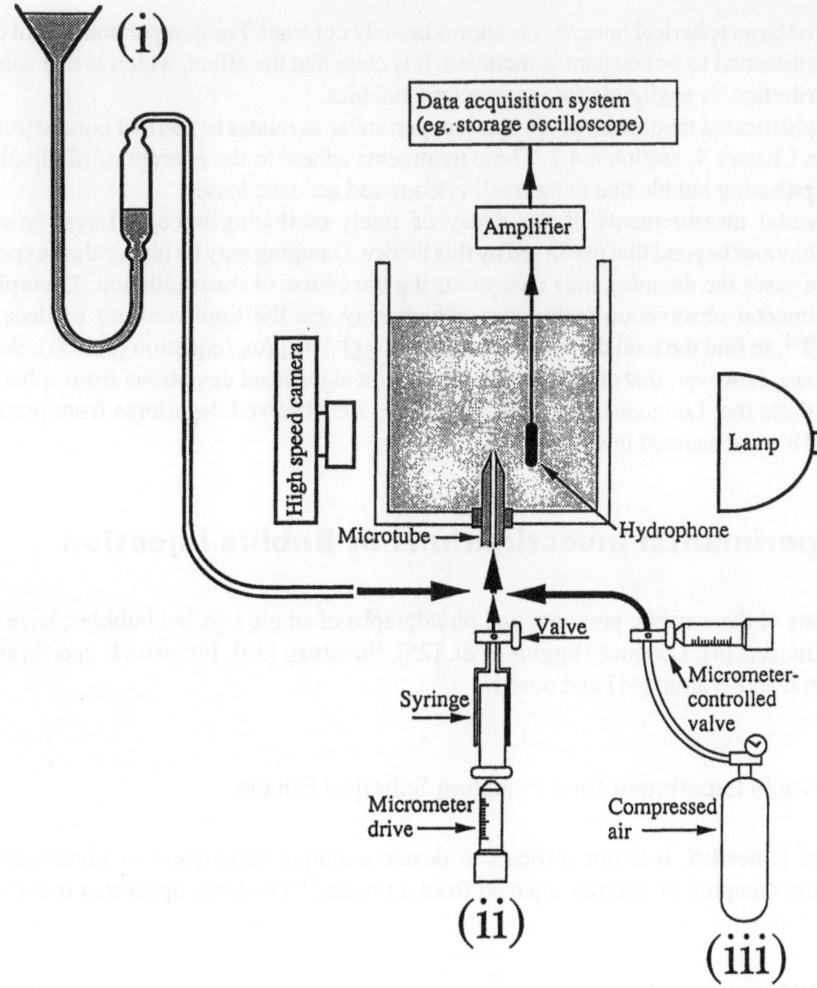

Figure 3.21 The apparatus to detect the acoustic emissions of an injected bubble. Three gas flow techniques are illustrated: (i) Minnaert [4], (ii) Leighton and Walton [41], (iii) Longuet-Higgins *et al.* [25].

As an interesting historical note, in 1933 Minnaert [4] listened, by ear, to the noise generated by injected bubbles and determined the fundamental frequency of the signal by comparing the emissions with the pitch of a standard tuning fork, a feat of considerable skill, particularly when one considers that the 'ping' emitted by the injected bubble lasts only of the order of milliseconds.

(ii) Measurement of Bubble Size. There are many non-acoustic techniques for the measurement of bubble size, the sophistication and accuracy of which vary greatly. An average bubble size produced by a given nozzle can be found by collecting a known number of bubbles in an inverted liquid-filled measuring cylinder [41]. It is, of course, necessary to correct for the hydrostatic head (through, for example, equation (1.16)) to find the size that the bubble assumed at the relevant depth. The size of the individual bubbles can be found by flash photography,

though the larger the bubble, the more it deviates from the spherical shape[23] and so the less useful this technique. One technique suitable for large bubbles is to use a density bottle of the type shown in Figure 3.22. The bottle is designed to contain a highly reproducible volume of liquid: it is filled, the top inserted in place using the ground-glass joint, and excess liquid displaced through the capillary. The difference in mass between the bottle when full, and when containing a bubble, gives the mass of water displaced by the bubble, and hence its volume. Such techniques, which measure the bubble *volume* (or mass) are intrinsically more accurate than direct measurements of the bubble *radius*. This is because differentiation of the volume $V_0 = 4\pi R_0^3/3$ yields $dV_0/dR_0 = V_0/3R_0$, which implies that if a radius measurement R_0 is deduced from a volume measurement V_0, the percentage error $\Delta R_0/R_0$ in the radius is only 1/3 that of the volume measurement $\Delta V_0/V_0$. In practice, however, the accuracy is controlled by details of the measurement technique, and the errors involved in the weighing technique (which arise from evaporation, dryness of the outside of the bottle, temperature instability etc.) tend to make this method less accurate as the bubble size decreases. As before, it is necessary to correct for the hydrostatic head.

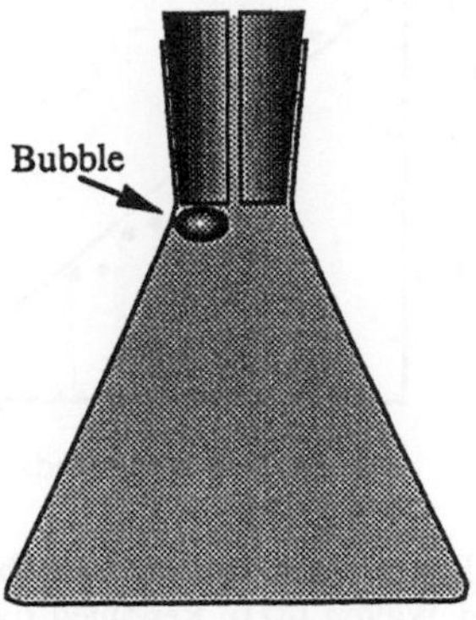

Figure 3.22 Use of a density bottle to find volume of large bubbles (after Leighton *et al.* [42]).

The bubble can be caught in a precisely rectangular thin-walled glass vessel, and the length of its principal axes measured [42, 43]. Minnaert [4] obtained the gas volume by catching each bubble and introducing it into a capillary tube: the length of the air column, measured at a controlled temperature, gave the bubble volume. Other techniques of bubble size measurement are discussed later.[24]

(b) Acoustic Measurements

(i) The Minnaert Frequency. Figure 3.23 shows the typical hydrophone trace of the emission from a single injected bubble [41]. The emission is a sinusoid which decays exponentially (as demonstrated in Figure 3.24), indicating that the source is a lightly damped oscillator.[25] Such

---

[23] See Chapter 2, sections 2.1.1 and 2.3.2(b).

[24] See sections 3.8.1(b) and 3.8.1(c). Acoustic techniques are particularly discussed in Chapter 5, section 5.1. The behaviour and size of the bubble, and the characteristics of the fluid medium and its enclosure, very much determine the appropriate technique to employ.

[25] See section 3.1.3.

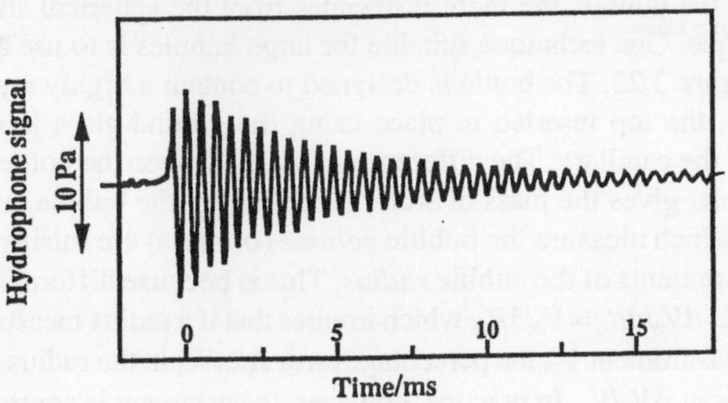

Figure 3.23 Typical hydrophone trace of the emission from a single injected bubble (after Leighton and Walton [41]). Reprinted by permission from *European Journal of Physics*, vol. 8, pp. 99–104; Copyright © 1987 IOP Publishing Ltd.

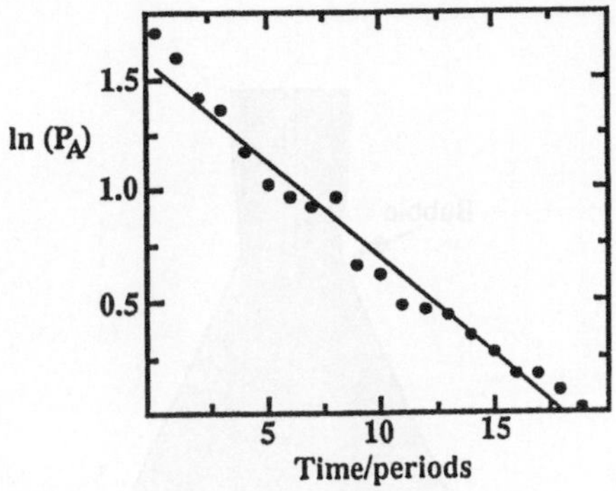

Figure 3.24 Replot of data in Figure 3.23, showing that the emission is a sinusoid which decays exponentially (after Leighton and Walton [41]). Reprinted by permission from *European Journal of Physics*, vol. 8, pp. 99–104; Copyright © 1987 IOP Publishing Ltd.

traces contain information on the value of the polytropic index during these free oscillations, the decay times and Q-factors of the bubbles, and the amplitude of the wall oscillation [41]. The data from such traces are used in Figure 3.25 to test Minnaert's theoretical prediction of the oscillation frequency.

Figure 3.25(a) shows a plot of $\nu_o/\sqrt{p_0}$ against R_0^{-1} for data from seven helium bubbles in water, the radius of the bubble being measured independently. Over the course of the experiment the atmospheric pressure varied by 4%, which would introduce a systematic 2% variation into the results if ν_o only were plotted against R_0. Therefore it is important to incorporate it into such plots. From equation (3.37), the result should be a line of gradient $(1/2\pi)\sqrt{3\kappa/\rho}$ if, as suggested by the preceding section for bubbles of this size (Figure 3.17(a)) surface tension has negligible effect on the resonance. Solid lines, corresponding to the adiabatic ($\kappa = \gamma = 5/3$) and the isothermal case ($\kappa = 1$) are shown. In Figure 3.25(b) the variation of the natural frequency with liquid density is tested. The two commonly available liquids with the greatest range in densities at room temperature are petrol (ρ = 660–690 kg m^{-3}) and saturated $NaClO_3$ solution (ρ = 1376 kg m^{-3}). The two liquids are hazardous, the former being flammable with explosive vapour, and the latter toxic, so that care must be taken. Figure 3.25(b) shows data for air bubbles in water and in saturated $NaClO_3$ solution, the lines again showing the range of polytropic index of air.

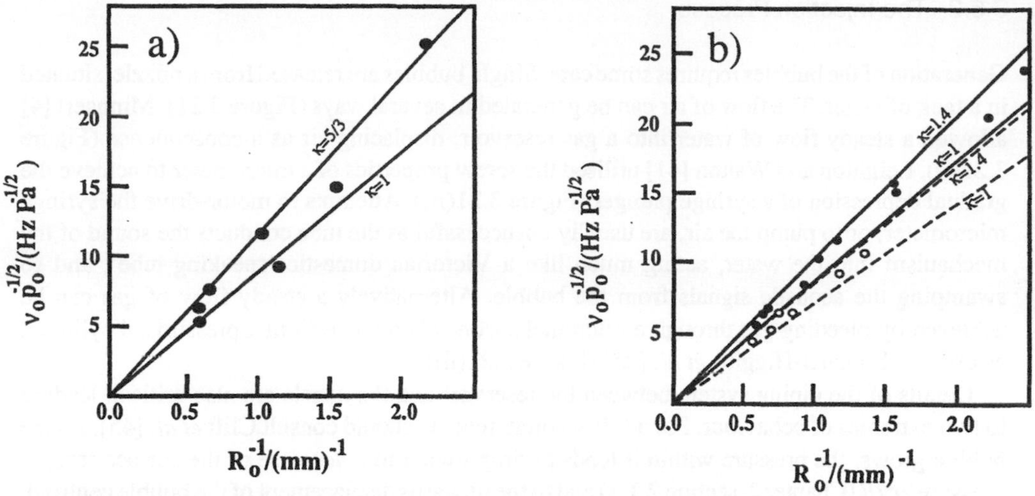

Figure 3.25 (a) A plot of $\nu_o/\sqrt{p_0}$ against R_0^{-1} for data from seven helium bubbles in water, the radius of the bubbles being measured independently. The prediction of equation (3.37) is shown by solid lines, corresponding to the adiabatic ($\kappa = \gamma = 5/3$) and the isothermal case ($\kappa = 1$) are shown. In (b) a plot of $\nu_o/\sqrt{p_0}$ against R_0^{-1} is made for air bubbles in water (solid lines and filled circles) and in saturated $NaClO_3$ solution (dashed lines and open circles), the lines again showing the range of polytropic index of air. (After Leighton and Walton [41]. Reprinted by permission from *European Journal of Physics*, vol. 8, pp. 99–104; Copyright © 1987 IOP Publishing Ltd.)

Figures 3.25(a) and 3.25(b) conform with equation (3.37), and though the value of the polytropic index is not readily apparent from them, there does seem to be an adiabatic, rather than an isothermal tendency, in agreement with the findings of the preceding section.

(ii) Amplitude of Oscillation. It is possible to use the formulation of this chapter to gain an estimate of the amplitude of oscillation of the bubble wall. Hydrophone traces, such as that illustrated in Figure 3.23, give the acoustic pressure amplitude (which decays exponentially, since the bubble is behaving as a lightly damped oscillator – see section 3.1.3). If, to a first approximation, the bubble is assumed to be a linear oscillator, then the velocity of the bubble wall, $U = U_0 e^{i\omega t} = \dot{R}$ is simple harmonic, and $U_0 = i\omega_0 R_{\varepsilon 0}$. If the bubble is approximated to be a rigid sphere, as Leighton and Walton [41] assumed, then the amplitude of the wall oscillation is given by equation (3.105). Pumphrey and Crum [44], using the rigid sphere model and fitting their data to the dipole radiation from bubbles entrained through the impact of liquid drops on a body of water, obtain an amplitude of oscillation of 3.3 μm. However, the acoustic pressure in the liquid around a bubble is not simply determined by interface motion, but by the instantaneous gas pressure within the bubble which, as shown in section 3.3.1(b), dominates in the case of small-amplitude linear pulsations. It is therefore more correct to apply equation (3.129) to the data. Doing this for the data of Leighton and Walton, Vokurka [private communication] obtained an oscillation amplitude of 10^{-7} m.

It is interesting to note that both theories assume a pulsating bubble which remains spherical at all times. However, when the injection process is studied in detail, it will be seen that this is not commonly the case. The shape oscillations which the bubble can undergo will be discussed further in this section and also in section 3.6.

3.5.2 The Injection Process

Generation of the bubbles requires some care. Single bubbles are released from a nozzle, situated in a tank of water. The flow of air can be generated in several ways (Figure 3.21). Minnaert [4] allowed a steady flow of water into a gas reservoir, displacing air as a consequence (Figure 3.21(i)). Leighton and Walton [41] utilised the screw properties of a micrometer to achieve the gradual depression of a syringe plunger (Figure 3.21(ii)). Attempts to motor-drive the syringe micrometer, or to pump the air, are usually unsuccessful as the tube conducts the sound of the mechanism into the water, acting much like a Victorian domestic 'speaking-tube', and so swamping the acoustic signals from the bubble. Alternatively a steady flow of gas can be achieved by bleeding gas through a micrometer-controlled valve from a pressurised cylinder, as used by Longuet-Higgins *et al.* [25] (Figure 3.21(iii)).

Details of the piping system between the reservoir and the nozzle are also critical, leading to two extremes of behaviour. For a full account, readers should consult Clift *et al.* [45]. As the bubble grows, the pressure within it tends to drop owing to a decrease of the surface tension pressure $2\sigma/R$ (Chapter 2, section 2.1.1) and to the upwards displacement of the bubble centroid. In one situation there may be some high-pressure drop restriction separating the nozzle orifice from the gas reservoir, such as a constriction behind the nozzle, or a long capillary (of much narrower bore than the orifice) between it and the reservoir. In these circumstances, the pressure changes within the bubble resulting from its growth are negligible compared with the pressure drop that occurs between the reservoir and the nozzle, i.e. across the restriction. Thus bubbles are produced under conditions of approximately constant flow. The formation of bubbles in a low-viscosity liquid, such as water, under conditions of constant flow is referred to as 'constant volume formation' if the flow is low, and 'constant frequency formation' if the constant gas flow is high, where frequency refers to the rate at which bubbles are formed. There is an intermediate regime where neither term is applicable. As the flow rate increases the formation is less predictable, since additional factors complicate the theory. These include the flow resulting from the release of the previous bubble. As will shortly be shown, at higher flows inter-bubble effects can become even more pronounced. Constant-volume conditions have proved most useful for producing single bubbles of equal size repeatably [25].

At the other extreme, if there is negligible pressure drop between the reservoir and the nozzle, constant pressure conditions are obtained. This regime is applicable to several industrial procedures. Bubbles produced under conditions of constant pressure tend to be larger than those produced at the same mean flow rate under conditions of constant flow.

It is important to note that the relevant 'reservoir' may be an unintentional result of careless experimental design. Any chamber of gas having a volume that is very much larger than the volume of any bubble produced will act as such a reservoir: attempts to generate constant flow conditions can be undermined by wide-bore tubing leading from the orifice and acting as a reservoir, and if no high-pressure drop restriction is included between tube and orifice. The small size of such 'accidental' reservoirs often causes both the rate of flow through the orifice and the pressure in the chamber to vary in time.

The nozzle can be made from commercial steel tubing, or by drawing glass tubing. It is better to sharpen the tip of the nozzle as shown in Figure 3.21, since this prevents irregular bubble formation resulting from the adhesion of the meniscus to the material of the nozzle between the bore and the outer diameter if that material is imperfectly wetted. With the larger bores, water pockets can enter the tubing and form 'plugs', hindering the reproducible production of single bubbles.

Longuet-Higgins *et al.* [25] undertook a detailed study of the mechanism for sound generation during the detachment of bubbles from underwater nozzles. The theory models a nozzle with very thin walls: bubble attachment can occur at any angle to the rim of the nozzle. The air flow is so slow that the growth of the bubble is quasi-static up until the onset of any instability. Pitts [46] discusses the stability of a bubble attached to a nozzle under conditions of both prescribed gas pressure and prescribed volume, though only the latter is considered by Longuet-Higgins *et al.* [25].

Theoretical calculations are used to deduce the sequence of bubble shapes assumed by the attached bubble at the nozzle (Figure 3.26). The scales are dimensionless, and the number associated with each curve gives the radius of curvature of the meniscus at the top of the bubble. As the bubble initially grows, the radius of curvature decreases from 1 to just under 0.5, whilst the volume of gas above the nozzle tip increases steadily. Near the point at which the tangent to the meniscus at the point of attachment to the nozzle becomes vertical, there is a large increase in volume, but the radius of curvature remains roughly constant. The volume and radius of curvature both then increase, until the latter equals about 0.655. At this point, the tangent to the meniscus at the point of attachment is again vertical, and there is a second sharp increase in volume. Soon after this point, the volume of gas above the top of the nozzle is a maximum. Longuet-Higgins *et al.* found that, for each nozzle, the theory predicts the existence of such a shape for the attached bubble where it has maximum volume. This volume cannot be exceeded whilst stability is maintained. Longuet-Higgins *et al.* take this point to be the moment when detachment occurs: if further air is forced in, the bubble must detach, the break occurring at the narrowest point in the bubble profile (the 'neck'). Fixing the break point in this manner allows the mass of gas within the bubble to be calculated, which is generally less than the maximum volume previously mentioned. Thus the main acoustic frequency generated by the bubble can

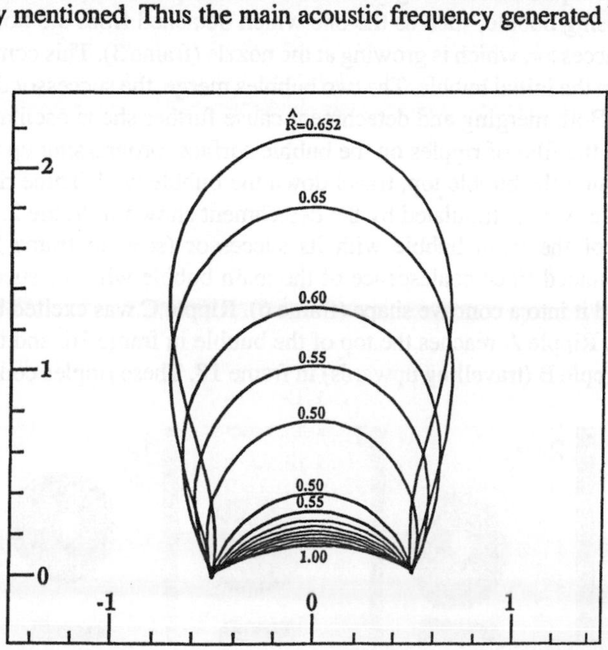

Figure 3.26 Theoretical calculations of the sequence of bubble shapes assumed by the attached bubble at the nozzle. The scales are dimensionless, and the number associated with each curve gives the radius of curvature of the meniscus at the top of the bubble (after Longuet-Higgins *et al.* [25]).

be predicted for each nozzle size. This is found to correspond well with experimental measurements taken in the limiting case of quasi-static bubble generation and thin-walled nozzles. Slow flow was obtained by using a micrometer-controlled valve to regulate the release of air from a pressurised reservoir, giving bubble production rates in the range of 0.01–5 s^{-1}. The outer walls of the larger nozzles tapered inwards towards the end to produce very thin edges at the tip.

Longuet-Higgins *et al.* [25] show that, as the rate of production of bubbles is increased, the amplitude of the acoustic pulse increased, and the acoustic frequency decreased, both dramatically, for a pressure-controlled system. At the lowest rates, the values tended to a limit. This limit is roughly the acoustic pressure and frequency for volume-controlled release, values which were far more stable with respect to increasing bubble production rate. The values did change in the same direction as for pressure-controlled release, but were very much smaller and more gradual. 'Gentle' release of bubbles corresponds to the case when the acoustic output has the limiting characteristic demonstrated at very low bubbling rates, and it is on these releases that quasi-static measurements should be taken.

In the experiment of Longuet-Higgins *et al.* [25], care was taken to use only slow rates of gas flow, so that the bubbles are generated singly in a quasi-static situation. Leighton *et al.* [47] experimentally studied the generation of bubbles as the gas flow through the nozzle is increased, using acoustic and high-speed photographic techniques.

Figure 3.27 is a series of consecutive frames with an inter-frame time of 0.71 ms, sampled from a film, showing an air bubble being blown from the metal nozzle. The initial shape distortion of the bubble, caused by the buoyancy and adhesion forces acting on the bubble, results in subsequent shape oscillations. As stated earlier, if the rate of production of bubbles is increased, the formation of one bubble can be influenced by the previous one through the 'updraft' and other liquid currents it generated. When the production rate is increased further (Figure 3.28), a rising bubble, such as the one which detached from the nozzle in frame 2, is contacted by its successor, which is growing at the nozzle (frame 3). This contact causes further shape oscillation in the initial bubble. The two bubbles merge, the successor detaching from the nozzle (frame 7). Both merging and detachment cause further shape oscillations, which have the appearance on the film of ripples on the bubble surface, progressing up the bubble. These ripples, after reaching the bubble top, travel down the bubble wall. Three ripples can be seen in frame 11. Ripple A was stimulated by the detachment shown in frame 2. Ripple B was the result of contact of the main bubble with its successor (seen in frame 3). This ripple is particularly pronounced since coalescence of the main bubble with the successor distorts the bubble wall behind it into a concave shape (frame 6). Ripple C was excited by the detachment shown in frame 8. Ripple A reaches the top of the bubble in frame 16, and then travels down, to interfere with ripple B (travelling upwards) in frame 18. These ripples could be expressed in

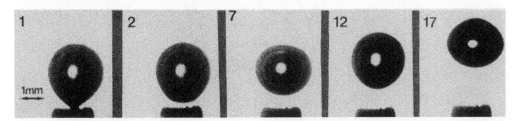

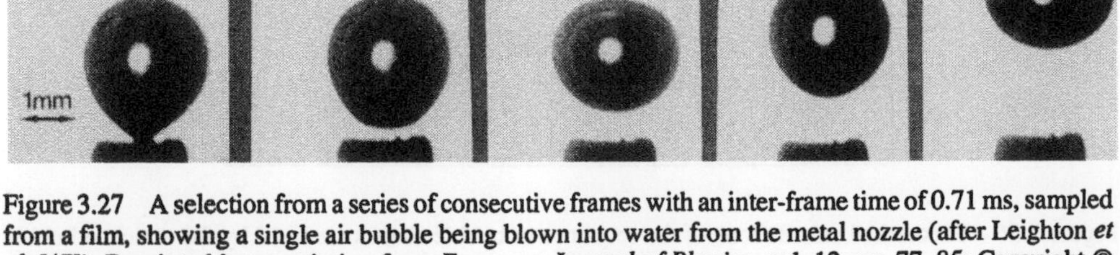

Figure 3.27 A selection from a series of consecutive frames with an inter-frame time of 0.71 ms, sampled from a film, showing a single air bubble being blown into water from the metal nozzle (after Leighton *et al.* [47]). Reprinted by permission from *European Journal of Physics*, vol. 12, pp. 77–85; Copyright © 1991, IOP Publishing Ltd.

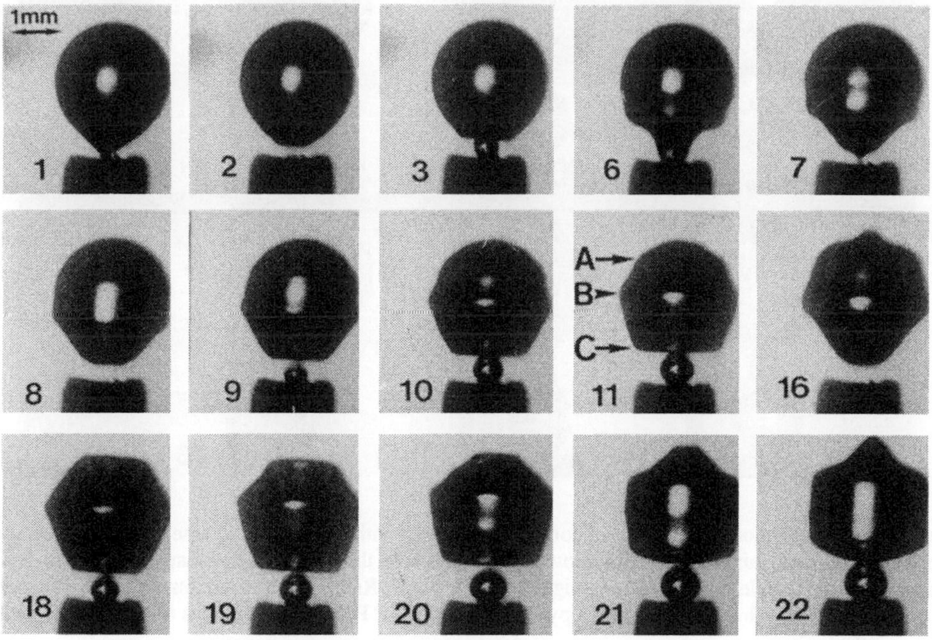

Figure 3.28 A selection from a series of consecutive frames with an inter-frame time of 0.24 ms, sampled from a film, showing air bubbles being blown into water from the metal nozzle, shape oscillations being excited by inter-bubble contact (after Leighton *et al.* [47]). Reprinted by permission from *European Journal of Physics*, vol. 12, pp. 77–85; Copyright © 1991, IOP Publishing Ltd. Gas flow rate ≈ 0.1 ml/s.

terms of spherical harmonics, as discussed in Chapter 2 section 2.2.4(c). The shape oscillations engendered by these inter-bubble contacts are more pronounced than those excited as a result of detachment from the nozzle, shown in Figure 3.27.

These processes continued after this sequence of frames, the main bubble absorbing in total four successors, and each time its volume increased. When the separation was large enough, the main bubble was observed to touch its successor without coalescence: such contact excited shape oscillation.

If the gas pressure in the nozzle is further increased, the rate of growth of the successors is increased. This means that more successors can be absorbed (five in the sequence from which Figure 3.29 was selected). The main bubble will therefore be more frequently excited, and will in the end have larger volume, owing to the absorption of a greater number of successors. Thus the oscillation frequency will change, and the departures from sphericity will be more pronounced. This, coupled with the fact that the successors grow more rapidly, means that even when the main bubble is far from the nozzle (see, for example, frame 1, which was exposed 13.7 ms after the initial release), a pronounced shape oscillation (frame 7) can lead to absorption of the successor (frame 16).

The shape seen in frames 7 to 15 of Figure 3.29 becomes the characteristic form as the air flow through the nozzle is increased. Figure 3.30, taken at a higher flow rate, shows a large bubble mass (the 'superior', labelled A) absorbing the newly released 'intermediate' (B) to grow

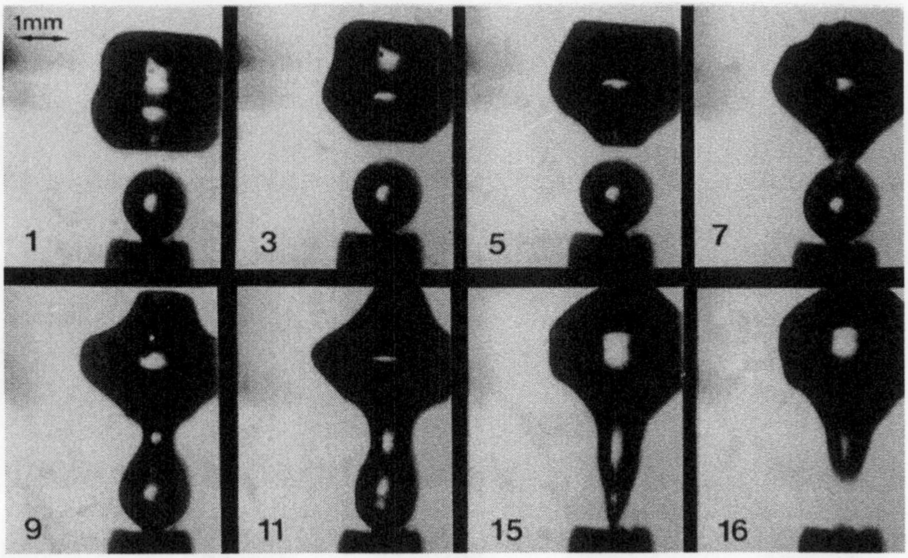

Figure 3.29 A selection from a series of consecutive frames with an inter-frame time of 0.22 ms, sampled from a film, showing air bubbles being blown into water from the metal nozzle, shape oscillations being excited by inter-bubble contact (after Leighton *et al.* [47]). Reprinted by permission from *European Journal of Physics*, vol. 12, pp. 77–85; Copyright © 1991, IOP Publishing Ltd. Gas flow rate ≈ 0.2 ml/s.

even larger (frames 1 to 3). The successor, C, contacts the intermediate and grows to itself take the intermediate position in frame 8 itself, whilst a new successor (D) is growing at the nozzle. In this way, gas is continually pumped into the superior which, as it grows larger, becomes susceptible to fragmentation (fragments are labelled E in frame 12). The superior can undergo an $n = 2$ form of oscillation and divide, generating very large bubbles (Figure 3.31). This may be a periodic event, since it occurs at the start of the sequence of frames with the release of the bubble labelled S, and a similar perturbation that would tend to divide the superior (A) is excited by the absorption of bubble B by A. Thus at the higher gas flow rates, the range of bubble sizes produced from a nozzle can be very wide. It is interesting to compare this form with that generated by the injection of large bubbles from wide bores, shown in Figure 2.17.

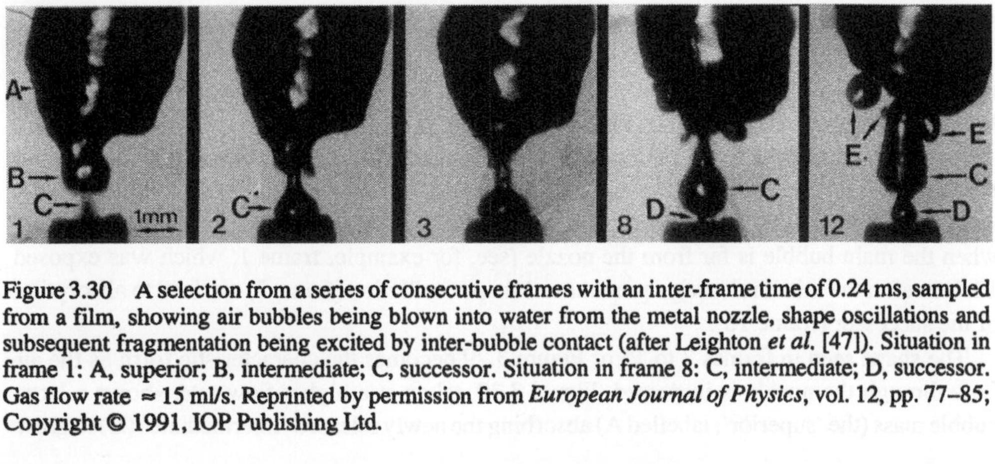

Figure 3.30 A selection from a series of consecutive frames with an inter-frame time of 0.24 ms, sampled from a film, showing air bubbles being blown into water from the metal nozzle, shape oscillations and subsequent fragmentation being excited by inter-bubble contact (after Leighton *et al.* [47]). Situation in frame 1: A, superior; B, intermediate; C, successor. Situation in frame 8: C, intermediate; D, successor. Gas flow rate ≈ 15 ml/s. Reprinted by permission from *European Journal of Physics*, vol. 12, pp. 77–85; Copyright © 1991, IOP Publishing Ltd.

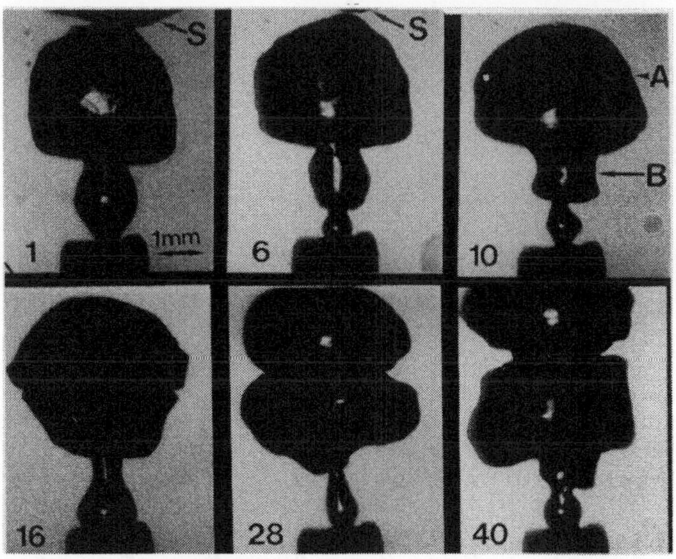

Figure 3.31 A selection from a series of consecutive frames with an inter-frame time of 0.15 ms, sampled from a film, showing air bubbles being blown into water from the metal nozzle (after Leighton *et al.* [47]). Reprinted by permission from *European Journal of Physics*, vol. 12, pp. 77–85; Copyright © 1991, IOP Publishing Ltd. Gas flow rate ≈ 30 ml/s.

To summarise, the acoustic emission from the release of a single bubble is familiar (Figure 3.23). However, when multibubble interactions occur, the emission can be considerably affected. If the inter-contact time is short compared with the decay time, then such contacts produce overlapping excitations, increasing the overall amplitude of the acoustic output [47]. However, if the inter-contact time is longer than the decay time, the emission shown in Figure 3.32 is observed. In general, excitations due to bubble–bubble contact tend to be larger than those generated by the initial detachment of the bubble from the nozzle.

At the higher flow rates the existence of a wide range of bubble sizes, both in the bubble chain growing from the nozzle and also those bubbles generated by fragmentation, increases the frequency spread. Detachment and fragmentation tend to occur unpredictably, so that the acoustic signal loses much of its periodicity and structure.

Inter-bubble contact can occur in several ways. Least intimate is a mere proximity effect, where one bubble can be thought of as being affected by the sound field or pressure gradients or fluid motions generated by another. Alternatively, the bubbles might touch, and for some time the bubbles will share a septum that is a common region of bubble wall. The bubbles might then separate, or a closer contact might occur in which the septum breaks down, so that there is no physical barrier preventing the mixture of the contents of the two bubbles. Subsequently there will be a tendency for these two bubbles to merge into one as, in order to minimise the surface energy and therefore the surface area, the gas pocket endeavours to become spherical. However, the liquid inertia invested in surface modes, and the growth of one bubble through an injection of gas, will tend to act against the merging of the two bubbles, and they may separate. Note should be made of the studies of the coalescence of pairs of growing bubbles, generated from adjacent nozzles, as observed using high speed photography by Yang and Maa [48]. The coalescence times were measured using optical techniques by Chuang *et al.* [49], for

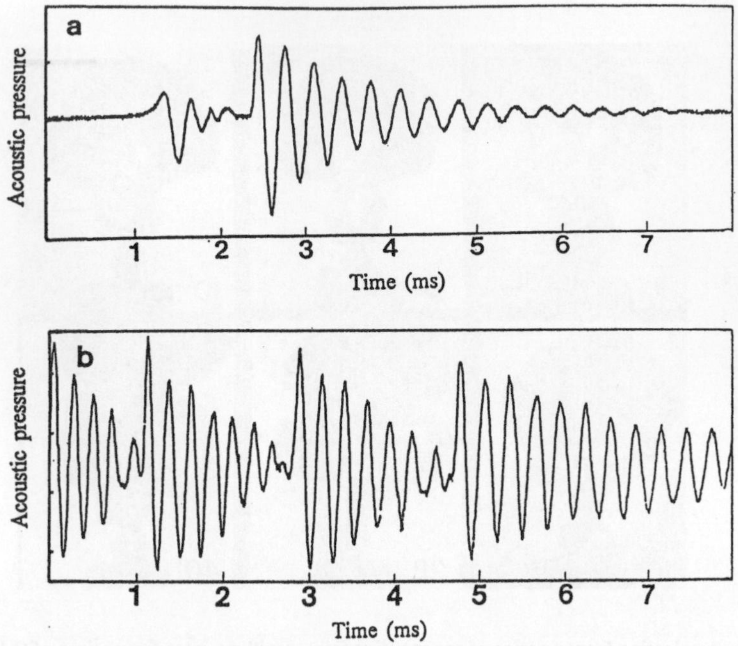

Figure 3.32 The acoustic trace following bubble injection from a glass nozzle of internal diameter 0.12 mm. Multiple excitations are seen, the excitations being (a) larger than, or (b) of the order of the initial excitation (Leighton *et al.* [47]). Reprinted by permission from *European Journal of Physics*, vol. 12, pp. 77–85; Copyright © 1991, IOP Publishing Ltd.

bubbles growing under constant flow conditions from orifices of diameter 2.8 mm separated centre-to-centre by 3.45 mm in an aluminium plate. They found that at high gas flow rates, the bubbles detached from the orifice before coalescence; and that the coalescence time was independent of flow rate over the range 0.42–2.7 bubble pairs per second.

At higher gas flow rates, the features of fragmentation, of coalescence and constriction still occur (Figure 3.33). However, when the flow velocity of injected gas becomes greater than the speed of sound at room temperature, the gas no longer undergoes expansion *at the orifice*. The diameter of the jet at the orifice equals the diameter of the orifice, and the transition is made from a 'bubbling' regime to a 'jetting' regime [50]. Though constriction of the gas column can still occur, at these higher injection rates the consequences can be much more violent: the gas jet may contain an initial supersonic core, and constriction of the jet can cause the air to be temporarily blown back against the nozzle region. This process is known as *back-attack* [51, 52]. It can occur several times a second, and repeated events can cause erosion of the nozzle. This can be a problem industrially when air is injected into liquid at high rates, for example, during the refining of steel. The injection of air into molten iron promotes mixing and is used to increase the rate of refinement. Replacement of the submerged nozzles (or *tuyeres*) if they become eroded necessitates shutdown of the furnace, and is therefore expensive. An understanding of the relation between the acoustic emission of bubbles and the corresponding flow patterns could prove useful in the study of cavitation phenomenon in opaque or hostile environments such as this.

Most of the theory of the bubble so far has assumed spherical symmetry. The above discussion makes it clear that, in many important situations, real bubbles can deviate from sphericity. Even if the bubble is considered as able to radiate sound only at its pulsation

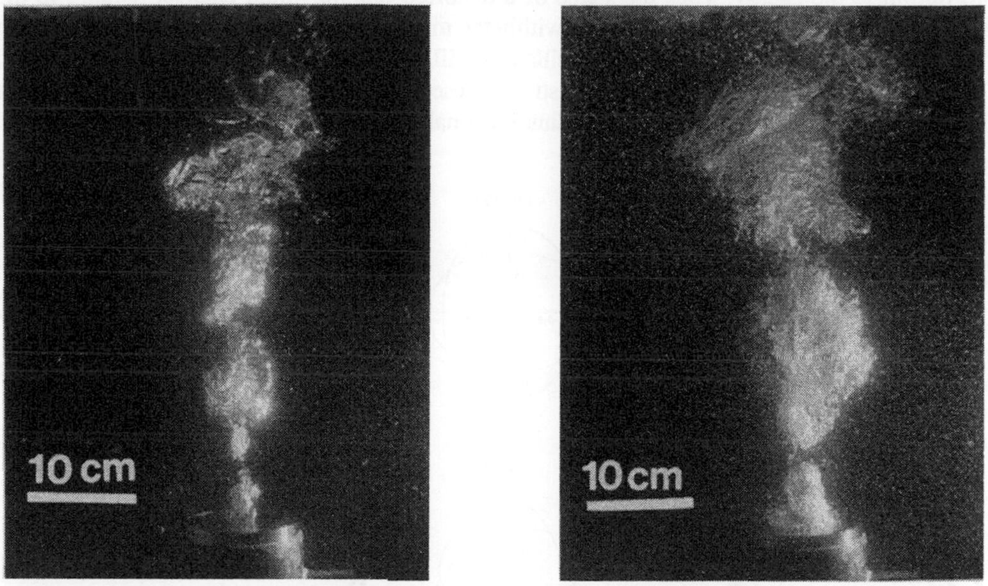

Figure 3.33 Injection of air into water at high flow rates through a nozzle of internal diameter 3 mm, set flush in a 10 cm diameter aluminium plate. (Photograph: TG Leighton, NA Jaques and K Fagan.)

frequency, ω_0, the shape oscillations will contribute indirectly to this emission by (a) bringing about excitation as observed in the above section, (b) increasing the volume of a bubble, and therefore lowering ω_0, through bringing about coalescence, and by (c) fragmentation, generating small bubbles (emitting higher frequencies) from larger ones. The following section will therefore discuss the nature of these shape oscillations, and investigate whether in addition to exciting and altering the ω_0 signal, they can contribute an acoustic emission of their own.

3.6 Nonspherical Bubble Oscillations

Minnaert [4] detected sound emission from freely-oscillating injected bubbles, the frequency of which agreed with the calculations from his theory of a spherical pulsating bubble. However, subsequent investigations [24, 25, 39–41] have demonstrated that such injected bubbles will commonly assume shapes significantly distorted from the spherical, as discussed in the previous section. The question of why such distorted bubbles should emit what appear to be acoustic signals at the breathing-mode frequency is a subject related to discussions dating from the last century, and is still controversial.

In Chapter 2, section 2.2.4(c) it was demonstrated how changes in shape of a bubble could be described in terms of a perturbation of appropriately summed spherical harmonics, superimposed on the spherical form. As an illustration, the perturbed shape was constructed from two-dimensional polar plots of the spherical harmonic. It is clear from the resultant shapes that an isolated bubble of that form would not be stable. It should be remembered that the surface tension of a liquid is numerically equal to the energy per unit area of surface, so that by enclosing the gas pocket within as small an area of bubble wall as possible, the system is tending

to minimise the energy. Thus the shape of a distorted bubble would change, tending to the spherical in order to encapsulate the gas within the minimum wall area. However, as with most systems of this type, overshoot and oscillation will occur.[26] In this way the shape-distorted bubble will oscillate. Figure 3.34 demonstrates these motions as predicted by linear theory for the superimposition upon a sphere of radius 1 a zonal spherical harmonic of amplitude 0.3, i.e.

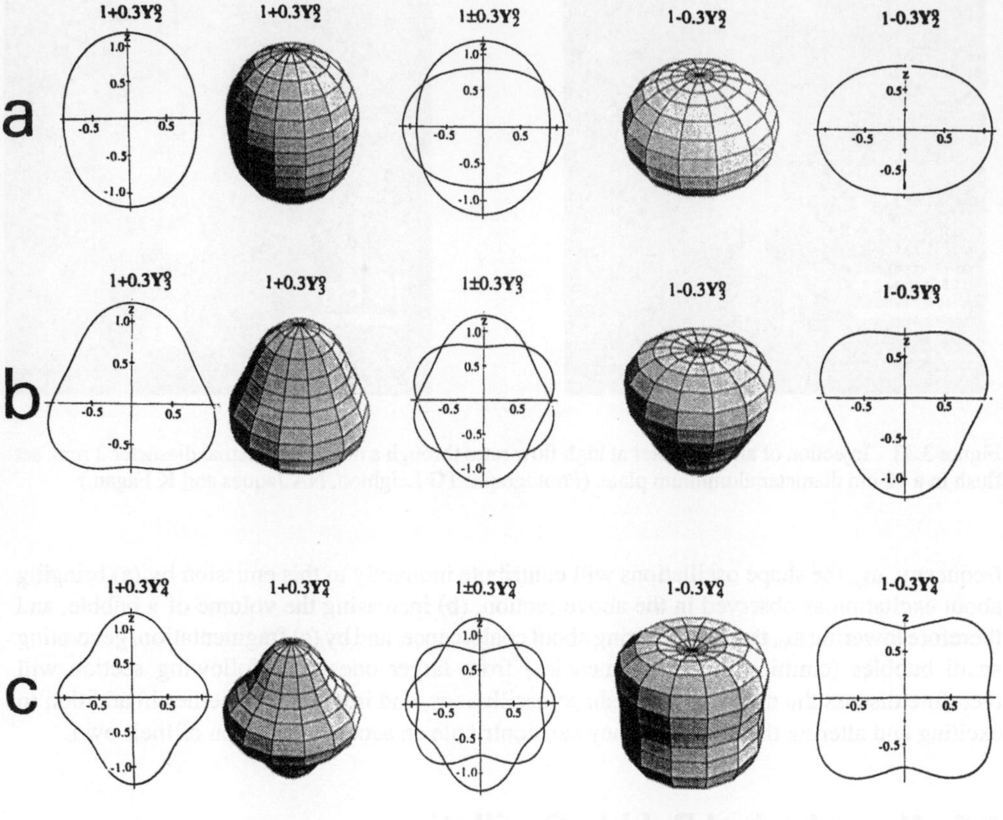

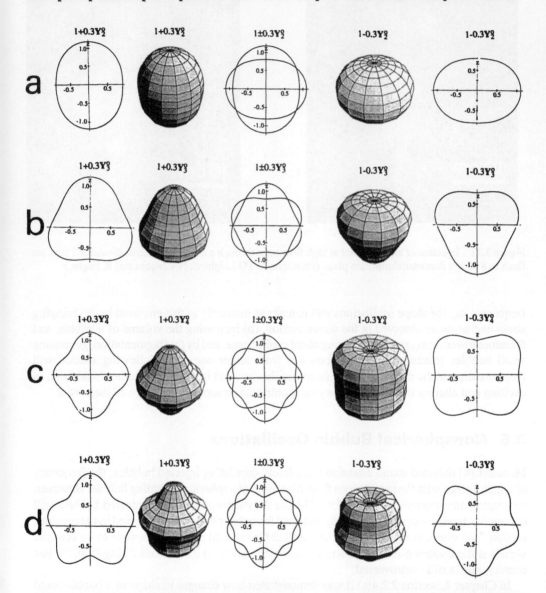

Figure 3.34 Shapes resulting from the linear superimposition upon a sphere of radius 1 unit of a zonal spherical harmonic of amplitude 0.3, of order n = (a) 2, (b) 3, (c) 4, (d) 5, (e) 6, (f) 7, (g) 8, (h) 9. For a given n the row has the two extremes of the oscillation shown in 3-D, $1 + 0.3Y_n^0$ on the left and $1 - 0.3Y_n^0$ on the right, with a 2-D polar diagram to accompany each on the outside. In the centre, the two extremes are superimposed.

26 A ball rolling down a valley will not stay at the bottom, the position of minimum energy to which it tends, until its kinetic energy has been dissipated.

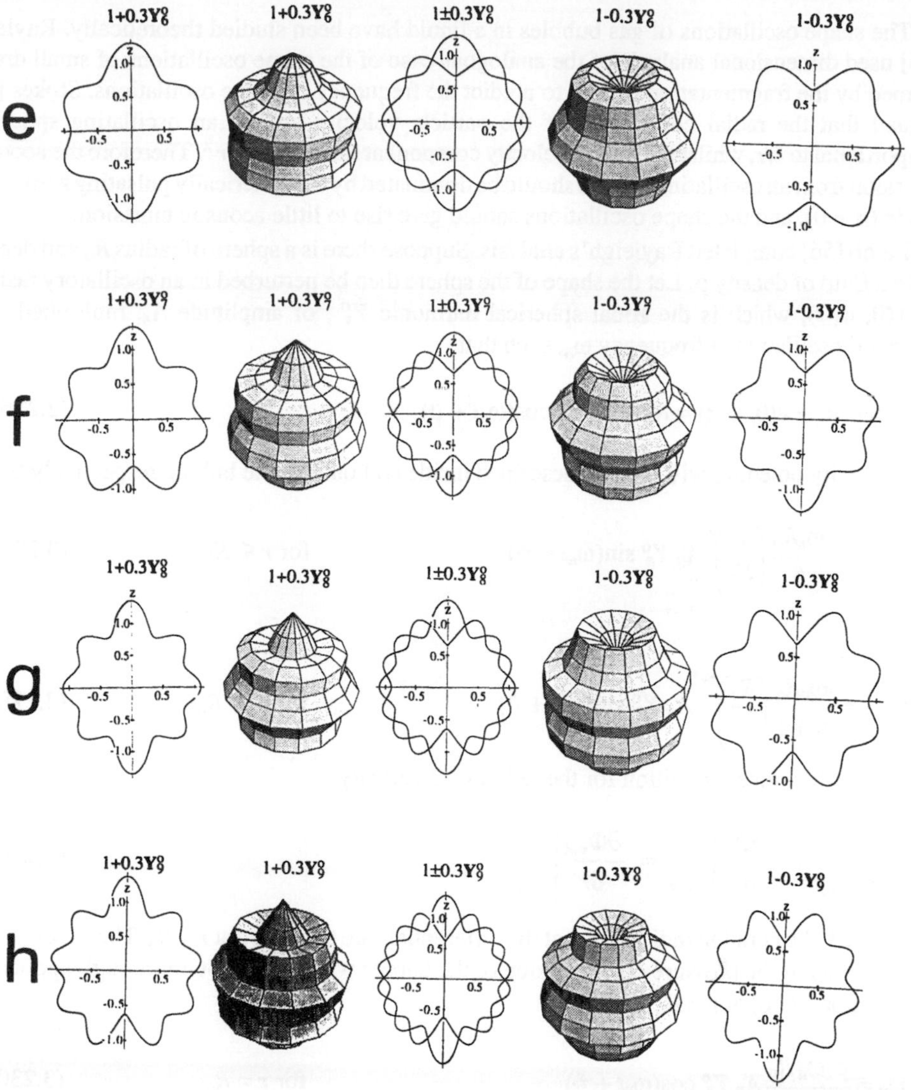

Figure 3.34 *continued*

$1 + 0.3Y_n^o\cos\omega t$. The order of the harmonic increases from $n = 2$ to $n = 9$ down the figure. For a given n, the row has the two extremes of the oscillation shown in 3-D, $1 + 0.3Y_n^o$ on the left and $1 - 0.3Y_n^o$ on the right, with a 2-D polar diagram to accompany each on the outside. In the centre, the two extremes are superimposed. In the figure, the harmonics have been superimposed individually upon separate spheres. The distortions thus described are the normal modes[27] of the bubble. In practice, any number of these modes of small amplitude can exist independently on a given sphere.

---

[27] A system will often be capable of oscillating in several different ways. The different forms of oscillation have different frequencies. These can be illustrated by holding a flexible chain or rope vertically, and letting the bottom end hang free. As the hand is oscillated back and forth in the horizontal direction, the chain will assume a number of discrete oscillatory forms, which have a vertical sine-wave-like appearance to the eye. The higher the frequency with which the hand drives the oscillation, the shorter the 'wavelength' of this sine wave. Each discrete form of oscillation is a mode. An arbitrary oscillation can be described by linear superimposition of normal modes [53].

The shape oscillations of gas bubbles in a liquid have been studied theoretically. Rayleigh [54] used dimensional analysis of the analogous case of the shape oscillations of small drops, formed by the fragmentation of jets, to predict the frequency of shape oscillations. Stokes [55] argued that the radial component of the particle velocity around an oscillating sphere is proportional to $1/r$, whilst the lateral velocity component falls off as $1/r^2$. Therefore the acoustic emission from an oscillating bubble should be dominated by the spherically pulsating zero-order mode ($n = 0$), and the shape oscillations should give rise to little acoustic emission.

Lamb [56] completed Rayleigh's analysis. Suppose there is a sphere of radius R_0 and density ρ_1 in a fluid of density ρ. Let the shape of the sphere then be perturbed in an oscillatory fashion by $\varepsilon(\theta, \varphi, t)$, which is the zonal spherical harmonic Y_n^o, of amplitude A_n, multiplied by a temporal oscillation at frequency ω_n, such that

$$\hat{R}(t) = R_0 + \varepsilon(\theta, \varphi, t) = R_0 + A_n Y_n^o \cos(\omega_n t + \vartheta) \tag{3.226}$$

The velocity potential will then be described inside and outside the bubble respectively by

$$\Phi_{in} = -\frac{\omega_n R_0}{n}\left(\frac{r}{R_0}\right)^n A_n Y_n^o \sin(\omega_n t + \vartheta) \qquad \text{for } r \leqslant R_0 \tag{3.227}$$

and

$$\Phi_{out} = \frac{\omega_n R_0}{n+1}\left(\frac{R_0}{r}\right)^{n+1} A_n Y_n^o \sin(\omega_n t + \vartheta) \qquad \text{for } r \geqslant R_0 \tag{3.228}$$

so that the boundary condition for the radial wall velocity

$$\left.\frac{\partial \varepsilon}{\partial t}\right|_{r=R_0} = \left.\frac{\partial \Phi_{in}}{\partial r}\right|_{r=R_0} = \left.\frac{\partial \Phi_{out}}{\partial r}\right|_{r=R_0} \tag{3.229}$$

is satisfied, the notation indicating that the differentials are evaluated at $r = R_0$. The time-varying components of the pressure on either side of the bubble wall can be found through application of equation (2.80) at $r = R_0$, giving

$$p_{in} = \frac{\rho_1 \omega_n^2 R_0}{n} A_n Y_n^o \cos(\omega_n t + \vartheta) \qquad \text{for } r = R_0 \tag{3.230}$$

and

$$p_{out} = -\frac{\rho \omega_n^2 R_0}{n+1} A_n Y_n^o \cos(\omega_n t + \vartheta) \qquad \text{for } r = R_0 \tag{3.231}$$

As we have seen in Chapter 2, section 2.1.1 (equation (2.5)), the difference in pressures across an interface that is perturbed from the spherical, and which therefore has two principal radii of curvature ($\hat{R}_1$ and $\hat{R}_2$) is given by

$$p_{in} - p_{out} = \sigma\left(\frac{1}{\hat{R}_1} + \frac{1}{\hat{R}_2}\right) \tag{3.232}$$

Lamb [56] finds the sum of the reciprocals of the principal radii to be

$$\left(\frac{1}{\hat{R}_1}+\frac{1}{\hat{R}_2}\right)=\frac{2}{R_0}+\frac{(n-1)(n+2)}{R_0^2}A_n Y_n^o \sin(\omega_n t+\vartheta) \tag{3.233}$$

(Lamb takes A_n to equal unity throughout.) Substitution of equations (3.230), (3.231) and (3.233) into equation (3.232) gives

$$\omega_n^2 = n(n-1)(n+1)(n+2)\frac{\sigma}{\{(n+1)\rho_1+n\rho\}R_0^3} \qquad (n>1) \tag{3.234}$$

To obtain the oscillations of a liquid drop in air, set $\rho_1 \gg \rho$ in equation (3.234) to obtain

$$\omega_n^2 \approx n(n-1)(n+2)\frac{\sigma}{\rho_1 R_0^3} \qquad (n>1) \tag{3.235}$$

To obtain the frequency of oscillation of a gas bubble in water, set $\rho \gg \rho_1$ in equation (3.234) to obtain

$$\boxed{\omega_n^2 \approx (n-1)(n+1)(n+2)\frac{\sigma}{\rho R_0^3} \qquad (n>1)} \tag{3.236}$$

The zero-order mode of the gas bubble has a frequency ω_o which from equation (3.204) is given by

$$\omega_0^2 = \frac{3\kappa p_o}{R_0^2\rho}\left(1+\frac{2\sigma}{p_o R_o}\right)-\frac{2\sigma}{\rho R_0^3}=\frac{3\kappa p_{i,e}}{R_0^2\rho}-\frac{2\sigma}{R_0^3\rho} \tag{3.237}$$

which, when the surface tension pressure is negligible in comparison with the hydrostatic pressure, reduces to equation (3.36).

The acoustic pressure field at a distance r can be found by applying equation (2.80) to equation (3.228), to give

$$P=-\frac{\rho\omega_n^2 R_o}{n+1}\left(\frac{R_o}{r}\right)^{n+1}A_n Y_n^o \sin(\omega_n t+\vartheta) \qquad \text{for } r \geqslant R_o \tag{3.238}$$

The acoustic pressure fields decay as $r^{-(n+1)}$. Since $n>1$, Lamb's linearised theory predicts that the acoustic pressure from the shape oscillation of bubbles would decay rapidly with radial distance r.

This agrees with the linearised theory of Chapter 2, section 2.2.4(c) which predicts that the velocity potential and pressure field decay as $r^{-(n+1)}$ for bubbles whose oscillations can be described by assigning spherical harmonic solutions to the velocity potential (the $n = 1$ mode excepted as being not pertinent to such oscillations). Thus although the $n = 0$ pulsation mode can emit sound strongly, the higher modes are by this theory much weaker emitters (the $n = 2$ 'pancake-needle' mode, for example, producing a pressure field which decays as r^{-3}). Another interpretation of this is given by Neppiras [57], who states that these oscillations of order $n \geqslant 2$ are pure distortion modes, not involving any area change, and are therefore weakly coupled to the liquid. Strasberg [39] also concluded from linear theory that there would be negligible acoustic emission from the shape modes, and that the acoustic output from a freely-oscillating bubble resulted solely from the breathing mode.

Other researchers have studied the shape oscillations in bubbles both in response to impulsive pressures [58] and periodic driving forces.[28] Of particular interest is the analysis of Longuet-Higgins [59], who has shown that nonlinear theory predicts that the distortion modes ($n \geq 2$) generate at second order a monopole acoustic radiation (that is, one where the pressure decays as r^{-1} as with the $n = 0$ mode). The frequency of this radiation is twice the basic frequency of the distortion mode, and the pressure amplitude is proportional to the square of the distortion amplitude.

Longuet-Higgins gives a solution for the $n \geq 2$ oscillations in the limit of an incompressible, irrotational fluid of negligible viscosity, applicable for small-perturbation approximations to the velocity potential and the bubble wall radius function. If the external velocity potential for a given mode is Φ_n, then Longuet-Higgins expands it for perturbations to first order ($\Phi_n^{(1)}$) and second order ($\Phi_n^{(2)}$), so that

$$\Phi_n = \Phi_n^{(1)} + \Phi_n^{(2)} + \ldots \tag{3.239}$$

The first-order terms are evaluated to give

$$\Phi_n^{(1)} \approx A_n \left(\frac{R_o}{r}\right)^{n+1} Y_n^m(\theta,\varphi) \frac{R_o \omega_n}{n+1} \sin\omega_n t \tag{3.240}$$

and

$$R_n^{(1)} \approx R_o + A_n Y_n^m(\theta,\varphi) \cos\omega_n t \tag{3.241}$$

where A_n is an amplitude reflecting the magnitude of the perturbation. The emitted pressure for such perturbations can be obtained through application of equation (2.80)

$$P_n = -\rho\dot{\Phi}_n \tag{3.242}$$

Thus, if

$$P_n = P_n^{(1)} + P_n^{(2)} + \ldots \tag{3.243}$$

then to first order the radiated pressure is

$$P_n^{(1)} \approx -\rho\dot{\Phi}_n^{(1)} = -A_n \left(\frac{R_o}{r}\right)^{n+1} Y_n^m(\theta,\varphi) \frac{\rho R_o \omega_n^2}{n+1} \cos\omega_n t \tag{3.244}$$

This clearly decays as $r^{-(n+1)}$. However, Longuet-Higgins then takes the formulation to second order, yielding additional terms for the velocity potential and radiated pressure

$$\Phi_n^{(2)} = \left\{\frac{(n-1)(n+2)(4n-1)}{4(2n+1)} \frac{\sigma A_n^2}{R_o^2}\right\}\left\{\frac{2\omega_n}{4\omega_n^2 - \omega_o^2}\right\} \frac{1}{r} \sin 2\omega_n t \tag{3.245}$$

and

$$P_n^{(2)} = -\left\{\frac{(n-1)(n+2)(4n-1)}{(2n+1)} \frac{\rho\sigma A_n^2}{R_o^2}\right\}\left\{\frac{\omega_n^2}{4\omega_n^2 - \omega_o^2}\right\} \frac{1}{r} \cos 2\omega_n t \tag{3.246}$$

[28]See Chapter 4, sections 4.3.3 and 4.4.6.

Longuet-Higgins notes that for long-distance propagation where the phase change in pressure resulting from finite travel times is not negligible, the $2\omega_n t$ term in pressure should be replaced by $2\omega_n(t - r/c)$ for $r \gg R_0$.

Equations (3.245) and (3.246) demonstrate that, when taken to second order, there will be terms in the velocity potential, wall radius function and radiated pressure that are independent of θ and φ, and which decay as r^{-1}, i.e. are monopole. Within the limitations of the assumption of no damping, the magnitude of these terms clearly depends on the factor

$$\left\{\frac{(n-1)(n+2)(4n-1)}{(2n+1)}\frac{\sigma}{R_0^3}A_n^2\right\}\left\{\frac{\omega_n^2}{4\omega_n^2 - \omega_0^2}\right\}R_0 \tag{3.247}$$

that is, on the square of the first-order surface wave amplitude A_n.

However, there is also an important resonance-like effect: the amplitude becomes increasingly large for those modes having frequency ω_n approaching $\omega_0/2$. Figure 3.35 shows the relationship between $2\omega_n$ and ω_0 for bubbles in the size range $1\ \mu\text{m} \leqslant R_0 \leqslant 10$ cm. For a few discrete bubble sizes, $\omega_n = \omega_0/2$ for one particular mode ($n = n$crit) and strong 'resonances' can be expected: one mode will emit at the frequency $2\omega_{n\text{crit}} = \omega_0$. Modes close to the critical one will emit less strongly, the amplitude decreasing with increasing difference between the mode number n and the critical one $n$crit. If the bubble size is one of those where no particular node has $\omega_n = \omega_0/2$, then those nodes having ω_n closest to $\omega_0/2$ will emit most strongly.

An important point to note is that whenever any mode produces this second-order monopole emission, the signal is at the frequency $2\omega_n$. Therefore if the excitation is strong because ω_n is close to, but not exactly equal to, $\omega_0/2$, the emission will not be at ω_0 but close to it (at $2\omega_n$). If several nodes are excited in this manner, one will expect a spread of frequencies around ω_0. The frequency spectrum will also contain terms in ω_n due to the first-order emission terms described in equation (3.244). Though these first-order emissions (at ω_n) will initially be larger than the second-order monopole emissions (at $2\omega_n$) at the bubble wall, they are *not* monopole and will therefore rapidly decay with increasing distance from the bubble.

Longuet-Higgins draws an interesting analogy between surface oscillations on the bubble and standing waves in deep water. In the latter case, second-order theory predicts a pressure fluctuation at twice the oscillatory frequency at some distance below the surface which, unlike

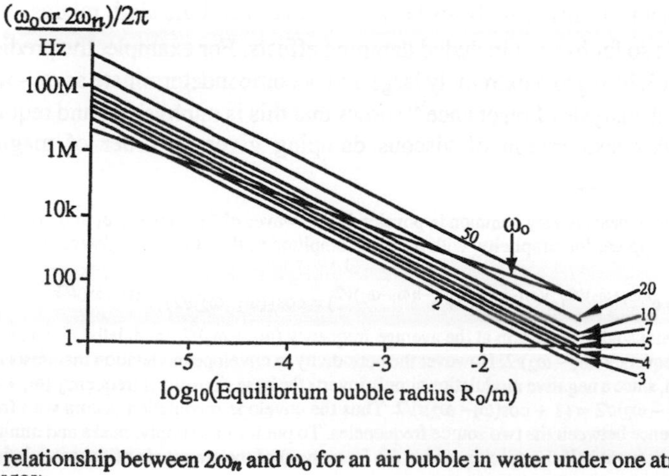

Figure 3.35 The relationship between $2\omega_n$ and ω_0 for an air bubble in water under one atmosphere (after Longuet-Higgins [59]).

the first-order result, does not decay with depth. Progressive waves give no such fluctuations. The root cause of the fluctuation is the change in height of the centre of mass of the fluid, an oscillation which occurs at twice the wave frequency. In a similar manner, the shape distortion modes can be likened to standing waves which are formed on the surface of the bubble through the superposition of two progressive surface waves, generating monopole emissions.

The excitation of such modes was investigated by Longuet-Higgins [60]. The initially undamped model considers a bubble which is distorted, but has identical volume to that of a spherical bubble containing the same amount of gas at equilibrium. The initial shape of a bubble is resolved into the normal modes and can be expressed as the sum of spherical harmonics. At times $t > 0$ these modes are shown to radiate independently at second order. By averaging over the sphere, Longuet-Higgins shows that the total contribution to the monopole radiation at second order is the sum of the contribution from the individual modes, without interaction. As expected, emission will be particularly strong from the harmonic possessing the linear mode frequency $\omega_n \approx \omega_0/2$, that acoustic emission of course being at $2\omega_n$. In addition, despite the fact that initially there is no volume change, the analysis predicts that the zero-order breathing mode will be excited.

Consider the simplest case, where only one of the distortion modes radiates strongly in monopole at second order. Then, since the two most potent emissions (that of the breathing mode ω_0, and that at twice the frequency of the resonating higher mode, i.e. at $2\omega_n$) are not exactly equal, beats will be set up at the difference frequency.[29] Thus the expected output of the bubble will be some sort of sinusoid at approximately $(\omega_0+2\omega_n)/2$, in an envelope modulated by the beat frequency. It is in fact likely that there will always be not just one, but several modes possessing a linear mode frequency ω_n close enough to $\omega_0/2$ to generate significant excitation. Thus, in general, one would expect a sinusoid at around ω_0 under an envelope which is modulated in some manner.

Longuet-Higgins illustrates that case of a bubble initially distorted from the spherical by a single peak, giving a bubble shape similar to that of bubbles just released from nozzles. The resulting modular expansion gives the appearance of waves of disturbance propagating around the bubble, and when the passing disturbances coincide at the poles of the bubble a high-amplitude perturbation is seen at that point. Since second-order radiated pressure tends as the square of the perturbation amplitude, this periodic occurrence generates frequencies that can also be seen in the predicted acoustic emission, modulating the envelope. This modulation is smaller than that resulting from beats between the zero and the sixth modes.

The analysis so far has not included damping effects. For example, the predicted amplitudes in (3.245) and (3.246) grow extremely large and become indeterminate as $\omega_n \to \omega_0$. Comparison with the general analysis of resonance[30] shows that this is unphysical, and requires the addition of damping. A consideration of viscous damping gives an order of magnitude estimate.

---

[29] The phenomenon of beats is very common in physics. If two waves of frequencies ω_1 and ω_2 are present in the same waveform (let us suppose, for simplicity, with the same amplitude), then the signal detected at a fixed point in space varies in time as

$$e^{i\omega_1 t} + e^{i\omega_2 t} = e^{i(\omega_1+\omega_2)t/2}(e^{i(\omega_1-\omega_2)t/2} + e^{-i(\omega_1-\omega_2)t/2}) = \cos\{(\omega_1-\omega_2)t/2)\}e^{i(\omega_1+\omega_2)t/2} \quad (3.f1)$$

This represents a sinusoidal waveform at the average frequency $(\omega_1 + \omega_2)/2$, modulated by an envelope that varies in amplitude with frequency $(\omega_1 - \omega_2)/2$. However the periodicity in envelope modulation manifests itself in the intensity of that modulation, since a negative modulation simply inverts the base sinusoid of frequency $(\omega_1 + \omega_2)/2$. The intensity varies as $\cos^2(\omega_1 - \omega_2)t/2 = (1 + \cos(\omega_1 - \omega_2)t)/2$. Thus the envelope modulation occurs with frequency $(\omega_1 - \omega_2)$, equal to the difference between the two source frequencies. To put it more simply, peaks and minima in a sinusoidally varying envelope occur at half the spacing (twice the frequency) of the amplitude of the modulation.

[30] See Chapter 4, sections 4.1.1 and 4.1.2(a).

However, clearly with these monopole emissions and volume changes dissipating energy, radiation and thermal damping must also be considered. As expected,[31] the analysis shows that damping causes the envelope to decay in time, introduces phase shifts, and changes the resonance frequency slightly, to ω_b. Thus consideration of such mode interactions must include the damping factors, since both the interaction itself and the acoustic output are frequency-dependent (the beat frequency, for example, is sensitive to small changes in the frequency of the resonance). The term $4\omega_n^2 - \omega_0^2$ in equations (3.245) and (3.246) is replaced by

$$\sqrt{(4\omega_n^2 - \omega_b^2)^2 + (4\beta_n\omega_n)^2} \tag{3.248}$$

which ensures finite amplitude response at resonance. The term β_n is the appropriate form of the general amplitude damping constant $\beta = (b/2m)$, where β_n^{-1} represents a time constant of decay for the amplitude of that mode.[32] Estimates of the β_n can be obtained through analysis of the rates of decay of energy in the modes [60], which depend on both linear (first-order) and nonlinear (second-order) terms [61], through

$$\beta_n = -\frac{1}{2A_n^2}\frac{dA_n^2}{dt} \tag{3.249}$$

Since the addition of damping modifies the theory through equation (3.248), the denominator is always finite in equations (3.245) and (3.246), so that the theory is uniformly valid with respect to the mode frequency ω_n provided that the distortion parameter (say, A_n/R_0) is very much less than the ratio β_n/ω_n. Thus when damping is included, the monopole pressure radiated at second order is

$$P_n^{(2)} = \frac{-1}{r}\left\{\frac{(n-1)(n+2)(4n-1)}{(2n+1)}\frac{\rho\sigma A_n^2}{R_0^2}\right\} \times \left\{\frac{\omega_n^2}{\sqrt{(4\omega_n^2 - \omega_b^2)^2 + (4\beta_n\omega_n)^2}}\right\}\cos 2\omega_n t \tag{3.250}$$

The effects of damping in such shape oscillations are further discussed in section 3.8.2(a) [38].

The total contribution to the monopole radiation at second order is the sum of the contributions from the individual modes [60]. At second order there is a volume change associated with these distortion modes. Volume pulsations are an essential accompaniment of shape oscillations at second order [61]. These volume changes are not, however, essentially the cause of the radiation, since equation (3.228) shows that monopole radiation would still occur even if the radial mode frequency were to vanish. Thus the radiation is not caused by second-order volume change.

One interesting prediction of the theory is that the initial magnitude of the monopole pressure fluctuation depends on the bubble radius only through the term $(A_n/R_0)^2$, that is, it depends on the *shape* of the bubble, not its absolute size (see equation (3.250)). However, after this initial moment, the pulse from the smaller bubble decays more rapidly as a result of the greatly increased damping.

It is clear from the analysis that higher-order interactions could occur generating, for example, the frequency of the distortion mode, ω_n [60]. Expansions of this sort would be

[31] See section 3.1.3.
[32] See section 3.1.3.

inadequate to describe the highly nonlinear behaviour of which a severely distorted bubble is capable.

Ffowcs Williams and Guo [62] suggest that the excitation of monopole emissions from shape oscillations, as proposed by Longuet-Higgins, is in fact only feasible if there is a driving system which can supply the energy for the breathing mode. If this is not present, Ffowcs Williams and Guo suggest that as a result of the sum of energies in the modes being unable to exceed the initial bubble energy, the energy of the excited breathing mode will be low. They suggest that the perturbation approach of Longuet-Higgins fails to consider accurately the long-term coupling (e.g. for times longer than around 4 to 5 periods of bubble oscillation), there being a gradual energy transfer between the shape to the breathing mode, which reduces the amplitude of the shape oscillation and so makes the perturbation analysis invalid. They conclude that monopole emissions from shape oscillations are unlikely to be acoustically significant as the breathing mode will never grow sufficiently. Ffowcs Williams and Guo do not include radiation and dissipation effects in their formulation. In reply Longuet-Higgins [61] points to this as being likely to lead to misleading results: through consideration of the $n = 6$ mode discussed by Ffowcs Williams and Guo, Longuet-Higgins illustrates that when damping is included, the slow transfer of energy from a shape to the breathing mode is insignificant. The decay of the shape oscillation is rapid, which means that the growth of the breathing mode, and its direct contribution to the pressure pulse, will be very small. In fact the excitation, considered as an initial value problem by Longuet-Higgins [60], does not come about through gradual transfer of energy, but through resonance during the first few cycles of the shape oscillation. It is through this mechanism that the 'resonance' amplification arises, which manifests itself in the formulation as the $((\omega_0^2 - 4\omega_n^2)^2 + (4\beta_n\omega_n^2))^{-1/2}$ term in equation (3.250). In the excitation of damped bubbles, Longuet-Higgins calculates that the initial energy of the shape oscillations significantly exceeds that required to produce the volume pulsations. Almost all of that energy is dissipated, mainly through thermal and radiation damping. Longuet-Higgins maintains that the effective energy balance is between the oscillation energies and their rates of dissipation, not between the forms of oscillation.

In summary of the mechanism therefore, Longuet-Higgins proposes that shape oscillations emit sound in monopole fashion at second order in the distortion parameter which measures the perturbation of the bubble from its equilibrium spherical shape, the pressure being proportional to the square of the first-order wave amplitude, A_n. It will be amplified when $2\omega_n$ approaches ω_0, the 'resonance' being substantially limited by acoustic radiation and thermal damping. The evidence suggests that the presence of energy in these modes may significantly affect the observed damping and pulse shapes[33] [38]. The contribution of this emission to ocean ambient noise is discussed by Longuet-Higgins [63], who uses the damped theory to predict that any broad spectrum of bubble-generated sound might be expected to contain peaks corresponding to these mode resonances. It should be noted that freshly entrained, excited bubbles in the sea will in general be close to the free boundary of the ocean surface, and will experience effectively less damping (by a factor of around 2). This will lead to an increase in the amplification and sharpness of the resonance over the damping of such bubbles in an infinite fluid [61]. Though a preliminary examination of ocean data by McConnell [64] and Farmer and Vagle [65, 66] enabled Longuet-Higgins [63] to speculate that there might be evidence of such peaks, particularly those relating to the even-numbered modes, subsequent research has revealed other mechanisms by which these peaks might arise.[34]

---

[33] See section 3.8.2(a).
[34] These are discussed in section 3.8.2(b)(ii).

We have seen that injected bubbles can be excited into passive emissions (section 3.5.1), that injection can generate extreme shape oscillations (section 3.5.2), and that shape oscillations may be associated with monopole emissions (section 3.6). However, the precise excitation mechanism is yet to be discussed in any depth. Since such a theme has wider implications than the simple injection process, it is appropriate to firstly introduce the environment where the entrainment and excitation of bubbles are most numerous, contributing to 'the murmur of the brook, the roar of the cataract, or the humming of the sea' [4]. That environment is the natural world.

3.7 Bubble Entrainment in the Natural World

3.7.1 Babbling Brooks, Rain and the Sea: Situations and Applications

The passive acoustic emissions of freely-oscillating bubbles were first used to size a natural bubble population by Leighton and Walton [41]. Figure 3.36 shows sections of the hydrophone record from a brook. Each exponentially decaying sinusoid can be identified with the excitation of a bubble whose size can be determined by application of equation (3.36) to the measured frequency of the sinusoid. The appropriate value of κ may be determined from plots similar to Figure 3.25(b) for air bubbles in water, or from predicted values:[35] the majority of bubbles in

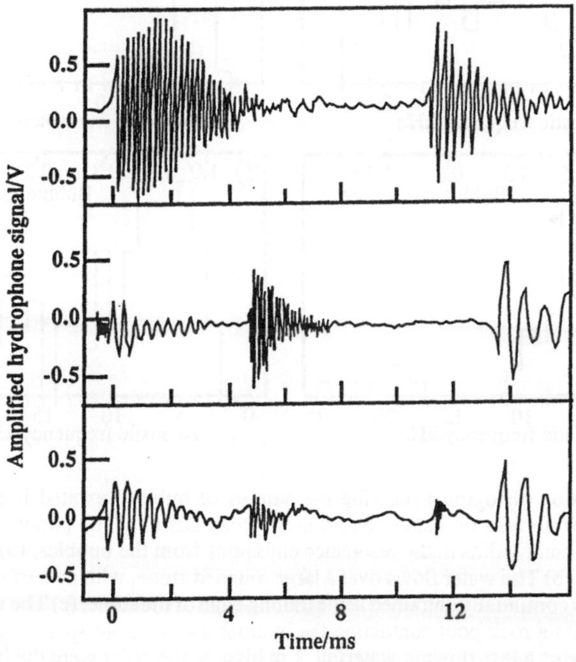

Figure 3.36 Sections of the hydrophone record from a babbling brook (after Leighton and Walton [41]).

[35]See section 3.4.2(a) and Chapter 4, section 4.4.2.

this size range will tend to the adiabatic. In this way Leighton and Walton were able to construct population histograms showing the number of bubbles excited in a given size range for brooks, streams and waterfalls on Kinder Scout in the Peak District (Figure 3.37). Flow over a large rounded stone, with no surface splashing, generated a narrow stream of bubbles at the trailing edge, the bubbles tending to be large with a relatively narrow size distribution (Figure 3.37(b)). Wider distributions are created when the water drops over a 30 cm natural weir (Figure 3.37(a)) or a 1 m waterfall (Figure 3.37(d)). There is a suggestion in these two of bimodal distribution, which is clearer in Figure 3.37(c), where the water flowed smoothly over a 30 cm drop into a rock pool without splashing.

Similar histograms can readily[36] be produced for other processes where bubbles are entrained and excited. Figure 3.38 shows histograms as a function of (i) the resonance frequency and (ii) the bubble size, determined from the passive underwater acoustic emission. Figure 3.38(a) shows the results for bubble entrainment through the action of breaking waves by the shore at Brighton on a calm day. There is an indication that 0.45 mm < R_0 < 0.47 mm is the preferred size range. Bubbles can also be entrained by the impact of liquid drops onto bodies of water.

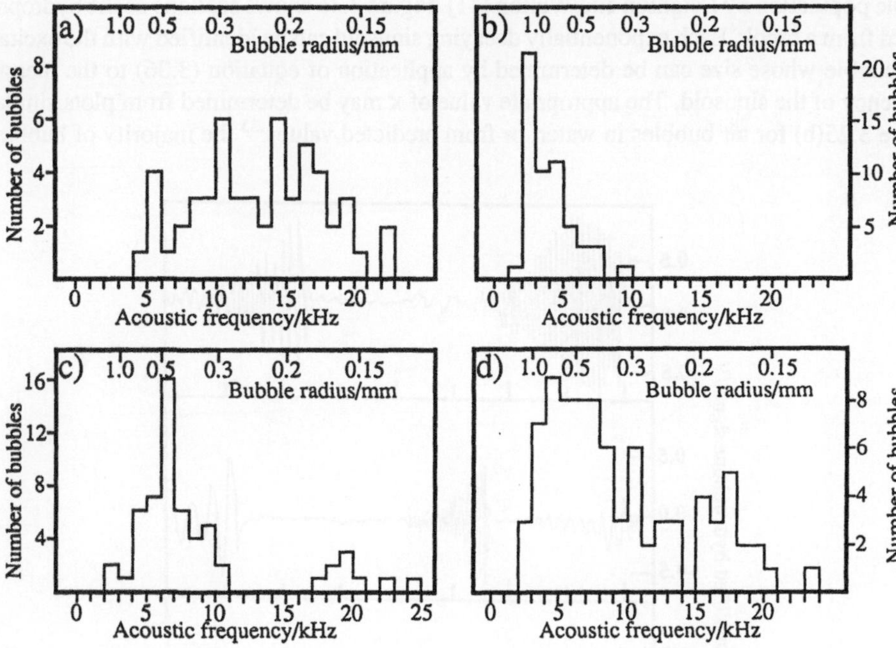

Figure 3.37 Population histograms showing the number of bubbles excited in a given size range for brooks, streams and waterfalls on Kinder Scout in the Peak District. The population was calculated by distinguishing 1 kHz bandwidths in the resonance emissions from the bubbles. (a) The water drops over a 30-cm natural weir. (b) The water flows over a large rounded stone, with no surface splashing; a narrow stream of bubbles was continually entrained at the trailing edge of the stone. (c) The water flowed smoothly over a 30-cm drop into a rock pool containing the hydrophone, without splashing. (d) The hydrophone was placed at the base of a fast-flowing waterfall, 1 m high, at the point were the bubbles emerged (after Leighton and Walton [41]).

[36] These light-hearted histograms are taken for small numbers of bubbles, the bubble traces being counted manually on the oscilloscope. For a serious survey of entrainment, the operator should use a recognition routine to automate the process and so count a much larger sample population. One such is the Gabor transform [67].

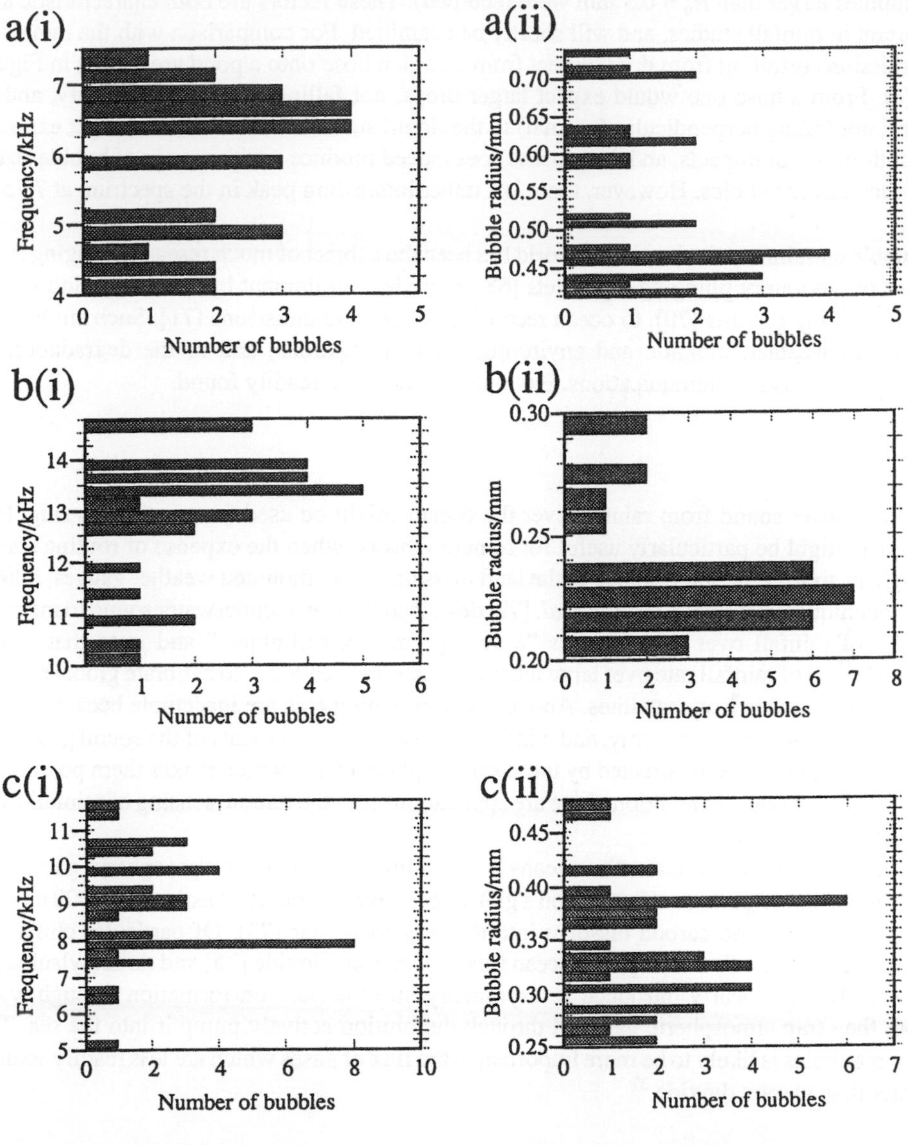

Figure 3.38 Bubble population histograms as a function of (i) the resonance frequency and (ii) the bubble size, determined from the passive underwater acoustic emission. (a) Bubble entrainment through the action of breaking waves by the shore at Brighton on a calm day. The underwater sound of the impact of (b) natural light rain on the fish pond at the Cavendish Laboratory, Cambridge, and (c) spray from a garden hose, impacting a garden pond.

Anyone who has dived into a swimming pool will be familiar with the entrainment of bubbles through the impact of a solid body onto a flat liquid surface. When a liquid drop impacts a body of water, the processes are significantly different: however, bubble entrainment may occur, and will be discussed later. Figure 3.38(b) shows the histograms pertaining to the impact of natural light rain on the fish pond at the Cavendish Laboratory, Cambridge. A peak in the acoustic spectrum is indicated at around 13–14 kHz, and there is a tendency only to entrain small bubbles (no bubbles larger than $R_0 = 0.3$ mm were recorded). These factors are both characteristic and important in rainfall studies, and will shortly be examined. For comparison with the rain data, the emissions resulting from drop impact from a garden hose onto a pond are shown in Figure 3.38(c). From a hose one would expect larger drops, not falling at terminal velocity, and in general not falling perpendicular (normal) to the liquid surface. Therefore one might expect a less uniform set of impacts, and the process does indeed produce a wider range of bubble sizes, and much larger bubbles. However, there is a rather interesting peak in the spectrum at 7.75–8 kHz.

Bubble entrainment in the natural world has been the subject of much research, ranging from the use of laboratory plunging liquid jets [68] to model entrainment by breaking wind waves [69] and falling streams [70], to ocean recordings of passive emissions [71]. Such studies are relevant to weather, climatic and environmental investigations, and to the degradation of underwater acoustic communications. Examples of these are readily found.

Environment

The underwater sound from rainfall over the oceans might be used for weather sensing: this technique might be particularly useful for remote sensors, when the expense of routing a ship through an area, or the interference of the land mass on island-mounted weather gauges, makes those techniques unsuitable. Laville *et al.* [72] described the use of underwater sound to measure the rate of rainfall over the ocean as "a very promising technique," and state that "local measurements of rainfall rate over large lakes or oceans are needed ... to calibrate global rainfall measurements made from satellites. Above water measurements are inaccurate because of the direct effect of wave action, spray, and wind. Underwater measurements of the sound generated by rainfall are not directly affected by these surface phenomena, which makes them potentially much more accurate." In addition, there are applications for "the remote sensing of rainfall from an underwater location."

On an environmental scale, the oceans act as a huge reservoir of atmospheric gases into which atmospheric gases can dissolve on a global scale. By conservative estimates, 1000 million tonnes of atmospheric carbon dissolve into the seas each year [73]. Of particular climatical significance [74] are the atmosphere/ocean fluxes of carbon dioxide [75] and dimethylsulphide [76, 77]. Bubbles clearly introduce an asymmetry into the flux: on formation through wave action, they trap atmospheric gas, and through dissolution actively pump it into the sea. The effect of bubbles is likely to be more important in the flux of gases which are less readily soluble in water than carbon dioxide.[37]

Ambient Sound in the Oceans

Bubbles entrained by rainfall or breaking waves contribute to the ambient sound in the oceans, which is the background noise for underwater acoustic communications. Since radio waves are evanescent in seawater,[38] to communicate underwater it is necessary to employ other means, such as acoustic signals. These, however, must be distinguishable from the background ambient noise. Urick [78] gives examples of mechanisms which contribute to this background. Tides

---

[37] See section 3.8.1(d).

[38] Electromagnetic energy was shown in Chapter 1, section 1.1.7 to decay far more rapidly than acoustic energy in seawater at comparable frequencies.

and the direct hydrostatic effect of waves on hydrophones can give rise to very low-frequency effects. Turbulence and nonlinear wave interactions can generate low-frequency pressure signals. Seismic effects are found below 1 Hz, and their importance may well extend from 10 Hz to 100 Hz if ship traffic is low. The latter is dominant in the region 50–500 Hz, and may even be detectable in excess of 1000 miles from the source. Ambient noise has been measured in both deep and shallow water [79–82], and since propagation is poorer over a continental shelf, shipping noise contributes significantly in shallow water. The noise from distant storms can also be important in this frequency range of 50–500 Hz. Knudsen *et al.* [83] summarised much World War II data by characterising the undersea noise in the frequency range 100 Hz to 25 kHz in the so-called 'Knudsen spectra'. Wenz [81] confirmed these spectra, and commented that the wind-dependence of the spectra between 50 Hz and 10 kHz could be caused by air bubbles. The Knudsen spectrum can conveniently be divided into three parts [84]. There is a dominant low-frequency component below 50 Hz, a shipping component at intermediate frequencies, and above 100 Hz one which peaks at 500 Hz, then falls off at ~ 6 dB per octave. At windspeeds greater than ~5 m/s, whitecaps are extensively present and the Knudsen spectrum originates from breaking waves [65, 85, 86]. However, the Knudsen spectrum can still be detected at windspeeds as low as 1 m/s. Suggestions that the 'popping' of surface bubbles may be responsible [87] have been dismissed by Prosperetti and Lu [88] as incapable of radiating sufficient acoustic energy into the water. Updegraaf and Anderson [89] observed a mode of wavebreaking that does not leave observable whitecaps, termed 'microbreaking' [90] which, through excitation of bubbles, may be the noise source at low windspeeds [84]. In a discussion published in 1983, Urick [78] lists the processes known to that date which might contribute to the Knudsen spectrum, including flow noise produced by the transmission into the sea of turbulent pressure changes generated as the wind blows over the rough sea surface, and the sound radiated from wind-induced sea surface waves of very long wavelength which propagate over the sea surface more rapidly than soundspeed in water. Since that publication, much original work has been done. Though the mechanism through which the Knudsen spectrum is generated is not fully understood [44], breaking waves undoubtedly contribute a very significant part in a number of ways, not least through the entrainment of bubbles. These emit sound upon entrainment, may generate low-frequencies through collective oscillations, and modify the propagation characteristics of sound through the ocean.

One fascinating application of the incoherent ambient noise in the oceans has been proposed by Buckingham [91]. Buckingham [private communication] uses the following analogy. For years, objects have been detected in the ocean using passive and active sonar. In the former, a submarine is, for example, detected by passively monitoring the noise it itself emits, for example, from the engines, or propeller noise. In active sonar, the searcher sends out an interrogating sound pulse, which bounces off the submarine (or fish shoal, sea bottom etc.), and on reception gives the direction and, from the time-of-flight, the distance of the target. A visual analogy would be to imagine how we see an aircraft in the sky: passive detection would correspond to our seeing the lights attached to its wings, and active detection would correspond to our illuminating it with the beam from a searchlight. This nicely illustrates the problem with active detection, in that it alerts the target to the presence of the detector.

However, we most usually see an aircraft (or anything else) when our eyes simply pick up the incoherent ambient daylight levels that are scattered from it. In the same way, Buckingham proposes that image of undersea objects might be obtained through detection and focusing of the incoherent ambient noise that is scattered from it. Buckingham [91] confirmed this principle by examining the change in the ambient noise detected when a previously uninterupted field is presented with 0.9×0.77 m^2 reflecting plywood targets. The geometry of the experiment, which took place off the pier of Scripps Institution of Oceanography, La Jolla, California, is shown in Figure 3.39. A 1.22 m diameter parabolic acoustic reflector with a hydrophone sensor at the focus acting as an acoustic lens, was used as the detector. Approximately 7 m distant from this, the plywood targets could be rotated such that they presented either a face or an edge to the detector, termed 'on' and 'off' respectively. Figure 3.40 shows the effect on the ambient noise resulting from rotation of the targets. In Figure 3.40(a) the noise detected signal is seen to

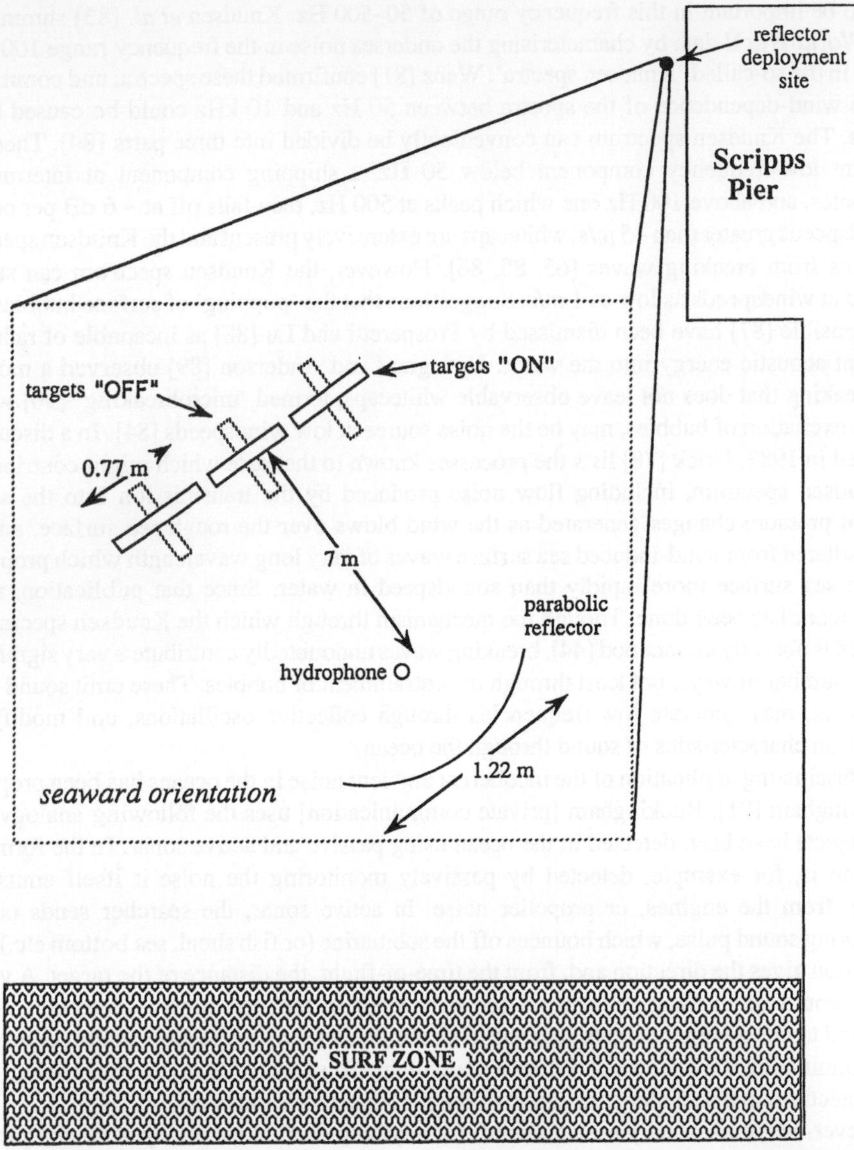

Figure 3.39 Plan view of the seaward configuration of the parabolic reflector and targets in the sea off Scripps pier. The deployment site is about 200 m out from the shore; at the end of the pier the water depth is about 7 m. The weather conditions during the three-day experiment were calm (sea-state 1 or less). (After Buckingham [91]. Reprinted by permission from *Nature* vol. 356, pp. 327–329; Copyright © 1992 Macmillan Magazines Limited.)

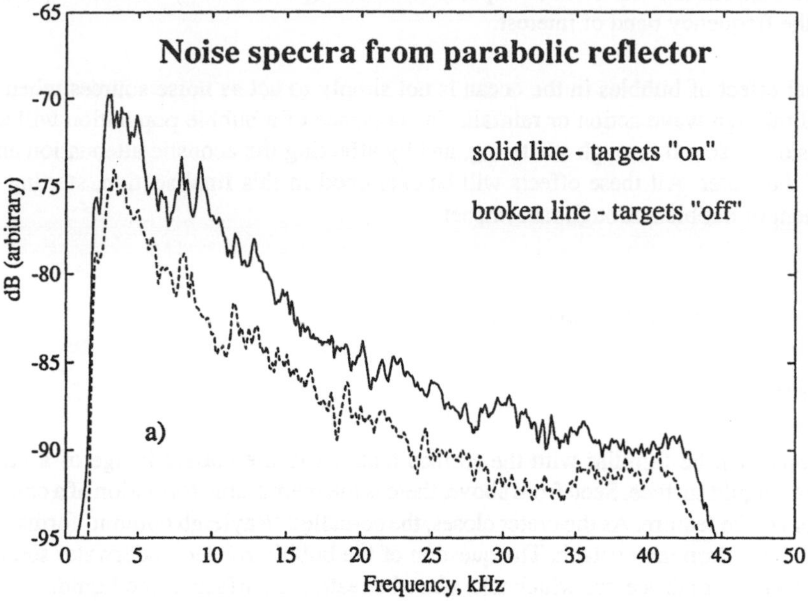

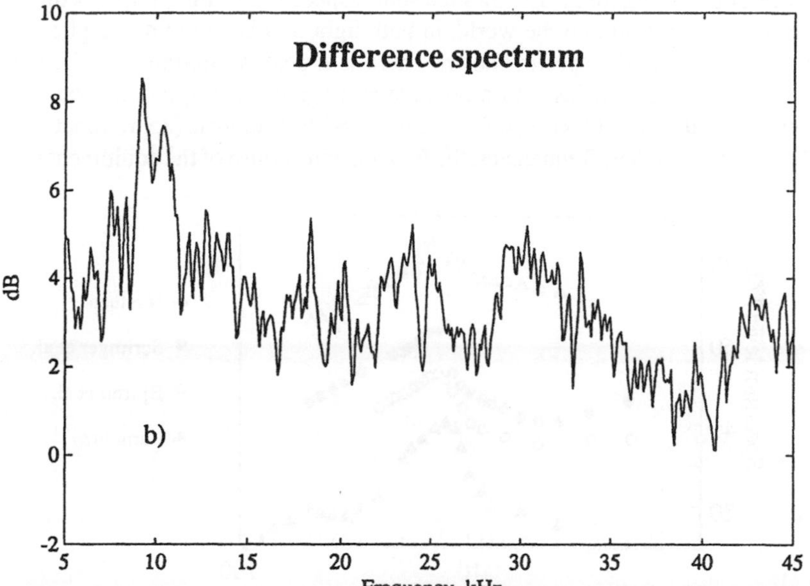

Figure 3.40 (a) Noise spectra obtained with the beam of the parabolic reflector orientated seaward and the targets at a range of 7 m. (The rapid roll-off above 43 kHz is an effect of the anti-alias filtering.) (b) The difference spectrum between the 'off' and 'on' states. (After Buckingham [91]. Reprinted by permission from *Nature* vol. 356, pp. 327–329; Copyright © 1992 Macmillan Magazines Limited.)

increase in the frequency range 5–50 kHz when the targets are 'on': Figure 3.40(b) plots the difference between the 'on' and 'off' spectra, illustrating a nominally uniform separation over most of the frequency band of interest.

The actual effect of bubbles in the ocean is not simply to act as noise sources when they are entrained through wave action or rainfall. The presence of a bubble population will affect the transmission of sound through scattering, and by affecting the acoustic attenuation and sound speed of the water. All these effects will be examined in this final section, starting with the entrainment of bubbles through rain impact.

3.7.2 The Underwater Sound of Rain

(a) Introduction

Most people will be familiar with the surface features in the classic image of a liquid drop impacting a liquid surface. Seen from above, there is the memorable formation of a crater, ringed by a coronet-like feature. As the crater closes, the so-called 'Rayleigh column' forms, and from the tip of this a drop may detach. The question of the bubble-related underwater sound of rain relates, however, to processes which take place beneath the surface of the liquid.

(i) Characteristics. Figure 3.41 shows the sound spectrum of rain, taken at different times and at various locations around the world, in both light and heavy rain. Despite this, there is remarkable similarity in the spectra shown. There is a peak at around 14 kHz, with a sharp fall-off to the lower frequencies, and a possible rise again starting below around 3 kHz. At frequencies higher than the 14-kHz peak there is a less steep fall-off (approximately –9 dB per octave) though at the highest frequencies, the frequency response of the equipment may become limiting.

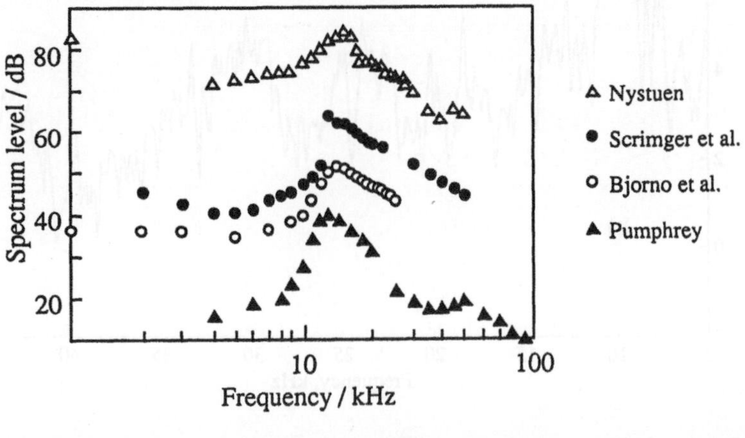

Figure 3.41 The sound spectrum of rain, taken at different times and at various locations around the world, in both light and heavy rain (after Pumphrey [92]). The data of Scrimger and Nystuen have a reference of 1 $\mu Pa^2/Hz$; the other data sets have arbitrary reference levels. Nystuen's data were taken in Clinton Lake, Illinois, during heavy rain. The other data are for light rain, recorded in Cowichan Lake, Vancouver Island (Scrimger), and in land-based water tanks at Lyngby, Denmark (Bjørnø) and Oxford, Mississippi (Pumphrey).

When a solid sphere impacts a body of water, bubbles are always entrained provided the impact speed exceeds a certain threshold [93, 94]. In contrast the entrainment of bubbles is far more complicated, and it is only within a limited regime of impact velocity u_d and drop size that bubbles are entrained in a repeatable manner. The term *regular entrainment* was coined for the process where a single bubble is captured as the crater formed by the impact develops in a repeatable manner. Such an event is shown in Figure 3.42.

Frames from a high-speed sequence, taken at around 950 frames per second (f.p.s.), are split to show the image on the left and the simultaneous hydrophone trace vertically on the right, increasing time being measured down the trace. The frames are in order but not necessarily consecutive, the tip of the hydrophone being visible at the lower right of each image. A drop of diameter $L_d = 3.2$ mm impacts the liquid surface (frame (a)), and the crater develops, a bubble of measured diameter 0.89±0.08 mm being pinched off from its base in frame (g). At this moment, the exponentially decaying sinusoid typical of bubble entrainment (compare with Figure 3.23) is detected by the hydrophone. At the end of the figure, in frame (j), the bubble is in the body of the liquid and the ageing crater develops the familiar vertical 'Rayleigh column'. Detailed study of the hydrophone trace from regular entrainment reveals several interesting features [96, 97]. An impression of total emitted pressure field can be gained from Figure 3.43, taken from Medwin *et al.* [96], which shows that there are two distinct signals: the characteristic bubble-generated sinusoid is emitted some time after a pressure signal which results from the initial impact of the drop. There are clearly high frequencies associated wih this initial impact, and as the signal in Figure 3.43 was filtered through an 8–50-kHz bandpass filter, there will be some distortion, particularly to the initial impact pulse. The results should therefore not be taken to mean that the initial impact signal undergoes one or more cycles of oscillation [98]. The initial sharp acoustic pulse, which has a duration of order μs (measurements suggest 10–40 μs, though instrumentation limits the temporal resolution), is a far-field emission which falls off as r^{-1}. This is followed by near-field pressure fluctuations which fall off as r^{-2}, as was elegantly demonstrated by Pumphrey [98], who measured the signal at various depths and plotted r^2p against time (Figure 3.44). The r^2p plot from the far-field initial pulse becomes larger with increasing depth, since the pressure falls off as r^{-1}; in contrast, the r^2p plot of the near-field component remains constant, proving that it decays as r^{-2}. The near-field pressure fluctuations may have their source in the dipole nature of the field imparted by the free surface [98] or in incompressible effects [99–101].

Whilst a pressure emission from the initial impact is ubiquitous, the bubble signal is, of course, only generated if entrainment occurs. Measurements of the time between the initial impact and the bubble emission for regular entrainment have given intervals of 17.7 ms [96] and 24 ms [97].

(ii) Measurement of Drop Size. The size of the incident liquid drops can be measured through a variety of techniques. One of the most common is by photography, using either high-speed or flash, triggered, for example, when the drop interrupts a light beam. However, distortions from the spherical shape can mean that estimation of the drop size from images may be slow and involved. There are several quick and convenient methods which can be employed and are particularly useful if there are a large number of raindrops involved. In principle, the percentage error in measurement will be reduced by a factor of 3 if the radius is calculated from direct observation of the volume of the drop.[39] One very quick method of measuring the mass,

[39]Following the discussion of bubble size in section 3.5.1(a)(ii), since the raindrop volume V_d is proportional to the cube of the drop diameter L_d, then differentiation of the natural logs shows that $\Delta L_d/L_d = (1/3)\Delta V_d/V_d$.

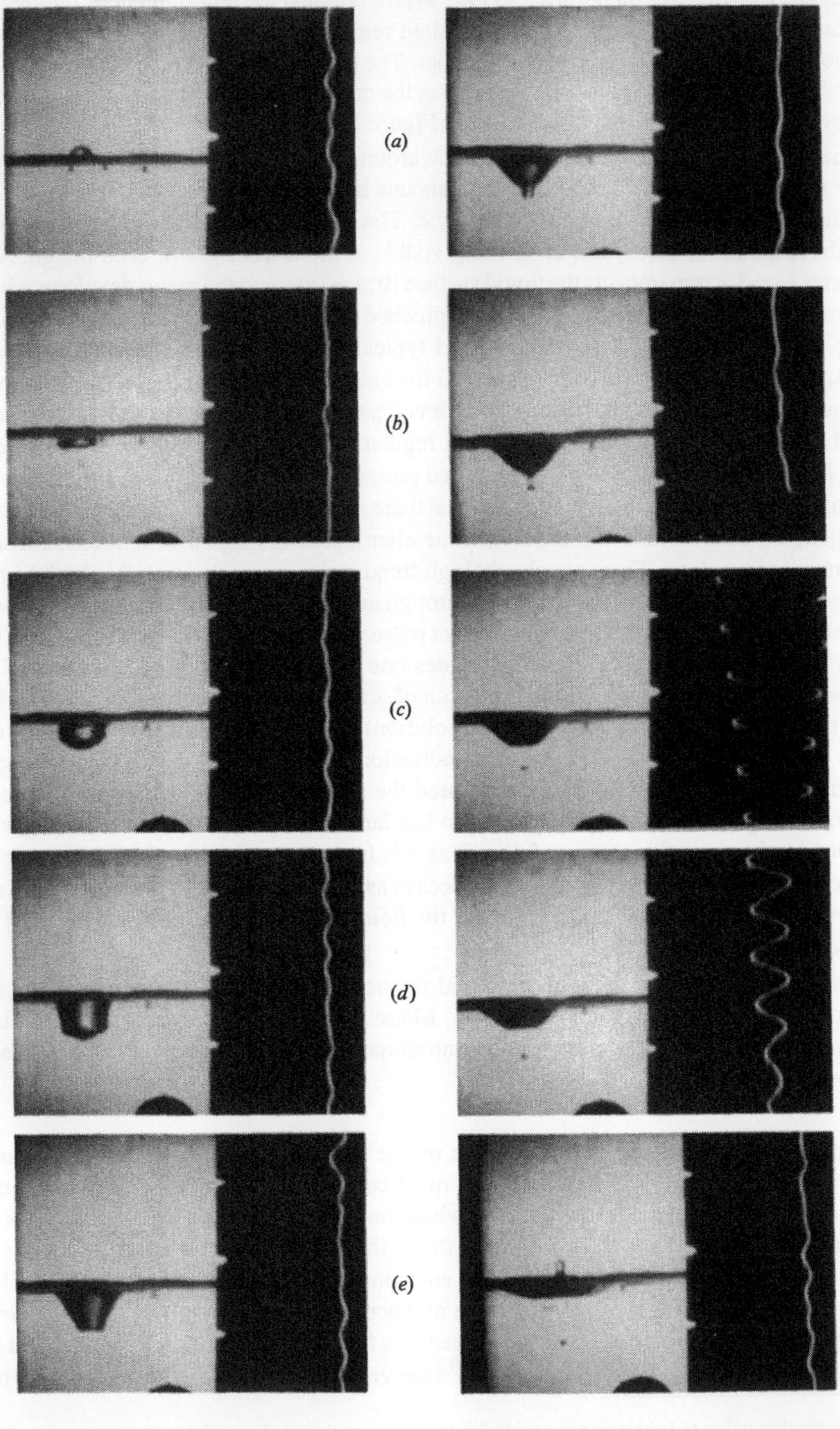
(a)
(b)
(c)
(d)
(e)
(f)
(g)
(h)
(i)
(j)

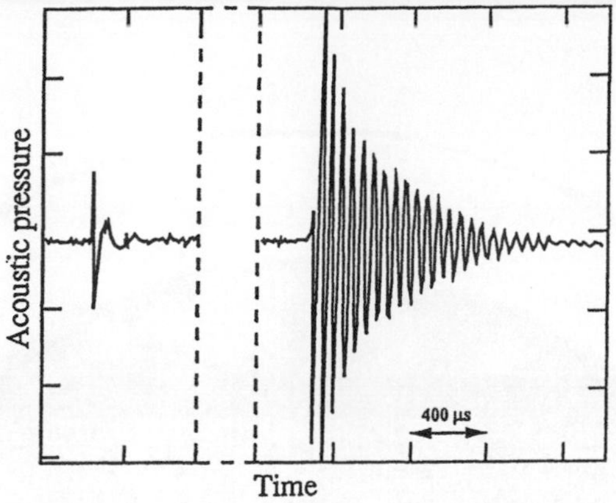

Figure 3.43 The pressure signal from the impact of a 0.83-mm diameter drop travelling at terminal velocity. The signal on the left is associated with the initial impact. The signal on the right, which begins 17.7 ms after the initial impact signal (the time axis is discontinuous between the two signals), corresponds to the bubble emission. The signal was filtered through an 8–50 kHz bandpass filter (after Medwin *et al.* [96]).

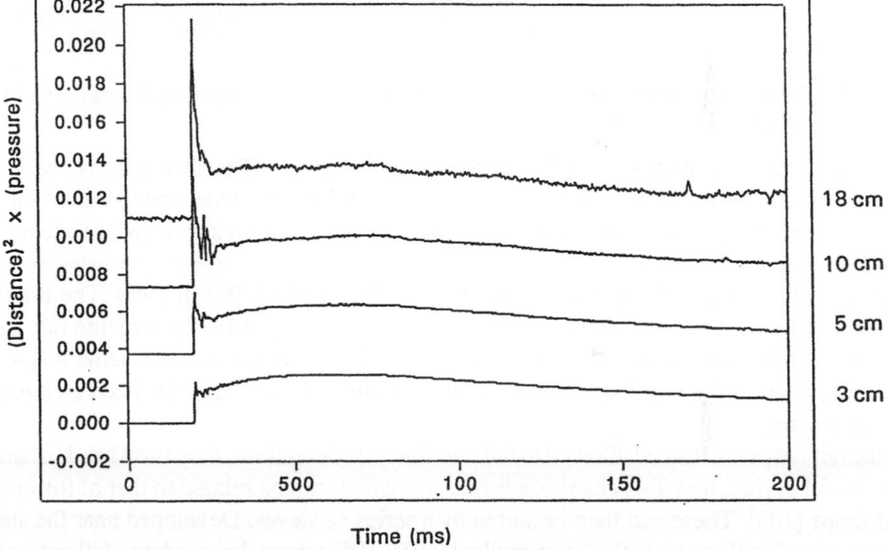

Figure 3.44 The pressure perturbation below the 4.6-m/s impact of a 2.93-mm diameter drop, measured at depths of 3, 5, 10 and 18 cm (from bottom trace to top). Time on the abscissa is measured in milliseconds. The ordinate has measures (distance)^2 × (pressure) in arbitrary but consistent units (Pumphrey, private communication, after [98]).

Figure 3.42 High-speed photography (950 f.p.s.) of regular entrainment through the 1.5-m/s impact of a 3.2-mm diameter laboratory drop, with simultaneous hydrophone trace imaged with each frame (time increasing vertically down the trace). The frames are in order but not necessarily consecutive, the start time of the frames being as follows: (a) 0, (b) 3.0, (c) 6.5, (d) 11.5, (e) 17.0, (f) 19.0, (g) 20.0, (h) 21.0, (i) 22.0 (j) 31.5 ms to an accuracy of 0.5 ms. The tip of the inverted hydrophone (of the type shown in Figure 1.18(a)) can be seen at the base of several of the frames. Coincident with the appearance of an exponentially decaying sinusoid of frequency 7.4 ± 0.6 kHz on the hydrophone trace, a bubble of measured diameter 0.89 ± 0.08 mm is entrained (after Pumphrey and Elmore [95]).

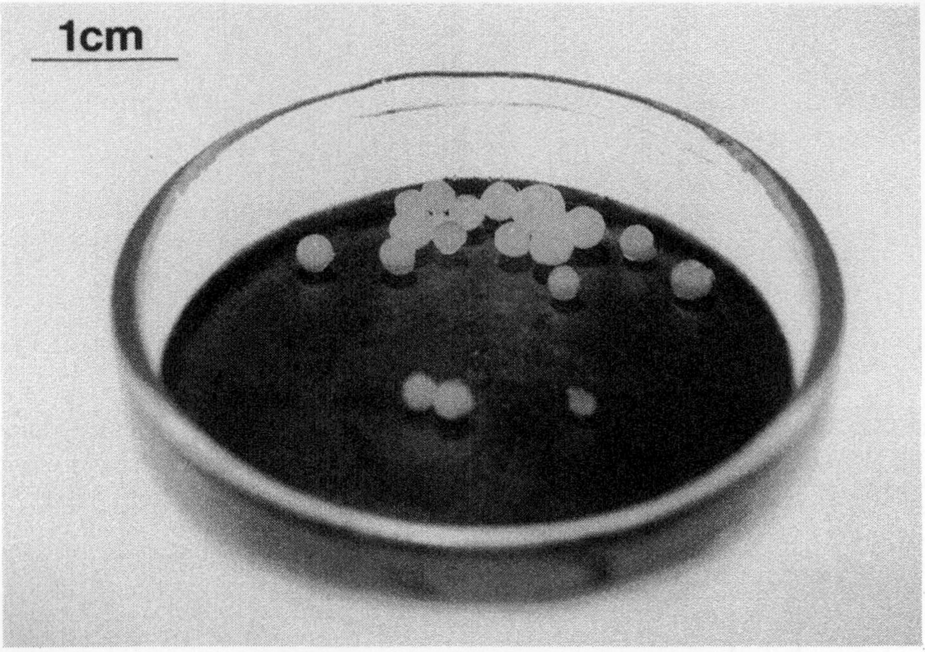

Figure 3.45 Ice drops produced by catching liquid drops in a suitably cushioned container of liquid nitrogen. (Photograph: TG Leighton.)

and so the volume, is by capturing the drops in liquid nitrogen. The frozen drops, which are shown in Figure 3.45, can be weighed in a *dry* atmosphere (otherwise condensation on their surface can lead to measurement errors). This measurement of drop size is independent of the shape.

Another interesting technique was first introduced by Wahl [102] in 1950. The drops are captured in a Petri dish containing two layers of oil. A light oil, say 10W petroleum lubricating oil, floats on a layer of denser silicone hydraulic oil. The captured drops assume a spherical shape and persist at the interface of the two oils (Figure 3.46), and may be sized *en masse* by image processing.

If the raindrops are instead allowed to fall into finely divided flour, they form beads of dough. After careful drying, they form hard beads of a size which can be related to that of the original liquid drops [103]. These can then be sorted by a series of sieves. Developed near the start of this century and still in use, is the 'stain method' [104, 105] where the raindrops fall onto a filter paper coated with a trace of solid dye (Rhodamine B, for example [106]). The size of the stain formed by each drop can be related to the size of that drop.

The above methods are applicable to drops in the size range of 0.25–7 mm diameter. Drops in clouds and fogs have diameters in the range 5–250 μm, and though not of direct relevance to bubble entrainment can usually be sized through the reduction in intensity of a light beam, through photography, or through collection on a slide covered with magnesium oxide, soot or oil, which can then be analysed by, for example, photomicrography [106].

A commercially available system[40] can be used to find the drop size distribution and mass median diameter [107].

[40] Optical Array Spectrometer Probe, and Two Dimensional Precipitation Imaging Probe (Particle Measuring Systems Inc, Colorado, USA).

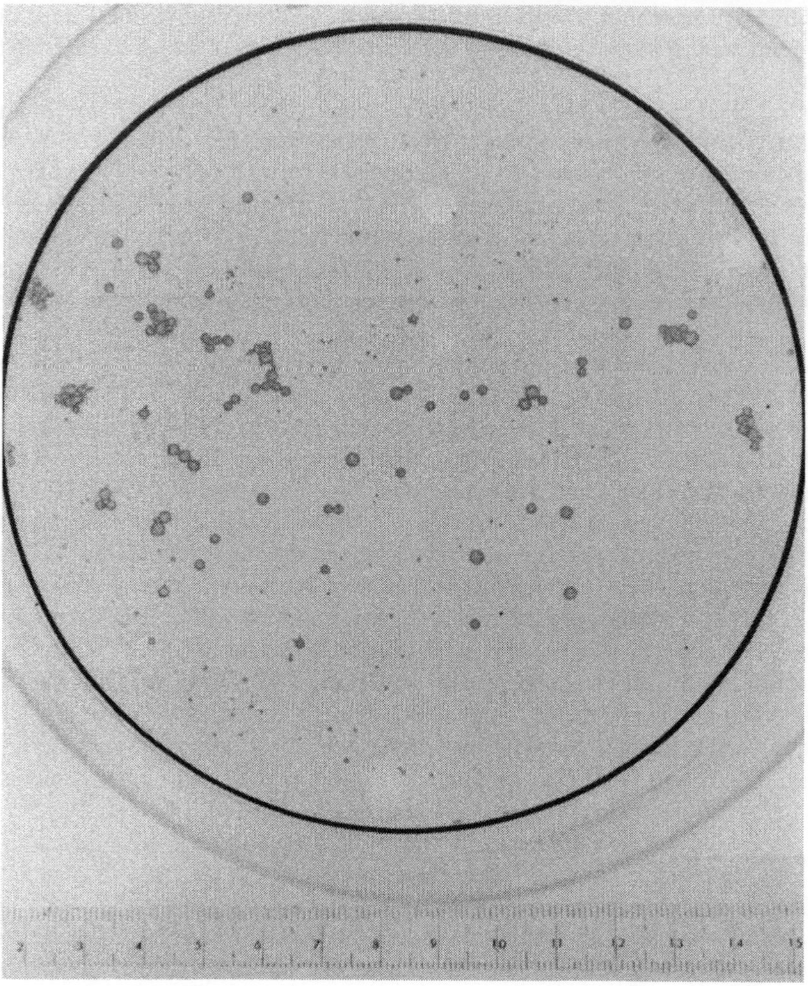

Figure 3.46 Rain drops caught between two layers of oil. (Photograph courtesy of P Tattershall and K Minter, RAE, Farnborough, private communication.)

(b) History

The first real attempt to investigate the sound produced when liquid drops impact a body of liquid under laboratory conditions was made by Franz [108], using high-speed photography and acoustic records. He showed that the impact always generated a sharp pulse of sound as the drop initially contacts the surface. In addition, a bubble could sometimes form within the liquid as a result of the drop impact some time after the initial contact, and when this happens an exponentially decaying sinusoid acoustic signal typical of bubble entrainment was produced (very similar to that shown in Figure 3.23 for an injected bubble). Since bubbles were not always generated by the impact, Franz was unsure of their importance. In the absence of good field measurements of rain noise on bodies of water, Franz concluded that the pulse caused by the initial impact was of more importance to the sound spectrum than the seemingly sporadic entrainment of bubbles. He thus predicted a very broad power spectrum for rain, peaking at around 3 kHz, which does not agree with later measurements. The reason for the discrepancy is that Franz studied the sound made by a shower of drops, all of which were much larger than raindrops.

The first field measurements to show the presence of the 14-kHz peak were not made until more than twenty years later (an earlier investigation, made in a shallow lake by Bom [109], was only for frequencies below 9.6 kHz). Scrimger [110] and Nystuen [99] made the observation, which has since been confirmed [111, 112]. Nystuen and Farmer [99, 100] made predictions of the shape of the initial short acoustic pulse, and nonacoustic dynamic pressure resulting from the flow field, through numerical modelling of the drop splash flow field. On the basis of this, Nystuen suggested that the initial impact is responsible for the 14-kHz peak [99, 100].

However, Pumphrey [92] showed that, although every impact might not entrain a bubble, drops of a given size and velocity could yield repeatable results. In this regime of *regular entrainment*, a given impact either would or would not entrain a bubble consistently. In fact, not only are the regular entrainment events repeatable, they are capable of numerical simulation (Figure 3.50). Three other classes of bubble generation could be defined [95], the regimes being dependent on the drop size and the impact velocity. *Irregular entrainment* involves the production of bubbles through the complex details of the splash. Pumphrey and Elmore [95] also name this 'Franz-type' entrainment because, for a given drop diameter and impact velocity, in this regime, one cannot predict whether or not a bubble will be entrained. *Large bubble entrainment* confines most of the volume of the crater within the bubble. This entrainment might be influenced by surface oscillations on the falling drop [95]. However, Pumphrey and Elmore [95] dismiss it as a likely source of rain noise as only a small proportion of drops possess the correct size and velocity to cause it, and because it never occurs for drops at terminal velocity (see Figure 3.47(a)). What we now know to be both regular and large bubble entrainment events feature in an article by Jones [113] who, in 1915, observed drops falling into a body of water. He noticed that there were critical ranges of distances through which the drop would fall and produce a characteristic sharp 'click' (which would now be classed as regular entrainment). Occasionally a "softer, duller sound" occurred, and a bubble was afterwards observed at the point where impact had occurred. This would now be ascribed to large bubble entrainment. Jones observed no bubble remaining behind after the 'click' events, presumably because the regularly entrained bubbles were too small or burst too rapidly for him to detect.

Regular, irregular and large bubble entrainment all generate significant sound. *Mesler entrainment* [114, 115] is generally a low-velocity unpredictable phenomenon which occurs when a multitude of small bubbles ($R_0 \sim 50$ μm) are trapped in the initial stages of impact. It does not produce significant acoustic emission. Oğuz and Prosperetti [116] suggested that the gas pockets may perhaps be trapped in the troughs of capillary waves which spread out from the point of contact on both the drop and the surface of the water body. The peaks of the waves on the spreading drop meet those on the liquid, trapping toroidal gas pockets, which through instability break up to generate small bubbles.

(c) Regular Entrainment at Normal Incidence

(i) Results. Clearly the simplest form of bubble generation by rain impact is regular entrainment at normal incidence. In the attempt to determine the contribution of acoustic bubbles to the underwater sound of rain, Pumphrey and Walton [117] correlated hydrophone data of the underwater sound with flash photography for liquid drop impacts in laboratory conditions, but did not realise at the time that the emissions from bubble entrainment were fundamentally

different from the mechanism favoured by Franz. They showed that the exponentially decaying sinusoid in the acoustic signal occurred at the resonance frequency of a bubble of the size entrained (equation (3.38)). Once it was known that bubbles are responsible for that part of the acoustic signal, to demonstrate that bubble entrainment contributes significantly to the sound spectrum of natural rain it is firstly necessary to show that natural rain included significant numbers of drops of a size and velocity to produce bubbles. Pumphrey *et al.* [97] did this, and in addition showed that these bubbles are of a size such that their resonant frequencies correspond to the peak in the spectrum.

Figure 3.47(a) shows that regular entrainment occurs when a drop of a certain size impacts a body of water at a known velocity, measured under laboratory conditions. The region for large bubble entrainment is indicated. Mesler entrainment may occur anywhere on the diagram, but is most noticeable in the low-velocity regime owing to the absence of any other type of entrainment. Irregular entrainment may occur anywhere above the upper limit for regular entrainment: the area indicated represents conditions where the high energy of the splash almost always succeeds in entraining bubbles.

The area shaded to signify regular entrainment corresponds to impact parameters that repeatably generate a single bubble, and it is this that is given greatest consideration in explaining rainfall spectrum. Figure 3.47(a) corresponds to impact at normal incidence only. The region is terminated by a line corresponding to the terminal velocities of drops in air [118]. Natural rain drops will fall at this velocity, entrainment data at this border having been taken by Medwin *et al.* [96]. Therefore if regular bubble entrainment is to occur during natural rainfall, that rain must include a significant number of drops of a size corresponding to the intersection of the terminal velocity plot with the shaded region. This corresponds to drop diameters of 0.8–1.1 mm. Figure 3.47(b) shows drop size distributions measured by Scrimger *et al.* [111] for three rain showers. All three have significant numbers of drops occurring in the desired range. Thus the two diagrams (3.42(a) and (b)) taken together demonstrate that natural rainfall can produce significant amounts of repeatable bubble entrainment.

Pumphrey *et al.* [119] showed that these raindrops will entrain bubbles of a size commensurate with an appropriate contribution to the acoustic spectrum. Figure 3.48 details the shaded area of repeatable bubble entrainment shown in Figure 3.47(a), and subdivides it to illustrate the bubble size entrained by the impact. The contour plot shows the natural frequencies of these bubbles. The intersection of this plot with the terminal velocity line shows the frequencies present in natural rainfall as a result of repeatable bubble entrainment. It is clear that there will be a large contribution in the region above 12 kHz, but not much at lower frequencies. The evidence suggests that repeatable bubble entrainment contributes significantly to the 14-kHz peak in the sound spectrum, and to the associated rapid fall-off at lower frequencies.

Evidence of the importance of such bubble entrainment to the acoustic spectrum is provided from the observation that addition of surfactant to the water can suppress the 14-kHz spectral peak for both laboratory [120] and real rain [97], and laboratory observation revealed that surfactants also suppressed regular bubble entrainment [120]. The evidence therefore that regular entrainment of bubbles is responsible for the 14-kHz peak in the acoustic spectrum from real rain is very convincing.

(ii) Mechanism of Regular Entrainment. In an attempt to explain why regular entrainment occurs, and its restriction to a few combinations of drop diameter L_d and impact velocity u_d, Pumphrey and Elmore [95] replotted the boundaries of the regular entrainment regime in terms of the Froude number

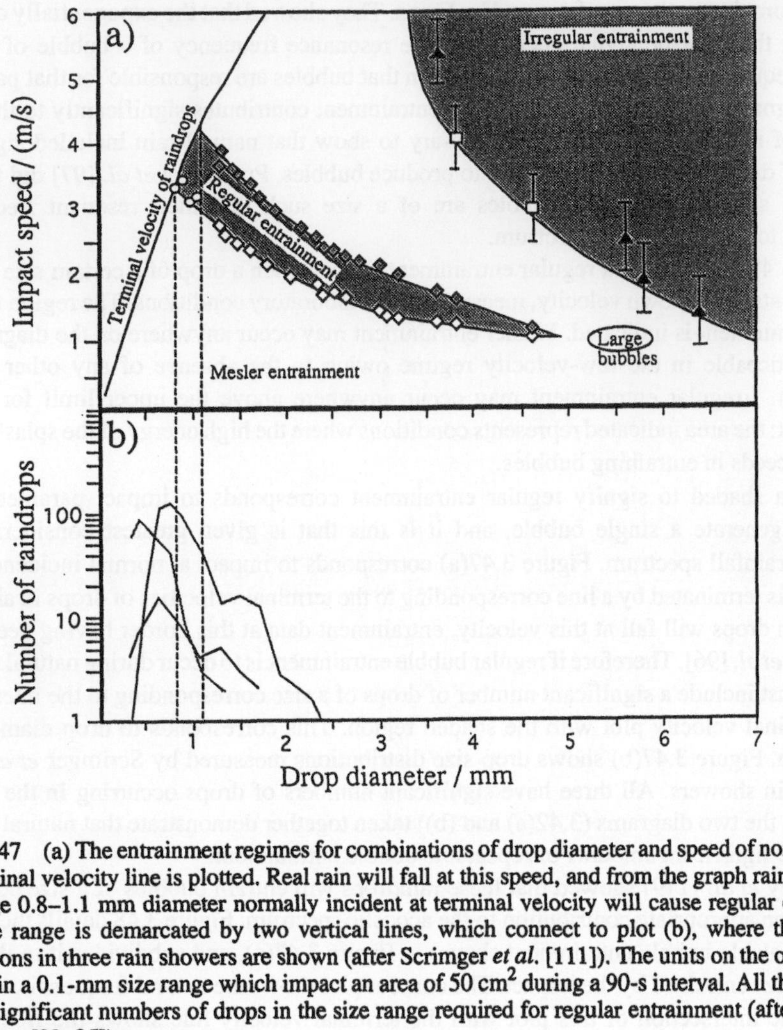

Figure 3.47 (a) The entrainment regimes for combinations of drop diameter and speed of normal impact. The terminal velocity line is plotted. Real rain will fall at this speed, and from the graph raindrops in the size range 0.8–1.1 mm diameter normally incident at terminal velocity will cause regular entrainment. This size range is demarcated by two vertical lines, which connect to plot (b), where the drop size distributions in three rain showers are shown (after Scrimger *et al.* [111]). The units on the ordinate refer to drops in a 0.1-mm size range which impact an area of 50 cm^2 during a 90-s interval. All three showers contain significant numbers of drops in the size range required for regular entrainment (after Pumphrey and Elmore [92, 95]).

Irregular entrainment
□ Pumphrey [92]
▲ Pumphrey and Elmore [95]
Regular entrainment
Upper limit ◈ Pumphrey [92]
Lower limit ◇ Pumphrey [92]
○ Pumphrey and Elmore [95]

$$\{\mathcal{Fr}\} = \frac{u_d^2}{L_d g} \tag{3.251}$$

and the Weber number

$$\{\mathcal{We}\} = \frac{\rho L_d u_d^2}{\sigma} \tag{3.252}$$

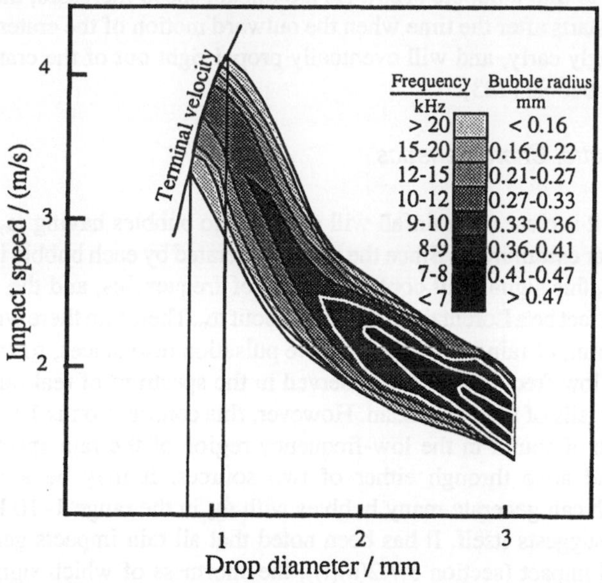

Figure 3.48 Detail of the shaded area of repeatable bubble entrainment shown in Figure 3.47(a), subdivided to illustrate the bubble size entrained and resonance frequency emitted by the impact (after Pumphrey and Elmore [95]).

Gravity and surface tension are the two most important forces acting in regular entrainment, and use of these dimensionless parameters allows the two forces to be expressed separately in terms of the drop diameter and the impact velocity. If $\{\mathcal{F}r\}$ is plotted on a logarithmic scale, and $\{\mathcal{W}e\}$ on a linear one, the regular entrainment regime is bounded by straight lines. This suggests that the upper and lower limits of the regime might be expressed as

$$\{\mathcal{F}r\} = \wp_1 e^{\wp_2 \{\mathcal{W}e\}} \tag{3.253}$$

where $\wp_1$ and $\wp_2$ are dimensionless parameters that take different values for the upper and lower boundaries. However, as the authors point out, this interesting observation does not immediately help explain the mechanism of regular entrainment.

Several mechanisms have been postulated. The numerical simulations of the drop formation process [121] do enable the theory that regular entrainment is influenced by shape changes in the falling drop (which are known to be a factor in determining whether a drop forms a vortex ring on impact – [122]) to be dismissed [95]. More likely theories for regular entrainment have been based on the dynamics of the crater, rather than those of the drop. Oğuz and Prosperetti [121], commenting how the different positions on the wall will reverse their motion at different moments as the cavity created by the impact first expands and then closes again, suggest that the relative timing of these reversals might determine whether or not a bubble is entrained. Pumphrey [92] speculates that regular entrainment might occur when the peaks of a capillary wave, travelling down the crater, meet at the base and pinch off a bubble. The notion of a bubble entrapment by a capillary wave which originates at the crater rim and grows in amplitude through geometrical focusing as it travels downwards has received support from recent numerical work [123, 124]. In addition, such simulations show that as the impact speed increases an upwards axial jet develops from the base of the crater, and this can fill the crater before the sides can close up to trap a bubble [123]. As the impact speed increases, the formation of the jet, which always starts after the time when the outward motion of the crater walls is reversed, becomes increasingly early, and will eventually propel right out of the crater, preventing the entrainment of the bubble [123].

(d) Off-peak Spectral Characteristics

From Figure 3.48 it is clear that rainfall will generate no bubbles having ω_o less than about 12 kHz through regular entrainment. Since the signal generated by each bubble is an exponentially decaying sinusoid, this signal will contain a range of frequencies, and the spectral profile of each bubble will in fact be a Lorentzian,[41] centred about ω_o. Therefore there will be contributions to the sound spectrum of rain beyond the bubble pulsation resonances, so that, for example, a component of the low-frequency noise observed in the spectrum of real rain will result from the low-frequency tails of the Lorentzian. However, this contribution is 10–15 dB too quiet to be the only source of sound in the low-frequency region of the rain spectrum [126]. Other contributions could arise through either of two sources. It may be a result of irregular entrainment, which can generate many bubbles with ω_o in the range 1–10 kHz [95]. Another source, however, suggests itself. It has been noted that all rain impacts generate an acoustic spike at the initial impact (section 3.7.2(a)(i)), the shortness of which signifies a broadband frequency spectrum. Analysis of the initial pulse is complicated by nonacoustic near-field pressure variations resulting from flow. These decay as r^{-2}, that is, the pressure follows an inverse square law with distance from the splash. The actual acoustic component of the pressure, which corresponds to only the very initial spike (measured to be of duration $\lesssim$ 10 μs [97, 98, 126], decays as r^{-1}. Although, in theory, one therefore merely needs to examine the hydrophone record at large distances from the impact to distinguish the acoustic component, in practice contamination of the far-field signal by noise and reflections makes this difficult. Franz [108] employed a high-pass filter. Pumphrey [98] adopts a different technique, measuring the pressure pulse at various depths below the impact site (as for Figure 3.44), and then combining the data to produce a representation of the far-field pulse. Thus analysis of the pressure at various distances from the impact site can enable the acoustic and nonacoustic components to be distinguished. Laboratory measurements have suggested that the pressure amplitude associated with this pulse is proportional to the impact velocity u_d raised to the power of between 2.5 and 3 [92, 97], and proportional to the drop diameter L_d raised to between around 1.6 and 1.8 [92]. Other estimates are that the amplitude of the initial pressure spike varies as $u_d^{2.8\pm0.2}$ and $L_d^{1.5\pm0.2}$ [98]. These make interesting comparison with the predictions of Franz [108] and Oğuz and Prosperetti [123] that the far-field pressure pulse should go as u_d^3 and L_d^1:

$$p_{\text{impact}} = \frac{\rho u_d^3 L_d}{2c} \frac{\cos\theta}{r} u \tag{3.254}$$

where the pressure radiated by the initial impact p_{impact} depends on the dimensionless function u, which is itself a function of the dimensionless parameters $u_d t/L_d$ and u_d/c, the impact Mach number. Simple arguments based upon a water-hammer[42] model for the initial impact suggest that the pressure from this should be proportional to u_d^1 [99, 101], though this assumes that the

[41] See reference [125]. The Lorentzian imparts a superficially 'bell-shaped' profile to a plot of the amplitude component as a function of frequency. In this case, the Lorentzian will be centred at the bubble pulsation resonance, with higher and lower frequencies contributing to the spectrum in progressively decreasing amounts.
[42] See Chapter 5, section 5.4.1.

liquid can be modelled by an arrested water column: the effect of the curvature at the front of the drop, as reviewed by Field [127], is discussed in Chapter 5, section 5.4.1. Guo and Ffowcs Williams [128, 129] formulate the compressive waves emitted when a spherical drop impacts an initially quiescent water surface in order to predict the initial far-field pressure pulse, the solutions suggesting that the radiated wavepacket will carry with it acoustic energy proportional to the kinetic energy of the falling drop and to the cube of the impact Mach number, u_d/c. The researchers note how this cubic result is to be expected from the dipole nature of the source [130], and has been confirmed by Franz [108]. Adaptation of the analysis for the impact of ellipsoidal drops predicts that these are more efficient far-field radiators than spherical ones [128]. It should be noted that the pressure pulse emitted by these water-hammer-like sources are of very short duration.

Pumphrey [98] compares the pulse shapes and power spectra from the initial impact studies of Nystuen and Farmer [100, 101, 131], and Medwin *et al.* [96] and concludes that the power spectrum decreases monotonically with frequency for frequencies greater than 1 kHz, and therefore contributes little to the underwater noise of rain above 14 kHz. From their numerical simulations Oğuz and Prosperetti [123] dismiss the initial impact as being incapable of explaining the underwater sound of rain impacting a liquid body. In the general case, which includes bubble generation by breaking waves (where in addition to drop impact, bubbles might also be produced by enhanced microbreaking and local disruption of the surface vortical layer [123]), they conclude that entrainment of air would probably dominate the acoustic emission, certainly in conditions of drop impact at normal incidence. Oblique impact can arise if there is a wind, and in deciding the relative importance of impact and bubble-generated sounds in nature, the effect of wind must be considered.

(e) The Effect of the Wind

To recap, rainfall onto liquid bodies generates impact noise. In addition, drops of diameter between 0.8 and 0.1 mm travelling at terminal velocities and normal incidence to a smooth water surface always entrain bubbles. Acoustic radiation from such bubbles is in this case stronger than the impact noise [96]. Such drops are almost always present in rainfall.

However, normally incident drops are far more effective at entraining bubbles than ones that come in obliquely [96]. The velocity of real rain is rarely perpendicular to the surface, both through wind action altering the drop path from the vertical, and through tilting of the liquid surface due to wind, wave and ripple action.

Nystuen and Farmer [101] studied the effect of wind on rain-generated noise in a lake. As the wind increased from 0.6 to 3.3 m/s, the 15-kHz peak became less prominent and broader, the peak shifting up by a few kHz. Medwin *et al.* [96] studied the impact at normal and oblique incidence of single raindrops travelling at terminal velocities, for acoustic radiation properties and for the probability of bubble entrainment. For normal incidence and terminal velocity, drops of diameter 0.83, 0.91 and 0.98 mm (100 drops each) generated peak initial pressures through bubble radiation of 0.55 ± 0.2, 0.4 ± 0.2 and 0.4 ± 0.1 Pa respectively at a depth of 1 metre. In contrast, the peak pressure due to the initial impact event was only around 0.14 Pa at the same depth. Bubbles entrained by rain impact are clearly in a geometry to generate a dipole-like field [44]. Any roughness of the liquid surface will tend to 'squash' the radiation pattern [96]. When the incidence is oblique, the radiation patterns for laboratory drops tested by Medwin *et al.* [96] remain 'dipole-like', though the expected asymmetry sets in. As a typical angle, Medwin *et al.* [96] discuss 20°, which requires a windspeed of only 1.3 m/s for a drop of diameter 1 mm. The experimenters found that a 20° deviation causes a 30% decrease in the energy of the emission associated with the initial drop impact, but a negligible effect on the energy of the bubble radiation *if a bubble is entrained.* Whilst for the appropriate drop size there is always bubble entrainment for normal impacts, at 20° only 10% of impacts generate bubbles. This factor means that for the rain field taken as a whole, the dominance of bubble noise over impact noise is reduced by a factor of 10, so that the acoustic energy produced from bubbles compared with impacts is now in the ratio of only 20:1 [96]. This explains the observed reduction in the 15-kHz peak compared with the broadband noise produced by the initial impact.

(f) Final Comments

(i) Summary of Rain Impact Experiments. Thus the acoustic spectrum of rain impacting a body of water can be thought of as containing a broadband feature due to the initial impact, and a peak resulting from a summation of resonance spectra of regularly entrained bubbles. Such a description allows many aspects of measured spectra to be interpreted, including the peak-suppression effects of wind and surfactants. However, these real spectra may in addition be complicated by forms of entrainment other than regular, by reverberation[43] from reflective lake bottoms, and through surface coverings of bubbles which could absorb sound [126].

Laville *et al.* [72] corroborated the finding that the 13–15-kHz peak is bubble-induced, whilst the broadband component to the acoustic spectrum arises through the initial impact. They explain lack of correlation in the literature as arising from the sensitivity of bubble generation to climatic and surface conditions.

Pumphrey and Crum [44] counted the bubbles entrained by a spray of drops generated as water was pumped through a hyperdermic needle. The distribution tended to show that the number of bubbles entrained varied as an inverse power of their resonance frequency where that power was roughly equal to 3 (i.e. number of bubbles proportional to ω_0^{-a} where $a \approx 2.9$), valid for the approximate frequency range 13–100 kHz. Data for real rain showed more scatter, but generally agreed with this finding.

(ii) An Analytical Formulation of Crater Development. Longuet-Higgins [132] presents an analytical expression for the fluid flow around the conical cavity that is often generated by liquid drop impact on water. Longuet-Higgins uses potential flow, satisfying Laplace's equation, and thus supposes an incompressible and irrotational liquid. Since the model assumes a conical cavity, the solutions must be axisymmetric. Examples of such solutions to Laplace's equation are discussed in Chapter 2, section 2.2.4(c). Once the potential has been found, the flow pressure can be deduced from equation (2.71). Since the fluid particle accelerations are large (greater than around 100 m/s^2) the effect of the gravitational acceleration is negligible. Therefore equation (2.71) reduces to

$$p = -\rho\dot{\Phi} - \tfrac{1}{2}\rho(\vec{\nabla}\Phi)^2 \qquad (3.255)$$

If, in the first instance, surface tension is neglected, the dynamic boundary condition is that $p = 0$. Longuet-Higgins also gives the kinematic boundary condition

---

43 As shown in Chapter 1, sound can be reflected from interfaces between media of different acoustic impedance. The sum total of the sound scattered in this way is called *reverberation*. The scatterers may be many small inhomogeneities, which can scatter or reradiate sound.

$$\left(\frac{\partial}{\partial t} + (\vec{\nabla}\Phi).\vec{\nabla}\right)p = 0 \tag{3.256}$$

Before solving exactly for full conical flow, Longuet-Higgins gives a simple formulation for the instantaneous flow. Realising that axisymmetric solutions to Laplace's equation exist in the form

$$\Phi_{\text{cone}} = \hat{A}(t)r^n P_n(\cos\theta) \tag{3.257}$$

where $\hat{A}(t)$ is a function of time, he takes the $n = 2$ solution to Laplace's equation, with an amplitude proportional to the time t. Though this fails to satisfy the kinematic boundary condition (equation (3.256)), at $t = 0$ the potential (being proportional to t) also equals zero. Substitution of this potential, at $t = 0$, into (3.255) shows that the dynamical boundary condition, $p = 0$, is satisfied at this time when, for this $n = 2$ spherical harmonic, $(3\cos^2\theta - 1) = 0$. This corresponds to the free surface taking an angle $\theta = \arccos(1/\sqrt{3}) \approx 54.7°$. This simple instantaneous solution, applicable only at $t = 0$, therefore describes a cone, the vertex at the origin and all points on the conical liquid surface at an angle of 54.7° to the vertical axis.

Longuet-Higgins does, however, find a solution which satisfies both boundary conditions for the general time, of the form

$$\Phi_{\text{cone}} = \frac{1}{2}\hat{A}(t)r^2(3\cos^2\theta - 1) \tag{3.258}$$

so that the free surface of the cavity is indeed always a cone. The function $\hat{A}(t)$ can be obtained, this $n = 2$ potential being one of a class of hyperbolic flows used by Longuet-Higgins [133] in discussion of the jets formed by the disintegration of a bubble film.

By transforming the potential $\Phi_{\text{cone}} = \hat{A}(2z^2 - x^2 - y^2)/2$ into cartesian coordinates, Longuet-Higgins shows that the particle velocity in the same frame, $\nabla\Phi_{\text{cone}} = \hat{A}(-x, -y, 2z)$, depends linearly on x, y and z. This means that any liquid particles which initially lie on a straight line will remain in a straight line at all future times, though the direction of that line will be time-dependent. Therefore particles on the surface of the cone will remain there, the cone angle varying in time. A typical particle path is shown in Figure 3.49 and such trajectories are driven by the pressure gradient which, having no component parallel to the inviscid liquid surface, is normal to it.

The particle path curves away from the axis, so that some component of this pressure gradient must be normal to the particle velocity. However, the pressure gradient must be normal to the liquid surface, as stated above, and so when the particle velocity is also normal to that surface, the pressure gradient and the particle velocity are parallel. In this situation, which occurs at some critical angle $\theta_{\text{cone}}^{\text{perp}}$, there is no component of force perpendicular to the particle velocity. Thus if the liquid surface is to remain conical, the flow must have a singularity in time. Longuet-Higgins concludes that physically it would seem difficult to force the flow beyond this critical configuration. The angle $\theta_{\text{cone}}^{\text{perp}}$ is attained when the particle velocity $\nabla\Phi_{\text{cone}} = \hat{A}(-x, -y, 2z)$, is perpendicular to the surface, which is described by the vector (x, y, z). Thus at the critical configuration

$$(x, y, z).(x, y, -2z) = x^2 + y^2 - 2z^2 = 0 \tag{3.259}$$

which, in spherical coordinates, occurs when $3\cos^2\theta_{\text{cone}}^{\text{perp}} - 1 = 0$ giving $\theta_{\text{cone}}^{\text{perp}} \approx 54.7°$. Longuet-Higgins finds favourable comparison with measured cone angles from the numerically predicted flow shapes of Oğuz and Prosperetti [121]. Figure 3.50 shows an example of the simulations

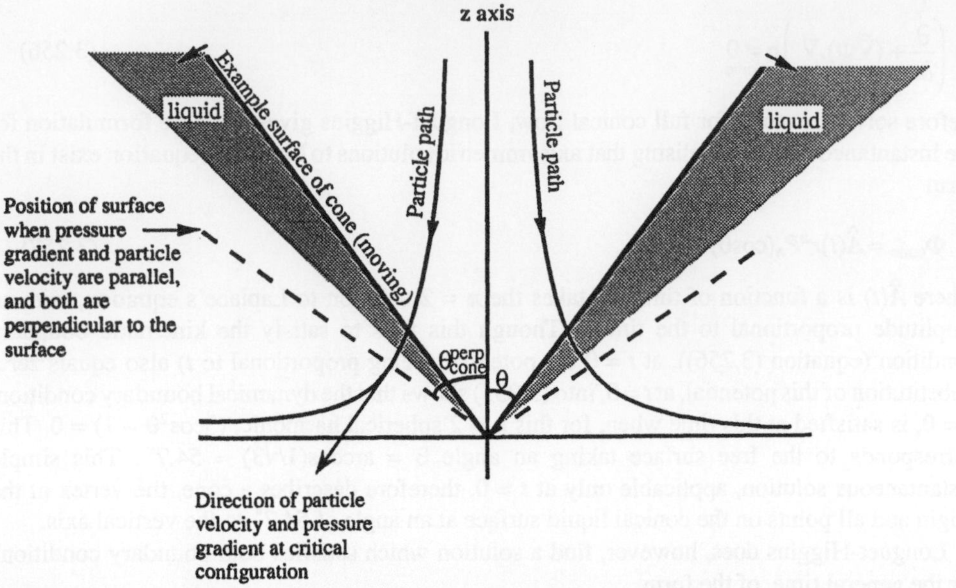

Figure 3.49 A cross-section of the flow analytically predicted for drop impact (after Longuet-Higgins [132]).

of Oğuz and Prosperetti [121] where the developing ideal cone is superimposed in cross-section on the simulation with the angle at the vertex being twice the half-angle. The measure of time given on each frame is in units of the ratio of the drop radius to the impact velocity. In Figure 3.50(a) the impact velocity in this case is too low to entrain a bubble, but the outer wall of the cavity does remain approximately conical, and in the last frame has attained a half-angle of $109.5°/2 = 54.7°$. It must, however, be stressed that on the whole the degree of agreement was not as precise, the average maximum cone angle being $113 \pm 8°$, which is nevertheless very encouraging. In Figure 3.50(b), where the impact speed is higher, a bubble is entrained. The physical reason for the singularity at the critical angle is that the flow must be forced through the critical configuration, so that there is a weak shock, when $\hat{A}(t) \propto t^{-3}$.

The analysis can be modified to account partly for the effect of surface tension by the addition of a sink, the strength of which varies linearly with time, at the cone vertex. This is made possible because in fact the vertex is surrounded by a small cavity on the free surface. In this way, the formulation can model bubble entrainment occurring as a result of the convergence at the vertex of a circular ripple which travels down the cone. The ripple is considered as a small perturbation, of capillary lengthscale, on the cone.

3.7.3 Bubble Entrainment from a Free Surface: the Excitation Mechanism

Introduction

One critical issue that has so far not been properly discussed is the method by which the excitation is delivered to the bubble. In section 3.5 it was shown how bubbles injected through nozzles are excited to oscillate and acoustically emit. That section ended with evidence of the extreme shape oscillations a bubble could undergo, whilst the very next section (3.6) introduced

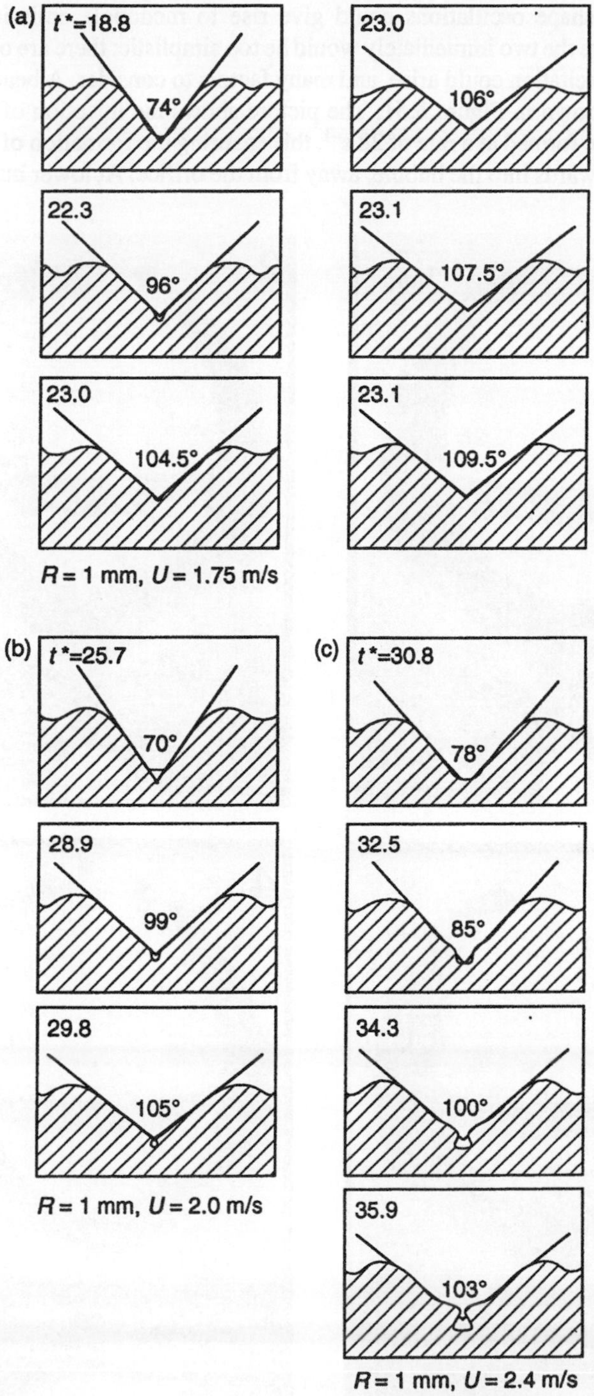

Figure 3.50 Axial section of the liquid surface numerically simulated from the impact of a 1-mm radius drop at a speed of (a) 1.75 m/s, (b) 2.0 m/s, (c) 2.4 m/s (after Oğuz and Prosperetti [121] and Longuet-Higgins [132]).

the theory of how shape oscillations could give rise to monopole emissions close to ω_0. However, to associate the two immediately would be too simplistic: there are other mechanisms through which the excitation could arise, and many factors to consider. A beautiful example of one such factor is shown in Figure 3.51. The picture shows the injection of a bubble from a nozzle. Bubbles were formed at a rate of 28 s^{-1}, this ensuring the formation of a strong axial jet which is directed upwards into the bubble, away from the orifice. At lower bubbling rates (e.g.

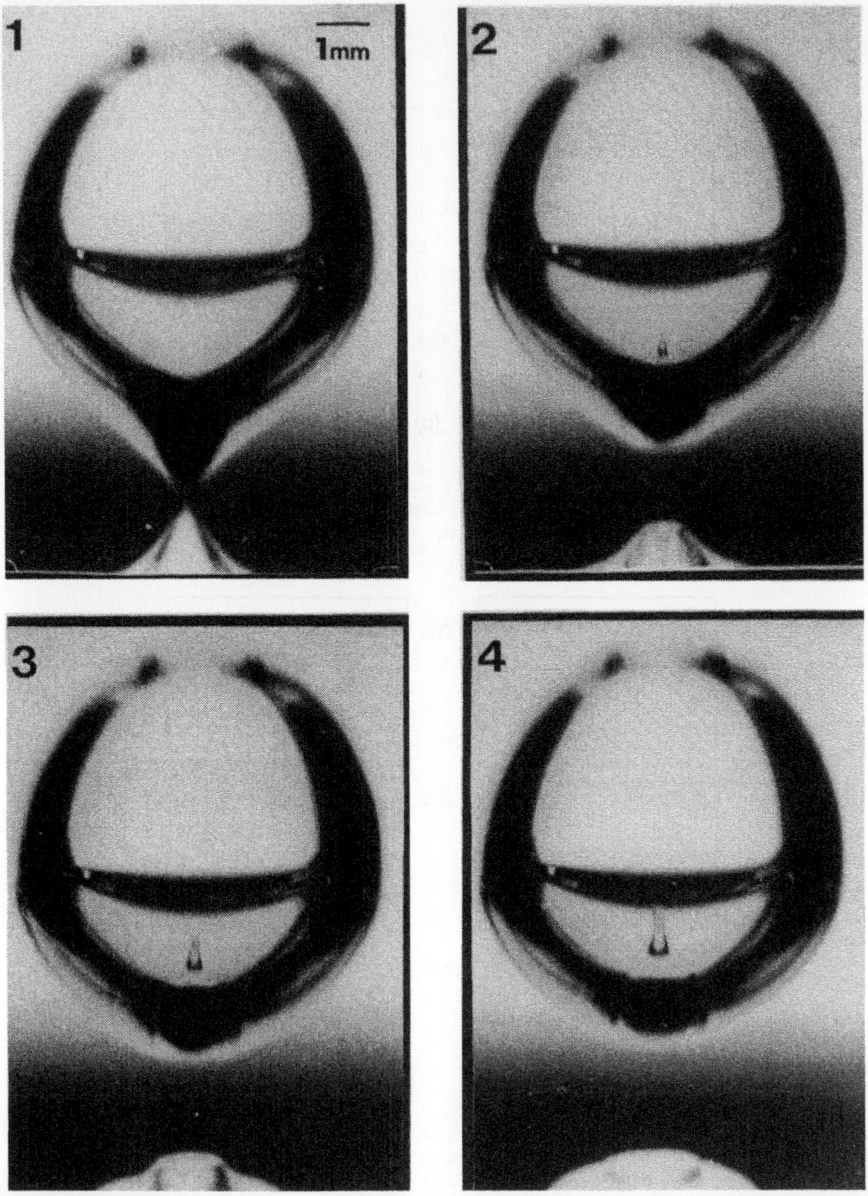

Figure 3.51 An upwards axial jet within a newly injected bubble. (Photograph courtesy of K. Lunde, private communication.)

1 s^{-1}), the jets are very small and short-lived. The fact that the bubble wall is optically reflective, so that to the usual observer it acts as a spherical mirror, imaging on its surface its surroundings, means that the bubble will tend to hide away internal details such as the jet. Lunde [134] used special optical techniques to see inside the bubble, viewing it at an angle of 15° above the horizontal, and placing a mirror on the floor of the water tank. Where this mirror meets the lit back wall of the tank, the unlit corner is the source of the horizontal dark line image crossing the bubble interior. Another optical effect results from the 15° angling: this causes the jet, which grows from the base of the bubble, to appear in the middle. Such jets may play a part in stimulating the acoustic emission [25].

The mechanism of excitation is clearly dependent on the geometry and details of entrainment. Such details have been elucidated for bubble production through injection and rain impact. Most bubbles in the natural world are, however, entrained and excited through wave action. Several types of breaker can be identified. In addition to *surging breakers*, which develop as a wave runs up on a steep beach, other forms of breaker can also be distinguished [135, 136]. Mason [137] discriminates between two types of breakers. Firstly there are *spilling breakers*, where the broken water seems to develop more gently from an instability at the sharp crest and forms a quasi-steady whitecap on the forward slope [138]. Because of the presence of a significant number of air bubbles, trapped as the wave breaks gently at the crest, the whitecap is lighter than the water below. The density difference inhibits mixing with the face of the wave, and the whitecap rides on top of the sloping sea surface, retaining its identity. Longuet-Higgins and Turner [138] proposed that a spilling breaker can be regarded as a turbulent gravity current riding down the forward slope of the wave. The component of gravity in the relevant direction draws on the distinct turbulent flow of the whitecap which, particularly at its front edge, entrains water from the laminar wave surface below and entrains air from above, so maintaining the density difference (Figure 3.52). Secondly there are *plunging breakers*, where the wave crest curls forward and plunges deeply into the slope of the wave some distance from the crest. Whether enveloped by plunging breakers (where gravity is the restoring force for the wave motion), or trapped in troughs in steep capillary (where surface tension acts as the restoring force) or capillary–gravity waves, one might envisage the resulting pockets of air to resemble horizontal cylinders. Longuet-Higgins [63] describes how such cylinders might develop.

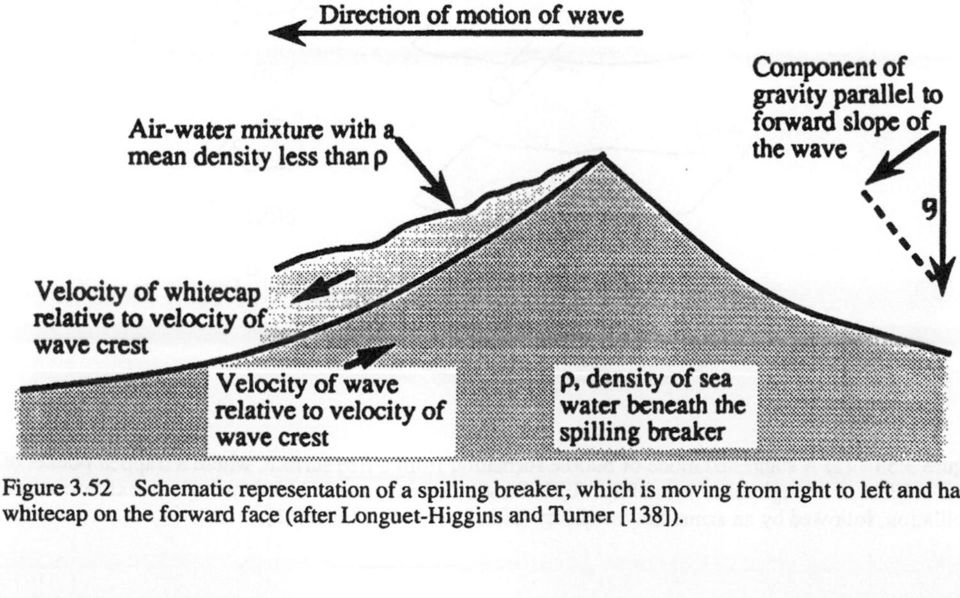

Figure 3.52 Schematic representation of a spilling breaker, which is moving from right to left and has a whitecap on the forward face (after Longuet-Higgins and Turner [138]).

Longitudinal instabilities would cause the cylinders to fragment into 'sausage-like' gas pockets, which becoming detached from their neighbours might further contract into spherical 'meatball' bubbles [63] (Figure 3.53(a)). Kolaini *et al.* [139] generated capillary–gravity waves, having wavelengths of the order of 1 mm (capillary) and centimetres (gravity) by blowing air over a laboratory water tank, and demonstrated the generation of sound through bubbles pinched off from the troughs of elongated capillary waves. Bubble entrainment by energetic gravity waves became prominent at high wind speeds. Longuet-Higgins [63] hypotheses a second entrainment geometry, illustrated in Figure 3.53(b). Here similar lengths of bubble, which through instability could break up into spherical fragments, might be formed through the generation of a toroidal gas pocket, the limiting case of a severe $n = 2$ shape oscillation. Imagine, for example, the bubble in Figure 2.14(a)(iv) oscillating so far that the bubble walls meet up in the middle, generating a 'ring'- or 'doughnut'-like gas pocket which then breaks up into smaller spherical bubbles.

It is useful at this point to consider what features the entrainment processes that can excite passive emissions considered so far have in common. Specifically, examples of excitation discussed in section 3.5.2 occurred through the processes of injection and coalescence, and in this section through rain impact, and through the simplified illustration of cylindrical and toroidal gas pockets. In all cases, one ends with an approximately spherical bubble. Since the passive emissions which result from free oscillations are characteristically driven by a brief impulsive force, we must look to the moments just prior to the formation of the final bubble for the excitation. In all the above cases, those moments are characterised by (i) the closing of a surface, effectively separating the entrapped gas from a much larger volume of gas, and (ii) the bubble passing through a severe distortion from the spherical form. In addition, there is a form of local fluid flow which produces closure.

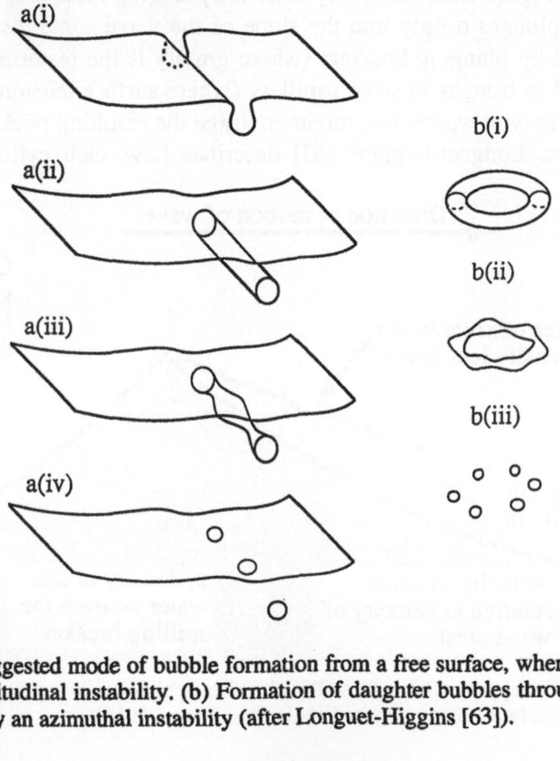

Figure 3.53 (a) A suggested mode of bubble formation from a free surface, where a trapped pocket of air is subject to a longitudinal instability. (b) Formation of daughter bubbles through an extreme $n = 2$ oscillation, followed by an azimuthal instability (after Longuet-Higgins [63]).

Other observations on the excitation process can now be viewed in the light of these common features. Flow can, for example, set up shape distortions and bring about fragmentation [140]. Kerman [85] suggests that turbulent pressure fluctuations are the source of bubble excitation. Turbulent flow, as discussed by Crighton and Ffowcs Williams [141], induces pressure fluctuations which are not coherent over the bubble surface and are therefore unlikely to excite the breathing mode directly. Pumphrey and Ffowcs Williams [142] excited an injected bubble by allowing it to rise into the stream of a horizontal under-water turbulent water jet. Excitation occurs when shear forces fragment the bubble, the level of the ω_0 emission being comparable with that generated when the bubble first leaves the nozzle. If the bubble does not fragment in the jet (i.e. there is no closing of a surface), emissions at ω_0 might still be excited, though the signal is much weaker (~20 dB). Pumphrey and Ffowcs Williams [142] concluded that turbulent excitation without fragmentation endows a far lower level of excitation than either injection or fragmentation. Acknowledging that their experiments are inexact models of the breaking wave, Pumphrey and Ffowcs Williams [142] tentatively point out that any turbulent pressure fluctuations that have frequencies similar to ω_0 will be on length scales smaller than R_0, and so be ineffective at directly exciting a volume pulsation. That is, the pressure changes at the bubbles surface will not occur uniformly so as to excite the breathing mode in the bubble and by engendering a spherical compression or dilation. However, though the turbulent pressure differentials discussed above could not cause spherical compression or expansion of the bubble, they may generate shape oscillations in the bubble, which might radiate in the monopole manner at second order at frequencies close to the pulsation resonance.

Medwin and Beaky [23], in examining the emission from bubbles in a laboratory wave-breaking facility, describe the bubble oscillations as 'shock excited', implying a single initial impulsive, rather than a periodic, forcing term. Though they were aware of the proposed mechanisms [26, 59, 60, 143, 144], they do not specify the nature of the shock.

The proposed methods of excitation divide into two groups. Firstly there are those which postulate excitation of the breathing mode, either because at the moment of closure the liquid at the wall has an initial radial velocity, or through a pressure change which is uniform over the bubble surface, arising from the Laplace pressure [132] or the hydrostatic head [26, 143]. The second mechanism excites a shape mode, either through the initial distortion of the bubble from the spherical form at the moment of closure, or through a deviation from sphericity in the fluid motion at the bubble wall at the instant of closure [145]. It is likely that the relative importance of the mechanisms is case-specific: the situation where the bubble has an initial distortion and, at that moment, has a stationary wall as discussed earlier for the injection process, is quite different from the case where the fluid at the wall has a finite velocity at the instant of bubble closure. Similarly, the Laplace pressure will have increasing influence as the bubble size decreases.

(a) Spherically Symmetric Excitation Through the Laplace Pressure and the Hydrostatic Head

It has been proposed that the excitation of the bubbles is due to the addition of a sudden ambient pressure as the liquid surface closes when a bubble is entrained. Crowther [26] and Hollett and Heitmeyer [143] found that a pressure change corresponding to only a 0.1-m head of water would account for the acoustic emissions from bubbles entrained by breaking waves. In addition, before the bubble wall closes completely around the bubble, there will be some orifice in the wall/free surface which will ensure that the gas that is about to be encapsulated will be at the pressure of the atmosphere just above the free surface, p_{atm}. However, after closure, the

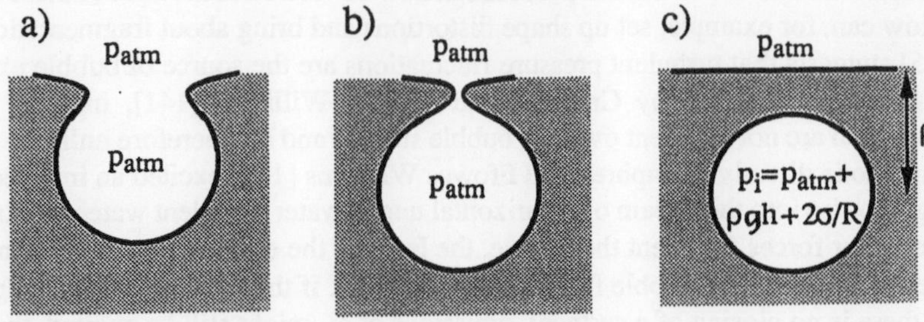

Figure 3.54 The gas pressures at which a bubble is formed (a) by trapping air at initially atmospheric pressure, and (b) through deformation of a free surface. At the moment of closure there is an initial excess pressure across the bubble wall due to the hydrostatic and Laplace pressures. Consequently the bubble contracts, and its volume will oscillate about an equilibrium position where the gas pressure inside the bubble will equal the sum of atmospheric, hydrostatic and Laplace pressures, as shown in (c).

encapsulated gas will no longer have communication with the atmospheric reservoir. At the very instant of closure, the gas will still be at a pressure p_{atm}, but the liquid pressure at the wall will be $p_{atm} + \rho g h$, and there will also be the added surface tension pressure (Figure 3.54).

It is not difficult to examine the effect of the surface tension and hydrostatic pressure terms together [95, 142]. Consider the gas that is to be trapped within the bubble. In the atmosphere, it is at a pressure of p_{atm} (approximately 1 bar). It remains at roughly this pressure until the liquid closes up around it, isolating it from the rest of the atmosphere. If this closure is assumed to occur instantaneously with the placement of the bubble at a depth h below the liquid surface, then rather than experiencing an ambient p_{atm} due to the atmosphere, the gas pocket is now subjected that p_{atm} *plus* an excess pressure due to the surface tension (p_σ) and the liquid head ($\rho g h$). There is clearly an imbalance, which will cause the gas within the bubble to be compressed. Assuming linear small-amplitude pulsation, the bubble therefore has energy to collapse from this initial maximum, through some equilibrium radius (where the internal pressure equals $p_{atm} + p_\sigma + \rho g h$) and be carried by the inertia of the liquid to some minimum, where the internal pressure will be approximately $p_{atm} + 2(p_\sigma + \rho g h)$. If no energy were lost, the bubble would expand back to its initial radius, where the internal gas pressure would be once again p_{atm}. Thus initially the amplitude of the internal pressure oscillation is $(p_\sigma + \rho g h)$. Damping will gradually reduce this.

In an infinite liquid, the initial far-field pressure radiated in the liquid, if this were the mechanism of excitation, would be[44]

$$P(r,t) = P(R_o,t)\frac{R_o}{r}\,e^{-\beta(t-r/c)} \tag{3.260}$$

where $P(R_o,t)$ is the acoustic pressure amplitude at the wall when the bubble has equilibrium radius, equal to

$$P(R_o,t) = \left(\frac{2\sigma}{R_o} + \rho g h\right)e^{i(\omega t-kr)} \tag{3.261}$$

[44]See section 3.3.1.

However, if the entrainment is by wave or rain action, the bubble will be near a water surface which, if planar, will cause the emission to take dipole form,[45] now having additionally a dependence on the polar angle θ

$$P(r,\theta,t) = \left(\frac{2\sigma + \rho g h R_0}{r}\right) 2hk\cos\theta e^{-\beta(t-r/c)} e^{i(\omega t-kr)} \tag{3.262}$$

so that the initial strength of the dipole is

$$D_{\mathrm{dip,i}} = (2\sigma + \rho g h R_0)2hk \tag{3.263}$$

which can be expressed as a function of the Minnaert approximation ω_M for the resonance frequency ω_0

$$D_{\mathrm{dip,i}} = \frac{2gh^2\sqrt{3\kappa p_0 \rho}}{c} + \frac{4\sigma h}{c}\omega_M \tag{3.264}$$

(there is a typographical error in reference [142]). The contribution to the dipole strength from the hydrostatic pressure is independent of frequency, but varies as h^2, whilst that resulting from the surface tension overpressure is proportional to both the depth and the bubble resonance frequency. Using the approximation $2\pi\nu_0 = \omega_0 \approx \omega_M$, the two terms in equation (3.242) are compared with the experimentally observed values of Pumphrey, Crum and Elmore [44, 95] in Figure 3.55 for a bubble at a depth of 3.2 mm (suggested by Pumphrey and Elmore [95] following observations from high-speed sequences of the type shown in Figure 3.42). For comparison, the result of theory for $h = 1$ cm is also shown also. The contribution from the hydrostatic pressure is constant for a given h at all frequencies, and is therefore given by the interception of the plots at zero hertz. It is therefore not a major excitation mechanism, for these values add a constant contribution of only around 0.003 N/m ($h = 3.2$ mm) and 0.03 N/m

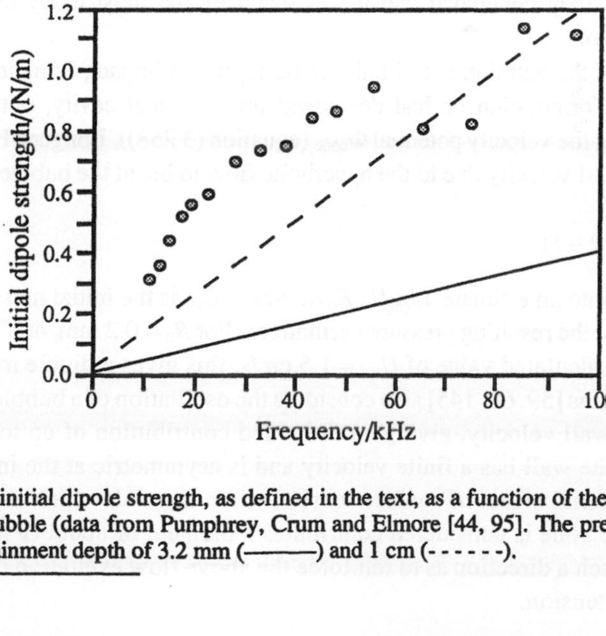

Figure 3.55 The initial dipole strength, as defined in the text, as a function of the resonance frequency of the entrained bubble (data from Pumphrey, Crum and Elmore [44, 95]. The predicted dependence is shown for an entrainment depth of 3.2 mm (———) and 1 cm (- - - - - -).

[45]See section 3.3.2.

($h = 1$ cm) to the initial dipole strength. Bearing in mind that the experimental data are taken from rain impact, for which a bubble depth of 3.2 mm was considered relevant by Pumphrey and Elmore [95], it is clear that another excitation mechanism must have contributed to the excitation, the contribution from the Laplace pressure increasing with decreasing bubble size, as one might expect from a surface tension effect.

Longuet-Higgins [145], estimating that the initial pressure in the bubble of order $2\sigma/R_o$ will introduce a pressure oscillation of $2\sigma/r$ in the liquid, suggests a contribution to the dipole moment equal to $2\sigma = 0.15$ N/m.

(b) Excitation Through Radial Liquid Flow at Moment of Closure, and Through Static and Flow Asymmetries

When the bubble is formed, the liquid flow that results in encapsulation may contain radial components, which obviously represent a force in the right direction to excite the zero-order mode. Pumphrey and Elmore apply the pulsating sphere formulation for dipole strength (equation (3.146)) to the bubble to obtain the initial dipole strength that results from a liquid flow of velocity amplitude $U_{0}^{R_o}$ at the bubble wall:

$$D_{\mathrm{dip,i}} = 2\rho c(kR_o)^2 hU^{R_o}_{0,\mathrm{i}} = \frac{6\gamma p_o}{c} hU^{R_o}_{0,\mathrm{i}} \tag{3.265}$$

$U^{R_o}_{0,\mathrm{i}}$ is the value of $U_{0}^{R_o}$ at the moment of closure, as discussed in section 3.3.2(b)(ii). Though Pumphrey and Elmore [95] were unable to estimate $U^{R_o}_{0,\mathrm{i}}$ from high-speed film records of bubble entrainment through the impact of a liquid drop, Pumphrey and Ffowcs Williams [142] were able to do so for bubble entrainment in a laboratory situation where a shallow stream of water flows into deeper water over a horizontal cylindrical obstacle, the axis of the cylinder being perpendicular to the flow. This estimated wall velocity gives them a dipole strength of $D_{\mathrm{dip,i}} = 1.3 \pm 0.6$ N/m, and they concluded that it was therefore the most likely candidate for the excitation mechanism.

In a discussion of the entrainment of bubbles through rain impact, Longuet-Higgins [145] used the analytical formulation he had developed for a conical cavity, outlined in section 3.7.2(f)(ii). Knowing the velocity potential Φ_{cone} (equation (3.258)), Longuet-Higgins finds the dipole-like radial fluid velocity due to the hyperbolic flow to be, at the bubble wall

$$U = \hat{A}(t)R_o(3\cos^2\theta - 1) \tag{3.266}$$

and substitutes this into an estimate, $\rho\omega_o U_{o,\mathrm{i}} R_o^2/r$, where $U_{o,\mathrm{i}}$ is the initial mean radial velocity at the bubble wall, of the resulting pressure oscillations. For $R_o = 0.2$ mm, $\omega_o/2\pi = 14$ kHz and Longuet-Higgins' calculated value of $U_{o,\mathrm{i}} = 1.5$ cm/s, this gives a dipole moment of 0.053 N/m. Longuet-Higgins [59, 60, 145] also considers the oscillation of a bubble that is initially distorted, with no wall velocity, giving an estimated contribution of up to 0.15 N/m, and where the fluid at the wall has a finite velocity and is asymmetric at the instant of bubble closure. The additional effect of surface tension when the sink which is associated with the curvature of the cone is considered contributes a moment of about $\sigma/\sqrt{2} = 0.052$ N/m, the flow being in such a direction as to reinforce the above flow excitation calculated in the absence of surface tension.

(c) Conclusions: the Excitation Mechanism

Measured initial peak dipole moments for entrainment by normally incident raindrops include observations in the approximate ranges 0.4–0.5 N/m [96], 0.3–1.1 N/m [44], and beneath laboratory breakers (no wind) 0.1–1.0 N/m (see Section 3.8.2(a)). Estimated excitations, as described above, are summarised below:

(i) For hydrostatic head: 0.003 N/m (h = 3.2 mm) and 0.03 N/m (h = 1 cm) (see above). In section 3.8.2(b)(ii), it will be shown that results by Buckingham [146] suggest entrainment depths of h = 1.5 m occur in the ocean, which would in such circumstances give an excitation by the hydrostatic head of 600 N/m (equation (3.242)).

(ii) For the Laplace pressure: 0.15 N/m [132]. As shown in Figure 3.55 for h = 3.2 mm (0.04 N/m at 10 kHz, 0.4 N/m at 100 kHz) and h = 1 cm (0.12 N/m at 10 kHz, 1.2 N/m at 100 kHz). For h = 1.5 m, as discussed above, the excitations are 18 N/m at 10 kHz, 180 N/m at 100 kHz.

(iii) Radial flow: $D_{dip,i}$ = 1.3±0.6 N/m [142]; 0.053 N/m (R_0 = 0.2 mm, ν_o = 14 kHz [132]).

(iv) Asymmetric excitation: 0.052 N/m [132].

Given both the disparity in some estimates, and the dependence on entrainment depth and bubble size, it is therefore not possible to say conclusively that any one mechanism dominates under a given set of circumstances. The depth is clearly important, and estimates of depth to be found in the literature include: 3.2 mm for regular entrainment through liquid drop impact [95]; 8.7 mm for entrainment in a laboratory facility where waves spill as a result of the water slope (there being no wind) [38]; and 1.5 m for oceanic breakers [146]. The estimate for the hydrostatic effect under 1.5 m of water is intriguingly large, though of course the model shown in Figure 3.54 of a single bubble in an otherwise bubble-free liquid, entrained from a flat surface, is too simple to account for the complex dynamics of oceanic breakers. The final section of this chapter will examine by far the most common, though probably the most difficult to study, populations of bubbles entrained from a free surface.

3.8 Oceanic Bubbles

In the previous section, the importance of knowledge of the details of bubble entrainment was illustrated in examination of the question of the mechanism through which the bubble is excited to emit. Therefore this section, which examines the acoustic effects of oceanic bubbles, will open with a discussion of the structure and dynamics of the bubble population in the sea.

3.8.1 Characteristics of Oceanic Bubble Populations

Clear and full accounts of oceanic bubble populations can be found in papers by Thorpe [147,148] and Monahan and Lu [149], and it is upon these that this introduction to the subject is based.

Wind can cause surface gravity waves in deep water to break as whitecaps (also known as 'whitehorses'). Large numbers of bubbles are entrained by this process, and their subsequent behaviour depends on their size and the local currents. *Foam* consists of bubbles *on* the sea surface, most of which are in physical contact with their neighbours, if not overlying them [148]. It is probably made up of the larger bubbles, which rise by buoyancy after a wave-breaking event and can persist in the foam layer for roughly 10–60 s [148]. Foam exists in patches, and covers a fraction of the sea surface that can be measured from above by photography. Monahan and Muircheartaigh [150] give the following approximate expression for the fraction of surface covered by foam, relating it to w_{10}, the windspeed at a height 10 m above the surface:

$$\text{Fractional foam coverage} = 3.84 \times 10^{-6} \left[\frac{w_{10}}{\text{m/s}} \right]^{3.41} \qquad (3.267)$$

One would not expect bubbles in foam to be as acoustically active as the bubbles discussed so far in this section. The latter, surrounded by a body of liquid, couple well with sound fields because of the great difference in compressibility between the gas and the liquid. This is not the case with bubbles in foam, owing to the proximity of contacting neighbours and the atmosphere. However, 'acoustic bubbles' do result from the wave-breaking action: at windspeeds of greater than around 7 m/s, there is just *below* the surface a continuous layer of bubbles. At greater depths, the bubbles tend to be grouped into clouds, which persist for 60–300 s, and are generally of a smaller size than those bubbles found in foam [148]. When considering oceanic bubbles, both in foam and subsurface, it should be noted that the sea is effectively a 'dirty' environment, and so as discussed earlier, will tend to stabilise bubbles[46] and immobilise the interface,[47] and impart to it the translational dynamics of a rigid body.

Aleksandrov and Vaindruk [151] recognised that the orbital motion of surface waves would not be capable of transporting or sustaining a bubble population to the observed depths of several metres, and that turbulence is in fact responsible. Though buoyancy would tend to cause the bubble to rise upwards at speed u_b, if the surrounding water is moving downwards with a vertical velocity component u_w, then the net upwards velocity of the bubble is $u_b - u_w$, which will be negative if $u_w > u_b$, signifying net downwards motion of the bubble. In 1963, Kanwisher [152], using an echo sounder submerged at a depth of 30 m, described 'foam' being swept down to a depth of 20 m. Thorpe [147] points out that, since Blanchard and Woodcock [153] have shown that bubbles thought to be typical of those entrained by breaking waves persist for only a few minutes, then to be observed at these depths, individual bubbles must have downward velocities of many centimetres per second. Kanwisher observed the positive correlation between the windspeed and number of bubbles entrained, and concluded that atmospherical gas exchange with the sea may be significantly influenced by bubble mechanisms at higher windspeeds. The asymmetry which bubbles introduce to the atmosphere/ocean gas exchange system is explored in section 3.8.1(d).

Prior to a discussion of the acoustical effects of the sub-surface bubble population, this section will examine these aspects, specifically: (i) the formation and evolution of bubble clouds; (ii) measurement of the cloud dimension; (iii) measurement of the bubble size distribution of oceanic bubbles; and (iv) the contribution of bubbles to gas exchange between atmosphere and oceans.

(a) Formation and Evolution of Oceanic Bubble Clouds

Figure 3.56 shows a representation by Monahan and Lu [149] of many surface and subsurface features, which items can be thought of as, in the first instance, two- and three-dimensional respectively. The diagram includes the stages of development of a bubble plume, illustrating the accompanying surface features. Features from earlier graphic representations by Thorpe

[46] See Chapter 2, section 2.1.2(b).
[47] See Chapter 2, section 2.3.2(b)(ii).

[154, 155] and Monahan [156, 157] are incorporated. The near-surface continuous-layer of bubbles, which occurs for windspeeds in excess of 7 m/s, is not shown.

Figure 3.56 shows a volume of sea relevant to a windspeed of about 10 m/s. The below-surface features are rotated forward through a right-angle, the hinge line being coincident with the front edge of the sea surface shown. The features are to scale (note the arrows measuring 10 m on surface and in depth, at the right of the figure). Anchors and anchor chains are added to clarify the direction. In this way it is possible on one figure to represent the subsurface features relevant to the commonly observed surface features. Surface feature A is a spilling wave crest or active whitecap, the surface accompaniment to the subsurface concentrated bubble plume (type α). Estimates of the void fraction go as high as 30% at the crest of the spilling wave [138],

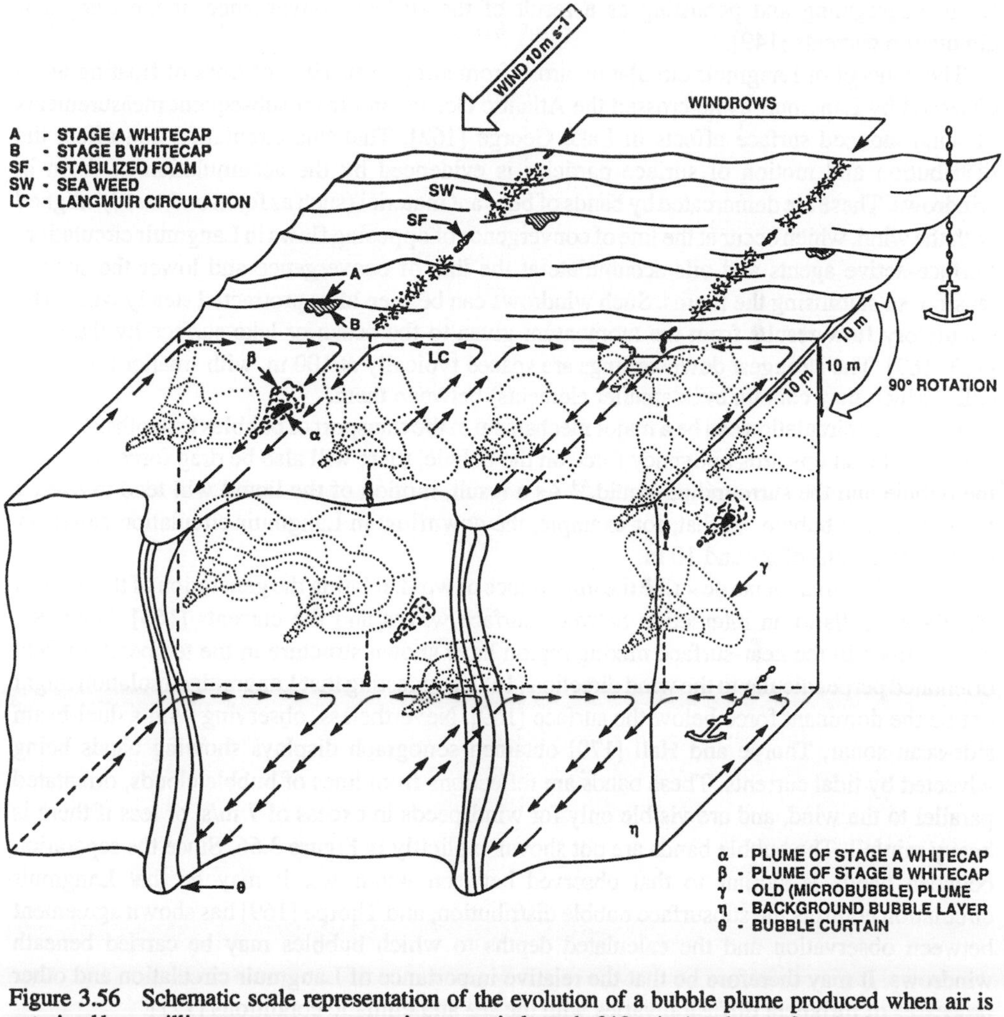

Figure 3.56 Schematic scale representation of the evolution of a bubble plume produced when air is entrained by a spilling wave crest, appropriate to a windspeed of 10 m/s. A volume of sea is shown, though hinged forward at the upper front end so that whilst the sea surface at the rear of the diagram is viewed from above, it is separated from the subsurface features, which are shown on their side and which the reader (looking through the front of the diagram) views as though from below. Anchors are drawn on these sections to clarify the direction. Arrow bars at the right of the hinge give a 10-m scale. Langmuir circulation is represented by subsurface arrows (after Monahan and Lu [149]). © 1990 IEEE

though this decreases with depth, and so is acoustically very important. These features are transient events, only persisting for longer than 1 s if they are regenerated by the parent spilling wave. Otherwise they immediately decay into the mature whitecap, or hazy foam patch, on the surface (type B features in Figure 3.56), and a type β-plume below. Bubble concentration and size spectrum are probably quite constant with depth [157, 158]. The β-plumes are columns of aerated water, attenuating with depth. The cross-sectional area decays exponentially, with decay length ≈0.5 m [149], and a void fraction estimates from 0.01–0.02% [159] to 0.1–0.2% [160]. Collective oscillations within β-plumes may act as sources of low-frequency sound.[48] The B and β features coexist for several seconds, dissipating together: the area of a mature whitecap decreases approximately exponentially [161] with an 'e-folding time' of 3.5–4.3 s. After a β-plume is dissipated, no subsurface feature exists unless it is as the underside of stabilised foam, aggregating and persisting as a result of the surface convergence of the Langmuir circulation currents [149].

The concept of Langmuir circulation arose from a report in 1923 of lines of floating weed observed by Langmuir as he crossed the Atlantic Ocean, and from subsequent measurements of wind-induced surface effects in Lake George [162]. That this circulation can affect the distribution and motion of surface particles is evidenced by the accumulation of foam in windrows. These are demarcated by bands of buoyant material (such as foam and weed), aligned with the wind, which occur at the line of convergence of opposing flows in Langmuir circulation. Surface-active agents and oils accumulate at the line of convergence and lower the surface tension, so stabilising the foams. Such windrows can be seen after protracted steady wind. The circulatory flow results from the momentum given to the ocean or lake surface by the wind [163–167]. The strongest downwellings are spaced typically at 100 m, with smaller Langmuir cells (which may contain even smaller elements) between them.

Langmuir circulation can be a major mechanism in the transport of bubbles to depths. Though there will be an upwards buoyancy force on the bubble, there will also be drag forces between the bubble and the surrounding liquid.[49] As a result, motion of the liquid will tend to exert a force upon the bubble, so that, for example, the downflow in Langmuir circulation can carry bubbles to depths of around 10 m.

Theoretical treatments describe the importance of wind, relating the generation of the vertical circulation cells to an interaction between surface waves and the currents [163]. However, observations in the near-surface mixing region have shown structure in the temperature field orientated perpendicular to the wind direction [168], suggesting that Langmuir circulation might not be the dominant force below the surface [169]. Nevertheless, observing with a dual-beam side-scan sonar, Thorpe and Hall [170] obtained sonograph displays showing bands being advected by tidal currents. These bands are reflections from lines of bubble clouds, orientated parallel to the wind, and are visible only for windspeeds in excess of 7 m/s, or less if there is heavy rainfall. The bubble bands are not shown explicitly in Figure 3.56. Since the separation between bands is similar to that observed between windrows, it may be that Langmuir circulation controls the subsurface bubble distribution, and Thorpe [169] has shown agreement between observation and the calculated depths to which bubbles may be carried beneath windrows. It may therefore be that the relative importance of Langmuir circulation and other processes of turbulent diffusion varies with the sea and climatic conditions [171].

Thorpe [169] describes a simple formulation which accommodates gas exchange from bubbles, the motion of which is governed by buoyant forces, turbulent diffusion of bubbles from the surface, and advection by Langmuir circulation. The analytical prediction for motion

[48] See section 3.8.2(c)(i).
[49] See Chapter 2, section 2.3.

agrees with a random-walk calculation, which can be applied to regimes where the analytical formulation is invalid, where the bubble population contains a wide range of radii and bubbles are composed of several gases. The random-walk model allows quantitative estimations of the effect of circulation on the acoustic scattering cross-section.[50] Calculations also estimate the heat flux resulting from Langmuir circulation. Thorpe concludes that the contribution to the acoustic scattering of an incident acoustic beam resulting from Langmuir circulation is a significant fraction of the net value. Provided the entrainment process is uniform (e.g. in heavy rain) or frequent (e.g. in strong winds), Langmuir circulation should induce a perturbation to the bubble distribution detectable by side-scan sonar.

Large subsurface γ-plumes which persist for hundreds of seconds [148] are acoustically detectable after a spilling wave has vanished. As a result of the longer lifetimes, any sample of ocean like that illustrated in Figure 3.56 will contain many more of these γ-type plumes than the β-type, the former being larger owing to the spread of bubble by turbulent diffusion [149]. Thorpe [147] used an inverted upward-pointing echo sounder, located at a depth of 30 m, for below-surface observation of these clouds of small bubbles. The bubbles collected in two types of cloud distinguishable on sonograph records. *Columnar clouds* form during unstable or convective conditions when the air is cooler than the surface water temperature, presumably influenced by the vertical convection and low shear. *Billow clouds* conversely are tilted roll-like structures, appearing in stable conditions when the air is warmer than the surface water temperature, the heat flux through the surface stabilising the water column. The depth of penetration of the bubble clouds increases with increasing windspeed, and for a given windspeed is greater in convective than in stable conditions (equation (3.270)). Since small bubbles have dissolved and large ones have been removed by buoyancy, the range of bubble radii present in γ-plumes is narrower than those in α- or β-plumes. Estimates of the void fraction suggest values of $10^{-4}\%$ to $10^{-5}\%$ [159]. The concentration decreases with depth. The γ-plumes can be billowy or columnar, though in Figure 3.56 Monahan and Lu [149] have illustrated idealised clouds, having a horizontal cross-sectional area which narrows exponentially with depth, a representation which may not be appropriate to the general case.

Figure 3.57 [147] illustrates billow clouds, and Figure 3.58 [147] columnar clouds. Plotted against a common time axis (time increasing to the left) is (from the top) the sonograph display with the mean level of the surface arrowed, a contour plot of $\log_{10}\Omega_c$, the values of $\log_{10}\Omega_c$, and finally the windspeed. Thorpe [147] describes a delay of only about 2 minutes in the appearance or deepening of clouds in response to a squall (Figure 3.57) or an increase in windspeed over Loch Ness, in accordance with the rapid response of the wavefield and the increase in whitecaps. After the squall has passed, the bubbles persist for a much longer period ($\gtrsim$ 10 minutes).

The positions of bubbles which have persisted for a long time[51] become randomised, and they can be thought of as a background population, the concentration of which is uniform in the horizontal plane and falls off exponentially with depth. Thus if some given cut-off concentration is taken to mark the base of this background population, then it can be labelled as the item η on Figure 3.56.

Persistent features, such as windrows, γ-plumes and the background bubble population, are influenced by several factors. The stability of bubble constituents will be influenced by local concentrations of dissolved atmospheric gases in the water, and by any stabilising factors (such as surfactants) [149]. Considering the discussion on bubble stabilisation by surface-active substances and motes given in Chapter 2, section 2.1.2(b), it is not surprising that bubbles

[50] The acoustic scattering cross-section is a measure of the ability of the target to scatter incident sound waves. Generally speaking, this increases with the bubble population. See section 3.8.2(c)(ii) and Chapter 4, section 4.1.2(d).
[51] And which may be stabilised by surface-active material in a manner similar to the nuclei discussed in Chapter 2, section 2.1.2(b).

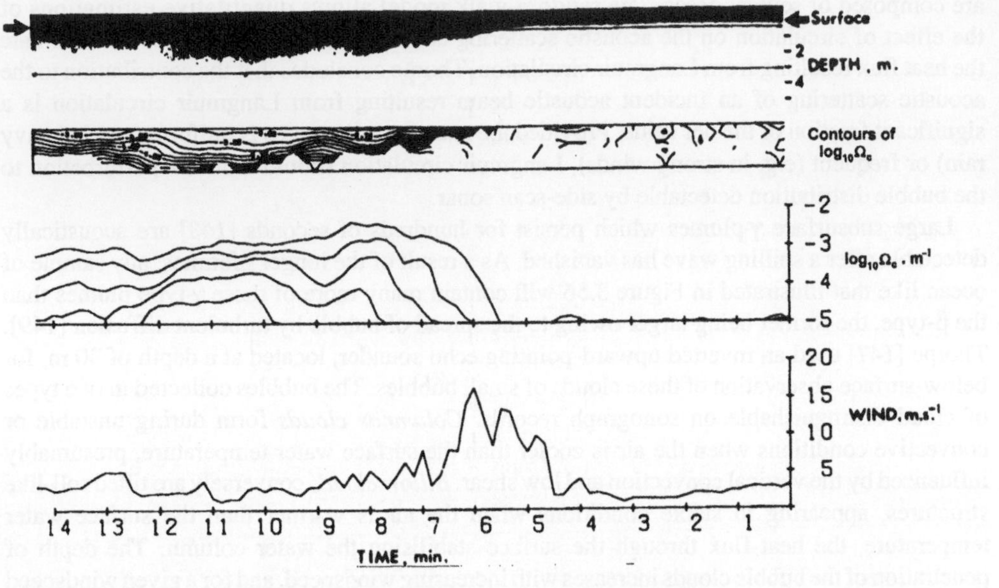

Figure 3.57 A sonograph display of billow clouds in Loch Ness. Plotted against a common time axis (time increasing to the left) is (from the top) the sonograph display with the mean level of the surface arrowed, a contour plot of $\log_{10}\Omega_c$, the values of $\log_{10}\Omega_c$, and finally the windspeed (after Thorpe [147]). The time history illustrates the response of the bubble clouds to a brief squall.

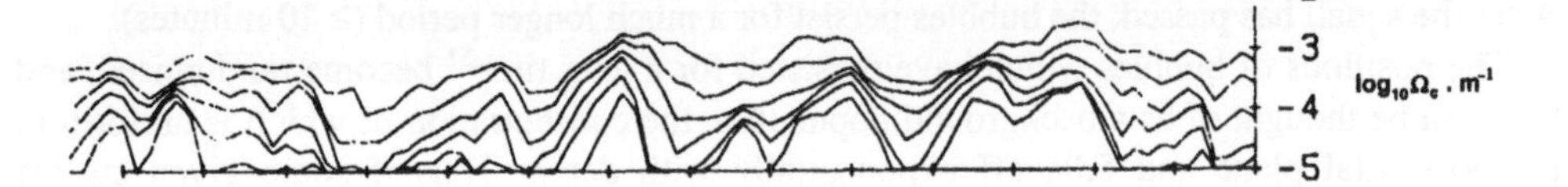

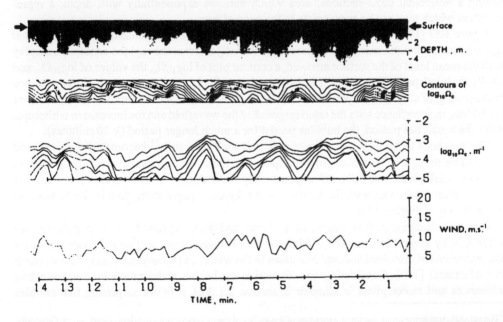

Figure 3.58 A sonograph display of columnar clouds in Loch Ness, the plots as for Figure 3.57 (after Thorpe [147]).

produced by seawater waves tending to persist longer than those from fresh water waves [172]). In addition, bubbles will be moved by prevailing water motions. Half of the γ-plumes represented by Monahan and Lu in Figure 3.56 are aligned along the surface convergence zones of the Langmuir circulation, where the material forming the windrows also accumulates. The down-welling component of the circulation will distort the background bubble population, generating curtains or walls where the concentration of bubbles is increased above the norm for the otherwise horizontally uniform background. Monahan and Lu illustrate two such curtains (marked θ in Figure 3.56).

(b) Bubble Counting in the Sea

Research into the characteristics of the oceanic bubble population has been underway for several years. Attempts have been made to analyse the entrainment and ambient populations of bubbles in the sea (Toba [173], using optical methods; Updegraff and Anderson [89, 174]). Acoustic techniques, including the use of passive acoustic emissions from bubbles entrained through laboratory-generated spilling breakers, have been utilised to determine the bubble population and entrainment rate, and the volume of trapped gas [172]. Oceanic passive emissions have similarly been studied, and related to the Knudsen spectra [174]. Acoustic techniques for bubble detection are reviewed in Chapter 5, section 5.1.

The bubble population in the sea tends in the main to increase with windspeed and decrease with depth below the surface. A discussion and review can be found in Thorpe [147]. Optical investigations by Blanchard and Woodcock [153] have shown that most of the bubbles in the near-shore zone have radii less than 100 μm. Kolovayev [175] was probably the first to catch, photograph and count bubbles below breaking waves in the open sea at windspeeds of up to 13 m/s, by allowing them to rise onto a transparent plate. The bubbles may, however, have dissolved on rising [147]. At depths between 1.5 and 8 m the most common bubbles were those possessing a radius of around 70 μm, and very few bubbles were greater than $R_0 \approx 300$ m. Johnson and Cooke [176] photographed bubbles *in situ* in the sea, using a camera suspended from a surface float at depths of 0.5–4.0 m, and windspeed of 8–13 m/s. They observed that the distribution of bubble size narrowed with increasing depth, the larger bubbles disappearing. The number of bubbles greater than the minimum size they could detect ($R_0 \approx 17$ μm) also decreased with increasing depth, decaying roughly exponentially over depthscales of the order of 1 m at windspeeds of 11–13 m/s, such that at a depth of 1.8 m, the density of bubbles of a detectable size was 1.56×10^5 m^{-3}. They obtained a modal bubble radius in the size distribution of 40–50 μm. However, Walsh and Mulhearn [177] suggest that the photographic observations lack the resolution to count the smallest bubbles accurately. MacIntyre [178] suggests that in addition to the lower limit imposed by resolution, there is an upper size limit on the reliable data resulting from poor statistics, so that only the data for bubbles in the range 60 μm $\leqslant R_0 \leqslant$ 150 μm are reliable. Medwin [179], making acoustic observations *in situ* through examination of attenuation at various frequencies, suggests that there is a higher proportion of the much smaller bubbles. Medwin and Breitz [180] confirm that the peak in the size spectrum occurs at a radius of less than 30 μm, and Su *et al.* [181] suggest that the peak is around 20 μm. Farmer and Vagle [66] deployed an upwardly-pointing four-frequency echo sounder (28, 50, 88 and 200 kHz) to investigate the bubble size distribution. Time-of-flight of reflected acoustic pulses gave a measure of distance, with a 10-cm resolution for vertical samples. The backscattered intensities at the four frequencies for each vertical sample gave point measurements of the bubble population at four specific radii in the range $15 \leqslant R_0 \leqslant 100$ μm. These were then employed in an iterative calculation to give the size distribution of the whole population. They confirmed that an upper bound on the size spectral peak occurs at $R_o = 22\ \mu m$, the true peak being closer to 20 μm.

One of the few oceanic observations which involved measurements made close to breaking waves (i.e. at depths less than 0.5 m) was by Medwin and Breitz [180], who used the variation in Q of several modes of a floating acoustical resonator to determine the bubble spectral size distribution at a depth of 0.25 m below a spilling oceanic breaker. The one-dimensional resonator consisted of a flat transducer facing a reflective plate 126 mm away, so that modes could be set up in the water between these. Bubbly water could flow in readily between the plates from the environment. From resonance broadening measurements for nine specific bubble sizes in the range $30\ \mu m < R_o < 270\ \mu m$, Breitz and Medwin [182] found an average bubble density of

$$n_b^{\mu}(R_o) = 7.8 \times 10^8 \left[\frac{R_o}{1\ \mu m}\right]^{-2.7} \quad (\text{for } 30\ \mu m < R_o < 270\ \mu m) \tag{3.268}$$

where n_b^{μ} is the number of bubbles per cubic metre per micrometre increment in radius. In the same radius range, the maximum bubble density detected was $n_b^{\mu}(R_o) = 1.6 \times 10^9\ [R_o/1\ \mu m]^{-2.7}$. Medwin and Breitz [180], however, found that only the larger bubbles in the range $60\ \mu m < R_o < 240\ \mu m$ followed a $n_b^{\mu}(R_o) \propto [R_o/1\ \mu m]^{-2.6}$ distribution: the population of smaller bubbles ($30\ \mu m < R_o < 60\ \mu m$) decayed with depth as $n_b^{\mu}(R_o) \propto [R_o/1\ \mu m]^{-4}$. A $[R_o/1\ \mu m]^{-4}$ model distribution fits most of the data obtained by bubble counting reasonably well [183].

After entrainment has occurred, buoyancy, gas diffusion, stabilisation and water motion will control the population to be found in a given volume of the ocean. However, the bubble population below surface ultimately begins with the population of bubbles entrained from the surface. Since in entrainment the bubbles emit sound, acoustic measurements at the surface will give information regarding the genesis of the bubble population in the sea. Medwin and Daniel [172], studying the generation of bubbles at the surface in a laboratory wave-breaking facility, compared the arrival times of single-bubble signals (similar to Figure 3.23) received at two pairs of hydrophones to determine the position of the bubble when it began to emit sound. From the spatial distribution of 500 bubbles, generated by 10 breakers, Medwin and Daniel tentatively suggest that the data might sometimes show lines of simultaneous bubble generation on the surface, and speculate that this might result if capillary waves were involved in the entrainment process.[52]

Most of the bubbles were generated in the first 500 ms after the passage of the breaker. For bubbles of all sizes, Medwin and Daniel suggest as a best-fit to their data, a bubble generation rate of

$$\frac{dN_b}{dt} = 3.13e^{-0.0070t} \tag{3.269}$$

Toba [173] studied bubbles in the volume of water below the surface, and extrapolated to deduce the surface bubble count. Medwin and Daniel compare their surface bubble data with that of Toba. The smallest bubble radii detected were, respectively, 0.05 mm and 0.15 mm, and the largest radii were 7.4 mm and 6 mm. Thus Medwin and Daniel detected a larger range of bubble size. Their maximum bubble density was five times that of Toba. Thorpe [184] makes the general point, however, that whilst laboratory experiments may adequately generate conditions of wind and turbulence in the air, their limited size prohibits the reproduction of the

[52] See section 3.7.3.

larger surface waves. As a result, the turbulence in the water in laboratory facilities may seriously fail to simulate oceanic conditions, about which little is known. He illustrates this by noting that Broeker [185], generating bubbles at windspeeds of 11–17 m/s in one of the largest laboratory facilities used in this field, measured a depthscale of 11 cm over which the subsurface bubble population exponentially decayed. This is very much less than those found by Johnson and Cooke [176] and Thorpe [147] in the ocean, which were of the order of 1 m. This suggests that the laboratory facility failed to simulate the high levels of subsea turbulence encountered in the ocean. Thorpe [184] notes that laboratory simulations of bubble entrainment may be similarly in error, in particular in studies of the flux of atmospheric gases from bubbles into the sea: bubbles carried to great depths by oceanic turbulence experience a significant hydrostatic overpressure, which drives the gas into solution, a phenomenon which laboratory facilities of lesser depth and turbulence will not duplicate.

(c) Dimensions of the Bubble Cloud

In view of the above comments, a distinction will be made in this section between observations of oceanic clouds, and of laboratory measurements.

(i) Oceanic Observations. Thorpe and Stubbs [186] used 248-kHz sonar to detect bubbles in Loch Ness. If w_{10} is the windspeed as measured at 10 m height, and $\Delta T^{\text{air}}_{\text{water}}$ the temperature of the air minus that of the water at the surface, then the average penetration depth of the bubble cloud, L_c, is given approximately by

$$\left[\frac{L_c}{\text{m}}\right] \approx 0.31\left(1-\left[\frac{\Delta T^{\text{air}}_{\text{water}}}{\text{K}}\right]\right)\left(\left[\frac{w_{10}}{\text{m/s}}\right]-2.5\right) \tag{3.270}$$

where depth, temperature and windspeed are normalised by 1 m, 1 kelvin and 1 m/s respectively, applicable for $|\Delta T^{\text{air}}_{\text{water}}| \lesssim 6$ K [147]. In the sea off Oban, the average depths were greater, given approximately by

$$\left[\frac{L_c}{\text{m}}\right] \approx 0.4\left(\left[\frac{w_{10}}{\text{m/s}}\right]-2.5\right) \tag{3.271}$$

where the data were obtained for south-westerly and westerly winds, with $-2\text{ K} < \Delta T^{\text{air}}_{\text{water}} < 2\text{ K}$. It should be noted that the measured data fit equations (3.270) and (3.271) only with a significant degree of scatter. The average depth of bubble clouds was greater in the Irish Sea than off Oban, and greater on the edge of the continental shelf off England than in the Irish Sea [148], though seasonal and climatic differences pertain to these experimental data.

Direct sonar measurements in Loch Ness by Thorpe [147], and his analysis of Johnson and Cooke's [176] sea results, suggest that Ω_c decays approximately exponentially with depth (with an 'e-folding depth' of metre order), and increases with windspeed. A comparison of sonar measurements suggest that, whilst penetration depths are similar in both salt and fresh water for $w_{10} \lesssim 7$ m/s, at large windspeeds ($\gtrsim 10$ m/s) clouds in the sea off Oban penetrate deeper than the fresh water ones in Loch Ness, and the average values of Ω_c tend to be greater [147]. Figure 3.59 shows the form of bubbles clouds, as measured in the Irish Sea by an upwardly pointing sonar. The logarithm of the acoustic scattering cross-section is contoured for different values of the windspeed w_{10}, and for the ratio of w_{10} to the dominant surface wavespeed. At low values of this ratio, the deepest clouds reach to about five times the significant wave height H_s (or about 0.16 times the dominant surface wavelength). An increase in windspeed results in deeper penetration, so that when the ratio is 1.22, the clouds reach 5.2 times the wave height and 0.25 times the wave length. The form of the contour plot is intriguing, and does suggest a subsurface layer of bubbles, with clouds penetrating to greater depths. However, it should be borne in mind that such plots represent the time-variation in the scattering of an upwardly pointing sonar beam, which will vary both because of the cloud shape and because of the advection of the cloud through the sonar beam. As a result, it is not possible to unambiguously determine the shape of these bubble clouds.

The observations of Thorpe [148] regarding the scale of the bubble clouds are comparable with the depth to which bubbles were carried in the laboratory studies of Toba *et al.* [187], i.e. 2.5–3 times the wave height. Thorpe [148] analyses the dynamics of a model cloud as it is generated by wave action and advected over the sonar. The results indicate that relatively few of the breaking waves contribute to the bubble cloud population at a depth of 10.3 m: most events presumably contribute to the near-surface bubble layer. The horizontal area of the bubble clouds at a depth of 10.3 m is about an order of magnitude greater than the whitecap area, whilst the area of whitecaps containing waves which are in the process of breaking and producing clouds which will extend to 10.3 m or deeper is an order of magnitude less than the total whitecap area.

Efforts were made to by Monahan and Lu [149] to relate the early, acoustically relevant stages in the development of the subsurface bubble plumes, to the whitecaps which are the

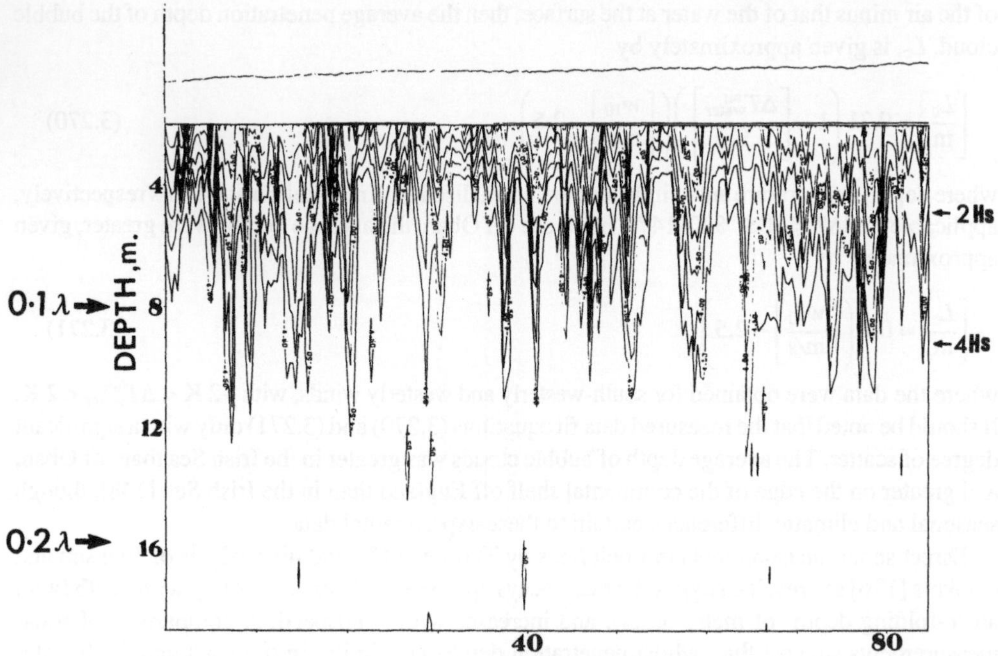

Figure 3.59 Bubbles clouds, as measured in the Irish Sea by an upwardly pointing sonar, shown as contours of the logarithm of the acoustic scattering cross-section, as a function of time. The windspeed w_{10}, and the ratio of w_{10} to the speed of the dominant surface wave, are respectively: (a) 6.9 m/s, 0.63; (b) 10.1 m/s, 0.86; (c) 13.3 m/s, 1.22. The depth is shown on the vertical axis, in metres, in multiples of the wave length of the dominant surface wave (on the left), and in multiples of the significant wave height H_s (on the right) (after Thorpe [148, 171]).

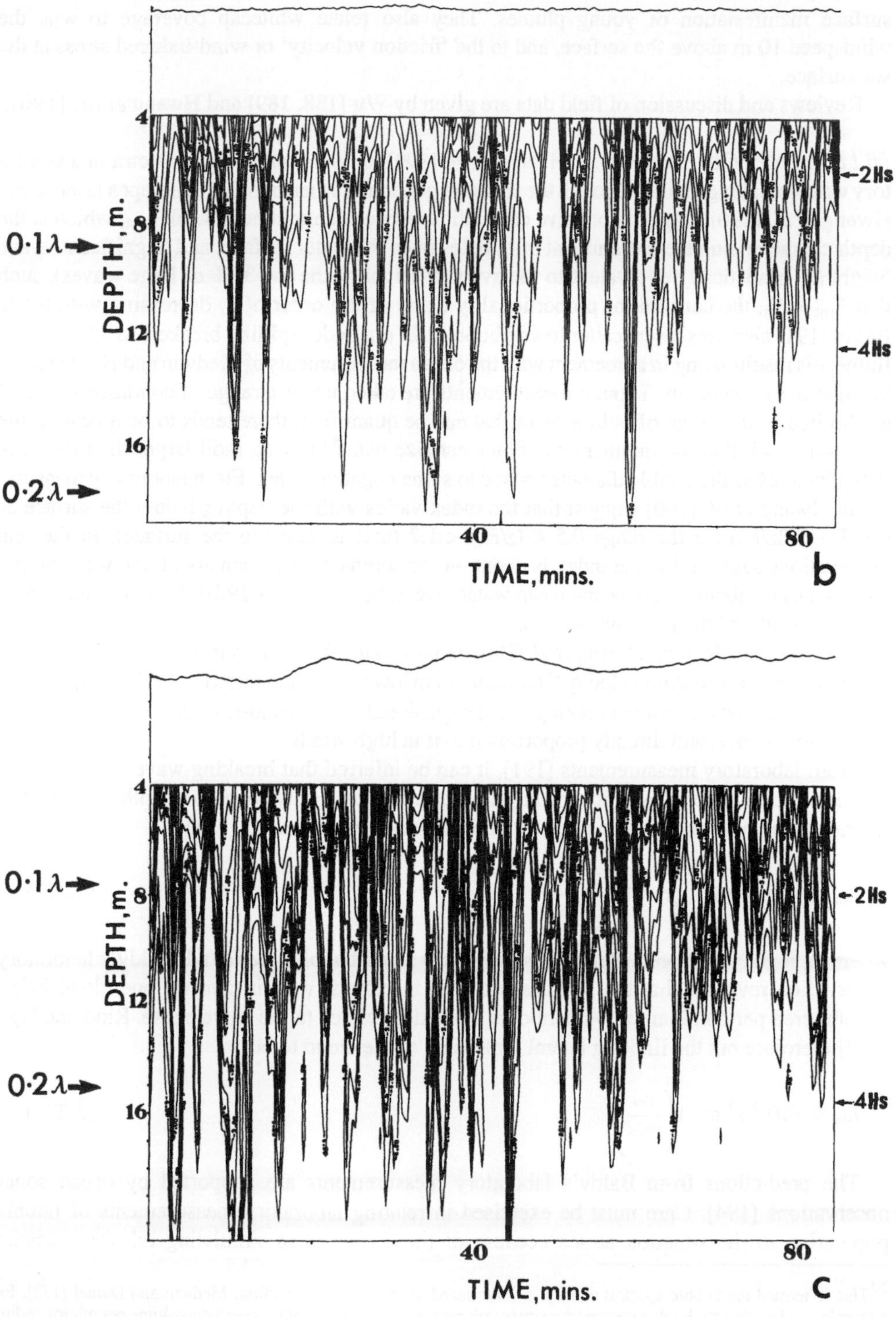

Figure 3.59 *continued*

surface manifestation of young plumes. They also relate whitecap coverage to w_{10}, the windspeed 10 m above the surface, and to the 'friction velocity' or wind-induced stress at the sea surface.

Reviews and discussion of field data are given by Wu [188, 189] and Hwang *et al.* [190].

(ii) Laboratory Observations. Hwang *et al.* [190] measure bubble size spectra in a laboratory wave-breaking facility through the shadows generated when bubbles intercept a laser beam. Hwang *et al.* [190] suggest a positive correlation between entrainment depth h_{ent}, which is the depth characterising the entrainment of bubbles from the water surface, and 'significant' wave height H_s (statistically equivalent to the average height of the top 33% of large waves), such that $h_{ent} \propto H_s$, the constant of proportionality being of the order of 1, decreasing with fetch. Baldy [191] also uses laser scatter to size bubbles 25 cm under spilling breakers in a laboratory flume, his results being in agreement with the ocean measurements of Medwin and Breitz [180], taken at the same depth. Though measurements are taken under a range of conditions, and all are limited in the range of bubble sizes that can be quantified, there tends to be a peak in the spectrum, such that the number of bubbles per size band[53] having radii larger than the peak value falls off as the bubble diameter raised to some negative index. From laboratory measurement, Hwang *et al.* [190] suggest that the index varies with the depth z below the surface as $-(0.7 + 2.8z/H_s)$ for the range $0.5 < (z/H_s) < 1.2$ (that is, close to the surface). In the sea, observations suggest that the index is about −4 for depths greater than about one wave height ($z/H_s > 1$), and about −2 above the mean water level ($z/H_s < 0$) [190–192]). This is in agreement with the results of the previous section.

In a laboratory facility, Hwang *et al.* [190] demonstrated that, if the surface bubble population varies as the wind-friction velocity[54] raised to ~5 in low winds, and raised to ~2.5 in high winds, then the frequency of wave breaking was proportional to the wind-friction velocity raised to ~2.2 in low winds, and directly proportional to it in high winds.

From laboratory measurements [191], it can be inferred that breaking waves will produce clouds of small bubbles in which the largest ones are present only down to a depth of about half the r.m.s. height of the breaking wave, that is down to approximately [193]

$$L_{depth} = (1.5 \times 10^{-2}\ \mathrm{s^2\ m^{-1}}) \left(\frac{v_w}{\mathrm{m/s}}\right)^2 \tag{3.272}$$

where v_w is the windspeed. Below this, only very small bubbles are found. Baldy's laboratory observation revealed that the bubbles occurred in clouds, with in general one cloud being encountered per dominant wave period, at a position related to the wave phase. Rino and Ngo [193] therefore put the limiting lateral dimension of the cloud to be

$$L_{lat} \leqslant (10^{-2}\ \mathrm{s^2\ m^{-1}}) \left(\frac{v_w}{\mathrm{m/s}}\right)^2 \tag{3.273}$$

The predictions from Baldy's laboratory measurements are supported by ocean sonar observations [194]. Care must be exercised in relating laboratory measurements of bubble population to the situation at sea because of the difficulty in simulating relevant oceanic

[53] This is termed the bubble spectral density, as introduced in the preceding section. Medwin and Daniel [172], for example, explain that the bubble spectral density in volume is the number of bubbles per unit volume per micron radius increment, which is a function of the bubble radius. They contrast this with the surface bubble spectral density per unit area.

[54] The square of the wind-friction velocity equals the ratio of the wind stress to the air density.

conditions. For example, in the above laboratory studies the age of the wave must be considered, and in the previous section limitations in depth and turbulence in laboratory facilities were discussed, and their effect on gas transfer studies considered. This is the topic of the next section.

(d) Gas Flux Through Oceanic Bubbles

Gas flux between the atmosphere into the oceans can occur either through direct transfer across the sea surface [183, 195], or through the dissolution of bubbles which, trapping gas at atmospheric pressure, dissolve as a result of the Laplace[55] and hydrostatic overpressures. As a result, unlike direct transfer, bubbles can force gas even into saturated water, and so produce supersaturation. Oxygen is on average about 3% supersaturated in near-surface ocean waters [196]. All else being equal, one would expect the flux of a very soluble gas (e.g. CO_2), which rapidly reaches equilibrium conditions, to be dominated by direct transfer across the sea surface, whilst bubble-mediated flux would be important for much less soluble gases.

Several of the studies of the flux of gas between atmosphere and ocean employ laboratory facilities and Thorpe [184] comments on the problems inherent in these experiments. The inability to simulate oceanic turbulence, and to carry bubbles to great depth where the hydrostatic overpressure will force the gas into solution, have already been mentioned. Details of the entrainment process are also important: if the wave fails to inject large bubbles to significant depths, and their vertical distribution depends only on buoyancy, turbulent diffusion and Langmuir circulation, then few large bubbles will be carried to depths of greater than 1 m, so that only smaller bubbles will contribute significantly to gas flux [183]. Since the presence of surface-active entities in the water can significantly alter the behaviour of the gas within the bubble (for example, by preventing interface motion which in turn causes internal gas circulation to be damped or eliminated;[56] and by stabilising the bubble against dissolution[57]), studies relevant to the ocean should simulate the complex chemical character of seawater. The cleanliness of the bubble is very important to the gas flux [183, 197], dirty bubbles being generally more efficient at transferring gas into the ocean. Thorpe [184] notes that some laboratory experiments were undertaken in fresh water. Bubbles formed in salt water tend to be more numerous, particular regarding the smallest bubbles [198], and are less prone to coalesce than bubbles in fresh water [199]. Both effects are suggestive of a stabilising skin at the bubble wall. Thorpe [184] notes that other gas flux studies have excluded gas bubbles by keeping to windspeeds low enough to ensure that wave breaking did not occur.

Thorpe [147] discusses the diffusion of gas between the sea and bubbles, both clean and dirty, moving vertically in oceanic bubble clouds. He concludes that diffusive processes and not Boyle's Law will determine the rate of change of size of these bubbles everywhere except very close to the surface, and that it is diffusion which is likely to control the size distribution of bubbles in the sea.

Theoretically there exists a steady-state position when $u_b = u_w$, so that there is no net vertical motion of the bubble, and where the partial pressures of the gas inside the bubble equal those outside. Following the analysis of Thorpe [147], equilibrium for a bubble containing only nitrogen will occur at a depth h_o where

[55] See Chapter 2, section 2.1.1.
[56] See Chapter 2, section 2.3.2(b)(ii).
[57] See Chapter 2, section 2.1.2(b).

$$p_0 + \rho g h_0 + \frac{2\sigma}{R_0} = p_{N_2} \qquad (3.274)$$

where p_{N_2} is the partial pressure of nitrogen in the water far from the bubble, presumed to be constant. If s_{N_2} is the percent oversaturation of nitrogen in the water, then

$$p_{N_2} = p_0(1 + s_{N_2}/100) \qquad (3.275)$$

Substitution into equation (3.274) gives the depth below the surface at which equilibrium occurs to be

$$h_0 = \left(\frac{s_{N_2} p_0}{100} - \frac{2\sigma}{R_0}\right)\left(\frac{1}{\rho g}\right) \qquad (3.276)$$

This equilibrium depth only exists if the value $h_0 > 0$. From equation (3.276), h_0 is positive only if $s_{N_2} > 0$, that is if the water is supersaturated. This is obvious, since if it were not, gas would dissolve out from the bubble. For the bubble in addition to be stationary, the upwards velocity component imparted by buoyancy must just balance the local downwards velocity of the water. The larger the bubble, the greater that buoyant speed, so that there will be a single critical radius at which the bubble is stationary. Larger bubbles will tend to rise, and smaller bubbles will tend to be carried downwards by the descending water. The rise speeds of buoyant bubbles are complicated. However, dirty bubbles of radius less than 80 μm approximate to rigid spheres undergoing Stokes drag. Therefore the buoyant and drag forces, $4\pi R_0^3 \rho g/3$ and $6\pi\eta R_0 u_b$ respectively, can be equated to give the bubble rise speed of $u_b \approx 2R_0^2 \rho g/9\eta = 2R_0^2 g/9\eta_k$, where η_k is the kinematic viscosity, equal to the ratio of the shear viscosity η to the fluid density. Equating this to the downward speed of the water, u_w, gives the critical bubble radius $R_0 \approx 3\sqrt{u_w \eta_k / 2g}$.

Thus equilibrium for this pure nitrogen bubble can only be attained at a given depth for a critical bubble radius, provided the water is supersaturated. However, the position of equilibrium is unstable. Perturbations in position upwards or downwards will cause the bubble to undergo continuing motion in that direction, as a result of the bubble size increasing/decreasing in respect to the decreased/increased hydrostatic pressure encountered at lesser/greater depths, a size change which increases/decreases u_b, and so makes the difference between it and u_w finite.

If a bubble of critical radius is at a depth less than the critical one, there will be a flux of gas into the bubble. The flux is out of the bubble if it is deeper than the equilibrium position. In practice, it is unlikely that a bubble will attain the equilibrium depth with the critical radius and correct gas composition to remain static at that position, without changing site or size. The process can perhaps be more usefully described by defining a 'saturation depth', h_s, such that

$$h_s = p_0 s_{N_2} / (100 \rho g) \qquad (3.277)$$

so that the supersaturation condition can be expressed through

$$p_{N_2} = p_0(1 + s_{N_2}/100) = p_0 + \rho g h_s \qquad (3.278)$$

This enables the equilibrium depth to be related to the saturation depth

$$h_0 = h_s - \frac{2\sigma}{R_0} \qquad (3.279)$$

Given that the fluid is indeed supersaturated (i.e. $s_{N_2} > 0$, $h_s > 0$), then there will be a flux of gas into the bubble if that bubble is at a depth *less* than h_s, and in possession of a radius greater than $2\sigma/(h_s - h_o)$. Gas flux will occur out of a bubble if it is smaller than this radius. Thorpe [147] calculates the temporal variation in depth for bubbles, both dirty and clean, of varying initial radii or vertical water speeds, and containing either nitrogen or air. Thorpe obtains the following empirical rules, where depth, speed and time are normalised by 1 metre, 1 cm/s and 1 minute respectively. The maximum depth to which dirty bubbles are carried is given by

$$\left[\frac{h_{max}}{\text{m}}\right] \approx 1.9\left[\frac{u_w}{\text{cm/s}}\right] \qquad \text{for } h_{max} \leqslant 8 \text{ m} \tag{3.280}$$

with a maximum error within 10%. The maximum lifetime, t_{max}, is related to the maximum depth by

$$\left[\frac{t_{max}}{\text{minutes}}\right] \approx \sqrt{25\left[\frac{h_{max}}{\text{m}}\right]} \qquad \text{for } h_{max} < 6 \text{ m} \tag{3.281}$$

The lifetime exceeds this estimate if $h_{max} > 6$ m. Thorpe and Humphries [200], studying the somewhat random breaking of waves on Loch Ness, are able to fit their data with some scatter to an equation relating the frequency of wave breaking, ν_{break}, to the windspeed w_{10}

$$\left[\frac{\nu_{break}}{\text{Hz}}\right] \approx 2.2 \times 10^{-3}\left(\left[\frac{w_{10}}{\text{m/s}}\right] - 2.5\right) \tag{3.282}$$

Comparison of equations (3.281) and (3.282), using equation (3.270) to estimate the depth, shows that bubbles will persist from one wave-break to the next (i.e. $t_{max} \geqslant 2\pi/\nu_{break}$) if the windspeed w_{10} is in excess of 6.5 m/s [147]. Thus whilst at lower windspeeds the clouds exist discretely, they will overlap at higher ones.

Thorpe [147] concludes that gas flux from the bubbles occurs mainly in the upper 2 m of the water column. Measurements taken in the fresh water of Loch Ness suggest that the contribution made by bubbles to the overall gas flux is small at windspeeds of up to 12 m/s. However, at sea (near Oban, Scotland) the gas flux from bubbles is significant at windspeed of 12 m/s. The question of whether, for example, at higher windspeed, bubbles may dominate gas transfer between atmosphere and ocean is examined by Thorpe [184], who compares the contributions to the gas flux from bubbles with that due to direct transfer across the sea surface, as expressed by Ariyel *et al.* [201] in collating the results of several laboratory experimenters. The degree to which the water must be supersaturated in order to stop the dissolution of O_2 and N_2 from bubbles is determined, as a function of the windspeed. If the degree of supersaturation is less than this, gas will be forced into solution until this level is reached. If the degree of supersaturation exceeds this level, then the dominant factor will be the direct exsolution of gas across the sea surface, which continues until the supersaturation is reduced to the critical level. Thus this degree of supersaturation is a stable point, all other conditions (such as water temperature, biological effects etc.) remaining equal. The response time of the system is the timescale over which it will change towards this equilibrium supersaturation. For a near-surface mixing layer depth of 20 m and a windspeed w_{10} of 12 m/s, the timescale over which equilibrium is approached is about 2.3 days. The conclusion reached by Thorpe [184] is that the supersaturation of O_2, as suggested by the data available, cannot be explained by gas flux through bubbles. This formulation is applicable only to gases whose concentration will not change significantly over the lifetime of the bubble, which in most circumstances limits it to O_2 and N_2 from the common gases.

Woolf and Thorpe [183] use a numerical Monte-Carlo model which incorporates features of bubble dynamics and upper ocean mixing to determine the transfer of nitrogen, oxygen, carbon dioxide and argon through the development of the bubble population following the initial entrainment of air by breaking waves. The various parameters are interdependent: as described above, the bubble rises at a speed assumed to be the terminal velocity relative to the local water, which in turn may be moving and so advecting the bubble. The resulting depth-changes alter the hydrostatic component of the overpressure imposed upon the bubble contents, and so affect the gas flux and the bubble size. Changes in the bubble size alter the Laplace contribution to the overpressure,[58] and the terminal rise speed. The composition of a bubble is a complicated function of its past history. More surprisingly, the gases do not behave independently, but in contrast to direct transfer at the sea surface, the fluxes of different gases through a bubble are coupled: the exchange of a particular gas depends in part on the saturation level of N_2 and O_2 in the upper ocean [183]. Direct transfer across the sea surface is incorporated into the model. The water motions included in the model consist of a downward water jet which results from wave breaking, the orbital motion of surface waves, structured cellular motion (due to Langmuir circulation or convection), and turbulence and motion of low spatial and temporal coherence. Incorporating the effects of the supersaturation of each gas and of windspeed allows the flux of each gas to be determined, and reveals that whilst bubbles can induce a 1–2% supersaturation of poorly soluble gases such as Ar, O_2 and N_2, a globally significant supersaturation of a very soluble gas, such as CO_2, cannot be supported by the bubble population.

3.8.2 Acoustic Effects of Oceanic Bubbles

In the ocean, bubbles have two main acoustic effects. Firstly they act as sources of sound, giving out characteristic acoustic emissions upon entrainment of the type discussed extensively in this chapter. Such emissions will be discussed in section 3.8.2(a). The second effect is that the presence of bubbles modifies the propagation characteristics of sound through the medium. One way to view this is to consider the bubbly liquid as a continuum with bulk properties (such as sound speed, attenuation etc.) that are different from those of bubble-free liquid. The general effect on the sound speed is discussed in section 3.8.2(b)). Since bubbles are not dispersed evenly throughout the ocean, but are in general contained to within a few metres of the surface, they will therefore cause a sound speed difference in those first few metres, which can generate a waveguide, the implications of which will be discussed in section 3.8.2(c). In envisaging a horizontal waveguide formed from a bubbly continuum, one is perceiving the bubble population in the sea to consist of layers, the population density of bubbles varying only in the vertical direction. However, as was shown in section 3.8.1, distinct variations can occur in the horizontal direction when a proportion of the bubble population is structured into clouds. Section 3.8.2(d) examines some of the acoustic effects associated with clouds, specifically: (i) the low-frequency emissions resulting from oscillation modes in bubble clouds and (ii) the interaction between incident sound beams and bubble clouds, with particular attention to the problem of acoustic backscatter.

---

[58]See Chapter 2, section 2.1.1.

(a) Types of Single-bubble Passive Emission from Laboratory Breakers

As seen in section 3.5.2 in discussion of the injection of bubbles from nozzles, details of the entrainment and excitation process can produce very characteristic hydrophone signals. Medwin and Beaky [23] identified several species of bubbles which, upon entrainment in laboratory-generated spilling breakers, emitted characteristic single-bubble acoustic signals. The entrainment process was empowered by the slope of the wave: there was no wind. The bubbles in general acted as transient dipoles as a result of the presence of the free surface which was up to a few millimetres from the position of the sound source. Experimental measurements of the initial dipole strength gave values of $D_{\mathrm{dip,i}} = 0.1\text{–}1.0$ N/m under laboratory breaking waves in the frequency range 2–30 kHz.

Medwin and Beaky examined over 2000 bubble emissions, from a single bubble to combinations of bubbles and other events ('nondescript' bubbles). They developed a classification scheme to cover the variety of emissions observed in their laboratory spilling breaker system.

Type A events accounted for about 65% of the emissions. Type A1 is the familiar output from a lightly damped single spherical bubble, of the type illustrated for bubble production from a nozzle in Figure 3.23. The first cycle reaches a lower positive pressure, and is of shorter duration than the succeeding cycles, where the sinusoid has a single frequency, and where there is a single time constant associated with the decay. In type A2 emissions (Figure 3.60(a)) there are two time constants associated with the decay: after an initial phase where the damping is almost twice that predicted by theory, the time constant finally agrees with the theoretical prediction. Medwin and Beaky [23] suggest that the increased damping at the start results from an addition nonlinear dissipation mechanism (e.g. streaming, interaction with sea surface). Longuet-Higgins [38] analyses the effect of energy transfer between the breathing mode ($\boldsymbol{n} = 0$) and a shape oscillation when $2\omega_n$ is close to ω_0, as discussed in section 3.6, and concludes that this mechanism may be responsible for the nonmonotonic decay in amplitude.

If a large bubble which is oscillating sheds a much smaller bubble fragment, the excited high-frequency emission from the smaller bubble will be superimposed on the lower-frequency (and generally larger amplitude) emission from the large parent, leading to type B emissions (Figure 3.60(b)). A large bubble, oscillating at around 3.1 kHz, produces first a daughter which emits at about 50 kHz, and then one emitting at 32.3 kHz. These emissions can be seen as small-amplitude high-frequency perturbations on the hydrophone signal shown in Figure 3.60(b). Longuet-Higgins [38] notes that after separation occurs between the parent and the second daughter, the amplitude of the lower-frequency emission decreases after about five periods to roughly one-quarter of its value prior to separation, and then increases to about one-half the initial value. (As seen in section 3.5.2, it is very likely that bubble fragmentation will be associated with significant shape oscillations, both in that these may lead to fragmentation, and that they may be excited as a result of the departures from sphericity during separation.) Longuet-Higgins [38] demonstrates that this could result from energy transfer between the 3.1-kHz breathing mode and the $\boldsymbol{n} = 11$ shape mode[59] (resonance at 2.93 kHz).

An effect to which Longuet-Higgins refers in relation to type B emissions [38], and to which Medwin and Beaky ascribe the form of type C emissions (Figure 3.60(c)), is the small depth of the oscillating bubble below a free surface. As a result, the near-surface bubble is part of a dipole system, through the presence of an image,[60] and type C emissions are generated if the bubble

[59] See Chapter 3, section 3.6.
[60] See Chapter 3, section 3.3.2.

is initially excited at a distance less than six radii from the surface, and at first travels away from it.[61] Thus at the start of the pulse the distance between the bubble and the image is increasing, and so the dipole strength D_{dip} and consequently the pressure output increases. Damped decay then occurs, accompanied by a change to a lower frequency.

Medwin and Beaky identified a type D emission (Figure 3.60(d)), characterised by a low-frequency amplitude modulation of the basic sinusoid, and took this to be a 'beats'

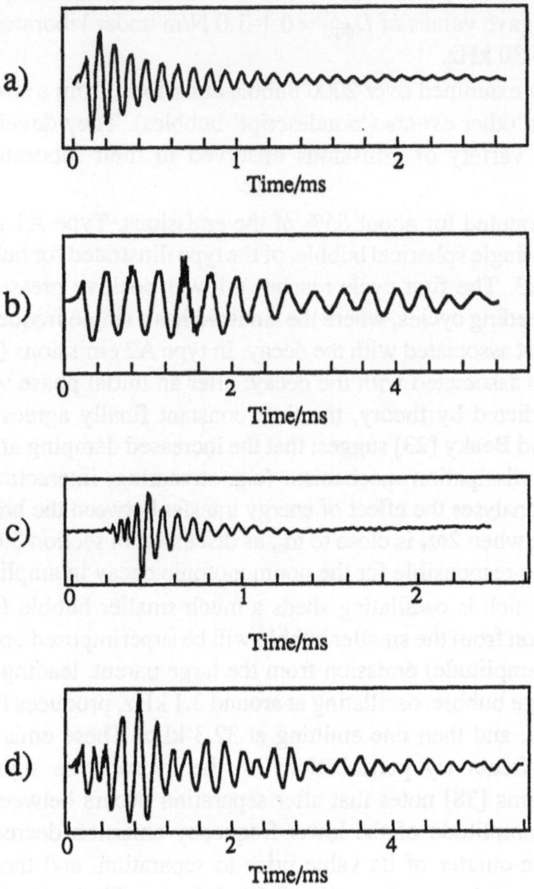

Figure 3.60 Passive acoustic emissions recorded in a laboratory wave-breaking facility (no wind). (a) Type A2 emission, resonance frequency 10.4 kHz, corresponding to $R_0 = 312$ μm. (b) Type B emission: basic frequency = 3.1 kHz ($R_0 = 1.05$ mm); first daughter at 50 kHz ($R_0 = 65$ μm); second daughter at 32.3 kHz ($R_0 = 101$ μm). (c) Type C emission: basic frequency is initially 25.6 kHz, final frequency 11.5 kHz. (d) Type D emission: basic frequency = 4.6 kHz, amplitude modulation frequency ~1 Hz (after Medwin and Beaky [23]).

[61] Though there will be repulsive forces between the free-surface image (section 3.3.2(d)), the motion of the bubble is likely to be dominated by more potent translatory forces, such as buoyancy and impressed forces associated with drag between the bubble and local water currents which may advect it.

phenomenon that occurs when two bubbles of similar size are excited close to each other in both time and space, as might happen, for example, if a larger bubble fragments into two similar daughters.

The fragmentation of a bubble into two daughters of similar size, so that ω_1 and ω_2 are similar, but not identical, could give rise to beats. In addition, shape oscillations on a single bubble can generate a monopole emission at second order, as described in section 3.6, which will be particularly strong if this $2\omega_n$ emission is close to the pulsation resonance ω_o, and beats can be set up between these to generate envelope modulations. Longuet-Higgins [38] notes that the basic frequency of 4.6 kHz lies midway between the two frequencies where the breathing mode will resonate with the $n = 9$ (5.2-kHz) and $n = 10$ (3.85-kHz) modes.

Longuet-Higgins [38] has demonstrated the potential for the undoubtedly often pronounced shape oscillations to affect the emission of newly entrained bubbles. The analysis also enables deduction of the distance below the free surface of the emitting bubble (8.7 mm for the trace shown in Figure 3.60(a)), which as shown in section 3.7.3 is an important consideration when discussing the excitation mechanism. In analysis of the combined frequency spectra of Medwin and Beaky's bubble pulses, Longuet-Higgins [63] detects peaks which, he speculates, could correspond to mode resonances.

However, when one considers the spectra of noise at sea, other perhaps dominant mechanisms come into play. In such circumstances the interaction of bubbles with sound manifests itself not simply in their ability to generate their passive acoustic emissions, but to scatter and modulate sound that originates from other sources. The response of a bubble to sound impressed upon it by an exterior source is the subject of the next chapter, and much of the basic physics behind the phenomena discussed in the remainder of this chapter must necessarily wait until Chapter 4. One effect of the presence of bubbles manifests itself as changes in the sound speed, and to understand several of the mechanisms behind recent interpretations on the spectra of noise in the sea, one must firstly appreciate the way in which bubbles affect the near-surface sound speed profile.

(b) The Effect of Bubbles on Sound Speed

A simple impression of how the presence of bubbles affects the sound speed can be attained by appealing to the basic arguments of Chapter 1, section 1.1.1(b). Since the speed of sound depends on the inertia and stiffness of the transmitting medium, the speed in a bubbly liquid, c_c, will differ from that in bubble-free liquid, since both parameters are changed. As shown in Chapter 1, in general the speed of sound in a medium is given by $\sqrt{B/\rho}$ (equation (1.10)), where B is the bulk modulus. The reciprocal of the bulk modulus, $B^{-1} = (-1/V)(dV/dp)$, is often termed the 'compressibility'. It refers to the change in volume that results from a change in the impressed pressure. Clearly, if a liquid sample contains bubbles which can respond readily to pressure perturbations by significant volume changes, and which though they take up volume are less massy than the liquid, both the compressibility and the density of the bubbly liquid will differ from those of the pure liquid, and so therefore will the sound speed.

(i) The Speed of Sound in Bubbly Liquids. The volume changes in a bubble will not usually be in phase with the driving pressure,[62] and since the compressibility arises from the ratio of the volume to the pressure changes, the compressibility of a bubbly liquid will in general be complex. The complex compressibility of the bubbly component of the mixture can be added

[62]See Chapter 4, section 4.1.

to the compressibility of the pure liquid phase to give a complex summed compressibility. This can be substituted into the expression $\sqrt{B/\rho}$ to give the ratio ω/k_c^{comp}, where k_c^{comp} is the complex wavenumber of the bubbly liquid, the real component of which yields the physical sound speed of the bubbly mixture.

In just such a way, effective analogues of key pure liquid parameters are employed to describe propagation in bubbly mixtures [202, 203], the predictions being in good agreement with experiment [204]. Lu *et al.* [205] present a linearised version of the model, approximating the bubble pulsations to the simple harmonic, which gives results applicable to void fractions of up to a few per cent.

In Chapter 4, section 4.1.2(e) the complex wavenumber k_c^{comp} of sound in a medium containing a uniform distribution of bubbles, all of radius R_0, is found using linearised theory in the limit of small-amplitude oscillations in bubble radius and pressure. Following the discussions of Chapter 1, section 1.1.7, whilst the imaginary part of the wavenumber describes the attenuation, the real part of k_c^{comp} gives the speed of sound in the bubbly medium, $c_c = \omega/\mathrm{Re}\{k_c^{comp}\}$. The result is that in general the speed of sound in a bubbly liquid depends on the size and number of bubbles, and on the frequency of the sound. The presence of bubbles therefore will make the liquid dispersive.[63] The formulation reduces to simple forms when the acoustic frequency and number density of bubbles take extreme values. In the limit of low acoustic frequencies and high number densities, the sound speed in a bubble cloud is (from equation (4.50))

$$c_c \approx \sqrt{\frac{\omega_0^2}{4\pi n_b R_0}} \tag{3.283}$$

which, if ω_M from equation (3.38) is used to approximate for ω_0, reduces to

$$c_c \approx \sqrt{\frac{3\kappa p_0}{\rho n_b(4\pi R_0^3/3)}} \approx \sqrt{\frac{\kappa p_0}{\rho\{VF\}}} \tag{3.284}$$

giving in the isothermal case

$$c_c \approx \sqrt{\frac{p_0}{\rho\{VF\}}} \tag{3.285}$$

where {VF} is the 'void fraction' (strictly the gas-volume fraction) [206–208].

In the opposite limit of low number densities, the sound speed is given by equation (4.53):

$$c_c \approx c\left\{1 - (2\pi c^2 n_b R_0)\left(\frac{\omega_0^2 - \omega^2}{(\omega_0^2 - \omega^2)^2 + 4\beta^2\omega^2}\right)\right\} \tag{3.286}$$

where β represents the dissipation.[64] Substitution can be made for a dimensionless damping constant for a single bubble $d = 2\beta/\omega$, which is the off-resonance equivalent of $\delta = 2\beta/\omega_0$. This gives the sound speed in a cloud of bubbles, all of radius R_0, number density n_b, to be

[63] See Chapter 1, sections 1.1.1(a) and 1.2.3(c).
[64] See Chapter 4, section 4.1.2.

$$c_c = c\left\{1 - \left(\frac{2\pi R_o n_b c^2}{\omega^2}\right)\left(\frac{(\omega_o/\omega)^2 - 1}{\{(\omega_o/\omega)^2 - 1\}^2 + d^2}\right)\right\} \tag{3.287}$$

in the limit of low number density. If there is a distribution of bubble sizes within the cloud, such that $n_b^{gr}(z,R_o)\mathrm{d}R_o$ is the number of bubbles per unit volume at depth z having radii between R_o and $R_o + \mathrm{d}R_o$, the speed of sound is a function of both the depth and the acoustic frequency [27]:

$$c_c(z,\omega) = c\left\{1 - (2\pi c^2)\int_{R_o=0}^{\infty} \frac{R_o}{\omega^2}\left(\frac{(\omega_o/\omega)^2 - 1}{\{(\omega_o/\omega)^2 - 1\}^2 + d^2}\right) n_b^{gr}(z,R_o)\mathrm{d}R_o\right\} \tag{3.288}$$

For low insonation frequencies ($\omega \ll \omega_o$), equation (3.287) reduces to

$$c_c = c\left\{1 - \left(\frac{2\pi R_o n_b c^2}{\omega_o^2}\right)\right\} \approx c\left\{1 - \tfrac{1}{2}\{\mathrm{VF}\}\frac{\rho c^2}{\kappa p_o}\right\} \tag{3.289}$$

using the approximation $\omega_M \approx \omega_o$.

(ii) The Effect of Vertical Variations in Bubble Population on Sound Speed. The above discussion illustrates how the speed of sound in the ocean can depend on the bubble population, in general the presence of bubbles decreasing the sound speed. Because of this effect, an alternative explanation to that of Longuet-Higgins for the presence of peaks in the acoustic spectrum in the sea has been proposed by Farmer and Vagle [66]. They suggest that waveguide propagation may occur in the ocean-surface bubble layer. Bubbles are entrained from the free surface of the ocean, so as the concentration of bubbles decreases with depth in the bubble-rich first few metres below the surface, the sound speed increases. If, for whatever reason, the sound speed increases with depth, sound waves propagating downwards will tend to turn, propagating at angles closer to the horizontal. This can be seen by using a construction of Huygen's wavelets for the sound, much as was done in Figure 1.3 for water waves approaching a beach, to show how a wavefront will turn as it progresses. Figure 3.61 shows how, if the sound speed increases with depth, sound initially propagating down from the surface may be refracted upwards towards the surface, from which it will reflect downwards. Repetition of this cycle can trap acoustic energy in a near-surface region [209, 210]. A wave of frequency greater than around 2 kHz might propagate through such a waveguide or duct, though scattering and absorption by bubbles, and the presence of bubble clouds and an irregular sea surface, might cause attenuation [146]. Farmer and Vagle [66] postulated, as a result of their analysing the frequency components of ambient ocean noise, that the trapping of sound in such a waveguide might influence the ambient acoustic spectra.

The breaking of a wave will emit sound [26]. Farmer and Vagle [66] video-recorded wave-breaking events at sea, and recorded the resulting sound. The spectrum of this acoustic emission does not exhibit a continuous frequency content, but instead is found to contain fine structure which is coherent over the duration of the event. In addition, the spectral fine structure is similar between one wave-breaking event and the next. The fine structure does vary during a given storm, and between storms. Farmer and Vagle report having observed such structure in a range of spectra measured in different geographic environments, including Georgia Strait (British Columbia), on La Perouse Bank (west of Vancouver Island), and in the FASINEX

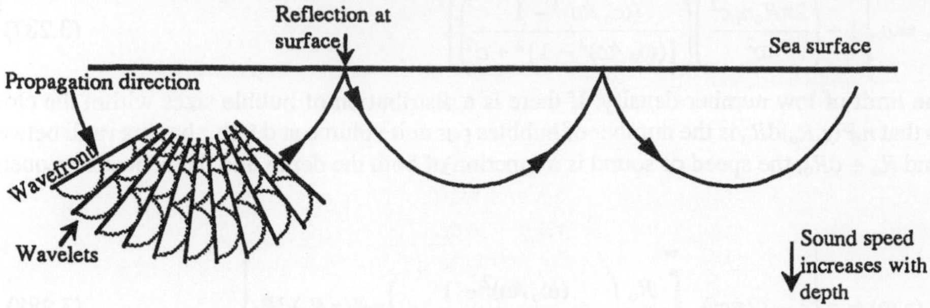

Figure 3.61 If the sound speed increases with depth, sound initially propagating down from the surface may be refracted upwards towards the surface, from which it will reflect downwards. The cycle may be repeated.

experiment, made 200 miles south-west of Bermuda. Spectra of the sound at the La Perouse site contained well-defined peaks at roughly 3-kHz separation in the range 3–20 kHz, but at FASINEX the spectrum contained broad bands, between which there are clear nulls, the separation being around 4 kHz. From sonar measurements at four frequencies (28, 50, 88 and 200 kHz) Farmer and Vagle iteratively reconstructed the bubble population $n_b^{\mathrm{gr}}(z,R_0)$, and by substituting this into equation (3.288) calculated the sound speed as a function of depth and frequency, $c_c(z,\omega)$. Farmer and Vagle decided that this sound speed profile resembles[65] an exponential law, where the sound speed increases with depth to approach in the limit the speed of sound in the bubble-free water:

$$c_c(z,\omega) = c - c_\omega e^{z/L_\omega} \qquad (z < 0) \tag{3.290}$$

where the depth z is a negative, being measured downwards from the surface. The parameters c_ω and L_ω are functions of the acoustic frequency ω with the dimensions of speed and length respectively. The modes found by Farmer and Vagle have a near-surface oscillatory component, which below a certain depth (the extinction depth) becomes evanescent, and decays with increasing depth. For each mode there is a 'cut-off' frequency: the mode cannot be excited at frequencies less than this. At the cut-off frequency the horizontal wavenumber, which was real, becomes imaginary, and the mode can no longer propagate horizontally. If the sound speed profile is taken to be exponential, an interesting effect is predicted. This is that as the cut-off frequency is approached from above, the extinction depth of any given mode increases, and the oscillatory part of the eigenfunction extends to greater depths. The pressure signal detected by a receiver position at some fixed depth in the decay region will, for a given modal amplitude, increase as the cut-off frequency is approached from above [66]. In the situation where all modes have the same energy, the acoustic spectrum detected by a hydrophone placed at depth beneath a strongly refractive region of the water column will contain peaks at frequencies approaching the cut-off relevant to each mode. This, suggest Farmer and Vagle, is the source of the peaks in the sound spectra they measured. Since the eigenfunctions decay exponentially with depth, and the higher the mode the greater the penetration depth of the eigenfunction, their theory predicts that the deeper the position of the hydrophone the less pronounced will be the peaks.

[65] It should be noted that, with the same data, Buckingham [146] favours an inverse-square profile, as discussed later. There is no basic discrepancy between these two profiles, since by judicious choice of parameters the exponential and inverse-square profiles can be made to look almost identical.

The peaks relevant to the lower modes will decay with increasing depth more rapidly than those pertaining to higher modes.

Farmer and Vagle suggested that information on the bubble population may be inferred from the drop-out frequencies detected as peaks in the acoustic spectrum. Buckingham [146] supports the idea that the distortion of signal as it passes through the subsurface layers might be used to obtain information if the process were understood sufficiently to allow interpretation. However, in application of his theories to the data and waveguide mechanism of Farmer and Vagle, he concludes that the loss of modes from the detected signal, such as is described by Farmer and Vagle, is not on its own sufficient to explain the dominant features in the sound spectra. He employs so-called inverse-square model profiles for the sound speed with depth, both upwardly and downwardly refracting, though the latter is not immediately applicable to the transmission of sound in normal modes through these bubble-induced ducts. Buckingham's theory predicts modes which are oscillatory beneath the surface down to an extinction depth, below which they are evanescent. The extinction depth increases exponentially with mode number, and with ω^{-1}, so that as the acoustic frequency increases, the mode is contained within a shallower duct. No mode is wholly evanescent (that is, all modes will propagate in the horizontal direction in a shallow region, the depth of which depends on the mode), and there is no mode cut-off. Most of the energy of the mode is contained above the extinction depth, and, if the detector is positioned at depths greater than this, the mode will 'drop out' of the detected signal. It is also shown that the pressure node distribution is identical in most of the modes, so that particular choices of the receiver depth will bring about multi-mode suppression [146]. In addition to the receiver being placed potentially at a depth coincident with zeros in the oscillatory parts of many modes, clearly as the acoustic frequency increases more modes will 'drop out', in that their extinction depth will pass into regions shallower than the depth of the receiver. The 'extinction' or 'drop-out' frequency of a mode is defined as being that frequency for which the extinction depth equals the depth of the receiver. Buckingham finds that the mode strength equals the ratio of the acoustic frequency to the normalised extinction depth, and decays exponentially with mode number. Therefore in a given situation, where source and receiver are at fixed depths and a single fixed frequency is investigated, there will be a 'window' of modes which contribute significantly to the acoustic field. Lower modes will have too shallow an extinction depth, so that the receiver is in the evanescent region. If the mode number is too high, the mode strength is weak, and when the field is expressed as a summation over modes it is heavily attenuated [146]. As the frequency increases, more modes contribute significantly to the summation. Interference between them can occur, complicating the sound field.

Buckingham fits an upwardly refracting sound speed profile to the measured data points on the variation of sound speed with depth taken by Farmer and Vagle [66]. Comparing this with the results from the exponential sound speed model profile employed by Farmer and Vagle, he suggests that mode drop-out is not sufficient to explain the structure of the sound spectra. Spectral peaks and nulls at La Perouse are instead generally attributed to interference between modes, the detailed structure resulting from the size of the contributing mode window. The latter increases with increasing frequency, and as more and more modes are involved, the results of inter-mode interference become more complex. The duct at FASINEX is 2.5 times deeper than that at La Perouse, so that more modes can interfere and the expected spectra will be more complex. The fine structure resulting from interference peaks is barely discernible in the FASINEX spectrum, the main feature being a modulation, where the nulls between the broad bands correspond to ones in the modulation envelope. (The modulation at La Perouse is not rapid enough to affect the spectrum significantly, as a result of the shallower duct.) Thus Buckingham was by his theory able to attribute the qualitative differences in the two spectra mentioned above to differences in the sound speed profile. In addition, since the modulation observed at FASINEX is dependent on the depth of the acoustic source, and since the sources are entrained bubbles, emitting for some few milliseconds after entrainment,[66] then from the width of these modulation peaks in the FASINEX data Buckingham was able to estimate the depth of the source. He found them to be at a depth of about 1.5 m below the surface, roughly coincident with the bottom of the bubble clouds. This is an extremely interesting observation, firstly in that it is surprisingly deep, and secondly, as illustrated in the discussion of Figure 3.55, estimates of the potency of the various mechanisms for the excitation of passive emissions can be very dependent on the depth of the bubble at the instant of excitation:[67] the difference between the observed 3.2 mm for raindrop entrainment and 1.5 m for Buckingham's results are excitingly different.

Ducts Generated by Nonbubble Sources. Though bubbles may be introduced at depth into the sea, for example, through methane seeps, the near-surface duct described above results from the fact that there is a significant concentration of bubbles which are confined to near-surface regions. However, ducts caused by sound-speed variations resulting from other sources can occur throughout the oceans. These are extremely useful, and in confining the sound they prevent losses that would occur through simple spherical spreading of the acoustic energy. Ducts can be formed which have a depth of several thousand metres, and which can therefore transmit very low frequencies (1–50 Hz) for several hundred kilometres with no bottom interactions. The *deep sound channel* (or *sofar channel*) results from a minimum in the sound speed profile which varies in depth from about 1.2 km at mid-latitudes to near the surface in polar waters. Urick [211] describes one remarkable utilisation of the channel by the so-called SOFAR (*so*und *f*ixing *a*nd *r*anging) system to enable the rescue of downed aviators at sea. A small explosive charge detonated in the proximity of the aviator emits an acoustic signal which is channelled by the duct, to be received at stations up to several thousands of miles away. The position of the aviator can be found from the arrival times of the signal at two or more stations [146, 211].

Over the range of the other critical parameters which can affect sound speed in the oceans, the speed of sound increases with increasing temperature (Figure 1.27), salinity and hydrostatic pressure. Though hydrostatic pressure increases linearly with depth, the water temperature may decrease or increase with depth: an example of the latter is found beneath ice in the Arctic Ocean, where at the surface the water temperature can be −2° C [146]. In fact, wind-powered water circulation can, in the absence of prolonged calm periods, give rise to a *mixed layer* of isothermal water, which can extend to depths of the order of 100 m. The sound speed increases linearly with depth as a result of the increase in hydrostatic pressure, and so is upwardly refracting. If there is insufficient mixing, near-surface regions can be prone to comparatively rapid changes, such as diurnal temperature variations.

As a result of these dependences, the sound speed profile (i.e. variation of c with depth) can take various forms, though at the greatest depths one would expect the effect of hydrostatic pressure to cause a linear increase. This can lead to interesting effects. Figure 3.62 shows the ray paths of sound emitted from a source for a given profile, where the sound speed first decreases with increasing depth, then, as expected, finally increases. Two paths are illustrated. As one would expect from the Huygens construction of Figure 3.61, the closer the ray path is

---

[66] See section 3.5.

[67] However, a classical dipole model is inappropriate for oceanic bubbles separated from the surface by 1.5 m of water containing such a dense, probably inhomogeneous bubble population.

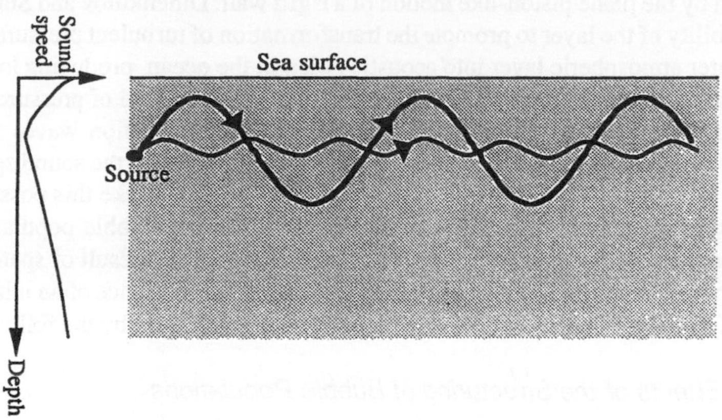

Figure 3.62 The ray paths of sound emitted from a source for a given profile, where the sound speed first decreases with increasing depth, then eventually increases. Two paths are illustrated.

to the horizontal, the more rapidly it bends.[68] In the ocean, ray paths between source and receiver can conceivably include bending features such as this, and also reflections from the surface (of the type shown in Figure 3.61) and the bottom. The bending of rays in this manner can cause interesting effects [212], such as the sonar 'shadow zones': the paths followed by sonar pulses emitted from a given source may be deflected away from certain zones, so that one submarine may hide there undetected by another. In contrast, close raypaths may converge, the envelope formed by their intersection being known as a *caustic*: if it occurs near the surface it is called a *convergence zone*. A detector placed at a convergence zone will hear greatly increased acoustic levels from the source. Such zones have approximately 50 km spacing in the ocean. If the source were a short pulse, a detector would receive a series of pulses corresponding to the arrival of signals along these and similar paths. (Which arrives first depends on the details of the profile: though the ray that takes wider excursion travels a longer path, it goes through water which has in general a higher sound speed.) Such techniques are being researched in an attempt to determine, for example, the feasibility of global temperature monitoring [256, 257]. Spiesberger and Metzger [131] describe measurements of the travel time of pulses, centred at 250 Hz and emitted at two-hourly intervals one day in four. These pulses travelled about 3000 km in the North Pacific Ocean before detection, and the results can be used to test various algorithms for the speed of sound in water. Useful introductions to the transmission of sound in the ocean are given by Spindel [213] and Urick [214].

Other Acoustic Effects of a Subsurface Bubble Layer. The subsurface bubble layer can have other acoustic effects, several of which are discussed in the following section. The sound emitted from single oscillating bubbles upon entrainment has been discussed at some length in this chapter. In addition to this effect, and that of waveguiding, the bubbles in the near-surface layer can have other acoustic implications. Didenkulov and Sutin [215] consider the contribution of the nonlinear generation of the second harmonic to the high-frequency noise spectrum in the ocean. Kumar and Brennen [216] model the nonlinear harmonic generation when a bubble

[68] In fact, at the other extreme, a vertical ray path would not bend at all if the sound speed varied in depth only. This can be seen by considering what would happen if a wavefront of the type shown in Figure 3.61 were horizontal: all the wavelet circle would be of the same radius, and the new wavefront would be parallel to the old one.

cloud is driven by the plane piston-like motion of a rigid wall. Didenkulov and Sutin [215] also consider the ability of the layer to promote the transformation of turbulent pressure fluctuations in the near-water atmospheric layer into acoustic noise in the ocean, producing low-frequency sound (up to several tens of hertz) [217]. Since the propagation speed of pressure fluctuations cannot exceed the speed of sound in the medium, pressure fluctuation waves in air cannot normally be communicated into water to radiate acoustic waves since the sound speed in water is higher than that in air [215]. However, the presence of bubbles make this possible, through two mechanisms: (i) nonlinear transformation in the subsurface bubble population; and (ii) scattering in regions in the layer where the sound speed varies as a result of spatial variations in the bubble population density. This example illustrates the importance of an inhomogeneous distribution of bubbles to ocean acoustics, a theme which is taken up by the following section.

(c) Acoustic Effects of the Structuring of Bubble Populations

In the previous section, the effect of a vertical variation in the bubble population was to cause a change in sound speed in the same direction, which generated a waveguide effect. However, the net direction of interest is often horizontal, since in many cases this is the direction source-to-receiver. Though, as we have seen, the vertical variation in bubble population can affect the horizontal propagation, so too of course could horizontal variations in the population. This can be illustrated at a most basic level by considering a situation where the horizontal variation in bubble population in a given region divides it into two well-defined regions, one being more densely populated than the other. The interface between these two regions of different bubble density will be approximately a vertical plane, and a horizontal sound signal will strike this interface at some angle to the normal. We have seen that the presence of bubbles in a region of liquid can be accommodated by assigning to that region of liquid macroscopic properties of sound speed, compressibility, density and effective acoustic impedance, and in Chapter 1 the reflection and transmission characteristics of a beam striking an impedance-mismatch interface were discussed. Even this simple model, where the sound speed in the horizontal direction varies as a single step function, illustrates reflection and transmission phenomena. In the ocean, where graduations in bubble population will be more complicated, and can occur in both the horizontal and vertical directions so that bubbles are confined in clusters and other regions of the liquid are much less densely populated, that medium will be far richer in acoustic phenomena.

(i) Collective Oscillations of Bubbles in a Cloud. The ambient noise in the ocean shows marked wind-dependence for frequencies from several tens of kilohertz down to at least a few hundred hertz. Since the windspeed affects the breaking of waves, and since the bubbles so produced can be acoustically active, it is reasonable to suggest that this sound might be associated with the bubble population. However, bubbles acting as individual oscillators are not responsible, since the resonance bubble radius for frequencies of 500 Hz and 100 Hz are respectively ~ 6 mm and ~ 3 cm (equation (3.39)). It is unlikely that a significant population of bubbles of this size could be generated and persist against the effects of fragmentation and buoyancy. If one assumes that a typical oceanic bubble has a radius of 1 mm and a resonance frequency ν_0 of 3 kHz, then in order to obtain acoustic emissions at 100 Hz it might be thought reasonable to envisage low-frequency oscillations of some larger body associated with bubbles. It was independently suggested by Prosperetti [144, 218, 219] and Carey [220–222] that the collective oscillations of bubble clouds might be the source of these low-frequency emissions. Later theoretical treatments of the phenomenon are given by Omta [223] and by d'Agostina

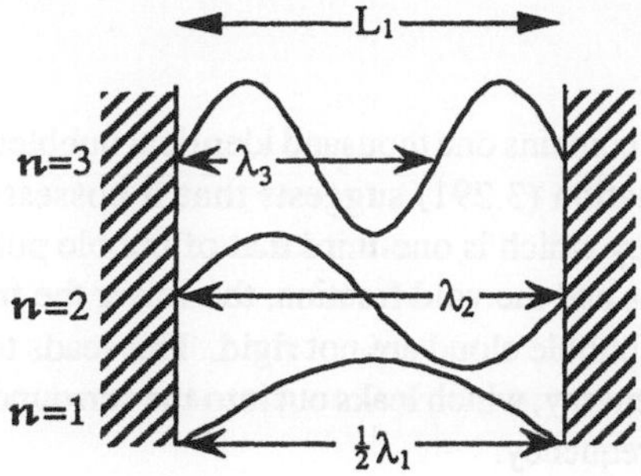

Figure 3.63 The three lowest modes of oscillation for a one-dimensional space of length L_1, having perfectly rigid walls.

and Brennen [224]. The bubbles in a cloud are coupled by mutual acoustic and hydrodynamic interactions, and such a system of coupled oscillators can have normal modes at much lower frequencies than the resonance frequencies of the individual oscillators.[69] Any arbitrary motion of the system can be analysed as a summation of various contributions from the different modes.

An alternative argument uses the fact that, on a macroscopic scale, the cloud can be seen as some medium possessing an effective acoustic impedance and reduced sound speed, as discussed earlier. If the walls were rigid, it would possess modes with frequencies dependent on the cloud dimensions. To illustrate in one dimension, since there must be zero displacement at rigid walls, then if there is dimension L_1 between those walls the appropriate mode must have a wavelength λ_n such that $n\lambda_n/2 = L_1$ (Figure 3.63). Since the wavevector of the mode has magnitude $k_n = 2\pi/\lambda_n$, then $k_n = n\pi/L_1$. If c_c is the speed of sound in the cloud, then the mode frequency $\varpi_n = c_c k_n$, so that $\varpi_n = n\pi c_c/L_1$. For a real cloud, the geometry is of higher dimension. If the symmetry is cylindrical, one could expect to employ Bessel functions,[70] rather than linear waveforms, to describe the modes. However, the principle is the same.

As outlined in section 3.8.2(b)(i), in the limits of high bubble number densities and $\omega \ll \omega_0$, the effective sound speed in the bubbly medium in the isothermal limit is given by equation (3.285) [206–208]. If the cloud has linear dimension L_c, and contains a *total* of N_b identical bubbles of radius R_0, then $\{VF\} \approx N_b(R_0/L_c)^3$. For some general geometry in a cloud of linear dimension L_c, Lu *et al.* [205] obtain to a first approximation the mode frequency[71] ϖ_n for a cloud of identical bubbles possessing individual pulsation resonances of ω_0:

---

[69] This effect can be nicely demonstrated by hanging two identical pendulums (called 'left' and 'right') from a short washing line. The resonance frequency of the pendulum is determined by its length (see footnote to section 3.1.1). The coupling arises through the transfer of energy from one swinging pendulum to another down the line. If one pendulum (left) is initially stationary, and the other (right) set into motion, a normal mode ensues. Over several swings, energy leaves the right pendulum, which oscillates with smaller amplitude, and excites the left pendulum. This mode has a period of many swings of the pendulum, a period which can be timed between the moments when the right pendulum is stationary and vertical. After this position, it is again excited into motion as energy is transferred back to it down the line from the high-amplitude oscillations of the left pendulum. The mode has a longer period, and a lower frequency, than the individual pendulum. The more pendulums that are hung on a line, the more modes there are.

[70] See Chapter 1, section 1.1.4.

[71] The index n labels the mode. Though the same symbol is used to express the labelling integer in both cases, it should be noted that these modes are completely different from the shape modes of a single bubble discussed in section 3.6. These modes, which were described in terms of spherical harmonics, had the lowest mode $n = 0$ ('breathing mode'), and shape oscillations described by perturbation of the sphere by the $n \geqslant 2$ spherical harmonics. The $n = 1$ mode has no physical significance. The $n \geqslant 2$ bubble shape modes have frequencies labelled ω_n (equation (3.236)). In contrast, the modes of collective oscillation of a bubble cloud are more like linear sinusoids (for rectangular geometries) described in Figure 3.63, or Bessel functions for cylindrical geometries. The lowest mode of collective oscillation of any cloud is labelled $n = 1$. To clarify the difference, cloud modes have frequencies labelled ϖ_n.

$$\varpi_n = \omega_0 \frac{n}{N_b^{1/3}\{\text{VF}\}^{1/6}} \tag{3.291}$$

If cloud of length scale 10 cm contains one thousand identical bubbles each having an individual pulsation resonance ω_0, equation (3.291) suggests that it possesses a mode corresponding to $n = 1$ which has a frequency which is one-third that of bubble pulsation, i.e. $\varpi_1 = \omega_0/3$. The greater the number of bubbles and the void fraction, the lower the frequency.

In reality, the walls of the bubble cloud are not rigid. This leads to a loss of energy from the oscillation, and this acoustic energy, which leaks out into the surrounding liquid, can be detected as sound at the appropriate frequency.

Having characterised the medium in terms of effective parameters, all that remains to model the modes in the bubble cloud is to satisfy continuity relations and boundary conditions when the cloud is of limited size, surrounded by bubble-free liquid. Lu *et al.* [205] do this for three types of cloud geometry:

(a) layers of bubbles, which if vertical represent a screen of bubbles generated through Langmuir circulation,[72] and if horizontal and adjacent to a pressure-release surface represents a layer of bubbles on the ocean surface;

(b) a hemispherical cloud adjacent to a pressure-release surface, which could be used to model the bubble plume generated by an isolated wave-breaking event;

(c) a cylindrical cloud.

In all cases, modes of significantly lower frequency than that of the bubble pulsation resonance were predicted. Both the mode frequency and the decay rate (the imaginary component of the eigenfrequencies) decrease with increasing void fraction and size of the cloud.

The theoretical results are relatively insensitive to the rigidity of the walls. However, the fact that the walls are not perfectly rigid is vital to the fact that sound energy can leak out of the cloud and so contribute to the ambient noise in the surrounding fluid.

Yoon *et al.* [225] describe a controlled laboratory test of the proposed mechanism and theory of the collective oscillations of bubble clouds. They injected steady-state clouds of cylindrical shape between tank base and water surface using a multi-needle device, with up to forty-nine needles being arranged in concentric circles. The radii of the circles were 2, 4 and 6 cm, possessing 8, 16 and 24 needles respectively, with a single needle at the centre. Spreading through the zig-zag rise path[73] of bubbles caused these circles to generate clouds of radii 4.6, 5.4 and 7.0 cm. Modal waves were excited in a column through the broadband noise produced as the bubbles were generated at the nozzles. The acoustic signal detected by a hydrophone was frequency-analysed to determine the lowest mode frequency, which was identified as a prominent low-frequency peak in the acoustic spectrum. The frequency dependence on the void fraction and the radius of the columnar cloud were measured and found to be in good agreement with theory (Figure 3.64). A weak dependence on bubble size was predicted for radii R_0 of 1.5, 2.0 and 2.5 mm, and demonstrated experimentally for $R_0 = 1.6$ mm and 2.0 mm. Theory predicts a decay rate for the oscillations of a few tens of s^{-1}, giving relatively short decay times. The fact that, despite this, the spectral peak corresponding to the lowest mode was so prominent, coupled with the relatively low energy of the exciting detachment process, suggest that these collective oscillations are readily excited. One key difference between this laboratory cloud and oceanic clouds when considering modal excitation is that the former had two facing planar surfaces, the

[72] See section 3.7.4(d)(i)
[73] See Chapter 2, section 2.3.2(b)(iv)

water surface and the tank base, which would encourage the excitation of modes: no such encouragement would occur in the sea.

Roy *et al.* [226] eliminated such boundary effects when they investigated the low-frequency (<2 kHz) scattering of incident sound from roughly cylindrical bubble clouds injected at 91.4-m depth into Seneca Lake, New York (130 m of fresh water). Both conventional and parametric sources were used to generate the sound. They obtained apparent low-frequency resonance peaks at 1.3 kHz, ~950 Hz and 450 Hz (which would correspond to single bubble resonances for R_0 = 8.2, 10 and 21 mm respectively, bubble sizes which Roy *et al.* judge to be highly unlikely). Quantitative agreement was found between the target strength data near 450 Hz and cloud resonance scattering calculations.

(ii) The Response of Bubble Clouds to Incident Sound Beams. One of the most powerful tools for probing the structure of oceanic clouds has been the effect of the bubble population on an incident sound beam. An outline of some of the effects is given in this section.

Analyses of Propagation of Sound Beams Through Structured Populations. Research into the propagation of sound through bubbly clouds or layers illustrates some of the variety

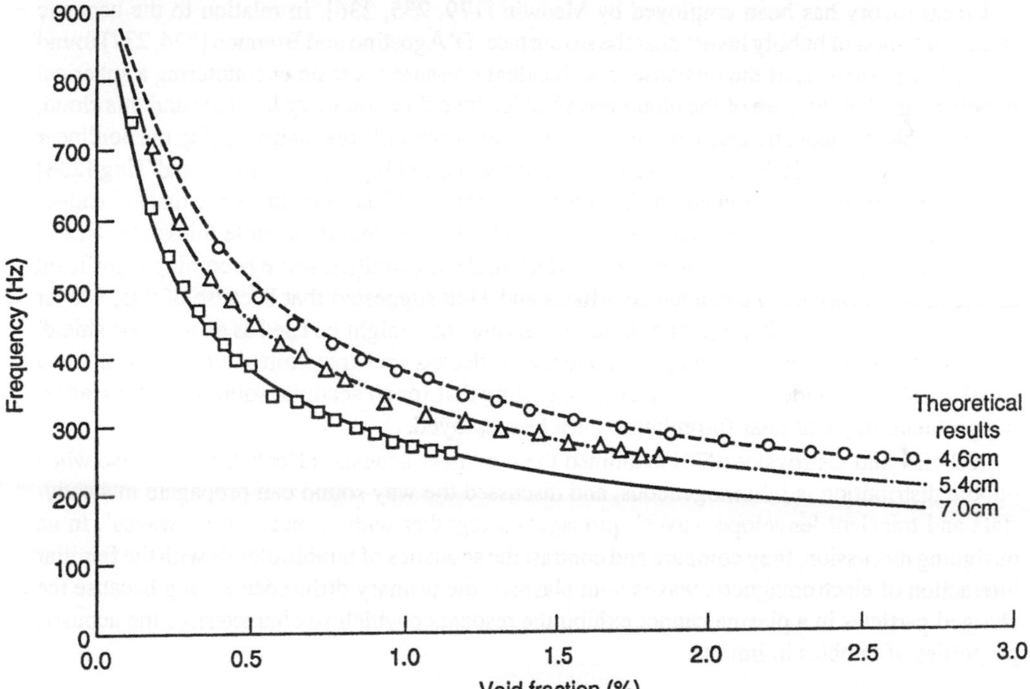

Figure 3.64 Variation of the lowest mode of an injected cloud with void fraction: comparison of theory and experiment (after Yoon *et al.* [225]). Results are presented for three column radii:

Column radius	Theory	Experiment
7.0 cm	———	□
5.4 cm	- - - - -	◇
4.6 cm	—·—	△

Column height = 0.82 m; average bubble radius = 2.0 mm; distance from cloud centre to tank wall (assumed in theory) = 0.75 mm.

mentioned earlier. Multiple scattering will occur, and should be accounted for [227]. Bubbly layers may enhance the efficiency of forward parametric radiation [228], and difference-frequency[74] waves can be backscattered from bubble layers subjected to waves of two acoustic frequencies [229].

Ostrovsky and Sutin [230] expand from the case of a single nonlinearly-oscillating bubble in a sound field, emitting scattered signals at, for example, the second harmonic (or difference frequencies, if given biharmonic insonation), to that of the bubble cloud. In the latter case, the nonlinear signals from individual bubbles are summed to yield a scattered acoustic field containing coherent and incoherent components. Observed effects include parametric radiation, cross-beam interactions, and phase conjugation. As sound propagates through a bubbly liquid it can change the distribution of bubbles in that liquid.[75] This effect can lead to self-transparency (where wave attenuation decreases with increasing acoustic beam power [231, 232]), self-diaphragming (the acoustic beam becomes narrower as it propagates owing to stronger attenuation of the weak field at the periphery of the beam [233]), and self-focusing of the acoustic beam (where an increase in the concentration of small bubbles on the beam axis causes a reduction in sound speed there, and causes the beam to bend inwards towards the axis, focusing itself [234]).

Linear theory has been employed by Medwin [179, 235, 236], in relation to the acoustic characterisation of bubbly layers near the sea surface. D'Agostino and Brennen [224, 237] found from a linear analysis of the response to an incident pressure wave on encountering a spherical bubble cloud that the core of the cloud was shielded by a thin boundary layer around this cloud, provided that the acoustic frequency exceeded the single-bubble resonance. Using the nonlinear model equations of Caflisch *et al.* [203] to describe the bubbly liquid, Miksis and Ting [238] investigated the transmission and reflection characteristics of the acoustic wave that is incident on a bubble layer. There are multiple resonances for bubbly layers of finite thickness. Near a resonance frequency, most of the wave is reflected, the transmitted wave becoming significant as one moves away from a resonance. Miksis and Ting suggested that because of this, a layer containing a random collection of bubbles of varying sizes might be used as an acoustic shield. The results show that there may be nonlinear reflected and transmitted waves even for a small-amplitude incident pressure wave, suggesting that for an accurate solution to the scattering problem, the nonlinear formulation must be employed.

Kobelev and Ostrovsky [234] examined the nonlinear acoustic effects that can arise when bubble distribution is inhomogeneous, and discussed the way sound can propagate in steady-state and transient 'envelope waves', propagating together with 'concentration waves'. In an intriguing discussion, they compare and contrast the acoustics of bubble clouds with the familiar interaction of electromagnetic waves with plasmas, the primary difference arising because the charged particles in a plasma cannot exhibit the resonance which so characterises the acoustic properties of bubbles in liquids.

Acoustic Backscatter. If an acoustic wave is directed at the sea surface from below, at low incidence (e.g. direction of propagation being typically 15° to the horizontal), the signal can be backscattered. In general, the magnitude of backscattering increases with the windspeed. Backscatter from the ocean surface at low frequencies (less than a few kHz) was thought to be the result of Bragg scattering from waves on the ocean surface [239]. The theory gives good agreement at low windspeeds (< 5 m/s).

---

[74] See Chapter 5, section 5.1.1(d).
[75] See Chapter 4, section 4.4.1.

High levels of reverberation of underwater sound associated with the ocean surface at windspeed in excess of about 10 m/s have been observed for some time [240]. Initially this was thought to be due to scattering from surface waves, but recent advances in scattering theory [241–243] enabled a 15–20 dB discrepancy between this model and experimental measurements of grazing angle scattering in high sea states to be deduced [244, 245]. Bubbles being known to be very acoustically active, the dependence of both the bubble population and the scatter discrepancy on windspeed is suggestive. Therefore bubbles were suggested as a source of the additional scatter.

If sound is propagating through a medium containing a small bubble, that bubble will be forced into pulsation against thermal and viscous dissipative mechanisms. The energy required to overcome this dissipation will be absorbed from the incident acoustic wave. In addition, the pulsating bubble is itself a source of sound, so that as the driven bubble oscillates it will scatter acoustic energy from the incident wave. This loss of energy from the acoustic wave resulting from the presence of a bubble is usually expressed as the ratio of the time-averaged power loss per bubble, $<\dot{W}>$, to the intensity of the incident acoustic beam, I. The quantity has the dimensions of area, and is called the *extinction cross-section*, Ω_b^{ext}, of the bubble so that[76]

$$\Omega_b^{ext} = \frac{<\dot{W}>}{I} \tag{3.292}$$

A plane wave of intensity I, travelling a distance Δz through a population of n_b bubbles per unit volume, each having an extinction cross-section Ω_b^{ext}, has its intensity reduced by $\Delta I = -n_b \, \Omega_b^{ext} \Delta z$. Integration gives

$$I = I_0 \exp(-n_b \, \Omega_b^{ext} z) \tag{3.293}$$

where I_0 is the intensity at $z = 0$. Since the intensity in plane waves is proportional to the square of the acoustic pressure, the latter varies as

$$P \propto \exp(-n_b \, \Omega_b^{ext} z/2) \tag{3.294}$$

It is often of interest to know specifically how much of the energy of an incident acoustic beam is scattered by the bubble, as opposed to being dissipated through thermal and viscous mechanisms. The ratio of the time-averaged power loss per bubble that results from scattering alone, to the intensity of the incident acoustic beam is called the *scattering cross-section*, Ω_b^{scat}, of the bubble.

At resonance, the scattering and extinction cross-sections of an oceanic bubble are about three orders of magnitude greater than its geometrical cross-section, which can be a considerable advantage when using acoustic spectroscopy, rather than optical techniques, to measure bubble populations.

The cross-sections can still be formulated if there is a range of bubble sizes present by defining a function describing the number density size distribution of bubbles in a cloud. The bubble density at a certain depth is given by $n_b^{gr}(R_0)$ for increment[77] dR_0 such that $n_b^{gr}(R_0)dR_0$ is the number of bubbles per unit volume with radius between R_0 and $R_0 + dR_0$. Thus the extinction cross-section for sound propagating through a bubble cloud with population described by $n_b^{gr}(R_0)$ is simply

---

[76] See Chapter 4 section 4.1.2(d).

[77] If, as is usually the case for oceanic bubbles, the increment dR_0 is fixed to equal 1 μm, then $n_b^{gr}(R_0) = n_b^{\mu}$.

$$\Omega_c^{ext} = \int_{R_o=0}^{\infty} \Omega_b^{ext} n_b^{gr}(R_o) dR_o \tag{3.295}$$

This parameter, which has units of [length]$^{-1}$, describes that component of attenuation of an acoustic beam propagating through a cloud which is due to the bubbles, since the intensity of the beam decays with distance of propagation z as

$$I \propto e^{-\Omega_c^{ext} z} \tag{3.296}$$

or

$$\frac{dI}{I} = -\Omega_c^{ext} dz \tag{3.297}$$

Knowing the cross-section for scatter from a bubble, the backscatter cross-section is simply $\Omega_b^{b\text{-}s} = \Omega_b^{scat}/4\pi$, the solid angle ratio arising since the backscatter is omnidirectional. When insonating a bubble cloud one examines the backscattering cross-section per unit volume, that is $\Omega_{cV}^{b\text{-}s}$ (the volume backscattering coefficient):

$$\Omega_{cV}^{b\text{-}s} = \frac{1}{4\pi} \int_{R_o=0}^{\infty} \Omega_b^{scat} n_b^{gr}(R_o) dR_o \tag{3.298}$$

Thus the backscatter and attenuation that results from the presence of bubbles, all of which have $R_o \ll \lambda$, can in principle be related to the bubble size spectrum of the bubble cloud [246].

To explain high levels of acoustic reverberation and backscatter for windspeeds in excess of 10 m/s, Henyey [240] presents a hypothesis in which it is necessary to classify oceanic bubbles into two types. When air is entrained through the action of a breaking wave, the resulting bubble population will contain a significant number of large bubbles. These, which Henyey distinguishes as 'macrobubbles', in his model rapidly rise to the surface through buoyancy. The smaller bubbles, which are not rapidly lost through buoyancy, he terms 'microbubbles'.[78] In this model, therefore, newly formed clouds have a high void fraction, and evolve rapidly owing to the buoyant loss of macrobubbles. Older clouds ('middle-aged' or 'fossil'), with ages up to a few minutes, will contain smaller bubbles. As discussed earlier, small bubbles can be stabilised by surface-active agents. The persistent population is that of microbubbles, which are to a first approximation mainly removed from the population through dissolution rather than buoyancy. Buoyancy plays a secondary role in the motion of microbubbles to advective flows, turbulent diffusion etc: they may, for example, be carried to depths in excess of 10 m by downwelling currents, the most notable of which are those resulting from Langmuir circulation, as described earlier.

As stated above, the observed acoustic reverberation levels are 15–20 dB greater than can be accounted for through scattering from surface waves alone in high sea states. Henyey considers that macrobubble scattering is unlikely to account for much of the observed extra

---

[78] It should be noted that this distinction in no way relates to the scheme of Vokurka, outlined in section 3.3.1(b), where the macro/microbubble distinction depended not simply on the equilibrium bubble size but also on the pulsation amplitude.

scattering. As scatters, individual microbubbles are too small to have a great effect at the frequencies concerned (of the order of 100 Hz). Rino and Ngo [193] demonstrate this by consideration of the wavenumber of a coherent acoustic field within a sparse distribution of scattering bubbles (k_c), which is related to the wavenumber in seawater (k_w) through

$$k_c \approx k_w + \frac{2\pi}{k_w} < f(\vec{a}_s, \vec{a}_i) > n_b \tag{3.299}$$

where n_b is the number density of bubbles, f is the complex scattering function of the bubble, which has arguments $\vec{a}_i$ and $\vec{a}_s$, the unit vectors along the incident and scattered directions respectively. The ensemble average of $f(a_s, a_i)$ is taken (indicated by '< >') over bubble characteristics (mainly over R_0, the bubble radius). If the bubbles are small, the scattering is nearly isotropic, and referring to Morse and Feschback [247], Rino and Ngo [193] give a result for f similar to the form derived later in Chapter 4, equation (4.51), for a medium of bubbles of identical size. They state that, since at the lower acoustic frequencies ($\omega \ll \omega_0$) thermal losses are negligible, in this microbubble regime the extinction of the coherent wave field is mainly due to this scattering (which will convert the energy from coherent to incoherent wave field). They further state that even the scattering losses are small.

However, as discussed in section 3.8.2(b), though the cumulative effect from individual microbubbles is very small, if the insonation frequency is very much less than any bubble resonance, the cloud can be viewed as a uniform body with an effective acoustic impedance, and an internal sound speed of

$$c_c = c\left(1 + \frac{\{\mathrm{VF}\}}{2}\frac{c^2\rho}{c_g^2\rho_1}\right)^{-1} \tag{3.300}$$

where c_g and ρ_1 are the sound speed in, and density of, the gas (air) respectively. If the void fraction is low, equation (3.300) can be expanded binomially to obtain agreement with equation (3.289) for a uniform cloud of low number density, when one remembers that $c_g = \sqrt{\kappa p_0/\rho_1}$ (equation (1.19)). Rino and Ngo [193] state that for $2\times10^{-4} \leqslant \{\mathrm{VF}\} \leqslant 2\times10^{-2}$, the discrepancy in sound speed can be significant, potentially giving backscatter 20 dB greater than that from the surface models alone. Henyey [240] shows that the scattering is very dependent on the plume dimension, the acoustic cross-section having a dependence on the fourth power of the plume depth.

Henyey considers in his model the scattering from the curtains of bubbles that occur in the downwelling regions of Langmuir circulation, and which are marked on the surface by the appearance of windrows. The concentration of bubbles will vary across the curtain, the regions of concern in Henyey's model being those which penetrate most deeply, since his calculation shows that the scattering cross-section of the region is proportional to the fourth power of the depth. It is convenient to refer to these regions as 'curtain plumes'.[79] To simplify the calculation, Henyey [240] models these plumes as cylinders of radius 1–2 m. A further simplification assumes that above a certain depth no bubble dissolution occurs, and below it all bubbles dissolve instantaneously: this approximation gives the penetration depth of the column, which is different for different columns. Because the acoustic scattering cross-section of the curtain plumes depends on the plume depth raised to the fourth power, shallow plumes are assumed to

---

[79] Henyey himself refers to them simply as 'plumes'. However, in his model they are specifically regions of high bubble concentration and deep penetration in the curtain, in contrast to the more general term 'plume' used in section 3.8.1(a).

have negligible contribution: the depths of the curtain plumes which do contribute significantly are assumed to be considerably larger than any deviations in flatness of the wavy water surface. The data of Farmer and Vagle [66] are used for information on the variation of the average bubble population with depth and bubble size. These measurements were made for the bubble range $15 \leqslant R_0 \leqslant 100$ μm. Henyey had to employ a power-law extrapolation after Thorpe [147] for information on larger bubble sizes. Henyey's model discusses variations in frequency and grazing angle, but not windspeed, within the limitations of experimental data. The assumption of a simple cylindrical geometry for the curtain plume limits the validity of the model to frequencies below about 2 kHz.

McDonald [248] calculates the backscattering from the microbubbles to use in conjunction with the scattering from larger bubbles which, through resonance, scatter strongly at the lower frequencies. McDonald uses a mechanism of weak scattering from clouds of these smaller, subresonant bubbles to explain the low-frequency backscatter. Unlike Henyey, MacDonald does not begin by explicitly assuming a mechanism for the formation of the clouds important to scattering, but instead appeals to experimental acoustic results which indicate the presence of clouds having filamentary or sheetlike shape, 'hanging' from the sea surface in the manner of stalactites (to use the analogy of McDonald [248]). McDonald models the subresonant microbubble cloud volume scatterers as vertical cylinders of elliptical cross-section horizontally, the population and sound speed defect decaying exponentially with depth (inside the clouds, the sound speed is reduced). Adjustments to the size and ellipticity enable the model to describe filaments or sheets. Figure 3.65 shows backscatter through ray geometries, emphasising corner reflection from a number of individual scattering cylinders, showing how volume scatter may occur before or after the incident acoustic ray has contacted the sea surface. Scatter properties are obtained through the Born approximation for acoustic frequencies between 200 Hz and 20 kHz, above which the resonance scatter mechanism should produce a significant increase in backscatter. The model predicts that at frequencies below which resonance does not dominate, but high enough for $\lambda/4$ to be less than the semi-minor axis of the ellipse, specular reflection should exceed volume backscatter. Reasonable agreement with experimental data is demonstrated for variation with incident angle and windspeed [249–251], McDonald recommends that the weak scattering model be used in addition to that of the resonant scatterers.

McDonald's model predicts low backscatter in the limit of low frequencies and small grazing angles. However, additional backscatter could come through interaction of the incident wave

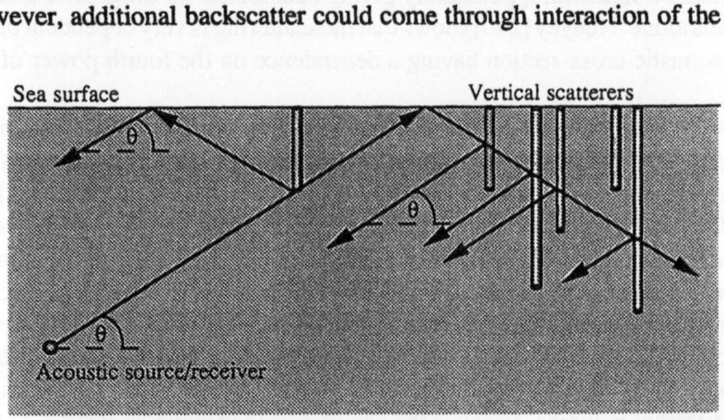

Figure 3.65 Backscatter ray diagram, emphasising corner reflection from a number of individual scattering cylinders, showing how volume scatter may occur before or after the incident acoustic ray has contacted the sea surface (after MacDonald [248]).

with the vertical dimension of the cloud (~15 m). The bubble population at this depth is probably the result of Langmuir circulation. McDonald recommends measurements on cloud properties (e.g. sound speed) *in situ* at this depth. In addition, he points to the scarcity of data for scattering at low frequencies.

Rino and Ngo [193] model the interaction between the surface and the bubble cloud. Originally Rino *et al.* [252] numerically simulated the backscatter from evolving nonlinear sea-surface perturbations. The cross-section of the surface to acoustic backscatter was reduced by top–bottom asymmetries in the surface waves, and the Doppler spectra were predicted. For windspeeds in excess of a small threshold, and low frequencies ($\lambda \geq 1$ m) in high sea states, first-order theory predicts an approximately constant reverberation noise level. A refinement to theory, which encompasses tilting of the Bragg 'carpets' due to large-scale waves, increases the cross-section only slightly. The inclusion of corrections for nonlinearities still does not enable surface roughness models to account for the experimentally observed acoustic backscatter.

In a private communication to Rino, Ellinthorpe suggested that subsurface bubble clouds would dominate the surface scatter for frequencies where the wavelength is comparable with the size of the cloud. Rino *et al.* [252] combine the interactive calculated scatter components from the bubble cloud and the rough surface in high sea states in an idealised model of the system. Were the sea a plane surface and the cloud very close to it, scatter from the cloud would be accompanied by scatter from an image cloud in antiphase, since the sea surface from below is approximately a free boundary. This would give reduced backscatter. However, as the cloud moves away from the surface, the source and image can interfere constructively. In addition, the sea surface is in fact not planar but rough. Thus bubble clouds breaking away from the surface can be effective scatterers. The results suggest that the model, idealised in several aspects (for example, in having a constant cloud cross-section independent of wind speed), nevertheless could account for observed acoustic surface reverberation in high sea states. Rino and Ngo call for more experimental information on the characteristics of subsurface bubble clouds, particularly their windspeed dependence.

Jessup *et al.* [253] observed sizeable events in the time series of the radar cross-section, which they associated with breaking waves. These events, or 'sea spikes', travel in the ocean at approximately the phase speed of the dominant waves [254]. There is some uncertainty as to the origin of sea spikes [255]. Loewen and Melville [255] compare the microwave backscatter with the acoustic emissions from breaking waves. The frequencies of these laboratory water waves were 0.88, 1.08 and 1.28 Hz, corresponding to wavelengths of 1.76, 1.28 and 0.94 m. The average backscattered microwave power and the mean-square acoustic pressure correlate with each other, and with the wave slope and dissipation, for waves of moderate slope. Both of the detected radiations increase with the slope and wavelength of the wavepacket. However, temporally significant microwave backscatter is seen before either acoustic emission or visible wave breaking commence, suggesting that it is the surface geometry of the wave prior to breaking that dominates the microwave backscatter.

The above examples of the acoustic effects of oceanic bubbles illustrate the variety of sources for the sound that is modified by the presence of bubble populations. Though by no means a rigid categorisation, waveguiding ducts both artificial signals, and also ambient noise in the sea that results from storms, shipping, turbulence etc, and also from bubble generation. As has been examined in section 3.7, the act of entrainment can excite individual bubble oscillations: in laboratory investigations into the low-frequency emissions of bubble clouds, the entrainment of bubbles from nozzles no doubt contributed to the excitation of the detected modes.

Conclusions

At the beginning of this chapter, it was explained that the bubbles under consideration were not 'undriven', but were excited in an impulsive, rather then a periodic, manner. In this way, the natural oscillations of single bubbles were investigated, and their resonance and damping properties expounded. With this information it was possible to examine the natural (resonance) emission of bubbles in the natural world, where in response to an impulsive excitation, they emit at the resonance frequency. That impulse came about through the entrainment mechanism, or sometimes through coalescence and fragmentation of bubbles. Bubbles in brooks and waterfall, in rainfall, and in the sea, were investigated. However, towards the end of the chapter, in discussion of bubbles in the oceans, it became necessary to introduce the action of bubbles when sound was directed at them from an external source, both because acoustic probes have been a major tool in examining the bubble population of the seas, and because of the need to know how the population modulates the sound propagating through the ocean. Concepts such as the acoustic cross-section and the sound speed in bubbly liquids were introduced. It is therefore now appropriate and necessary to examine the interaction between a bubble and the sound wave that can drive it into oscillation. This important and complex phenomenon is examined in the next chapter.

References

[1] French AP. Vibrations and Waves. Nelson, London, 1979, pp 43–45

[2] Nelson PA and Elliott SJ. Active Control of Sound. Academic Press, London, 1992, pp 38–42

[3] Vokurka K. On Rayleigh's model of a freely oscillating bubble. II. Results. Czech J Phys 1985; B35: 110–120

[4] Minnaert M. On musical air-bubbles and sounds of running water. Phil Mag 1933; 16: 235–248

[5] Bragg Sir WH. The World of Sound. Bell, London, 1921, pp 69–74

[6] Worthington AM and Cole RS. Impact with a liquid surface, studied by the aid of instantaneous photography. Phil Trans Roy Soc 1897; A189: 137–148

[7] Worthington AM. A Study of Splashes. Longmans, Green, London, 1908

[8] Kinsler LE, Frey AR, Coppens AB and Sanders JV. Fundamentals of Acoustics. Wiley, New York, 3rd edn, 1982, pp 112–115, 228

[9] Wood A. Acoustics. Blackie, London, 1940. References to 1960 re-issue, p 105

[10] Vokurka K. On Rayleigh's model of a freely oscillating bubble. I. Basic relations. Czech J Phys 1985; B35: 28–40

[11] Cole RH. Underwater Explosions. Princeton University Press, Princeton, New Jersey, 1948

[12] Strasberg M. Gas bubbles as sources of sound in liquids. J Acoust Soc Am 1956; 28: 20–26

[13] Vokurka K. Amplitudes of free bubble oscillations in liquids. J Sound Vib 1990; 141: 259–275

[14] Bjørnø L and Levin P. Underwater explosion research using small amounts of chemical explosives. Ultrasonics 1976; 14: 263–267

[15] Pritchett JW. Incompressible calculations of underwater explosion phenomena. Proc 2nd Int Conf Numerical Methods Fluid Dynamics. Springer, Berlin, 1971, pp 422–428

[16] Urick RJ. Principles of Underwater Sound. McGraw-Hill, New York, 3rd edn, 1983, p 87

[17] Vokurka K. The scaling law for free oscillations of gas bubbles. Acustica 1986; 60: 269–276

[18] Samek L. Acoustic waves emitted by radial oscillations of a spherical bubble in viscous compressible heat conductive liquids. Czech J Phys 1983; B33: 1108–1114

[19] Kinsler LE, Frey AR, Coppens AB and Sanders JV. Fundamentals of Acoustics. Wiley, New York, 3rd edn, 1982, pp 163–165

[20] Nelson PA and Elliott SJ. Active Control of Sound. Academic Press, London, 1992, pp 26–27

[21] Morse PM and Ingard KU. Linear acoustic theory. In: Handbuch der Physik (Flügge S, ed.), Vol. 11, Part 1. Springer, Berlin, 1961

[22] Morse PM and Ingard KU. Linear acoustic theory. In: Handbuch der Physik (Flügge S, ed.), Vol. 11, Part 1. Springer, Berlin, 1961, p 313
[23] Medwin H and Beaky MW. Bubble sources of the Knudsen sea noise spectra. J Acoust Soc Am 1989; 86: 1124–1130
[24] Strasberg M. The pulsation frequency of non-spherical gas bubbles in liquids. J Acoust Soc Am 1953; 25: 536–537
[25] Longuet-Higgins MS, Kerman BR and Lunde K. The release of air bubbles from an underwater nozzle. J Fluid Mech 1991; 230: 365–390
[26] Crowther PA. Bubble noise creation mechanisms. Proc NATO Adv Workshop on Natural Mechanisms of Surface Generated Noise in the Ocean, Lerici, Italy, 15–19 June 1987. In: Sea Surface Sound (Kerman BR, ed.). Reidel, Dortrecht, The Netherlands, 1988, pp 131–150
[27] Clay CS and Medwin H. Acoustical Oceanography: Principles and Applications. Wiley, New York, 1977
[28] Kinsler LE, Frey AR, Coppens AB and Sanders JV. Fundamentals of Acoustics. Wiley, New York, 3rd edn, 1982, p 409
[29] Longuet-Higgins MS. The sound field due to an oscillating bubble near an indented free surface. J Fluid Mech 1990; 221: pp 675–683
[30] Oğuz HN and Prosperetti A. Bubble oscillations in the vicinity of a nearby plane free surface. J Acoust Soc Am 1990; 87: 2085–2092
[31] Devin C Jr. Survey of thermal, radiation, and viscous damping of pulsating air bubbles in water. J Acoust Soc Am 1959; 31: 1654
[32] Eller AI. Damping constants of pulsating bubbles. J Acoust Soc Am 1970; 47: 1469–1470
[33] Chapman RB and Plesset MS. Thermal effects in the free oscillations of gas bubbles. Trans ASME D 1972; 94: 142–145
[34] Pfriem H. Akust Zh 1940; 5: 202–207
[35] Walton AJ. Three phases of matter. Clarendon, Oxford, 2nd edn, 1983, p 226
[36] Walton AJ. Three phases of matter. Clarendon, Oxford, 2nd edn, 1983, p 156
[37] Mallock A. The damping of sound by frothy liquids. Proc Roy Soc 1910; A84: 391–395
[38] Longuet-Higgins MS. Nonlinear damping of bubble oscillations by resonant interaction. J Acoust Soc Am 1992; 91: 1414–1422
[39] Strasberg M. Gas bubbles as sources of sound in liquids. J Acoust Soc Am 1956; 28: 20–26
[40] Fitzpatrick HM and Strasberg M. Hydrodynamic sources of sound. In: Proc 1st Symp Naval Hydrodynamics, Washington DC. NAS-NRC Publ 515, US Printing Office, Washington DC, 1957, pp 241–280
[41] Leighton TG and Walton AJ. An experimental study of the sound emitted from gas bubbles in a liquid. Eur J Phys 1987; 8: 98–104
[42] Leighton TG, Lingard RJ, Walton AJ and Field JE. Bubble sizing by the nonlinear scattering of two acoustic frequencies. In: Natural Physical Sources of Underwater Sound (Kerman BR, ed.). Kluwer, Dordrecht, The Netherlands, 1992
[43] Leighton TG, Lingard RJ, Walton AJ and Field JE. Acoustic bubble sizing by the combination of subharmonic emissions with an imaging frequency. Ultrasonics 1991; 29: 319–323
[44] Pumphrey HC and Crum LA. Free oscillations of near-surface bubbles as a source of the underwater noise of rain. J Acoust Soc Am 1990; 87(1): 142–148
[45] Clift R, Grace JR and Weber ME. Bubbles, Drops and Particles. Academic Press, New York, 1978, pp 321–330
[46] Pitts E. The stability of pendant drops. Part 2: Axial symmetry. J Fluid Mech 1974; 63: 487–508
[47] Leighton TG, Fagan KJ and Field JE. Acoustic and photographic studies of injected bubbles. Eur J Phys 1991; 12: 77–85
[48] Yang Y-M and Maa J-R. A photographic study of bubble coalescence. Journal of Chinese Institute of Engineers 1983; 6(4): 257–263
[49] Chuang KT, Stirling AJ and Baker JC. An optical/electronic technique for measuring bubble coalescence times. I&EC Fundamentals 1984; 23: 109
[50] Ozawa Y and Mori K. Characteristics of jetting observed in gas injection into liquid. Trans ISIJ 1983; 23: 764–768
[51] Aoki T, Masuda S, Hatono A and Taga M. Proc Int Conf Injection Phenomena in Extraction and Refining (Wraith A, ed.), Vol. 1. Department of Metallurgy and Engineering Materials, University of Newcastle-upon-Tyne, UK, 1982

[52] Ozawa Y and Mori K. Effect of physical properties of gas and liquid on bubble-jetting phenomena in gas injection into liquid. Trans ISIJ 1986; 26: 291–295
[53] French, AP. Vibrations and waves, Nelson, London, 1979, chapters 5 and 6
[54] Rayleigh Lord. On the capillary phenomena of jets. Proc Roy Soc 1879; 29: 71–97
[55] Stokes GG. On the communication of vibration from a vibrating body to surrounding gas. In: Mathematical and Physical Papers, Vol. 4. Cambridge University Press, London, pp 299–317 (originally Phil Trans Roy Soc 1868)
[56] Lamb H. Hydrodynamics. Cambridge University Press, London, 5th edn, 1924, pp 448–450 (Lamb uses the alternative convention $\vec{v} = -\nabla\Phi$)
[57] Neppiras EA. Acoustic cavitation. Phys Rep 1980; 61: 159–251
[58] Herman WAHJ. On the instability of a translating gas bubble under the influence of a pressure step. Thesis, Eindhoven Univ Technol, The Netherlands, 1973
[59] Longuet-Higgins MS. Monopole emission of sound by asymmetric bubble oscillations. Part 1. Normal modes. J Fluid Mech 1989; 201: 525–541
[60] Longuet-Higgins MS. Monopole emission of sound by asymmetric bubble oscillations. Part 2. An initial value problem. J Fluid Mech 1989; 201: 543–565
[61] Longuet-Higgins MS. Resonance in nonlinear bubble oscillations. J Fluid Mech 1991; 224: 531–549
[62] Ffowcs Williams JE and Guo YP. On resonant nonlinear bubble oscillations. J Fluid Mech 1991; 224: 507–529
[63] Longuet-Higgins MS. Bubble noise spectra. J Acoust Soc Am 1990; 87(2): 652–661
[64] McConnell SO. Remote sensing of the air–sea interface using microwave acoustics. IEEE Proc Oceans 1983; 85–92
[65] Farmer DM and Vagle S. Observations of high-frequency ambient sound generated by wind. In: Sea Surface Sound (Kerman BR, ed.). Kluwer, Boston, 1988, pp 403–415.
[66] Farmer DM and Vagle S. Waveguide propagation of ambient sound in the ocean-surface bubble layer. J Acoust Soc Am 1989; 86: 1897–1908
[67] Friedlander B and Porat B. Detection of transient signals by the Gabor Representation. Trans IEEE 1989; ASSP37: 169–180
[68] Detsch RM and Sharma RN. The critical angle for gas bubble entrainment by plunging liquid jets. Chem Eng J 1990; 44: 157–166
[69] Koga M. Bubble entrainment in breaking wind waves. Tellus 1982; 34: 481–489
[70] Lin TJ and Donnelly HG. Gas bubble entrainment by plunging laminar liquid jets. AIChE J 1966; 12: 563–571
[71] Updegraff GE and Anderson VC. Bubble noise and wavelet spills recorded 1 m below the ocean surface. J Acoust Soc Am 1991; 89: 2264
[72] Laville F, Abbott GD and Miller MJ. Underwater sound generation by rainfall. J Acoust Soc Am 1991; 89: 715–721
[73] Anderson I and Bowler S. Oceans spring surprise on climate modellers. New Scientist 1990; 125: 1707
[74] Watson AJ, Upstill-Goddard RC and Liss PS. Air–sea gas exchange in rough and stormy seas measured by a dual-tracer technique. Nature 1991; 349: 145
[75] Etcheto J and Merlivat L. Satellite determination of the carbon dioxide exchange coefficient at the ocean–atmosphere interface: a first step. J Geophys Res 1988; 93(C12): 15669–15678
[76] Andreae MO. In: The Role of Air–Sea Exchange in Geochemical Cycling (Buat-Menard P, ed.). Reidel, Dordrecht, The Netherlands, 1986, pp 331–363
[77] Charlson RJ, Lovelock JE, Andreae MO and Warren SG. Oceanic phytoplankton, atmospheric sulphur, cloud albedo and climate. Nature 1987; 326: 655–661
[78] Urick RJ. Principles of Underwater Sound. McGraw-Hill, New York, 3rd edn, 1983, pp 202–209
[79] Urick RJ. Ambient Noise in the Sea. Naval Sea Systems Command, Washington DC, 1983
[80] Ross D. Mechanics of underwater noise. Pergamon, New York, 1976
[81] Wenz GM. Acoustic ambient noise in the ocean: spectra and sources. J Acoust Soc Am 1962; 34: 1936–1956
[82] Zakarauskas P, Chapman DMF and Staal PR. Underwater acoustic ambient noise levels on the eastern Canadian continental shelf. J Acoust Soc Am 1990; 87: 2064–2071
[83] Knudsen VO, Alforf RS and Emling JW. Underwater ambient noise. J Marine Res 1948; 7: 410–429

[84] Pumphrey HC and Ffowcs Williams JE. Bubbles as sources of ambient noise. J Ocean Eng 1990; 15(4): 268–274
[85] Kerman BR. Underwater sound generation by breaking wind waves. J Acoust Soc Am 1984; 75: 149–165
[86] Kerman BR. Audio signature of a laboratory breaking wave. In: Sea Surface Sound (Kerman BR, ed.). Kluwer, Boston, 1988
[87] Shang EC and Anderson VO. Surface generated noise under low speed at kilohertz frequencies. J Acoust Soc Am 1986; 79: 964–971
[88] Prosperetti A and Lu NQ. Cavitation and bubble bursting as sources of oceanic ambient noise. J Acoust Soc Am 1988; 84: 1037–1097
[89] Updegraff GE and Anderson VC. *In situ* acoustic signature of low sea-state microbreaking. J Acoust Soc Am 1989; 85(Suppl 1): S146
[90] Rohr J, Glass R and Castille B. Effect of monomolecular films on the underlying ocean ambient noise field. J Acoust Soc Am 1989; 85: 1148–1157
[91] Buckingham M. Acoustic daylight: imaging the ocean with ambient noise. Nature 1992; 356: 327–329
[92] Pumphrey HC. Sources of ambient noise in the ocean: an experimental investigation. PhD dissertation, University of Mississippi, 1989
[93] Richardson EG. The impact of a solid on a liquid surface. Proc Phys Soc 1948; 61: 352–367
[94] Richardson EG. The sounds of impact of a solid on a liquid surface. Proc Phys Soc 1955; 68: 541–547
[95] Pumphrey HC and Elmore PA. The entrainment of bubbles by drop impacts. J Fluid Mech 1990; 220: 539–567
[96] Medwin H, Kurgan A and Nystuen JA. Impact and bubble sound from raindrops at normal and oblique incidence. J Acoust Soc Am 1990; 88: 413–418
[97] Pumphrey HC, Crum LA and Bjørnø L. Underwater sound produced by individual drop impacts and rainfall. J Acoust Soc Am 1989; 85: 1518–1526
[98] Pumphrey HC. Underwater rain noise – the initial impact component. Proc Inst Acoust 1991; 13: 192–200
[99] Nystuen JA. Rainfall measurements using underwater ambient noise. J Acoust Soc Am 1986; 79: 972–982
[100] Nystuen JA and Farmer DM. The sound generated by precipitation striking the ocean surface. In: Sea Surface Sound (Kerman BR, ed.). Kluwer, Boston, 1988, p 485
[101] Nystuen JA and Farmer DM. The influence of wind on the underwater sound generated by light rain. J Acoust Soc Am 1987; 82: 270–274
[102] Wahl NE. Rain erosion properties of plastic materials. AFTR-5686(Suppl 1), February 1950
[103] Law JC and Parsons DA. Relation of raindrop size to intensity. Trans Am Geophys Union 1943; part ii: 457
[104] Lenard P. Met Zeit 1904; XXI: 248–262
[105] Defant A. Met Zeit 1905; XXII
[106] Fyall A and Strain RNC. A 'whirling-arm' test rig for the assessment of the rain erosion of materials. Report CHEM.509, Royal Aircraft Establishment, Farnborough, Hants, UK, 1956
[107] The High Speed Test Track, Facilities and Capabilities. Munitions Systems Division, 3246th Test Wing, 6585th Test Group, Test Track Division, Holloman Air Force Base, New Mexico, April 1989, p 48
[108] Franz GJ. Splashes as sources of sound in liquids. J Acoust Soc Am 1959; 31: 1080–1104
[109] Bom N. Effect of rain on underwater noise level. J Acoust Soc Am 1969; 45: 150–156
[110] Scrimger JA. Underwater noise caused by precipitation. Nature 1985; 318: 647–649
[111] Scrimger JA, Evans DJ, McBean GA, Farmer DM and Kerman BR. Underwater noise due to rain, hail and snow. J Acoust Soc Am 1987; 81: 79–86
[112] Scrimger JA, Evans DJ and Yee W. Underwater noise due to rain – open ocean measurements. J Acoust Soc Am 1989; 85: 726–731
[113] Jones AT. The sounds of splashes. Science 1920; 295–296
[114] Carrol K and Mesler R. Bubble nucleation studies. Part II: Bubble entrainment by drop-formed vortex rings. AIChE J 1981; 27: 853–856
[115] Esmailizadeh L and Mesler R. Bubble entrainment with drops. J Colloid Interface Sci 1986; 110: 561–574

[116] Oğuz HN and Prosperetti A. Surface tension effects in the contact of liquid surfaces. J Fluid Mech 1989; 219: 149–171
[117] Pumphrey HC and Walton AJ. Experimental study of the sound emitted by water drops impacting on a water surface. Eur J Phys 1988; 9: 225–231
[118] Dingle AN and Lee Y. Terminal fallspeeds of raindrops. J Appl Met 1972; 11: 877–879
[119] Pumphrey HC and Elmore PA. The entrainment of bubbles by drop impacts. J Fluid Mech 1990; 220: 539–567
[120] Pumphrey HC and Crum LA. Acoustic emissions associated with drop impacts. In: Sea Surface Sound (Kerman BR, ed.). Kluwer, Boston, 1988
[121] Oğuz HN and Prosperetti A. Bubble entrainment by the impact of drops on liquid surfaces. J Fluid Mech 1990; 219: 143–179
[122] Chapman DS and Critchlow PR. Formation of vortex rings from falling drops. J Fluid Mech 1967; 29: 177–185
[123] Oğuz HN and Prosperetti A. Numerical calculations of the underwater noise of rain. J Fluid Mech 1991; 228: 417–442
[124] Oğuz HN and Prosperetti A. Bubble entrapment by axisymmetric capillary waves. In: Engineering Science, Fluid Dynamics, A Symposium to Honor Theordore Yao-Tsu Wu (Yates GT, ed.). World Scientific, 1990, pp 191–202
[125] Leighton TG. The frequency analysis of transients. Eur J Phys 1988; 9: 69–70
[126] Pumphrey HC. Sources of underwater rain noise. In: Natural Physical Sources of Underwater Sound. (Kerman BR, ed.). Kluwer, Dordrecht, The Netherlands, 1992
[127] Field JE. The physics of liquid impact, shock wave interactions with cavities, and the implications to shock wave lithotripsy. Phys Med Biol 1991; 36: 1475–1484
[128] Guo YP. Impact sound from water drops. Proc Inst Acoust 1991; 13: 183–191
[129] Guo YP and Ffowcs Williams J. A theoretical study on drop impact sound and rain noise. J Fluid Mech 1991; 227: 345–355.
[130] Lighthill MJ. On sound generated aerodynamically: I. General theory. Proc Roy Soc 1952; A211: 564–587
[131] Spiesberger JL and Metzger K. New estimates of sound speed in water. J Acoust Soc Am 1991; 89: 1697–1700
[132] Longuet-Higgins MS. An analytical model of sound production by rain-drops. J Fluid Mech 1990; 214: 395–410
[133] Longuet-Higgins MS. Bubbles, breaking waves and hyperbolic jets at a free surface. J Fluid Mech 1983; 127: 103–121
[134] Lunde K. PhD thesis. Cambridge University, 1992
[135] Kjeldsen SP and Olsen GB. Breaking waves (16 mm film). Technical University of Copenhagen, Denmark, 1971
[136] Galvin CJ. Wave breaking in shallow water. Waves on Beaches (Meyer RR, ed.). Academic Press, New York, 1972, pp 413–456
[137] Mason MA. Some observations of breaking waves. In: Gravity Waves. US Nat Bur Standard Circular 521, 1952, pp 215–220
[138] Longuet-Higgins MS and Turner JS. An 'entraining plume' model of a spilling breaker. J Fluid Mech 1974; 63: 1–20
[139] Kolaini A, Roy R and Crum LA. The production of high-frequency ambient noise by capillary waves. In: Natural Physical Sources of Underwater Sound (Kerman BR, ed.). Kluwer, Dordrecht, The Netherlands, 1992
[140] Ryskin G and Leal LG. Numerical solution of free-boundary problems in fluid mechanics. Part 3. Bubble deformation in an axisymmetric straining flow. J Fluid Mech 1984; 148: 37–43
[141] Crighton DG and Ffowcs Williams JE. Sound generated by turbulent two-phase gas flows. J Fluid Mech 1969; 36: 585–603
[142] Pumphrey HC and Ffowcs Williams JE. Bubbles as sources of ambient noise. IEEE J Ocean Eng 1990; 15: 268–274
[143] Hollett R and Heitmeyer R. Noise generation by bubbles formed in breaking waves. Proc NATO Adv Workshop Natural Mechanisms of Surface Generated Noise in the Ocean, Lerici, Italy, 15–19 June 1987. In: Sea Surface Sound (Kerman BR ed.). Reidel, Dortrecht, The Netherlands, 1987, pp 449–462

[144] Prosperetti A. Bubble dynamics in oceanic ambient noise. In: Sea Surface Sound (Kerman BR, ed.). Kluwer, Boston, 1988, pp 151–172

[145] Longuet-Higgins MS. An analytical model of sound production by raindrops. J Fluid Mech 1990; 214: 395–410

[146] Buckingham MJ. On acoustic transmission in ocean-surface waveguides. Phil Trans Roy Soc 1991; A335: 513–555

[147] Thorpe S. On the clouds of bubbles formed by breaking wind-waves in deep water, and their role in air–sea gas transfer. Phil Trans Roy Soc 1982; A304: 155–210

[148] Thorpe SA. Measurements with an automatically recording inverted echo sounder; ARIES and the bubble clouds. J Phys Ocean 1986; 16: 1462–1478

[149] Monahan EC and Lu NQ. Acoustically relevent bubble assemblages and their dependence on meteorological parameters. IEEE J Ocean Eng 1990; 15: 340–345

[150] Monahan EC and Muircheartaigh IO. Optimal power-law description of oceanic whitecap coverage dependence on wind speed. J Phys Ocean 1980; 10: 2094–2099

[151] Aleksandrov AP and Vaindruk ES. In: The Investigation of the Variablity of Hydrophysical Fields in the Ocean (Ozmidov R, ed.). Nauka, Moscow, 1974, pp 122–128

[152] Kanwisher J. On the exchange of gases between the atmosphere and the sea. Deep Sea Res 1963; 10: 195–207

[153] Blanchard DC and Woodcock AH. Bubble formation and modification in the sea and its meteorological significance. Tellus 1957; 9: 145–158

[154] Thorpe SA. Small-scale processes in the upper ocean boundary. Nature 1985; 318: 519–522

[155] Thorpe SA. The horizontal structure and distribution of bubble clouds. In: Sea Surface Sound (Kerman BR, ed.). Kluwer, Boston, 1988, pp 173–183

[156] Monahan EC. Whitecap coverage as a remotely monitorable indication of the rate of bubble injection into the ocean mixed layer. In: Sea Surface Sound (Kerman BR, ed.). Kluwer, Boston, 1988, pp 85–96

[157] Monahan EC. Near-surface bubble concentration and oceanic whitecap coverage. In: Proc 7th Conf Ocean–Atmos Interaction (Anaheim, California). American Meteorological Society, Boston, Massachusetts, 1988, pp 178–181

[158] Monahan EC. From the laboratory tank to the global ocean. In: Climate and Health Implications of Bubble-mediated Sea–Air Exchange (Monahan EC and Van Patten MA, eds.). CT Sea Grant College Program, Groton, Connecticut, 1989, pp 43–63

[159] Monahan EC, Kin JP, Wilson MB and Woolf DK. Oceanic whitecaps and the marine microlayer spanning the boundary separating the sub-surface bubble clouds from the aerosol laden marine atmosphere. Whitecap Report 3, University of Connecticut, Avery Point, 1987, pp 1–108

[160] Bezzabotnov VZ. Some results on the changes of the structure of sea foam formations in the field. Fiz Atmosfery Okeana 1985; 21: 101–104 (in Russian)

[161] Monahan EC, Davidson KL and Spiel DE. Whitecap aerosol productivity deduced from simulation tank measurements. J Geophys Res 1982; 87: 8898–8904

[162] Langmuir I. Surface water motion induced by wind. Science 1938; 87: 119–123

[163] Leibovich S. The form and dynamics of Langmuir circulations. Ann Rev Fluid Mech 1983; 15: 391–427

[164] Weller R, Dean J, Marra J, Price J, Francis E and Boardman D. Three-dimensional flow in the upper ocean. Science 1985; 227: 1552–1556

[165] Smith J, Pinkel R and Weller R. Velocity structure in the mixed layer during MILDEX. J Phys Ocean 1987; 17: 425–439

[166] Weller RA and Price JF. Langmuir circulation within the oceanic mixed layer. Deep Sea Res 1988; 35: 711–747

[167] Zedel L and Farmer DM. Organised structures in subsurface bubble clouds: Langmuir circulation in the open ocean. J Geophys Res 1991; 96: 8889–8900

[168] Thorpe SA and Hall AJ. The mixing layer of Loch Ness. J Fluid Mech 1980; 101: 687–703

[169] Thorpe S. The effect of Langmuir circulation on the distribution of submerged bubbles caused by breaking wind waves. J Fluid Mech 1984; 142: 151–170

[170] Thorpe SA and Hall AJ. The characteristics of breaking waves, bubble clouds and near-surface currents observed using side-scan sonar. Continental Shelf Res 1983; 1: 353–384

[171] Thorpe S. Bubble clouds and the dynamics of the upper ocean. Quart J Meteorol Soc 1992; 118: 1–22

[172] Medwin H and Daniel AC Jr. Acoustical measurements on bubble production by spilling breakers. J Acoust Soc Am 1990; 88: 408–412

[173] Toba Y. Memoirs of the college of science. University of Kyoto, Series A, 1961; XXIX(3): Art 4, 313–344

[174] Updegraff GE and Anderson VC. Bubble noise and wavelet spills recorded 1 m below the ocean surface. J Acoust Soc Am 1991; 89: 2264–2279

[175] Kolovayev PA. Investigation of the concentration and statistical size distribution of wind produced bubbles in the near-surface ocean layer. Oceanology 1976; 15: 659–661

[176] Johnson BD and Cooke RC. Bubble populations and spectra in coastal waters: a photographic approach. J Geophys Res 1979; 84(C7): 3761–3766

[177] Walsh AL and Mulhearn PJ. Photographic measurements of bubble populations from breaking waves at sea. J Geophys Res 1987; 92: 14553–14656

[178] MacIntyre F. On reconciling optical and acoustic bubble spectra in the mixed layer. In: Oceanic Whitecaps and Their Role in Air–Sea Exchange Processes (Monanhan EC and O'Muircheartaigh I, eds.). Reidel, Dordrecht, The Netherlands, 1986; pp 95–100

[179] Medwin H. *In situ* acoustic measurements of microbubbles at sea. J Geophys Res 1977; 82: 971–976

[180] Medwin H and Breitz ND. Ambient and transient bubble spectral densities in quiescent seas and under spilling breakers. J Geophys Res 1989; 94: 12751–12759

[181] Su MY, Ling SC and Cartmill J. Optical measurements in the North Sea. In: Sea Surface Sound (Kerman B, ed.). Kluwer, Boston, 1988

[182] Breitz N and Medwin H. Intrumentation for *in situ* acoustical measurements of bubble spectra under breaking waves. J Acoust Soc Am 1989; 86: 739–743

[183] Woolf DK and Thorpe SA. Bubbles and the air–sea exchange of gases in near-saturation conditions. J Marine Res 1991; 49: 435–466

[184] Thorpe SA. The role of bubbles produced by breaking waves in super-saturating the near-surface ocean mixing layer with oxygen. Ann Geophys 1984; 2: 53–56

[185] Broeker HC. Effects of bubbles upon the gas exchange between the atmosphere and ocean. In: Symposium on Capillary Waves and Gas Exchange. 1980; Heft Nr 17: 127–139

[186] Thorpe SA and Stubbs AR. Bubbles in a freshwater lake. Nature 1979; 279: 403–405

[187] Toba Y, Tokuda M, Okuda K and Kawai S. Forced convection accompanying wind waves. J Ocean Soc Japan 1975; 31: 192–198

[188] Wu J. Bubble populations and spectra in near-surface ocean: summary and review of field measurements. J Geophys Res 1981; 86: 457–463

[189] Wu J. Bubbles in near-surface ocean: a general description. J Geophys Res 1988; 93: 587–590

[190] Hwang PA, Hsu YHL and Wu J. Air bubbles produced by breaking wind waves: a laboratory study. J Phys Ocean 1990; 20: 19–28

[191] Baldy S. Bubbles in the close vicinity of breaking waves: statistical characteristics of the generation and dispersion mechanism. J Geophys Res 1988; 93: 8239–8248

[192] Baldy S and Bourguel M. Bubbles between wave trough and wave crest levels. J Geophys Res 1987; 92: 2919–2929

[193] Rino CL and Ngo HD. Low-frequency acoustic scatter from subsurface bubble clouds. J Acoust Soc Am 1991; 90: 406–415

[194] Crawford GB and Farmer DM. On the spatial distribution of ocean bubbles. J Geophys Res 1987; 92: 8231–8243

[195] Liss PS and Merlivat L. Air–sea gas exchange rates: introduction and synthesis. In: The Role of Air–Sea Exchange in Geochemical Cycling (Buat-Menard P, ed.). Reidel, Dordrecht, The Netherlands, 1986, pp 113–127

[196] Broeker WS and Peng T-H. Tracers in the Sea. Eldigio Press, Palisades, NY, 1982

[197] Memery L and Merlivat L. Modelling of gas flux through bubbles at the air–water interface. Tellus 1985; 37B: 272–285

[198] Scott JC. The role of salt in whitecap persistence. Deep Sea Res 1975; 22: 653–657

[199] Zieminski SA, Hume RM and Durham R. Rates of oxygen transfer from air bubbles to aqueous NaCl solutions at various temperatures. Mar Chem 1976; 4: 333–346

[200] Thorpe SA and Humphries PN. Bubbles and breaking waves. Nature 1980; 283: 463–465

[201] Ariyel NZ, Byutner EK and Strokina LA. Estimates of the rate of gas exchange across the ocean–atmosphere interface. Izvestiya Atmosph Oceanic Phys 1981; 17: 782–787

[202] Foldy LL. The multiple scattering of waves. Phys Rev 1945; 67: 107–119

[203] Caflisch RE, Miksis MJ, Papanicolaou GC and Ting L. Effective equations for wave propogation in bubbly liquids. J Fluid Mech 1985; 153: 259–273

[204] Commander KW and Prosperetti A. Linear pressure waves in bubbly liquids: comparison between theory and experiments. J Acoust Soc Am 1989; 85: 732–746

[205] Lu NQ, Prosperetti A and Yoon SW. Underwater noise emissions from bubble clouds. IEEE J Ocean Eng 1990; 15: 275–285

[206] Wood A. Acoustics. Blackie, London, 1940. References to 1960 re-issue, p 364

[207] Van Wijngaarden L. One-dimensional flow of liquids containing small gas bubbles. Ann Rev Fluid Mech 1972; 4: 369–394

[208] Crighton DG and Ffowcs Williams JE. Sound generated by turbulent two-phase gas flows. J Fluid Mech 1969; 36: 585–603

[209] Kinsler LE, Frey AR, Coppens AB and Sanders JV. Fundamentals of Acoustics. Wiley, New York, 3rd edn, 1982, p 403

[210] Urick RJ. Principles of Underwater Sound. McGraw-Hill, New York, 3rd edn, 1983, p 148

[211] Urick RJ. Principles of Underwater Sound. McGraw-Hill, New York, 3rd edn, 1983, p 159

[212] Urick RJ. Principles of Underwater Sound. McGraw-Hill, New York, 3rd edn, 1983, pp 135–136 and 159–182

[213] Spindel RC. Sound transmission in the ocean. Ann Rev Fluid Mech 1985; 17: 217–237

[214] Urick RJ. Principles of Underwater Sound. McGraw-Hill, New York, 3rd edn, 1983

[215] Didenkulov IN and Sutin AM. The influence of a subsurface bubble layer on wind ambient noise generation. In: Natural Physical Sources of Underwater Sound (Kerman BR, ed.). Kluwer, Dordrecht, The Netherlands, 1992

[216] Kumar S and Brennen CE. Nonlinear effects of the dynamics of clouds of bubbles. J Acoust Soc Am 1991; 89: 707–714

[217] Isakovitch MA and Kur'yanov BF. To the theory of low-frequency noises in the ocean. Akust Zh 1970; 16: 62–74

[218] Prosperetti A. Bubble-related ambient noise in the ocean. J Acoust Soc Am 1985; 78(Suppl 1): S2

[219] Prosperetti A. Bubble-related ambient noise in the ocean. J Acoust Soc Am 1988; 84: 1042–1054

[220] Carey WM. Low-frequency ocean surface noise sources. J Acoust Soc Am 1985; 78(Suppl): S1–S2

[221] Carey WM. Low-frequency noise and bubble plume oscillations. J Acoust Soc Am 1987; 82(Suppl): S62

[222] Carey WM and Browning D. Low-frequency ocean ambient noise: measurements and theory. In: Sea Surface Sound (Kerman BR, ed.). Kluwer, Boston, 1988

[223] Omta R. Oscillations of a cloud of bubbles of small and not so small amplitude. J Acoust Soc Am 1987; 82: 1018–1033

[224] d'Agostino L and Brennen CE. Acoustical absorption and scattering cross sections of spherical bubble clouds. J Acoust Soc Am 1988; 84: 2126–2134

[225] Yoon SW, Crum LA, Prosperetti A and Lu NQ. An investigation of the collective oscillations of a bubble cloud. J Acoust Soc Am 1991; 89: 700–706

[226] Roy RA, Carey W, Nicholas M and Crum LA. Low-frequency scattering from bubble clouds. Proc 14th Int Congr Acoustics, Beijing, China, 1992

[227] Skelton EA and Fitzgerald WJ. An invariant imbedding approach to the scattering of sound from a two-phase fluid. J Acoust Soc Am 1988; 84: 742

[228] Kustov LM, Nazarov VE, Ostrovskii LA *et al.* Acoust Lett 1982; 6: 15

[229] Donskoi DM, Zamolin SV, Kustov LM and Sutin AM. Nonlinear backscattering of acoustic waves in a bubble layer. Acoust Lett 1984; 7: 131

[230] Ostrovsky LA and Sutin AM. Nonlinear sound scattering from subsurface bubble layers. In: Natural Physical Sources of Underwater Sound (Kerman BR, ed.). Kluwer, Dordrecht, The Netherlands, 1992

[231] Kobelev Yu, Ostrovsky L and Sutin A. Self-illumination effects for acoustic waves in a liquid with gas bubbles. Pis'ma v ZhETF, 1979; 30(7): 423–425

[232] Kobelev Yu, Ostrovsky L and Sutin A. Self-induced transparency and frequency conversion effects for acoustic waves in water containing gas bubbles. In: Cavitation and Inhomogeneities in Underwater Acoustics (Lauterborn W, ed.). Springer, Berlin, 1980, pp 151–156

[233] Kustov L, Nazarov V and Sutin A. Narrowing of directivity pattern of acoustic radiation in bubble layer. Sov Phys Acoust 1987; 33(3)
[234] Kobelev YuA and Ostrovsky LA. Nonlinear acoustic phenomena due to bubble drift in a gas–liquid mixture. J Acoust Soc Am 1989; 85: 621–629
[235] Medwin H. Acoustic fluctuations due to microbubbles in the near-surface ocean. J Acoust Soc Am 1974; 56: 1100–1104
[236] Medwin H. Counting bubbles acoustically: a review. Ultrasonics 1977; 15: 7–14
[237] d'Agostino L and Brennen CE. On the acoustical dynamics of bubble clouds. ASME Cavitation and Multiphase Forum, American Society of Mechanical Engineers, New York, 1983, p 72
[238] Miksis MJ and Ting L. Effects of bubbly layers on wave propagation. J Acoust Soc Am 1989; 86: 2349–2358
[239] McDaniel ST. High-frequency sea surface scattering: recent progress. J Acoust Soc Am 1988; 84(Suppl 1): S121
[240] Henyey FS. Acoustic scattering from ocean microbubble plumes in the 100 Hz to 2 kHz region. J Acoust Soc Am 1991; 90: 399–405
[241] Dashen R, Henyey FS and Wurmser D. Calculations of acoustic scattering from the ocean surface. J Acoust Soc Am 1990; 88: 310–323
[242] Thorsos E. Acoustic scattering from a 'Pierson–Moskowitz' sea surface. J Acoust Soc Am 1990; 88: 335–349
[243] Milder M. An improved formalism for wave scattering from rough surfaces. J Acoust Soc Am 1991; 529–541
[244] Chapman R and Harris J. Surface backscattering strength measured with explosive sound sources. J Acoust Soc Am 1962; 34: 1592–1597
[245] Chapman R and Scott H. Surface backscattering strengths measured over an extended range of frequencies and grazing angles. J Acoust Soc Am 1964; 36: 1735–1737
[246] Medwin H. Acoustical bubble spectrometry at sea. Cavitation and Inhomogeneities in Underwater Acoustics, Vol. 4. Springer, Berlin, 1980, pp 187–193
[247] Morse PM and Feschback H. Methods in Theoretical Physics – Part II. McGraw Hill, New York, 1953, pp 1498–1501
[248] McDonald BE. Echoes from vertically striated subresonant bubble clouds: a model for ocean surface reverberation. J Acoust Soc 1991; 89: 617–622
[249] Bachmann W. A theoretical model for the backscattering strength of a composite roughness sea. J Acoust Soc Am 1973; 54: 712
[250] Crowther PA. Acoustic scattering from near-surface bubble layers. In: Cavitation and Inhomogeneities in Underwater Acoustics (Lauterborn W, ed.). Springer, New York, 1979, pp 194–204
[251] Nutzel B, Herwig H, Monti JM and Koenigs PD. The Influence of Surface Roughness and Bubbles on Sea Surface Acoustic Back-scattering. NUSC Tech Rep 7955, Naval Underwater Systems Center, New London, Connecticut, 1987
[252] Rino CL, Crystal TL, Kiode AK, Ngo HD and Guthart H. Numerical simulation of acoustic and EM scattering from linear and nonlinear ocean surfaces. Radio Sci 1991; 26: 51–57
[253] Jessup AT, Keller WC and Melville WK. Measurements of sea spikes in microwave backscatter at moderate incidence angles. J Geophys Res 1990; 95: 9679–9688
[254] Ewell GW, Tuley MT and Horne WF. Temporal and spatial behaviour of high resolution sea clutter 'spikes'. IEEE Radar 84 Conf, 1984, pp 100–104
[255] Loewen MR and Melville WK. Microwave backscatter and acoustic radiation from breaking waves. J Fluid Mech 1991; 224: 601–623
[256] Munk WH and Forbes AMG. Global Ocean Warming: An Acoustic Measure? J Phys Oceanography 1989; 19: 1765–1778
[257] Baggeroer A and Munk W. The Heard Island Feasibility Test. Physics Today 1992; 45: 22–30

4
The Forced Bubble

An entrepreneur submitted to the Department of Trade and Industry a proposal for collaboration with academia in the prediction of racehorse winners. Enthusing on the value of such a project, the Department financed three years of research, whereby the entrepreneur provided field data, which was subsequently analysed by the Statistics group of a mathematics department at a certain university. At the end of three years, the university submitted a written report to the entrepreneur and the DTI which they stated was '100% certain'. Passing rapidly over the pages of formulae, the entrepreneur read the one-line conclusion at the end: "One horse always wins." Somewhat dismayed, but still feeling the project had worth, the entrepreneur approached the Science and Engineering Research Council. They suggested that, since the problem of a fast racehorse might reduce to the drag factor of the horse, the circulation of the blood etc., for a more practical input the entrepreneur should collaborate with the Fluid Dynamics group at the university. The project went ahead along these lines, and at the end of three years the entrepreneur turned up to take possession of the report, which he would personally pass on to the SERC. "Is it 100% certain to predict the result?" he asked, as the researchers handed over the thick bound copy. "No, no," they replied, "This is a practical science, and nothing is ever certain. But we have 95% confidence in the winner that the theory predicts for any given race." The entrepreneur beamed. "That's good enough for me!" he smiled, and opening the report, read the first line: *Assume a spherically-symmetric horse ...*

4.1 The Forced Linear Oscillator

In Chapter 3, it was convenient to discuss the unforced linear oscillator in terms of a bob on a spring, which acts as an analogue for the small-amplitude pulsation of a gas bubble in a liquid. It is worthwhile using the same analogy when discussing the driven linear oscillator. As before, the displacement of the bob models the motion of the bubble wall, and the spring constant k acts as the restoring force (that is, the pressure changes in the gas compressed or rarefied within the bubble). The driving force, which in the case of the acoustic bubble is the acoustic pressure changes within the liquid, may be represented in the analogue by a periodic displacement of the point of attachment of the spring. In this manner, a person holding a spring would drive a bob at the remote end by raising and lowering the hand. As before, the inertia associated with the bubble system arises mainly from the motion of the surrounding fluid, in contrast to the simple spring–bob analogue, where the inertia resides in the mass of the bob.

4.1.1 The Undamped Driven Oscillator

(a) Transient and Steady-state Responses

In the absence of damping, the equation of motion of the bob can be found by applying Newton's Second Law to the one-dimensional spring–bob system discussed in Chapter 3, section 3.1, with the addition of a driving force. Against this force, $F_0 e^{i\omega t}$, the spring exerts a force of $k\varepsilon$, and the difference between these forces equals the product of the mass and the acceleration:

$$m\ddot{\varepsilon} + k\varepsilon = F_0 e^{i\omega t} \tag{4.1}$$

where ω is the frequency of the driving force, of amplitude F_0. The zero of time is chosen so that F_0 is real.

The behaviour of the oscillator can be understood qualitatively. Assume at some time $t = t_0$ the driving force begins, and continues for all time. There are two main times of interest: at $t = t_0$, and at $t = \infty$.

Consider when the oscillator has been subjected to the driving force for a very long time ($t \rightarrow \infty$). For all times before and after that moment, the oscillator can 'see' one continuous sinusoid. The only temporal information received is the frequency and phase of the driving force. Therefore the oscillator can only respond at that frequency. This *steady-state* response of the oscillator is therefore a sinusoidal oscillation at the frequency ω of the driving force, and at some definite phase relation to it.

However, at the moment when the force begins ($t = t_0$), the oscillator 'sees' a continuous sinusoid from $t = t_0$ to $t = \infty$, but zero driving force from $t = -\infty$ to $t = t_0$. This information contains many frequencies, as Fourier analysis of the time-series data containing $t = t_0$ would show. Obviously, a major component would be at the driving frequency ω, and so this would be expected to feature in the response of the oscillator. However, the oscillator is especially sensitive to its resonance frequency ω_0. Therefore if it detects that frequency component among the frequencies present at $t = t_0$, it will respond appropriately. There will therefore be a component of the resonance frequency ω_0 in the bubble response at times $t > t_0$. A schematic representation of an oscillator's possible response is shown in Figure 4.1. At times $t < t_0$, there is no oscillation. At large times after the onset of the driving force, the response is steady state

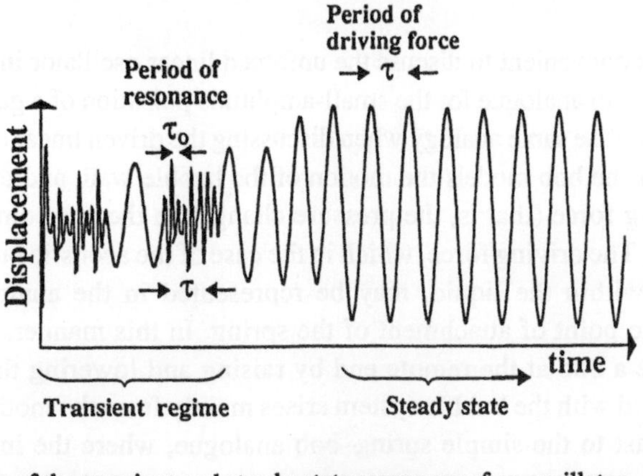

Figure 4.1 Schematic of the transient and steady-state responses of an oscillator.

(i.e. at ω). For finite times after the onset, the response is complicated by contributions of other frequencies, and the oscillator resonance frequency, $\omega_0 = 2\pi/\tau_0$, is present in the response. This region is termed the *transient response*.

Mathematical arguments demonstrate a similar frequency response. The equation of motion is given by equation (4.1), which division through by the system inertia, m, gives

$$\ddot{\varepsilon} + \frac{k}{m}\varepsilon = \frac{F_0}{m}e^{i\omega t} \tag{4.2}$$

As was stated earlier, shortly after the onset of the driving force, the sensitivity of the bubble to its resonance frequency ω_0 is important. As was shown in Chapter 3, section 3.1.2, to examine the response to the resonance frequency ω_0 at finite times $t > 0$ it is necessary to look at the undriven form of the equation:

$$\ddot{\varepsilon} + \frac{k}{m}\varepsilon = 0 \tag{4.3}$$

Substitution of a trial solution in the manner of Chapter 3, section 3.1.2(a) gives a general solution of the form

$$\varepsilon = \Xi_1 e^{i\omega_0 t} + \Xi_2 e^{-i\omega_0 t} \tag{4.4}$$

where $\omega_0 = \sqrt{k/m}$, and where the complex conjugates Ξ_1 and $\Xi_2 = \Xi_1^*$ may accommodate phase angles. However, this cannot describe the oscillation when the system is subjected to a driving force: if the solution given by equation (4.4) is substituted into the left-hand side of the equation of motion (equation (4.2)), it gives zero, since it is in fact a solution to equation (4.3). Since zero cannot equal $(F_0/m)e^{i\omega t}$ except in the absence of a driving force, then another term must be added to the left-hand side of equation (4.4) so that when this solution goes through the 'sausage machine' which is the right-hand side of equation (4.2), the result is equal to $(F_0/m)e^{i\omega t}$. This term must obviously contain the driving frequency ω. Therefore the general solution to equation (4.2) is

$$\varepsilon = \Xi_1 e^{i\omega_0 t} + \Xi_2 e^{-i\omega_0 t} + Ae^{i\omega t} \tag{4.5}$$

where A is real in cases of zero damping. Equation (4.4) is called the *complementary function*, and the additional term in equation (4.5) (here equal to $Ae^{i\omega t}$) the *particular integral*. The former contains the frequency information at ω_0 which provides transients in the response after the onset of the driving force, and the latter describes the steady-state response at frequency ω.

(b) Amplitude and Phase Response in the Steady State

In the steady state, the effects of the complementary function are negligible, as discussed above. The response as $t \to \infty$ is therefore given by $\varepsilon = Ae^{i\omega t}$. Substitution of this into the equation of motion of the forced undamped oscillator (equation (4.2)) gives the amplitude

$$A = \frac{F_0}{m(\omega_0^2 - \omega^2)} \tag{4.6}$$

The variation of the amplitude A with driving frequency ω is shown in Figure 4.2. At low frequencies the amplitude A tends to $(F_0/m\omega_0^2)$ and is positive for all $\omega \ll \omega_0$. As $\omega \to \infty$, the

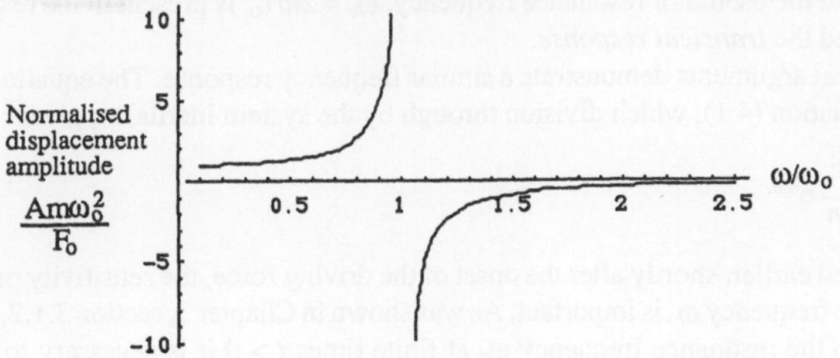

Figure 4.2 Variation in amplitude with driving frequency for an undamped oscillator.

amplitude A tends to zero, and is negative for all $\omega \ll \omega_0$. The amplitude becomes undefined at $\omega = \omega_0$.

This result tallies with physical experience, though the amplitude at $\omega = \omega_0$ is in practice large in magnitude, but finite (which would result from the addition of some small but finite damping). This can be seen by considering the spring–bob system. Say that the spring is very flexible, and has a large mass on the end (consider a billiard ball suspended on the end of a 'slinky™'). The resonance frequency $\omega_0 = \sqrt{k/m}$ will be low. If the hand holding the slinky™ is not oscillated, but instead raised uniformly, the ball will follow uniformly (after initial transients have damped out). Here $\omega = 0$ and $A = (F_0/m\omega_0^2)$. If the hand is moved up and down extremely slowly, the ball will move in phase with the hand. This corresponds to $\omega \ll \omega_0$. However, if an effort is made to oscillate the system at a higher frequency, the response is quite different (Figure 4.3). Consider the movement required of the hand to generate high-frequency oscillation of the billiard ball. When the ball is moving down, the hand rises up in order to extend the spring and so increase the upwards force on the ball, arresting and then reversing its motion. However, as soon as the ball begins to move upwards, the hand is lowered, compressing the spring. This increases the downwards force on the bob, stopping and then reversing its upwards motion. In this case $\omega \gg \omega_0$ and the displacement of the oscillator is in antiphase with the driving force.

In contrast, one might envisage a very stiff spring with a light mass on it. An extreme case of this might be a pencil: the stiff wood gives a high k and the eraser on the remote end represents a very small mass m. This rather extreme example of a spring–bob system has a very high resonance frequency of $\omega_0 = \sqrt{k/m}$! If this system is subjected to vertical oscillations by hand, the motion of the bob is exactly in phase with the driving force (when the hand is raised, the eraser rises!). Therefore when $\omega \ll \omega_0$, the driving force and the displacement response will be in phase. This correlates with Figure 4.2, where A is real and positive.

So at $\omega \ll \omega_0$ the driving force and the oscillator displacement are in phase (called *stiffness controlled*), and at $\omega \gg \omega_0$ they are in antiphase (*inertia controlled*) for the undamped forced harmonic oscillator. Note that the velocity is $\dot{\varepsilon} = i\omega\varepsilon = \omega\varepsilon e^{i\pi/2}$, that is the velocity leads the displacement by $\pi/2$. Therefore the velocity also undergoes a phase change of magnitude π at resonance, going from a phase of where it leads the force by $\pi/2$ when $\omega \ll \omega_0$, to one where it lags the force by $\pi/2$ for $\omega \gg \omega_0$. At resonance, the velocity is in phase with the driving force.

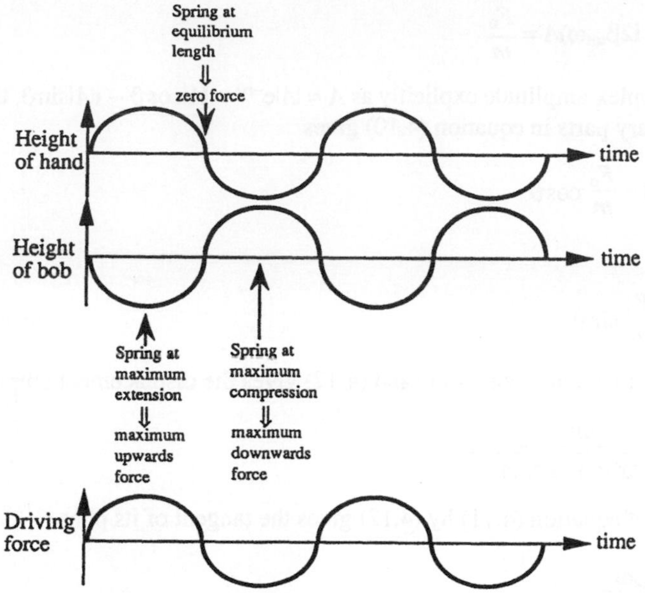

Figure 4.3 The response of an inertia-controlled oscillator, as illustrated by the displacement of a heavy mass which is supported from the hand by a 'floppy' spring.

4.1.2 The Damped Forced Oscillator

(a) Amplitude and Phase in the Steady State

When damping is added to the driven oscillator, the equation of motion becomes

$$m\ddot{\varepsilon} + b_{tot}\dot{\varepsilon} + k\varepsilon = F_0 e^{i\omega t} \tag{4.7}$$

which is found through the addition of the driving force to equation (3.16). The dissipative constant is written as b_{tot} to clarify that it must incorporate *all* the dissipative processes that occur. With the usual substitutions of $\omega_0 = \sqrt{k/m}$ and $\beta_{tot} = b_{tot}/(2m)$, division of equation (4.7) by m gives

$$\ddot{\varepsilon} + 2\beta_{tot}\dot{\varepsilon} + \omega_0^2\varepsilon = \frac{F_0}{m} e^{i\omega t} \tag{4.8}$$

If the steady-state response is assumed (i.e. $t \to \infty$), the solution takes the form of the particular integral only. To give oscillatory solutions, the damping must be light. Substitution of the particular integral solution $\varepsilon = Ae^{i\omega t}$ (where A may be complex to allow for a finite phase between displacement and driving force) gives

$$((\omega_0^2 - \omega^2) + i2\beta_{tot}\omega)\varepsilon = \frac{F_0}{m} e^{i\omega t} \tag{4.9}$$

or

$$((\omega_0^2 - \omega^2) + i2\beta_{tot}\omega)A = \frac{F_o}{m} \tag{4.10}$$

Writing the complex amplitude explicitly as $A = |A|e^{-i\vartheta} = |A|\cos\vartheta - i|A|\sin\vartheta$, then equating the real and imaginary parts in equation (4.10) gives

$$(\omega_0^2 - \omega^2)\,|A| = \frac{F_o}{m}\cos\vartheta \tag{4.11}$$

and

$$2\beta_{tot}\omega\,|A| = \frac{F_o}{m}\sin\vartheta \tag{4.12}$$

Squaring and adding equations (4.11) and (4.12) gives the displacement amplitude

$$|A| = \frac{F_o/m}{\sqrt{(\omega_0^2 - \omega^2)^2 + (2\beta_{tot}\omega)^2}} \tag{4.13}$$

whilst division of equation (4.11) by (4.12) gives the tangent of its phase

$$\tan\vartheta = \frac{2\beta_{tot}\omega}{(\omega_0^2 - \omega^2)} \tag{4.14}$$

For a simply linear oscillator, where the damping is independent of frequency, these expressions are conveniently written in terms of the quality factor, $Q = 1/\delta_{tot}$. Making use of the approximation $Q^{-1} = 2\beta_{tot}/\omega_b \approx 2\beta_{tot}/\omega_0$ (equation (3.157)), equations (4.13) and (4.14) can be rewritten

$$|A| = \frac{F_o}{m\omega_0^2} \quad \frac{\omega_0/\omega}{\sqrt{\left(\dfrac{\omega_0}{\omega} - \dfrac{\omega}{\omega_0}\right)^2 + \dfrac{1}{Q^2}}} \tag{4.15}$$

noting that $m\omega_0^2 = k$ [1], and

$$\tan\vartheta = \frac{1/Q}{\left(\dfrac{\omega_0}{\omega} - \dfrac{\omega}{\omega_0}\right)} \tag{4.16}$$

Plots of $|A|$ and ϑ as a function of the driving frequency ω can be seen in Figure 4.4, for different values of the damping. Comparison with Figure 4.2 shows that the addition of damping to the system limits the amplitude at resonance and reduces the frequency of maximum response. The phase change at resonance is no longer discontinuous. If the oscillator is driven at frequencies well below resonance (stiffness-controlled), it is in phase with the driver ($\vartheta = 0$). The amplitude tends to $F_o/(m\omega_0^2)$. As the frequency is increased towards resonance, the displacement tends to lag behind the driver, so that at resonance the displacement lags the driver by $\pi/2$ (i.e. $\vartheta = \pi/2$). When the oscillator is driven at frequencies much greater than resonance (inertia-controlled), the displacement is in antiphase with the driver ($\vartheta = \pi$). The amplitude tends to zero as $\omega \to \infty$, since the response time of the oscillator is much slower than the timescales over which the force reverses direction: the oscillator therefore responds to the net force in a cycle, which is zero. The addition of damping tends to reduce the frequency at which

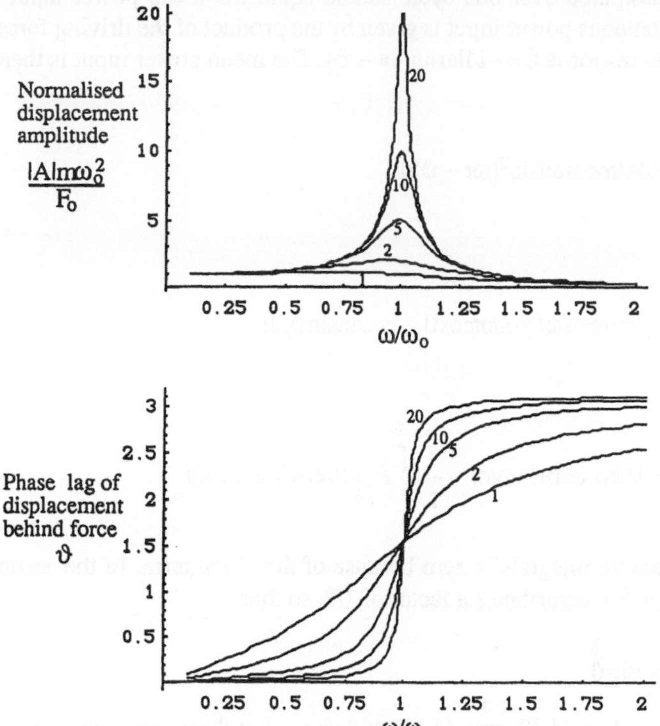

Figure 4.4 (a) The amplitude and (b) the phase response of a damped oscillator as a function of the driving frequency.

the response is a maximum to a frequency less than ω_0. By equating to zero, the differential of equation (4.15) with respect to ω shows that the maximum occurs at a frequency $\omega_0\sqrt{1-1/(2Q^2)}$, whereupon the displacement has amplitude $F_0Q/(m\omega_0^2\sqrt{1-1/(4Q^2)})$.

(b) Energy Dissipation in the Forced Damped Oscillator

The energy dissipated over one cycle in the forced damped oscillator is given by the time-average of the product of the damping constant b with the square of the velocity. This is because, as seen in Chapter 3, section 3.1.3, to overcome damping a force of $b_{tot}\dot{\varepsilon}$ must be supplied by the oscillator, and the power this requires is the product of this force with $\dot{\varepsilon}$. This gives the instantaneous rate at which energy is lost from the oscillator. Since $\varepsilon = |A|\cos(\omega t - \vartheta)$ then $\dot{\varepsilon} = -|A|\omega\sin(\omega t - \vartheta)$. Thus the energy dissipated over one cycle is the integral of rate of energy loss over that cycle

$$\int_0^{\tau_v} b_{tot}|A|^2\omega^2\sin^2(\omega t - \vartheta)dt = \frac{\omega^2|A|^2 b_{tot}}{2} = \omega^2|A|^2\beta_{tot}\, m \qquad (4.17)$$

since the average of $\{\sin^2\omega t = (1-\cos 2\omega t)/2\}$ over one cycle is 1/2.

The energy dissipated over one cycle should equal the mean power input to the system, $<\dot{W}>$. The instantaneous power input is given by the product of the driving force $F_0\cos\omega t$ with the velocity of the response $\dot{\varepsilon} = -|A|\omega\sin(\omega t - \vartheta)$. The mean power input is therefore given by

$$<\dot{W}> = \frac{1}{\tau_v}\int_0^{\tau_v} F_0|A|\omega\cos\omega t\sin^2(\omega t - \vartheta)dt$$

$$= \frac{1}{\tau_v}\int_0^{\tau_v} F_0|A|\omega\cos\omega t(\sin\omega t\cos\vartheta - \cos\omega t\sin\vartheta)dt$$

$$= \frac{1}{\tau_v}\int_0^{\tau_v} F_0|A|\omega\cos\vartheta\sin\omega t dt - \frac{1}{\tau_v}\int_0^{\tau_v} F_0|A|\omega\cos^2\omega t\sin\vartheta dt \tag{4.18}$$

The first of the above integrals is zero because of the $\sin\omega t$ term. In the second integral, the presence of the $\cos^2\omega t$ contributes a factor of 1/2, so that

$$<\dot{W}> = \tfrac{1}{2}F_0|A|\omega\sin\vartheta \tag{4.19}$$

Substitution of equation (4.12) into (4.19) will show that the energy dissipated over one cycle (equation (4.17)) equals the mean power input to the oscillator. Note that if there were no dissipation, that is, b were zero, then the force would be in phase or antiphase with the displacement, and ϑ would equal zero or π (equation (4.14)). The mean power input would be zero, as much energy entering the system each quarter-cycle as leaves it the next.

(c) The Pulsating Bubble as a Linear Forced Damped Oscillator

The small-amplitude spherically symmetric linear response of a bubble in the RF frame is expounded by the equation of motion[1]

$$m_{RF}^{rad}\ddot{R}_\varepsilon + b_{RF}\dot{R}_\varepsilon + k_{RF}R_\varepsilon = -P_A.4\pi R_o^2\cos\omega t \tag{4.20}$$

the negative sign being used to ensure a displacement in the direction of increasing coordinate (r) is taken as the positive (a *reduction* in pressure is required for an increase in bubble radius). The oscillation of the pulsating bubble of equilibrium radius R_0 may be characterised as

$$R(t) = R_o + R_\varepsilon \tag{4.21}$$

Comparison with the system set up in section 4.1 enables the following equivalences to be substituted into the formulation: $R_\varepsilon \equiv \varepsilon$, $F_o = -4\pi R_o^2 P_A$.

Introducing the linearly oscillating solution

$$R_\varepsilon = \{-R_{\varepsilon o}\}\cos(\omega t - \vartheta) \tag{4.22}$$

[1]See Chapter 3, section 3.2.1(a).

is the particular equivalent of $Ae^{i\omega t} = |A|e^{-i\vartheta}e^{i\omega t}$. The particular case for the bubble of ε is R_ε, and the particular case of $|A|$ is the real *negative* number $\{-R_{\varepsilon o}\}$, the negative sign arising again because a positive acoustic pressure causes a displacement of the bubble wall in the direction of *decreasing* radial coordinate r. The relation between the real positive constant $R_{\varepsilon o}$ and the acoustic pressure is found by applying this particular case of the bubble to equation (4.13):

$$R_{\varepsilon o} = \frac{P_A}{R_o \rho} \frac{1}{\sqrt{(\omega_o^2 - \omega^2)^2 + (2\beta_{tot}\omega)^2}} = \frac{P_A}{R_o \rho \Pi} \tag{4.23}$$

using the explicit form for the radiation mass in this frame, $m_{RF}^{rad} = 4\pi R_o^3 \rho$, and the substitution

$$\Pi = \sqrt{(\omega_o^2 - \omega^2)^2 + (2\beta_{tot}\omega)^2} \tag{4.24}$$

From the analysis given in section 4.1.2(a), the phase lag ϑ is governed by the following relations:

$$\cos\vartheta = \frac{\omega_o^2 - \omega^2}{\Pi} \tag{4.25}$$

$$\sin\vartheta = \frac{2\beta_{tot}\omega}{\Pi} \tag{4.26}$$

and by equation (4.14). If the damping of an oscillator were independent of frequency, then these expressions might be simplified by employing the approximation $Q = \omega_o/2\beta_{tot} \approx \omega_o/2\beta_{tot}$ (equation (3.157)) to write $\Pi/\omega^2 = \sqrt{[(\omega_o/\omega)^2 - 1]^2 + \omega_o^2/(\omega Q)^2}$ or $\Pi/\omega_o^2 = \sqrt{[1-(\omega/\omega_o)^2]^2 + \omega^2/(\omega_o Q)^2}$. However, with the acoustic bubble, the damping does depend on the frequency of oscillation, so such substitutions are not strictly valid. If such changes in damping with frequency were negligible, then one might expect a bubble undergoing small-amplitude oscillations to exhibit response curves of the type shown in Figure 4.4. The frequency dependence of b_{tot} is one important way in which the real bubble differs from the simple linear oscillators discussed earlier. Other differences are introduced at the end of this section.

(d) The Absorption, Scattering and Extinction Cross-sections of a Bubble

When driven into small amplitude pulsation by an acoustic field, the response of the bubble approximates to the forced, damped linear pulsation discussed above. As seen in Chapter 3, section 3.4.2, energy will be subtracted from a plane acoustic wave travelling through a medium containing a bubble, as the bubble, driven into pulsation, will lose energy through thermal and viscous dissipation. Additional energy is subtracted from the incident plane wave to supply energy for the sound scattered or re-radiated by the pulsating bubble. This loss of energy from the incident wave can be characterised by assigning an extinction cross-section Ω_b^{ext} to the bubble. This cross-section is given by the ratio of the time-averaged power subtracted from the wave as a result of the presence of the bubble, to the intensity of the beam (equation (3.292)). The proportions of the energy from the incident beam which are taken by the two contributing processes are represented by the absorption cross-section, Ω_b^{abs} (for thermal and viscous dissipation), and the scattering cross-section Ω_b^{scat}. Assuming the bubble pulsations are approximately linear, the above theory can be applied to calculate these cross-sections.

Consider a plane wave of amplitude P_A and frequency ω which drives a bubble into pulsation, the relevant equation of motion being any of (3.40) to (3.43), depending on the frame chosen. From section 4.1.2(b), the time-average rate of energy dissipation by the bubble through thermal losses is $\omega^2|A|^2 b^{\text{th}}/2$, and through viscous losses is $\omega^2|A|^2 b^{\text{vis}}/2$, where the dissipative constant for the relevant frame is used. The time-average rate at which energy from the plane wave is re-radiated by the scattering bubble is $\omega^2|A|^2 b^{\text{rad}}/2$. The cross-section is given by the ratio of the relevant dissipation rate to the incident intensity, where the dissipative constants $b = 2m\beta$ and $b_{\text{tot}} = 2m\beta_{\text{tot}}$ that are employed are relevant to the frame (RF, RP, VF or VP), and where b is specific for the mechanism of energy loss (b^{rad}, b^{th} or b^{vis}). In the RF frame, where $|A| = R_{\varepsilon 0}$ and where the plane wave intensity is $I = P_A^2/(2\rho c)$ (equation (1.39)), the scattering cross-section of a bubble is therefore found by employing $b_{\text{RF}}^{\text{rad}}$ and equation (4.23) in the ratio:

$$\Omega_b^{\text{scat}} = \frac{\omega^2 R_{\varepsilon 0}^2\, b_{\text{RF}}^{\text{rad}}\, \rho c}{P_A^2} = \frac{\omega^4 4\pi R_o^2}{\Pi^2} = \frac{4\pi R_o^2}{((\omega_o/\omega)^2 - 1)^2 + (2\beta_{\text{tot}}/\omega)^2} \qquad (kR_o \ll 1) \qquad (4.27)$$

using $b_{\text{RF}}^{\text{rad}} = 4\pi R_o^2 \rho_o c (kR_o)^2$ for $kR_o \ll 1$ (equation (3.87)). In this linear regime it is independent of $R_{\varepsilon 0}$. This cross-section is plotted in Figure 4.5(a), with the approximation that damping is invariant with bubble radius ($Q = 10$). The cross-section peaks at the bubble radius resonant with the insonation frequency, and then slowly increases with bubble size. At the other extreme, as the bubble size becomes much less than resonance, the cross-section tends to zero. At resonance, the acoustic scattering cross-section is about five hundred times the geometric cross-section of the resonant bubbles. The effect of damping, and an indication of the degree to which the assumption of frequency-invariant damping is appropriate, is seen in Figure 4.5(b), where the cross-section of a bubble with $Q = 30$ is compared with that of the bubble in Figure 4.5(a) for identical insonation conditions. The peak cross-section is larger, being now roughly four thousand times the geometric cross-section of resonant bubbles. If the damping does change significantly near resonance, such plots need to be interpreted carefully. The degree to which the dimensionless damping constant of a given bubble varies with the insonation frequency is discussed in section 4.4.2. In Figure 4.5(c), the acoustic scattering cross-section for insonation at a higher frequency (248 kHz) is shown: it is clear that the peak at resonance is only a local minimum, and that bubbles at much larger than resonance can scatter to a greater degree than those at resonance. The scattering by resonant bubbles is due to the strong coupling with the incident wave, as manifested by the large amplitude of wall pulsation. Much larger bubbles, in contrast, pulsate to a negligible degree: in the limit of the bubble size being much larger than the acoustic wavelength, the process is geometric, the bubbles generating acoustic shadows.

A similar process to the above could readily be employed to obtain the formulations of the cross-sections associated with thermal and viscous dissipation. However, they are obtained more simply by noting that the only difference in the formulation arises from the choice of b when calculating the time-average rate of energy dissipation by the bubble, $\omega^2|A|^2 b/2$. The choice of the forms of F_o and m depend only on the frame, and not the process. Therefore the ratio of the time-average rate of energy dissipation through thermal losses to that through radiation losses is $(\omega^2|A|^2 b^{\text{th}}/2)/(\omega^2|A|^2 b^{\text{rad}}/2) = b^{\text{th}}/b^{\text{rad}} = d_{\text{th}}/d_{\text{rad}}$ regardless of the frame. Here the off-resonance damping is described by dimensionless damping constants d_{rad}, d_{th} and d_{vis} (and their sum d_{tot}) which at resonance takes the values δ_{rad}, δ_{th} and δ_{vis} (and their sum δ_{tot}). These constants are discussed in section 4.4.2. Since the ratio of the energy losses are in the ratio of the damping constants, then the cross-section associated with thermal dissipation is

$$\Omega_b^{\text{th}} = \frac{d_{\text{th}}}{d_{\text{rad}}} \Omega_b^{\text{scat}} \qquad (4.28)$$

Similarly the cross-section associated with viscous dissipation is

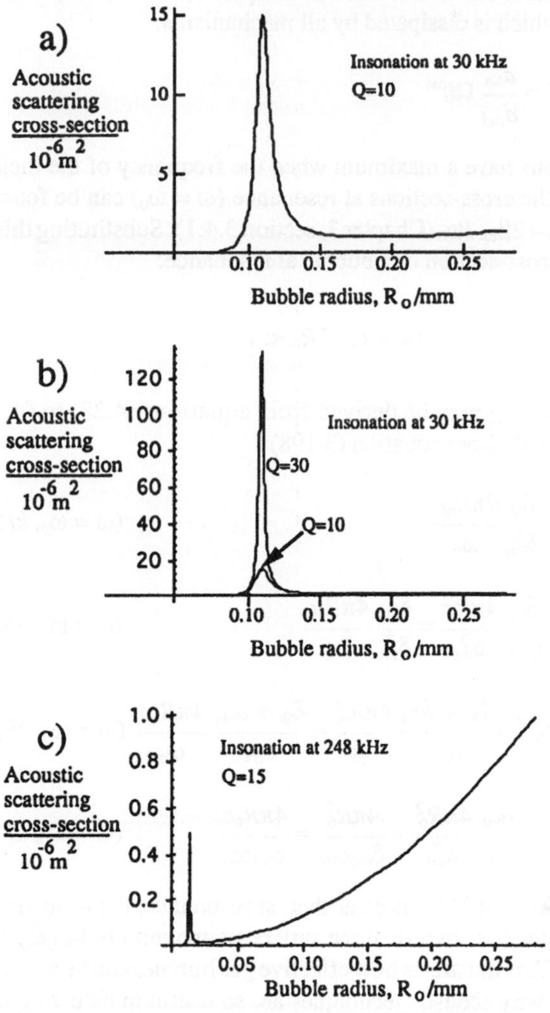

Figure 4.5 The scattering cross-section of a bubble, for which the damping is assumed to be constant with frequency. Linear harmonic bubble pulsations assumed. (a) Insonation at 30 kHz at $\delta_{tot} = 0.1$ ($\equiv Q = 10$). (b) Insonation at 30 kHz for bubbles having $Q = 10$ (as in (a)) and $Q = 30$ ($\equiv \delta_{tot} \approx 0.033$). (c) Insonation at 248 kHz for $Q = 15$ ($\equiv \delta_{tot} \approx 0.066$).

$$\Omega_b^{vis} = \frac{d_{vis}}{d_{rad}} \Omega_b^{scat} \tag{4.29}$$

It is usual to group together the thermal and viscous losses to generate the absorption cross-section of the bubble:

$$\Omega_b^{abs} = \Omega_b^{th} + \Omega_b^{vis} = \frac{d_{th} + d_{vis}}{d_{rad}} \Omega_b^{scat} \tag{4.30}$$

The extinction cross-section, as discussed in Chapter 3, section 3.7, reflects the proportion of the incident energy which is dissipated by all mechanisms:

$$\Omega_b^{ext} = \Omega_b^{scat} + \Omega_b^{abs} = \frac{d_{tot}}{d_{rad}} \Omega_b^{scat} \tag{4.31}$$

These cross-sections have a maximum when the frequency of the incident sound ω equals ω_0. Useful forms of the cross-sections at resonance ($\omega = \omega_0$) can be found using the approximation $\delta_{tot} = 2\beta_{tot}/\omega_b \approx 2\beta_{tot}/\omega_0$ (Chapter 3, section 3.4.1). Substituting this into equation (4.26) gives the scattering cross-section of a bubble at resonance:

$$\Omega_b^{scat} = \frac{4\pi R_o^2}{\delta_{tot}^2} \qquad (\omega = \omega_0,\ kR_o \ll 1) \tag{4.32}$$

from which the following may be derived from equations (4.28) to (4.31), the substitution $\delta_{rad} = \omega_0 R_o/c$ being made from equation (3.198):

$$\Omega_b^{th} = \frac{\delta_{th}}{\delta_{rad}} \frac{4\pi R_o^2}{\delta_{tot}^2} = \frac{\delta_{th}}{\delta_{tot}^2} \frac{4\pi R_o c}{\omega_0} \qquad (\omega = \omega_0,\ kR_o \ll 1) \tag{4.33}$$

$$\Omega_b^{vis} = \frac{\delta_{vis}}{\delta_{rad}} \Omega_b^{scat} = \frac{\delta_{th}}{\delta_{rad}} \frac{4\pi R_o^2}{\delta_{tot}^2} = \frac{\delta_{vis}}{\delta_{tot}^2} \frac{4\pi R_o c}{\omega_0} \qquad (\omega = \omega_0,\ kR_o \ll 1) \tag{4.34}$$

$$\Omega_b^{abs} = \frac{\delta_{th} + \delta_{vis}}{\delta_{rad}} \Omega_b^{scat} = \frac{\delta_{th} + \delta_{vis}}{\delta_{rad}} \frac{4\pi R_o^2}{\delta_{tot}^2} = \frac{\delta_{th} + \delta_{vis}}{\delta_{tot}^2} \frac{4\pi R_o c}{\omega_0} \qquad (\omega = \omega_0,\ kR_o \ll 1) \tag{4.35}$$

$$\Omega_b^{ext} = \Omega_b^{scat} + \Omega_b^{abs} = \frac{\delta_{tot}}{\delta_{rad}} \frac{4\pi R_o^2}{\delta_{tot}^2} = \frac{4\pi R_o^2}{\delta_{rad}\delta_{tot}} = \frac{4\pi R_o c}{\delta_{tot}\omega_0} \qquad (\omega = \omega_0,\ kR_o \ll 1) \tag{4.36}$$

From equations (4.32) to (4.36) it is clear that, at resonance, the acoustic cross-sections are of order δ^{-2} times greater (i.e. two or three orders of magnitude larger) than the geometrical cross-section, $4\pi R_o^2$. This illustrates how effective gas bubbles can be at scattering and absorbing acoustic energy, and why acoustic techniques are so useful in detecting bubbles. Fanelli et al. [2] have derived the acoustic scattering cross-section of a gas–vapour bubble forced into linear shape oscillations.

(e) Compressibility and Sound Speed in Bubbly Liquids

As shown in Chapter 1, section 1.1.1(b)(i), the speed of sound in a compressible continuum is given by the bulk modulus $B = -V(dp/dV)$ (equation (1.3)), such that the sound speed is given by $c = \sqrt{B/\rho}$ (equation (1.10)). When a liquid contains gas bubbles, its bulk modulus decreases, since for any given change in pressure, the change in volume is greater than in the bubble-free liquid: In simple terms, the bubbles expand significantly in response to a pressure reduction, and are contracted more than the surrounding liquid during compression, though since the bubbly liquid is being treated as a homogeneous continuum in this model, consideration of the response of individual bubbles should not be taken too far. Nevertheless it is convenient to write the bulk modulus of the bubbly liquid (B_c) in terms of the bulk modulus of the bubble-free liquid (B_w), which is due to the change in volume of the bubble-free liquid in response to pressure change, plus a term B_{bub}, which is due to the pressure-mediated volume change of the entire bubble population:

$$B_c = B_w + B_{bub} \tag{4.37}$$

where

$$c_w = \sqrt{\frac{B_w}{\rho_w}} \tag{4.38}$$

is the speed of sound in the bubble-free liquid, which has density ρ_w. To calculate the change in the compressibility, we simply have to consider how one bubble responds to a pressure change, and integrate up for all bubbles. To a first approximation, consider that the bubble population contains n_b identical bubbles per unit volume, where each bubble has a volume $V_1(t)$ which can change in time in response to a variation in the applied pressure. When the whole population is affected by such a change, the resulting response is characterised by

$$B_{bub} = -\frac{1}{n_b \Delta V_1} \Delta p \tag{4.39}$$

Since we have assumed small-amplitude linear oscillations throughout this derivation, then $R_{\varepsilon o} \ll R_o$. Therefore the bubble volume $V_1(t) = 4\pi R(t)^3/3 = 4\pi R_o^3(1+R_\varepsilon/R_o)^3/3$ may be approximated to first order as

$$V_1(t) \approx V_o(1 + 3R_\varepsilon/R_o) \tag{4.40}$$

where $V_o = 4\pi R_o^3/3$, so that

$$\Delta V_1(t) \approx \frac{3V_o R_\varepsilon}{R_o} \tag{4.41}$$

Substituting this into equation (4.39), the bubble-mediated bulk modulus is therefore

$$B_{bub} = -\frac{R_o \Delta p}{3 n_b V_o R_\varepsilon} \tag{4.42}$$

The value of R_ε may be found through the methods of section 4.1.2(c). Comparing equations (4.7) and (4.20), since the particular case for the bubble of ε is R_ε, and of F_o is $(-P_A 4\pi R_o^2)$ and of m is $m_{RF}^{rad} = 4\pi R_o^3 \rho_w$, equation (4.9) becomes

$$R_\varepsilon = -\frac{P_A e^{i\omega t}}{R_o \rho_w} \frac{1}{((\omega_o^2 - \omega^2) + i2\beta_{tot}\omega)} \tag{4.43}$$

Noting that $\Delta p = -P_A e^{i\omega t}$, substitution of equation (4.43) into equation (4.42) yields

$$B_{bub} \approx \frac{R_o^2 \rho_w((\omega_o^2 - \omega^2) + i2\beta\omega)}{3 n_b V_o} \tag{4.44}$$

The nature of the dissipation will be discussed later. This, B_{bub}, is a complex entity, and so the wavenumber of sound in a bubbly liquid will also be complex:

$$k_c^{comp} = \frac{\omega}{c_c} = \omega\sqrt{\frac{\rho}{B_c}} \tag{4.45}$$

Substituting from equations (4.37) and (4.38), with the approximation $\rho_c \approx \rho_w$, yields

$$k_c^{comp} = \omega\sqrt{\frac{\rho_c}{B_w + B_{bub}}} \approx \omega\sqrt{\frac{\rho_w}{B_w + B_{bub}}} = \frac{\omega}{c_w}\sqrt{1 + \frac{c_w^2 \rho_w}{B_{bub}}} \tag{4.46}$$

Substitution for $V_o = 4\pi R_o^3/3$, and for B_{bub} from equation (4.44) gives

$$k_c^{comp} \approx \frac{\omega}{c_w}\sqrt{1 + \frac{4\pi n_b c_w^2 R_o}{((\omega_o^2 - \omega^2) + i2\beta\omega)}} \tag{4.47}$$

Clearly, k_c^{comp} contains a real part, which gives the effective wavenumber of propagation, and an imaginary part which gives the attenuation.[2] Though viscous and thermal dissipation will subtract energy from the wave and convert it into heat, radiation losses correspond to the scattering of sound by bubbles, which does not subtract energy from the wave as a whole but merely converts it from coherent to incoherent radiation [3]. It is interesting to examine the limiting forms.

If the unity term is negligible in comparison with the other, then equation (4.47) reduces to

$$k_c^{comp} = \sqrt{4\pi\omega^2 n_b R_o}\sqrt{\frac{\omega_o^2 - \omega^2 - 2i\beta\omega}{((\omega_o^2 - \omega^2)^2 + 4\beta^2\omega^2)}} \tag{4.48}$$

If the propagating frequencies are very low then $\omega \ll \omega_o$ and equation (4.48) gives the wavenumber of the propagation

$$k_c = \mathrm{Re}\{k_c^{comp}\} = \sqrt{\frac{4\pi\omega^2 n_b R_o}{\omega_o^2}} \tag{4.49}$$

The speed of sound in the cloud in this limit of low acoustic frequencies and high number densities is therefore

$$c_c = \frac{\omega}{k_c} \approx \sqrt{\frac{\omega_o^2}{4\pi n_b R_o}} \tag{4.50}$$

If the number density is low so that the fractional term is very much less than unity, equation (4.47) can be expanded binomially giving, to first order

$$\begin{aligned} k_c^{comp} &= \left(\frac{\omega}{c}\right)\left(1 + \frac{2\pi c^2 n_b R_o}{\omega_o^2 - \omega^2 + 2i\beta\omega}\right) \\ &= \left(\frac{\omega}{c}\right)\left(1 + (2\pi c^2 n_b R_o)\frac{\omega_o^2 - \omega^2 - 2i\beta\omega}{(\omega_o^2 - \omega^2)^2 + 4\beta^2\omega^2}\right) \end{aligned} \tag{4.51}$$

The propagation wavenumber is therefore

[2] See Chapter 1, section 1.1.7.

$$k_c = \mathrm{Re}\{k_c^{\mathrm{comp}}\} = \left(\frac{\omega}{c}\right)\left\{1 + (2\pi c^2 n_b R_o)\left(\frac{\omega_o^2 - \omega^2}{(\omega_o^2 - \omega^2)^2 + 4\beta^2\omega^2}\right)\right\} \tag{4.52}$$

and the sound speed by

$$c_c = \frac{\omega}{k_c} \approx c\left\{1 - (2\pi c^2 n_b R_o)\left(\frac{\omega_o^2 - \omega^2}{(\omega_o^2 - \omega^2)^2 + 4\beta^2\omega^2}\right)\right\} \tag{4.53}$$

If there is a distribution of bubble sizes within the cloud, such that $n_b^{gr}(z,R_o)dR_o$ is the number of bubbles per unit volume at depth z having radii between R_o and $R_o + dR_o$, the speed of sound is a function of both the depth and the acoustic frequency [4]:

$$c_c(z,\omega) = c\left\{1 - (2\pi c^2)\int_{R_o=0}^{\infty} \frac{R_o}{\omega^2}\left(\frac{(\omega_o/\omega)^2 - 1}{\{(\omega_o/\omega)^2 - 1\}^2 + d^2}\right) n_b^{gr}(z,R_o)\, dR_o\right\} \tag{4.54}$$

Limiting forms derived from equations (4.50) and (4.53) are given in section 3.8.2(b)(i).

The discussion so far in this chapter has been applicable to linear oscillators. By its very nature, the bubble would be expected to oscillate nonlinearly at finite amplitudes: expansion and compression are not symmetrical, and whilst the bubble could expand without limit, upon compression the radial displacement cannot exceed the size of the equilibrium bubble radius. The real bubble may potentially differ from the ideal linear oscillator discussed here in a second way, in that as the oscillation proceeds, the system itself is generally changing. The equilibrium bubble radius may decrease as the bubble dissolves, or increase through rectified diffusion (see section 4.4.3). As seen in Chapters 2 and 3, the bubble is often not spherical (more sophisticated treatments have, for example, discussed the acoustic scattering cross-section of a bubble undergoing shape oscillations [2]). The oscillator may change in extremely dramatic ways: the bubble may, for example, fragment. The remainder of this chapter discusses observations of real bubbles, and some of the models for their behaviour, which improve upon that of a sphere undergoing linear harmonic oscillations by incorporating to a greater or lesser extent the features mentioned above.

4.2 Nonlinear Equations of Motion for the Spherical Pulsating Bubble

In the spring–bob illustration of a *linear* oscillator[3] it was shown that the system response (the displacement of the bob) is directly proportional to the driving force (equation (3.2)). One of the simplest examples of a *nonlinear* system can be expressed by stating that the system response is related to the driving force through a power series expansion:

$$¥(t) = s_0 + s_1.\hbar(t) + s_2.\hbar^2(t) + s_3.\hbar^3(t) + s_4.\hbar^4(t)... \tag{4.55}$$

where ¥ is the general response of the oscillator (for example, it might represent the displacement) and $\hbar$ represents the driving force. The parameters s_0, s_1, s_2 etc. are coefficients. If the

[3] See Chapter 3, section 3.1, and also section 4.1.

oscillator were linear, s_2, s_3 and higher coefficients would equal zero, and the oscillator would behave in the simple manner described in section 4.1.

The bubble is a nonlinear oscillator. By examining the physics of the oscillation process, it is possible to obtain relationships between the driving acoustic pressure and the response of the bubble, and so examine the nature of the nonlinearity. Though such formulations are always subject to simplifying approximations, they are often quite complicated expressions. One of the simplest is described in the next section.

4.2.1 The Rayleigh–Plesset Equation

(a) A Simple Equation of Motion for the Spherical Pulsating Bubble

In order to examine the dynamics of the bubble at finite amplitudes, we will now develop an approximate, nonlinear equation of motion. This equation describes the response of a spherical bubble to a time-varying pressure field in an incompressible liquid. At a time $t < 0$, a bubble of radius R_0 is at rest in an incompressible, viscous liquid. The hydrostatic pressure is p_0, a constant. At time $t > 0$, a pressure $P(t)$ which varies with time is superimposed on p_0, so that the liquid pressure at a point remote from the bubble is $p_\infty = p_0 + P(t)$. Consequently the bubble radius will change to some new value $R(t)$. During this process, the liquid will acquire a kinetic energy of

$$\frac{1}{2}\rho\int_R^\infty \dot{r}^2 4\pi r^2 \mathrm{d}r \tag{4.56}$$

which (using the liquid incompressibility condition $\dot{r}/\dot{R} = R^2/r^2$ given in equation (2.25)) can be integrated to give $2\pi\rho R^3\dot{R}^2$. Equating this to the difference between the work done remote from the bubble by p_∞ and the work done by the pressure p_L in the liquid just outside the bubble wall gives

$$\int_{R_0}^{R} (p_L - p_\infty)\, 4\pi R^2 \mathrm{d}R = 2\pi R^3 \dot{R}^2 \rho \tag{4.57}$$

which is just a re-expression of equation (2.29). Differentiation of equation (4.57) with respect to R, noting that

$$\frac{\partial(\dot{R}^2)}{\partial R} = \frac{1}{\dot{R}}\frac{\partial(\dot{R}^2)}{\partial t} = 2\ddot{R} \tag{4.58}$$

gives

$$\frac{p_L - p_\infty}{\rho} = \frac{3\dot{R}^2}{2} + R\ddot{R} \tag{4.59}$$

The liquid pressure p_L for a pulsating bubble containing gas and vapour was found in Chapter 2, section 2.1.3(a), though in the discussion of the response of a bubble to a quasi-static pressure change. Bearing in mind this restriction, equation (2.18) can be substituted into equation (4.39), and p_∞ expanded to $p_0 + P(t)$, to give

$$R\ddot{R} + \frac{3\dot{R}^2}{2} = \frac{1}{\rho}\left\{\left(p_0 + \frac{2\sigma}{R_0} - p_v\right)\left(\frac{R_0}{R}\right)^{3\kappa} + p_v - \frac{2\sigma}{R} - p_0 - P(t)\right\} \tag{4.60}$$

This initial treatment can be improved to incorporate the effects of viscosity. To achieve this, it is simplest to start by consideration of the viscous equation of motion and the boundary conditions, and substitute the results into Bernoulli's equation. This yields the same equation of motion as equation (4.60), but with viscous effects included.

(b) The Effect of Viscosity

The effect of viscosity on the equation of motion of a bubble in an incompressible fluid was considered by Poritsky [5]. The inclusion initially presents a paradox. The Eulerian equation of motion is given by equation (2.133)

$$\rho\vec{a} = \rho\Sigma\vec{F}_{ext} - \vec{\nabla}p' + \frac{\eta}{3}\vec{\nabla}(\vec{\nabla}.\vec{v}) + \eta\nabla^2\vec{v} \tag{4.61}$$

where η is the shear viscosity, $\Sigma\vec{F}_{ext}$ is the sum of the external forces (volume and surface) per unit mass of fluid, p' is the pressure existing at some boundary within the liquid, and the fluid acceleration $\vec{a}$ is given[4] by equation (2.60):

$$\vec{a} = \frac{D\vec{v}}{Dt} = \left(\frac{\partial}{\partial t} + \vec{v}.\vec{\nabla}\right)\vec{v}$$

If the liquid is incompressible, then $\vec{\nabla}.\vec{v} = 0$ (equation (2.81)) and $r^2\dot{r} = R^2\dot{R}$ (equation (2.25)). This implies that in the spherically symmetric case considered here

$$\vec{v} = \dot{r} = (R^2\dot{R})\frac{1}{r^2} \tag{4.62}$$

Thus $\vec{v}$, which is in the radial direction only, is proportional to r^{-2}, the constant of proportionality being a function of time only. Therefore in conditions of potential flow, $\vec{v}$ is the gradient of a velocity potential Φ which is proportional to r^{-1},

$$\Phi = -(R^2\dot{R})\frac{1}{r} \tag{4.63}$$

Since $\nabla^2\Phi = 0$, then

$$\nabla^2\vec{v} = \nabla^2(\vec{\nabla}\Phi) = \vec{\nabla}(\nabla^2\Phi) = 0 \tag{4.64}$$

Substitution of $\vec{\nabla}.\vec{v}$ and equation (4.64) into equation (4.61) gives

$$\rho\vec{a} = \rho\Sigma\vec{F}_{ext} - \vec{\nabla}p' \tag{4.65}$$

---

[4]See Chapter 2, section 2.2.2 and also Chapter 3, section 3.4.2(d).

showing that all the terms, including the viscosity, are zero (see Chapter 3, section 3.4.2(d)). This demonstrates the paradox, that apparently no difference exists between the equations governing the pulsation of a spherical bubble in a viscous fluid, and those in a nonviscous fluid. In fact, this result implies that the resultant net effect of viscosity stresses per unit volume at any point internal to the fluid is zero, but the stresses themselves may be nonzero. Viscous effects therefore manifest themselves through the boundary condition, and not through the Navier–Stokes equation, thus resolving the paradox.

The three principal stresses, taken to be in the spherical coordinate directions, are p_r, p_θ and p_φ, and are related through the standard equations [6]

$$p_r = -p' - \frac{2\eta}{3}(\varepsilon_r' + \varepsilon_\theta' + \varepsilon_\varphi') + 2\eta\varepsilon_r' \tag{4.66}$$

$$p_\theta = -p' - \frac{2\eta}{3}(\varepsilon_r' + \varepsilon_\theta' + \varepsilon_\varphi') + 2\eta\varepsilon_\theta' \tag{4.67}$$

$$p_\varphi = -p' - \frac{2\eta}{3}(\varepsilon_r' + \varepsilon_\theta' + \varepsilon_\varphi') + 2\eta\varepsilon_\varphi' \tag{4.68}$$

where ε_r', ε_θ' and ε_φ' are the three principal rates of strain, and where the pressure p', existing at a boundary within the fluid, is given by

$$p' = -\tfrac{1}{3}(p_r + p_\theta + p_\varphi) \tag{4.69}$$

Consider the bubble. As a result of the spherical symmetry, the θ and φ directions are equivalent, and since the fluid is incompressible

$$\varepsilon_r' + \varepsilon_\theta' + \varepsilon_\varphi' = 0 \tag{4.70}$$

Substitution of these simplifications reduces equations (4.66), (4.67) and (4.68) to

$$p_r = -p' + 2\eta\varepsilon_r' \tag{4.71}$$

and

$$p_\theta = p_\varphi = -p' - \eta\varepsilon_r' \tag{4.72}$$

The pressure within the bubble balances the liquid pressure at the bubble wall, p_L. Since therefore $p_L = -p_r$, then equations (4.69), (4.71) and (4.72) yield

$$p_L = p' - 2\eta\varepsilon_r' \tag{4.73}$$

The principal rate of strain ε_r' is simply $\partial v/\partial r$, which from equation (4.62) gives

$$\varepsilon_r' = \frac{\partial v}{\partial r} = -2(R^2\dot{R})\frac{1}{r^3} \tag{4.74}$$

so that equation (4.73) becomes

$$p_L = p' + 4\eta(R^2\dot{R})\frac{1}{r^3} \tag{4.75}$$

Within the fluid, the Navier–Stokes equation reduces to equation (4.65). Assuming $\sum\vec{F}_{ext} = 0$, and noting the spherical symmetry within the fluid, equation (4.65) reduces to

$$-\frac{1}{\rho}\frac{\partial p'}{\partial r} = \left(\frac{\partial}{\partial t} + v\frac{\partial}{\partial r}\right)v \tag{4.76}$$

Integrating this with respect to r, noting that $\vec{v} = \vec{\nabla}\Phi$, gives a result identical in form to Bernoulli's equation (equation (2.73)):

$$\frac{p'}{\rho} = \frac{p_\infty}{\rho} - \frac{\partial \Phi}{\partial t} - \frac{v^2}{2} \tag{4.77}$$

where the constant of integration is found by noting that at $r = \infty$ the liquid velocity v is zero and the liquid pressure is p_∞. Substitution for the velocity and velocity potential from equations (4.62) and (4.63) respectively gives

$$\frac{p'}{\rho} = \frac{p_\infty}{\rho} + \frac{1}{r}\frac{\partial(R^2\dot{R})}{\partial t} - \frac{R^4\dot{R}^2}{2r^4} \tag{4.78}$$

This can then be evaluated at the bubble wall to give

$$\frac{p'}{\rho} = \frac{p_\infty}{\rho} + R\ddot{R} + \tfrac{3}{2}\dot{R}^2 \tag{4.79}$$

Eliminating p' from equations (4.75) and (4.79) at $r = R$ yields

$$\frac{1}{\rho}\left(p_L - p_\infty - \frac{4\eta\dot{R}}{R}\right) = R\ddot{R} + \tfrac{3}{2}\dot{R}^2 \tag{4.80}$$

Substitution from equation (2.18), and writing $p_\infty = p_0 + P(t)$, gives

$$R\ddot{R} + \frac{3\dot{R}^2}{2} = \frac{1}{\rho}\left\{\left(p_0 + \frac{2\sigma}{R_0} - p_v\right)\left(\frac{R_0}{R}\right)^{3\kappa} + p_v - \frac{2\sigma}{R} - \frac{4\eta\dot{R}}{R} - p_0 - P(t)\right\} \tag{4.81}$$

Lauterborn [7] suggested that this equation be referred to as the RPNNP equation in tribute to the workers who contributed to its formulation: Rayleigh [8], Plesset [9], Noltingk and Neppiras [10], Neppiras and Noltingk [11] and Poritsky [5], though nowadays it is more commonly called the 'Rayleigh–Plesset' equation.

The Rayleigh–Plesset equation is nonlinear, as one would expect from a system where, except at low amplitude, the expansion cannot be symmetric with the compression. However, just as the bubble itself approximates to a linear resonator at low amplitudes of oscillation, so should the equation of motion. This is now demonstrated.

(c) Small-amplitude Behaviour

Assume the time-varying pressure has the form of a sinusoidal sound wave of amplitude P_A and circular frequency ω, i.e. $P(t) = -P_A e^{i\omega t}$. Assume also that vapour pressure and viscosity are negligible, and there are small-amplitude variations in bubble radius about the limiting value (so that $R(t) = R_o + R_\varepsilon(t)$, as discussed in section 4.1.2(c)). If $R_\varepsilon \ll R_o$, then in an expansion to first order in powers of R_o^{-1}, equation (4.81) becomes

$$\ddot{R}_\varepsilon + \omega_0^2 R_\varepsilon = \frac{P_A}{\rho R_o} e^{i\omega t} \tag{4.82}$$

where ω_0 is the resonant frequency, and is given by

$$\omega_0^2 = \frac{1}{\rho R_o^2}\left\{3\kappa\left(p_0 + \frac{2\sigma}{R_o}\right) - \frac{2\sigma}{R_o}\right\} \tag{4.83}$$

If surface tension terms are negligible, this reduces to

$$\omega_0 = \frac{1}{R_o}\sqrt{\frac{3\kappa p_0}{\rho}} \tag{4.84}$$

in agreement with equation (3.36), which was derived through consideration of a bubble undergoing small-amplitude, simple harmonic pulsations. At the other extreme, where surface tension dominates (e.g. for very small bubbles), the resonant frequency is

$$\omega_0^2 = \frac{2\sigma}{\rho R_o^3}(3\kappa - 1) \tag{4.85}$$

The Rayleigh–Plesset equation (equation (4.81)) gives the direct small-amplitude approximation to the resonant frequency of a bubble as [12]

$$\omega_0 = \frac{1}{R_o\sqrt{\rho}}\sqrt{\left\{3\kappa\left(p_0 + \frac{2\sigma}{R_o} - p_v\right) - \frac{2\sigma}{R_o} + p_v - \frac{4\eta^2}{\rho R_o^2}\right\}} \tag{4.86}$$

Examination of the Rayleigh–Plesset equation therefore demonstrates how, in the small-amplitude limit, the bubble can tend to behave as a forced linear oscillator.

4.2.2 Other Equations of Motion

Certain assumptions are inherent in the derivation of the Rayleigh–Plesset equation [12]. These are that: (a) there is a single bubble in an infinite medium, (b) at all times the bubble remains spherical, (c) spatially uniform conditions exist within the bubble, (d) the bubble radius is small compared with the acoustic wavelength, (e) no body forces (e.g. gravitational) are present, (f) bulk viscous effects can be ignored, (g) the density of the liquid is large, and its compressibility is infinitesimally small, compared with the values for the gas within the bubble, (h) the gas content of the bubble is constant and (i) the vapour pressure is constant during the oscillation.

The approximation which is perhaps most easily improved upon is that the liquid is incompressible. As a consequence of the density of the liquid being assumed to be constant, the speed of sound becomes infinite.[5] The approximation limits applicability of the Rayleigh–Plesset equation.

---

[5]See Chapter 2, section 2.1.3(b)(i).

There are three other common approximations which take into account, to a greater or lesser extent, the compressibility of the liquid [13]. They are here listed in order of increasing sophistication. The *acoustic approximation* assumes the speed of sound to be a finite constant equal to $\sqrt{\partial p/\partial \rho}$, where p/ρ is a constant. A finite, constant speed of sound in the liquid is also assumed in the *Herring approximation*. However, the latter takes fuller account of the energy stored through compression of the liquid. As a result, whilst the acoustic approximation is accurate in predicting stable cavitation, it is limited to the descriptions of only sound radiation dissipation during transient cavitation. The incompressible and acoustic approximations therefore in essence relate to incompressible fluids. The Herring approximation accurately describes transient collapse at the lower expansion ratios (R_{max} being only a few times greater than R_0). Higher expansion ratios require the *Kirkwood–Bethe* approximation, which assumes the speed of sound is a function of the motion. The Kirkwood–Bethe Hypothesis [14] states that for spherical waves of finite amplitude, the quantity $r\Phi$ (Φ is the velocity potential in the liquid, and r the radial coordinate) propagates with a velocity equal to the sum of the fluid velocity and the local velocity of sound. Cavitation in compressible fluids requires either the use of the Herring or the Kirkwood–Bethe approximation.

The Herring–Trilling equation is given by

$$R\dot{R}\frac{\partial \dot{R}}{\partial R}(1-2M)+\frac{3\dot{R}^2}{2}(1-4M)=\frac{R\dot{R}}{\rho_0 c}\frac{\mathrm{d}p_L}{\mathrm{d}R}+\frac{p_\infty - p_L}{\rho_0} \tag{4.87}$$

where M is the acoustic Mach number ($M = \dot{R}/c$). Note that for $\dot{R} \ll c$, we revert to the incompressible formulations. For example, the collapse of an empty cavity, neglecting surface tension, setting $p_L = 0$ reduces equation (4.87) to

$$R\dot{R}\frac{\partial \dot{R}}{\partial R}(1-2M)+\frac{3\dot{R}^2}{2}(1-4M)=\frac{p_\infty}{\rho_0} \tag{4.88}$$

which, as $M \to 0$, gives Rayleigh's formulation for the collapse of an empty cavity. This suggests why the incompressible approximation is unsuitable for the description of violent transient collapse, when the speed of the bubble wall may not be insignificant in comparison with the speed of sound in the liquid.

It is therefore interesting to express Gilmore's equation [15] in terms of $\dot{R}$, the velocity of the bubble wall:

$$\left(1-\frac{\dot{R}}{c_L}\right)\frac{\mathrm{d}\dot{R}}{\mathrm{d}t}+\frac{3}{2}\frac{\dot{R}^2}{R}\left(1-\frac{\dot{R}}{3c_L}\right)=\left(1+\frac{\dot{R}}{c_L}\right)\frac{H}{R}+\frac{1}{c_L}\frac{\mathrm{d}H}{\mathrm{d}t}\left(1-\frac{\dot{R}}{c_L}\right) \tag{4.89}$$

where H is the difference in the liquid enthalpy between the bubble wall and infinity, and $c_L(t)$ is the time-dependent speed of sound in the liquid at the bubble wall. Both H and c_L are functions of $p_L(t)$.

Keller and Miksis [16] produced a radial equation suitable for describing large-amplitude forced oscillations, incorporating the effects of acoustic radiation by the bubble and using the approximation of a linear polytropic index. This was modified by Prosperetti *et al.* [17] to incorporate a more exact formulation for the internal pressure. These two formulations, and that of Flynn [18], are compared with experimental measurements by Gaitan *et al.* [19].[6] All three

[6] See Chapter 5, section 5.2.3.

analyses assume that the internal pressure remains uniform throughout the bubble, and that the bubble contents obey the ideal gas law. Church [20] considers the effect of using a van der Waals equation of state for the gas.

In general, it is not possible to obtain full analytical solutions to these equations of motion. However, in later sections it will be useful to discuss their small-amplitude limiting forms in order to generate equations of motion which approximate to the general equation of motion of the forced damped oscillator. In this way we can produce approximate analytical solutions, which enable trends of behaviour to be discerned.

It was through examination of full numerical solutions of the Rayleigh–Plesset and other equations that led to the suggestion that it might be possible to classify the response of bubbles to an acoustic driver into two broad types: *stable* and *transient*. This concept will now be discussed.

4.2.3 Numerical Solutions of the Rayleigh–Plesset Equation: Stable and Unstable Cavitation

Figures 4.6 to 4.8 show numerical solutions to the Rayleigh–Plesset equation for the starting conditions that at $t = 0$, the bubble has equilibrium radius and is motionless ($R = R_0$ and $\dot{R} = 0$). The solutions are for air bubbles in water of temperature 20° C. Though, as shown above, at low amplitude a bubble will respond in a manner similar to a linear damped oscillator, when driven at higher amplitudes, nonlinear effects are apparent. Figure 4.6 shows an air bubble having $R_0 = 2$ mm driven by a 10-kHz sound field. Though the acoustic pressure amplitude is high (0.27 MPa), the bubble is much larger than resonance size, and the response is of low amplitude. There are two main frequencies present in the radial response. One is a slower oscillation (period 0.6 ms) associated with the bubble resonance frequency of 1.7 kHz (equation (3.39)). The other, more rapid oscillation, has a period of 0.1 ms, and is therefore associated with the 10-kHz driving frequency. On the same time axis, the wall velocity is shown. Also shown are the pressure and temperature of the gas within the bubble, though it should be emphasised that the latter two are for indication only. They are simply derived from a calculation that assumes spatially uniform adiabatic conditions within the bubble i.e. *no heat flow across the bubble wall*.

Figure 4.7 shows the radius/time plot for a bubble of $R_0 = 0.10$ mm, that is just less than resonance size, in a 10-kHz sound field, where $P_A = 0.24$ MPa. The oscillation is high amplitude and nonlinear. The transient response can be seen, and the rapid oscillations at the start, which occur in the first 0.04 ms, have a period of around 0.014 ms and so an associated frequency of around 70 kHz, which is twice the resonance frequency of the bubble (equation (3.39)). The steady-state response can be seen in the periodic unit that occurs between the times 0.43 ms and 0.87 ms: the interval of 0.44 ms is associated with a frequency of 2.3 kHz. The driving frequency would be manifest as a period of around 0.1 ms, which is roughly that of the feature that occurs twice between 0.68 ms and 0.88 ms. Thus the response is nonlinear, but after the transients have died out, the solution becomes periodic. The form of oscillation shown in Figures 4.6 and 4.7 is known as *stable cavitation*.

If the bubble were simply a linear oscillator, we would expect that smaller bubbles in this sound field would respond with lower amplitude, as they are further from resonance. However, the effect of decreasing the equilibrium bubble size in this sound field is shown in Figure 4.8, which gives the Rayleigh–Plesset radius–time predictions for four bubbles of progressively smaller size ($R_0 = 60$ μm, 50 μm, 10 μm and 1 μm) in this 10-kHz sound field, where $P_A =$ 0.24 MPa. The response is dramatically different from that shown in the previous figures. The bubble expands to a maximum size, and then rapidly collapses. The smaller the bubble, the

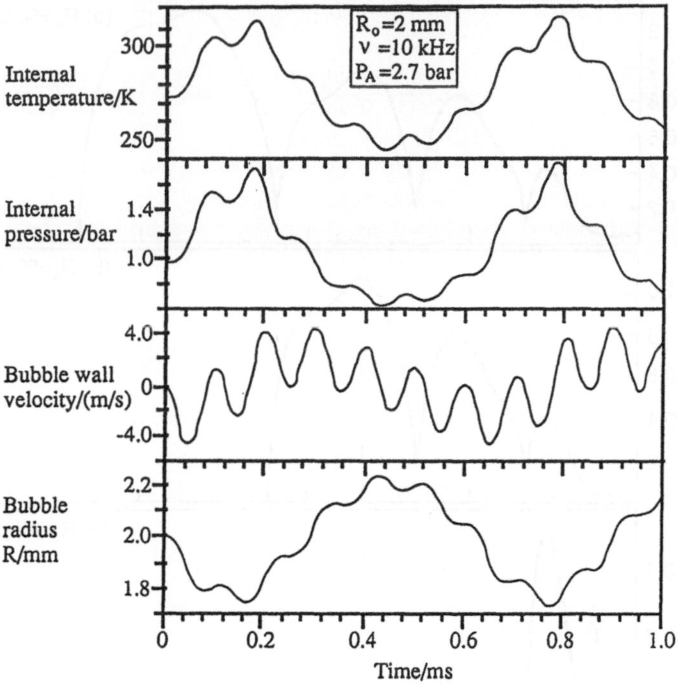

Figure 4.6 An air bubble of equilibrium radius 2 mm in water of temperature 20°C and under one atmosphere static pressure, is subjected to a 10-kHz sound field, of pressure amplitude 2.7 bar. Insonation begins at $t = 0$. Bubble radius and radial wall speed as calculated from the Rayleigh–Plesset equation are plotted against a common time axis, as are the pressure and temperature of the gas (adiabatic gas behaviour and constant vapour pressure are assumed).

fewer oscillations it undertakes before undergoing this violent collapse. Such collapses are high-energy events, and can generate a wide range of potentially destructive effects (erosion, cell disruption, sonoluminescence etc.). This type of event is known as *unstable cavitation.* A more common name is *transient cavitation.*[7]

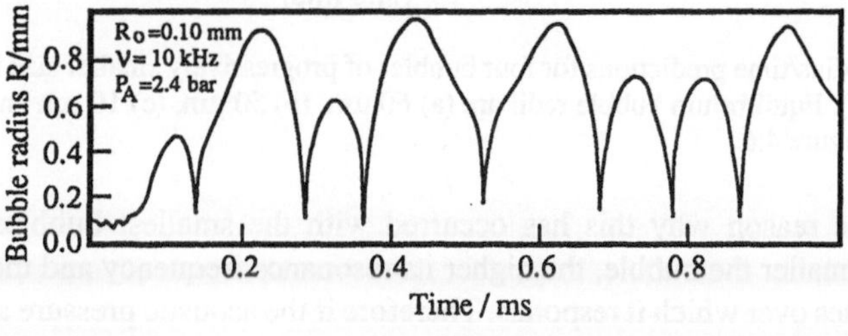

Figure 4.7 The radius/time plot of an air bubble in water ($R_0 = 0.10$ mm) in a 10-kHz sound field, where $P_A = 2.4$ bar. Other conditions as for Figure 4.6.

[7]Such cavities are 'transient' in that the cavities have a finite lifetime. The term should not be confused with the transients that distinguish the initial behaviour of an oscillator from the steady state, as discussed earlier in this chapter.

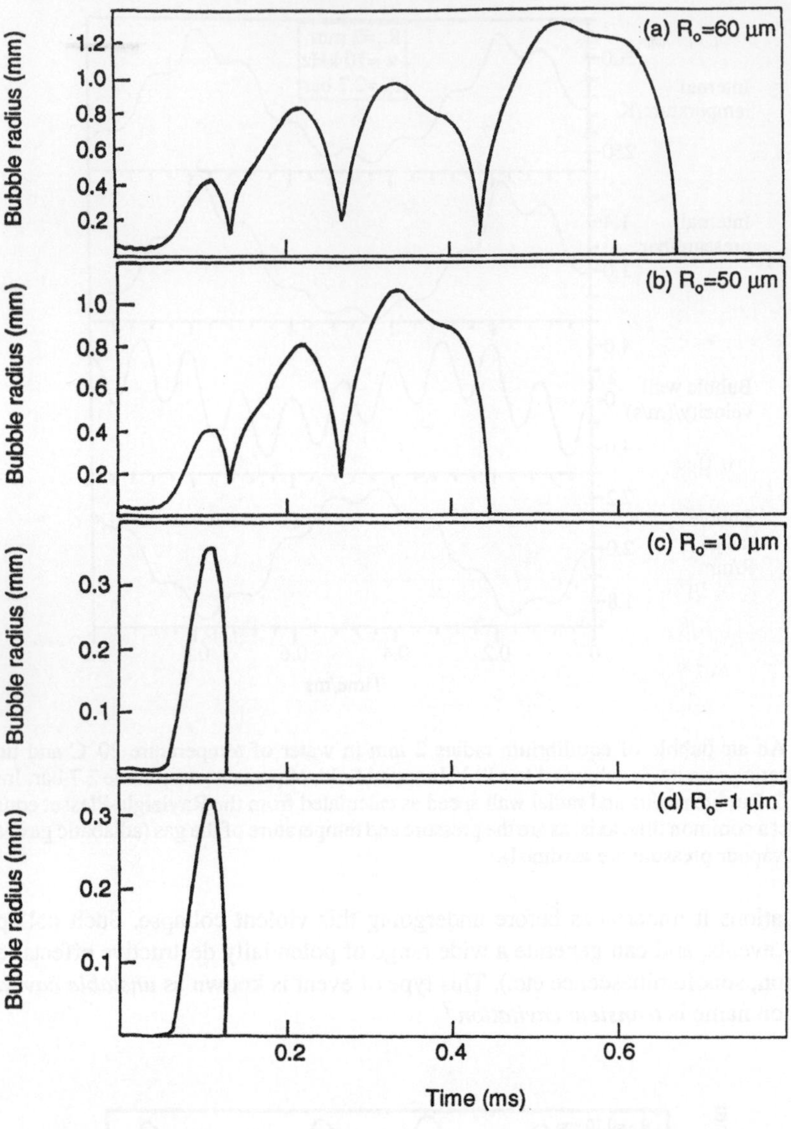

Figure 4.8 The radius/time predictions for four bubbles of progressively smaller size in a 10-kHz sound field ($P_A = 2.4$ bar). Equilibrium bubble radii are (a) 60 μm, (b) 50 μm, (c) 10 μm and (d) 1 μm. Other conditions as for Figure 4.6.

Physically, the reason why this has occurred with the smallest bubbles is not hard to appreciate. The smaller the bubble, the higher its resonance frequency and therefore the more rapid the timescales over which it responds. Therefore if the acoustic pressure amplitude is high enough to put the liquid into tension, from the point of view of a very small bubble there is ample time to grow whilst the tension is maintained, before the phase of the sound field turns compressional. Thus the bubble can grow very large, in a manner similar to that discussed in Chapter 2, section 2.1.3(a). By the time the sound wave does exert a compressional influence, there is a great deal of 'elastic' energy stored within the expanded bubble, so that its collapse during the compressional phase of the sound field is very violent. Larger bubbles have a lower resonance frequency, and therefore a slower response time, and do not have time to grow as much during a single tension cycle: it may take several cycles before the bubble reaches a large stationary maximum simultaneously with the onset of a compressional force, if at all. At the other extreme, in the very smallest bubbles the surface tension forces inhibit growth.[8] Therefore one might expect a range of bubbles of intermediate size to undergo transient cavitation, the remaining bubbles being too large or too small to undergo anything other than stable cavitation.

The discovery that a small change in certain parameters (P_A, R_0, ω) could bring about a dramatic change in the radius–time curve of solutions to the Rayleigh–Plesset equation was first observed in the early 1950s by Noltingk and Neppiras [10, 11]. They also observed that, during the final stages of violent collapse, the speed of the bubble wall could exceed that of sound in the liquid, thereby invalidating their incompressible theory. They nevertheless suggested that even when the wall velocities were supersonic, deductions from their theory could still give an indication of the relative importance of the bubble and acoustic parameters in estimating the intensity of cavitation.

Flynn [13], recognising the need to introduce simple models to describe the extremes of bubble behaviour, proposed in 1964 the terms 'transient' and 'stable'. The former "grow to some maximum size under the influence of a sound field and ... [then] ... collapse violently," whilst the latter "pulsate over relatively long intervals of time." It is important to note that this distinction in the first instance reflects the violence of the collapse above all else. Even so, a temporal component has been introduced, and the use of the word 'transient', with its temporal connotations, is perhaps unfortunate. One should remember that the two classes of cavitation, stable and transient, are simple models for extreme forms of behaviour: a plethora of intermediate forms, some seeming to fall into both classes and some into neither, exist. From the original distinction, much work has been undertaken to define transient cavitation rigorously, partly to put some order to the intermediate cases (generally by the experimentalists), and partly to enable the threshold to be defined (generally by the theorists).

The concept of the threshold is an attractive one. As can be seen from Figures 4.7 and 4.8, in that particular sound field bubbles of large equilibrium radius undergo stable cavitation, whilst the smaller bubbles shown undergo transient collapse. There must exist some *threshold* equilibrium radius for a bubble in this sound field that denotes the transition from stable to transient behaviour. Obviously there must also be a similar threshold in acoustic pressure, since we have already seen that in low-amplitude pressure fields the bubble approximates to a linear oscillator, and so must undergo low-amplitude stable cavitation. This must also apply to the bubbles of the radii used in Figure 4.8 ($R_0 = 60$ μm, 50 μm, 10 μm and 1 μm) in a 10-kHz sound field when the acoustic pressure amplitude is small: although they are transient in the figure, when P_A = 0.24 MPa, they must undergo stable cavitation at very low driving pressures. Therefore for these bubbles there must exist some *threshold* acoustic pressure that marks their transition from transient to stable cavitation.[9]

However, it is worth saying at the outset that the search for a rigorous definition of transient cavitation has engendered some confusion which still exists today. Though perhaps the best work has been done by approaching the problem from a thermodynamic point of view (a consideration based on the 'violence' of the collapse), considerations of the temporal aspects

[8] See Chapter 2, section 2.1.3(a).

[9] This simple illustration of the threshold acoustic pressure for transient cavitation does not imply that a threshold pressure always exists. That a bubble which would be transient could, if subjected to a reduced acoustic pressure amplitude, undergo stable cavitation does not prove the inverse, that stable cavities will go transient in a more powerful field.

(or life-cycle) of the bubble are appealing in their immediacy: it is simpler to understand a radius–time curve than a formulation of how energy concentrates within the bubble. Indeed the two overlap when, from the thermodynamic definitions, daughter definitions which are one step further from the basis but easier to understand, have been given in terms of the radii attained during the life-cycle of a bubble.

In the following sections the main features of transient and stable cavitation, and the concept of thresholds, will be explored. It should be borne in mind that these are extreme and idealised forms of the behaviour, and discussion of the intermediate forms that can be found experimentally will be included.

4.3 Transient Cavitation

4.3.1 Transient Cavitation Threshold

There is clearly a fundamental difference between the concepts of stable and transient cavitation. If there is a clear transition from the one type of cavitation to the other, it occurs at the transient cavitation threshold. In section 4.2.3 it was shown how, if all other parameters remain the same, a variation in equilibrium bubble radius could result in a transition across that threshold. Similarly, the discussion indicated how a change in the acoustic pressure could, under certain circumstances, also bring about transition across the threshold, all other parameters remaining the same. If one does assume that a sharp distinction exists, and defines regimes of stable and transient cavitation, variation in other parameters, such as acoustic frequency, might also result in the transition.

The details of transient collapse are similarly dependent on these parameters. Figure 4.9 [12,13,21] shows the effect of increasing the pressure amplitude on bubbles in water undergoing transient collapse. All the bubbles show the characteristic initial explosive growth, and the final rapid collapse. For a bubble of this or any size, as the pressure is increased there is a transition from the situation of growth to a single maximum prior to collapse, to one in which the bubble grows to two maxima before collapse. The reason is that, as higher acoustic pressure amplitudes are employed, the bubble is initially subjected to greater tensions during the rarefaction phase of the sound wave, and so the initial growth occurs to a greater extent. The bubble can attain a maximum radius so large that it does not have time to collapse during the compression phase of the sound cycle. Before the collapse is complete, it is subjected to another tension, which halts the collapse and causes growth to a second maximum. The final, transient collapse occurs when the contraction coincides with the acoustic compression phase. Nuclei which cavitate after only one acoustic cycle are called 'prompt' nuclei [22].

Regardless of the details of the behaviour, there are two important phases which must occur in the life of a transient bubble. It must firstly nucleate, undergoing extensive growth which is initially explosive in character. Secondly, given that such growth has occurred, the bubble must collapse in such a manner as to concentrate energy significantly. There are thresholds in bubble radius and acoustic pressure[10] associated with each of these phases. For example, in order to grow in this manner, the initial bubble must be *larger* than some critical radius. In order to concentrate energy in spite of dissipation losses, the ratio R_{max}/R_{min} must exceed a certain critical

[10] And in other parameters too, such as the acoustic frequency, though radius and pressure are usually the most useful in discussion.

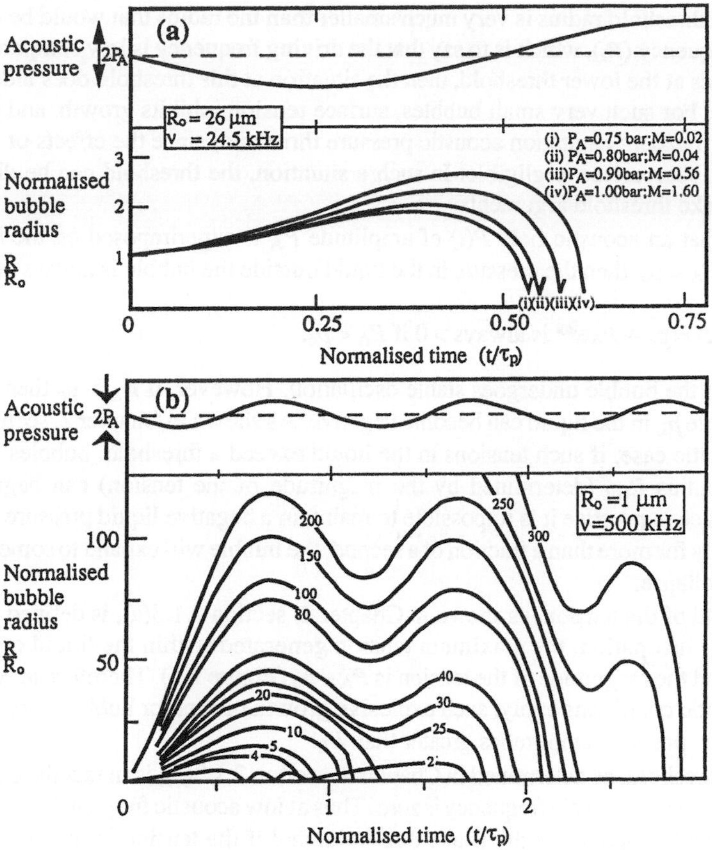

Figure 4.9 Radius/time plots showing the effect of increasing the pressure amplitude on an air bubble undergoing transient collapse in water (after Walton and Reynolds [12]). Time is measured in multiples of the acoustic period. (a) $R_0 = 26$ μm, ν = 24.5 kHz (after Flynn [13]). (b) $R_0 = 1$ μm, ν = 500 kHz. The small numbers against each curve are P_A/p_0. (After Neppiras [21].)

value, and for this to occur the initial bubble must be *smaller* than *Flynn's dynamical radius*. These two thresholds are now formulated.

(a) The Initial Explosive Growth

There is a clear similarity between the initial, growth stage of the transient collapse and the explosive growth of a bubble in the quasi-static situation, approached in Chapter 2, section 2.1.3(a) through discussion of the Blake threshold. The latter is effectively a nucleation threshold, where only the effects of surface tension are considered. The Laplace pressure, $p_\sigma = 2\sigma/R$, tends to contain the gas and it is against this that the tension in the liquid must work in order to bring about explosive growth. Since $p_\sigma \propto R^{-1}$, then as the initial radius of the bubble under consideration becomes very small, the containing influence of the Laplace pressure becomes very large indeed. Thus it is clear that for a given liquid tension there will be a critical bubble radius such that bubbles smaller than this will not grow explosively.

If this lower threshold radius is very much smaller than the radius that would be resonant at the acoustic frequency (R_r), which is to say that the driving frequency is low compared with the ω_0 of the bubbles at the lower threshold, then the situation at this threshold does indeed tend to the quasi-static. For such very small bubbles, surface tension inhibits growth, and effectively determines the transient cavitation acoustic pressure threshold, since the effects of inertia and viscosity are in comparison negligible. In such a situation, the threshold can be discussed in terms of the Blake threshold arguments.

Let us say that an acoustic field $P(t)$ of amplitude P_A is superimposed on the hydrostatic pressure p_o. If $P_A < p_o$, then the pressure in the liquid outside the bubble is always positive, i.e.

$$p_L = p_o + P(t) = p_o + P_A e^{i\omega t} \text{ is always} > 0 \text{ if } P_A < p_o.$$

In this situation the bubble undergoes stable oscillation. However, if $P_A > p_o$ then during the cycle the pressure p_L in the liquid can become negative. As shown in Chapter 2, section 2.1.3(a) for the quasi-static case, if such tensions in the liquid exceed a threshold, bubbles larger than some critical radius R_{crit} (determined by the magnitude of the tension) can begin to grow explosively. Since in practice it is impossible to maintain a negative liquid pressure around the expanding cavity for more than a fraction of a second, the bubble will expand to some maximum size and then collapse.

The threshold of the tension, as shown in Chapter 2, section 2.1.3(a), is defined as equal to $p_o - P_B$. During insonation, the maximum tension generated within the liquid occurs when $p_L = p_o - P_A$, and the magnitude of the tension is $P_A - p_o$ (Figure 2.5). Therefore in sound fields where quasi-static conditions apply, such explosive growth occurs for bubbles larger than R_{crit} only for acoustic pressure amplitudes greater than P_B.

The Blake threshold, as determined in Chapter 2, section 2.1.3(a), is in fact the extreme case of the above, when the acoustic frequency is zero. Thus at low acoustic frequencies, this transient cavitation threshold approaches the Blake threshold, and if the tension in the liquid could be maintained, the bubble so nucleated would not collapse.

It is possible to use the above analysis as a starting point from which to investigate the regime in which transient cavitation will occur, in particular in investigating the threshold size which a bubble must exceed if it is to undergo explosive initial growth in response to a tension. That is to say, if the pressure and frequency conditions are appropriate to cause transient cavitation in *some* bubbles, then there is a lower limit to the initial size of the bubble which will go transient. As the quasi-static limit is approached, the approximate criterion for transient cavitation for this lower radius limit is that the bubble radius must exceed R_{crit}, and the acoustic pressure amplitude must exceed P_B.

All these effects are illustrated in Figure 4.10 from Apfel and Holland [23]. The plots demarcate the thresholds for insonation by a single-cycle sinusoidal pulse: bubbles initially in the situation that places them geometrically above the threshold will undergo transient collapse. The thresholds are derived through an approximate analytical consideration of the response of a bubble nucleus to a single acoustic pulse [24], which will be discussed in section 4.3.1(c)(iii). For the moment it is sufficient to say that, as illustrated in the discussion of Figure 4.8, crucial to the question of whether a bubble will grow sufficiently is the response time of the bubble relative to the acoustic period. The response time of the bubble can be characterised by time constants associated with viscous and inertial effects, and with surface tension. The criterion for transient collapse used by Holland and Apfel in deriving Figure 4.10 was that the compressed

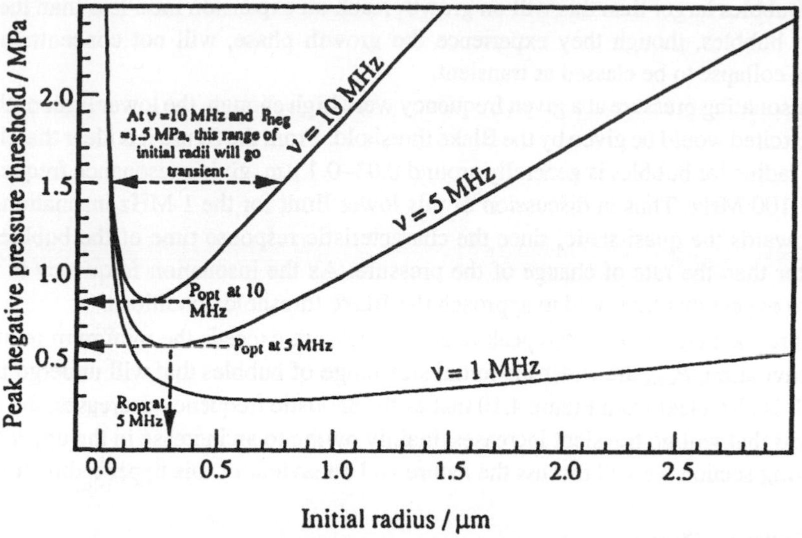

Figure 4.10 Plot of the cavitation threshold in water as a function of initial nucleus radius for three frequencies of insonification: 1, 5 and 10 MHz. Nuclei consist of air bubbles initially at 300 K that undergo growth in a single cycle of ultrasound and collapse adiabatically to a temperature of 5000 K. Surface tension, viscosity and inertia of the host fluid are included in this analytical model. For 5 MHz, the optimal nucleus radius is $R_{opt} = 0.3$ μm with a corresponding cavitation threshold, P_{opt}, of 0.58 MPa peak negative pressure. At ν = 10 MHz, the minimum peak negative acoustic pressure has increased in magnitude to around $P_{opt} = 0.85$ MPa, and the optimum radius reduced to 0.2 μm, Note that at a peak negative pressure greater in magnitude than the minimum magnitude for a given frequency, a broader size range of nuclei cavitate. This is illustrated in the figure for ν = 10 MHz. (After Apfel and Holland [23]. Reprinted by permission from *Ultrasound in Medicine and Biology*, vol. 17, pp. 179–185; Copyright © 1991 Pergamon Press Ltd.)

gas within the bubble should be heated[11] to temperatures in excess of 5000 K. In most practical situations the frequency will be fixed: for illustrative purposes we will assume it to be 5 MHz. Clearly, if the insonation pressure amplitude is such that the magnitude of the peak negative pressure P_{neg} is less than that at the minimum in the graph ($P_{neg} \leq P_{opt}$, where $P_{opt} = 0.58$ MPa at ν = 5 MHz), the model predicts that no bubbles will ever collapse transiently. When $P_{neg} = P_{opt}$, only optimally sized bubbles having radius R_{opt} (which is frequency-dependent) will undergo transient collapse. At pressure amplitudes greater than this, there will in general be a range of bubble sizes which will collapse transiently. For example, if a peak negative pressure of 0.7 MPa is applied at 5 MHz, there will be a lower limit radius threshold: bubbles initially having radii smaller than about 0.11 μm will not collapse because, as we have seen, surface tension will prevent the initial explosive growth required. The upper limit in radius, as mentioned earlier, arises because the ratio of the maximum radius reached at the end of the expansion phase (R_{max}) to the initial radius (R_o) must exceed a certain value so that energy is concentrated sufficiently on the collapse. This process will be discussed in the next section, but suffice to say for now that those bubbles which do in fact grow explosively tend to attain roughly similar maximum radii. Therefore the expansion ratio R_{max}/R_o really depends mostly upon the value of R_o, and if the ratio must exceed a critical value then there will be an initial radius R_o

[11] After equation (1.f3), one may write $TV^{\kappa-1}$ = constant. This implies that if $\kappa \neq 1$, that is if the process is not isothermal, then a gas will heat up on compression and cool on expansion, as discussed in Chapter 1, section 1.1.1(b)(ii).

such that bubbles larger than this will on growth reach an expansion ratio less than the critical one. Such bubbles, though they experience the growth phase, will not concentrate energy enough on collapse to be classed as transient.

If the insonating pressure at a given frequency were high enough, the lower limit of the range of nuclei excited would be given by the Blake threshold. From the figure it is clear that the lower threshold radius for bubbles is generally around 0.03–0.1 μm, giving resonance frequencies ν_o of around 100 MHz. Thus in discussion of this lower limit for the 1-MHz insonation, we are tending towards the quasi-static, since the characteristic response time of the bubble is very much faster than the rate of change of the pressure. As the insonation frequency is reduced further, we expect this threshold to approach the Blake threshold conditions.

The more the magnitude of the peak negative pressure exceeds the minimum required for acoustic cavitation, P_{opt}, the wider the initial size range of bubbles that will undergo transient collapse. It is also clear from Figure 4.10 that as the acoustic frequency decreases, the range of bubble radii that will go transient increases, mainly owing to an increase in the upper limit. In the following section we will discuss the nature and behaviour of this upper radius limit.

(b) The Collapse Phase

After initial explosive growth, the expansion of the bubble prevents the tension in water being maintained. The growth slows and stops, the bubble reaching a maximum radius. After this maximum, the bubble collapses. Work is done on the bubble, supplied through the kinetic energy of the spherical convergence of the surrounding liquid. Acting against this supply of energy to the bubble are dissipative mechanisms: for example, energy may be conducted away from the bubble as heat and effectively lost to the thermal reservoir of the surrounding liquid. If a transient collapse is thought of as being one where energy is concentrated significantly in the bubble, the effects of energy supply must outweigh the effects of dissipation. From this balance we obtain the upper size limit on a bubble that can 'go transient'.

(i) Dissipation Effects. In order to examine the dynamics of expansion and collapse, Flynn [18] derived an equation of motion which resembles the Rayleigh–Plesset equation corrected for the compressibility of the liquid when the motions are finite. Such correction factors had already been derived by Herring and Trilling, though the resulting formulation was not amenable to Flynn's purpose. He therefore derived the correction factors through analogy of their form at infinitesimal amplitude $(1 - \dot{R})$, with the Gilmore equation.[12]

Flynn [18, 25] included the effects of thermal conduction, viscosity, surface tension and liquid compressibility (to a limited extent) in his equation of motion. He then used it to examine the response of an argon bubble of initial radius $R_0 = 5$ μm, following the passage of a short negative pressure pulse, of half-width 0.83 μs, through water. The general form of the response is shown in Figure 4.11. By an appropriate choice of time constants, the pulse is so short that its amplitude is negligible by the time the bubble has expanded to a maximum radius R_{max}. The subsequent motion of the cavity is therefore free oscillation, rebounding and expanding to maximum radii that decrease with time owing to energy loss through damping. Flynn used the ratio of two successive maxima, R^n_{max} and R^{n+1}_{max}, to define a nonlinear logarithmic damping decrement Δ^{nonlin}_{log}:

[12] See section 4.2.2.

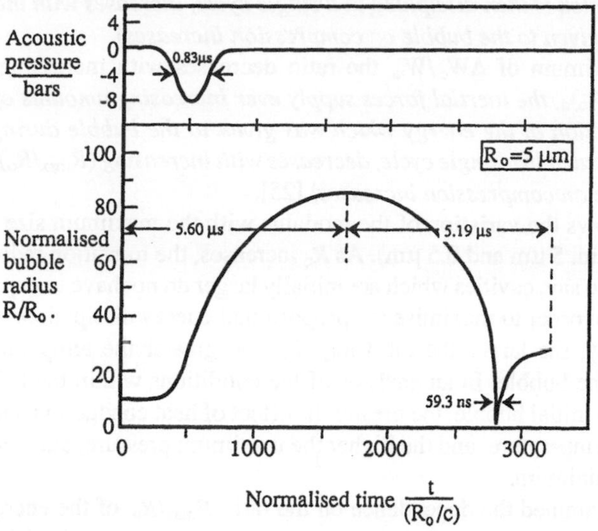

Figure 4.11 The response of an argon bubble ($R_0 = 5$ µm) to the pressure pulse plotted on a common normalised time axis (after Flynn [25]).

$$\Delta_{\log}^{\text{nonlin}} = \log_{10}\left(\frac{R_{\max}^{n} - R_0}{R_{\max}^{n+1} - R_0}\right) \tag{4.90}$$

Since the majority of the bubble motion shown in Figure 4.11 has the bubble at a size greater than equilibrium, heat flow will be into the bubble. Flynn showed that the heat influx was so effective that the motion was isothermal around $R = R_{\max}$, the temperature within the bubble equalling that of the surrounding liquid.

During compression, work is done on the bubble by the pressure forces in the liquid (when the work is taken as negative), and work is done by the bubble on expansion using the energy of the compressed bubble contents (in which case the work is taken as positive). Flynn defines an energy dissipation modulus as being the ratio of ΔW_c, the mechanical work done on or by a gas bubble in a complete cycle, to W_n, the work done on a bubble in its compression from the initial maximum at the start of the nth cycle to its next minimum radius. The quantity ΔW_c represents energy lost through dissipation, whilst W_n is the energy supplied to the bubble through mechanical work. Flynn's energy dissipation modulus, $\Delta W_c / W_n$, represents the fraction of the energy which was given to the bubble during contraction which is then dissipated in a single cycle. Flynn found that for a given cavity of equilibrium size R_0, the modulus $\Delta W_c / W_n$ showed a maximum at some value of $R_{\max}$. He termed this value the transition threshold radius ratio, $(R_{\max}/R_0)_d$, which is the value of $R_{\max}/R_0$ for which

$$\frac{\text{Energy dissipated per cycle}}{\text{Mechanical work done on the cavity by the liquid during contraction}}$$

is a maximum. This ratio represents the fraction of the available energy which is dissipated in a cycle; this must be small for an energetic collapse.

Below the maximum of $\Delta W_c / W_n$, the ratio increases with increasing $R_{\max}/R_0$. *Thus if* $(R_{\max}/R_0) < (R_{\max}/R_0)_d$, *the fraction of the energy which was given to the bubble during contraction and which is then dissipated in a single cycle, increases with increasing* (R_{max}/R_0) *(i.e. as the energy given to the bubble on compression increases).*

Above the maximum of $\Delta W_c/W_n$, the ratio decreases with increasing R_{max}/R_0. *Thus if* $(R_{max}/R_0) > (R_{max}/R_0)_d$, *the inertial forces supply ever increasing amounts of kinetic energy to the cavity, the fraction of the energy which was given to the bubble during contraction and which is then dissipated in a single cycle, decreases with increasing* (R_{max}/R_0) *(i.e. as the energy given to the bubble on compression increases)* [25].

Figure 4.12 shows the variation of the modulus with the maximum size reached, for three cavities (R_0 = 50 μm, 5 μm and 0.5 μm). As R_0 increases, the transition radius ratio $(R_{max}/R_0)_d$ decreases. That is to say, cavities which are initially larger do not have to expand to such a large maximum radius in order to maximise the proportional energy dissipation.

Not surprisingly, the larger the ratio R_{max}/R_0, the greater the temperature and pressures generated within the bubble. In an analysis of the conditions within the bubble, Flynn found that the smaller the initial bubble, the greater the effect of heat conduction and so the lower will be the maximum temperature, and the higher the maximum pressure, attained when the bubble radius reached a minimum.

In all, Flynn examined the dependence on the ratio R_{max}/R_0, of the energy dissipation and the damping moduli, the maximum temperature and pressures obtained in collapse and the resonance frequency of the bubble. He discovered that each of these quantities undergo a transition in the region $2R_0 \leq R_{max} \leq 3R_0$. Flynn found that, in addition to the dependence on bubble size, above the transition (at large amplitudes), the compressibility of the liquid controls

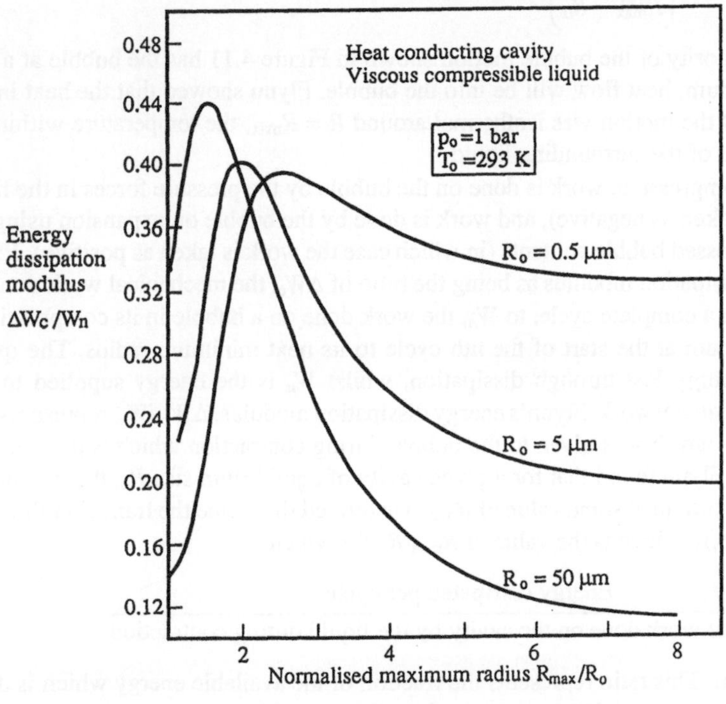

Figure 4.12 The variation of the modulus with the maximum size reached, for three cavities (R_0 = 50 μm, 5 μm and 0.5 μm) (after Flynn [25]).

these quantities, whilst at small amplitudes, behaviour is dominated by heat conduction and, to a lesser extent, viscous effects.

(ii) Inertial Effects. To investigate this dependence on R_{max}/R_0, and to analyse further the transition, Flynn [18, 25] decomposed the equation of motion of the bubble to express the radial acceleration as the sum of two terms, the inertial function (IF) and the pressure function (PF), such that

$$\ddot{R} = \mathrm{IF} + \mathrm{PF} \tag{4.91}$$

For example, treating the Rayleigh–Plesset equation in this manner[13] would yield

$$\mathrm{IF} = -\frac{3\dot{R}^2}{2} \tag{4.92}$$

and

$$\mathrm{PF} = \frac{1}{\rho}\left\{\left(p_0 + \frac{2\sigma}{R_0} - p_v\right)\left(\frac{R_0}{R}\right)^{3\kappa} - \frac{2\sigma}{R} - \frac{4\eta\dot{R}}{R} - p_0 - P(t)\right\} \tag{4.93}$$

The inertial function is so-called since it is reminiscent of a kinetic energy gradient, in that it represents the effect of inertial forces in the liquid. For a free collapse, for example, it would be negative, representing the spherical convergence of the liquid. The pressure function represents the acceleration which is due to the summed pressure forces.

Figure 4.13 shows $\ddot{R}$, IF and PF for the first contraction phase of an argon-filled bubble. For increasing time, the IF and PF curves should be read from right to left, since the figures represent contraction. From equation (4.92), IF will always be negative, except at the start and the end of the collapse (when $R = R_{max}$, R_{min} respectively) at which point it will be zero (since $\dot{R} = 0$). It therefore passes through a minimum, which is a negative value of maximum magnitude, corresponding to the instant that the wall velocity is the greatest, before compression of the gas within the bubble slows the collapse. At the start of the collapse (extreme right, not shown on either plot), the PF will be negative, will decrease to a minimum, and will then increase.

In Figure 4.13(a), the pulse is of relatively low amplitude (peaking at $P_A = 5.03$ bar). As described above, the contraction phase occurs well after the pulse has passed, so that $P(t) \approx 0$ for the interval described in the figure. The maximum size reached by the bubble, R_{max}, equals 7.3 μm (i.e. $R_{max} = 1.46R_0$). From the figure it is clear that during this contraction the acceleration $\ddot{R}$ is dominated by the pressure function PF. In contrast, Figure 4.13(b) shows the response of the same bubble to a much larger pulse, of magnitude 41.2 bar. The bubble expands to a larger maximum radius of 10.05 μm (i.e. $R_{max} = 2.01R_0$). The radial acceleration $\ddot{R}$ is now dominated by the inertial function IF. Flynn found that in many cases, $\ddot{R}$ is determined almost completely by either PF alone during the contraction of the bubble, or by IF alone during the complete oscillatory cycle.

At its minimum, PF changes only slowly, so that if IF falls below the PF at the latter's minimum, the PF cannot change rapidly enough to prevent the IF from controlling the collapse until R is close to R_{min}. Thus if the IF is more negative than the PF at the minimum of the PF, inertial forces will dominate during most of the collapse.

---

[13] In fact, Flynn started with the more complex equation of motion outlined previously.

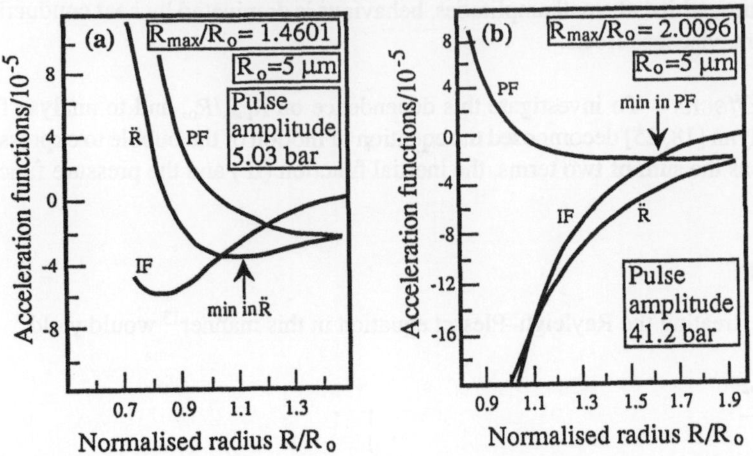

Figure 4.13 Showing how $\ddot{R}$, the inertial function (IF) and the pressure function (PF) vary during the contraction of an argon-filled bubble of initial radius 5 μm in water which has been subjected to a negative Gaussian pressure pulse, as shown in Figure 4.11. The peak negative pressure of the Gaussian pulse, and expansion ratio R_{max}/R_0, are respectively (a) 5.03 bar and 1.4601; (b) 41.2 bar and 2.0096. The acceleration functions are in nondimensional form in which each is divided by c^2/R_0, where c is the speed of sound in the liquid. (After Flynn [25].)

Whilst the position of the minimum in PF is almost independent of R_{max}/R_0, the point of intersection of PF and IF depends almost solely on that ratio. There will therefore be a critical value of the ratio, termed the critical normalised maximum radius $(R_{max}/R_0)_c$, where IF intersects PF at its minimum, so that bubbles with an expansion ratio R_{max}/R_0 greater than the critical value have collapses controlled by the inertial forces. Bubbles with an expansion ratio R_{max}/R_0 less than the critical value have collapses controlled by either the pressure forces alone (small-amplitude motions), or in conjunction with the inertial forces.

If on expansion R_{max}/R_0 has exceeded the critical ratio $(R_{max}/R_0)_c$, inertial forces dominate the subsequent collapse, and the spherical convergence of the surrounding liquid transfers ever increasing quantities of kinetic energy to the contracting bubble [25].

(iii) Dynamical Cavitation Threshold. As described earlier if $(R_{max}/R_0) > (R_{max}/R_0)_d$, the fraction of the energy which was given to the bubble during contraction and which is then dissipated in a single cycle, decreases with increasing (R_{max}/R_0). In the preceding section it was found that if $(R_{max}/R_0) > (R_{max}/R_0)_c$, inertial forces dominate the subsequent collapse, and the spherical convergence of the surrounding liquid transfers ever increasing quantities of kinetic energy to the contracting bubble.

Thus if (R_{max}/R_0) exceeds *both* $(R_{max}/R_0)_d$ and $(R_{max}/R_0)_c$, the inertial forces supply an ever increasing amount of energy to the collapse, an ever decreasing fraction of which is lost through dissipation. Flynn hypothesised that this rapid concentration of energy is the basis of the energetic phenomena (high temperatures, sonoluminescence, shocks, erosion etc.) that we associate with transient cavitation, and proposed the criterion that R_{max}/R_0 should exceed *both* $(R_{max}/R_0)_d$ and $(R_{max}/R_0)_c$ as a possible threshold for transient cavitation. Figure 4.14 shows $(R_{max}/R_0)_d$ and $(R_{max}/R_0)_c$ plotted against R_0. Flynn's dynamical cavitation threshold, $(R_{max}/R_0)_t$, is shown to lie against whichever curve is uppermost, so that both critical ratios are exceeded.

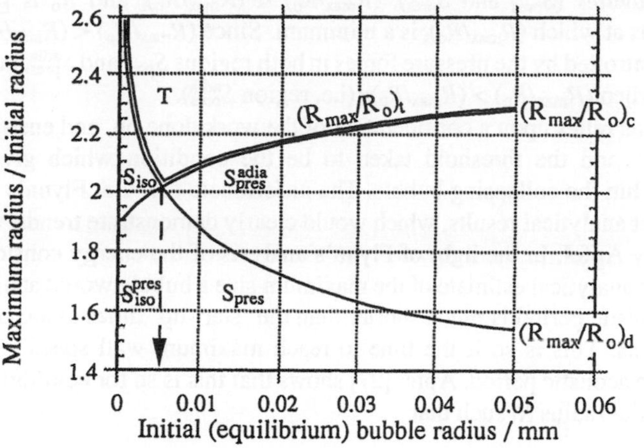

Figure 4.14 The transition radius $(R_{max}/R_o)_d$, critical radius $(R_{max}/R_o)_c$, and dynamical threshold $(R_{max}/R_o)_t$, all normalised to the equilibrium bubble radius, are plotted as functions of the equilibrium bubble radius (after Flynn [25]). The domains are explained in the text.

There is a minimum in the dynamical cavitation threshold $(R_{max}/R_o)_t$ when $(R_{max}/R_o)_d$ equals $(R_{max}/R_o)_c$ which in the conditions of the figure occurs at $R_o \approx 5$ µm (arrowed in Figure 4.14).

The dissipation criterion $(R_{max}/R_o)_d$ dominates $(R_{max}/R_o)_t$ for bubbles with $R_o \lesssim 5$ µm. If in this regime R_{max}/R_o increases in the range $(R_{max}/R_o)_c < (R_{max}/R_o) < (R_{max}/R_o)_d$ (region S_{iso} on the figure), both the energy supplied to, and dissipated from, the bubble increase rapidly, and the collapse tends to the isothermal, leading to higher maximum pressures as $(R_{max}/R_o)_t$ is approached.

The inertial criterion $(R_{max}/R_o)_c$ dominates $(R_{max}/R_o)_t$ for bubbles with $R_o \gtrsim 5$ µm. If in this regime R_{max}/R_o increases in the range $(R_{max}/R_o)_d < (R_{max}/R_o) < (R_{max}/R_o)_c$ (region S_{pres}^{adia} on the figure), the energy supplied to the bubble increases slowly, but the relative heat loss decreases rapidly. The collapse tends to the adiabatic, leading to higher maximum temperatures as $(R_{max}/R_o)_t$ is approached [25].

Flynn [25] suggested that, in view of this, $(R_{max}/R_o)_t$ might be linked with the onset of the pressure-related phenomena that we associate with transient cavitation when $R_o \lesssim 5$ µm, and with the onset of the temperature-related phenomena that we associate with transient cavitation when $R_o \gtrsim 5$ µm. Flynn went on to suggest that transient cavitation be defined as occurring when the bubble expands to a maximum radius such that R_{max}/R_o is greater than its dynamical threshold $(R_{max}/R_o)_t$; and stable cavitation be defined as occurring when a pulsating cavity never expands to a size such that R_{max}/R_o is greater than its dynamical threshold radius $(R_{max}/R_o)_t$.

Therefore Figure 4.14 can be divided into regions of a characteristic cavitation type:

Transient cavities: $(R_{max}/R_o) > (R_{max}/R_o)_t$ – region T. The dominance of inertial forces during the collapse is similar to the situation described in Chapter 2, section 2.1.3(b), the collapse of a Rayleigh cavity. The rapid collapse will be adiabatic, whilst the slower growth phase will tend to the isothermal [12].

'Small' stable cavities (regions S_{iso} and S_{iso}^{pres}). $(R_{max}/R_o) < (R_{max}/R_o)_t$ and R_o is less than the equilibrium radius at which $(R_{max}/R_o)_t$ is a minimum. Since $(R_{max}/R_o) < (R_{max}/R_o)_d$, the motion will tend to the isothermal in both regions S_{iso} and S_{iso}^{pres}), but only when $(R_{max}/R_o) < (R_{max}/R_o)_c$ (i.e. in region S_{iso}^{pres}) will the pressure forces completely control the collapse.

'Large' stable cavities **(S_{pres} and S_{pres}^{adia}).** $(R_{max}/R_o) < (R_{max}/R_o)_t$ and R_o is greater than the equilibrium radius at which $(R_{max}/R_o)_t$ is a minimum. Since $(R_{max}/R_o) < (R_{max}/R_o)_c$, the motion will tend to be controlled by the pressure forces in both regions S_{pres} and S_{pres}^{adia}, but will only tend to the adiabatic when $(R_{max}/R_o) > (R_{max}/R_o)_d$ (i.e. region S_{pres}^{adia}).

Flynn's analysis relies upon a consideration of the work done on, and energy lost from, the collapsing cavity, and the threshold taken to be the condition which gives high energy concentration within the collapsing bubble. The unfortunate fact that Flynn's theory does not lend itself to exact analytical results, which would clearly demonstrate trends, was dealt with a few years later by Apfel. In the light of Flynn's analysis of the energy considerations, Apfel [26] made a basic analytical estimate of the maximum size a bubble would attain. The estimate is valid if the acoustic period is much shorter than the 'start-up' times associated with inertial and viscous effects. This is so if the time to reach maximum wall speed is less that about one-quarter of the acoustic period. Apfel [27] shows that this is so for equilibrium bubble radii less than the inertial radius R_I such that

$$R_I \approx \frac{\pi}{2\omega}\sqrt{\frac{P_A - p_o}{\rho}} \tag{4.94}$$

Within this approximation, Apfel's analysis is as follows. The bubble grows while the liquid is under tension. Then, when the pressure becomes positive, the growth slows, ceases, and a Rayleigh-like[14] collapse follows. By application of the Rayleigh model, Apfel argued that R_{max} could be estimated to be

$$R_{max} = \frac{4}{3\omega}(P_A - p_o)\sqrt{\frac{2}{\rho P_A}}\left(1 + \frac{2}{3p_o}(P_A - p_o)\right)^{1/3} \tag{4.95}$$

This estimate predicts that R_{max} should be independent of R_o during transient cavitation, for bubbles which are initially much smaller than resonance. Noltingk and Neppiras [10] had given support for this idea by solving the RPNNP equation numerically, and it is evident in the Rayleigh–Plesset solutions of Figures 4.8(c) and 4.8(d). It is possible to use this fact to convert (in a manner that should not be taken too seriously) Figure 4.14, so that one might obtain a physical appreciation of the relevance of Flynn's energy-mediated dynamical threshold for transient cavitation to the initial bubble size. Since, in any given set of conditions of P_A, p_o, ρ and ω, the value of R_{max} is roughly independent of R_o for bubbles of much less than resonance size, Figure 4.14 might be reciprocated and re-plotted as $(R_o/R_{max})_c$, $(R_o/R_{max})_d$ and $(R_o/R_{max})_t$ against R_o (Figure 4.15), to suggest how bubbles of different initial bubble radii (on the abscissa) lie within different regimes of behaviour, as given by the locus of the straight-line plot corresponding to the value of R_{max}. For given insonation conditions, if the bubble is to go transient, R_o must be small enough to place R_o/R_{max} within region T, which in Figure 4.15 is *under* the $(R_o/R_{max})_t$ curve. If a known sound field is applied to a given liquid, then P_A, p_o, ρ and ω are fixed, and therefore so (approximately) is the value of R_{max}. Thus the spectrum of behaviour of the bubbles in the liquid, all of which attain the same R_{max}, is given by a straight line of gradient R_{max}^{-1}. Plots of R_o/R_{max} for $R_{max} = 0.01, 0.02, 0.03, 0.05$ and 0.10 mm are shown.[15] The graph indicates the trend of behaviour. If all bubbles achieve, for example, a maximum radius of 0.05 mm, then only bubbles that were initially smaller than 0.0225 mm in radius will

[14] See Chapter 2, section 2.1.3(b)(ii).

[15] High values of R_{max} tend to signify high insonation pressures, long growth times etc. If bubbles of different R_o attain different R_{max}, the trend of behaviour can be seen by moving between different (- - -) curves.

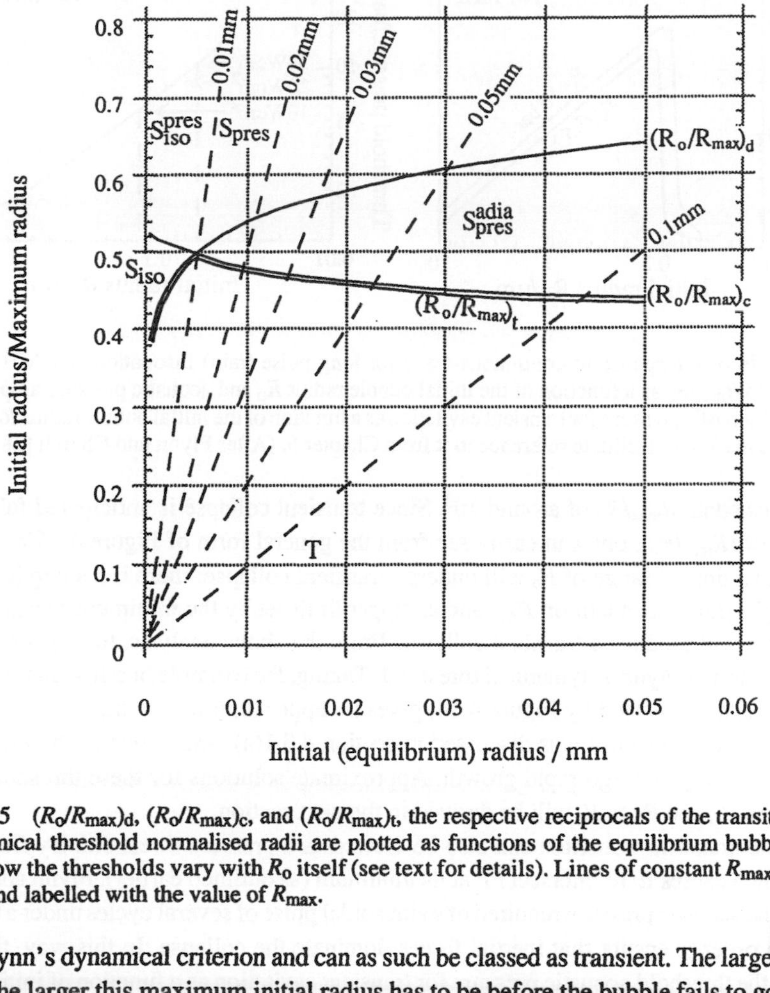

Figure 4.15 $(R_o/R_{max})_d$, $(R_o/R_{max})_c$, and $(R_o/R_{max})_t$, the respective reciprocals of the transition, critical and dynamical threshold normalised radii are plotted as functions of the equilibrium bubble radius to suggest how the thresholds vary with R_o itself (see text for details). Lines of constant R_{max} are plotted (- - - - -) and labelled with the value of R_{max}.

satisfy Flynn's dynamical criterion and can as such be classed as transient. The larger the value of R_{max}, the larger this maximum initial radius has to be before the bubble fails to go transient. Even if all bubbles do not attain the same R_{max}, one can determine whether or not the bubble will go transient by locating the position of the point on the appropriate R_{max} curve corresponding to the initial bubble radius, as indicated on the horizontal axis. Thus we have shown that there is an upper limit on the initial, equilibrium size of a cavity that, given the insonation conditions, will collapse transiently.

(c) Upper and Lower Thresholds for Transient Cavitation

It is clear from Figure 4.14 that the dynamical threshold occurs at around $(R_{max}/R_o) \approx 2$ for the value of R_o where it is a minimum. Figure 4.16 [28] demonstrates the idea of thresholds for transient cavitation. In Figure 4.16(a), the expansion ratio (R_{max}/R_o) is plotted as a function of the initial, equilibrium radius of the bubble, for a 1-MHz sound field of amplitudes 7 bar and 10 bar. The form of Figure 4.16(a) is typical of the phenomenon: Flynn [13], utilising conditions from Noltingk and Neppiras [10, 11] and Flynn [29], generated similar plots in which optimally

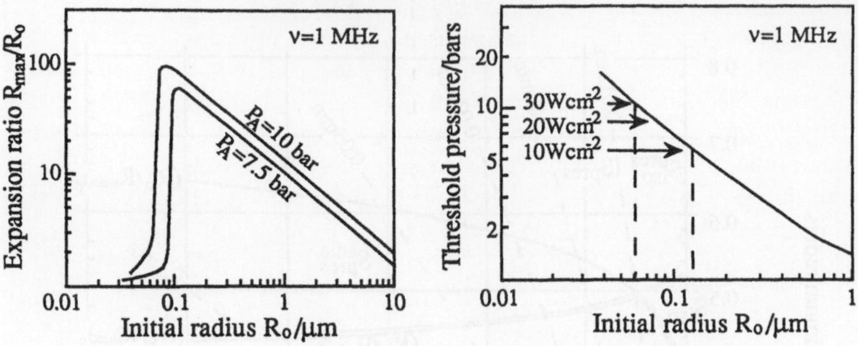

Figure 4.16 Bubble response to continuous-wave (or long pulse train) insonation at 1 MHz. (a) The expansion ratio R_{max}/R_o as a function of the initial bubble radius R_o and acoustic pressure amplitude P_A. (b) Threshold acoustic pressure for transient cavitation as a function of the initial bubble radius R_o. Specific intensities are marked to facilitate reference to it from Chapter 5. (After Flynn and Church [28].)

sized bubbles attain R_{max}/R_o of around 10^3. Since transient collapse is anticipated following a large value of (R_{max}/R_o), one can easily see from the general form of Figure 4.16(a) that only bubbles with a limited range of R_o will undergo transient collapse: there is a sharp lower limit at around 0.07 μm (dependent on P_A), and an upper limit set by the minimum expansion ratio (R_{max}/R_o) that will produce a transient collapse. From the above analysis, this upper threshold is clearly a result of Flynn's dynamical threshold. Taking, for example, a criterion threshold of $(R_{max}/R_o)_t \approx 2$, as suggested by Figure 4.14, gives an upper limit on the transient cavity to be about 7 μm. The lower limit was discussed in section 4.3.1(a), and is due to the fact that the cavity must initially undergo rapid growth. Approximate solutions for these thresholds in the limits of $R_o \ll R_r$, and $R_o \gg R_r$ will be derived in the next section.

The stable–transient transition can also be viewed as a pressure threshold. By finding that pressure which causes IF to intersect PF at its minimum (as outlined earlier), Flynn and Church calculated the acoustic pressure required of a sinusoidal pulse of several cycles under a Gaussian envelope in order to ensure that inertial forces dominate the collapse. In this way, they were able to plot the threshold acoustic pressure for transient cavitation as a function of initial bubble radius. This is shown in Figure 4.16(b).

(i) Approximation for the Lower Threshold Radius. This lower threshold radius of 0.07 μm occurs well below the resonance radius, which at 1 MHz is roughly 3 μm. Thus the pressure function dominates, and the Blake threshold is an appropriate approximation. The acoustic pressure required to generate transient cavitation at the lower limit is

$$P_B = p_o + \frac{8\sigma}{9}\sqrt{\frac{3\sigma}{2R_B^3\,(p_o + (2\sigma/R_B))}} \qquad \text{(valid for } R_o \ll R_r\text{)} \qquad (4.96)$$

In this equation the threshold value for R_o is termed R_B, the Blake radius. Bubbles smaller than R_B will not grow explosively. P_B is the minimum acoustic pressure at which a gas bubble of radius R_B will grow in the manner described in Chapter 2, section 2.1.3(a), under quasi-static conditions (i.e. $R_B \ll R_r$).

(ii) Approximation for the Upper Threshold Radius (applicable for $R_o \gg R_r$). Apfel's analytical estimate (equation (4.95)) can be used to relate a pressure threshold P_t at Flynn's dynamical limit to the upper limit threshold value of R_o, namely R_t. This is done by relating R_{max} to R_o as enabled through Flynn's requirement that on expansion R_{max}/R_o has exceeded the critical ratio $(R_{max}/R_o)_c$ so that inertial forces dominate the subsequent collapse, and the spherical convergence of the surrounding liquid transfers ever increasing quantities of kinetic energy to the contracting bubble. Using $(R_{max}/R_o)_c = 2$, Walton and Reynolds [12] obtain

$$\frac{R_t}{R_r} = \frac{2}{3}\frac{P_t - p_0}{p_0}\sqrt{\frac{2p_0}{3P_t}}\left(1 + \frac{2}{3p_0}(P_t - p_0)\right)^{1/3} \quad \text{(valid for } R_r \ll R_o < R_I\text{)} \tag{4.97}$$

where R_r is simply the resonant equilibrium radius of the bubble, found from equation (3.36) to be $R_r = \sqrt{(3p_0/\rho\omega^2)}$ where ω is the frequency of the sound wave and κ is set equal to unity. Deriving as it does from Flynn's dynamical threshold, equation (4.97) is valid only for $R_o \gg R_r$ (when the inertial function dominates). Even when the inertial function does dominate, equation (4.97) is only valid if equation (4.95), from which it derives, is also valid. This is so if the acoustic period is much shorter than the 'start-up' times associated with inertial and viscous effects, which implies that the bubble must be smaller than the inertial radius R_I, which is given approximately by equation (4.94).

(iii) Approximate Analytical Formulation Incorporating Both Upper and Lower Radius Thresholds. Holland and Apfel [24] published an extension to the arguments of Apfel [30] providing an improved analytical formulation for *both* the growth and collapse phases together. The model transient collapse is shown in Figure 4.17 for the transient response of a bubble to a single acoustic cycle. The negative phase of the sound pressure precedes the positive, so that the liquid initially goes into tension. There is a time lag before the bubble responds with rapid growth. This delay is caused by the inertia and viscosity of the liquid, and by the initially large Laplace pressure p_σ. If these time delays comprise a significant fraction of the acoustic period, much higher acoustic pressures will be required to induce transient cavitation.

Even when the acoustic cycle enters the positive phase, and there are compressive pressures applied by the sound field, the liquid momentum causes the bubble to continue to expand until it reaches a stationary position at some maximum radius. It then begins the characteristic collapse.

Though the actual characteristics of the growth may be complicated, attributing a simplified version to the bubble allowed Holland and Apfel to produce an approximate formulation. The negative pulse begins at time $t = 0$. At time $t = t_1$, the magnitude of the negative acoustic pressure in the liquid exceeds P_B, the Blake threshold pressure, where

$$t_1 \approx \left(\frac{1}{\omega}\right)\left\{\frac{\pi}{2} - \sqrt{2\left(1 - \frac{P_B}{P_A}\right)}\right\} \tag{4.98}$$

Holland and Apfel [24] also estimate the additional delay, Δt_σ, caused by the surface tension pressure, by comparison of p_σ with the driving terms in the Rayleigh–Plesset equation:

$$\Delta t_\sigma \approx \frac{2\sigma}{P_A - P_B}\sqrt{\frac{3\rho}{2(P_A - P_B)}} \tag{4.99}$$

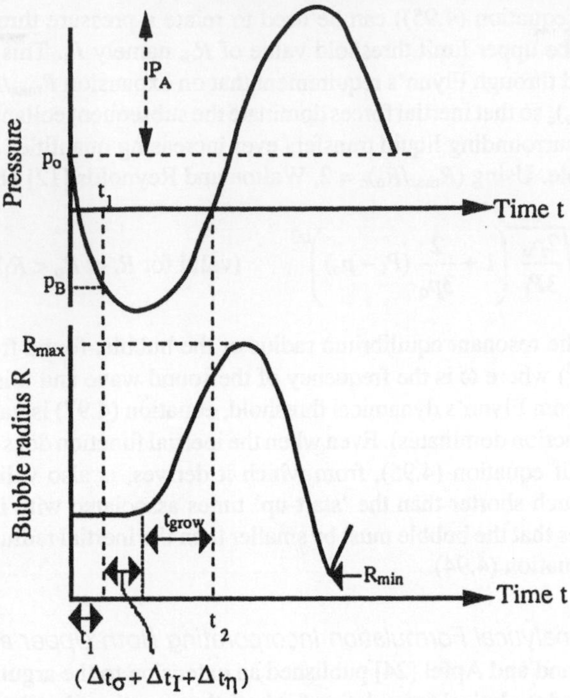

Figure 4.17 The model transient response of a bubble to a single acoustic cycle, where the negative phase of the sound pressure precedes the positive (after Holland and Apfel [31]). © 1989 IEEE.

Another approximation gives t_2, the time when the liquid ceases to be in tension

$$t_2 \approx \left(\frac{1}{\omega}\right)\left\{\frac{\pi}{2} + \sqrt{2\left(1 - \frac{p_0}{P_A}\right)}\right\} \tag{4.100}$$

The difference $t_2 - t_1$ represents the duration of the 'force' that causes the bubble to grow.[16] This force that drives the growth can in estimation be quantified through the time-averaged pressure difference across the bubble wall, ΔP_{wall}:

$$\Delta P_{wall} = \frac{1}{t_2 - t_1}\int_{t_1}^{t_2} (P_A \sin\omega t - p_0)\,dt \tag{4.101}$$

the gas pressure within the bubble being omitted from ΔP_{wall} as it falls rapidly. Equation (4.101) can be evaluated to give ΔP_{wall} to second order

[16]Though in the approximation the tension must exceed P_B to initiate growth, clearly once growth is initiated it will continue to be driven when the tension falls to less than P_B, which was simply the R_0-dependent starting condition.

$$\Delta P_{wall} \approx \tfrac{1}{3}(P_A + P_B - 2p_0 + \sqrt{(P_A - p_0)(P_A - P_B)}\,) \tag{4.102}$$

In addition to Δt_σ, there are similar time delays resulting from inertia and viscosity, which sum to give [24, 30]:

$$\Delta t_I + \Delta t_\eta \approx \frac{2R_0}{3}\sqrt{\frac{\rho}{\Delta P_{wall}}} + \frac{4\eta}{\Delta P_{wall}} \tag{4.103}$$

Using the above scales as a basis, Holland and Apfel estimate the maximum radius attained by the bubble to be

$$R_{max} = \left(R_0 + \sqrt{\frac{2\Delta P_{wall}}{3\rho}}\, t_{grow}\right)\left(\frac{\Delta P_{wall}}{p_0} + 1\right)^{1/3} \tag{4.104}$$

where

$$t_{grow} = (t_2 - t_1) - (\Delta t_\sigma + \Delta t_I + \Delta t_\eta) \tag{4.105}$$

is the net time for bubble growth. If an adiabatic collapse is assumed, then following Apfel [30], equation (4.84) can be used to relate the maximum temperature attained during the collapse, T_{max}, to the insonation conditions:

$$\nu = \frac{1}{3\pi R_0}\sqrt{\frac{3\Delta P_{wall}}{\rho}}\left\{\sqrt{\left(1 - \frac{p_0}{P_A}\right)} + \sqrt{\left(1 - \frac{P_B}{P_A}\right)}\right\} \div$$

$$\left\{\left(\frac{T_{max}}{T_0(\gamma - 1)\,(1 + [\Delta P_{wall}/p_0])}\right)^{1/3} - 0.46 + \frac{4\eta}{R_0}\sqrt{\frac{2}{3\rho\Delta P_{wall}}} + \frac{2\sigma}{p_0 R_0}\sqrt{\frac{\Delta P_{wall}}{p_0}}\left(\frac{p_0}{P_A - P_B}\right)^{3/2}\right\} \tag{4.106}$$

where $\nu = \omega/2\pi$ is the frequency of insonation, and T_0 the initial temperature of the bubble interior. Having obtained this formulation, Holland and Apfel solve equation (4.106) numerically to find the locus of points corresponding to the achievement of a specific condition during these collapses: that the internal temperature of the bubble should reach 5000 K. They take this as their criterion for transient collapse, citing the sonochemical results obtained using comparative rate thermometry by Suslick *et al.* [32].[17] The achievement of this temperature Holland and Apfel chose as the threshold of transient cavitation, and thus use the formulation to obtain plots of the threshold, similar to the one shown in Figure 4.10.

Having obtained an analytical formulation for the transient cavitation threshold based on the collapsing bubble achieving temperatures in excess of some chosen value, Holland and Apfel compare their transient cavitation predictions with those of Flynn and Church [22], who employed the Gilmore–Akulichev model of bubble motion. However, Flynn and Church used the definition that the expansion ratio R_{max}/R_0 equals 2 at the threshold. The temperature T_{max} clearly will depend on the expansion ratio R_{max}/R_0, and since they are not independent, then in

[17] See Chapter 5, section 5.2.1(a)(ii).

order to compare the threshold of Holland and Apfel with that of Flynn and Church, it is necessary to formulate the dependence to ensure that like is compared with like. Apfel [30] uses equation (2.39), $T_{max} \approx T_0(\gamma - 1)(R_{max}/R_0)^3$ where T_0 is the initial gas temperature, taken to be 300 K. This relationship was derived in Chapter 2, section 2.1.3(b)(ii) in discussion of the Rayleigh-like collapse of a gas-filled cavity, where the cavity about to undergo transient collapse is assumed to have just undergone isothermal growth. It is accurate only for bubbles having $R_0 \gtrsim 40$ µm, and is simply a reduction of the relationship given by Noltingk and Neppiras [10], which assumes negligible surface tension and viscosity:

$$T_{max} \approx T_0(\gamma - 1)(R_{max}/R_0)^3 (p_{L_0}/p_0) \tag{4.107}$$

the reduction clearly occurring when the pressure outside the cavity as it begins to collapse, p_{L_0} is assumed to equal p_0.

Walton and Reynolds [12] include surface tension in estimating the temperatures and pressures attained in transient collapse. The use of κ instead of γ in equations (2.35) and (2.37) takes a limited account of the effects of heat conduction, to give the final pressure and temperature inside a gas-filled cavity that has collapsed from an initial volume to a minimum, V_{min}, under the action of an unchanging hydrostatic pressure p_∞. Assume that the bubble has initially expanded slowly and isothermally from radius R_0 to R_{max}. The collapse then occurs rapidly. The approximation is made that this collapse occurs so rapidly that the external liquid pressure remains effectively constant for the duration of the collapse, and that it occurs when the acoustic pressure reaches its maximum (so that p_∞ in the previous formulation is replaced by $P_A + p_0$). This contrasts with the assumption of Apfel who, in obtaining equation (2.39), assumed that $p_{L_0} = p_0$, which is to say that the acoustic pressure was zero for the duration of the collapse. To obtain the maximum pressure and temperature attained upon collapse within the limitations of this derivation, it is therefore only necessary to incorporate surface tension into the relationship between $p_{g,m}$ and p_∞ in equations (2.35) and (2.37). Assuming that the liquid hydrostatic pressure outside the bubble is p_0, then when $R = R_0$ the gas pressure within the bubble is $p_0 + 2\sigma/R_0$. Applying equation (1.16) to the isothermal growth phase gives

$$R_0^3(p_0 + 2\sigma/R_0) = R_{max}^3\, p_{g,m} \tag{4.108}$$

With a polytropic approximation, substitution of this into equations (2.35) and (2.37), where the bubble attains a maximum radius $R_{max} = R_m$, yields the maximum gas pressure and temperature attained in the transient collapse, $p_{max,t}$ and $T_{max,t}$, (which are equivalent to $p_{g,max}$ and T_{max} used in the discussion of the Rayleigh-like collapse of a gas-filled cavity):

$$p_{max,t} \approx \left(\frac{R_{max}}{R_0}\right)^{\frac{3}{(\kappa-1)}} \left(p_0 + \frac{2\sigma}{R_0}\right)^{\frac{-1}{(\kappa-1)}} \left(\kappa - 1\right)^{\frac{\kappa}{(\kappa-1)}} \left(P_A + p_0\right)^{\frac{\kappa}{(\kappa-1)}} \tag{4.109}$$

$$T_{max,t} = T_0 \left(\frac{R_{max}}{R_0}\right)^3 \left(p_0 + \frac{2\sigma}{R_0}\right)^{-1} (\kappa - 1)(P_A + p_0) \tag{4.110}$$

In order to compare like threshold prediction with like, Holland and Apfel used equation (2.39) to adjust their definition of transient cavitation to occur when the bubble attained a temperature of 960 K. Agreement between the two sets of predictions (Holland and Apfel [24], and Flynn and Church [22]) is good for small nuclei and low frequencies. There are differences, however. For example, the ranges of nuclei size that are predicted to undergo transient collapse are in general broader in Flynn and Church's results than in Holland and Apfel's. The latter discuss the effects of the approximations inherent in their threshold derivation. For example, equation (2.39) is accurate for large bubbles, as outlined, and for very small bubbles will not lead to significant error since the surface tension term in the denominator of equation (4.106) dominates. Šponer *et al.* [33–35], in attempting to determine the thresholds for sonoluminescence, model the adiabatic collapse of an isolated gas bubble in a viscous compressible liquid, and set the cavitation threshold to be the acoustic pressure required to attain a collapse temperature of 1550 K.

(d) The Mechanical Index

Apfel and Holland [23] derive a *mechanical index*, which represents the likelihood that transient cavitation will occur, applicable to microbubble growth in the limit of short-pulse, low duty cycle insonation. This limitation arises simply because the theory used in the derivation is that outlined in section 4.3.1(c)(iii), pertinent to prompt cavitation resulting from a single cycle of ultrasound. Clearly, effects relating to longer insonation periods, such as the growth by rectified diffusion[18] of stable bubbles to the transient cavitation threshold, are not covered. However, the index could be used to gauge the probability of prompt transient cavitation resulting from diagnostic ultrasound fields.

The advantage of the formulation given by Apfel [30] and Holland and Apfel [24] is that, though less exact than numerical solutions, it enables the key parameters, and their relationships, to be identified. From the above analysis it is clear that surface tension, fluid inertia and viscosity delay the growth of a bubble subjected to tension, and as the time for growth t_{grow} is reduced, so is the likelihood of violent cavitation. From solutions of equation (4.106) by Holland and Apfel [24], the relationship between the acoustic pressure P_A, the maximum temperature attained during the collapse T_{max}, and the initial bubble radius R_0 can be deduced. In order for the bubble to grow significantly (a phase they assume to be isothermal), then to reach a given T_{max} on collapse (as outlined earlier, Holland and Apfel chose 5000 K), the relevant contributions to the minimum acoustic pressure that will achieve this (a) increase with decreasing R_0 owing to surface tension, and (b) increase with increasing R_0 as a result of inertial and viscous effects [23]. Thus their analytic model allows the causes of the upper and lower limits on the initial size of a bubble which can go transient to become apparent. Figure 4.10 demonstrates this clearly, Apfel and Holland plotting the transient cavitation threshold as a function of the peak negative pressure and the initial bubble radius. Each curve shows a minimum. For a given insonation pressure P_A at a fixed frequency, the locus of bubble behaviour is shown by a horizontal line at that pressure. If that line does not intersect the threshold plot, the formulation predicts that there will be no transient collapse. If it does, it will generally do so at two points, which demarcate the intermediate range of bubble sizes which will go transient *in response to a single acoustic cycle*. There is a minimum. As the insonation frequency increases, the bubble radius which requires minimum pressure to go transient decreases, since inertial and viscous forces increase with increasing frequency. For the same reason, the acoustic pressure required to cause transient cavitation in all but the smallest bubbles increases with increasing frequency. Surface tension dominates the response of the smallest bubbles. As the frequency is reduced, one will expect the threshold for small bubbles to tend to the quasi-static case described as the Blake threshold, where surface tension dominates. As the pressure increases, stimulating greater bubble response, even the higher-frequency transient thresholds tend to a limit for the lower

[18] See section 4.4.3.

radius threshold. At a given insonation frequency, if the acoustic pressure were high enough, the lower radius limit would tend to the Blake threshold.

If one is interested in a worst-case assessment of the likelihood that transient cavitation will occur when a liquid is insonated, clearly one must assume that the bubble population contains bubbles at the radius corresponding to the minimum in the threshold curve. At a given frequency, it is bubbles of this radius, R_{opt}, which the analysis predicts will require the smallest peak negative pressure, P_{opt}, to undergo prompt transient cavitation in response to a single acoustic cycle. For example, in Figure 4.10 insonation at 5 MHz gives a minimum in the threshold corresponding to the coordinates $P_{opt} = 0.58$ MPa and $R_{opt} = 0.3$ μm. Apfel and Holland [23] generate a plot of P_{opt} against frequency for water and whole blood, using pure fluid bulk property values for the σ, ρ and η relevant to the two fluids. The liquids are assumed to contain the relevant nuclei at size R_{opt}. Apfel and Holland employ a two parameter least-squares fit to these plots in order to obtain a relationship between P_{opt} and insonation frequency $\nu = \omega/2\pi$. They find that

$$\frac{(P_{opt})^{a_1}}{\nu} = a_2 \tag{4.111}$$

where if P_{opt} is measured in MPa and ν in MHz, the constant a_1 takes values of 2.10 for water and 1.67 for blood, and a_2 has values of 0.06 for water and 0.13 for blood. For a given sound field with a maximum negative pressure of P_{neg}, then by taking a value of $a_1 \approx 2$ to approximate the appropriate physiologically relevant liquid, an index I_{index} can be defined for the sound in that liquid:

$$I_{index} = \frac{\left[\dfrac{P_{neg}}{\text{MPa}}\right]^2}{\left[\dfrac{\nu}{\text{MHz}}\right]} \tag{4.112}$$

This index is roughly proportional to the mechanical work that can be performed on a bubble in the negative phase of the acoustic cycle, and it is interesting to note that in the dynamic strength testing of solids, the response of a material subjected to a stress pulse of amplitude P_{str} and duration t_{str} depends critically upon the product ($P_{str}^2 t_{str}$) [36]. Defect extension, for example, occurs once this product exceeds a critical value [37]. This too is a 'prompt' phenomenon.

The index I_{index} for prompt cavitation represents an approximate measure of the worst-case likelihood of transient cavitation in response to a single acoustic cycle and as such, remembering the single-pulse limitation, could be used as an index to estimate the potential for transient cavitation resulting from insonation by diagnostic ultrasound of low duty cycle and short pulse lengths (no more than a few acoustic cycles). Clearly, it would be inappropriate to apply this index to insonations not approximating to a single acoustic cycle. In addition, the index is strictly relevant only for cavitation in the liquid where P_{neg} was measured. Apfel and Holland [23] outline the protocol that should be followed in estimating I_{index} for a given ultrasonic device. The peak negative pressure output from the device, as measured in water, must be derated to give the appropriate peak negative pressure that would be attained *in vivo*, to account for attenuation [38] and nonlinear propagation [39]. The centre frequency (which for accuracy is expected to be of the order MHz) is used for ν. They recommend that the pulse length should not exceed 10 cycles, nor the duty cycle 1:100.

Apfel and Holland [23] suggest that a value of I_{index} below 0.5 would indicate that, even in the presence of a broad size distribution of nuclei, the conditions are not sufficient to allow significant bubble expansion (provided that the pulse approximates to an isolated single-cycle, and that rectified diffusion can be neglected). If $I_{index} \geqslant 0.5$, Apfel and Holland suggest that "the user should be advised of the potential for bubble activity." Though this does not mean that sound fields with I_{index} greater than 0.5 will produce a bioeffect, it does mean that the degree by which the calculated index exceeds 0.5 will reflect the amount of bubble activity that will generate collapse temperatures of 5000 K or more, and so may correlate with the probability of a bioeffect. By examining Figure 4.10 it is clear that one reason for this is because, as the pressure increases, the range of bubble radii that can go transient increases. Another reason would be that, in general, a given bubble would be expected to respond to a greater extent. However, care must be taken when trying to predict, for example, the maximum collapse pressure attained within a bubble when it is excited by a microsecond pulse of ultrasound at pressure amplitudes greater than the transient cavitation threshold [40]. An increase in the acoustic pressure amplitude in the transient regime may in fact bring about a decrease in the amplitude of oscillation, and in the magnitude of the collapse pressure, as a result of the nonlinearity of the bubble response, and in particular the phase difference between the acoustic field and the bubble response.

Based on this work, the AIUM, NEMA and FDA have adopted a *mechanical index*, weighted for the frequency response, of $I_{MI} = (P_{neg}/\sqrt{\nu})$. This is in order to allow estimation, from a real-time output display, of the potential for cavitation *in vivo* during diagnostic ultrasound scanning [229]. Since $I_{MI} = \sqrt{I_{index}}$, the critical value for I_{MI} is $\sqrt{0.5} \approx 0.7$.

Devices are currently being produced which generate pulsed waveforms significantly different from the modelled sinusoidal pulse of Apfel and Holland. In particular shocks, arising either through nonlinear propagation, or engineered into the system (for example, in lithotripsy), bear consideration when the choice of measurement technique for, and interpretation of, the pressure is made. The pulses are asymmetrically distorted, the peak-positive usually exceeding the peak-negative pressure. The acoustic energy is invested in a range of frequencies. Models by Aymé and Carstensen [41–43], based on the response of a free-floating spherical nucleus to diagnostic-type pressure signals, suggest that the peak-positive pressure is a very poor indicator of the bubble response. The peak-negative pressure generally underestimates the bubble response [42]. A better predictor of bubble response is the magnitude of the acoustic pressure amplitude of the fundamental in the Fourier series expansion of the pulse [42], a parameter on which depends the strength of the pressure spikes radiated out from the rebounding bubble into the liquid [43]. If these spikes are the mechanism for damage to occur in the medium surrounding the bubble,[19] then it is this fundamental component which is of importance. This is consistent with the killing of *Drosophilia* larvae exposed to pulsed, symmetric sinusoidal fields, and to pulsed, asymmetric distorted fields [41]. Put very loosely, the reason for the predominant importance of the amplitude of the fundamental is that, just as there is a resonant period associated with every bubble, so too is there a timescale over which it responds: in general, small bubbles can respond quickly, whilst large bubbles are 'sluggish'. As demonstrated in Chapter 1, the form of the shock is associated with the investment of energy into higher frequencies, and there is not enough time for a micrometre-sized bubble in a diagnostic field to respond significantly to these shorter periods. Aymé and Carstensen demonstrate this, stating that: "In the 1–10 MHz frequency range, the distorted pulse is approximately

[19] See Chapter 5, section 5.4 – of particular relevance is the biological damage that can arise in response to microsecond pulses of ultrasound, as discussed in Chapter 5, section 5.4.2(b)(iii).

viewed by a microbubble with initial size $R_o \leqslant 3\ \mu m$ as a sinusoidal pulse with peak pressure P_A equal to the pressure amplitude of the fundamental." To put it another way, Figure 4.4 demonstrates the response of an oscillator to driving forces of various frequencies. Though the oscillator has finite response even at frequencies well below resonance, as the frequency of the driving force increases to much greater than the resonance frequency of the oscillator, the amplitude of the response falls to zero.

Two final points should be noted. Firstly, the mechanical index gauges the likelihood of prompt cavitation, and nothing more: the main clinical bioeffect may be related to some other mechanism. Secondly, the model for the index is based on the assumption of a free-floating spherical nucleus of optimum size. In certain circumstances it may be that the nucleus is of a different type. These problems will be illustrated later[20] following further examination of the effect of short pulses on *Drosophilia*. Results will be presented which suggest that when the nuclei are cylindrical air pockets in larval trachae, and these are subjected to lithotripsy pulses, the positive peak pressure seems to be the important parameter (rather than the negative pressure, which one would expect to excite nuclei).

4.3.2 Definition of Transient Cavitation

The attempt by Holland and Apfel to correlate their predictions to those of Flynn and Church illustrates one feature of transient cavitation that can be a continual problem, and that is the various definitions that are employed to distinguish its occurrence. In that comparison, the workers were fortunate that the threshold predicted by Holland and Apfel is somewhat insensitive to the exact value of the temperature T_{max} chosen. In fact, these two definitions are in essence not dissimilar, both being based on the ability of the collapse to concentrate energy sufficiently. As such, these definitions of transient collapse have their foundation in the mechanistic energetics, as formulated by Flynn, where the collapse is controlled by inertial forces generated by the spherical compression of the liquid, such that the work done on the bubble outweighs the dissipative losses. In order to achieve this energy concentration, Flynn has calculated that on expansion R_{max}/R_0 must exceed some critical value (around 2), and that the greater R_{max}/R_0 is above this value, the greater the concentration of energy in the bubble. This has lead to secondary definition of transient cavitation, which is that R_{max}/R_0 must exceed about 2.

An alternative approach leads to a similar definition. Following the discussion by Noltingk and Neppiras [10] of the Rayleigh-like collapse of a gas-filled cavity introduced in Chapter 2, section 2.1.3(b)(ii), Apfel [26] suggests that the threshold be $R_{max}/R_0 \approx 2.3$. This expansion ratio causes the bubble wall just to reach supersonic collapse speeds in water [44]. This can be shown through differentiation of equation (2.32) to find when $\dot{R}$ is a maximum, and then setting that maximum equal to the speed of sound in the liquid. For expansion ratios greater than 2.3 the bubble wall will, on collapse, exceed the velocity of sound in water in the incompressible limit [10]. In such a situation the assumption of incompressibility is invalid: strictly it is appropriate only up to $M = \dot{R}/c \approx 0.2$ [21]. Apfel [45], in using the conditions which cause the bubble wall just to go supersonic to define the transient cavitation threshold, notes that the formulation employed assumes that inertial and viscous effects are negligible. Thus the threshold as found in this way is only appropriate if the threshold radius predicted with this formulation is less than R_I. Apfel's approximation for the threshold is found by substitution of the criterion $R_{max}/R_0 \approx 2.3$ into equation (4.95). It shows that in a sound field of acoustic pressure amplitude

[20] See Chapter 5, sections 5.4.2(b)(iii), 5.4.2(b)(iv) and 5.4.2(c).

P_A, for a bubble to undergo transient collapse the initial equilibrium bubble radius should be less than R_t, where

$$R_t = \frac{1}{2.3}\left(\frac{4}{3\omega}\right)(P_A - p_o)\sqrt{\frac{2}{\rho P_A}}\left(1 + \frac{2}{3p_o}(P_A - p_o)\right)^{1/3} \Rightarrow$$

$$R_t = \frac{0.82}{\omega}\frac{1}{\sqrt{P_A\rho}}(P_A - p_o)\left(1 + \frac{2}{3}\frac{(P_A - p_o)}{p_o}\right)^{1/3} \quad \text{(valid only if } R_t < R_I\text{).} \qquad (4.113)$$

An alternative way of viewing this threshold is to recast equation (4.113) to say that a bubble of radius R_o will only go transient if the acoustic pressure amplitude P_A exceeds a threshold value P_t where they are related by

$$R_o = \frac{0.82}{\omega}\frac{1}{\sqrt{P_t\rho}}(P_t - p_o)\left(1 + \frac{2}{3}\frac{(P_t - p_o)}{p_o}\right)^{1/3} \quad \text{(valid only if } R_o < R_I\text{).} \qquad (4.114)$$

To within the approximation for the resonance given by equation (3.36) for $\kappa = 1$, equation (4.113) predicts a value for R_t which is 2.3/2 times smaller than that predicted by equation (4.97), because the ratio chosen for $(R_{max}/R_o)_c$ is now 2.3 (based on the wall speed definition of transient cavitation) rather than 2 (based on Flynn's energy consideration).

Neppiras [21] notes that, whatever additional complications are in evidence, all transient cavities begin their collapse in the manner of the Rayleigh cavity, and suggests that this 'Rayleigh-like' behaviour could be included in the definition of transient cavitation.

Thus the foundation for the above definitions of transient cavitation is the energetics. The primary definition, based upon the dominance of the inertial function during the collapse, arises from an insight into the mechanistic energetics. However, the definition derived from this, $R_{max}/R_o \gtrsim 2$, though less fundamental, is generally more useful in that it is easier to apply to the results of bubble dynamics. It is comparable to the definition based upon the wall attaining on collapse the speed of sound in the liquid, which gives $R_{max}/R_o \gtrsim 2.3$.

For comparison, Soviet workers [46, 47] have solved equation (4.81) and plotted dR/dt against R on a phase-plane. The resulting curves may be closed (indicating stable cavitation) or show discontinuities, the curve running off to the infinite regions of the plane (transient cavitation). By noting the form of the curve as the parameters are varied, transient cavitation thresholds have been found.

However, these treatments generally approach the phenomenon from the theoretical viewpoint, and are therefore limited by the assumption of an isolated bubble, which remains spherical at all times, in an infinite liquid. For this reason, the energetic definition is sometimes inadequate. If the collapse is extremely violent, then the bubble may fragment. This process has been studied for pressure change in a shock tube by Matsumoto and Aoki [48], who found that the number of remaining nuclei/fragments from the collapse increased with increasing gas content in the liquid, and also increased when the shape of the bubble just prior to collapse is aspherical. Fragmentation obviously is incompatible with assumptions of everlasting spherical symmetry, yet it is very important in describing the process by which a medium can be re-seeded with nuclei during very violent cavitation.

Therefore an alternative way to describe violent cavitation is through consideration of bubble life-cycles. As shown in section 4.2.3, the bubble expands to many times its initial radius, and then collapses to a minimum size. This collapse will be violent. Irregularities in the spherical shape of the bubble may develop towards the end of the collapse, and may become more pronounced on rebound. As a result, the bubble may fragment shortly after rebound.

Clearly, bubble fragmentation is incompatible with any assumptions of spherical symmetry, and in this respect one might consider the life-cycle description of violent cavitation to be superior to those based on theoretical energetics. In addition, fragmentation is an important process, enabling a medium to become re-seeded with bubble nuclei. Apfel [26] states that "The physical significance of the transient cavitation threshold is a much discussed point. It may well be that it closely corresponds to the conditions for surface instabilities and thus bubble break up in low viscosity liquids." However, to define transient cavitation on the grounds of fragmentation is wholly inadequate, since it is not a reliable indicator of the energetics. This can be seen in two cases. Firstly, fragmentation often results from asymmetries in the environment, and need not be associated with a particularly violent collapse. A strong asymmetry, such as the proximity of a boundary or another bubble, can cause an oscillating bubble to collapse without the generation of high temperatures or other violent effects. Indeed, if the model allows a bubble to experience shape asymmetries and fragmentation, then during collapse the degree of compression and temperature attained might be reduced from those achieved in the spherical case. Secondly, high-energy spherical oscillations, satisfying all the energetic definitions of transient cavitation, may in some circumstances be achieved without fragmentation.[21]

Flynn [13] describes a transient cavity as one where the initial collapse approximates to that of the Rayleigh cavity, and states that this model of transient cavity is most useful in representing physical bubbles which collapse violently, and then fragment or rebound to repeat the cycle.

In the real world, one is interested on the *effects* of cavitation. Definitions of the above type are useful, but in the practical situation one often really wishes to be able to make a statement about the cavitation, to say whether it is violent enough to produce a given extreme condition (e.g. attain a sufficiently high temperature, generate free radicals, radiate a strong shock wave etc.) or not. The terms 'stable' and 'transient' are perhaps unfortunate, since they suggest a temporal differentiation, when in fact they are most useful in describing the energy and effects of the collapse.

Clearly, in a low-amplitude sound field a bubble will pulsate at low amplitude, a process which is undeniably stable cavitation. As the driving force increases, the amplitude of pulsation in general also increases. A pulsating bubble, which cycle after cycle collapses in a manner which satisfies the energetic definitions of transient cavitation (e.g. inertial function dominates on collapse, $R_{max}/R_0 \geq 2$ or 2.3, $T_{max} \geq 5000$ K, the wall velocity exceeds c at some point), clearly is a form of transient cavitation if, as seems reasonable and useful, one wishes to use this class of cavitation to make a statement about the violence of the effects. Since it is not transient in the temporal sense, for clarification it would perhaps be re-classed as 'repetitive transient cavitation'.

The subdivision of transient cavitation might be completed by defining 'fragmentary transient cavitation', where the collapse satisfies the energetic requirements of a transient collapse and in addition the bubble breaks up. To subdivide further would perhaps give a cumbersome classification. However, for completeness there would be 'cyclic fragmentary transient cavitation', where at least some of the bubble fragments are capable of growth by rectified diffusion or coalescence to seed further transient collapses; and 'non-cyclic fragmentary transient cavitation', where all the bubble fragments dissolve away.

Key to this perception of transient cavitation has been the energetic effects, and it should be noted that initial explosive growth is not necessarily always succeeded by a high-energy

---

[21] See Chapter 5, section 5.2.3.

collapse. For example, a bubble might undergo initial explosive growth to more than twice its initial radius, but if the liquid is saturated with a gas (such as carbon dioxide) which will rapidly come out of solution once a concentration gradient has been set up, then during the prolonged expansion phase the exsolution of gas may significantly reduce the violence of the subsequent collapse. It is important to distinguish whether the acoustic field is continuous-wave or pulsed: for example, following a 'cushioned collapse' of the type just previously mentioned, a bubble would simply pulsate nonlinearly, and the oscillations would damp out, if the insonation were a pulse. If, however, it were continuous-wave, the dynamics of this driven bubble would be completely different. Also, the effect of transients in the motion of the type shown in Figure 4.1 must be considered, as must the fact that it may be inappropriate to discuss a bubble in terms of 'resonance' if the sound is a short pulse [40]. This scheme for subdividing transient cavitation is summarised in Figure 4.18, and bears comparison with the classification illustrated in section 4.4.8, where forms of stable cavitation are similarly analysed.

Even this classification of transient collapse is incomplete. Other behaviour may be observed: gross asymmetries in the environment can lead to such phenomena as jetting, discussed in Chapter 5, section 5.4.1. The question of the surface instabilities, which can be the source of fragmentation, is discussed in the next section.

4.3.3 Surface Instabilities

The above description of transient cavitation refers to its fragmentation through surface instabilities. The bubble expands to many times its initial radius, and then collapses to a minimum size. Either during this first, or a subsequent, collapse, irregularities in the spherical shape of the bubble may develop. Sources of these irregularities include asymmetries in the environment such as the proximity of other bubbles or surfaces (e.g. container walls, upper surface of the liquid body etc.), or gravity. Asymmetries may also arise if there are significant pressure gradients across the bubble, for example, due to focused acoustic fields, of if the bubble is in the near-field of a transducer, or if the acoustic wave length is not very much greater than

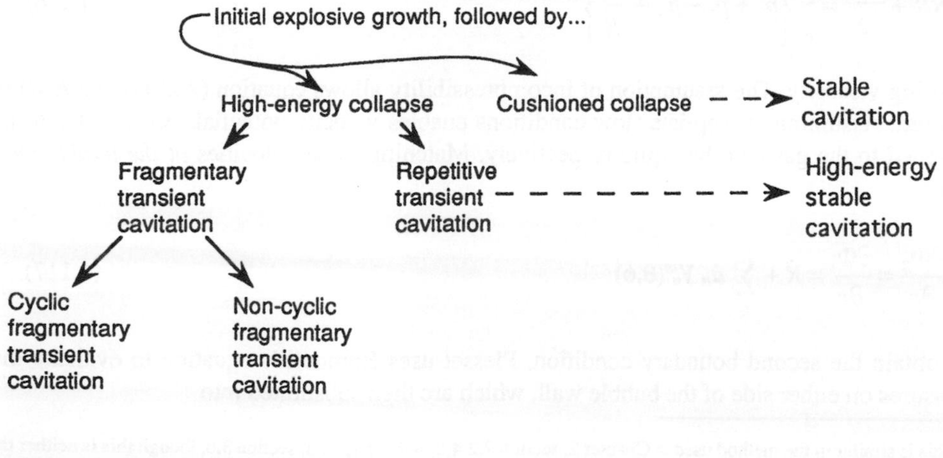

Figure 4.18 Possible classes of transient cavitation.

the bubble radius. The asymmetries in the bubble shape can become more pronounced on rebound, and as a result the bubble fragments shortly afterwards.

This is an important phenomenon. It will set a limit on the maximum pressure and temperatures that can be generated through transient collapse; in addition, fragmentation may re-seed a liquid medium with bubble nuclei.

The first theoretical treatment of the stability of the surface separating two fluids was given by Taylor [49]. He demonstrated that, for a plane surface, the interface is stable only when the acceleration is directed from the liquid to the gas phase. On reworking the theory, this conclusion was verified by Plesset [50], and agrees with the experimental results of Lewis [51]. However, observations showed that the opposite conclusion was true for a spherical surface: the interface becomes unstable on collapse, when the acceleration is directed from liquid to gas. Birkhoff [52, 53] analysed the instability of the collapsing Rayleigh cavity, his results suggesting that the interface is only stable when the acceleration is negative, and is increasing in magnitude as the bubble radius changes at a rate greater than R^{-5}.

(a) Stability of a Spherical Surface

Plesset [50] studied the stability of a spherical surface using spherical harmonics to describe the perturbations.[22] Instability manifests itself in the growth of initially small perturbations. Superimposing a summation of spherical harmonic perturbations on the spherical bubble [54] gives as the equation of the bubble wall

$$\hat{R} = R(t) + \sum_{nm} a(t)_{nm} Y_n^m(\theta,\phi) \tag{4.115}$$

where $Y_n^m(\theta,\phi)$ is a spherical harmonic, of amplitude a_{nm}. The harmonics are assumed to be uncoupled. As in Chapter 2, we are interested in zonal harmonics only, so that $m = 0$ and will not be indicated explicitly in the formulation. The variable R represents the position the unperturbed spherical wall would occupy as, for example, given by equation (4.59), where substitution of $p_L = (p_g + p_v - 2\sigma/R)$ (equations (2.1) and (2.2)) gives

$$R\ddot{R} + \frac{3\dot{R}^2}{2} = \frac{1}{\rho}\left\{p_g + p_v - p_\infty - \frac{2\sigma}{R}\right\} \tag{4.116}$$

ignoring viscosity. The assumption of incompressibility allows equation (2.25) ($r^2\dot{r} = R^2\dot{R}$) to be used. Assuming appropriate flow conditions enables velocity potentials, Φ_g and Φ_l, to be assigned to the gas and the liquid respectively. Matching these velocities *at the bubble wall* gives

$$\frac{\partial \Phi_g}{\partial r} = \frac{\partial \Phi_l}{\partial r} = \dot{R} + \sum_n \dot{a}_n Y_n^m(\theta,\phi) \tag{4.117}$$

To obtain the second boundary condition, Plesset uses Bernoulli's equation to evaluate the pressures on either side of the bubble wall, which are then substituted into

[22] This is similar to the method used in Chapter 2, section 2.2.4(c) and Chapter 3, section 3.6, though this is neither the only, nor sometimes the most convenient way, to describe the perturbation.

$$(p_g + p_v - p_L) = \sigma\left(\frac{1}{\hat{R}_1} + \frac{1}{\hat{R}_2}\right) \tag{4.118}$$

where $\hat{R}_1$ and $\hat{R}_2$ are the principal radii of curvature at the bubble.[23] Plesset applies equation (4.116) and linearises to obtain the stability condition for $n > 0$:

$$\ddot{a}_n + \frac{3\dot{R}\dot{a}_n}{R} + a_n\left\{\frac{\ddot{R}\,[n(n-1)\rho - (n+1)(n+2)\rho_g] - (n-1)n(n+1)(n+2)\sigma/R^2}{R[n\rho + (n+1)\rho_g]}\right\} = 0 \tag{4.119}$$

Introducing viscous effects for the limit when the viscous–diffusion length is very much less than the bubble radius, gives in the limit when the gas density is very much less than that of the liquid ($\rho_g \ll \rho$):

$$\ddot{a}_n + \dot{a}_n\left\{\frac{3\dot{R}}{R} + \frac{2(n+2)(2n+1)\eta_k}{R^2}\right\} + a_n(n-1)\left\{\frac{(n+1)(n+2)\sigma}{\rho R^3} + \frac{2(n+2)\eta_k\dot{R}}{R^3} - \frac{\ddot{R}}{R}\right\} = 0 \tag{4.120}$$

where $\eta_k = \eta/\rho$ is the kinematic viscosity [54]. Use of the substitution $a_n' = a_n R^{2/3}$, reduces equation (4.120) to

$$\ddot{a}_n' + a_n'\left\{(n-1)(n+1)(n+2)\frac{\sigma}{\rho R^3} - \frac{3\dot{R}^2}{4R^2} - \frac{(2n+1)\ddot{R}}{2R}\right\} = 0 \tag{4.121}$$

[21]. It is clear from the above equations that the stability of a spherical surface depends not just on its acceleration, but on its velocity. This is in contrast to the familiar Rayleigh–Taylor instability of a plane surface, which is dependent only on the acceleration. The dependence on velocity is a natural consequence of the spherical geometry [54]: during bubble growth, the streamlines diverge and produce a stabilising effect, whilst the reverse happens during contraction. This stability condition is equally applicable to stable and transient cavities. Shape perturbations during stable cavitation will be discussed later.[24]

(b) Stability of the Surface of a Transient Cavity

Simple interpretations of the above formulation qualitatively illustrate the stability of the cavity. Assume the bubble is growing very much larger than its initial radius (see, for example, Figures 4.8(c) and 4.8(d)). The growth phase is slow, and there is time for sufficient vapour to enter the bubble to maintain constant vapour pressure p_v. When the bubble is large, p_v is likely to be much greater than p_g, since the mass of the original permanent gas phase is small, and the mass flux of dissolved gas out of the liquid is assumed to be much slower than the formation of vapour

[23] See Chapter 2, section 2.1.1.
[24] See section 4.4.6.

through evaporation. Thus if $R \gg R_0$ and the ambient pressure is assumed to be time-independent and the surface tension negligible, equation (4.116) can be integrated once with respect to time in this approximation to give the radial velocity [54]

$$\dot{R} = \sqrt{\frac{2}{3}\frac{p_g + p_v - p_\infty}{\rho}} \tag{4.122}$$

Substitution of this into equation (4.120) shows that a_n tends to a constant value as $R \to \infty$. Thus *the expanding bubble is stable*, any destabilising influence resulting from the acceleration being countered by the stabilisation produced as the streamlines diverge and the surface stretches.

During collapse

$$\dot{R}^2 \approx -\frac{2(p_i - p_\infty)}{(3\rho)}\left(\frac{R_{max}}{R}\right)^3 \tag{4.123}$$

Thus towards the end of the collapse a similar analysis [54] shows that wall velocity $\dot{R}$ goes as $R^{-3/2}$, as indeed does the Rayleigh collapse velocity (equation (2.26)). Substitution of this into equation (4.121) at the limit of $R \to 0$ shows that a_n' has the form $-nw^2R^{-5}$, where w is a real constant. From this, Plesset and Prosperetti [54] find

$$a_n \sim R^{-1/4} \exp\left(\pm iw\sqrt{n} \int R^{-5/2} dt\right) \tag{4.124}$$

Since a_n therefore increases with decreasing radius, and oscillates with increasing frequency as $R \to 0$ [52, 53, 55], the surface is unstable. When a_n reaches a size of order R, the bubble breaks up. The full nonlinear calculation of this phenomenon has been done by Chapman and Plesset [56].

Plesset and Mitchell [55] found that a perturbation of a spherical surface would remain small for $1 \geqslant (R/R_{max}) \gtrsim 0.2$. The instability becomes violent for $R \lesssim R_{max}/10$, that is before the effects of viscosity, surface tension, or liquid compressibility become considerable. Therefore, a collapsing transient cavity initially containing gas at low pressure may expect to become unstable before reaching minimum size, and before the Rayleigh–Plesset equation (equation (4.60)) becomes invalid. However, relatively small bubbles containing gas at higher pressures will probably remain stable.

Benjamin [57] undertook an independent investigation, and formulated the stability criterion in terms of R_{min}, the *minimum* radius attained by the collapsing transient, as opposed to the maximum at the start of the collapse as used in the above formulation. Benjamin found that a large disturbance would not occur until

$$\left(\frac{R}{R_{min}}\right)^{3(\gamma-1)} < \gamma + \frac{\gamma - 1}{2n} \tag{4.125}$$

where n is again the order of the spherical harmonic.

4.3.4 Experimental Observation of Cavitation

Figures 4.8(c) and 4.8(d) indicate the response of a bubble of much less than resonance size to a 10-kHz high intensity sound wave ($P_A = 0.24$ MPa). Following sudden explosive growth, the bubble pauses at some maximum size which is insensitive to its initial size, and then collapses rapidly and irreversibly (i.e. without rebound). This behaviour has been photographed at high speed (8000 f.p.s.[25]) in Figure 4.19, the cavitation being viewed from above on axis in the cylindrical transducer described in Chapter 1, section 1.2.2(b)(i). The films capture the onset of insonation with a 10-kHz sinusoidal sound wave[26] of amplitude 2.4 bar.

Figure 4.19 shows eight frames from a section of film. Insonation begins between frames 1 and 2. Two bubbles of just less than resonance size (labelled A and B, with $R_o \approx 0.10 \pm 0.05$ mm) can be seen quiescent in frame 1. They begin to oscillate in frame 2. In frame 4, further bubbles, which were initially too small to be seen, grow to visible size. These bubbles have collapsed unstably and have all gone by frame 6.

It is, however, interesting to note the behaviour of the stably oscillating bubbles. Once insonation begins, bubbles A and B are driven together, and coalesce in frame 4 to form a single bubble (labelled C). Bubble C exhibits violent surface oscillations which in frame 5 almost break it up again. A third bubble (labelled D), also oscillating stably, travels into the depth of field in frame 3. It continues travelling perpendicularly to the picture, and by frame 8 is leaving the optical focus. The cause of attraction between bubbles A and B are radiation forces. Whilst a transient cavity usually does not persist long enough for these forces to influence it significantly, radiation forces can clearly be powerful influences on the motion of a stable cavity.

In his review paper, Neppiras [21] cites two experimental observations of historical importance. If a sample of aerated water is subjected to a relatively low-intensity focused sound field, large bubbles may be seen to levitate. At these pressures the acoustic emissions of such bubbles are small. At higher pressures, surface oscillations may set in, and the bubble surface will appear to shimmer. As the sound pressure amplitude is increased, 'streamers' are generated, with the emission of a hissing sound. These misty, ribbon-like structures consist of stable bubbles travelling at high speed through the liquid to the focus under radiation forces.[27] Blake [59] termed the event 'gaseous cavitation' as they are observed only in aerated liquids, and found that there was a threshold acoustic pressure for their generation. He observed a second threshold at higher pressures, where single short-lived bubbles formed and collapsed intermittently, which he termed 'vaporous cavitation'. Such events were accompanied by a sharp 'snap' noise. As the pressure increased, the rate of these events increased (though sometimes Blake observed them to disappear completely). This form of cavitation could be observed in completely and partially degassed water. As Flynn [13] points out, the terms 'gaseous' and 'vaporous' are merely labels of convenience: Blake did not actually know anything about the contents of his bubbles. One might assume that, if a bubble grows from an initially stabilised gas nucleus, there would be some component of the bubble contents that is a permanent gas phase. Then again, if the expansion has been great, a vapour content would be expected. One can only say that such entities may contain, gas, vapour, or a mixture.

Blake observed a short-lived transient cavity. Willard [60] describes another kind. A nucleus is carried into a region of high-intensity pressure variations, and pulsates at very large amplitude. It collapses and rebounds, the shock wave it emits during rebound propagating radially outwards

[25] f.p.s. = frames per second.

[26] Since the acoustic frequency was 10 kHz, to satisfy the Nyquist criterion the sampling rate would have to be at least 20 kHz to avoid the possibility of misleading frequency information being present in a given sample. The maximum framing rate of the Hadland Hyspeed camera used to produce these full-frame images is 8000 f.p.s., so therefore care is required when interpreting the results. The possibility exists that what appears to be a one-off event in an isolated 8-frame sample may in fact be periodic: to ensure that it is transitory in nature and coincident with the onset of insonation, many films were taken. Since the inter-frame time is longer than the acoustic period, the stably oscillating bubbles (A, B, C and D) undergo considerable volume change during each frame, and their perimeters are indistinct.

[27] See section 4.4.1(b).

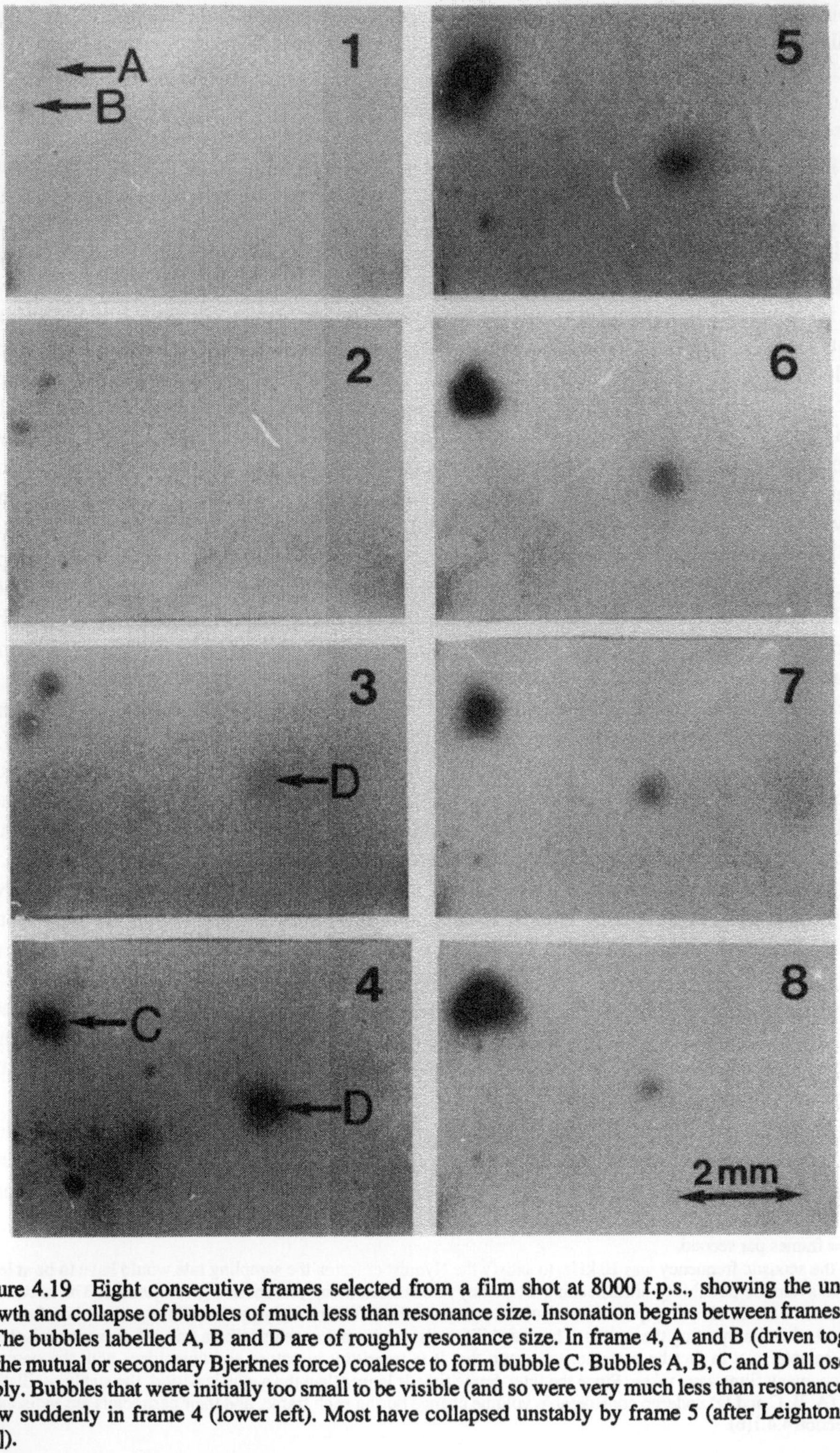

Figure 4.19 Eight consecutive frames selected from a film shot at 8000 f.p.s., showing the unstable growth and collapse of bubbles of much less than resonance size. Insonation begins between frames 1 and 2. The bubbles labelled A, B and D are of roughly resonance size. In frame 4, A and B (driven together by the mutual or secondary Bjerknes force) coalesce to form bubble C. Bubbles A, B, C and D all oscillate stably. Bubbles that were initially too small to be visible (and so were very much less than resonance size) grow suddenly in frame 4 (lower left). Most have collapsed unstably by frame 5 (after Leighton *et al.* [58]).

and causing the explosion of nuclei smaller than the original one. Though this occurs as a result of a periodic sound field, the active agent in this transient cavitation is a shock wave.

The radiation forces which are so evident in the observations described in this section will now be discussed in greater detail.

4.4 Stable Cavitation

In section 4.1, the ideal spring–bob system was used to illustrate the driven harmonic oscillator. Initially, the undamped condition was studied to bring out the main features of the behaviour (such as the phase change that occurs in the response as the system passes through resonance). The modifications to this behaviour resulting from damping were then investigated.

We have already seen from the solutions to the Rayleigh–Plesset equation how the bubble approximates to a linear oscillator when driven at low amplitude. Similarly the exponentially decaying acoustic sinusoid emitted by a freely oscillating bubble (Figure 3.23) tells us that such bubbles are lightly damped.[28] It is therefore reasonable to begin the investigation of the driven stably-pulsating bubble with the approximation of linear harmonic behaviour and no damping.

4.4.1 Radiation Forces

A pulsating bubble, driven by a sound field, will absorb and scatter the incident acoustic wave. As seen in Chapter 1, section 1.1.4, a travelling wave has a momentum associated with it, absorption of which can give rise to a radiation pressure. Therefore one would expect radiation pressure forces on a bubble which is being driven significantly in a travelling-wave field, which is to say the bubble is not very much greater than, or very much less than, its resonance. In a standing-wave field, which can be thought of as being two identical travelling-waves propagating along the same axis in opposite directions, one might at first sight think that there would be no net radiation pressure force, the contributions from the two component travelling-waves cancelling. This is only true if the bubbles are at certain key positions in the sound field: at other points in the standing-wave field the time-averaged radiation forces are unbalanced, and tend in fact to force the bubbles to those afore-mentioned key positions. Therefore, given an arbitrary starting position, a bubble in a standing-wave field will be driven by radiation forces to a certain location where, all other things being equal, they will remain. Bubbles do not simply react to radiation forces impressed by the incident driving wave: the pressure field radiated by one pulsating bubble can generate radiation forces on a second.

In order to demonstrate that a body of volume V in a pressure gradient $\vec{\nabla}p$ experiences a force $-V\vec{\nabla}p$, consider the cartesian volume element shown in Figure 4.20, which is stationary in the fluid. In the x-direction, the pressure on the face at $x = x_0$ is $p(x_0)$, and the pressure on the face at $x = x_0 + \Delta x$ is $p(x_0 + \Delta x) = p(x_0) + (\partial p/\partial x)\Delta x$. The area of the face perpendicular to the x-axis is $\Delta y \Delta z$, and the product of this area with the difference in pressure between the two faces gives the x-component of the force:

$$F_x = -\left.\frac{\partial p}{\partial x}\right|_{x_0} \Delta x\,\Delta y\,\Delta z \qquad (4.126)$$

[28]See Chapter 3, section 3.1.3, Figure 3.3.

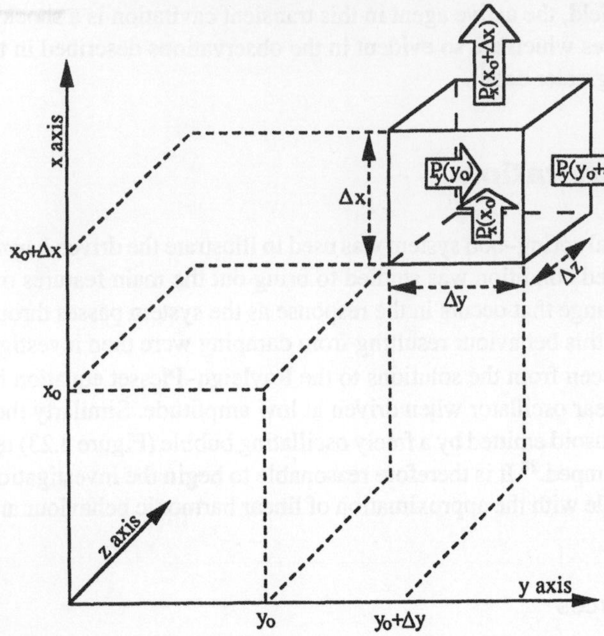

Figure 4.20 Consideration of the cartesian volume element which is stationary in the fluid demonstrates that a body of volume V in a pressure gradient $\vec{\nabla} p$, experiences a force $-V\vec{\nabla} p$.

Similarly the components of the force in y and z are $F_y = -(\partial p/\partial y)\Delta x \Delta y \Delta z$ and $F_z = -(\partial p/\partial z)\Delta x \Delta y \Delta z$ respectively. Since the product $\Delta x \Delta y \Delta z$ equals the volume of the element chosen, then this analysis gives the vector force to be $-V\vec{\nabla} p$. If this quantity varies in time, then the net force on the body is simply the time average of this, i.e.

$$<\vec{F}> = -<V\vec{\nabla} P> \tag{4.127}$$

where '< >' denotes the time-average. The general symbol for these bubble-related radiation forces will be $\bar{F}$, and implicit in this symbol is the idea that these are *time-averaged* forces, as is clear from equation (4.127). The treatments will be uni-axial, so that the vector nature of the force is incorporated into the sign of $\bar{F}$. The travelling-wave radiation force is $\bar{F}_{trav}$, and the radiation force exerted by a pulsating bubble on an particle is $\bar{F}_{part}$. The radiation force on a bubble in a standing-wave field ($\bar{F}_{B1}$), and exerted on one pulsating bubble by the sound field radiation by another, are termed the *primary* and *secondary* (or *mutual*) Bjerknes forces, respectively. These radiation forces will now be discussed.

(a) The Radiation Force on a Bubble in Travelling-wave Conditions

Consider a sinusoidal travelling acoustic plane wave. As the point of observation moves in the direction of propagation (the z-axis), the amplitude of the wave is constant but the phase varies, and the pressure is the sum of the acoustic and static terms:

$$p = p_0 + P_A \cos(\omega t - kz) \tag{4.128}$$

Since the only variation is in the z-direction, $\vec{\nabla} p = \frac{\partial p}{\partial z}$, giving

$$\vec{\nabla} p(z,t) = kP_A \sin(\omega t - kz) \tag{4.129}$$

If a bubble were placed at any point along the line of propagation, that is at any coordinate z, its pulsation would follow

$$R = R_0 - R_{\varepsilon 0}\cos(\omega t - kz - \vartheta) \tag{4.130}$$

Since we have assumed small-amplitude linear oscillations throughout this derivation, then $R_{\varepsilon 0} \ll R_0$. Therefore the bubble volume $V(t) = 4\pi R(t)^3/3 = 4\pi R_0^3(1+R_\varepsilon/R_0)^3/3$ may be approximated to first order as

$$V(t) \approx V_0(1 + 3R_\varepsilon/R_0) = V_0\{1 - (3R_{\varepsilon 0}/R_0)\cos(\omega t - kz - \vartheta)\} \tag{4.131}$$

where $V_0 = 4\pi R_0^3/3$, the equilibrium bubble volume. Substitution of equations (4.129) and (4.131) into equation (4.127) gives

$$\bar{F}_{trav} = -<V_o kP_A\{1 - (3R_{\varepsilon o}/R_o)\cos(\omega t - kz - \vartheta)\}\sin(kz - \omega t)>$$

$$= -V_o kP_A\left\{<\sin(kz - \omega t)> - \frac{3R_{\varepsilon 0}}{R_0}(<\sin^2(\omega t - kz)\sin\vartheta> + <\cos(\omega t - kz)\sin(\omega t - kz)>\cos\vartheta)\right\} \tag{4.132}$$

The $<\sin(kz - \omega t)>$ and the $<\cos(\omega t - kz)\sin(\omega t - kz)>$ terms average out to zero, and $<\sin^2(kz - \omega t)> = 1/2$, so that

$$\bar{F}_{trav} = V_0 kP_A \frac{3R_{\varepsilon 0}}{2R_0}\sin\vartheta \tag{4.133}$$

Substituting for $\sin\vartheta$ from equation (4.12), and for $|A|$ from equation (4.13), gives

$$\bar{F}_{trav} = V_0 kP_A \frac{3R_{\varepsilon 0}}{2R_0}\frac{2\beta_{tot}/\omega}{\sqrt{\{(\omega_0/\omega)^2 - 1\}^2 + (2\beta_{tot}/\omega)^2}} \tag{4.134}$$

The amplitude of oscillation $R_{\varepsilon 0}$ can be substituted from equation (4.23) to yield

$$\bar{F}_{trav} = \frac{3}{2}\frac{P_A^2 V_0 k}{R_0^2\rho\omega^2}\frac{2\beta_{tot}/\omega}{(\{(\omega_0/\omega)^2 - 1\}^2 + (2\beta_{tot}/\omega)^2)}$$

$$= \frac{3}{2}\frac{P_A^2 V_0 k}{R_0^2\rho\omega_0^2}\frac{2\beta_{tot}/\omega}{(\{1 - (\omega/\omega_0)^2\}^2 + (2\beta_{tot}\,\omega/\omega_0^2)^2)} \tag{4.135}$$

This might be simplified further by employing one of the approximate relationships for the product $(R_0\omega_0)$. For example, approximating ω_M for ω_0 (equation (3.38)), and using $\omega = ck$, reveals

$$\bar{F}_{trav} \approx \frac{1}{2}\frac{P_A^2 V_0}{\gamma p_0 c}\frac{2\beta_{tot}}{(\{1 - (\omega/\omega_0)^2\}^2 + (2\beta_{tot}\,\omega/\omega_0^2)^2)} \tag{4.136}$$

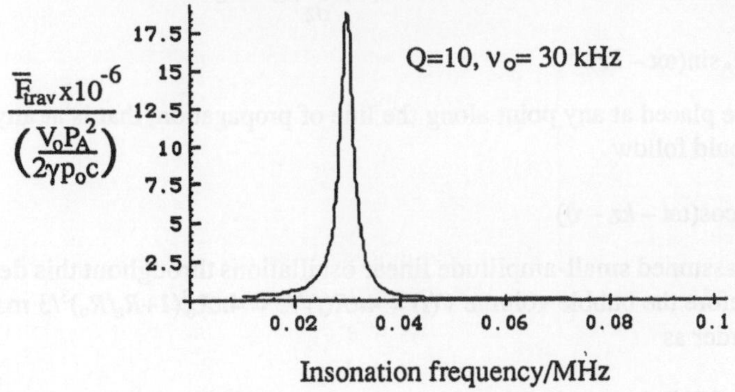

Figure 4.21 The time-average radiation force on a linearly oscillating bubble with resonance frequency $\nu_o = 30$ kHz in a travelling-wave field, shown as a function of the frequency of that field, where the bubble damping is assumed to be frequency-invariant ($\delta_{tot} = 0.1$, $Q = 10$).

The form of this is plotted in Figure 4.21, with the approximation that the damping is frequency-invariant, where $Q \approx \omega_o/2\beta_{tot}$.

The limiting forms of equation (4.136) for bubbles away from resonance are readily found in terms of $R_{\varepsilon o}$ through substitution for ϑ in equation (4.133). For bubbles significantly smaller than, or significantly larger than resonance, $\sin\vartheta = 0$ ($\vartheta = 0$ and π respectively). The force increases as the bubbles tend to resonance size. At resonance, when $\omega = \omega_o$, the force on a bubble in a travelling wave as given by equation (4.136) reduces to the ratio of the mean power input, as given by equation (4.19), to the speed of sound in the liquid [61]. The treatment for the general oscillator given in section 4.1 shows that the mean power input $<\dot{W}>$ equals the energy dissipated by the oscillator per cycle, which at resonance equals $\omega_o^2|A|^2\beta_{tot}m$ (equation (4.17)). For the case of the pulsating bubble, $|A| = -R_{\varepsilon o}$, and $m = 4\pi R_o^3\rho$. Therefore the ratio of the mean power input to the speed of sound at resonance is

$$\frac{<\dot{W}>}{c} = \frac{\omega_o^2 R_{\varepsilon o}^2\, \beta_{tot} 4\pi R_o^3 \rho}{c} \tag{4.137}$$

Equation (4.23) shows that at $\omega = \omega_o$, $R_{\varepsilon o}^2 = P_A^2/(4\rho^2 R_o^2 \beta_{tot}^2 \omega_o^2)$, and substituting this into equation (4.137) yields

$$\frac{<\dot{W}>}{c} = \frac{P_A^2 \pi R_o}{\rho \beta_{tot} c} \tag{4.138}$$

From equation (4.135), the force on a bubble resonating in a travelling-wave field is

$$\bar{F}_{trav} = \frac{3}{4}\frac{P_A^2 V_o k}{R_o^2 \rho \omega_o\, \beta_{tot}} \qquad (\omega = \omega_o) \tag{4.139}$$

Since $V_o = 4\pi R_o^3/3$, and at resonance $k = \omega_o/c$, comparison of equations (4.138) and (4.139) shows that

$$\bar{F}_{trav} = \frac{<\dot{W}>}{c} \quad \text{when } \omega = \omega_0 \text{ (resonance).} \tag{4.140}$$

From equation (3.292) this can be related to the extinction cross-section of the bubble:

$$\bar{F}_{trav} = \frac{I\Omega_b^{ext}}{c} \quad (\omega = \omega_0) \tag{4.141}$$

From equation (4.139), this force at resonance can be calculated explicitly by noting that at $\omega = \omega_0$ the damping constant may be related to $Q = 1/\delta_{tot} = \omega_o/2\beta_{tot} \approx \omega_0/2\beta_{tot}$ (Chapter 3, section 3.4.1). The dimensionless damping coefficient δ_{tot} equals the sum $\delta_{rad} + \delta_{th} + \delta_{vis}$. Substituting this into equation (4.139) and using the fact that $k = \omega_0/c$ at resonance yields

$$\bar{F}_{trav} \approx \frac{3}{2}\frac{P_A^2 V_0 k}{R_0^2 \rho \omega_0^2 \delta_{tot}} = \frac{2\pi P_A^2 R_0}{\rho c \omega_0 \delta_{tot}}$$

$$\approx \frac{2\pi P_A^2 R_0^2}{c\delta_{tot}\sqrt{3\kappa p_0 \rho}} \approx \frac{2\pi P_A^2}{c\delta_{tot}}\sqrt{\frac{3\kappa p_0}{\rho^3}}\,\frac{1}{\omega_0^2} \quad (\omega = \omega_0) \tag{4.142}$$

Here, equation (3.36) is used to approximately relate ω_0 to R_0, demonstrating that the radiation force at resonance is proportional to R_0^2 and ω_0^{-2}. Equating (4.140) to (4.142) and employing $I = P_A^2/(2\rho c)$ for plane waves (equation (1.39)) gives for the extinction cross-section of a resonating bubble $\Omega_b^{ext} = 4\pi R_0 c/(\omega_0 \delta_{tot})$, in agreement with equation (4.36). Yosioka and Kawasima [62] detail the radiation force on a bubble in a travelling-wave field.

(b) The Radiation Force on a Bubble in a Standing-wave Field: the Primary Bjerknes Force

Now consider a bubble in a standing-wave field. The pressure gradient oscillates, as does the bubble volume. As shown in section 4.1.1(b), a bubble of less than resonant size oscillates in-phase with the sound field, and bubbles larger than resonance oscillate π out-of-phase with the field. (It should be noted that since the positive driving pressure causes a reduction in volume, then the bubble volume will be a minimum when the pressure is a maximum if the two are oscillating in phase.) Therefore the quantity $-<V\vec{\nabla}P>$ will be in one direction for bubbles with $R < R_r$ and in the opposite direction for those with $R > R_r$, since obviously $<V\vec{\nabla}P>$ averages out to be in the direction that $\vec{\nabla}P$ takes when the bubble volume is largest. Therefore, in a standing-wave field, bubbles of less than resonant size travel up a pressure gradient towards the pressure antinodes, and those larger than resonance travel down the gradient to the node.

To formalise the argument, consider a bubble located at position y in the standing-wave field. In such a field, the amplitude of the pressure varies with position, but the phase of the wave does not.[29] Therefore the pressure experienced by the bubble, which is the sum of static and time-dependent terms, is

$$p(y,t) = p_0 + 2\hat{P}_A \sin(ky)\cos(\omega t) \tag{4.143}$$

where $\hat{P}_A$ is the acoustic pressure amplitude of the incident plane wave which, upon reflection from a free or rigid boundary, sets up the standing-wave field. From equation (4.143) then

[29]See Chapter 1, section 1.1.6.

$$\vec{\nabla} p(y,t) = 2k\hat{P}_A \cos(ky)\cos(\omega t) \tag{4.144}$$

where k is the wave vector and p_o is assumed to be constant. If $2\hat{P}_A \ll p_o$, then the bubble radius $R(t)$ will oscillate linearly as

$$R(t) = R_o - (R_{\varepsilon oa}\sin(ky))\cos(\omega t - \vartheta) \tag{4.145}$$

where $\vartheta = 0$ for bubbles much smaller than resonance, and $\vartheta = \pi$ for bubbles much larger than resonance (Figure 4.4(b)). Comparison of (4.145) with equations (4.21) and (4.22) shows that $-(R_{\varepsilon oa}\sin(ky))$ represents the radial amplitude of oscillation, the sin(ky) dependence arising since the acoustic pressure amplitude experienced by the bubble varies as sin(ky) in the standing-wave field. The negative sign in equation (4.145) follows from equation (4.21), since a positive acoustic pressure causes a reduction in bubble volume when the two are in phase.

It is possible to obtain the form of $R_{\varepsilon oa}$ explicitly by applying equation (4.23), and noting that the amplitude of the driving pressure is $2\hat{P}_A \sin ky$. Therefore replacing P_A in equation (4.23) by $2\hat{P}_A \sin ky$ yields

$$R_{\varepsilon o} = \frac{2\hat{P}_A \sin ky}{R_o \rho} \frac{1}{\sqrt{(\omega_o^2 - \omega^2)^2 + (2\beta_{tot}\omega)^2}} \tag{4.146}$$

Comparison with equation (4.145) therefore gives the amplitude of radial oscillation at the pressure antinode:

$$R_{\varepsilon oa} = \frac{2\hat{P}_A}{R_o \rho} \frac{1}{\sqrt{(\omega_o^2 - \omega^2)^2 + (2\beta_{tot}\omega)^2}} \tag{4.147}$$

Since we have assumed small-amplitude linear oscillations throughout this derivation, then $R_{\varepsilon oa} \ll R_o$. Therefore the bubble volume $V(t) = 4\pi R(t)^3/3 = 4\pi R_o^3(1 + R_\varepsilon/R_o)^3$ may be approximated to first order as

$$V(t) \approx V_o(1 + 3R_\varepsilon/R_o) = V_o\left\{1 - (3R_{\varepsilon oa}/R_o)\sin(ky)\cos(\omega t - \vartheta)\right\} \tag{4.148}$$

where $V_o = 4\pi R_o^3/3$, the equilibrium bubble volume. Substitution of (4.144) and (4.148) into (4.127) gives the primary Bjerknes force $\bar{F}_{B1}$ on the bubble to be

$$\bar{F}_{B1} = -\langle V \vec{\nabla} P \rangle$$

$$= -2k\hat{P}_A V_o\left(\cos ky \langle\cos\omega t\rangle - \frac{3R_{\varepsilon oa}}{R_o}(\sin ky \cos ky)\langle\cos\omega t \cos(\omega t - \vartheta)\rangle \right) \tag{4.149}$$

The time-averages evaluate as follows: $\langle\cos\omega t\rangle = 0$; and $\langle\cos\omega t\cos(\omega t - \vartheta)\rangle = \langle\cos^2\omega t\rangle\cos\vartheta + \langle\cos\omega t\sin\omega t\rangle\sin\vartheta$, the second term of which is zero. Since $\sin ky \cos ky = (\sin 2ky)/2$, then the primary Bjerknes force on the bubble is

$$\bar{F}_{B1} = \frac{3\hat{P}_A k R_{\varepsilon oa} V_o \sin(2ky)}{2R_o}\cos\vartheta \tag{4.150}$$

Substituting for $\cos\vartheta$ from equation (4.11) and for $|A|$ from (4.13), gives

$$\bar{F}_{B1} = \frac{3\hat{P}_A k R_{\varepsilon oa} V_0 \sin(2ky)}{2R_0} \frac{(\omega_0^2 - \omega^2)}{\sqrt{(\omega_0^2 - \omega^2)^2 + (2\beta_{tot}\omega)^2}} \tag{4.151}$$

Substitution for $R_{\varepsilon oa}$ from equation (4.147) enables formulation of the primary Bjerknes force explicitly in terms of $\hat{P}_A$, the acoustic pressure amplitude of the incident plane wave which, upon reflection, generated the standing-wave field:

$$\bar{F}_{B1} = \frac{4\pi R_0 \hat{P}_A^2 \omega \sin(2ky)}{\rho c} \frac{(\omega_0^2 - \omega^2)}{((\omega_0^2 - \omega^2)^2 + (2\beta_{tot}\omega)^2)}$$

$$= \frac{4\pi R_0 \hat{P}_A^2 \sin(2ky)}{\omega \rho c} \frac{(\omega_0/\omega)^2 - 1}{(((\omega_0/\omega)^2 - 1)^2 + (2\beta_{tot}/\omega)^2)} \tag{4.152}$$

incorporating $V_0 = 4\pi R_0^3/3$ and $\omega = ck$.

In order to appreciate the behaviour of the bubbles, consider first the limiting forms of equation (4.150). For bubbles smaller than resonance ($\omega_0 > \omega$) such that $\vartheta = 0$, equation (4.150) suggests that the force is

$$\bar{F}_{B1} = \frac{3\hat{P}_A k V_0 \sin(2ky)}{2R_0} R_{\varepsilon oa} \qquad (\omega_0 > \omega) \tag{4.153}$$

If the bubble is larger than resonance ($\omega_0 < \omega$) such that $\vartheta = \pi$, equation (4.150) suggests that the force is [63]

$$\bar{F}_{B1} = -\frac{3\hat{P}_A k V_0 \sin(2ky)}{2R_0} R_{\varepsilon oa} \qquad (\omega_0 < \omega) \tag{4.154}$$

Comparing this with the sin(ky) variation of the pressure field in equation (4.143) supports the statements made earlier, that if the bubble is smaller than resonance size it will tend to move to the pressure antinode, and if the bubble is larger than resonance size it will tend to move to the node.[30] Figure 4.22 gives a simple graphical explanation showing the response to a standing-wave field of bubbles of greater than, and less than, resonance size, such that $\vartheta = \pi$, 0 respectively. However, it is important to note that the forces, as given by equations (4.153) and (4.154), depend on the value of $R_{\varepsilon oa}$ at the relevant frequency. This is greatest close to resonance, and decreases away from resonance (Figure 4.4). The behaviour can be most readily shown by making the approximation that the bubble damping is independent of frequency, and equation (4.152) can be re-written using $2\beta_{tot}/\omega \approx \omega_0/(\omega Q)$, where Q is the quality factor. In the limit of this approximation, the primary Bjerknes force is given by

$$\bar{F}_{B1} \approx \frac{4\pi R_0 \hat{P}_A^2 \omega \sin(2ky)}{\rho c} \frac{(\omega_0/\omega)^2 - 1}{(((\omega_0/\omega)^2 - 1)^2 + (\omega_0/\omega)^2(1/Q)^2)} \tag{4.155}$$

This enables a graphical representation of the force experienced by bubbles for different values of ω_0/ω (which is small for large bubbles, and vice versa) at the various positions in the sound field. This is shown in Figure 4.23 as a surface plot, the two horizontal axes representing ky (for one wavelength) and ω_0/ω (in the range $\omega/2 \leqslant \omega_0 \leqslant 2\omega$). The height of the surface, as

---

[30] Versions of this proof appear in the literature [12, 64] where the variation of the amplitude of wall motion with position in the field is neglected: the resulting formulation is incorrect.

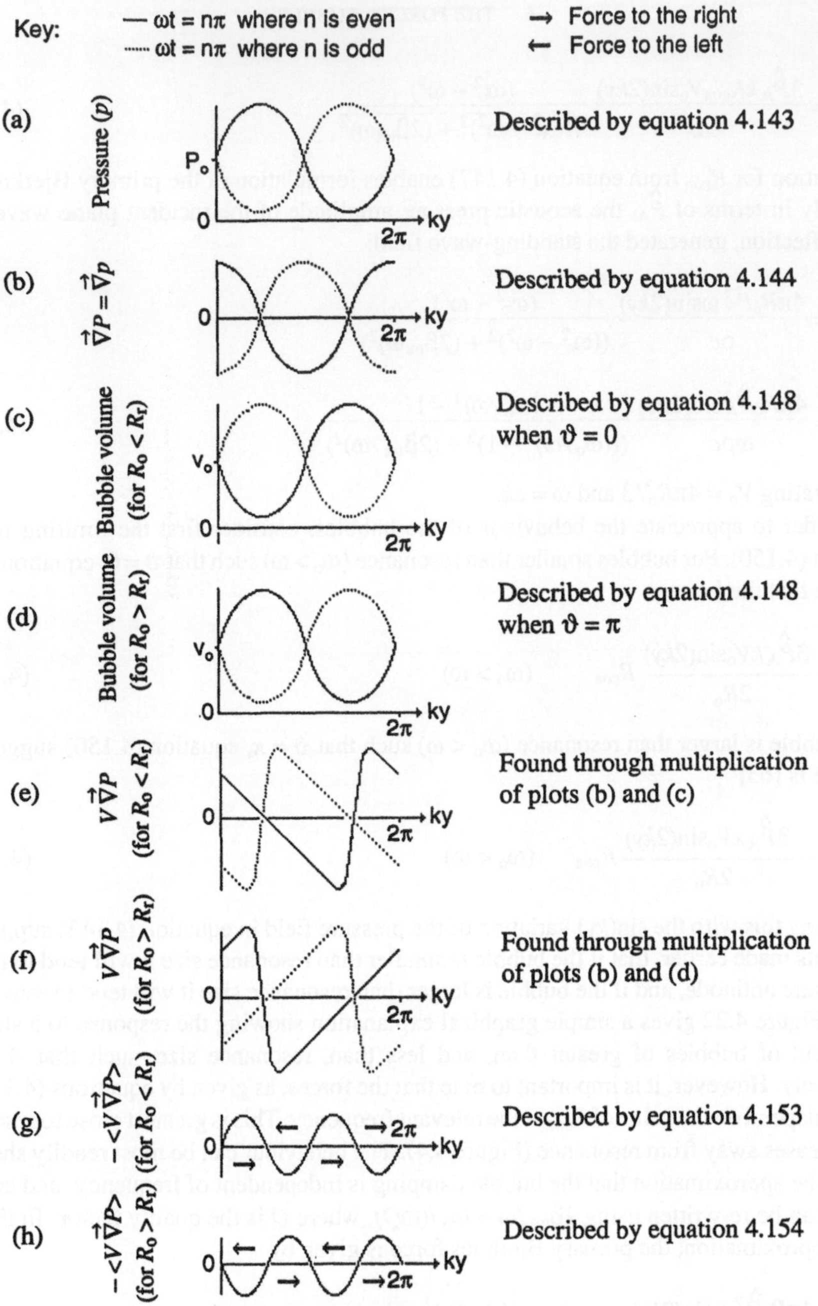

Figure 4.22 A simple graphical explanation showing the response to a standing-wave field of bubbles of greater than, and less than, resonance size, such that $\vartheta = \pi$, 0 respectively (after Leighton *et al.* [63]). Reprinted by permission from *European Journal of Physics*, vol 11, pp. 47–50; Copyright © 1990 IOP Publishing Ltd. The entire scheme can be derived qualitatively from the known pressure field (a) and the phase of the oscillating bubble ((c) and (d)). Plot (b) is obtained from the gradient of (a). The multiplication of the plots which generate (e) = (b) × (c) and (f) = (b) × (d) is readily done by noting the positions of the zeros in (b), (c) and (d), and the sign of the functions between those zeros. The sign of the functions in (g) and (h) is found through examination of the comparative magnitude of the functions in plots (e) and (f) in the positive and negative phases of the cycle. Comparison of the force (plots (g) and (h)) with the pressure field (a) yields the expected result.

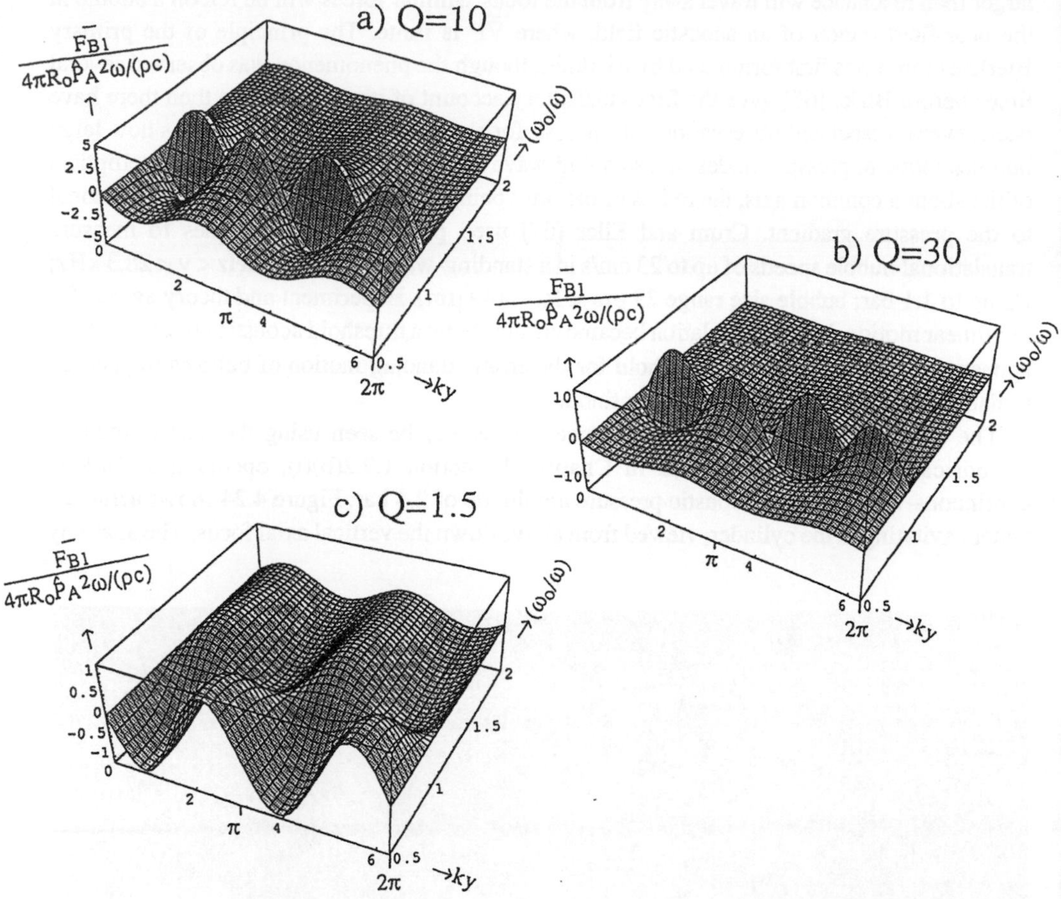

Figure 4.23 A surface plot of the normalised primary Bjerknes force, positive corresponding to a force in the direction of increasing *y*, and negative corresponding to decreasing *y*. The two horizontal axes represent *ky* (for one wavelength) and ω_0/ω (in the range $\omega/2 \leq \omega_0 \leq 2\omega$). Bubble oscillators damped such that: (a) $Q = 10$ (equivalent to $\delta_{tot} = 0.1$); (b) $Q = 30$ ($\equiv \delta_{tot} \approx 0.033$); and (c) $Q = 1.5$ ($\equiv \delta_{tot} \approx 0.67$).

measured against the vertical axis, gives the normalised primary Bjerknes force, positive corresponding to a force in the direction of increasing *y*, and negative corresponding to decreasing *y*. Bubble oscillators damped such that $Q = 10$ (equivalent to $\delta_{tot} = 0.1$) and $Q = 30$ ($\equiv \delta_{tot} \approx 0.033$) are shown in Figures 4.23(a) and 4.23(b) respectively. The amplitude of the force in both cases is large close to resonance, being larger for the higher *Q* i.e. lighter damping (note the change in scale), and undergoes a phase change there. This behaviour is expected, given the known behaviour of the damped driven oscillator (Figure 4.4). The decrease in amplitude as the bubble is increasingly off-resonance reduces the magnitude of the primary Bjerknes force on the bubble. For interest, the variation in the force for a bubble with relatively high dissipation ($Q = 1.5 \Rightarrow \delta_{tot} \approx 0.67$), seldom encountered in the physical world, is shown in Figure 4.23(c).

It should be remembered that the primary Bjerknes forces described above are active not just in standing-wave fields, but in any field containing a pressure gradient. Therefore in a focused acoustic field, bubbles below resonance will travel to the focal pressure antinode, and those larger than resonance will travel away from the focus. Similar forces will be felt on a bubble in the near-field region of an acoustic field, where $\vec{\nabla} P$ is finite. The principle of the primary Bjerknes force was first formulated by Bjerknes, though the phenomenon was observed several times before Blake [65] gave the first satisfactory account of its origin. Since then there have been several associated observations of interest: for example, Miller [66] describes how large bubbles move to pressure nodes in a standing-wave field. The bubbles then move in elliptical orbits about a common axis, the radius of the orbit being approximately inversely proportional to the pressure gradient. Crum and Eller [67] used photographic techniques to measure translational bubble speeds of up to 23 cm/s in a standing-wave field (23.6 kHz $< \nu <$ 28.3 kHz; P_A up to 1.1 bar; bubble size range 29 μm $< R_0 <$ 149 μm). Experiment and theory agreed for rectilinear motion, but the translation became erratic above a threshold acoustic pressure, which appeared to coincide with the threshold for the erratic dancing motion of bubbles trapped by radiation forces close to the pressure antinode.

The effect of the primary Bjerknes force can readily be seen using the axially focused cylindrical sound field described in Chapter 1, section 1.2.2(b)(i), operating at 10-kHz continuous-wave, with an acoustic pressure amplitude of 2.4 bar. Figure 4.24 shows aerated[31] water cavitating in the cylinder, viewed from above down the vertical axial focus. The anatomy

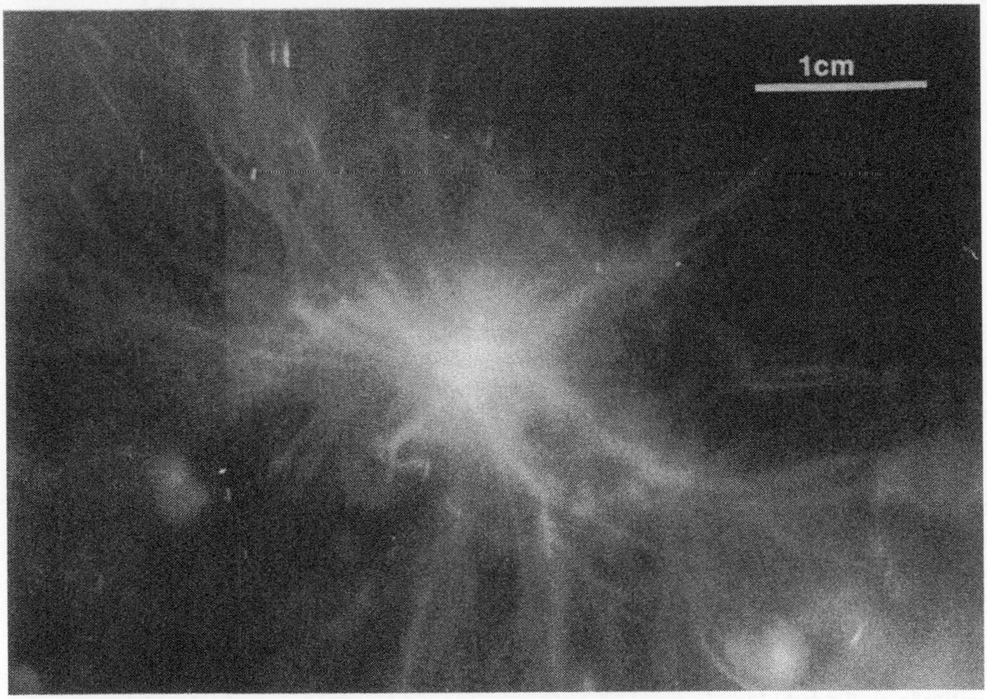

Figure 4.24 A photograph of aerated water cavitating in a cylindrically focused 10-kHz sound field (transducer shown in Figure 1.20), viewed from above down the vertical axial focus, where the pressure amplitude is 2.4 bar. Exposure time 1/30th s. Streamers are clearly visible. (Photograph: TG Leighton.)

[31] 'Aerated' is commonly taken to mean water freshly drawn from the tap. If the water has been treated in any way (e.g. distilled, degased, filtered of deionised) it is usual to state this explicitly. Thus, for example, the aerated water used here has, from the terminology, not been filtered, and so would be expected to contain many motes.

of this structure is explained in Figure 4.25. At the axial pressure focal antinode is a dense cloud of bubbles of less than resonance size, drawn there by the primary Bjerknes force from nucleation sites within the liquid. The bubbles travel to the focus in ribbon-like structures, of about 3-cm length, which are the 'streamers', following one another in a manner similar to bubbles rising in beer.[32] The nucleation sites from which they are continually generated, at the origin of each streamer, is relatively stationary within the liquid (the exposure of the frame was 1/30th s). Being unaffected by the primary Bjerknes force, they are probably sited on solid motes. In such an intense sound field, free-floating stabilised gas pockets, and nuclei created through cosmic ray action, would be driven to the pressure antinode. As a light-hearted illustration, a microtube containing air is inserted into the water. Two photographs (Figures 4.26 and 4.27), taken with a flash exposure of about 1 ms, show the result, with the important features arrowed: 1 – the uppermost rim of the piezoceramic; 2 – the microtube; 3 – the tip of the microtube; 4 – the inner surface of the upper piezoceramic; 5 – the focal pressure antinode; 6 – a line of bubbles travelling to the focus. Figure 4.26 shows intense activity at the tip of the tube, from which bubbles are sheared. Intense surface wave activity is visible on the large bubble, which is generating microstreaming circulation patterns,[33] revealed in the translations of small bubbles nearby. Figure 4.27 shows a line of bubbles from the microtube travelling to the focus in the same manner as other lines which have crevice sources. The use of the microtube is for illustration only, and is not meant to model a gas pocket in a mote accurately.

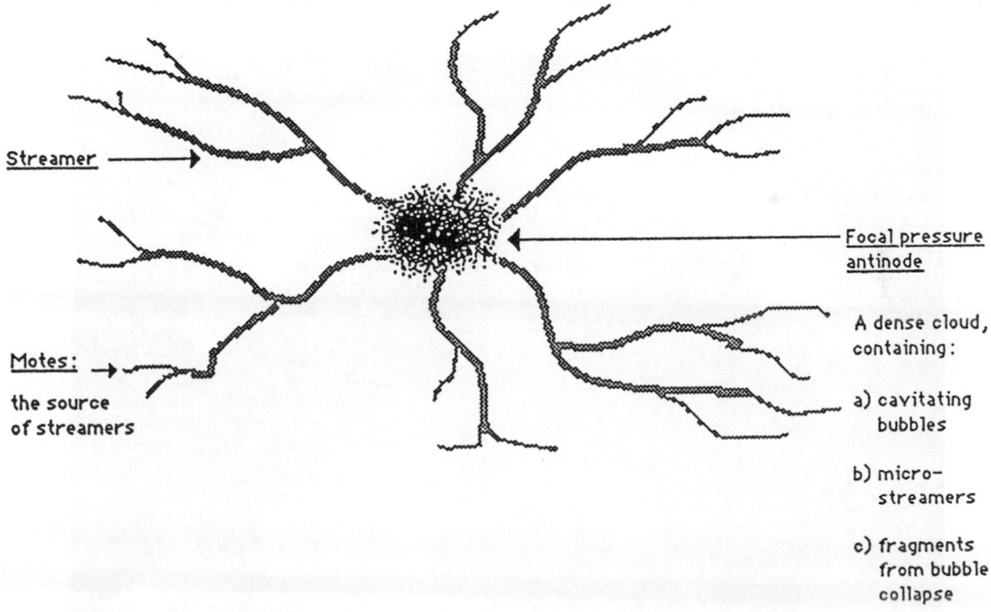

Figure 4.25 Schematic of the anatomy of the cavitation structure shown in Figure 4.24.

---

[32] See section 4.3.4.

[33] Microstreaming is discussed in Chapter 1, section 1.2.3(b), and in section 4.5. This latter discusses the correlation between microstreaming and surface oscillations on a bubble, which are the topic of section 4.6.

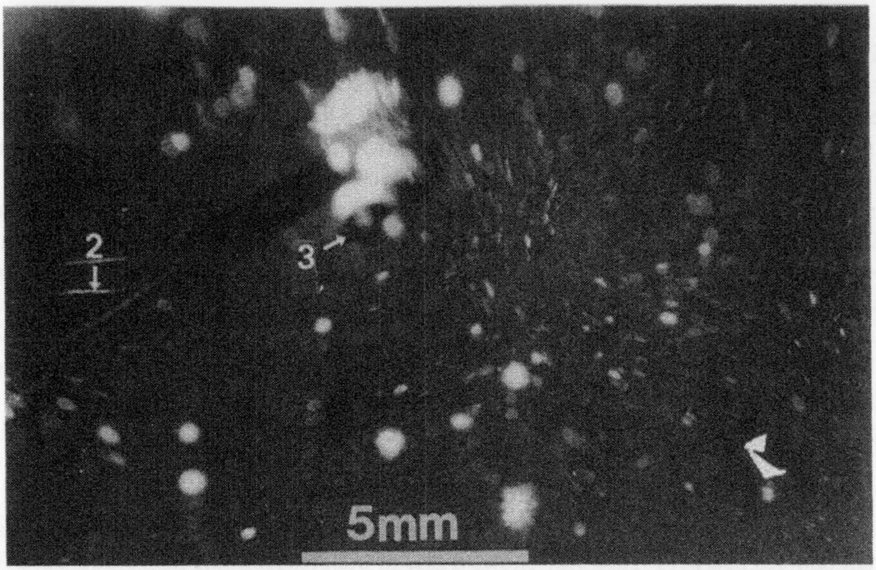

Figure 4.26 A photograph, taken in reflected illumination (a flash gun), of cavitation in aerated water, at the microtube tip. The experiment took place in the piezoceramic cylinder, with an acoustic frequency of 10 kHz (as for Figure 4.24, though exposure here is 1 ms). The tube tip is ~2 cm from the axial focus (see Figure 4.27). This photograph shows the tip of the microtubing. Bubbles are clearly visible, being sheared off the meniscus at the tip: the large bubble which has just been sheared off (above the tip) exhibits significant surface wave activity. Microstreaming circulatory patterns can be seen to the right of the tip. This is the origin of the streamer labelled 6 in Figure 4.27. 2. The microtube. 3. The tip of the microtube. After Leighton [228].

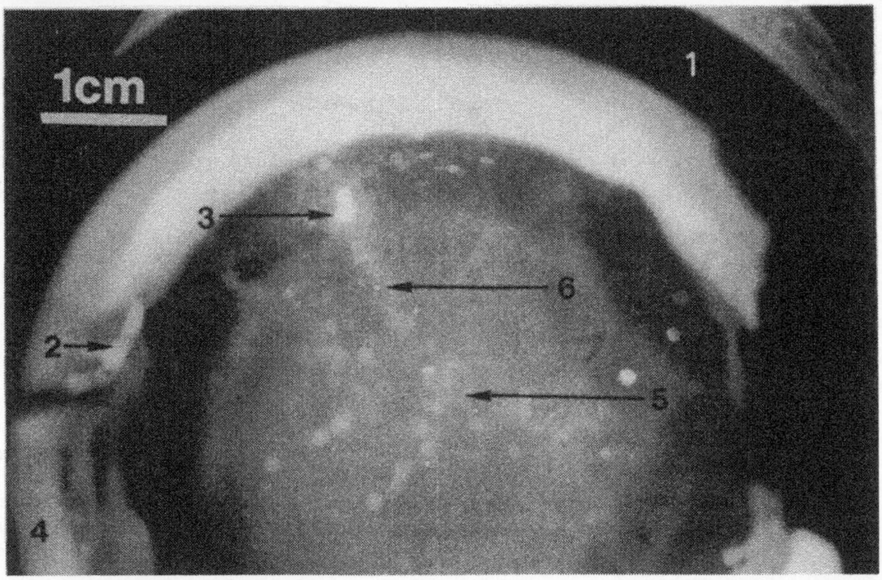

Figure 4.27 A flash photograph (exposure 1 ms), of cavitation in aerated water, using microtube seed bubbles as the source for a streamer. The experiment took place in the piezoceramic cylinder, with an acoustic frequency of 10 kHz, as for Figures 4.24 and 4.26. Several streamers are visible, reaching from the body of the liquid to the focal pressure antinode (where a cloud of bubbles can be seen). One of these streamers, however, originates at the tip of the microtubing. 1. The uppermost rim of the piezoceramic cylinder. 2. The microtube. 3. The tip of the microtube. 4. The inner surface of the upper piezoceramic. 5. The cloud of bubbles at the focal pressure antinode. 6. The streamer that originates at the microtube tip. Differences in the appearance of the streamers in Figures 4.24 and 4.27 arise through the difference in photographic exposures. After Leighton [228].

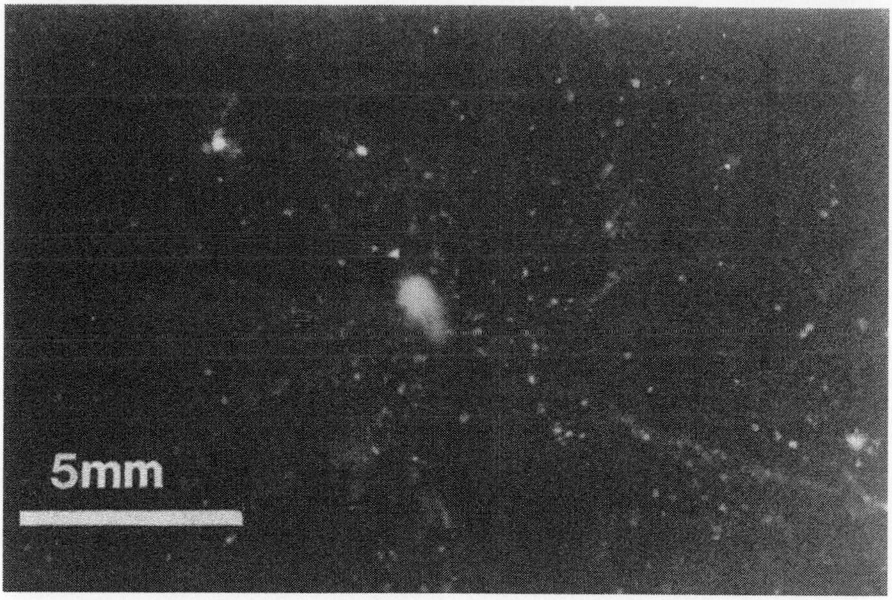

Figure 4.28 A photograph of the cavitation of the 90% glycerine/10% water mixture, occurring within the piezoceramic cylinder, taken by flash photography during insonation (exposure = 1 ms). Clearly visible is the bubble cloud at the focal pressure antinode, with its associated streamer formations, that also occurred in aerated water and are summarised for that liquid in Figure 4.25. Bubble dissolution and buoyant migration are much less than is the case in water, so that this structure will persist for several seconds after insonation has ceased. After Leighton [228].

A similar structure to that seen in water (Figure 4.24) is observed when the cylinder is filled with a 90% glycerine/10% water mixture (Figure 4.28), though the focal cloud is much more densely populated, and is about 1/10th the radius of its counterpart in aerated water.

The high population density at the axial focus suggests that there will be considerable interaction between bubbles. It is important to determine if in this experimental sound field this is so, as much of the theory presented so far in this chapter assumes a single bubble oscillating in an infinite medium.

Figures 4.29 to 4.31 are high-speed sequences (2000 f.p.s.[34]), taken at the antinodal focus. Figure 4.29 shows eight consecutive frames. A bubble of near-resonant size (labelled A) is oscillating stably in frame 1. It is approached by a second bubble (labelled B) which is similarly oscillating stably. In frame 3 this bubble is travelling perpendicularly to the plane of the picture, and between frames 3 and 4 it moves into optical depth of focus. The second bubble distorts the sound field experienced by bubble A, and sets up violent surface oscillations in the latter (in frame 5 these almost split bubble A in two!). The actual break-up occurs in frame 6. The

[34]In this section, the acoustic driving frequency is 10 kHz, and the framing rate 2000 f.p.s. The sampling rate for data is therefore well below the Nyquist frequency, and so these high-speed photographs have no use in analysing the oscillations of the bubbles. This is, for example, clear from the photographs, where the stably oscillating bubbles appear as ill-defined bodies: during the exposure time they have gone through approximately five oscillations. However, the low framing rate allows ciné films of sufficient duration to study the translations and life histories of bubbles.

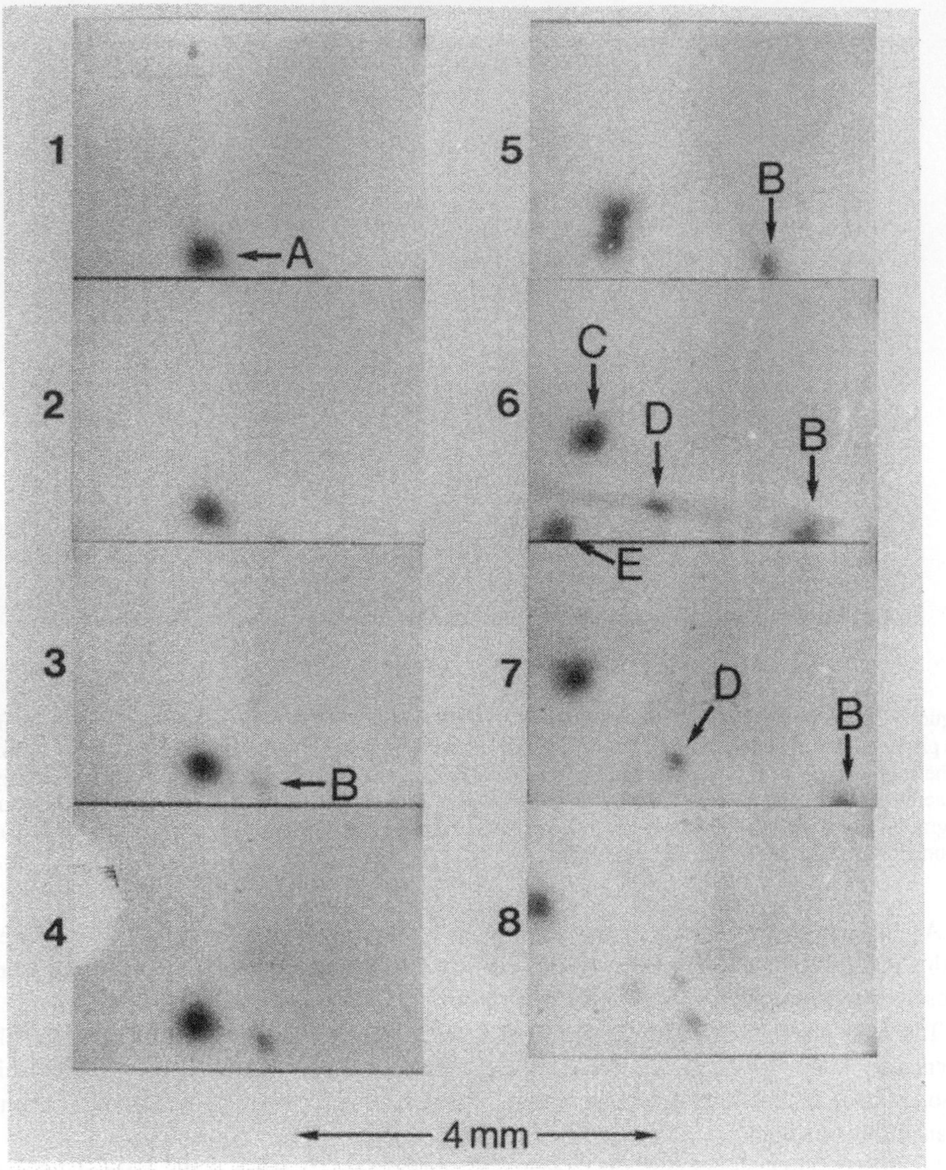

Figure 4.29 Eight frames selected from a film shot at 2000 f.p.s. Bubble A is initially oscillating stably. The approach of a similar bubble (B) disturbs the sound field around A. Violent shape oscillations in A, reminiscent in frame 5 of a second-order spherical harmonic perturbation, cause it to break up (frame 6) in bubble fragments, labelled C, D and E. These, and B, disperse rapidly. After Leighton [228].

product is three major bubble fragments (labelled C, D and E). The force of the event pushes bubbles B and E out of the field of view. By frame 7, bubble E has left, and bubble B is just leaving, the picture. Thus a bubble which would, in isolation, have undergone stable oscillation, in practice fragments owing to shape oscillations excited by the acoustic emissions from a second bubble. The stable cavitation is disturbed by an inhomogeneity in the sound field, thereby effectively ending the validity model theories given earlier in this chapter.

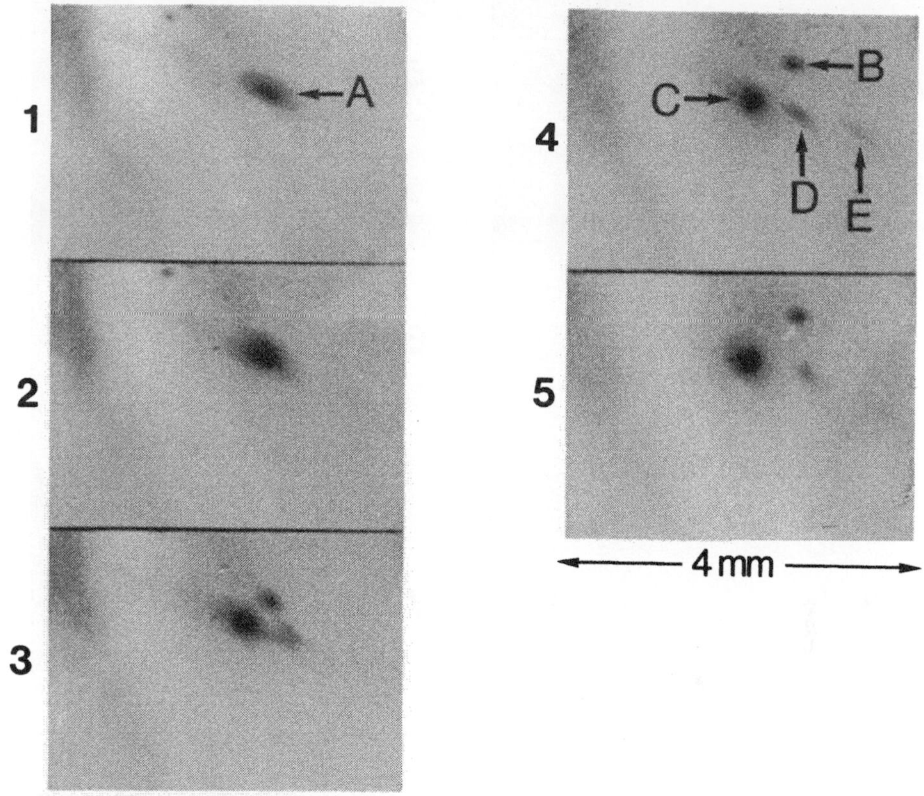

Figure 4.30 Five frames selected from a film shot at 2000 f.p.s. Bubble A is initially oscillating stably. In frame 3 it breaks up, the resulting bubble fragments (labelled B, C, D and E) dispersing rapidly. After Leighton [228].

In Figure 4.30, an oscillating bubble disintegrates into four fragments, bubble E being violently expelled. Thus the focal cloud consists of bubbles of resonant size or less, which may be growing by rectified diffusion, and which are oscillating violently. Some will disintegrate into fragments, and so the focal cloud will also contain such detritus in the form of bubble nuclei. Bubble fragments can also be generated, not just by a single collapse, but through the action of surface waves. Tiny bubbles can be sheared off the tips of surface waves on the bubble wall (Figure 4.31). These emission of such microbubbles can have important implications, as will be discussed later.[35] Though the exposure time of each frame is insufficient in Figure 4.31 to resolve the surface oscillations, it is possible to see a line of tiny bubble fragments move along a common path. Such fragments become separated from the parent bubble as a result of the intense surface wave activity [21]. In the figure, the fragments travel in a line out into the water. The bubbles follow each other, and are attracted to a second larger bubble (labelled B), with which they coalesce. The force of attraction is the mutual or secondary Bjerknes force, and is described in the next section.

[35]See sections 4.4.7 to 4.4.10.

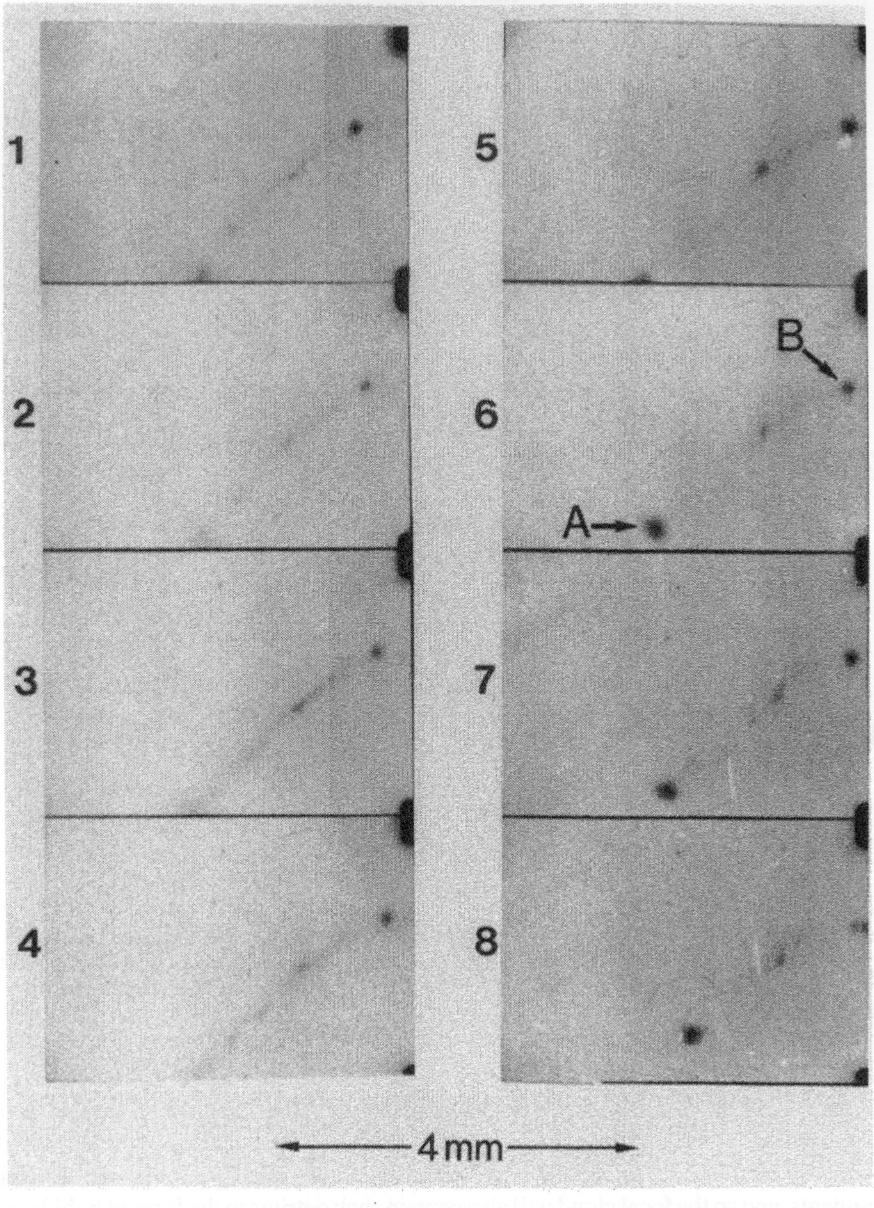

Figure 4.31 Eight frames selected from a film shot at 2000 f.p.s., showing a microstreamer within the focal cloud. The source is the bubble labelled A. Bubble fragments are sheared off from the surface of A; they travel in the microstreamer under the influence of the mutual Bjerknes force, to coalesce with the bubble marked B. (Taken in reflected illumination.) This may be compared with the coalescence of two bubbles seen in Figure 4.19. After Leighton [228].

(c) The Radiation Force Between Two Bubbles

In Chapter 3 a simple qualitative exploration of the attractive and repulsive forces between a freely-oscillating bubble and a boundary was made. It was shown how if the boundary is rigid, the bubble is attracted to it because the image oscillates in phase with the bubble. If the boundary is free, the image oscillates in antiphase, and the force is repulsive. The second bubble need not, of course, be an image but could be a real bubble oscillating with the appropriate phase. Firstly, that illustration is now formulated for the case of two spherical pulsators of the same frequency but with an arbitrary phase difference.

(i) The Force Between Two Spherical Pulsators. As was shown in Chapter 2, section 2.3, forces arise between a fluid and a body when there is relative acceleration between them, even if the fluid is frictionless. This results from the fluid that is 'carried along' some distance by the moving body, and can be accommodated in the formulation by the simple addition of an apparent mass to the original mass of the body. This apparent mass equals ρV_a, where ρ is (as usual) the liquid density and V_a equivalent to the volume of fluid carried along with the body.

When a bubble oscillates close to a boundary, there will be an oscillatory acceleration of the fluid $\vec{\dot{v}}$ felt by the bubble due to the pressure waves emitted by the image. This gives rise to a force between the bubble and the boundary.

Firstly we shall discuss the general case of a body of density ρ_b and volume V_b in an irrotational fluid which is accelerating uniformly at $\vec{\dot{v}}$ at all points, and which completely surrounds the body (the disturbance due to the body itself will be ignored), as outlined by Prandtl [68]. This acceleration corresponds to a fall in pressure p in the fluid of $\rho\vec{\dot{v}} = -\vec{\nabla}p$.

The body experiences a lift force as a result of the pressure gradient $\vec{\nabla}p$ across it, of magnitude $-V_b\vec{\nabla}p$ (Figure 4.20), giving a force of $\rho\vec{\dot{v}}V_b$ in the direction of the fluid acceleration. If $\rho \neq \rho_b$, then the acceleration of the body will not equal the acceleration of the fluid, and there will be a net acceleration of the body relative to the fluid of $\vec{\dot{u}}_b - \vec{\dot{v}}$, where $\vec{\dot{u}}_b$ is the acceleration of the body. Because of this relative acceleration, there will be a drag[36] proportional to the apparent mass ρV_a [68]. The total force on the body is therefore the lift force, $-V_b\vec{\nabla}p = \rho\vec{\dot{v}}V_b$ in the direction of the fluid acceleration, and the drag force $(\vec{\dot{u}}_b - \vec{\dot{v}})\rho V_a$, which opposes the lift and so acts in the direction opposite to the fluid acceleration. From Newton's Second Law

$$\rho(\vec{\dot{v}}V_b - (\vec{\dot{u}}_b - \vec{\dot{v}})V_a) = \rho_b\vec{\dot{u}}_bV_b \tag{4.156}$$

rearrangement of which gives

$$\vec{\dot{u}}_b = \frac{\vec{\dot{v}}\,(V_b + V_a)}{\dfrac{V_b\rho_b}{\rho} + V_a} \tag{4.157}$$

This formulation supports Figure 3.14. A solid body of greater density than water will undergo smaller acceleration, as one might expect (i.e. $\rho_b > \rho$ implies $\vec{\dot{u}}_b < \vec{\dot{v}}$). A body of equal density with the fluid will undergo equal acceleration (i.e. $\rho_b = \rho$ implies $\vec{\dot{u}}_b = \vec{\dot{v}}$). This is an important, if not unexpected, result, and enables researchers to study the flow properties of fluids by adding contrast agents of equal density.

The case of particular interest to bubble studies is when the body is less dense than the surrounding fluid. Since $\rho_b < \rho$, then $\vec{\dot{u}}_b > \vec{\dot{v}}$. Consider the flame in a lantern, which is lighter than the surrounding air. The glass of the lantern shields the flame from draughts as it moves in the air. The flame is therefore free to show accelerations relative to the body of air contained within the lantern. If the lantern is accelerated sideways from rest, the flame bends forwards, in the direction of motion of the lantern, it being subject to greater sideways acceleration than

[36] See Chapter 2, section 2.3.1(b).

the air within the lantern. If the sideways motion of the lantern is suddenly stopped, the air within the lantern is decelerated to a standstill, that is, it is subject to an acceleration in the direction opposing the former motion of the lantern. From equation (4.157), the flame must be subjected to a greater acceleration than the air, and so as the lantern is stopped the flame will bend in the direction opposing the former motion of the lantern. If the lantern is moving sideways with constant velocity, then the air contained is subject to no acceleration, and so the flame is unaffected and rises vertically. This phenomenon was recorded by V. Bjerknes [69], who observed the flame of a candle in a swinging stable lantern.

A pulsating bubble in an accelerating liquid will also be subject to a translational force, as given by equation (4.157). However, the volume of the bubble, and so the force exerted upon it, will be a variable. In Chapter 3, section 3.3.2(d) this was illustrated by considering the accelerations involved when a real bubble pulsates close to a rigid (Figure 3.15(a)) or a free (Figure 3.15(b)) plane boundary. That argument will be recapitulated to illustrate the accelerations and forces involved when two bubbles oscillate in phase or antiphase.

Consider the bubbles shown in Figure 3.15, though let both now be real bubbles[37] which pulsate at some frequency ω. The liquid in the region of the bubble on the right (labelled 2) undergoes planar oscillation in the z-direction at the same frequency. In Figure 3.15(a) the two bubbles pulsate in phase. In Figure 3.15(a)(i) both bubbles have reached maximum volume. The liquid surrounding bubble 2 will roughly follow the acceleration of the liquid in the detailed part of bubble 1, since they are colinear. As the bubbles are at maximum volume, and are about to contract, the acceleration of the wall of each bubble is maximally in towards the centre of the bubble. Thus the detailed part of the wall of bubble 1 has a maximum wall displacement to the right, and a maximum acceleration to the left. This reflects exactly the case of the pendulum discussed earlier. Thus when bubble 2 has maximum size and minimum density, the liquid around it is accelerating to the left, and so it will also accelerate to the left. Half a period later (Figure 3.15 (a)(ii)), the detailed part of the wall of bubble 1 will have maximum displacement to the left, and maximum acceleration to the right. The fluid there, and surrounding bubble 2, will similarly have maximum displacement to the left, and maximum acceleration to the right. Bubble 2 will therefore accelerate to the right. However, its volume is a minimum, and its density a maximum, so the acceleration of bubble 2 to the right is not as great as its former acceleration to the left. Therefore the net motion of bubble 2 will be to the left, towards bubble 1. A similar argument shows that if the bubbles are pulsating in antiphase, they will repel (Figure 3.15(b)).

Of course, the argument also holds for two bubbles driven into ocillation close to a boundary. If a bubble is close to a pressure-release surface, the image bubble will pulsate in antiphase, and the original bubble will be repelled by the surface. However, if the bubble is near a rigid surface, the image will pulsate in phase, and the bubble will be attracted to the interface. This can be seen in Figure 4.32. Bubbles injected into the base of the vertically vibrating cell discussed in Chapter 1, section 1.2.2(b)(ii) are attracted to the rigid cell wall as they pulsate in the 103-Hz sound field.

In Chapter 2, section 2.2.4(c) it was shown that the velocity field when a bubble pulsates in an infinite body of inviscid incompressible irrotational liquid under conservative fields falls off as r^{-2} (equation (2.101)). Allowing for the fact that such conditions are unlikely to be fully satisfied, the result does suggest through equation (4.157) that the mutual Bjerknes force follows

[37]So for 'image' in Figure 3.15, now read 'other'.

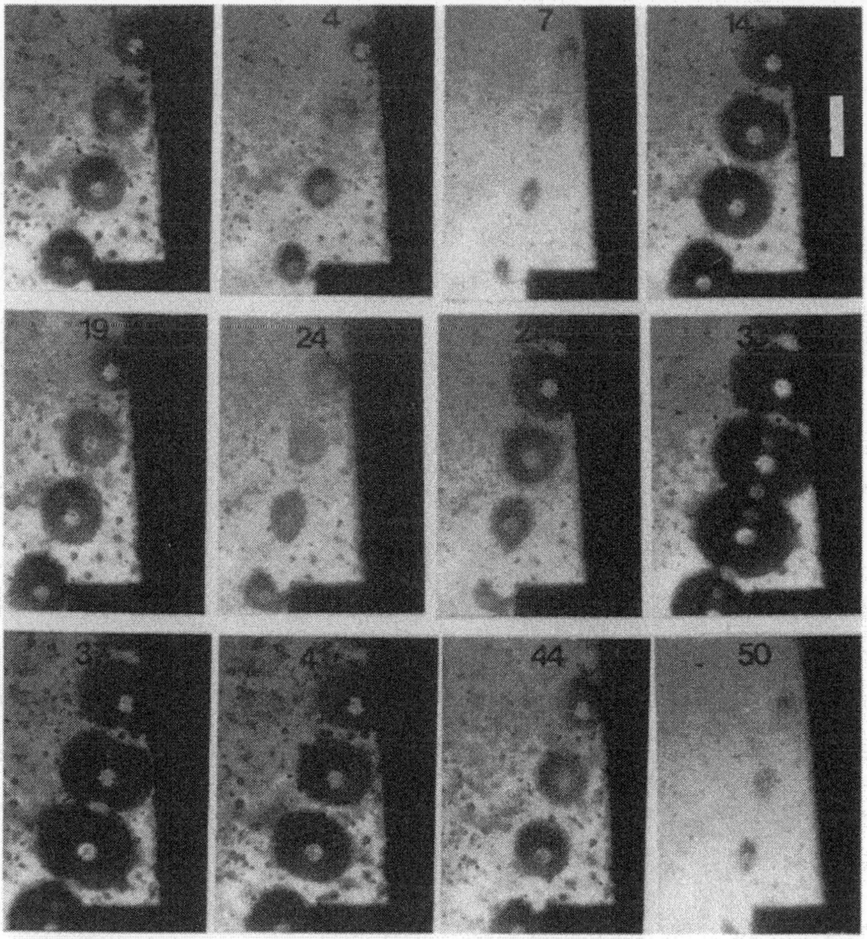

Figure 4.32 A line of air bubbles, filmed at 2000 f.p.s., is injected into the base of the vertically vibrating cell containing glycerol. The cell oscillates at 103 Hz, and the acoustic pressure amplitude is approximately 3700 Pa. The cell wall is to the right of the frame, but its position is not visible as the frame of the front window of the cell (seen in the photographs as a dark band to the right of each picture) just masks it. Visible at the bottom right of each frame, the dark rectangle is the silhouette of the punctured bung through which the air is injected. The line of rising bubbles departs from sphericity as a result of their close proximity (and the proximity of the image bubbles), and because of buoyancy. Many smaller bubbles, undergoing a similar pulsation, are also visible in the field of view. For scale, a white vertical bar of length 1 mm has been added to frame 14. (Photograph: TG Leighton, M Adlam and K Fagan.)

an inverse square law. The theory of hydrodynamical action at a distance was first published by C. A. Bjerknes in 1871, and expanded by his son, V. Bjerknes [70, 71].

Formulation of Force Between Two Spherical Pulsators. As shown above, a body of volume V_b and density ρ_b and subject to an additional apparent mass of ρV_a (where ρ is the liquid density) will, in a liquid accelerating at $\dot{v}$, be subject to a force resulting in an acceleration $\dot{u}_b$ of that body, as given by equation (4.157). It was further described how this interaction can give rise to the mutual Bjerknes force between two bubbles (and therefore, through images, between a bubble and a plane interface). In this section the magnitude of that force is quantified. It should be pointed out that the mutual Bjerknes force arises from the pressure field radiated from one pulsating bubble acting on a second, regardless of whether those pulsations are the result of free or forced oscillations.

The formulation is uniaxial, based upon the line between the bubble centres. Assume two spherically pulsating bubbles, 1 and 2, whose volume vary as $V_1 = V_{o1} - V_{\varepsilon o1}\cos\omega t$ and $V_2 = V_{o2} - V_{\varepsilon o2}\cos(\omega t + \vartheta_b)$, respectively, where V_{o1} is the equilibrium volume, and $V_{\varepsilon o1}$ the volume amplitude of pulsation, for bubble 1. If the bubbles are assumed to be spherical at all times with no internal circulation, then the apparent volume of fluid associated with each bubble is half the volume of the sphere (Chapter 2, section 2.3). Let $\dot{v}$ be the acceleration at the position of bubble 2 due to the oscillations of bubble 1 (the wavelength of sound is very much greater than the bubble radius, so that $\dot{v}$ is uniform over the bubble). If $\dot{u}_b$ is the acceleration of bubble 2, then from equation (4.157)

$$\frac{\dot{u}_b}{\dot{v}} = \frac{(V_2 + \frac{1}{2}V_2)}{\dfrac{V_2\rho_2}{\rho} + \frac{1}{2}V_2} \tag{4.158}$$

The density of bubble 2, ρ_2, varies in time, since V_2 also does. However, if the mass of gas and vapour within the bubble is assumed constant, then

$$\rho_2 V_2 = \rho_{2e} V_{o2} \tag{4.159}$$

(ρ_{2e} being the density of bubble 2 at equilibrium). Introducing the ratio $f = (\rho_{2e}/\rho)$, substitution of equation (4.159) into equation (4.158) gives

$$\frac{\dot{u}_b}{\dot{v}} = \frac{3V_2}{2fV_{o2} + V_2} = \frac{3(V_{o2} - V_{\varepsilon o2}\cos\omega t)}{(1+2f)V_{o2} - V_{\varepsilon o2}\cos\omega t} \tag{4.160}$$

Binomial expansion of the denominator gives

$$\left(1 - \frac{V_{\varepsilon o2}\cos\omega t}{(1+2f)V_{o2}}\right)^{-1} = 1 + \frac{V_{\varepsilon o2}\cos\omega t}{(1+2f)V_{o2}} \tag{4.161}$$

so that equation (4.160) reduces to

$$\frac{\dot{u}_b}{\dot{v}} = \left(\frac{3}{1+2f}\right)\left(1 - \frac{V_{\varepsilon o2}\cos\omega t}{V_{o2}}\right)\left(1 + \frac{V_{\varepsilon o2}\cos\omega t}{(1+2f)V_{o2}}\right) \tag{4.162}$$

To first order, that is taking terms of $(V_{\varepsilon o2}/V_{o2})^2$ and higher to be negligible,

$$\frac{\dot{u}_b}{\dot{v}} = \left(\frac{3}{1+2f}\right)\left\{1 - \left(\frac{2f}{1+2f}\right)\left(\frac{V_{\varepsilon o2}}{V_{o2}}\right)\cos\omega t\right\} \tag{4.163}$$

From Newton's Second Law, the force on the bubble must equal the product of the mass, $\rho_2 V_2$, and the bubble acceleration, $\dot{u}_b$, and from equation (4.159) the resulting term $\rho_2 V_2 \dot{u}_b$, equals $\rho_{2e} V_{o2} \dot{u}_b$. The *instantaneous* force on bubble 2, F_{rad2}, is therefore

$$F_{rad2} = \rho_{2e} V_{o2} \dot{u}_b = \left(\frac{3}{1+2f}\right) \dot{v} \rho_{2e} V_{o2} - \left(\frac{3}{1+2f}\right)\left(\frac{2f}{1+2f}\right)(V_{\varepsilon o2} \dot{v} \rho_{2e} \cos\omega t) \quad (4.164)$$

From Chapter 2, section 2.2.4(c), the fluid velocity at a distance r_b from a pulsating sphere of radius $R(t)$ is $v = \dot{R}R^2/r_b^2$ (equation (2.101)) in the radial direction, assuming liquid incompressibility. Since the volume of a sphere is $V = 4\pi R^3/3$, then $\dot{V} = 4\pi R^2 \dot{R}$. Therefore the liquid velocity a distance r_b from a sphere which pulsates with an instantaneous radius $R_1(t)$, instantaneous volume $V_1 = V_{o1} - V_{\varepsilon o1}\cos(\omega t + \vartheta_b)$, and for which $\dot{V}_1 = \omega V_{\varepsilon o1}\sin(\omega t + \vartheta_b)$, is

$$v = \frac{R_1^2 \dot{R}_1}{r_b^2} = \frac{-V_{\varepsilon o1} i\omega \cos(\omega t + \vartheta_b)}{4\pi r_b^2} \quad (4.165)$$

If bubble 2 is a distance r_b from the centre of bubble 1, it will therefore experience a liquid acceleration as a result of the pulsation of 1 given by

$$\dot{v} = \frac{V_{\varepsilon o1}\omega^2 \cos(\omega t + \vartheta_b)}{4\pi r_b^2} \quad (4.166)$$

Substitution of this into equation (4.164) gives the instantaneous force on bubble 2 resulting from the oscillation of bubble 1:

$$F_{rad2} = \left(\frac{V_{\varepsilon o1}\omega^2}{4\pi r_b^2}\right)\left(\frac{3}{1+2f}\right)\rho_{2e} V_{o2}\cos(\omega t + \vartheta_b)$$

$$-\left(\frac{3}{1+2f}\right)\left(\frac{2f}{1+2f}\right)\left(\frac{V_{\varepsilon o1}\omega^2}{4\pi r_b^2}\right)(V_{\varepsilon o2}\rho_{2e}\cos(\omega t + \vartheta_b)\cos\omega t) \quad (4.167)$$

This is the instantaneous force. The net radiation force is found by taking the time average, that is $\bar{F}_{rad2} = \langle F_{rad2} \rangle$, averaging, for example, over one cycle. The first term in equation (4.167) goes to zero, since $\langle \cos\omega t \rangle = 0$. The time average in the second term is $\langle \cos(\omega t + \vartheta_b)\cos\omega t \rangle = \langle \cos^2\omega t \cos\vartheta_b \rangle - \langle \sin\omega t \cos\omega t \sin\vartheta_b \rangle = \cos\vartheta_b/2$. Therefore the radiation force on bubble 2 exerted by bubble 1 is

$$\bar{F}_{rad2} = -\left(\frac{3}{1+2f}\right)\left(\frac{f}{1+2f}\right)\left(\frac{V_{\varepsilon o1}\omega^2}{4\pi r_b^2}\right)(V_{\varepsilon o2}\rho_{2e}\cos\vartheta_b) \quad (4.168)$$

The response of bubbles far from resonance follows immediately. If the two bubbles are oscillating in phase with one another (i.e. both greater than, or less than, or equal to resonance size), $\vartheta_b = 0$ and $\cos\vartheta_b = 1$ to give a force which has a negative sign, and so is attractive. This is also the force towards a rigid boundary experienced by a bubble a distance $h = r_b/2$ from that boundary. If the bubbles are oscillating in antiphase with one another (i.e. one is greater than, and the other less than, resonance size), $\vartheta_b = \pi$ and $\cos\vartheta_b = -1$ to give a repulsive force. This is also the force away from a pressure-release boundary experienced by a bubble a distance $h = r_b/2$ from that boundary. These results are in agreement with the qualitative explanation given earlier.

The variation of the force with the relative densities and with the phase difference ϑ_b is shown in Figure 4.33. The variation of the term $3f/(1+2f)^2$ is shown in Figure 4.33(a), from which it can be seen that the force is greatest when $f = 0.5$, i.e. when the density of the spheres is half

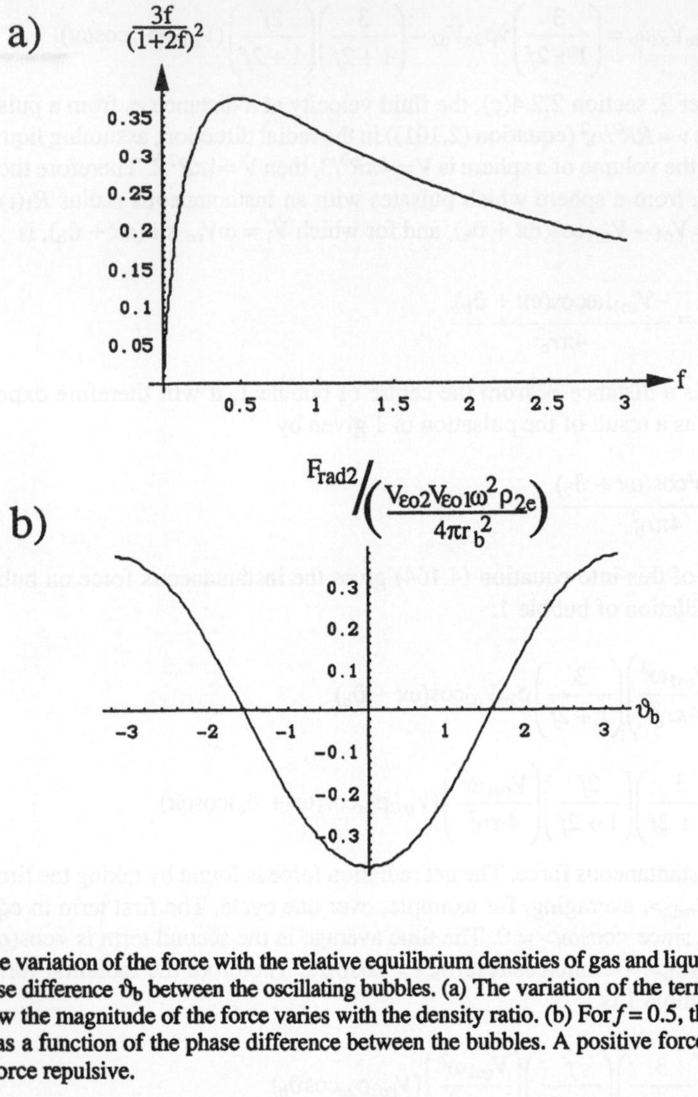

Figure 4.33 The variation of the force with the relative equilibrium densities of gas and liquid ($f = \rho_{1e}/\rho$) and with the phase difference ϑ_b between the oscillating bubbles. (a) The variation of the term $3f/(1+2f)^2$, which shows how the magnitude of the force varies with the density ratio. (b) For $f = 0.5$, the normalised force is plotted as a function of the phase difference between the bubbles. A positive force is attractive, and a negative force repulsive.

that of the surrounding liquid (as could be found through simple differentiation of equation (4.168) with respect to f). For this particular value of $f = 0.5$, the normalised force is plotted in Figure 4.33(b) against the phase difference between the bubbles. It can be seen that for $-\pi/2 < \vartheta_b < \pi/2$, the force is attractive (negative), it is zero when the bubbles pulsate such that $\vartheta_b = \pm\pi/2$, and is otherwise repulsive (positive).

Setting $f = 1$ in equation (4.168) gives the force on a sphere of equal density with the liquid. This situation was discussed by Prandtl [68].[38] Alternatively Batchelor [72] assumes the opposing condition that $2\rho_{2e} \ll \rho$, which is to say that the sphere is far less dense than the liquid, in a binomial expansion of equation (4.156) to give

[38]There is a minor typographical error in equation 10 of this reference – otherwise that formulation agrees with the reduced form of equation (4.168) when $f = 1$.

$$\ddot{u}_b \approx 3\left(1 - \frac{2\rho_{2e}}{\rho}\right)\dot{v} \qquad (4.169)$$

Batchelor then substitutes the volumes and fluid accelerations of two bubbles pulsating in phase into equation (4.169) to give an acceleration of bubble 2 towards bubble 1:

$$\ddot{u}_b \approx -\left(\frac{3\omega^2}{4\pi r_b^2 \rho}\right)(V_{o2}\rho_{2e})\left(\frac{V_{\varepsilon o1}V_{\varepsilon o2}}{V_{o2}^2}\right) \qquad (4.170)$$

This acceleration is equal to the ratio of the force on one bubble due to another pulsating in phase (as given by equation (4.168) for $\vartheta_b = 0$), to the bubble mass $\rho_2 V_2 = \rho_{2e} V_{o2}$, provided the liquid is much denser than the gas. As such, equation (4.170) agrees with equation (4.168).

(ii) Force Between Two Bubbles When There is an Incident Sound Field. When two bubbles are driven by an incident sound field, to a first approximation that driving field will have two effects. Firstly it will exert a radiation force on the two bubbles of the types discussed in sections 4.4.1(a) and 4.4.1(b), depending on whether it is travelling or standing wave, focused etc. Secondly it will tend to lock two bubbles to a certain phase difference: for example, if both bubbles are much less than, or much greater than, resonance size, they will tend to oscillate in phase; if one is greater than, and the other less than, resonance size, they will tend to oscillate in antiphase. This will tend to cause attractive and repulsive forces, respectively, between the two bubbles, in addition to the radiation force from the incident field. In practice this is a simplification: a full calculation would have to consider mutual scattering effects. However, Coakley and Nyborg [61] obtained a simplified physical picture by assuming that the bubble will simply respond to any gradient in the pressure field within the medium at its location[39] and by assuming that the pressure field is made up simply through a summation of the incident acoustic wave and the pressure field emitted by the other bubble. Assume the incident field is

$$p(t) = p_0 + P_A\cos(\omega t - kx) \qquad (4.171)$$

From section 4.1.2(c), the response of bubble 1, which is at $x = 0$, is

$$\begin{aligned} R_1(t) &= R_{o1} + R_{\varepsilon 1} = R_{o1} - R_{\varepsilon o1}\cos(\omega t - \vartheta_1) \\ &= R_{o1} - \frac{P_A}{R_o\rho}\frac{1}{\sqrt{(\omega_o^2-\omega^2)^2 + (2\beta_{tot}\omega)^2}}\cos(\omega t - \vartheta_1) \\ &\equiv R_{o1} - \frac{P_A}{R_o\rho}\frac{1}{\sqrt{(\omega_o^2-\omega^2)^2 + (2\beta_{tot}\omega)^2}}e^{i(\omega t - \vartheta_1)} \end{aligned} \qquad (4.172)$$

employing the complex notation, where the real part of the function represents the physical oscillation. In the same notation, assuming spherical symmetry, the pressure radiated by this bubble is found from equation (3.108) to be

$$P_{b1} = \frac{i\omega\rho_o R_o^2 \dot{R}_{\varepsilon 1}}{r} \qquad (4.173)$$

though strictly it would be more accurate to employ equation (3.126). Since equation (4.172) implies that

---

39 As outlined in section 4.4.1.

$$\dot{R}_{\varepsilon 1} = -\frac{P_A}{R_0 \rho} \frac{1}{\sqrt{(\omega_0^2 - \omega^2)^2 + (2\beta_{tot}\omega)^2}} i\omega e^{i(\omega t - \vartheta_1)} \tag{4.174}$$

substitution into equation (4.173) implies

$$P_{bl} = \frac{\omega^2 P_A R_0}{r} \frac{1}{\sqrt{(\omega_0^2 - \omega^2)^2 + (2\beta_{tot}\omega)^2}} e^{i(\omega t - \vartheta_1)}$$

$$= P_A \frac{R_0}{r} \frac{\left(\frac{\omega}{\omega_0}\right)^2}{\sqrt{\left\{1 - \left(\frac{\omega}{\omega_0}\right)^2\right\}^2 + \left(\frac{2\beta_{tot}}{\omega_0}\right)^2 \left(\frac{\omega}{\omega_0}\right)^2}} e^{i(\omega t - \vartheta_1)} \tag{4.175}$$

Making the approximation that the damping of the bubble is independent of frequency allows the substitution $Q = \omega_o/2\beta_{tot} \approx \omega_0/2\beta_{tot}$ (equation (3.157)):

$$P_{bl} = P_A \frac{R_0}{r} \frac{\left(\frac{\omega}{\omega_0}\right)^2}{\sqrt{\left\{1 - \left(\frac{\omega}{\omega_0}\right)^2\right\}^2 + \left(\frac{1}{Q}\right)^2 \left(\frac{\omega}{\omega_0}\right)^2}} e^{i(\omega t - \vartheta_1)} \tag{4.176}$$

Clearly, from this equation the phase between the radiated pressure field lags a phase factor ϑ_1 behind the oscillating component of the incident pressure field

$$P(t) = P_A \cos\omega t \equiv P_A e^{i\omega t} \tag{4.177}$$

Therefore when $\omega_{01} \gg \omega$, they will be in phase ($\vartheta_1 = 0$, from Figure 4.4) and will reinforce one another. However, when $\omega_{01} \ll \omega$ they will be in antiphase ($\vartheta_1 = \pi$), and will work to inexactly cancel one another. When the ratio ω/ω_{01} is at neither of these extremes, the phase lag is intermediate between 0 and π, and the resultant pressure amplitude in the medium is given by the sum of the real parts of $P(t)$ and $P_{bl}(t)$. One would expect this to cause interesting effects when ω_{01} is close to resonance, and at distances not far from the bubble wall. This is because the closer the bubble is to resonance, the more rapidly the phase lag ϑ_1 changes with respect to changes in ω_0/ω (Figure 4.4(b)); and because the incident acoustic pressure has constant value throughout the field, but the field radiated by this pulsating bubble (bubble 1) falls off with distance from the bubble centre, at the rate of r^{-1} (equation (4.176)). In Figure 4.34 the sum of the radiated pressure field P_{bl} (as given by equation (4.176)) and the incident sound field (as given by equation (4.171)) are plotted as a function of r/R_0 (clearly only $r \geqslant R_0$ has any physical significance), for difference values of the ratio ω/ω_0, for a bubble having $Q = 15$. The curves show the net normalised pressure amplitude, as given by the magnitude $|P_A + P_{bl}|/P_A$, as the ratio ω/ω_0 varies from zero (when bubble 1 is infinitesimally small) to $\omega/\omega_0 = 20$ (when bubble 1 is very much larger than the resonance radius).

When the insonation frequency is less than the resonance of bubble 1, the radiated pressure field from 1 adds an additional pressure onto the insonating pressure field, the amount increasing as one approaches the bubble, and as the insonation frequency approaches resonance ($\omega = \omega_0$). This is maximal at resonance. Above resonance ($\omega = 1.1\omega_0$ in the figure), the relative phases

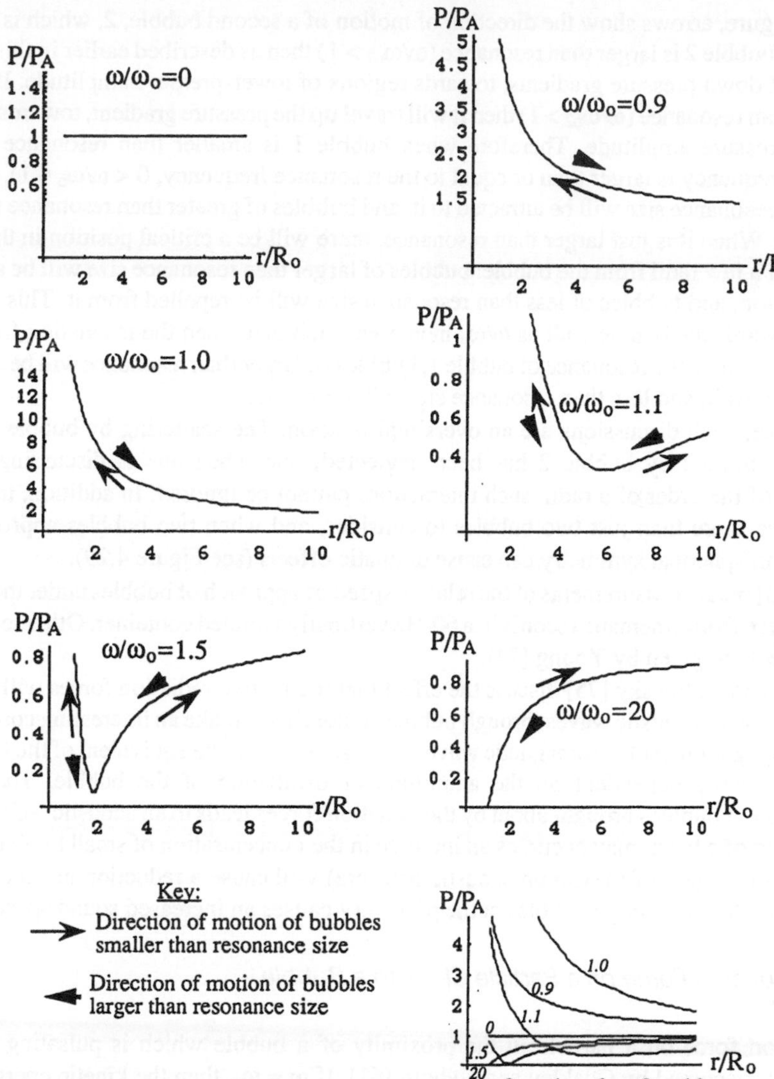

Figure 4.34 The sum of the radiated pressure field P_{bl} and the incident sound field, normalised to the acoustic pressure amplitude of the incident field, is plotted as a function of r/R_o (clearly, only $r \geqslant R_o$ has any physical significance), for a bubble having $Q = 15$. The curves show the net pressure amplitude, as given by the magnitude $|P_A + P_{bl}|/P_A$, for different values of ω/ω_o. The arrows indicate the direction of motion of the bubbles which are smaller and greater than resonance size. To enable comparison, the six pressure fields are plotted together at the base of the figure.

of the scattered and incident signals cause a minimum, which is sharper and occurs closer to the bubble wall as the insonation frequencies become increasingly greater than resonance ($\omega = 1.5\omega_o$ in the figure), such that in the limit, the scattered and incident fields are effectively in antiphase, and the pressure field decays towards the bubble wall where it is zero (as shown for $\omega = 20\omega_o$). The final plot in Figure 4.34 shows these curves superimposed to indicate the relative scale.

On the figure, arrows show the direction of motion of a second bubble, 2, which is close to bubble 1. If bubble 2 is larger than resonance ($\omega/\omega_{o2} > 1$) then as described earlier in this section it will travel down pressure gradients, towards regions of lower-pressure amplitude. If bubble 2 is larger than resonance ($\omega/\omega_{o2} > 1$) then it will travel up the pressure gradient, towards regions of higher-pressure amplitude. Therefore when bubble 1 is smaller than resonance (i.e. its resonance frequency is larger than or equal to the resonance frequency, $0 < \omega/\omega_o \leqslant 1$), bubbles of less than resonance size will be attracted to it, and bubbles of greater than resonance size will be repulsed. When it is just larger than resonance, there will be a critical position in the liquid a distance of a few radii from the bubble: bubbles of larger than resonance size will be attracted to this position, and bubbles of less than resonance size will be repelled from it. This position migrates towards the bubble wall as ω/ω_o increases, such that when the insonation frequency is much greater than the resonance of bubble 1, bubbles of larger than resonance will be attracted to the bubble wall, and less than resonance size will be repelled.

In practice, such discussions are an oversimplification. The scattering by bubble 1 of the sound field emitted by bubble 2 has been neglected, and when one is discussing bubble separations of the order of a radii, such interactions cannot be ignored. In addition, there will in general be more than just two bubbles to consider, and when two bubbles approach, the breakdown in spherical symmetry can cause dramatic effects (see Figure 4.29).

Crum [73] made measurements of the relative speed of approach of bubbles under the mutual Bjerknes force from cinematic records in a 60-Hz vertically vibrated container. Other references on the subject are given by Young [74].

Kobelev and Ostrovsky [75] discuss the effect that interactive radiation forces will have on the propagation of acoustic waves through bubble clouds. They make an interesting comparison with the propagation of electromagnetic waves through plasmas, the equivalent of the Coulomb charge now being dependent on the amplitude of oscillation of the bubble. The spatial redistribution of bubbles brought about by the radiation forces leads to an acoustic 'self-action'. Self-focusing of a beam may occur, as an increase in the concentration of small bubbles on the beam axis (the region of maximum acoustic pressure) will cause a reduction in sound speed, whilst accumulation of large bubbles at the periphery causes an increased sound speed there.

(d) The Radiation Force on a Particle Close to a Bubble

The radiation force on a particle in the proximity of a bubble which is pulsating close to resonance is discussed by Coakley and Nyborg [61]. If $\omega \approx \omega_o$, then the kinetic energy of the system dominates the potential, and the radiation force on a particle of volume V_p and density ρ_p is given by the energy gradient:

$$\bar{F}_{part} = \frac{3V_p(\rho_p - \rho_o)}{2\rho_p + \rho_o} \frac{\partial \langle \phi_K \rangle}{\partial r} \tag{4.178}$$

where $\langle \phi_K \rangle$ is the time-averaged kinetic energy density, equal to $\rho_o \omega^2 R_{\varepsilon o}^2 R_o^4 / (4r^4)$, so that

$$\bar{F}_{part} = -\frac{3V_p(\rho_p - \rho_o)}{2\rho_p + \rho_o} \frac{\rho_o \omega^2 R_{\varepsilon o}^2 R_o^4}{r^5} \tag{4.179}$$

This force is greatest at the bubble wall, and falls off rapidly with distance. Using this formulation, Coakley and Nyborg [61] show that in a sound field of $P_A = 0.05$ atm and $\nu = 1$ MHz, a bubble with $R_o = 3.3$ μm and $R_{\varepsilon o} = R_o/10$ will attract and collect from a saline solution all suspended platelets that were in the region $R_o < r < 2R_o$ to the bubble surface within 3 ms from the start of insonation. In practice, such transport features may be influenced by microstreaming (see section 4.4.5).

The above discussion of radiation forces has included the translational migration of bubbles through a sound field that is spatially nonuniform; and the interactive forces between several bubbles, and between bubbles and solid bodies (walls or particles). This simple linear discussion of radiation forces suggests the wealth of behaviour that can occur in a real sound field. However, in such fields the nonlinear aspects of forced oscillations may readily be observed. For example, Figure 4.35(a) shows the oscillation of a gas bubble in the simulated acoustic field described in Chapter 1, section 1.2.2(b)(ii). The 35-frame sequence shows the periodic repeating cycle of oscillation, which contains two minima (frames 6 and 19), the collapse to the second being very rapid: this is clearly not a linear harmonic motion. In addition to such spherically-symmetric oscillations about the equilibrium position, there may also be departures from the spherical shape (Figure 4.35(b)). Many of the phenomena of interest in stable cavitation do involve nonlinearities; however, linearisation of analysis is often necessary. The damping of driven bubbles is discussed in the next section; to make the problem tractable, the analyses assume for the most part single, isolated, spherical bubbles in a uniform sound field. Having in this section considered some common and simple departures from the ideal of an isolated spherical bubble, those analyses will now be discussed.

4.4.2 The Damping of Bubbles

The thermal behaviour of oscillating gas bubbles is still the subject of much work, a subject gaining recent impetus from the need for accurate models of damping required for the study of the chaotic behaviour of bubbles (see section 4.4.7(c)) [76, 77]. In this section, two illuminating models from the 1970s will be examined in depth, later work being summarised in section 4.4.2(c).

(a) Eller's 1970 Analysis

In Chapter 3, section 3.4 we examined the damping of a bubble oscillating at its resonance frequency, following the formulation of Devin [78]. Eller [79] extended these arguments to examine the damping of bubbles driven off-resonance, yet still for the limit when the wavelength of the driving sound field (both in liquid and in the gas) is much greater than the radius of the bubble.

Devin's method studied the response of a bubble to a driving frequency ω, and then found the behaviour at resonance with the reduction $\omega = \omega_0$. Eller's method relies upon this formulation, obtaining the identical form for the polytropic index

$$\kappa = \gamma\left\{1+\delta_{th}^2\right\}^{-1}\left[1+\frac{3(\gamma-1)}{R_o/l_D}\left\{\frac{\sinh(R_o/l_D)-\sin(R_o/l_D)}{\cosh(R_o/l_D)-\cos(R_o/l_D)}\right\}\right]^{-1} \tag{4.180}$$

as found by Devin (compare the above with equation (3.194), using $\kappa = \gamma/\alpha_{th}$, equation (3.197)). Eller's spring constant, derived with surface tension neglected, agrees with equation (3.193) (Devin's result before surface tension effects are included):

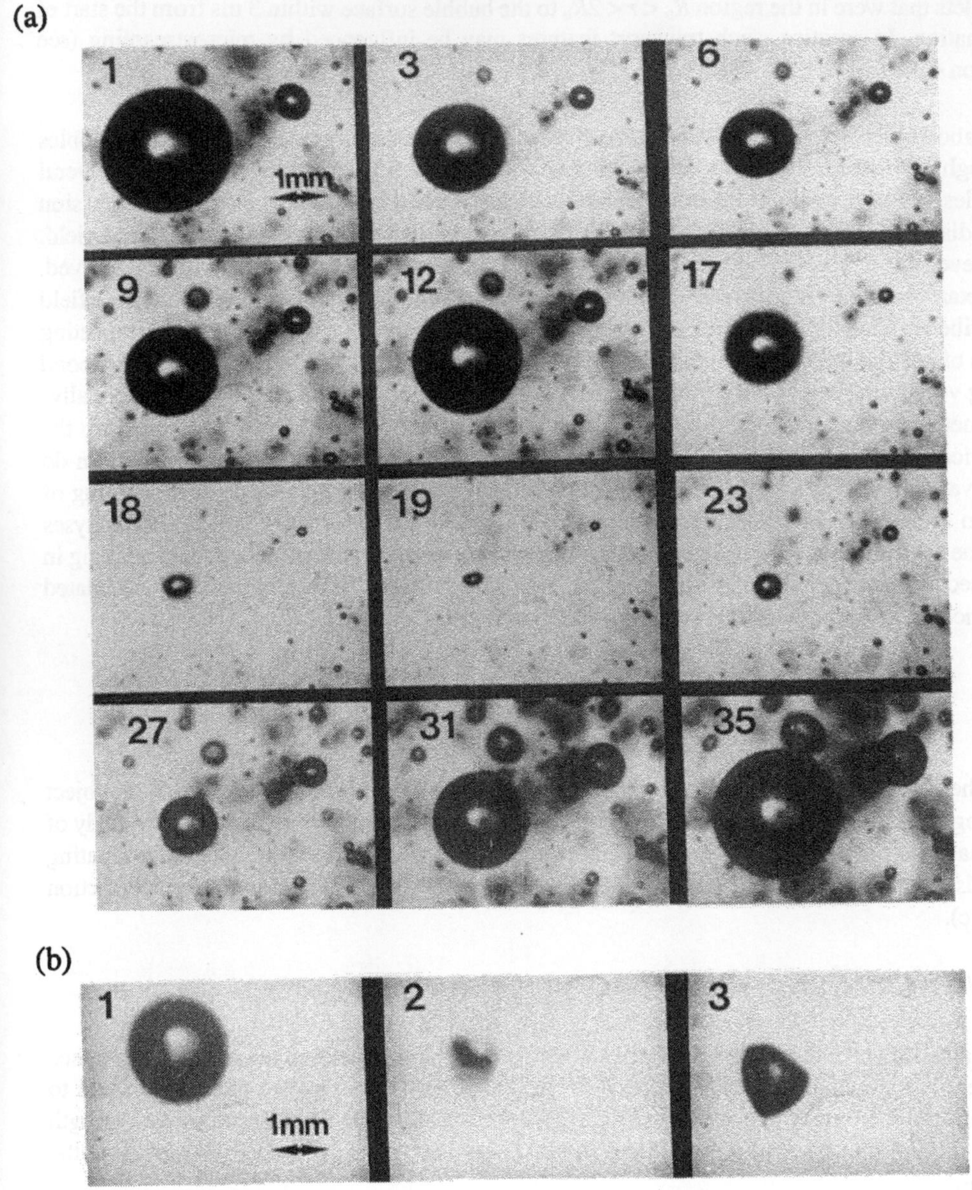

Figure 4.35 Two sequences, filmed at 2000 f.p.s., showing bubble collapses in the 100-Hz simulated sound field discussed in Chapter 1, section 1.2.2(b)(ii). (a) A selection of frames from the sequence of 35 consecutive frames which show the period unit in the motion of a stable air bubble oscillating in glycerol ($p_0 = 600$ Pa, $P_A = 3900$ Pa). The bubble contracts from maximum size in frame 1 to a minimum in frame 6, expands to a second maximum (frame 12), then collapses to a second minimum (frame 19) before expanding to the initial size. The second collapse is gradual up to frame 17, but then becomes very rapid. (b) The collapse and rebound of an air bubble in water ($p_0 = 780$ Pa, $P_A = 880$ Pa). The bubble loses its spherical shape (after Leighton *et al.* [227]). Reprinted by permission from *European Journal of Physics*, vol. 11, pp. 352–358; Copyright © 1990 IOP Publishing Ltd.

$$k_{VP} = \frac{\kappa p_0}{V_0} = \frac{\gamma}{\alpha_{th}}\frac{p_0}{V_0} \tag{4.181}$$

where of course $V_0 = 4\pi R_0^3/3$. From an equation of motion in the *volume–pressure* frame

$$m_{VP}^{rad}\ddot{V}_\varepsilon + b_{VP}\dot{V}_\varepsilon + k_{VP}V_\varepsilon = -P_A e^{i\omega t} \tag{4.182}$$

Eller found that the variation in *radius* $R(t) = R_0 + R_\varepsilon(t)$ can be expressed as

$$\frac{R_\varepsilon(t)}{R_0} = \frac{P_A e^{-i\omega t}}{3\kappa p_0}\left\{\frac{\rho R_0^2\omega^2}{3\kappa p_0} - 1 + id_{tot}\right\}\left\{\left(\frac{\rho R_0^2\omega^2}{3\kappa p_0} - 1\right)^2 + d_{tot}^2\right\}^{-1} \tag{4.183}$$

for $R_\varepsilon \ll R_0$, where $d_{tot} = d_{rad} + d_{th} + d_{vis}$ is the total (dimensionless) damping coefficient. Eller's resonance occurred when the critical quantity $(\rho R_0^2\omega^2)/(3\kappa p_0)$ equalled unity. Off-resonance, it was not quite equal to the ratio of the resonance to the driving frequency, the discrepancy arising through the frequency dependence of the polytropic index (see equation (4.180)), though it should be remembered that Eller's formulation takes no account of surface tension.

The quantity d_{tot} is defined as

$$d_{tot} = \frac{\omega b_{VP}}{k_{VP}} \tag{4.184}$$

For a bubble in resonance, this damping coefficient becomes $d_{tot} = d_{rad} + d_{th} + d_{vis} = \delta_{tot} = \delta_{rad} + \delta_{th} + \delta_{vis}$, as described by equation (3.165). Eller's dimensionless damping constants are

$$d_{th} = \frac{3(\gamma-1)\{(R_0/l_D)[\sinh(R_0/l_D) + \sin(R_0/l_D)] - 2[\cosh(R_0/l_D) - \cos(R_0/l_D)]\}}{(R_0/l_D)^2[\cosh(R_0/l_D) - \cos(R_0/l_D)] + 3(\gamma-1)(R_0/l_D)[\sinh(R_0/l_D) - \sin(R_0/l_D)]} \tag{4.185}$$

$$d_{rad} = \frac{\rho}{3\kappa p_0}\frac{(R_0\omega)^3}{c} \tag{4.186}$$

and

$$d_{vis} = \frac{4\eta\omega}{3\kappa p_0} \tag{4.187}$$

Figure 4.36 shows Eller's results for the damping constants of (a) bubbles in the size range 20 μm $\leqslant R_0 \leqslant$ 2 mm in a 10-kHz sound field (d_{vis} is too small to appear on the figure); and (b) a 100-μm radius bubble in a sound field in the frequency range 1 kHz $\leqslant \nu \leqslant$ 100 kHz. From such examinations Eller published the following generalised conclusions:

(a) For driving frequencies less than the bubble resonance, the thermal damping dominates.
(b) For driving frequencies greater than the bubble resonance, the radiation damping dominates.
(c) The damping tends to undergo a local minimum in the transition region between the dominance of thermal and radiation damping. This local minimum occurs at a driving frequency close to, but not exactly on, the bubble resonance.

It should be noted that the thermal diffusivity of the gas ($D_g = K_g/(\rho_{1e}C_p)$) may not be constant over the bubble pulsation if the diffusion rate is very rapid, or if mass exchange between the

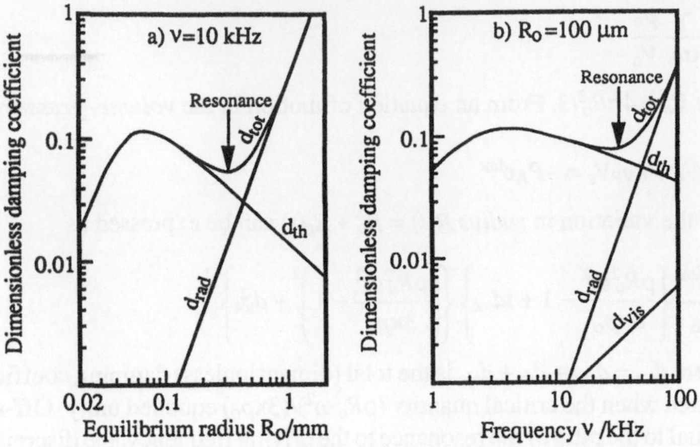

Figure 4.36 The damping constants of (a) bubbles in the size range 20 μm $\leqslant R_0 \leqslant$ 2 mm in a 10-kHz sound field (d_{vis} is too small to appear on the figure); and (b) a 100 μm radius bubble in sound fields in the frequency range 1 kHz $\leqslant \nu \leqslant$ 100 kHz (after Eller [79]).

contents of the bubble and the liquid is not negligible. In such a case the above expressions (e.g. equation (4.180)) may not be valid. The thermal diffusivity is discussed further in Prosperetti [80], whose damping analysis is discussed in the next section.

(b) Prosperetti's 1977 Analysis

In 1977, Prosperetti [80] presented a linearised theory for the small-amplitude forced pulsation of a bubble, describing the thermal effects in terms of the effective polytropic index and thermal damping constant. The analysis uses the conservation equations of continuum mechanics. In contrast to Eller's analysis, spatial uniformity of pressure within the bubble is no longer assumed, and for insonation at high frequency this is particularly pertinent. The results are presented in the form of tables, employing two dimensionless parameters made up from the insonation frequency ω, the equilibrium bubble radius R_0, and characteristic properties of the gas and the liquid.

(i) Basic Formulation. From consideration of the normal stresses at the bubble wall, and neglecting the viscosity of the gas, the pressures on either side of the bubble wall can be found through evaluation of equation (4.75) at $r = R$, with the usual incorporation of the Laplace pressure $p_\sigma = 2\sigma/R$:

$$p_i - p_L = \frac{2\sigma}{R} + \frac{4\eta\dot{R}}{R} \qquad (4.188)$$

which, when the bubble is stationary ($\dot{R}=0$) reduces to equation (2.2). As with Devin's theory, p_L is the sum of the static pressure, the acoustic pressure, and $P_{bl}(r = R_0)$, the acoustic pressure field radiated by the bubble evaluated at the bubble wall:

$$p_L = p_0 + P_A e^{i\omega t} + P_{bl}(r = R_0) \qquad (4.189)$$

The pressure field radiated from a pulsating bubble was discussed in Chapter 3, section 3.3.1. In a method equivalent to Prosperetti's 1977 analysis, substituting $\cos\chi_0$ from equation (3.102), and $\ddot{R} = i\omega U_o e^{i\omega t}$ from equation (3.98), into equation (3.105), and noting that in the long-wavelength limit $\chi_0 \to \pi/2$ so that $e^{i\chi_0} \to i$, yields

$$P_{bl}(r,t) \approx \frac{\rho_o R_o^2 \ddot{R}}{r\sqrt{1+(\omega R_o/c)^2}} e^{-ik(r-R_o)} \tag{4.190}$$

where $\omega R_o/c = kR_o$. (Prosperetti [80] finds P_{bl} by substituting the velocity potential for the fluid around a pulsating bubble

$$\Phi(r,t) = -\frac{\dot{R}R_o^2}{r}\frac{e^{-i\omega(r-R_o)/c}}{1+i\omega R_o/c} \tag{4.191}$$

as given by Landau and Lifshitz [81] for the case when viscous terms are negligible, into equation (2.80).)

When evaluated at the bubble wall ($r = R_o$), equation (4.190) can be substituted into equation (4.189) to give the pressure in the liquid just beyond the bubble wall:

$$p_L = p_o + P_A e^{i\omega t} + \frac{\rho R_o \ddot{R}}{1+i\omega R_o/c} \tag{4.192}$$

The small-amplitude pulsations, $R = R_o + R_\varepsilon$ (where $R_\varepsilon \ll R_o$), produce small-amplitude perturbations p_ε from the equilibrium internal pressure

$$p_i = p_{i,e} + p_\varepsilon(t) \tag{4.193}$$

Substitution of equation (4.192) into equation (4.188) recasts the pressure imbalance across the bubble wall in terms of the radial and pressure fluctuations:

$$\frac{\rho R_o^2}{1+i\omega R_o/c}\left[\frac{\ddot{R}_\varepsilon}{R_o}\right] + 4\eta\left[\frac{\dot{R}_\varepsilon}{R_o}\right] - \frac{2\sigma}{R_o}\left[\frac{R_\varepsilon}{R_o}\right] = p_\varepsilon - P_A e^{i\omega t} \tag{4.194}$$

To obtain the form of p_ε, Prosperetti employs a polytropic relation, and in addition uses a dissipative term, similar in form to an additional contribution to make up an effective liquid viscosity, which through analogy with equation (4.75) is

$$p_i = p_o\left(\frac{R_o}{R}\right)^{3\kappa} - 4\eta_{th}\left[\frac{\dot{R}_\varepsilon}{R_o}\right] \tag{4.195}$$

the term η_{th} being the additional thermal contribution to the effective viscosity. Thus from equation (4.193) we obtain

$$p_\varepsilon = p_{i,e}\left\{\left(\frac{R_o}{R}\right)^{3\kappa} - 1\right\} - 4\eta_{th}\left[\frac{\dot{R}_\varepsilon}{R_o}\right] \approx 3\kappa p_{i,e}\left[\frac{R_\varepsilon}{R_o}\right] - 4\eta_{th}\left[\frac{\dot{R}_\varepsilon}{R_o}\right] \tag{4.196}$$

Substitution for p_ε of equation (4.196) into equation (4.194) gives

$$(\rho R_o)\left[\ddot{R}_\varepsilon\right]+(\rho R_o)\left(4\frac{\eta+\eta_{th}}{\rho R_o^2}+\omega\frac{\omega R_o/c}{1+(\omega R_o/c)^2}\right)\left[\dot{R}_\varepsilon\right]$$

$$+(\rho R_o)\left(\frac{3\kappa p_{i,e}}{\rho R_o^2}-\frac{2\sigma}{\rho R_o^3}+\frac{\omega^2(\omega R_o/c)^2}{1+(\omega R_o/c)^2}\right)\left[R_\varepsilon\right]=-P_A e^{i\omega t} \tag{4.197}$$

We now have an equation of motion of the bubble in the radius–pressure frame, as discussed in Chapter 3, section 3.2.1(a): $m_{RP}^{rad}\ddot{R}_\varepsilon + b_{RP}\dot{R}_\varepsilon + k_{RP}R_\varepsilon = -P_A e^{i\omega t}$. The formulation gives $m_{RP}^{rad} = \rho R_o$, in agreement with equation (3.90), where the inertia of the bubble is assumed to be invested in the motion of the surrounding fluid. The stiffness is

$$k_{RP}=(\rho R_o)\left(\frac{3\kappa p_{i,e}}{\rho R_o^2}-\frac{2\sigma}{\rho R_o^3}+\frac{\omega^2(\omega R_o/c)^2}{1+(\omega R_o/c)^2}\right) \tag{4.198}$$

and the ratio of this to the mass gives

$$\omega_0^2=\frac{k_{RP}}{m_{RP}}=\left(\frac{3\kappa p_{i,e}}{\rho R_o^2}-\frac{2\sigma}{\rho R_o^3}+\frac{\omega^2(\omega R_o/c)^2}{1+(\omega R_o/c)^2}\right) \tag{4.199}$$

This gives the resonance frequency as a function of the equilibrium bubble radius and the driving frequency. The form is explained later. The final term in equation (4.199), $(\omega^2(\omega R_o/c)^2)/(1+(\omega R_o/c)^2)$, is the result of the acoustic radiation from the bubble which affects the stiffness of the system (equation (4.198)) through the momentum given to the moving liquid. If this is neglected, we obtain the approximate form of the resonance frequency

$$\omega_0^2\approx\frac{3\kappa p_{i,e}}{\rho R_o^2}-\frac{2\sigma}{\rho R_o^3} \tag{4.200}$$

where substitution for $p_{i,e}$ from equation (2.14) will give the resonance in terms of the hydrostatic pressure p_o in agreement with equation (4.83). This can be compared with Devin's result (equation (3.204)).

The thermal, radiation and viscous contributions to the dissipative constant can be seen from equation (4.197) to be:

$$b_{RP}^{th}=\frac{4\eta_{th}}{R_o} \tag{4.201}$$

the thermal dissipation constant in the radius–pressure frame,

$$b_{RP}^{rad}=(\rho\omega R_o)\left(\frac{\omega R_o/c}{1+(\omega R_o/c)^2}\right) \tag{4.202}$$

the radiation dissipative constant in the radius–pressure frame, and

$$b_{RP}^{vis}=\frac{4\eta}{R_o} \tag{4.203}$$

the viscous dissipation constant in the radius–pressure frame. The dimensionless damping constant is therefore

$$d_{\text{th}} = \frac{\omega b_{\text{RP}}^{\text{th}}}{k_{\text{RP}}} = \frac{4\eta_{\text{th}}}{R_o}\frac{\omega}{\rho R_o}\left(\frac{3\kappa p_{\text{i,e}}}{\rho R_o^2} - \frac{2\sigma}{\rho R_o^3} + \frac{\omega^2(\omega R_o/c)^2}{1+(\omega R_o/c)^2}\right)^{-1} \tag{4.204}$$

$$d_{\text{rad}} = \frac{\omega b_{\text{RP}}^{\text{rad}}}{k_{\text{RP}}} = \omega^2\left(\frac{\omega R_o/c}{1+(\omega R_o/c)^2}\right)\left(\frac{3\kappa p_{\text{i,e}}}{\rho R_o^2} - \frac{2\sigma}{\rho R_o^3} + \frac{\omega^2(\omega R_o/c)^2}{1+(\omega R_o/c)^2}\right)^{-1} \tag{4.205}$$

$$d_{\text{vis}} = \frac{\omega b_{\text{RP}}^{\text{vis}}}{k_{\text{RP}}} = \frac{4\eta}{R_o}\frac{\omega}{\rho R_o}\left(\frac{3\kappa p_{\text{i,e}}}{\rho R_o^2} - \frac{2\sigma}{\rho R_o^3} + \frac{\omega^2(\omega R_o/c)^2}{1+(\omega R_o/c)^2}\right)^{-1} \tag{4.206}$$

If the acoustic radiation term $(\omega^2(\omega R_o/c)^2)/(1+(\omega R_o/c)^2)$ is deemed negligible, equations (4.205) and (4.206) reduce to Eller's formulation for d_{rad} and d_{vis}, as given by equations (4.186) and (4.187) respectively.

To evaluate the formulation of equations (4.204) to (4.206) requires knowledge of the two unknown elements, η_{th} and κ. Their evaluation is involved, and Prosperetti's results are presented below.

(ii) Polytropic Index and 'Thermal Viscosity'. To evaluate η_{th} and κ, Prosperetti employs two dimensionless groups, G_1 and G_2:

$$G_1 = \frac{\rho_1 V}{N_{\text{m}}}\frac{D_{\text{g}}\omega}{\gamma R_{\text{g}} T_\infty} \tag{4.207}$$

where ρ_1 and D_{g} are the density and thermal diffusivity of the gas, and T_∞ the absolute temperature of the liquid thermal reservoir surrounding the bubble. The quantities $\rho_1 V$ and N_{m} are, respectively, the mass and the number of moles of gas contained within the bubble. Therefore $(\rho_1 V/N_{\text{m}})$ is the molar mass of the gas. Since the adiabatic speed of sound in the gas c_{g} is given by equation (1.19) for $\kappa = \gamma$, then if T_∞ is taken as the gas temperature, substitution of equation (1.15) gives

$$c_{\text{g}}^2 = \frac{\gamma p_{\text{g}}}{\rho_1} = \frac{\gamma}{\rho_1 V} N_{\text{m}} R_{\text{g}} T_\infty \tag{4.208}$$

Since the thermal diffusion length in the gas l_{D} equals $\sqrt{D_{\text{g}}/2\omega}$, then G_1 is approximately the square of the ratio of the thermal diffusion length to the acoustic wavelength in the gas, and would be expected to be very small ($\sim 10^{-5}$–10^{-9}).

The second dimensionless parameter is

$$G_2 = \frac{\omega R_o^2}{D_{\text{g}}} = \frac{1}{2}\left(\frac{R_o}{l_{\text{D}}}\right)^2 \tag{4.209}$$

and since the bubble radius would usually be much larger than the thermal penetration depth, G_2 would be expected to be larger ($\sim 10^{-1}$ to 10^9). Thermal diffusion in the liquid is characterised by the parameters

$$G_3 = \frac{\omega R_o^2}{D_1} \tag{4.210}$$

and

$$G_4 = \frac{K_l}{K_g}(1 + (1+\mathrm{i})(G_3/2)^{1/2}) \tag{4.211}$$

where D_l is the thermal diffusivity, and K_l the thermal conductivity, of the liquid. Defining further parameters

$$\Gamma_{1,2} = \mathrm{i} + G_1 \pm \sqrt{(\mathrm{i} - G_1)^2 + 4\mathrm{i}G_1/\gamma} \tag{4.212}$$

$$\Lambda_1 = \forall_1 \coth \forall_1 - 1 \tag{4.213}$$

and

$$\Lambda_2 = \forall_2 \coth \forall_2 - 1 \tag{4.214}$$

where

$$\forall_{1,2} = \sqrt{\frac{\gamma G_2}{2}\left(\mathrm{i} - G_1 \pm \sqrt{(\mathrm{i} - G_1)^2 + \frac{4\mathrm{i}G_1}{\gamma}}\right)} \tag{4.215}$$

these are substituted into

$$G_5 = \frac{G_4(\Gamma_2 - \Gamma_1) + \Lambda_2\Gamma_2 - \Lambda_1\Gamma_1}{G_4(\Lambda_2\Gamma_1 - \Lambda_1\Gamma_2) - \Lambda_2\Lambda_1(\Gamma_2 - \Gamma_1)} \tag{4.216}$$

From this complex entity, Prosperetti derives expressions for η_{th} and κ:

$$\eta_{th} = \frac{\omega\rho_g R_o^2}{4}\,\mathrm{Im}\{G_5\} \tag{4.217}$$

and

$$\kappa = \frac{\omega^2\rho_g R_o^2}{3p_{i,e}}\,\mathrm{Re}\{G_5\} \tag{4.218}$$

Figure 4.37 [80] shows the value of the polytropic index as a function of G_2 (the square of the ratio of the bubble radius to the thermal penetration depth), for a diatomic gas ($\gamma = 7/5$). Labelled curves show the behaviour for different values of G_1 (the square of the ratio of the thickness of the layer where conduction will cause significant temperature changes, to the acoustic wavelength in the gas).

To understand the form, simply consider an increase in G_2 to refer to an increase in the equilibrium bubble radius considered, all other conditions remaining fixed. The smallest bubbles ($G_2 \rightarrow 0$) behave isothermally, as expected since in this regime the thermal diffusion length in the gas is much greater than the bubble radius (or, from an alternative viewpoint, the thermal energy associated with changes in bubble volume is conducted away at almost the rate with which it is generated). For larger bubbles, as G_2 increases and the thermal penetration depth becomes negligible compared with the bubble radius, the process tends to the adiabatic limit. However, this only occurs so long as G_1 is not large. A large G_1 would correspond to a high driving frequency ω, which prevents κ reaching the adiabatic limit. In fact, the graph shows that for the highest values of G_2 (bubble radius), the polytropic index falls again. This effect is greater, the greater the value of G_1 (e.g. the higher ω).

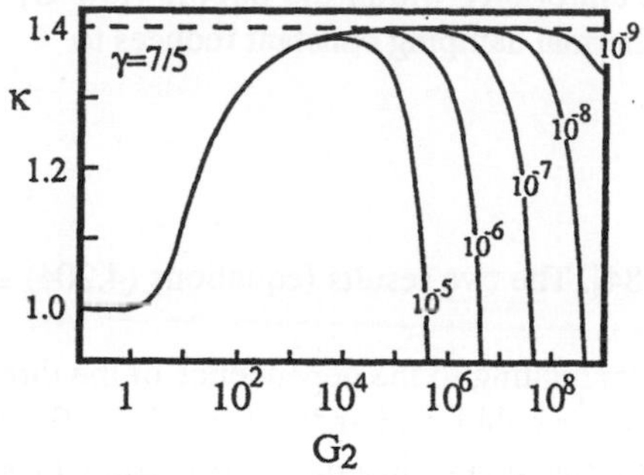

Figure 4.37 The value of the polytropic index as a function of G_2 (the square of the ratio of the bubble radius to the thermal penetration depth), for a diatomic gas ($\gamma = 7/5$). Labelled curves show the behaviour for different values of G_1 (the square of the ratio of the thickness of the layer where conduction will cause significant temperature changes, to the acoustic wavelength in the gas). (After Prosperetti [80].)

The physical cause of such effects, which manifest themselves in large bubbles at high insonation frequencies, are pressure nonuniformities within the bubble. The motion of the bubble wall, and the pressure–volume behaviour of the bubble (the source of the polytropic index), are governed by the pressures on either side of the interface. If the pressure is nonuniform within the bubble, the gas pressure of importance is the local pressure at the bubble wall. At that wall, there is an acoustic impedance mismatch (gas/liquid interface), and the pressure field within the bubble (i.e. within the gas) corresponds to a standing wave [80]. Therefore, at the bubble wall, the perturbation from equilibrium of the internal pressure within the bubble ($p_\varepsilon(t)$) can become positive during contraction of the bubble, depending on the spatial phase of the acoustic wave. (At low frequencies, of course, when the pressure within the bubble is spatially uniform, $p_\varepsilon(t) > 0$ during bubble contraction). It is the pressure at the wall which determines the effective pressure–volume relation, and a decrease in pressure there as the bubble volume decreases corresponds to a negative polytropic index. Such a result signifies that the oscillatory system is unstable, and the model invalid in this regime. It is in approaching this regime that the polytropic index falls in Figure 4.37. Prosperetti's model gives κ oscillating between positive and negative values for this regime (not shown in the figure).

Prosperetti points out that, when the spatial nonuniformity within the bubble is not negligible, though the polytropic index has no thermal meaning, it nevertheless still has mechanical relevance, in that it reflects the gas pressure at the bubble wall, and so the compressibility of the bubble. As a result, Prosperetti shows that his results do not contradict those of Plesset [82], who predicts that at high frequencies, the oscillations should tend to the isothermal, a statement referring to the thermodynamics of the bubble considered as a whole. In the high-frequency domain, the net internal energy variations are small.

Crum [83] demonstrates how Prosperetti's formulation for the polytropic index reduces to that of Eller (equation (4.180)) when (i) the heat capacity of the liquid is much larger than that of the gas; (ii) surface tension is neglected, (iii) $G_1G_2 \ll 1$, which is to say that the bubble radius is very much smaller than the wavelength of the sound in the gas; and (iv) $G_1 \ll 1$, i.e. the mean free path of the gas molecules is very much smaller than the wavelength of the sound in the gas, which, as discussed above, is the case except at the highest frequencies. Through observations of individual bubbles which were levitated by radiation forces at the pressure antinode in a cylindrical cell similar to that described in Chapter 1, section 1.2.2(b)(i), Crum made measurements of the polytropic index consistent with the above theory.

In the regime of spatial uniformity within the bubble (low G_1 (or ω), large G_2 (or R_o)), Prosperetti's result for the thermal damping constant reduces to

$$d_{th} = \frac{3(\gamma-1)}{R_o}\sqrt{\frac{D_g}{2\gamma\omega}} \tag{4.219}$$

in agreement with Pfriem [84]. The two results (equations (4.204) and (4.219), are compared in Figure 4.38.

Prosperetti presents results relating to the dependence of the three dissipation constants on the insonation frequency for air bubbles of radii 10^{-3}, 10^{-4}, 10^{-5} and 10^{-6} m in water. The variation of the damping as a function of insonation frequency ω is shown in Figure 4.39 for bubbles of equilibrium radius (a) 1 mm, (b) 100 μm, (c) 10 μm, (d) 1 μm. On the vertical axis, the damping constant b_{RP} is divided by $2\rho R_o$ (i.e. twice m_{RP}) to make it independent of the spatial dimension. The circle marks the position of damping at the bubble resonance, in agreement with the values given by Chapman and Plesset [85]. The trends are clear. Acoustic damping is negligible at low frequencies, whilst thermal damping is negligible at high frequencies. This does not contradict the earlier statement, that the bubble tends to the isothermal at *both* the lowest and the highest frequencies, since at high frequencies the pressure perturbations within the bubble are no longer spatially uniform, and the isothermal behaviour reflects the *net* energy changes. Viscous damping is negligible at the larger bubble radii, and (as can be seen from equation (4.203)) is independent of the insonation frequency.

The figures show a minimum in the total damping coefficient at around the bubble resonance, as a result of this being the transition region between thermally and acoustically dominated damping. Prosperetti points out that the fact that b_{RP}^{th} and b_{RP}^{rad} balance in the region of the bubble resonance is a coincidence peculiar to this particular system. In the figures, the minimum is at a higher frequency than the resonance for μm bubbles, and at a lower frequency than resonance for the mm size bubbles. However, as the bubble size decreases, the increasing contribution from the viscous damping causes the minimum to disappear.

It is therefore now possible to explain the dependence of the bubble resonance on the driving frequency and bubble radius, as given by equation (4.199), and plotted for illustrative purposes in Figure 4.40. The resonance takes the isothermal form for small bubbles and low driving

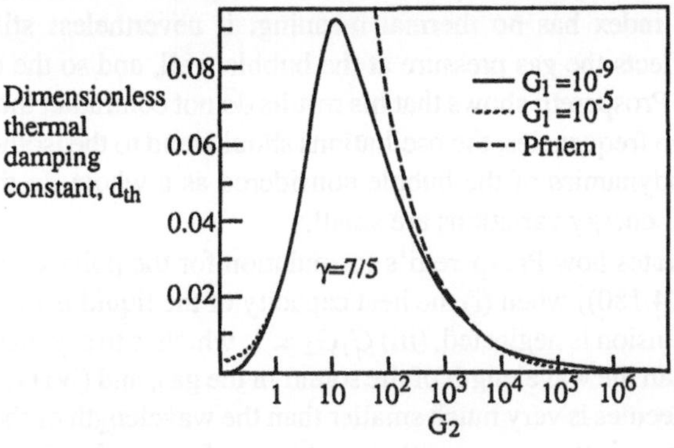

Figure 4.38 Comparison of the predictions for the thermal damping constant given by Prosperetti [80] and Pfriem [84], as a function of G_2 for a diatomic gas ($\gamma = 7/5$). The two curves correspond to $G_1 = 10^{-9}$ and $G_1 = 10^{-5}$ (After Prosperetti [80].)

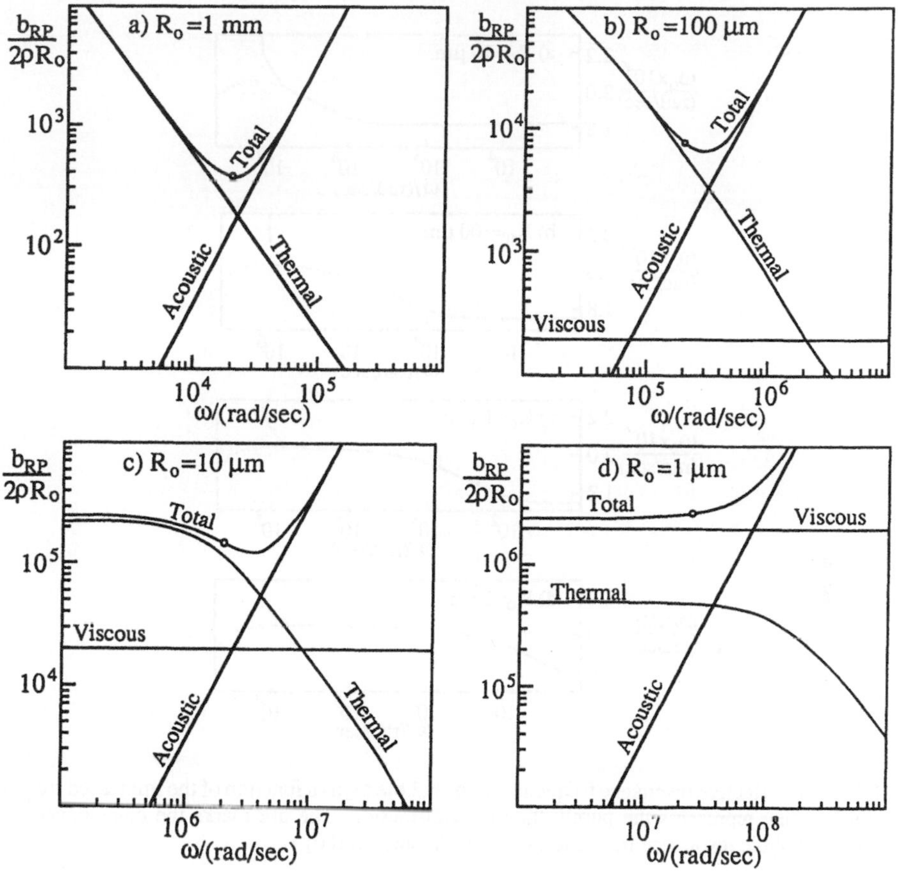

Figure 4.39 The variation of the damping as a function of insonation frequency ω is shown for bubbles of equilibrium radius (a) 1 mm, (b) 100 μm, (c) 10 μm, (d) 1 μm. On the vertical axis, the damping constant b_{RP} is divided by $2\rho R_0$ (i.e. twice m_{RP}) to make it independent of the spatial dimension. The circle marks the position of damping at the bubble resonance. (After Prosperetti [80].)

frequencies, as explained above. As both increase, the resonance departs from the isothermal form. However, at frequencies of about ten times the bubble resonance, the acoustic radiation terms become important and the value of the resonance increases dramatically. Such acoustic effects dominate, because at this point the thermal effects would tend to make the resonance decrease: it coincides with the fall in κ illustrated in Figure 4.37.

(c) Summary

Plesset and Prosperetti [54], Apfel [26] and Prosperetti [86, 87] review the damping of spherical pulsating bubbles. The use of a polytropic assumption to model the internal pressure within a bubble containing mainly noncondensible gas is clearly an approximation that will be inappropriate under some conditions. Through linearised conservation equations for mass, momentum and energy, the variation of the polytropic index (particularly with the frequency of insonation ω) and the significance of thermal (compared with viscous and radiation) losses have been demonstrated [54, 80, 84, 85, 88, 89].

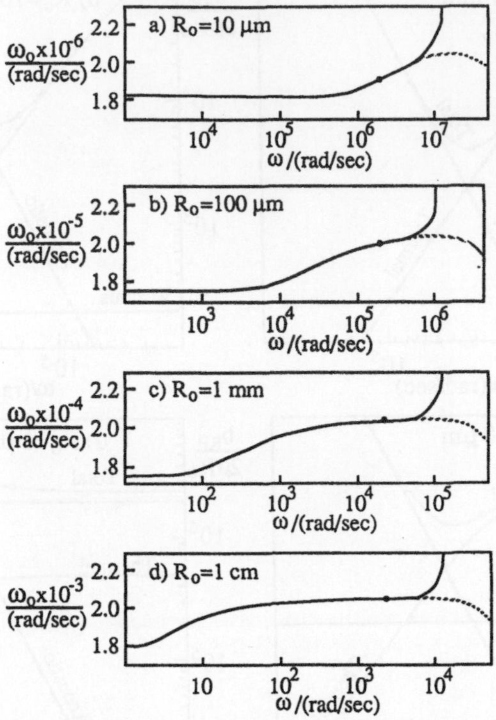

Figure 4.40 The effective resonance frequency of the bubble ω_0 as a function of the impressed frequency ω. The broken line represents the purely thermal contribution. The dot marks the coincidence of the insonation and bubble resonance frequencies. (After Prosperetti [80].)

As seen in section 4.4.2(b), Prosperetti [80] departed from the assumption of spatial uniformity within the bubble. Building on a model [90, 91] of small-amplitude forced spherical pulsations, which incorporates spatial nonuniformities in density, pressure, velocity, concentration and temperature both inside and outside the bubble, Fanelli, Prosperetti and Reali [2] give a theoretical treatment for a bubble containing gas and vapour which is forced into linear shape oscillations. Miksis and Ting [92] provide a set of nonlinear differential equations for the radial pulsations of a bubble, incorporating surface tension, thermal and viscous effects, assuming a thin thermal boundary layer in the gas, which when linearised reduces to the 1977 results of Prosperetti [80].

Nonlinear investigations into the damping of a forced bubble include the experimental [93], and a formulation based upon a direct evaluation of the internal pressure within the bubble [17]. A nonlinear model, originally proposed by Nigmatulin *et al.* who made the approximation of spatially uniform internal pressure, has proved fruitful [76, 77, 94–98].

The evaporation and condensation of vapour and mass flux at the bubble wall may be important to the damping of oscillating bubbles [90, 91, 99]. The question of such mass flux is an important one. The primary difference so far evident between the stable bubble and the ideal spring–bob system described in section 4.1 is that the bubble is a nonlinear oscillator with frequency-dependent damping. There are, however, other differences. Whilst the spring–bob system does not change in time, the bubble is perfectly able to do so. As seen in section 4.3, the transient bubble may fragment. The stable bubble can also undergo changes in its nature. One such, called *rectified diffusion*, can cause the equilibrium radius R_0 to increase in time as a result of mass flux across the bubble wall. This phenomenon is the topic of the next section.

4.4.3 Rectified Diffusion

(a) Introduction: the Shell and Area Effects

A gas bubble in a liquid, in the absence of a sound field, will slowly dissolve owing to the excess internal gas pressure required to balance the surface tension pressure of $2\sigma/R_0$ (Chapter 2, section 2.1.1). Thus, in the absence of a sound field and any stabilising mechanisms of the type discussed in Chapter 2, section 2.1.2(b), all bubbles gradually dissolve [100].

In the presence of a sound field, the situation is quite different. During stable cavitation, it is usually assumed that since evaporation and condensation take place so much more rapidly than the bubble dynamics, the vapour pressure within the bubble remains constant at the equilibrium value. However, this is not so for the gas content of the bubble, a gas which will also be dissolved in the liquid. Harvey *et al.* [101], studying the formation of bubbles in animals, suggested a mechanism by which bubbles undergoing stable cavitation in a sound field can experience a steady increase in R_0, their equilibrium radius. This inwardly directed *rectified diffusion* comes about through the active pumping of gas, initially dissolved in the liquid, into the bubble, using the energy of the sound field. There are two contributory elements to a full description of the processes: an 'area effect' and a 'shell effect'.

The Area Effect. Whilst the bubble radius is less than equilibrium, the gas inside is at a greater pressure than the equilibrium value, and thus diffuses out into the liquid. Conversely, when the bubble radius is greater than R_0 (except for when it is just greater than R_0, at which time there is an excess internal pressure of $(2\sigma/R)$ resulting from surface tension) the internal gas pressure is less than the equilibrium value, and so gas diffuses from the liquid into the bubble interior. The net flow rate of the gas, however, is not equal during the compressed and expanded phases of the bubble motion, because the area of the bubble wall (the transferal interface) is greater in the latter case than in the former. This process is illustrated in Figure 4.41.[40] Therefore, over a period of time, there will be a net influx of gas to the bubble interior.

The Shell Effect. This process occurs because the diffusion rate of a gas in a liquid is proportional to the concentration gradient of the dissolved gas.[41] As the bubble pulsates, a spherical shell of liquid surrounding the bubble will change volume, and so the concentration gradient will change. Consider the bubble shown at equilibrium in Figure 4.42(a). Two of the shells surrounding the bubble are illustrated schematically. When the bubble is expanded, each liquid shell contracts (Figure 4.42(b)). The concentration of dissolved gas in the liquid adjacent to the bubble wall is less than the equilibrium value (Henry's Law[42]), but the shell is thinner than when the bubble is at equilibrium radius, so that the gradient across the shell is higher. Thus the rate of diffusion of gas towards and into the bubble is high. When the bubble is contracted, the liquid shells surrounding the bubble are expanded (Figure 4.42(c)). Though the

[40] This figure is for illustrative purposes only – it does not fully take into account all relevant effects (e.g. surface tension).

[41] To put it in simpler terms, the greater the difference in concentration across a given distance, the greater the amount of substance that will diffuse across that distance in a given time.

[42] See Chapter 2, section 2.1.1.

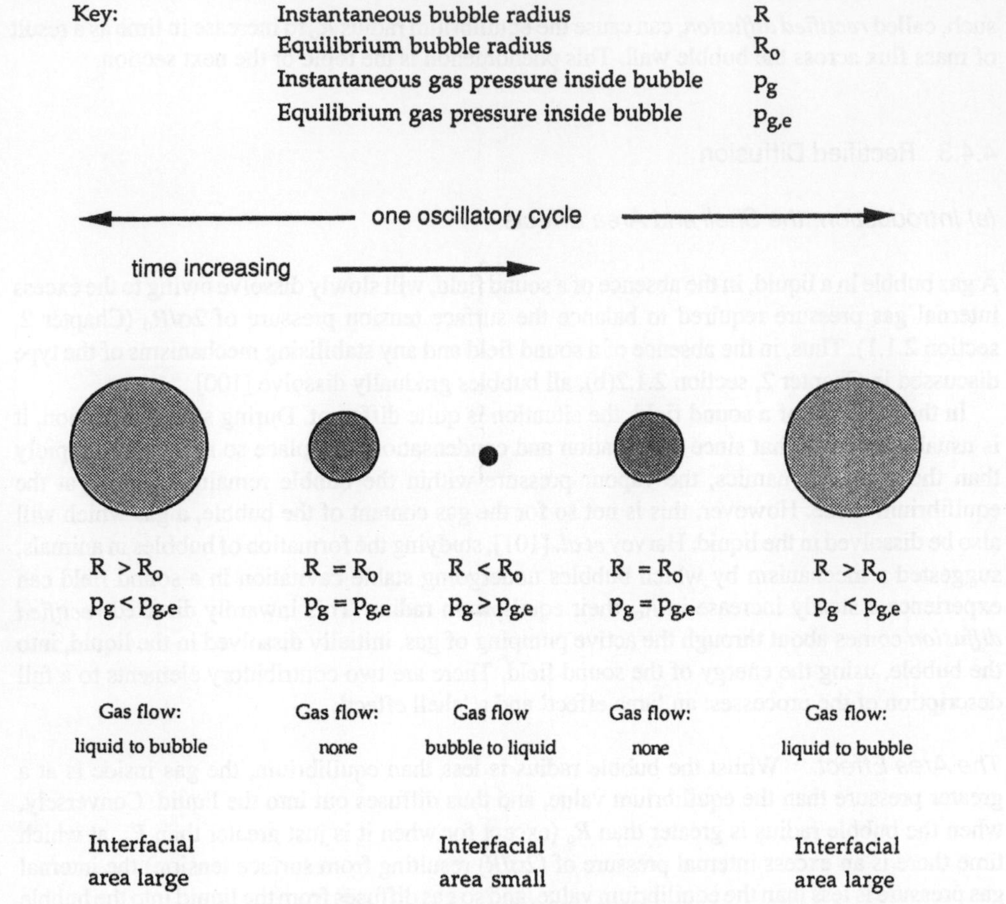

Figure 4.41 The 'area' effect. The pressure within a pulsating bubble, and the surface area of the transfer (the bubble wall), both oscillate about an equilibrium value, causing a flux imbalance.

concentration of gas near the bubble wall is higher than when the bubble is expanded (Henry's Law), the increased thickness of the shell means that the concentration gradient is not as great as when the bubble is expanded. The two factors (gas concentration at the bubble wall, and shell thickness) work together when the bubble is expanded, but against one another when the bubble is contacted: on expansion there is a large concentration gradient driving gas a short distance, and in the second case a lesser gradient driving gas a longer distance. The former effect is dominant.

Thus not only is the area asymmetrical in expansion and contractions, so is the diffusion rate: the two effects reinforce one another. The combined effect means that during stable cavitation in sufficiently intense fields (Crum [102] estimates for $P_A \gtrsim 0.01$ MPa) the time-average bubble radius R_o will increase.

Rectified diffusion of gas[43] has far-reaching consequences: bubble nuclei can grow to provide intense cavitational activity, and small bubbles may grow to a resonant or threshold size where more violent action is usually observed. For example, if a bubble fragments through

[43] It is interesting to note that, just as mass may be the subject of rectified diffusion, so may heat. Wang [103] provides an analysis of this effect based upon evaporation effects in an oscillating bubble.

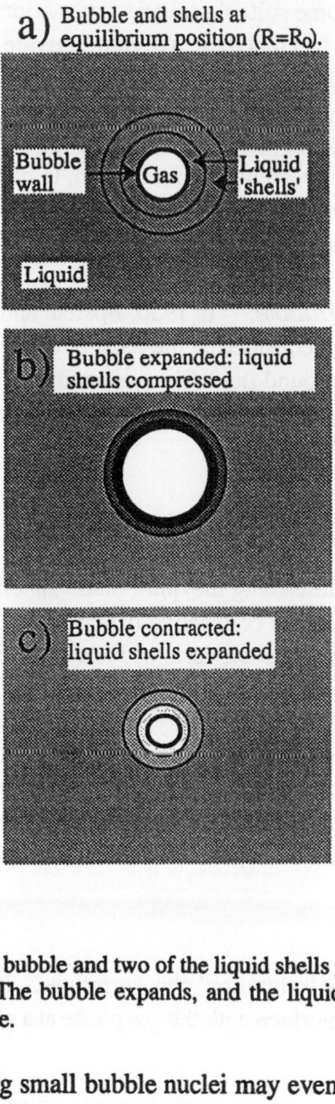

Figure 4.42 The 'shell' effect: the bubble and two of the liquid shells surrounding it are shown. (a) The bubble is at equilibrium size. (b) The bubble expands, and the liquid shells contract. (c) The bubble contracts, and the liquid shells dilate.

a transient collapse, the resulting small bubble nuclei may eventually grow to collapse themselves, so providing cavitation with a self-enhancing mechanism of positive feedback.

(b) Initial Formulations

(i) Mass Flux. The transport of dissolved gas within a liquid is governed by Fick's Law of Mass Transfer:

$$\frac{dC}{dt} = \frac{\partial C}{\partial t} + \vec{v}.\vec{\nabla}C = D\nabla^2 C \tag{4.220}$$

where D is the diffusion constant of dissolved gas within the liquid, C is the concentration of dissolved gas within the liquid, and $\vec{v}$ is the liquid velocity at a point.

To solve the mass transfer of gas between the liquid and a spherical bubble requires solution of both equation (4.220) and some suitable equation of motion for the bubble. However, such a solution is complicated by the fact that these two equations are coupled through the convective term $\vec{v}.\vec{\nabla}C$, and through the appearance of p_g in the equation of motion.[44] The gas pressure will vary in time as gas leaves the bubble in a manner governed by Fick's Law, so that the equation of motion is dependent on equation (4.220).

Solution of mass transfer between bubble and liquid, to describe either passive dissolution or rectified diffusion, therefore usually involves making approximations which reduce the coupling between Fick's Law and the equation of motion.

(ii) Passive Dissolution of a Bubble. In 1950, Epstein and Plesset [104] published a solution for the two equations when considering the passive dissolution of a bubble under a static pressure, in the absence of any sound field. The gas exchange is slow, as is the bubble motion and consequent liquid velocity, so that the convective term can be neglected. In this way

$$\frac{\partial C}{\partial t} = D\left(\frac{\partial^2 C}{\partial r^2} + \frac{2}{r}\frac{\partial C}{\partial r}\right) \tag{4.221}$$

approximates to the equation describing the quasi-static spherical diffusion of gas in a liquid. This frees equation (4.220) from its convective dependence on the equation of motion, and, being now independent, this approximate form of Fick's Law is amenable to solution.

It is assumed firstly that the gas concentration within the bubble is spatially uniform, and secondly that the system has spherical symmetry, so that the radial component of liquid velocity at distance r from the bubble centre in an incompressible fluid is $v_r = R^2\dot{R}/r^2$ (equation (2.25)).

Epstein and Plesset assumed that at time $t = 0$ there would be uniform gas concentration C_∞ throughout the liquid (measured here in moles per unit volume). At times $t > 0$, the dissolved gas concentration at the bubble wall, C_R, is determined through Henry's Law, where

$$C_R = H_D^{-1} p_g \tag{4.222}$$

H_D being Henry's constant. The 'saturation' concentration C_o is the dissolved gas concentration in equilibrium across a plane interface with the gas phase at a given partial pressure. In this case

$$C_o = H_D^{-1} p_o \tag{4.223}$$

so that division by equation (4.222) and substitution from equation (2.8) for the static case ($p_\infty = p_o$, $R = R_o$), with vapour pressure assumed to be negligible, gives

$$\frac{C_R}{C_o} = 1 + \frac{2\sigma}{p_o R_o} \tag{4.224}$$

The concentration at the wall, C_R, will change in time as R_o decreases, as shown by this equation. However, the change will generally be slow and was neglected by Epstein and Plesset. They obtained for the rate of change of the number of moles of gas N_m contained within the bubble

[44] For example, p_g appears in the Rayleigh–Plesset equation, through the analysis given in Chapter 2, section 2.2.1.

$$\frac{dN_m}{dt} = 4\pi D R^2 (C_\infty - C_R)\left(\frac{1}{R_0} + \sqrt{\frac{1}{\pi D t}}\right) \tag{4.225}$$

As expected, the bubble will dissolve if $C_\infty < C_R$. If $C_\infty > C_R$, the liquid is supersaturated with the gas, and the net mass transfer will be into the bubble. Neppiras [21] gives a modified form of the result of Epstein and Plesset:

$$\frac{\partial m_b}{\partial t} = 4\pi R_0^2 C_0 D\left(\frac{C_\infty}{C_0} - 1 - \frac{2\sigma}{p_0 R_0}\right)\left(\frac{1}{R_0} + \frac{1}{\sqrt{\pi D t}}\right) \tag{4.226}$$

where m_b is the mass of gas within the bubble, C_∞ is the mass-concentration of gas dissolved in liquid remote from the bubble, and C_0 is the saturation concentration. Replacing dm_b/dt by $4\pi\rho R_0^2 \dot{R}_0$ gives the rate of decrease of bubble size by diffusion:

$$\dot{R}_0 = \frac{C_0 D}{\rho}\left(\frac{C_\infty}{C_0} - 1 - \frac{2\sigma}{p_0 R_0}\right)\left(\frac{1}{R_0} + \frac{1}{\sqrt{\pi D t}}\right) \tag{4.227}$$

From equation (4.227) it is obvious that, even if the liquid is saturated with dissolved gas (i.e. $C_\infty = C_0$), the bubble will still decrease in size ($\dot{R}_0 < 0$) owing to the surface tension term. Neppiras gives the example of a bubble of radius 10 µm, which dissolves away completely in air-free water in about 1.17 s, but takes about 6.63 s in saturated water. If $C_\infty/C_0 = 0.5$, complete dissolution requires 2 s.

(iii) The Dynamic Problem. Epstein and Plesset [104] assumed quasi-static conditions for the formulation of passive dissolution. If, however, the bubble is being driven by a sound field, the problem is dynamic and the coupling of equation (4.220) with the equation of motion through the convective term cannot be neglected. The equation of motion may yet be treated as independent of the diffusion equation if the timescale for change in the equilibrium bubble radius is very much longer than the time for a single pulsation. Solutions to the equation of motion found in this manner can then be used to solve equation (4.220). However, the equation of motion is nonlinear, and solution of (4.220) involves the application of boundary conditions to a nonstationary bubble wall. Faced with these problems, Blake [105] assumed small-amplitude sinusoidal motions to eliminate the nonlinearity in the equation of motion, and assumed the bubble wall fixed in space, at which position the wall area and the concentration of gas would vary as if the wall were moving. However, such methods make no allowance for the 'shell' effect mentioned above. Thus Blake's consideration only accounted for the area effect, and disagreed with the first experimental measurements [106, 107]. Blake's treatment was modified [108, 109], with negligible change to the final result, and again this result disagreed widely with the experimental data. The cause was simply that the approximation of ignoring the convective term in the gas diffusion equation in order to uncouple the equations of motion and gas diffusion is inappropriate. When the mass diffusion length $\sqrt{4Dt}$ of gas in the liquid [110] is small compared with the bubble radius, the convective term is not negligible. Thus, because of the timescales involved, the approximation is justifiable with passive bubble dissolution, but not with rectified diffusion.

Realising this, Hsieh and Plesset [111] reinstated the convective term in the equation for mass transfer in a spherically symmetric frame:

$$\frac{\partial C}{\partial t}+V\frac{\partial C}{\partial r}=D\left(\frac{\partial^2 C}{\partial r^2}+\frac{2}{r}\frac{\partial C}{\partial r}\right) \tag{4.228}$$

They accounted for the shell effect by expanding the boundary condition in a Taylor series about the equilibrium bubble wall position, but they too were restricted to small-amplitude wall motions. Their analysis employed the model of a uniform spherical bubble situated in an infinite liquid saturated with dissolved gas at the gas pressure inside the bubble. Viscous and thermal effects were ignored. They found that if the bubble radius R_o is very much greater than the diffusion length $\sqrt{4Dt}$, but very much smaller than R_r, the resonant bubble radius for the frequency of the incident sound wave, then provided the passive dissolution of the bubble is negligible, the mass flow rate is

$$\frac{\partial m_b}{\partial t} = \tfrac{8}{3}\pi R_o C_o D\left(\frac{p_A}{p_{i,e}}\right)^2 \tag{4.229}$$

where p_A is the amplitude of the oscillation of the pressure of the *permanent gas phase* within the bubble,[45] and $p_{i,e}$ is the pressure within the bubble when $R = R_o$. Neppiras [21] modified equation (4.229) to account explicitly for the effects of surface tension, to give

$$\frac{\partial m_b}{\partial t} = \tfrac{8}{3}\pi R_o C_o D\left(\frac{p_A}{p_o}\right)^2\left(1+\frac{2\sigma}{p_o R_o}\right) \tag{4.230}$$

giving that a small gas bubble would double its radius in a time t_{RD}, where

$$t_{RD}=\frac{9R_o\rho_g}{4C_oD}\left(\frac{p_A}{p_o}\right)^{-2}\left(1+\frac{2\sigma}{p_oR_o}\right)^{-1} \tag{4.231}$$

Here ρ_g is the density of the permanent gas within the bubble. As stated above, this formulation is restricted to small-amplitude oscillations. Kapustina [112] and Safar [113] derived a multiplicative correction factor to equations (4.230) and (4.231) of

$$\left\{\left(1-\frac{\omega^2}{\omega_o^2}\right)^2+\frac{\omega}{\omega_o}\delta^2\right\}^{-1} \tag{4.232}$$

where δ^{-1} is the Q-factor of the resonance. This factor enables the formula to be applied to a larger range of bubble sizes, including bubbles growing through resonance size. Treatment by Skinner [114, 115] similarly allows for bubble growth through resonance.

If the liquid were not saturated with dissolved gas, the passive dissolution of the bubble (described by equation (4.226)) competes with the mass-flow from rectified diffusion (equations (4.230) and (4.232)) to give a threshold value for p_A/p_o, below which the bubbles will dissolve away, and above which they will grow by rectified diffusion. The value of the threshold as given by the above formulation, is found by simply obtaining from equations (4.226), (4.230) and (4.232) the oscillatory amplitude in the gas pressure at which the net mass flow across the bubble wall is zero, which is

---

[45]Not to be confused with P_A, the amplitude of oscillation of the insonating acoustic pressure field within the liquid.

$$\left(\frac{p_A}{p_0}\right)^2 = \frac{3}{2}\left(1 + \frac{2\sigma}{p_0 R_0} - \frac{C_\infty}{C_0}\right)\left(1 + \frac{2\sigma}{p_0 R_0}\right)^{-1}\left\{\left(1 - \frac{\omega^2}{\omega_0^2}\right)^2 + \frac{\omega}{\omega_0}\delta^2\right\} \tag{4.233}$$

as given by Neppiras [21] in his 1980 review article. It should be noted that as the bubble increases in size, the threshold falls until the bubble reaches resonance size. After this point, the threshold increases rapidly as the bubble size increases. However, as Eller [116] pointed out, bubbles may tend to aggregate owing to radiation forces (e.g. bubbles of larger than resonance size will collect at pressure nodes in a standing-wave field, and secondary Bjerknes forces may cause attraction): this can bring about coalescence, by which bubbles can attain sizes much larger than resonance.

Safar's treatment [113] gives a threshold for rectified diffusion such that the acoustic pressure amplitude P_A in the liquid must exceed P_d, where

$$\frac{P_d}{p_0} = \frac{[3\kappa\{1 + (2\sigma/R_0 p_0)\} - (2\sigma/R_0 p_0)]\,\{1 - (\omega/\omega_0)^2\}\,\sqrt{1 - (C_\infty/C_0) + (2\sigma/R_0 p_0)}}{\sqrt{6(1 + (2\sigma/R_0 p_0))}} \tag{4.234}$$

It is interesting to compare equations (4.233) and (4.234) with the value of the threshold acoustic pressure required for rectified diffusion found in 1961 by Strasberg [107], which is based on the theories of Hsieh and Plesset [111]:

$$P_d = p_0\sqrt{\left(\frac{3}{2}\right)\left(1 + \frac{2\sigma}{R_0 p_0} - \frac{C_i}{C_0}\right)} \tag{4.235}$$

Unlike equation (4.233), this formulation predicts no dependence on the resonance frequency of the bubble.

Key to much of the current work in rectified diffusion is the formulation for mass flux developed in 1965 by Eller and Flynn [117]. As explained above, a full treatment of rectified solution would require analysis of the coupled equations which describe the motion and heat transfer of the bubble–liquid system, and those continuity relations appropriate to the bubble wall. Eller and Flynn [117] simplified the problem in two ways, by assuming that temperature varies in neither time nor space (so eliminating heat transfer considerations), and secondly by assuming spatial uniformity within the bubble with regard to the thermodynamic variables (pressure, temperature, gas concentration etc.), parameters which may, however, vary in time. In this way they separated the solution of the equation of motion and of the diffusion equation into two individual problems.

Firstly, they found a solution for Fick's Law applicable to any general oscillation $R(t)$, provided that the oscillation was in some way periodic. Their result depends on known parameters, and also on time averages of $R(t)$ and $R(t)^4$. The second part of the solution involves solving an appropriate equation of motion to calculate these time averages, which are then substituted in Eller and Flynn's solution to Fick's Law. As will be shown later, several workers have adopted this solution to Fick's Law, completing the analyses by calculating the two time-averages by applying suitable analytical or numerical techniques to a chosen equation of motion. Since it therefore provides an important background, we will firstly discuss Eller and Flynn's solution to the mass diffusion equation.

(c) The Eller-Flynn Solution for Mass Diffusion

Following Plesset and Zwick [118], Eller and Flynn recast equation (4.220) in terms of the new variables h_R, U_C and ξ:

$$h_R = (r^3 - R^3)/3 \tag{4.236}$$

where h_R is a substitution parameter that is proportional to the volume of liquid enclosed between the bubble wall and a sphere at radius r;

$$\frac{\partial U_C}{\partial h_R} = C(h_R,t) - C_\infty \tag{4.237}$$

U_C representing the difference in the mass of dissolved gas (rather than the concentration) from the initial conditions; and

$$\xi = \int_0^t R^4(t')\mathrm{d}t' \tag{4.238}$$

where ξ is a temporal parameter through the time integral of R^4. Substitution of these parameters into equation (4.220) gives

$$\left(1 + \frac{3h_R}{R^3}\right)^{4/3} \frac{\partial^2 U_C}{\partial h_R^2} = \frac{1}{D}\frac{\partial U_C}{\partial \xi} \tag{4.239}$$

which may be simply expanded in the region $(3h_R R^3) \ll 1$ to give

$$\left\{1 + 4\frac{h_R}{R^3} + 2\left(\frac{h_R}{R^3}\right)^2 + \ldots\right\} \frac{\partial^2 U_C}{\partial h_R^2} = \frac{1}{D}\frac{\partial U_C}{\partial \xi} \tag{4.240}$$

The appropriate boundary conditions apply, one of these describing the concentration change at the bubble wall. In a manner analogous to Epstein and Plesset [104] mentioned above, Henry's Law is assumed to hold in the dynamic case as it did in the static case. If the gas within the bubble behaves as an isothermal perfect gas, then

$$p_g R^3 = \left(p_0 + \frac{2\sigma}{R_0}\right) R_0^3 \tag{4.241}$$

so that from equations (4.222), (4.223) and (4.224)

$$C_R = \frac{C_0}{p_0}\left(p_0 + \frac{2\sigma}{R_0}\right)\left(\frac{R_0}{R}\right)^3 \tag{4.242}$$

Thus the concentration of dissolved gas at the bubble wall at equilibrium C_{R_0} gives the boundary condition

$$\frac{\partial U_c}{\partial h_R} = C_{R_0}\left(\frac{R_0}{R}\right)^3 - C_\infty \equiv F(\xi) \tag{4.243}$$

The function $F(\xi)$ represents the difference between the concentration of dissolved gas at the bubble wall, and the concentration far from the bubble; it is clearly time-dependent. It should be noted that while Eller and Flynn assumed isothermal conditions in equation (4.241), Eller [119] later generalised $F(\xi)$ for non-isothermal conditions.

In the spherical frame, the flux of gas across the bubble wall is

$$J = -D \left. \frac{\partial C}{\partial r} \right|_{r=R} = -DR^2 \left. \frac{\partial^2 U_C}{\partial h_R^2} \right|_{h_R=0} \tag{4.244}$$

The rate of change of the total number of moles of gas within the bubble is therefore

$$\frac{dN_m}{dt} = \oint \vec{J}.d\vec{S} \tag{4.245}$$

where the integration of flux takes place over the entire bubble surface, to give

$$\frac{dN_m}{dt} = 4\pi DR^4 \left. \frac{\partial^2 U_C}{\partial h_R^2} \right|_{h_R=0} \tag{4.246}$$

the differential occurring at $h_R = 0$ (the bubble surface). Combination of equation (4.246) with equations (4.220) and (4.238) gives

$$\frac{dN_m}{dt} = 4\pi R^4 \left. \frac{\partial U_C}{\partial \xi} \right|_{h_R=0} \quad \Rightarrow$$

$$\frac{dN_m}{d\xi} = 4\pi \left. \frac{\partial U_C}{\partial \xi} \right|_{h_R=0} \tag{4.247}$$

which can be integrated with respect to ξ to give the total change in the number of moles within the bubble:

$$\Delta N_m = N_m - N_{m,i} = 4\pi U_c \Big|_{h_R=0} \tag{4.248}$$

the notation denoting the evaluation of U_C at $h_R = 0$ (the bubble wall), where N_m and $N_{m,i}$ are respectively the current and initial number of moles of gas within the bubble.

Zero-order Solution. Eller and Flynn decomposed equation (4.240) into a series of equations for the powers of (h_R/R^3), where U_C is similarly decomposed into a summation $U_C = U_{C0}+U_{C1}+U_{C2}+ \ldots U_{Cn}$, where n is the relevant power of (h_R/R^3). The zero-order equation

$$\frac{\partial^2 U_{C0}}{\partial h_R^2} - \frac{1}{D}\frac{\partial U_{C0}}{\partial \xi} = 0 \tag{4.249}$$

with boundary conditions

$$U_{C0}(h_R, t = 0) = 0 \tag{4.250}$$

an initial condition

$$\lim_{h_R \to \infty} \frac{\partial U_{C0}}{\partial h_R} = 0 \tag{4.251}$$

and

$$\left. \frac{\partial U_{C0}}{\partial h_R} \right|_{h_R=0} = C_R - C_\infty = F(\xi) \tag{4.252}$$

is solved by Laplace transform to give

$$\Delta_0 N_m = -4\sqrt{\pi D} \int_0^\xi \frac{F(\xi - \xi')}{\sqrt{\xi'}} \, d\xi' \tag{4.253}$$

The quantity $\Delta_0 N_m$ is the zero-order approximation to the change in the number of moles contained within the bubble, and will be a function of the temporal parameter ξ.

First-order Solution. The first-order equation

$$D \frac{\partial^2 U_{C1}}{\partial h_R^2} - \frac{\partial U_{C1}}{\partial \xi} = -4D \frac{h_R}{R^3} \frac{\partial^2 U_{C0}}{\partial h_R^2} \tag{4.254}$$

with the boundary conditions

$$U_{C1}(h_R, t = 0) = 0 \tag{4.255}$$

an initial condition

$$\lim_{h_R \to \infty} \frac{\partial U_{C0}}{\partial h_R} = 0 \tag{4.256}$$

and

$$\left. \frac{\partial U_{C1}}{\partial h_R} \right|_{h_R=0} = 0 \tag{4.257}$$

is solved by the use of a Green's function to give

$$\Delta_1 N_m = 32D \left\{ \int_0^\xi \frac{\sqrt{\xi - \xi'}}{R(\xi')^3} \, d\xi' \int_0^{\xi'} F(\xi'') \frac{d}{d\xi''} \left(\frac{\sqrt{\xi' - \xi''}}{\xi - \xi''} \right) d\xi'' \right\} \tag{4.258}$$

$\Delta_1 N_m$ is the first-order additive correction to $\Delta_0 N_m$.

The 'High-frequency' Approximation. Equations (4.253) and (4.258) provide more detail than is usually required. As the bubble pulsates, there will be a variation in the gas flux across the bubble wall on the timescale of the pulsation frequency, and this information is incorporated into the two equations.[46] However, with rectified diffusion we are physically only concerned with the net flow of gas into the bubble that occurs over a much longer timescale. Therefore the high-frequency information can be eliminated from equations (4.253) and (4.258), greatly simplifying the result, yet still retaining the pertinent features which describe rectified diffusion and net bubble growth over many oscillations. This 'high-frequency approximation' is valid for $\omega \gg D/R_o^2$.

In order to eliminate the high-frequency components, Eller and Flynn assume that the bubble motion, and therefore the related functions, are periodic, where this 'bubble period' τ_b is some small integral multiple of the acoustic period τ_v. Since $R(t)$ is periodic in t, then ξ will be so likewise, with a transformed period ξ_b. The interval ξ_b represents the value of the variable ξ for a time corresponding to one bubble period τ_b. There will therefore be a circular frequency associated with ξ of magnitude $\omega_\xi = (2\pi/\xi_b)$. This frequency will differ from the insonation frequency ω by $O(R_o^4)$. Eller and Flynn give the approximate relation $\omega_\xi \approx \omega/R_o^4$.

Since its argument is periodic, so too will $F(\xi)$ exhibit periodicity at frequency ω_ξ. Therefore $F(\xi)$ can be expanded as a Fourier series, with a time-invariant term F_o, and oscillating terms of the form $F_k e^{ik\omega_\xi}$. Evaluation of $F(\xi)$, it should be remembered, would yield information on the concentration of dissolved gas at the bubble wall (equation (4.243)).

F_o is the time average of $F(\xi)$ over a single period, i.e.

$$F_o = \frac{1}{\xi_b}\int_0^{\xi_b} F(\xi)\mathrm{d}\xi \tag{4.259}$$

Eller and Flynn change the variable to time t, and evaluate the integral to give

$$F_o = \frac{1}{\xi_b}\int_0^{\tau_b} R_o^4 F \mathrm{d}t = \frac{\tau_b R_o^4 C_{R_o}}{\xi_b} \langle (R/R_o)^4 \rangle \left(\frac{\langle (R/R_o) \rangle}{\langle (R/R_o)^4 \rangle} - \frac{C_\infty}{C_{R_o}} \right) \tag{4.260}$$

where as before C_{R_o} is the concentration of dissolved gas at the bubble wall at equilibrium, and the two time-averages (indicated by '< >') are

$$\langle (R/R_o) \rangle = \frac{1}{\tau_b}\int_0^{\tau_b} \frac{R}{R_o}\mathrm{d}t \tag{4.261}$$

and

$$\langle (R/R_o)^4 \rangle = \frac{1}{\tau_b}\int_0^{\tau_b} \left(\frac{R}{R_o}\right)^4 \mathrm{d}t \tag{4.262}$$

Having already explained the relation between ξ_b and τ_b, it is a simple matter to formulate that relation in terms of $\langle (R/R_o)^4 \rangle$:

---

[46] As Eller and Flynn point out, however, in order to uncouple the equation of motion from the diffusion equation at the onset of this derivation, and to express the boundary condition at the bubble wall, the diffusion was assumed to be a relatively slow process. Therefore it would be inappropriate to treat subsequent high-frequency diffusion information resulting from the formulation with any degree of confidence, and to eliminate it in the manner described in this section is a correct procedure.

$$\xi_b = \int_0^{\tau_b} R^4\,dt = R_0^4 \int_0^{\tau_b} \left(\frac{R}{R_0}\right)^4 dt = \langle (R/R_0)^4 \rangle \tau_b R_0^4 \tag{4.263}$$

which can be substituted into equation (4.260) to give

$$F_0 = C_{R_0}\left(\frac{\langle (R/R_0)\rangle}{\langle (R/R_0)^4\rangle} - \frac{C_\infty}{C_{R_0}}\right) \tag{4.264}$$

High-frequency Form of Zero-order Solution. Replacing $F(\xi)$ by its Fourier representation in the expression for $\Delta_0 N_m$ gives the high-frequency approximation. The contribution due to F_0 is $-8\sqrt{\pi D\xi}F_0$ (equation (4.253)). Examination of equation (4.263) reveals the approximate relation:

$$\xi \approx R_0^4 \langle (R/R_0)^4 \rangle t \tag{4.265}$$

Employing this in equation (4.264) approximates the F_0 contribution to a function of time t

$$8\sqrt{\pi D t \langle (R/R_0)^4\rangle}\, R_0^2 C_{R_0}\left\{\frac{C_\infty}{C_{R_0}} - \frac{\langle (R/R_0)\rangle}{\langle (R/R_0)^4\rangle}\right\} \tag{4.266}$$

In the high-frequency approximation, the contributions from the oscillating terms are negligible [117]. Therefore the high-frequency form of the zero-order solution $\Delta_0 N_m$ contains only contributions from F_0, the magnitude of which is given by equation (4.266).

High-frequency Form of First-order Solution. The solution to equation (4.258) is similarly found by substitution of Fourier series. In this way R^{-3} in the first integral is replaced, the time-invariant term being the average of the function over one period ξ_b

$$\frac{1}{\xi_b}\int_0^{\xi_b} R^{-3}\,d\xi = \frac{R_0}{\xi_b}\int_0^{\tau_b} \frac{R}{R_0}\,dt = \frac{1}{R_0^3}\frac{\langle (R/R_0)\rangle}{\langle (R/R_0)^4\rangle} \tag{4.267}$$

The contributions to the first-order solution $\Delta_1 N_m$ from the above static term in the R^{-3} series, and from F_0, are

$$\frac{-32DF_0}{R_0^3}\frac{\langle (R/R_0)\rangle}{\langle (R/R_0)^4\rangle}\frac{1}{\xi}\int_0^{\xi} \sqrt{\xi'(\xi-\xi')}\,d\xi' = \frac{-4\pi\xi D F_0}{R_0^3}\frac{\langle (R/R_0)\rangle}{\langle (R/R_0)^4\rangle} \tag{4.268}$$

Substitution for F_0 from equation (4.264) and for ξ from (4.265) gives the value of this contribution to be

$$4\pi D t R_0 C_{R_0} \langle (R/R_0)\rangle \left(\frac{C_\infty}{C_{R_0}} - \frac{\langle (R/R_0)\rangle}{\langle (R/R_0)^4\rangle}\right) \tag{4.269}$$

As for $\Delta_0 N_m$, the oscillating contributions to $\Delta_1 N_m$ are negligible in the high-frequency approximation. Therefore equation (4.269) gives the magnitude of $\Delta_1 N_m$ in the limit of the high-frequency approximation.

Gas Change Within the Bubble. It is now possible to determine the gas change within the bubble, in the limit of the high-frequency approximation. The change in the number of moles (equation (4.248)) is

$$\Delta N_m \approx \Delta_0 N_m + \Delta_1 N_m \tag{4.270}$$

$\Delta_0 N_m$ is given approximately by equation (4.266), and $\Delta_1 N_m$ is given by equation (4.269). Summation of these two equations therefore yields the change in the total number of moles of gas contained within the bubble:

$$\Delta N_m = N_m - N_{m,i} = C_{R_0}\left\{\frac{C_\infty}{C_{R_0}} - \frac{<(R/R_0)>}{<(R/R_0)^4>}\right\} \times\left[\left(8\sqrt{\pi D t<(R/R_0)^4>}\, R_0^2\right) + \left(4\pi D t R_0<(R/R_0)>\right)\right] \tag{4.271}$$

By differentiating equation (4.271) with respect to time, we obtain the rate of change of the number of moles of gas within the bubble, within the accuracy of the high-frequency approximation, and taking only the zero- and first-order terms in the expansion of ΔN_m

$$\frac{dN_m}{dt} = 4\pi D R_0 C_{R_0}\left(<R/R_o)> + R_0\sqrt{\frac{<(R/R_0)^4>}{\pi D t}}\right)\left(\frac{C_\infty}{C_{R_0}} - \frac{<(R/R_0)>}{<(R/R_0)^4>}\right) \tag{4.272}$$

The term C_{R_0} can be related to the saturation concentration C_0 by evaluating equation (4.242) at $R = R_0$:

$$C_{R_0} = \frac{C_0}{p_0}\left(p_0 + \frac{2\sigma}{R_0}\right) \tag{4.273}$$

and substituted in equation (4.272) to give

$$\frac{dN_m}{dt} = 4\pi D R_0 C_0\left(1 + \frac{2\sigma}{p_0 R_0}\right)\left(<R/R_o)> + R_0\sqrt{\frac{<(R/R_0)^4>}{\pi D t}}\right) \times\left(\frac{C_\infty}{C_0}\left[1 + \frac{2\sigma}{p_0 R_0}\right]^{-1} - \frac{<(R/R_0)>}{<(R/R_0)^4>}\right) \tag{4.274}$$

Having solved the diffusion equation, Eller and Flynn found the threshold condition by considering the transition point at which the bubble neither grows nor dissolves. The net flow of gas is zero, so that setting $(dN_m/dt) = 0$ in equation (4.274) gives the threshold condition

$$\frac{C_\infty}{C_{R_0}} = \frac{<(R/R_0)>}{<(R/R_0)^4>} \tag{4.275}$$

or, using equation (4.273)

$$\frac{C_\infty}{C_o}=\frac{<(R/R_o)>}{<(R/R_o)^4>}\left[1+\frac{2\sigma}{p_oR_o}\right] \tag{4.276}$$

The time-averages account for the dependence of the threshold on acoustic parameters such as P_A, ω etc. To quantify rectified diffusion, one needs to calculate $<(R/R_o)>$ and $<(R/R_o)^4>$.

(d) Growth Rates and Pressure Thresholds Through Calculation of the Time-averages

The technique adopted by Eller and Flynn and several other workers is to combine the above formulation for the description of the gas diffusion with particular solutions to $R(t)$ obtained from equations of motion. Examples of the various equations of motion that could be used are discussed by Lastman and Wentzell [120].

(i) The Eller–Flynn Time-averages. Eller and Flynn employed numerical solutions to an equation of motion which is effectively the Rayleigh–Plesset equation when conditions are assumed to be inviscid:

$$R\ddot{R}+\frac{3\dot{R}^2}{2}=\frac{1}{\rho}\left\{p_g-\frac{2\sigma}{R}-p_o+P_A\sin\omega t\right\} \tag{4.277}$$

which by comparison with the Rayleigh–Plesset equation (equation (4.81)) can be seen to omit vapour pressure and viscous effects. By using numerical solutions to this equation, Eller and Flynn departed from the small-amplitude treatment used by previous workers, and could incorporate nonlinear or large amplitude effects.

Eller and Flynn [117] compare their theoretical (numerical) results to the Hsieh–Plesset threshold, as expressed in Strasberg's prediction (equation (4.235)). The Hsieh and Plesset threshold [111], which predicts no effect of the bubble resonance, is only valid at frequencies far below the resonance (i.e. $R_o \ll R_r$). It is not valid at resonance or at frequencies greater than resonance. Through comparison with a single experimental result of Strasberg [107], Eller and Flynn demonstrated that their results were an improvement on the Hsieh–Plesset threshold.

Eller and Flynn compare their numerical results with an approximate analytical formulation, based on the linearised equation of motion for first-order perturbations about equilibrium:

$$\frac{R}{R_o}=1+\alpha(\sin\omega t-\sin\omega_o t) \tag{4.278}$$

where

$$\alpha=-\left\{\frac{P_A}{3p_o}\left(1+\frac{4\sigma}{3p_oR_o}\right)^{-1}\left(1-\frac{\omega^2}{\omega_o^2}\right)^{-1}\right\} \tag{4.279}$$

and

$$\omega_o^2=\frac{3p_o}{\rho R_o^2}\left(1+\frac{4\sigma}{3p_oR_o}\right) \tag{4.280}$$

giving

$$<R/R_o>=1 \tag{4.281}$$

and

$$\langle (R/R_0)^4 \rangle = 1 + 3\alpha^2 \left(1 + \frac{\omega^2}{\omega_0^2}\right) \tag{4.282}$$

as the first-order solution. In addition, second-order solutions were derived from an equation of motion containing terms in $(P_A/p_0)^2$:

$$\frac{R}{R_0} = 1 + \alpha \left(\sin\omega t - \frac{\omega}{\omega_0} \sin\omega_0 t\right) + \alpha^2 (K + \text{'oscillating terms'}) \tag{4.283}$$

where

$$K = \left(1 + \frac{5\sigma}{3p_0 R_0}\right)\left(1 + \frac{4\sigma}{3p_0 R_0}\right)^{-1}\left(1 - \frac{\omega^2}{\omega_0^2}\right) - \frac{\omega^2}{2\omega_0^2} \tag{4.284}$$

This gives, to second order in P_A/p_o,

$$\langle R/R_o \rangle = 1 + \alpha^2 K \tag{4.285}$$

and

$$\langle (R/R_0)^4 \rangle = 1 + 3\alpha^2 \left(1 + \frac{\omega^2}{\omega_0^2}\right) + 4\alpha^2 K \tag{4.286}$$

The contribution to the time-averages from the second-order terms in the expansion of $R(t)$ is of the same order as the contribution from the first-order terms, and so cannot be neglected. These results are closer to their numerical results than they are to the Hsieh–Plesset threshold. By the inclusion of inertial effects in the result of Hsieh and Plesset [111], Safar [113] showed that they were equivalent to the Eller–Flynn results.

Eller [121] measured the growth of bubbles trapped by radiation forces at a pressure antinode in a 26.6-kHz acoustic field, at acoustic pressures of up to about 0.3 bar. Primary Bjerknes forces held the bubble in place whilst measurements were made. The bubble radii ranged from 15 μm to 90 μm, measured through application of Stoke's Law to observations of the rise time once insonation had ceased.

The stated medium was air-saturated water, though as Church [122] points out, the changing hydrostatic pressure within the column of water would result in Eller underestimating the concentration of gas dissolved within the liquid. Air was bubbled into the base of Eller's liquid column, and the higher pressure there would allow more gas to dissolve. The bubbling, and the streaming resulting from a base-mounted transducer, would lead to a slight super-saturation of the water at the top of the column. Eller compared the observed pressure thresholds and growth rates with those based on the calculation of Eller and Flynn [117].

A solution to the undamped isothermal equation of motion

$$R\ddot{R} + \frac{3\dot{R}^2}{2} = \frac{1}{\rho}\left\{\left(p_0 + \frac{2\sigma}{R_0}\right)\left(\frac{R_0}{R}\right)^3 - \frac{2\sigma}{R} - p_0 - P_A \cos\omega t\right\} \tag{4.287}$$

is the expansion

$$\frac{R}{R_o} = 1 - \alpha \frac{P_A}{p_o} \cos\omega t + \alpha^2 K \left(\frac{P_A}{p_o}\right)^2 + \ldots \tag{4.288}$$

where

$$\alpha^{-1} = 3\left(1 + \frac{4\sigma}{3p_o R_o} - \frac{\rho\omega^2 R_o^2}{3p_o}\right) \tag{4.289}$$

and

$$K = \left(1 + \frac{5\sigma}{3p_o R_o} - \frac{\rho\omega^2 R_o^2}{12p_o}\right)\left(1 + \frac{4\sigma}{3p_o R_o}\right)^{-1} \tag{4.290}$$

The relevant time-averages are therefore given to second order by

$$<(R/R_o)> = 1 + \alpha^2 K \left(\frac{P_A}{p_o}\right)^2 \tag{4.291}$$

and

$$<(R/R_o)^4> = 1 + \alpha^2(3 + 4K)\left(\frac{P_A}{p_o}\right)^2 \tag{4.292}$$

Substitution of equations (4.291) and (4.292) into equation (4.276) and solving for P_A at the threshold (when the acoustic pressure amplitude P_A equals the threshold P_d) gives

$$P_d = \frac{p_o}{\alpha}\sqrt{\frac{1 + (2\sigma/R_o p_o) - (C_\infty/C_o)}{(3 + 4K)(C_\infty/C_o) - K(1 + (2\sigma/R_o p_o))}} \tag{4.293}$$

Equation (4.293) is similar, but not identical, to the results of Safar (mentioned previously). Eller [121] reduced equation (4.293) for the conditions of low surface tension effects ($\sigma \ll R_o p_o$) and saturation ($C_\infty = C_o$) to

$$P_d = p_o\left(1 - \frac{\rho\omega^2 R_o^2}{3p_o}\right)\sqrt{\left(\frac{3\sigma}{R_o p_o}\right)\left(1 - \frac{\rho\omega^2 R_o^2}{24p_o}\right)^{-1}} \tag{4.294}$$

The thresholds showed rough agreement, but the observed growth rates were much greater than those calculated, sometimes exceeding them by a factor of more than 20. In the 1969 paper Eller [121] suggested that acoustic microstreaming could account for the discrepancy. These small-scale circulation currents, which can occur around an oscillating bubble, were introduced in Chapter 1, section 1.2.3(b) and are more fully discussed in section 4.4.4. The presence of such currents would homogenise gas concentrations except in the thin boundary layer around the bubble. In the absence of microstreaming, the steady-state mass transfer of gas from the liquid to the bubble is given by

$$\frac{dN_m}{dt} = 4\pi D R_o (C_\infty - C_R) \tag{4.295}$$

However, if microstreaming were to occur, the gas concentration profile within the liquid would change. The concentration is C_∞ everywhere within the liquid, except for a thin shell of width L_{ms} around the bubble. Eller stated that the gas transport rate is increased by a factor of $(R_o+L_{ms})/L_{ms}$. Using Nyborg's [123] estimation that the thickness of the acoustic microstreaming boundary layer is $L_{ms} = \sqrt{\eta/\pi\rho\nu}$ (equation (1.96)), Eller estimated that his experimental conditions give a boundary layer of thickness 3.5 μm, which for a bubble of radius 60 μm would increase the gas diffusion rate by a factor of 18.

In a similar experiment in 1972, Eller [119] measured the growth of bubbles of radii from 50 μm to > 200 μm in an 11-kHz acoustic field, trapped by radiation forces at a pressure antinode, again with acoustic pressure of up to around 0.3 bar. The bubble radius was found by employing the measured rise velocity u_b in an iteration of the approximate equation of Langmuir and Blodgett [124, 125]

$$R_o^2 = \frac{9\eta_k}{2g} u_b \left\{1 + 0.197\left(\frac{2R_o u_b}{\eta_k}\right)^{0.63}\right\} \tag{4.296}$$

where g is the acceleration due to gravity, and η_k the kinematic viscosity of the liquid.

The analysis of Eller and Flynn [117] was restricted to the isothermal case through choice of boundary conditions. In 1972, Eller [119] modified the theory for nonisothermal boundary conditions, by rewriting F, which was obtained for the isothermal case from equations (4.241) to (4.243), as the difference between the instantaneous gas concentration dissolved in the liquid at the bubble wall C_R, and that far from the bubble C_∞. Neglecting any temperature-dependence of Henry's Law (equation (4.222)), $C_\infty/C_o = p_g/p_o$. Therefore the function F may be recast as

$$F(\xi) = C_R - C_\infty = C_o\left(\frac{p_g}{p_o} - \frac{C_\infty}{C_o}\right) \tag{4.297}$$

With this modification to the formulation of Eller and Flynn [117], Eller [119] gives the rate of change of the number of moles of gas within the bubble to be approximately:

$$\frac{dN_m}{dt} = 4\pi D R_o C_o \left(\langle R/R_o \rangle\right)\left(\frac{C_\infty}{C_o} - \frac{\langle (R/R_o)^4 (p_g/p_o) \rangle}{\langle (R/R_o)^4 \rangle}\right) \tag{4.298}$$

Eller [119] then applies to this solutions of the adiabatic form of the undamped equation of motion (the isothermal form of which is given in equation (4.287))

$$R\ddot{R} + \frac{3\dot{R}^2}{2} = \frac{1}{\rho}\left\{\left(p_o + \frac{2\sigma}{R_o}\right)\left(\frac{R_o}{R}\right)^{3\gamma} - \frac{2\sigma}{R} - P_o - P_A\cos\omega t\right\} \tag{4.299}$$

with the instantaneous gas pressure given by

$$\frac{p_g}{p_o} = \left(1 + \frac{2\sigma}{p_o R_o}\right)\left(\frac{R_o}{R}\right)^{3\gamma} \tag{4.300}$$

Following through the techniques of his 1969 paper, Eller [121] obtains for the experimental conditions $C_\infty = C_o$ and $\sigma/R_o \ll p_o$, the growth rate

$$\frac{\mathrm{d}R_o}{\mathrm{d}t} = \frac{D}{R_o}\frac{C_\infty}{C_0}\left\{\left(\frac{2\left[1-\dfrac{\rho\omega^2R_0^2}{24\gamma p_0}\right]}{3\gamma\left[1-\dfrac{\rho\omega^2R_0^2}{3\gamma p_0}\right]^2}\right)\left(\frac{P_A}{p_0}\right)^2 - \frac{2\sigma}{R_0 p_0}\right\} \tag{4.301}$$

Setting $\dot{R}_0 = 0$ gives the threshold acoustic pressure for growth by rectified diffusion

$$P_d = p_0\left(1-\frac{\rho\omega^2R_0^2}{3\gamma p_0}\right)\sqrt{\left(\frac{3\gamma\sigma}{R_0 p_0}\right)\left(1-\frac{\rho\omega^2R_0^2}{24\gamma p_0}\right)^{-1}} \tag{4.302}$$

which, if γ were replaced by unity, would give the isothermal condition, and so agree with equation (4.294). Eller concluded that for bubbles driven below resonance, the theoretical threshold acoustic pressure for growth by rectified diffusion was greater for adiabatic than for isothermal conditions. The results for growth by rectified diffusion were then compared with calculated values, for both isothermal and adiabatic conditions. As before, though the calculated threshold pressures were consistent with experiment, the calculated growth rates exceeded those observed by one to two orders of magnitude. Following Eller's suggestion in 1969 [121] that microstreaming could be affecting the growth rate by rectified diffusion, Gould [126] made microscopic measurements of the growth rate, and observed that surface oscillations on the bubble could enhance the growth. In the absence of such a phenomenon, Gould was able to obtain agreement with theory. Theoretical treatments which do examine the effects of microstreaming on the rates of growth by rectified diffusion are discussed in section 4.4.3(d)(iii).

Eller's analyses had not accounted for the effects of damping (note, for example, the absence of a phase factor in equation (4.288)). Since dissipation would be likely to have significant effects close to resonance, in 1975 Eller [116] published a modified formulation which included damping in an approximate form through the addition of a linear damping term $\omega_0 R_0 \dot{R}/Q$ to the equation of motion, where Q is the quality factor which is relevant to the linear resonance:

$$R\ddot{R} + \frac{3\dot{R}^2}{2} + \frac{\omega_0 R_0 \dot{R}}{Q} = \frac{1}{\rho}\left\{\left(p_0 + \frac{2\sigma}{R_0}\right)\left(\frac{R_0}{R}\right)^3 - \frac{2\sigma}{R} - P_0 - P_A\cos\omega t\right\} \tag{4.303}$$

where ω_0 is given by equation (4.280). Eller gives the solution

$$\frac{R}{R_0} = 1 - \alpha\frac{P_A}{p_0}\cos(\omega t - \vartheta) + \alpha^2 K\left(\frac{P_A}{p_0}\right)^2 + \ldots \tag{4.304}$$

where

$$\alpha^{-1} = \frac{\rho R_0^2}{p_0}\sqrt{(\omega^2-\omega_0^2)^2 + (\omega\omega_0 d_{tot})^2} \tag{4.305}$$

K is given by equation (4.290), and d_{tot} is the total dimensionless damping constant. The results he obtained are similar to those found in 1969 for growth rate and threshold (given by (4.293)), though using equation (4.305) for α.

(ii) The 1984 Crum Analysis. Crum [102] returned to approximate analytical solutions to the equation of motion in order to obtain the appropriate time-averages required to make use of

the mass-diffusion formulation of Eller and Flynn [117]. He considered the dependence, at any given moment, of the rate of diffusion of gas across the bubble wall on the *instantaneous* values of the gas pressure p_g within the bubble and the bubble radius R, as well the concentration of dissolved gas outside the bubble. The net flow depends on the time-average of these values (indicated by < >). Following Eller and Flynn [117], Crum obtains for the net rate of change in the number of moles of gas N_m within the bubble

$$\frac{dN_m}{dt} = 4\pi D R_0 C_o \left(\langle R/R_o \rangle + R_o \sqrt{\frac{\langle (R/R_o)^4 \rangle}{\pi D t}} \right) \left(\frac{C_\infty}{C_0} - \frac{\langle (R/R_0)^4 (p_g/p_0) \rangle}{\langle (R/R_0)^4 \rangle} \right) \tag{4.306}$$

The instantaneous gas pressure is p_g, which equals $p_{g,e} = p_0 + p_\sigma$ when $R = R_0$, and p_0 and p_σ are the liquid and surface tension pressures as discussed in Chapter 2, section 2.1.1 (vapour pressure terms being neglected). C_∞ is the concentration of dissolved gas in regions of the liquid far from the bubble, and C_0 is the saturation concentration of dissolved gas in the liquid. With one extra term, this is similar in form to that obtained by Eller in equation (4.298).

Simple thermodynamics gives the instantaneous gas pressure within the bubble to be

$$p_g = p_{g,e} \left(\frac{R_0}{R} \right)^{3\kappa} \tag{4.307}$$

and the equilibrium radius as a function of the number of moles of gas contained within the bubble

$$p_{g,e} \frac{4\pi R_0^3}{3} = N_m R_g T \tag{4.308}$$

Having an expression for dN_m/dt and the two elementary thermodynamic formulae given above, it is conceptually simple to see how the rate of change of equilibrium bubble radius $\dot{R}_0$ may be obtained. Combination of equations (4.306) and (4.308) [102] gives the rate of growth of the equilibrium radius:

$$\dot{R}_0 = \frac{D R_g T C_0}{p_0 R_0 \left(1 + \frac{4\sigma}{3 p_0 R_o} \right)}$$

$$\times \left(\langle R/R_o \rangle + R_o \sqrt{\frac{\langle (R/R_o)^4 \rangle}{\pi D t}} \right) \left(\frac{C_\infty}{C_0} - \frac{\langle (R/R_0)^4 (p_g/p_0) \rangle}{\langle (R/R_0)^4 \rangle} \right) \tag{4.309}$$

One only requires knowledge of the behaviour of R/R_0 (and therefore, from equation (4.307), of p_g/p_0) to make equation (4.309) of use. Crum [102] obtained his expression for R/R_0 as a nonlinear expansion in terms of the acoustic pressure, analogous to equation (4.303):

$$\frac{R}{R_0} = 1 + \alpha \cos(\omega t - \vartheta) \frac{P_A}{p_0} + \alpha^2 K \left(\frac{P_A}{p_0} \right)^2 + \ldots \tag{4.310}$$

where the coefficients are given by

$$\alpha^{-1} = \frac{\rho R_o^2}{p_o}\sqrt{(\omega^2-\omega_o^2)^2 + (\omega\omega_o d_{tot})^2} \tag{4.311}$$

$$K = \frac{(3\kappa + 1 - \rho\omega^2 R_o^2/3\kappa p_o)/4 + (\sigma/4R_o p_o)(6\kappa + 2 - 4/3\kappa)}{1 + (2\sigma/R_o p_o)(1 - 1/3\kappa)} \tag{4.312}$$

and

$$\vartheta = \tan^{-1}\frac{\omega\omega_o d_{tot}}{(\omega_o^2 - \omega^2)} \tag{4.313}$$

When conditions are such that the K-term in equation (4.310) is negligible, the formulation reduces to the small-amplitude linear form given in section 4.1 (equations (4.22) to (4.25)).

This expansion can be compared with those of Eller [119, 121], given in equations (4.283 ff) and (4.288 ff) respectively. Note that Crum has the parameter $\omega^2/\omega_M^2 = (\rho\omega^2 R_o^2/3\kappa p_o)$, in agreement with Eller's use of $(\rho\omega^2 R_o^2/3p_o)$ in the isothermal form (equation (4.294)) and $(\rho\omega^2 R_o^2/3\gamma p_o)$ in the adiabatic case (equation (4.302)).

The parameter d_{tot} is the off-resonance damping constant. Crum [102], using Eller's formulation[47] for the damped oscillation [79], evaluates the following time-averages:

$$\langle (R/R_o)\rangle = 1 + \alpha^2 K\left(\frac{P_A}{p_o}\right)^2 \tag{4.314}$$

$$\langle (R/R_o)^4\rangle = 1 + \alpha^2(3+4K)\left(\frac{P_A}{p_o}\right)^2 \tag{4.315}$$

and

$$\langle (R/R_o)^4(p_g/p_o)\rangle =$$

$$\left\{1 + \tfrac{3}{4}(\kappa-1)(3\kappa-4)\,\alpha^2\left(\frac{P_A}{p_o}\right)^2 + (4-3\kappa)K\alpha^2\left(\frac{P_A}{p_o}\right)^2\right\}\left(1 + \frac{2\sigma}{R_o p_o}\right) \tag{4.316}$$

By setting $\dot{R}_o = 0$ in equation (4.309) and substituting in equations (4.314), (4.315), and (4.316), Crum obtains the threshold condition by finding the acoustic pressure amplitude P_d for which $\dot{R}_o = 0$, where the net flux of gas through the wall of a bubble of initial equilibrium radius R_d is zero:

$$P_d^2 = \frac{(\rho R_d^2\omega_o^2)^2\left\{(1-\omega^2/\omega_o^2)^2 + d_{tot}^2(\omega^2/\omega_o^2)\right\}(1 + 2\sigma/R_d p_o - C_\infty/C_o)}{(3+4K)(C_\infty/C_o) - \left\{\frac{3}{4}(\kappa-1)(3\kappa-4) + (4-3\kappa)K\right\}(1 + 2\sigma/R_d p_o)} \tag{4.317}$$

Limiting Forms:

(a) Low-frequency Limit. Crum [102] reduces equation (4.317) into two approximate forms relevant to the kilohertz and megahertz frequency regimes. For frequencies of the order of a kilohertz

[47] See section 4.4.2(a).

$$d_{th} \approx \frac{3(\gamma-1)}{R_o / l_D} \tag{4.318}$$

where l_D is the thickness of a thermal boundary layer in the bubble.[48] In this regime, d_{th} is very much greater than either d_{rad} or d_{vis} (assumed negligible), so that $d_{tot} \approx d_{th}$. In the same limit,

$$\kappa \approx \frac{\gamma}{1+3(\gamma-1)R_o / l_D} \tag{4.319}$$

With surface tension deemed negligible

$$K \approx \frac{1}{4}\left(3\kappa + 1 - \frac{\rho\omega^2 R_o^2}{3\kappa p_o}\right) \tag{4.320}$$

and the resonance frequency reduces to the form given by equation (3.36). Using these approximations, Crum obtains

$$P_t^2 = \frac{(\rho R_d^2 \omega_o^2)^2 \left\{(1-\omega^2/\omega_o^2)^2 + d_{tot}^2\right\} (1 + 2\sigma/R_d p_o - C_\infty/C_o)}{(3+4K)\,(C_\infty/C_o)} \tag{4.321}$$

(b) High Frequency Limit. Owing to the small size of bubbles of interest in this regime, Crum [102] assumes isothermal behaviour. With the approximations $\kappa \approx 1$

$$d_{tot} \approx d_{vis} = \frac{4\omega\eta}{3p_o} \tag{4.322}$$

$$K \approx \frac{\left(1 - \dfrac{\rho\omega^2 R_o^2}{12\kappa p_o}\right) + \dfrac{5\sigma}{3p_o R_o}}{\left(1 + \dfrac{4\sigma}{3p_o R_o}\right)} \tag{4.323}$$

and

$$\omega_o^2 \approx \frac{1}{\rho R_o^2}\left(3p_o + \frac{4\sigma}{R_o}\right) \tag{4.324}$$

Crum reduces equation (4.317) to

$$P_t^2 = \frac{(\rho R_d^2 \omega_o^2)^2 \left\{(1-\omega^2/\omega_o^2)^2 + d_{tot}^2\right\} (1 + 2\sigma/R_d p_o - C_\infty/C_o)}{(3+4K)\,(C_\infty/C_o) - K(1 + 2\sigma/R_d p_o)} \tag{4.325}$$

This expression for the threshold is almost identical to that obtained by Eller in 1975 [116].

(iii) The Gilmore–Akulichev Solutions of Church. Obtaining analytical solutions such as was done by Crum allows approximate answers to be obtained quickly, and general trends to be elucidated. For example, the formulation suggests that the acoustic pressure threshold is

[48] See Chapter 3, section 3.4.2(a).

minimised, and the growth rate is maximised, near the bubble resonance ($R_o = R_r$).[49] However, fine details, such as the effect of subharmonic and harmonic contributions, require numerical solution of the equations of motion.

Church [122] combined solutions of the Gilmore–Akulichev equation [127] with the solution of Eller and Flynn [117] given in equation (4.274) for the mass flow to obtain values for the thresholds and growth rates, for insonation at 26.6 kHz, and at 1–10 MHz.

The quantities $\langle R/R_o \rangle$ and $\langle (R/R_o)^4 \rangle$, required for substitution into equation (4.274), are found by numerical integration of a fourth-order Runge–Kutta solution to the Gilmore–Akulichev equation, a combination of the Tait equation of state with the Gilmore equation (equation (4.70)). The speed of sound at the bubble wall, c_L, is related to c_o, the speed of sound at infinitesimal amplitudes, by

$$c_L = \sqrt{c_o^2 + (n-1)H} \tag{4.326}$$

where the enthalpy of the liquid, H, is

$$H = \int_{p_\infty}^{p(R)} \frac{dp}{\rho} \tag{4.327}$$

and p and ρ are the time-varying pressure and density in the liquid. Following Lastman and Wentzell [120], Church uses $n = 7$ and

$$p(t) = \frac{c_L^2 \rho}{n p_o}\left(\frac{\rho}{\rho_o}\right)^{\kappa} - \left(\frac{c_L^2 \rho}{n p_o} - 1\right) \tag{4.328}$$

with the liquid pressure at the bubble wall

$$p_L(R) = \left(p_o + \frac{2\sigma}{R_o}\right)\left(\frac{R_o}{R}\right)^{3\kappa} - \frac{2\sigma}{R} - \frac{4\eta \dot{R}}{R} \tag{4.329}$$

(compare with section 4.2.1(b)). Church calculates the polytropic index κ for each value of R_o, employing the method of Prosperetti [80] outlined in section 4.4.2(b) of this chapter. The pressure far from the bubble is

$$p_\infty = p_o - P_A \sin\omega t \tag{4.330}$$

Viscous losses (damping is not explicit in Gilmore's equation) are included in this formulation through the effect of η on $p(R)$ and H in equations (4.327) and (4.329). Radiation damping arises through the energy lost as the bubble compresses the liquid, which manifests itself in the formulation as the variation in the speed of sound at the bubble (equation (4.326)). Variation in the polytropic index could describe thermal damping, though for a complete analysis κ would have to vary continuously with R and $\dot{R}$: in Church's analysis, κ is fixed for a given R_o.

As before, the time-averages $\langle R/R_o \rangle$ and $\langle (R/R_o)^4 \rangle$ are numerically calculated, and the threshold obtained by substitution of the results into equation (4.275) (or (4.276)). The growth rate, $\dot{R}_o$, is found in a manner similar to that of Crum (compare equations (4.331) and (4.308)) by relating R_o to N_m using

---

[49] However, since one might expect large-amplitude oscillations as the bubble approaches resonance, the quantitative results from this theory might not be expected to have high accuracy.

$$\left(p_0 + \frac{2\sigma}{R_0}\right)\left(\frac{4\pi R_0^3}{3}\right) = N_m R_g T \qquad (4.331)$$

Church justifies the assumption of isothermal conditions, when the equations of motion he employs are polytropic, by pointing out that the growth rate equation he obtains is not a dynamic differential equation, but simply a description of the *net* change in equilibrium radius with time. Simple substitution of equation (4.331) into the result of Eller and Flynn (equation (4.272)) gives the growth rate of

$$\frac{dR_0}{dt} = \frac{DR_gT}{p_0R_0} C_{R_0} \left(1 + \frac{4\sigma}{3p_0R_0}\right)^{-1}$$

$$\times \left(\langle R/R_o \rangle + R_o \sqrt{\frac{\langle (R/R_0)^4 \rangle}{\pi D t}} \right) \left(\frac{C_\infty}{C_{R_0}} - \frac{\langle (R/R_0) \rangle}{\langle (R/R_0)^4 \rangle} \right) \qquad (4.332)$$

Church [128] observes that in equation (4.332), the moment from which time t is measured can be taken to be the instant when the acoustic pressure exceeds the thresholds for rectified diffusion [129]. Since this term appears as $\sqrt{1/t}$ in equation (4.332), its effect is large at small times, but it becomes increasingly insignificant as $t \to \infty$, and so the term $\sqrt{1/t}$ is often referred to as the 'transient term'. It may be interpreted as representing the initial depletion of gas from the shell surrounding a bubble growing by rectified diffusion (or, in the case of a dissolving bubble, the initial excess gas that has come from the bubble), $\sqrt{Dt}$ representing a diffusion thickness. Therefore that whole term may be interpreted for growing bubbles as the contribution to the growth from dissolved gas within a shell-like diffusion boundary layer just outside the bubble. The first term in that bracket might be considered physically to represent the contribution to growth from gas that resided in the liquid, outside the diffusion boundary layer. Once growth or dissolution begins, it will take a certain time to establish new equilibrium concentrations: once these are set up, there are steady-state conditions whereby there is a net gas flow into the diffusion boundary layer from the liquid reservoir outside, and from the diffusion boundary layer into the bubble. It is in the period up to the establishment of the new equilibrium that the so-called 'transient' term takes effect [128].

Church [122] calculates the time-averages numerically for air bubbles in water at a concentration equal to 90% and 100% of liquid saturation (the 90% mimicking tissue). His calculations are able to make allowances for deviations from periodicity in the R–t behaviour.

Figure 4.43 presents Church's results for the thresholds for rectified diffusion, compared with those for transient cavitation, at acoustic frequencies of 1, 3, 5 and 10 MHz. The threshold oscillates as R_0 changes, owing to harmonic and subharmonic resonances, with a global minimum at R_r.

Examination of Figure 4.43(b) (3 MHz) enables elucidation of bubble life-histories. The diagram is shown expanded for clarity in Figure 4.43(e). Consider the application of a fixed acoustic pressure (so that the history of an individual bubble will be represented by a horizontal line, growing bubbles travelling to the right and dissolving ones to the left).

Acoustic Pressure Fixed; Initial Equilibrium Radius Variable. If the initial radius/pressure conditions place a given bubble above the threshold for transient collapse, it will undergo

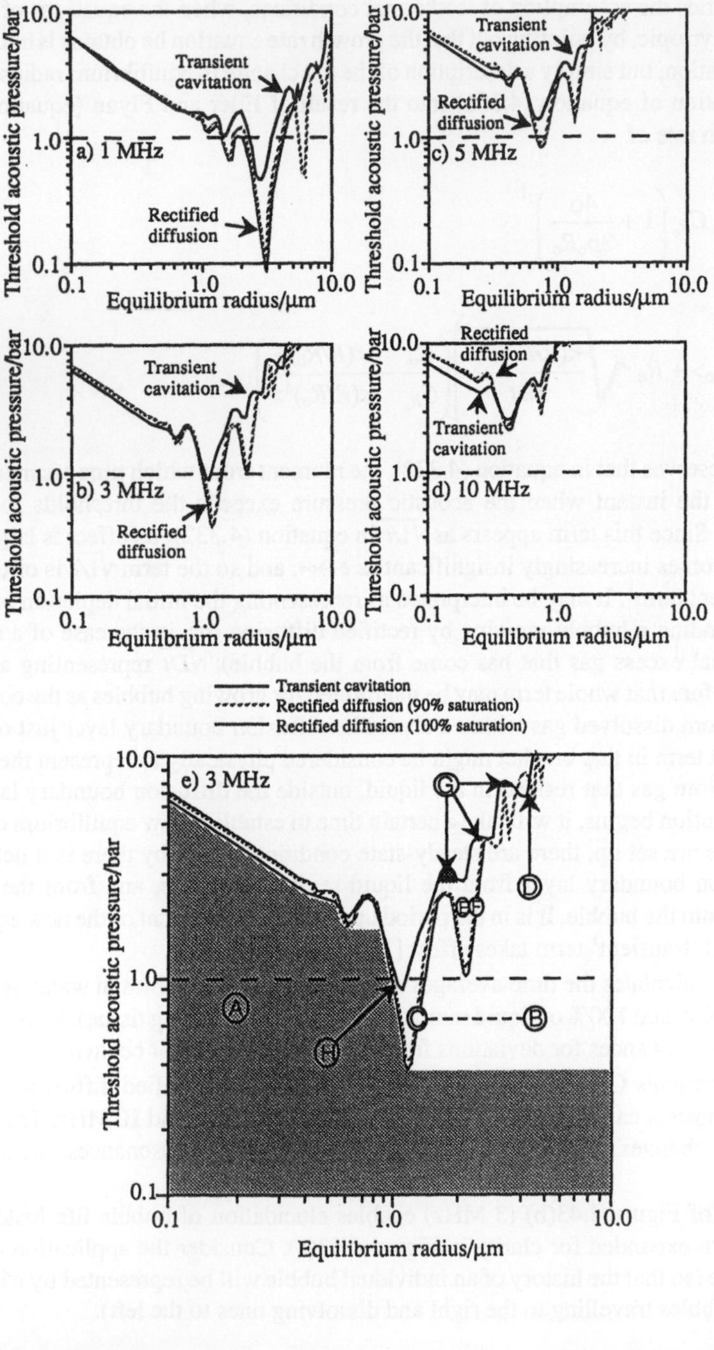

Figure 4.43 Thresholds for rectified diffusion at 90% and 100% saturation, and for transient cavitation, at (a) 1 MHz, (b) 3 MHz, (c) 5 MHz, (d) 10 MHz. The 3-MHz plot (b) is shown expanded in (e) with labelled regimes (see text). (After Church [122].)

transient collapse. If the bubble initially lies below that threshold, it will grow (if it lies above the threshold for rectified diffusion[50]) or dissolve (if it lies below it).

Bubbles Which Initially Lie Below Both Rectified Diffusion and Transient Thresholds. Such bubbles will initially dissolve, so that the equilibrium radius will gradually reduce. The history of the bubble will be represented on the diagram by a horizontal line, as indicated, travelling to the left. If this line does not intersect a threshold, the bubble will dissolve away. Such bubbles are those whose initial radius combines with the acoustic pressure to put them in region A (shaded) in Figure 4.43.

Other bubbles might initially be too large to grow by rectified diffusion, but on shrinking as their gas content dissolves may encounter a rectified diffusion threshold. If the latter has a positive gradient, then as soon as the bubble has grown slightly, the pressure threshold relevant to this slightly larger radius is greater than the applied acoustic pressure (assumed constant), so that the bubble will shrink. It will then re-intersect the threshold, and grow, trapped in this continual cycle of growth and shrinkage. An example of such is the bubble which starts at radius/pressure conditions labelled B on the figure, which ends up with an equilibrium radius undergoing slight changes around the radius at point C.

Alternatively, the bubble might as it dissolves encounter the transient cavitation threshold first, in which case it will undergo transient collapse. In the figure, this would apply to the bubble which under the initial conditions lies at the tip of the arrow D.

Bubbles Which Initially Lie Below the Transient Threshold But Above the Rectified Diffusion Threshold. Such a bubble will initially grow. Its history will therefore correspond to a horizontal line, initially moving to the right. The bubble can then do one of two things. If it intersects the transient cavitation threshold, it will undergo violent collapse. This occurs, for example, for bubbles which at the start of insonation lie in one of the regions labelled G (shaded).

Alternatively it may intersect the rectified diffusion threshold. Growing slightly beyond this threshold, it will be in a region where the pressure/radius conditions place it below the threshold. It will dissolve back to the point of intersection, and be trapped in a stable cyclic process at the point of intersection. An example of this history is given by the bubble which at the start of insonation is at the point labelled E, and becomes trapped at point F.

The dispositions of the stable and transient thresholds will vary with insonation frequency, as Church demonstrates in Figure 4.43 for insonation at 1, 3, 5 and 10 MHz. However, it is clear that the proportion of bubbles which grow by rectified diffusion to collapse upon reaching the transient threshold is not large (compare the shaded areas G to the total area of the plot). Such bubbles must at the start of insonation be in regions above the rectified diffusion threshold, and immediately to the left of a region of the transient cavitation threshold which has a negative gradient.

Church demonstrates that bubble histories which progress by rectified diffusion to end in transient collapse are not common in the biomedical frequency range of 1–10 MHz. At 3 MHz, for an insonation pressure of 1 bar, 80% of cavities that undergo transient collapse do so because they lie above the transient cavitation threshold immediately at the start of insonation (corresponding to initial bubble radii of 1.03–1.15 μm). Only 20% of all cavities that will go transient do so by growing there through rectified diffusion (corresponding to initial bubble radii of

[50]The threshold which bubbles may grow through rectified diffusion, and below which may not, will be called the 'rectified diffusion threshold'. However, in the literature this is sometimes termed the 'stable cavitation threshold', a terminology which will not be employed in this book, since a pulsating bubble which is dissolving is undergoing stable cavitation.

1.00–1.03 μm). Bubbles of initial radii 1.15–1.35 μm will grow by rectified diffusion to become trapped, stably cavitating, at a radius of 1.35 μm. Bubbles of initial radius greater than 1.35 μm will dissolve until they too become trapped at that radius. Bubbles of initial radius below 1.00 μm will dissolve away (Figure 4.43 (e)).

The likelihood of a bubble growing by rectified diffusion to transient collapse decreases with increasing frequency since the rectified diffusion threshold increases more rapidly with increasing frequency than does the transient cavitation threshold.

In a similar vein, there has been a popular belief in the contribution to violent cavitation of bubbles which grow by rectified diffusion to resonance radius R_r, and there collapse. Figure 4.43 illustrates that, again, this is uncommon at this particular frequency. The resonance radius for low-amplitude pulsation is around 1 μm, so that depending on the acoustic pressure, most bubbles below that size will either dissolve away or undergo transient collapse. Very few indeed that have $R_0 < R_r$ at the start of insonation lie above the rectified diffusion threshold, but are smaller than the minimum size required to be above the transient threshold, and will reach R_r by growth before reaching the transient threshold. Such bubbles are in the small region at the tip of the arrow H.

Church compares his results to the analytical predictions of Crum [102] and Crum and Hansen [129], with the expected differences discussed above. As a result of his theoretical investigation into rectified diffusion in the 1–10 MHz frequency range, Church was able to make the following general comments: (i) growth rates could be very considerable within this frequency range (1–10 MHz), quoting a change in equilibrium radius of around 20 μm s^{-1} for insonation at 1 MHz with an acoustic pressure of 1 bar; (ii) at the lower end of the frequency range, thresholds are sensitive to dissolved gas concentration, and are accurate over a wide range of bubble size. However, at the higher frequencies the threshold predictions are only accurate near R_r, the equilibrium radius in pulsation mode resonance with the insonation frequency. (iii) Most importantly, Church concluded that, in the 1–10 MHz range, most bubbles either undergo immediate transient collapse, or undergo stable cavitation and grow or shrink to become trapped at some stable size. At these frequencies, very few bubbles follow the popular tenet, that bubbles will undergo stable cavitation and grow by rectified diffusion from their original size to the transient threshold or to resonance size, at which point they will violently collapse.

Bubble Growth Rates and the Effect of Microstreaming. As outlined above, though predictions of the acoustic pressure threshold for growth by rectified diffusion agree with observation, in general the measured growth rates are often much greater than the predicted ones as a result of microstreaming, agreement only occurring when microstreaming is avoided. The reason for the increased growth rate is clear. As a bubble grows by rectified diffusion, the dissolved gas is taken from the liquid near the bubble. If there is no flow, then the rate at which the deficit is met depends on the rate at which dissolved gas can diffuse from regions farther out from the bubble. Since this is in general a slow process, the liquid outside the bubble wall will become depleted of dissolved gas. The resulting change in concentration gradient reduces the rate of further growth. However, microstreaming flows will tend to bring liquid from farther out close to the bubble wall. The convection of dissolved gas reduces the depletion, and increases the growth rate. Microstreaming will continually refresh the liquid at the bubble wall, giving it a dissolved gas concentration close to that found far from the bubble. The converse process is of course valid: if a bubble is dissolving, microstreaming will tend to remove from the region outside the bubble wall the excess dissolved gas concentration, so increasing the rate of dissolution.

Kapustina and Statnikov [130] considered the effect on the diffusion of microstreaming around a bubble that is fixed in space, whilst Davidson [131] separated the effects of rectified diffusion from the effect of acoustic microstreaming on mass transfer. As he did previously [122], Church [128] calculates the necessary time-averages from computed periodic radius–time solutions to the Gilmore–Akulichev equation. These are employed in a reinterpretation of the formulation of Eller and Flynn [117], where the $\sqrt{1/t}$ term (discussed following equation (4.332)) no longer represents the time since the acoustic pressure amplitude exceeded the threshold for growth, as it did in the absence of microstreaming, but is instead a function of the microstreaming velocity and the bubble radius. Referring back to the interpretation of the relevant terms in equation (4.332) as representing contributions to growth from gas outside and within the diffusion boundary layer, Church [128] notes that the microstreaming flow, which brings in liquid containing the gas concentration found in the reservoir, will affect the diffusion boundary layer. A spatially nonuniform, dynamic equilibrium gas concentration is established. This has a higher average value, and is established more rapidly, than would be the case in the absence of microstreaming. Church renames the $\sqrt{1/t}$ term the ‘decay’ term, since it will always give a finite contribution in the presence of microstreaming, and because t is a decay time regardless of streaming. To estimate its value, Church assumes a certain microstreaming pattern to be set up instantaneously as insonation commences. He considers the variation in time of the gas content of a package of fluid that comes up from the reservoir to the bubble wall, and moves around it before being taken away by the microstreaming flow. If the bubble is growing by rectified diffusion, gas will be withdrawn from the package when it is close to the bubble: key parameters are the difference in gas content from when it arrives at the wall to when it leaves, and the time interval between these events. Church [128] estimates the time required for the gas concentration in the liquid near the bubble to change to its steady-state value, which is the decay time. This is used in the formulation of Eller and Flynn [117]. In comparing his predictions to the measured growth rates of Eller [121] and Gould [126], Church finds reasonable agreement for the low kHz range. Predictions for the effect at biomedical frequencies are also presented. Figure 4.44 shows the predicted growth rates in a 26.6-kHz field ($P_A = 2 \times 10^4$ Pa) with and without microstreaming, as a function of the equilibrium bubble radius, compared with Eller's data. The theory suggests, as expected, that the microstreaming will enhance the growth rate. The data for $R_0 < 35$ µm fit the model of no microstreaming, whilst that for larger sizes agree with the prediction for microstreaming-assisted growth. As will be shown in the next section, microstreaming is associated with surface waves, and as Church points out, in this sound field, surface waves are generated for $R_0 \gtrsim 35$ µm.

(e) Conclusions for Formulations of Rectified Diffusion

The Hsieh and Plesset [111] threshold is only valid at insonation frequencies far below the resonance, and predicts no effect due to the bubble resonance. The Eller–Flynn threshold predicts a smooth minimum in the threshold pressure, and a smooth maximum in the growth rate, at the bubble resonance. This is a result of the pulsation amplitude being a maximum there. The results compare with the analytical solutions of Crum. However, the analytical solutions cannot predict the effect of harmonics and subharmonics, which Church shows to cause local minima in threshold, superimposed on the global minimum at radius corresponding to the low-amplitude pulsation resonance (when $R_0 = R_r$). In the absence of acoustic microstreaming and when there are no surface-active agents on the bubble surface (either of which could increase growth rates), equation (4.332) will afford predictions in good agreement with experiment.

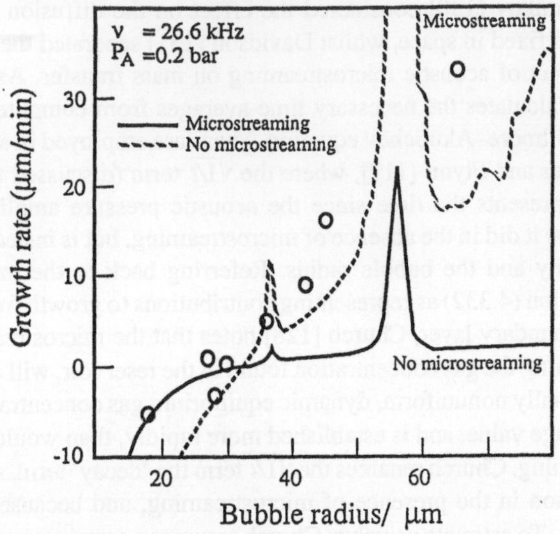

Figure 4.44 Bubble growth rates at ν = 26.6 kHz, P_A = 0.2 bar, with and without microstreaming, as predicted by Church [128]. These are compared with the data (o) of Eller [121]. (After Church [128].)

When microstreaming is present, the circulation may enhance both the rate of growth by rectified diffusion, and the rate of dissolution.

In 1981 Lewin and Bjørnø [132] made calculations based on Eller's 1975 analysis [116], which use equation (4.293) for the threshold, with equation (4.290) for *K*, but the damped form for α (equation (4.311)). They concluded that microbubbles are likely to exist in tissue (especially blood), that resonant ones could be significant in the interaction between tissue and sound, and that during clinical continuous-wave insonation, microbubbles might grow through rectified diffusion to a resonant size in the range 2–3.5 μm. At roughly the same time, ter Haar and Daniels and co-workers [133, 134] found evidence for ultrasonically induced cavitation *in vivo* in guinea pig legs irradiated at 0.75 MHz continuous-wave, generated by a commercial ultrasonic therapy device (Rank Sonacel Multiphon Mark III). The intensities were $80\ \text{mW cm}^{-2} \leq I_{SA} \leq 680\ \text{mW cm}^{-2}$, the peak being three times the spatial average at the position of the leg. The presence of bubbles with radii down to 5 μm was detected by a pulse-echo ultrasonic imager.[51] Following, this Crum and Hansen [135], assuming the existence of nuclei having $R_0 < 5$ μm, numerically integrated equations for the growth of bubbles by rectified diffusion to model the process under the experimental conditions of ter Haar and Daniels. Agreement was good, although the experiment had demonstrated growth to maximum size "within the first minute," while the theory predicted only a few seconds. In a similar experiment, Daniels *et al.* [136] observed the growth of bubbles in insonated agar-based gels to macroscopic size within 5 minutes. In an accompanying theoretical analysis, Crum *et al.* [137], noting that the treatment by Crum and Hansen [129, 135] would not predict the growth to sizes of the order of 1 mm by rectified diffusion alone, assumed that there were three key stages to the observed behaviour. Firstly, the previously stabilised micrometre-sized nucleus of the type discussed in Chapter 2, section 2.1.2 must destabilise, becoming a free-floating unstabilised gas bubble. Secondly, this bubble must grow to somewhat larger than resonance size through rectified

[51] See Chapter 5, section 5.1.1(a).

diffusion, at which no further growth by rectified diffusion is possible, as explained earlier. The third stage is peculiar to the details of the treatment of the gel. The gel was irradiated at a temperature (43° C or 37° C) higher than that at which it had come into equilibrium with respect to the dissolved gas (≈ 20° C). The liquid within the gel is therefore supersaturated with gas, so that the third stage is one of growth by ordinary diffusion. The theory enables the effects of the initial free-floating bubble size, temperature, acoustic pressure amplitude, frequency and pulsing regime to be predicted.

4.4.4 Thresholds and Regimes for the Spherical Pulsating Bubble

We have seen how in the theory of the spherical pulsating bubble there exist certain thresholds, which depend on the equilibrium bubble radius, the acoustic pressure amplitude, the insonation frequency etc. In 1981 Apfel [26, 27, 45] employed the available threshold formulations to draw up diagrammatic representations of cavitation regimes. He chose to plot the thresholds on graphs relating the equilibrium bubble radius to the acoustic pressure amplitude, making a different plot for each pertinent frequency.

Figure 4.45 shows two such plots, for (a) ν = 20 kHz and (b) ν = 1 MHz, for an air bubble in water saturated with dissolved gas. The static pressure p_0 is 1 bar, the water temperature 20° C. The thresholds are based on the equations at that time, and are:

The Blake threshold (Chapter 2, section 2.1.3(a); sections 4.3.1(a) and 4.3.1(c)(i)). Bubbles above this threshold will undergo explosive initial growth in response to a tension in the liquid.

The rectified diffusion threshold (section 4.4.3). Bubbles above this threshold can grow by rectified diffusion. The threshold is based on equation (4.234).

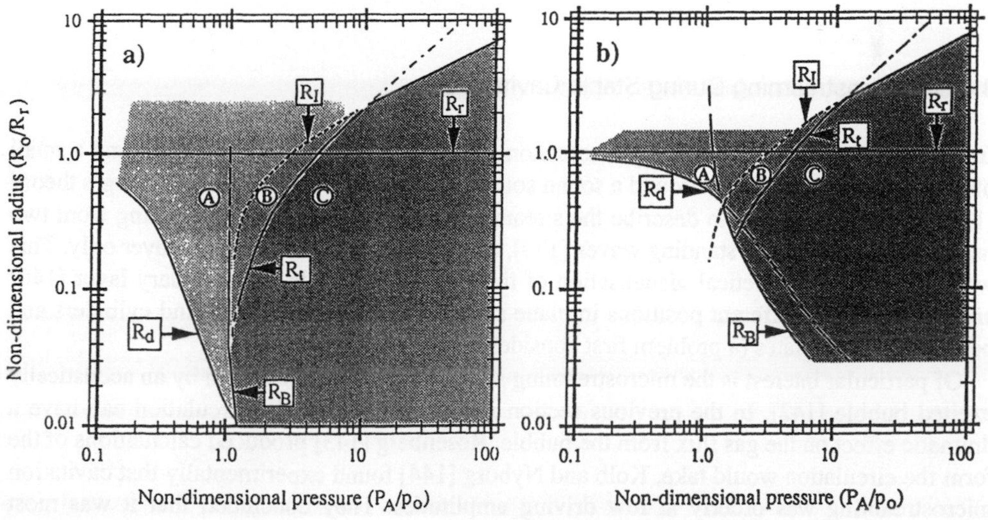

Figure 4.45 Acoustic cavitation prediction chart for air-saturated water at an acoustic frequency of (a) 20 kHz and (b) 1 MHz. The abscissa gives the threshold pressures, normalised to the ambient pressure p_0. The ordinate gives the appropriate threshold radius, normalised to the resonant radius R_r (After Apfel [26].)

The transient cavitation threshold (sections 4.3.1(b) and 4.3.1(c)(ii)). Strictly speaking, this is the threshold relevant to the upper limit on bubble radius, and to a first approximation a bubble will undergo transient collapse if it is larger than the Blake threshold, but smaller than the transient cavitation threshold. Equation (4.112) is derived from the Noltingk–Neppiras criterion that the expansion ratio R_{max}/R_0 at the threshold should be just sufficient to allow the bubble wall velocity to attain the speed of sound in the liquid on collapse. As explained in section 4.3.2, this formulation is only valid if the initial bubble size is smaller than the inertial radius R_I, as given by equation (4.94), for a given acoustic pressure.

Apfel could therefore divide the illustration into several regimes, bounded by these thresholds. In region A, bubbles may grow only by rectified diffusion. In region B, bubbles may grow by rectified diffusion or by direct mechanical means (i.e. with little gas transport). However, the bubble will not undergo transient collapse. In region C, bubbles will undergo transient collapse. The fragments from a transient collapse could clearly lie in another regime. Bubbles below the Blake and rectified diffusion thresholds will dissolve away (unless stabilised). Such divisions are clearly approximate, since they are based on approximate formulations. One of the most fundamental of these is the assumption that the bubble is spherical throughout the oscillation.

Apfel, acknowledging such limitations, included in his figures a line indicating the size of the linear resonance of the bubble ($R_0 = R_r$). This is because as a bubble approaches this size, several interesting phenomena may be observed. From the simple theory outlined in section 4.1, one would expect the oscillation amplitude to become large. However, a bubble undergoing large-amplitude stable oscillation close to resonance can begin to exhibit shape oscillations. These asphericities can lead to many important phenomena, such as the emission of tiny bubble fragments from the surface ripples on the bubble wall. Surface waves on bubble walls are also associated with the microstreaming circulation which, as has been noted in this section, can so influence rectified diffusion. Microstreaming and shape oscillations are discussed in the next two sections of this chapter.

4.4.5 Microstreaming During Stable Cavitation

In Chapter 1, section 1.2.3(b) the phenomenon of microstreaming in the liquid around a small obstacle in a sound field, or around a sound source, was introduced. Recently Nyborg's theory [138] has been modified to describe the streaming around a small sphere resulting from two out-of-phase orthogonal standing waves [139], applicable to a surface viscous layer only. This was followed by theoretical visualisation of the streaming outside the boundary layer [140] around spheres at different positions in plane standing waves, and also around cylinders and between parallel plates (a problem first considered by Rayleigh [141]).

Of particular interest is the microstreaming set up by the sound scattered by an acoustically excited bubble [142]. In the previous section it was shown how the circulation can have a dramatic effect on the gas flux from the bubble. Rosenberg [143] produced calculations of the form the circulation would take. Kolb and Nyborg [144] found experimentally that cavitation microstreaming was orderly at low driving amplitudes. They concluded that it was most pronounced when the insonation frequency drives the bubbles at their resonance frequency for pulsation, ω_0, and when the bubbles are situated on solid boundaries.[52] Through microscopic examination of the steady-state circulation patterns of a suspension of aluminium particles in

[52] This is probably due to the presence of the in-phase image discussed in Chapter 3, section 3.3.2.

water, Elder [145] was able to visualise the microstreaming patterns about a single bubble excited close to resonance. The bubble, which assumed a flattened shaped with maximum horizontal and vertical dimensions of around 0.3–0.5 mm, was sited upon the base of a tank containing a liquid of known viscosity (water, or a water/glycol mixture). The bubble adhered to the base because of the large contact angle which resulted from the 'soiled' nature of the metal, and through the action of the attractive acoustic forces which occur between a bubble and a rigid wall.[53] The tank, which had transparent walls, was firmly mounted on a vibrating steel piston, which was driven at 10 kHz by a magnetostrictive driver. Several different forms of circulation were observed: if, for example, the acoustic pressure amplitude was changed, all other conditions remaining the same, the circulation changed discontinuously through several stable regimes, which tended to have symmetry about the normal to the boundary. Giving ten illustrations of different circulation patterns, Elder summarises by distinguishing four characteristic regimes (Figure 4.46). In addition to the dependence on acoustic pressure amplitude, the pattern assumed in a given set of conditions was also found to be dependent on the liquid viscosity, as is to be expected since microstreaming is associated with the formation of a shear wave due to the retarding effect of the liquid viscosity[54] [21]. However, it is also affected by the amount of surface-active material present in the water: as was shown in Chapter 2,[55] such contaminants can collect on the bubble wall and dramatically affect the dynamics by retarding the free tangential flow of liquid past the surface. If a surface-contaminated bubble is driven at low amplitude in a low-viscosity liquid, the circulation illustrated in Figure 4.46(a) occurs. The pattern shown in Figure 4.46(b) was observed over a wide range of driving amplitude and liquid viscosities. The third regime, illustrated in Figure 4.46(c), usually occurred in low-viscosity liquids, though could be generated in high viscosity liquids if the driving pressure was great enough. In the figure, the bubble wall is drawn so as to represent schematically the orderly extremes of oscillation of a surface wave. This is because the onset of this microstreaming regime coincided with the initiation of the first surface mode. At higher amplitudes it was coincident with the cessation of the second regime. The circulation in the fourth regime was observed at high driving amplitudes and the lowest viscosities. These conditions allow more

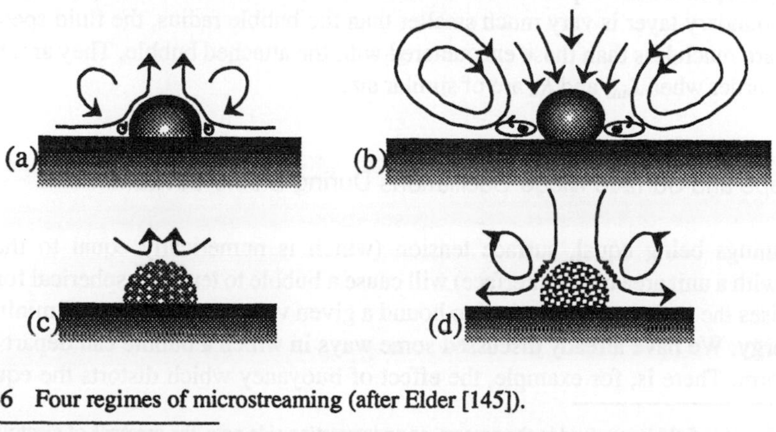

Figure 4.46 Four regimes of microstreaming (after Elder [145]).

[53] See Chapter 3, section 3.3.2(d) and Chapter 4, section 4.4.1(c).
[54] Surface waves and microstreaming are illustrated together in Figure 4.26.
[55] Sections 2.1.2(b) and 2.3.2(b).

than one mode to be excited upon the bubble wall, as represented in Figure 4.46(d). Surface modes on a stably oscillating bubble are discussed in the next section.

The microstreaming occurs within the acoustic microstreaming boundary layer of thickness $L_{ms} = \sqrt{2\eta / \rho\omega}$ (equation (1.96)), where η and ρ are the shear viscosity and density respectively of the liquid [146], over which the tangential component of the acoustic particle velocity falls to zero. Elder found that for large viscosities (η/ρ of the order of 10^{-4} m^2s^{-1}) the boundary layer was thick enough (approximately 100 μm) to be visible through the microscope, the measurements agreeing with the predicted value to within 10%. The existence of a boundary layer will result in velocity gradients and corresponding hydrodynamic shear stresses. This stress is given by the product of η with the velocity gradient, which Coakley and Nyborg [61] find to be $\eta\omega R_{\varepsilon o}^2/(R_o L_{ms})$. They give the example of a 5-μm radius bubble oscillating in water with wall amplitude $R_{\varepsilon o} = 0.5$ μm in a 1-MHz sound field. The boundary layer thickness will be 0.56 μm, the very small size resulting in a high velocity gradient (5.5×10^5 s^{-1}) and local shearing stress (550 Nm^{-2}). Thus the hydrodynamic shear stresses set up as a result of microstreaming could be capable of damaging and inducing changes in biological cells [147–151]. Bioeffects resulting from microstreaming are discussed in Chapter 5, section 5.4.2. The shear stresses that result when a gas bubble responds to continuous and pulsed acoustic fields typical of those used in diagnostic ultrasonics have been modelled by Lewin and Bjørnø [152].

In addition to generating shear stresses, microstreaming may affect processes in the medium through transport mechanisms associated with the circulating flow.[56] The circulation that occurs around the bubble can, for example, increase the bioeffect by transporting cells into the proximity of the active bubble [155], and so have an effect regardless of whether the bioeffect is brought about directly by the hydrodynamic shear found close to the bubble, or by some other cavitational mechanism. In another field, this transport effect might improve the quality of electrodeposits as the circulation causes dilution of the ion cloud [156], and may enhance ultrasonic cleaning [157]. Gould [126] observed greatly increased growth rates for bubbles through rectified diffusion following the onset of surface oscillations, with associated microstreaming. Microstreaming will affect mass-diffusion across the bubble wall through the convective term, by adding a steady streaming velocity to the oscillatory velocity [21].

The theory of Coakley and Nyborg [61] was developed for a bubble attached to a solid boundary. The microstreaming generated around a bubble which is pulsating and translating whilst freely suspended in a liquid has been formulated by Davidson and Riley [158]. They find that if the boundary layer is very much smaller than the bubble radius, the fluid speeds in the circulation are much less than those encountered with the attached bubble. They are, however, of the same order when L_{ms} and R_0 are of similar size.

4.4.6 Shape and Surface Mode Oscillations During Stable Cavitation

All other things being equal, surface tension (which is numerically equal to the energy associated with a unit area of the interface) will cause a bubble to tend to a spherical form, since that minimises the surface area required to bound a given volume of gas, and so minimises the surface energy. We have already discussed some ways in which a bubble can depart from the spherical form. There is, for example, the effect of buoyancy which distorts the equilibrium

[56] Though no acoustic field is involved in the process, as an interesting side note, the presence of electric fields can, like microstreaming, also affect heat and mass transfer processes, and generate circulation when bubbles of liquid drops are stationary or translating under gravity [153, 154], and so might through this mechanism bring about related observable effects.

shape. In addition, the bubble can be distorted by asymmetries in the environment, such as the proximity of other bubbles or boundary walls, gravity, shock fronts, and pressure gradients on scales small compared with the bubble radius.

In section 4.3 it was described how a small bubble nucleus might, when a tension is applied to the liquid, expand to a large sphere, and then undergo a transient collapse. In one scenario[57] it was shown how towards the end of the collapse, instabilities might generate surface irregularities which, on rebound, could be accentuated to such an extent that the bubble would break up. Such surface distortions set a limit to the maximum temperatures and pressures attained inside a bubble during transient collapse.

The stability of interfaces undergoing period motion has also been studied. Benjamin and Ursell [159] investigated the waves on the free surface of a vertically vibrated cylinder of liquid. These waves were at half the exciting frequency, a phenomenon first recognised by Faraday [160], and which will be discussed later.

Surface oscillations on bubbles undergoing stable cavitation were originally observed by Kornfeld and Suvarov [161]. They often manifest themselves as an added 'shimmer' seen on the bubble surface.

The resonance frequency of the nth spherical harmonic mode ($n \geqslant 2$) of a gas bubble in a liquid is given by equation (3.236) as $\omega_{on} = \sqrt{(n-1)(n+1)(n+2)\sigma/(\rho R_o^3)}$. This can also be readily found from either equation (4.119) or (4.120). Prosperetti's formulation (equation (4.120)) yields the viscous damping constant for these oscillations to be $(n+2)(2n+1)\eta_\kappa/R_o^2$ [54].

The small-amplitude behaviour of the surface oscillations on a pulsating bubble can be found by formulating the pulsation mode and retaining only linear terms [111, 162, 163]: $R = R_o - R_{\varepsilon 0}\sin\omega t$, where $R_\varepsilon \ll R_o$; $p = p_o + P_A \sin\omega t$, where $P_A \ll p_o$, and

$$p_L = \left(p_0 + \frac{2\sigma}{R_0}\right)\left(\frac{R_0}{R}\right)^3 - \frac{2\sigma}{R}$$

as given by equation (2.18), when vapour pressure is neglected, and isothermal conditions assumed. As a result, equation (4.121) reduces to a Mathieu relation

$$\ddot{a}_n' + a_n'\left(\omega_{on}^2 + \left\{\left(\frac{(2n+1)\omega^2}{2} - 3\omega_{on}^2\right)\left(\frac{R_\varepsilon}{R_0}\right)\right\}\sin\omega t\right) = 0 \tag{4.333}$$

Since equation (4.333) has Mathieu form, then depending on the magnitude of the terms ω_{on}^2 and the term in curly brackets '{ }', the solutions for the perturbation amplitude a_n' may have the form of modulated oscillations, the amplitude of which grow exponentially in time. Although, as Plesset and Prosperetti point out, equation (4.333) is only valid for $|a_n'| \ll R_o$, it is indicative that parametric excitation of the instability is possible, and is valid for the early stages of its formation. Neppiras [21] provides an order-of-magnitude stability criterion by noting that the solution to equation (4.333) will be unstable if the term in curly brackets { } is negative. Therefore if we consider a field of bubbles executing a low-amplitude pulsation mode such that R_ε/R_0 is a constant for all bubbles, then the mode n perturbation criterion is relevant to bubbles of equilibrium radii greater than

---

[57] See sections 4.3.2 to 4.3.4.

$$R_0 \approx \sqrt[3]{\frac{2(n-1)(n+1)(n+2)\sigma}{(2n+1)\rho\omega^2}\left[3+\left(\frac{R_\varepsilon}{R_0}\right)^{-1}\right]} \tag{4.334}$$

The stability of the solutions to the Mathieu equation for bubbles was first studied by Benjamin and Ursell [159], though with reference to a planar gas/liquid interface. Their results show that the modes most likely to grow will do so through parametric excitation at $\omega/2$. Benjamin and Strasberg [164] produced a theory similar to Plesset's, for the stability of waves on a spherical surface, which also gave the result of equation (4.119). They were also the first to report the erratic translatory motion of bubbles which is found to have roughly the same experimental threshold as the surface instability. It was to explain this 'bubble dancing' that Eller and Crum [165] carried out an approximation to equation (4.121) to higher order in R_ε/R_0.

When excited in a linear, non-parametric manner, the amplitude of the perturbation increases linearly with the acoustic pressure amplitude, and there exists a pressure threshold for parametric excitation with finite damping. Strong surface waves are parametrically excited at half the driving frequency, and are strongly coupled to the pulsation mode [21]. Sorokin [166] and Eisenmenger [167] evaluated the amplitude threshold for a plane surface to be

$$R_\varepsilon = \sqrt[3]{\frac{16\eta^3}{\omega\sigma\rho^2}} \tag{4.335}$$

If the order of the mode is high, this plane-surface result is applicable to perturbations on a sphere. Neppiras points out that for large bubbles in pulsation resonance, the amplitude of pulsation is approximately related to the acoustic pressure by

$$R_\varepsilon \approx \frac{P_A R_0}{3\gamma p_0 d_{tot}} \tag{4.336}$$

where d_{tot} is the dimensionless damping constant. Neppiras combines equations (4.335) and (4.336) to give the pressure threshold to excite surface waves on a resonant air bubble in water at 20 kHz to be only 0.0025 bar, and to be 0.037 bar at 500 kHz.

Neppiras [21] concluded that (a) surface waves are readily excitable on bubbles much larger that resonance; (b) strong coupling results whenever the frequency of a given perturbation mode comes within the bandwidth of the radial resonance; and (c) there exists a certain minimum bubble size below which surface modes cannot couple to the radial resonance.

Experimental attempts to find thresholds (in the acoustic pressure amplitude and frequency regimes) for exciting surface waves have been done by Strasberg and Benjamin [168], Gould [169] and Eller and Crum [165]. They show agreement with theory for the larger bubbles (though still below resonant size), but discrepancy for the smallest bubble sizes.

As seen in Chapter 3, section 3.6, the acoustic emissions from the shape oscillations on an undriven bubble (that is, one not subjected to a periodic sound field) are the cause of intense interest. No less are the possible emissions from the surface oscillations of bubbles in sound fields. As demonstrated in Chapter 3, shape oscillations generally have lower frequencies than that of the pulsation mode. However, the emission of low frequencies from an insonated bubble population are by no means the sole preserve of surface oscillations, as will be seen in the next section.

4.4.7 Acoustic Emissions from Driven Bubbles

As a bubble oscillates, it in turn emits acoustic waves. If the drive power is low, then the bubble simply pulsates in an approximately linear manner, and the emitted signal is simply at the insonation frequency. The spectrum of the acoustic signal generated by bubbles cavitating in more powerful acoustic fields may contain broadband signals, and also specific line-spectrum components at specific frequencies. The line-spectrum represents harmonics, ultraharmonics and subharmonics of the insonation frequency.

Esche [170] observed only the presence of harmonics at low insonation powers. However, at higher powers, he detected subharmonics (particularly at $\nu/2$, but also at $\nu/3$ and $\nu/4$) and a continuum, that is a broadband spectrum of acoustic frequencies. This was the first report of subharmonic emissions. Bohn [171] detected similar emissions.

In his 1980 review, Neppiras [21] summarised the available experimental data as a progression of emissions. If a liquid containing a bubble population is insonated at low power levels, continuous-wave at the fundamental frequency ν, the detected acoustic emissions are at ν only. At higher intensities, but below the transient cavitation threshold, harmonics are emitted at integer multiples of ν up to high order. The 2ν emission is prominent, its amplitude being proportional to the square of the fundamental. Low-level broadband continuum noise is present, which becomes very strong as the transient cavitation threshold is approached. The $\nu/2$ subharmonic appears intermittently, the duration of the emission being much shorter than the 'off-times'. Other subharmonics, and ultraharmonics at $(2n+1)\nu/2$ can be detected.

Numerous theories have been produced to explain the source of these emissions during both stable and transient cavitation. They are considered separately in the next two sections, though realistically there are types of cavitation that fall into both categories, and this is certainly evident in some of the mechanisms proposed for the emissions.

(a) Emissions During Transient Cavitation

The above, rather dry, statement that one encounters an increase in the broadband continuum as the transient cavitation threshold is approached fails to convey the often audible and distressing noise which is associated with the onset of violent cavitation.[58] This, in addition to the increase in harmonic and subharmonic emissions, can provide a form of audible marker for the onset of violent cavitation.

The continuum is attributed to the rapidly changing bubble radius. The subharmonic emissions might be generated by a prolonged expansion phase and a delayed collapse which can occur during transient cavitation (see Figure 4.9). Akulichev [172] demonstrated how as a result of inertial forces a bubble could continue to expand after the rarefaction half-period of the acoustic cycle becomes the compressive half-period. Similarly Apfel [26] points out that transient cavitation can be characterised by bubbles which survive for one, two or three acoustic cycles before collapse. Certainly if the bubble were to survive the transient collapse intact,[59] and repeat the form of motion illustrated in Figure 4.9, surviving several acoustic cycles before collapse, subharmonics emissions could be generated. Neppiras [21] also suggests that a form of periodic unstable oscillation of a bubble driven at twice its resonance near threshold might emit at subharmonic frequencies.

[58] See section 4.3.4.

[59] And therefore be classed as 'repetitive cavitation', as discussed in section 4.3.2, which in section 4.4.8 will also be classed as high energy stable cavitation.

The increase in the subharmonic prompted some workers in the past to use it as an indicator of the onset of transient cavitation. However, Walton and Reynolds [12] caution against this simplistic approach, recommending that one cannot say much more than the statement that "transient cavitation is characterised by a continuum in the sound spectrum." Suslick *et al.* [32], commenting on the work of Niemczewski [173], state that "clearly, white noise is not a good measure of the chemical effects of cavitation".[60] Vaughan and Leeman [174] state that "the generation of fractional harmonics, in particular the first and third half-harmonic, is a general characteristic of non-linear bubble pulsation and does not specifically indicate the occurrence of transient cavitation."

In a study of ultrasonic cavitation, Negishi [175] simultaneously monitored the acoustic and light emission. The latter, termed *sonoluminescence*, is an indicator of a particular type of violent collapse, and will be discussed in the next chapter. Negishi found a continuum in the acoustic spectrum occurred when he detected sonoluminescence. Kuttruff [176] confirmed these results by examining the circular shock waves produced in transient collapse using schlieren optics.

Other workers have demonstrated that the appearance of the half-harmonic is not correlated to the onset of sonoluminescent activity. Iernetti and Ceschia [177, 178], using pulsed 0.7-MHz ultrasound, showed that the appearance of the half-harmonic did not vary with the pulse length, repetition frequency and gas solubility in the same way as did the threshold for sonoluminescence. In test intervals ranging in duration from 2 to 10 minutes varying the acoustic power for 20-kHz insonation, Margulis and Grundel [179, 180] showed, by gradually increasing the power of the incident sound, that the sonoluminescent activity did not correlate with the strength of the $\omega/2$ subharmonic.

(b) During Stable Cavitation

If the bubble were a linear damped oscillator of the type discussed in section 4.4.1, one would expect to detect a variety of frequencies from bubbles subjected to acoustic pulses of only a very few cycles, since during such insonation, bubbles would not settle down to steady-state oscillation. However, a low-amplitude linear bubble pulsation emits only the insonation frequency once in the steady state. Nevertheless, as explained in section 4.2, the bubble is a nonlinear oscillator at finite amplitude. Even the simple power series expansion of the force/response relationship illustrates how frequencies other than the driving frequency can arise in the steady state. In that model, the general response ¥ of a bubble is a power series of the driving force ñ, for example, ¥$(t) = s_0 + s_1.$ñ$(t) + s_2.$ñ$^2(t) + s_3.$ñ$^3(t) + s_4.$ñ$^4(t)\ldots$, as given by equation (4.55). If the driving force were a single frequency, the acoustic pressure would be $P(t) = P_A\cos\omega t$. Substitution of this expression for $P(t)$ into equation (4.55), reveals how frequencies that are integer multiples of the driving frequency can arise. For example, the quadratic term ñ^2 will produce a harmonic at twice the driving frequency through $2\cos^2\omega t = 1 + \cos 2\omega t$; the cubic term ñ^3 will produce a harmonic at three times the driving frequency ($4\cos^3\omega t = \cos 3\omega t + 3\cos\omega t$); and so on. Thus at higher drive powers, the bubble undergoes nonlinear stable cavitation, and harmonics can be detected in the acoustic emission from the bubble [181]. The power series expansion given in equation (4.55) will not, however, predict subharmonics. Jordan and Smith [182] demonstrate how this may be done through Fourier series expansion of the response, or by perturbation methods. A comparison of formulations that can be used to

[60] See Chapter 5, section 5.2.1(a)(ii).

predict the frequency components present in nonlinear bubble oscillations is given by Du and Wu [183]. The nonlinear frequency response of clouds of bubbles has also been modelled [184].

Specific frequencies are characterised by two integers, n and m, such that the emission is at frequency $n\nu/m$, where ν is the driving frequency. Therefore the fundamental (also known as the first harmonic) has $n = m = 1$, and higher-order harmonics have $n = 2, 3, 4$ ... etc., and $m = 1$. Other frequencies, in addition to the harmonics, can in fact be detected from bubbles subjected to continuous-wave oscillation. These include subharmonics ($n = 1$ and $m = 2, 3, 4$ ... etc., so that, for example, the $\omega/2$ subharmonic was noted to be present intermittently), and ultraharmonics at $3\omega/2$, $5\omega/2$, $7\omega/2$ (i.e. $n = 2, 3, 4$ ... etc., and $m = 2, 3, 4$ ... etc.). In addition, a low-level noise (a continuum of frequencies) can be detected. In 1980, Neppiras [21] reviewed the current thinking. The noise was thought to be due to the emission of microbubbles from large-amplitude surface waves [185], such as shown in Figure 4.31, or to reversion of the oscillations of shocked bubbles out of the steady state, so that their own natural frequencies appear in the spectrum [21]. Such 'noise' would then be dependent on the bubble population, and have structure. Random frequencies, unconnected with ω, were thought to be the result of the shock excitation of large bubbles which would then oscillate at their own natural pulsation resonances [186, 187].

Neppiras [21] summarises three theories for the generation of the subharmonic. The first is parametric amplification by the liquid: if propagation through the medium itself is nonlinear, parametric excitation can occur in the absence of bubbles. The remaining two theories summarised by Neppiras are dependent on the bubble population: these are that the lower-frequency emissions might originate from surface waves on the bubble wall, or might be due to the excitation of bubbles larger than resonance.

(i) Surface Wave Theory. The stability of interfaces undergoing period motion was discussed in section 4.4.6. Faraday [160] was the first to report the appearance of waves at $\nu/2$, half the exciting frequency, on the free surface of vertically vibrated water, waves subsequently studied by Rayleigh [188] and Benjamin and Ursell [159]. Neppiras [186, 187] suggested that such Faraday waves on the surface of bubbles are generated at half the excitation frequency, and investigated the possibility that these surface oscillations might be responsible for subharmonic emissions. However, the observed intensity of the subharmonic signal is stronger than this hypothesis would predict because, as Strasberg [189] commented, current thinking was that bubbles executing shape oscillations are inefficient sound sources since the velocity potential decreases rapidly as one moves away from the bubble [54]. One of the general observations summarised by Neppiras did in fact suggest that modal oscillations having $n > 2$ may be responsible for some of the detected subharmonic emissions. The rapid fall-off in velocity potential of such modes would mean that this mechanism generates subharmonics detectable only very close to the bubble. Thus for the subharmonic to be detected the relevant bubble must approach to within close proximity of the hydrophone, which might have explained why the observed subharmonic in the stable regime is often intermittent. Indeed, Neppiras [21] discussed the attachment of pertinent bubbles to the hydrophone during the experiment cited above [186, 187]. In 1969 Neppiras [186] wrote "Although it has never been found possible to relate subharmonic signals with surface-wave activity on the bubbles, surface waves would be expected to contribute to the $\nu/2$ signals – the possibility is not yet ruled out." However, in the light of the nonlinear arguments introduced in 1988 by Longuet-Higgins and presented in Chapter 3, section 3.6, which suggest that shape oscillations involving no initial volume change might still generate monopole emissions at second order, this mechanism will have to be reviewed.

(ii) Large Bubble Theory. Another theory suggests that such emission might be the result of the acoustic field acting on a population of bubbles containing, either wholly or in part, bubbles with an equilibrium radius twice the size of the radius which would be the pulsation resonance of the acoustic field (i.e. bubbles with $R_o = 2R_r$) [190]. Similar mechanisms involving progressively larger bubbles to account for the lower subharmonics were proposed. However, bubbles are unlikely to grow by rectified diffusion to radii of $2R_r$, $3R_r$ etc., with consequent resonances at subharmonics of the driving frequency ($\omega/2$, $\omega/3$ etc.) since: (i) in growing through resonance there is significant chance of break-up through surface wave activity; and (ii) the acoustic pressure threshold for growth by rectified diffusion rapidly increases, and the growth rate rapidly decreases, once a bubble passes resonance size.[61] If they were pre-existing in the liquid, they would require stabilisation against buoyancy. Coalescence through mutual Bjerknes forces may generate such bubbles.

In 1969 Eller and Flynn [191] published a calculation of the threshold acoustic pressure required for bubbles to emit a signal at half the frequency of insonation. By analysing an equation of motion for a spherical bubble in an incompressible liquid, and testing for the stability of the solutions, they obtained the threshold acoustic pressure to generate the subharmonic at half the driving frequency:

$$P_{A2} = p_o \sqrt{\left(\frac{3}{2}\right)^2 \left\{\left(\frac{\omega}{\omega_o}\right)^2 - 4\right\}^2 + \left(\frac{6\Delta_{\log}}{\pi}\right)^2} \tag{4.337}$$

where $\Delta_{\log}$ is the logarithmic decrement, representing the damping of the pulsation. This pressure shows a pronounced minimum, for bubbles which are close to twice the size of those resonant with the insonating field: the $\omega/2$ signal corresponds roughly to the natural resonance ω_o of the bubbles in question. For such bubbles, the threshold can be found by setting $\omega_o = \omega/2$ in equation (4.337), giving a threshold of

$$P_{A2} = \frac{6p_o\Delta_{\log}}{\pi} \tag{4.338}$$

Neppiras notes that the subharmonics at $\omega/3$ etc. are less likely, owing to the inability of the bubble to grow to the larger sizes. Other formulations include those of Safar [192] and Nayfeh and Saric [193].

However, the theory that subharmonics are caused solely by emissions from bubbles of a size which were integer multiples larger than that of the bubble resonant with the sound field are inadequate. For example, in Figure 4.47 the largest bubble in the population is of resonance size: there are no other, larger bubbles present, yet subharmonics at $\nu/2$ and other frequencies were readily detected. Other arguments are outlined by Lauterborn and Holzfuss [195]: (i) it was found that bubbles larger than resonance can oscillate subharmonically; (ii) subharmonic emissions were observed from a standing-wave fields, and as described in section 4.4.1(b), bubbles larger than resonance would be forced away from the pressure antinodes by primary Bjerknes forces; (iii) following the numerical calculations from a simple (Rayleigh–Plesset) bubble model [196] an analysis by Lauterborn [7] revealed that there existed regimes characterised by a combination of certain parameter values where there were no usual steady-state, periodic solutions for the bubble motion. Analysis of a more sophisticated model showed that this behaviour was a genuine general result [127, 197–199]. Another notable investigation

[61] See section 4.4.3(d).

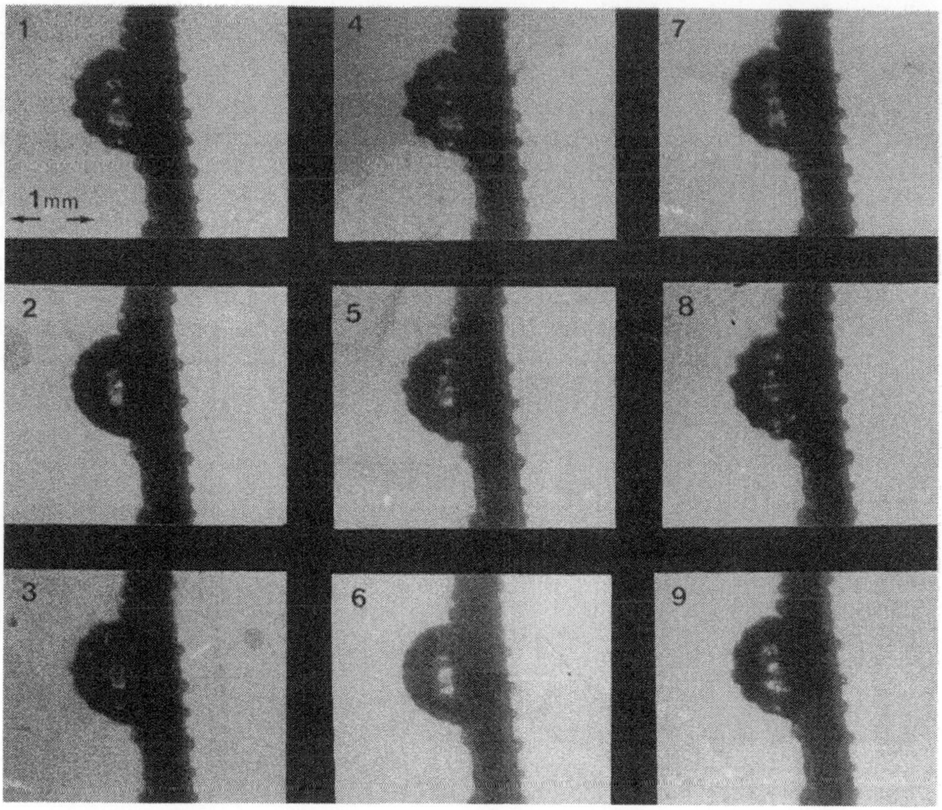

Figure 4.47 Nine consecutive frames, taken at 6000 f.p.s., showing an air bubble attached to a wire, in a 4.370-kHz sound field (after Leighton *et al.* [194]).

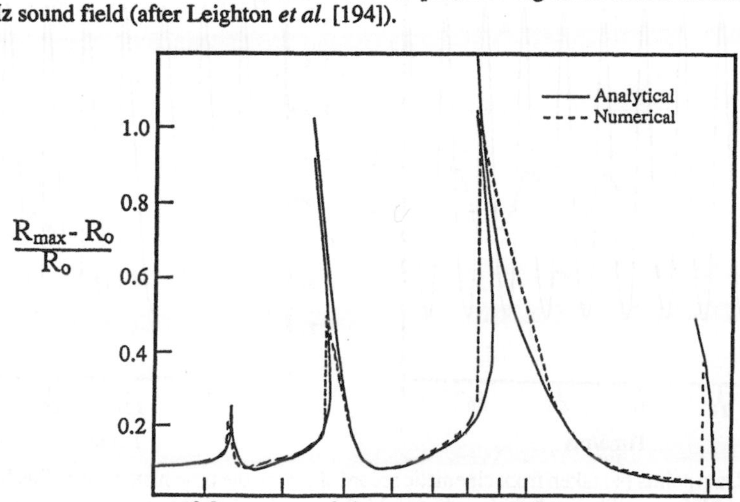

Figure 4.48 Comparison of the results of Lauterborn (numerical) and Prosperetti (analytical) for the variation with driving frequency in the response of a 10-μm (R_o) bubble where $\kappa = 1.33$, in water at 20 °C, where only viscous damping is considered. The ratio P_A/p_o equals 0.3. (After Plesset [206].)

predicting ultraharmonic and subharmonic resonances in a bubble was a theoretical treatment by Prosperetti [200] using perturbation methods. Later work included analyses by Samek [201, 202] and Francescutto *et al.* [203–205]. Figure 4.48 shows a historic comparison made by Plesset [206] between the results of Lauterborn (numerical) and Prosperetti (analytical) for the variation in the response with driving frequency. Clear resonances can be seen at multiples and fractions of the driving frequency.

It is worthwhile at this point visualising some subharmonic behaviour in an individual bubble using high-speed photographic records of air bubbles in glycerol, oscillating in the experimental low-frequency (103 Hz) sound field described in Chapter 1, section 1.2.2(b)(ii). The radius/time data are shown in Figure 4.49. The crosses are data points giving the instantaneous bubble radius, as measured from the cinematic record. For comparison, the solid curve indicates the

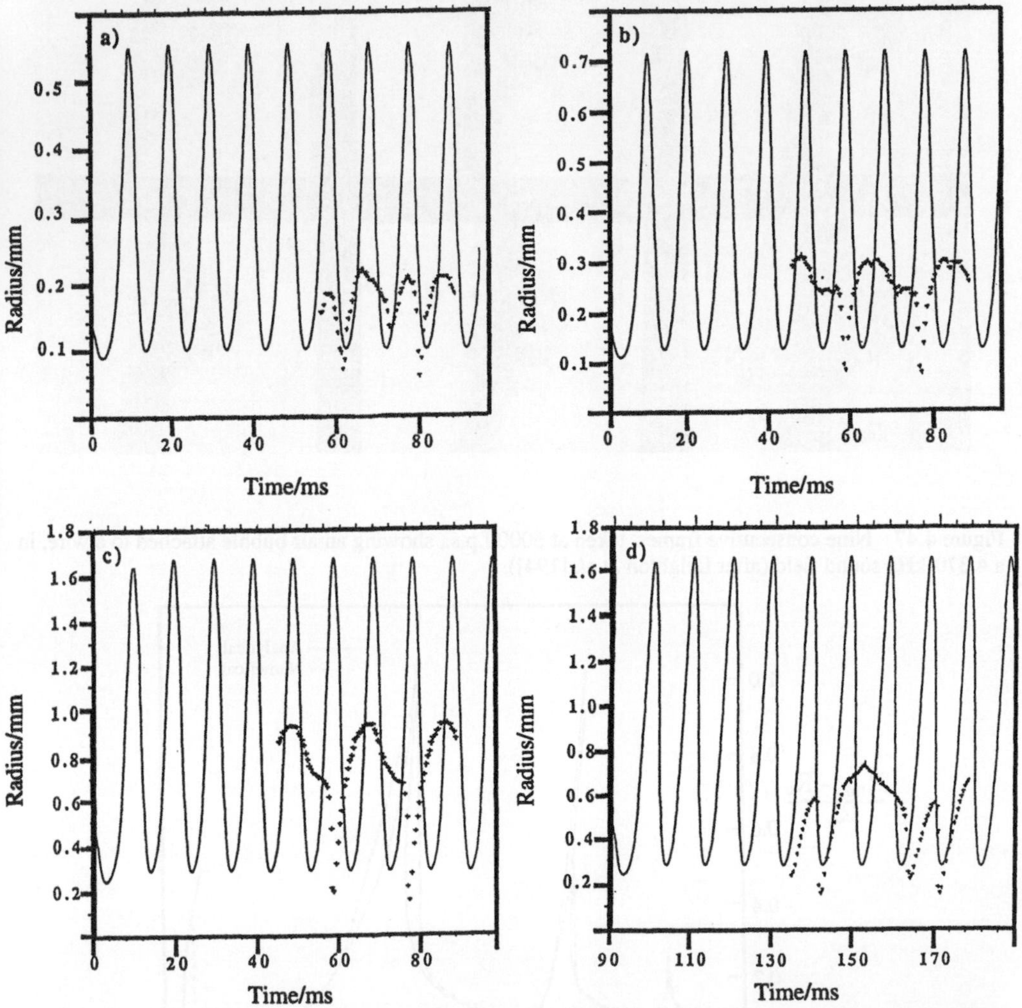

Figure 4.49 Data points (+) taken from cinematic record showing the time history of radius for a propane bubble in glycerol, for equilibrium bubble radii of (a) 0.14 mm, (b) 0.19 mm, (c) 0.47 mm. Plot (d) shows the behaviour of the bubble in (c) measured at a later time, after it has gone through a qualitative change in behaviour (corresponding to the change between Figures 4.50(a) and 4.50(b)). For comparison, the solid lines show the numerical Rayleigh–Plesset radius/time predictions, with the following parameters: $\rho = 1260$ kg/m^3, $p_0 = 333$ Pa, $\sigma = 62.96$ mN/m, $p_v = 100$ Pa, $\eta = 1.49$ Pa s, $\omega = 648.6$ rad/s, $P_A = 3469$ Pa, $\nu = 103$ Hz, $\kappa = 1$. (After Leighton *et al.* [207].)

Rayleigh–Plesset prediction (which in this case is not particularly successful at predicting the oscillation, but is nevertheless shown as it is at the same frequency as the insonating sound field, and so provides a basis for temporal comparison).[62] The technique used to obtain this was to photograph a field of bubbles, three of which were then chosen to give simultaneous measurements under identical conditions, which are recorded in Figures 4.49(a), 4.49(b) and 4.49(c) [207]. Thus we are viewing the response of the bubble as the parameter R_o (obtained from the photographic record prior to the start of the insonation at $t = 0$) is changed. In Figure 4.49(a), the data points show that for a bubble of $R_o = 0.14$ mm the oscillation is not sinusoidal. Allowing for the presence of transients and inhomogeneities in the surrounding medium, the motion appears to be periodic, which was confirmed by observation of later sections of the film, which lasted over 1 second. There are two clear frequencies present, one at twice the period of the sound field. For example, one can see from the minimum at $t = 81$ ms onwards, the beginning of the repeat of the cycle that occurred between 61 ms and 81 ms. However, that cycle is roughly divided into two by a smaller minimum at $t = 71$ ms: this indicates the presence in the oscillation of frequencies corresponding to that of the sound field, 103 Hz. However, as we keep the sound field the same and vary R_o, the form of the oscillation, and therefore the frequencies inherent in it, changes. In Figure 4.49(b) ($R_o = 0.19$ mm) the second peak is barely apparent, corresponding to a reduction in the component at the insonation frequency, 103 Hz. The minima at 52 ms and 71 ms almost merge with the maxima at 54 ms and 74 ms respectively. The subsequent collapses at 58 ms and 77 ms are correspondingly much sharper. Examination of later sections of the film confirmed that this motion was periodic. In Figure 4.49(c) ($R_o = 0.47$ mm) this form is further developed from that shown in Figure 4.49(b), so that the bubble oscillation is now predominantly at half the acoustic frequency, the only evidence of the second maximum being a perturbation in the rate of collapse at 56 ms and 75 ms. From the shape of the radius/time plot, frequencies are clearly still involved, though $\nu/2$ predominates. However, unlike the bubbles shown in Figures 4.49(a) and 4.49(b), the oscillation recorded by the bubble of radius $R_o = 0.47$ mm in Figure 4.49(c) is not periodic and stable. Figure 4.50 shows the photographic sequence of the bubble at (a) the time corresponding to Figure 4.49(c), and (b) at a later time. The form of the oscillation has changed, as can be seen from the corresponding data measurements (Figure 4.49(d)). Figure 4.49(d) shows experimental data from the same bubble at times of between 135 ms and 180 ms after the start of insonation. The form of the oscillation is similar to that shown in Figure 4.49(a), where two frequencies predominate. The plot suggests that the cycle evident between the minimum at 142 ms and 172 ms is repetitive, corresponding to a period of 30 ms, and therefore frequency of $\nu/3$ in the response. In addition, the interval between the minima at 142 ms and that at 162 ms suggests a period of twice that of the sound field, and that between the minima at 162 ms and 172 ms suggests an interval at the acoustic period.

This brief observation suggests (i) the presence of subharmonics in the oscillation of an individual bubble; (ii) that *qualitative* changes in the behaviour of the bubble can occur when one of the system parameters (in this case R_o) changes; and (iii) there exist certain combinations of parameter values which afford no usual steady-state, periodic solutions for the bubble motion, just as predicted in 1976 by Lauterborn [7].

Perhaps the most important feature of subharmonic oscillations was discovered in the late 1970s, and explained much of the rather esoteric observations made regarding the subharmonic and continuum emissions from ultrasonically excited bubbles. That feature is discussed in the next section.

---

[62] The finite accuracy of the measurement of the inter-frame time means that as the oscillation proceeds there is increasing systematic error in the time measurement.

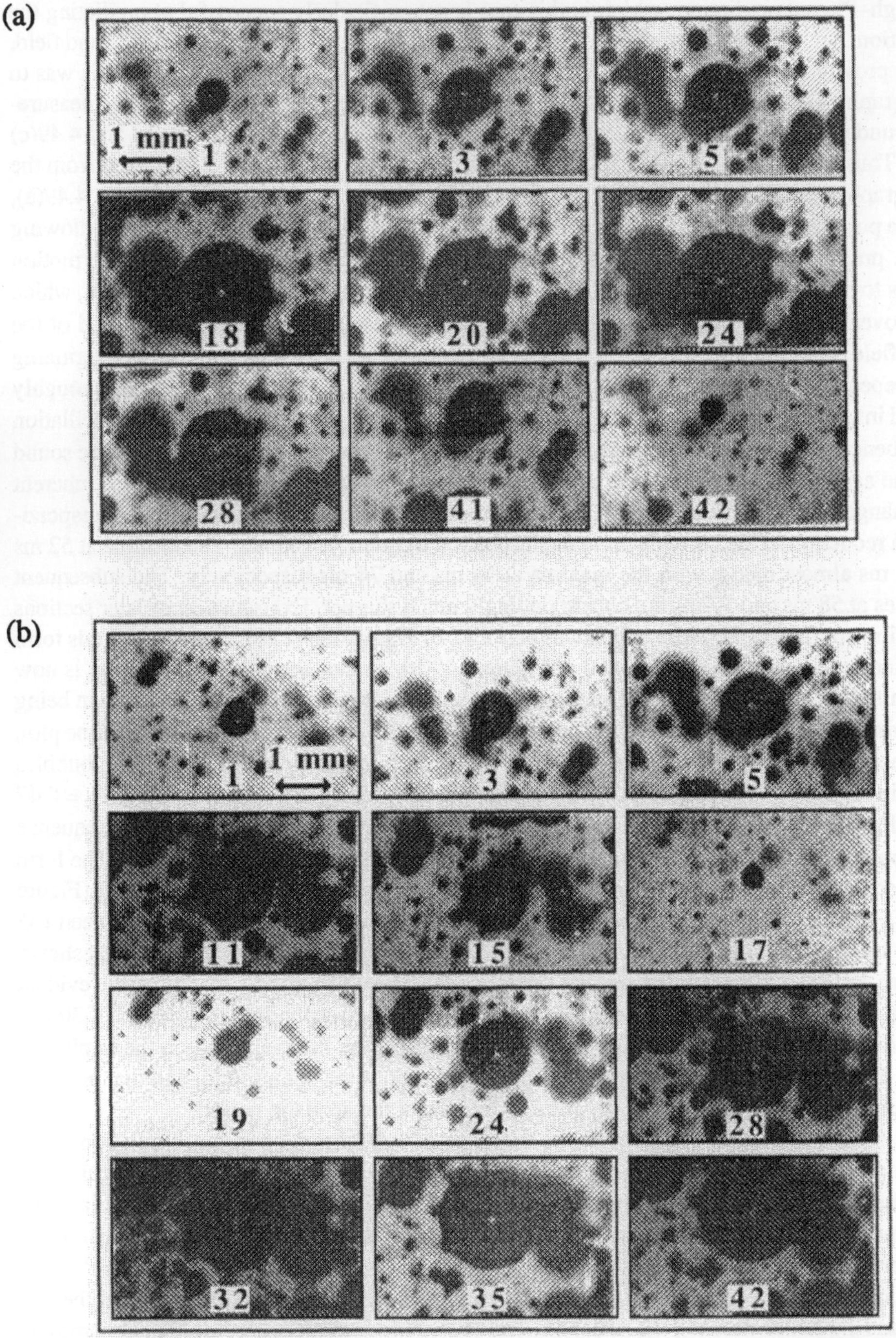

Figure 4.50 Selected frames showing propane bubbles in glycerol, for p_A = 3469 Pa, ν = 103 Hz. Inter-frame time = 0.429 ms. The radius of the largest bubble in the figure was used to obtain the data for Figures 4.49(c) and 4.49 (d). (a) The sequence from which the data for Figure 4.49(c) were taken. (b) The sequence from which the data for Figure 4.49(d) were taken (frame 1 corresponds to a time 135 ms after the start of insonation). (After Leighton *et al.* [207].)

(c) Chaotic Oscillations of Spherical Bubbles

We are perhaps used to considering the macroscopic world as being deterministic, of being governed by laws so that, if we were sufficiently knowledgeable and capable, we could completely predict how it will behave in the future. However, in recent years it has become clear that any sufficiently complicated system can, though mathematically determined, behave in a way that simulates randomness. Though it follows exact natural laws, it is unpredictable over all but the shortest timescales in the following sense. One might, for example, be able to formulate the dynamical problem but, being unable to solve it analytically, resort to a numerical solution, and find the latter to be such that a slight change in the starting conditions causes rapidly diverging errors. To put it another way, arbitrarily small perturbations or differences, which are so small as to be beyond the resolution of our measurement or computational techniques, that are present in the initial state of the system can grow exponentially with time to become of comparable size with the variables of the system: from that time onwards these unknowables are crucial in determining the dynamics of the system. Thus one would not be able to have confidence in the solution for anything but the smallest times after the start.[63] The irregular, unpredictable motion of deterministic system is termed *chaos*. It is a natural and typical ingredient of nonlinear systems of sufficiently high order.

Since, as we have discussed, an unchanging sinusoidal response is indicative of a single frequency, an irregular response may be thought of as containing many frequencies. This suggests the possibility of a connection between chaos and noise. An example of the occurrence of other frequencies is found in one of the commonest indications that a system possesses the aptitude for chaos, which is the appearance of subharmonics in the response when it is subjected to a periodic driving force of single frequency ν.

Subharmonics have been observed in a variety of common systems, including loudspeakers [208]. In 1981 Keolian *et al.* [209] observed subharmonics with frequency as low as $\nu/35$ in a variation of the experiment of Benjamin and Ursel [159]. They also observed chaos.

The regimes found by Lauterborn in 1976 [7], where there exist no usual steady-state, periodic solutions for the bubble motion, are in fact chaotic [210–214].

Lauterborn and Cramer [197] experimentally observed a sequence of emissions in water from period doubling to broadband noise as the acoustic pressure was increased, though other, less clear, sequences have also been observed [215]. Similar results have been found in liquid helium [216]. Period-doubling has been predicted by numerical results for bubbles assumed to be undergoing spherical pulsations [210] and "period doubling cascades to chaos and back" [217]. The points at which these qualitative changes occur at known as *bifurcations*. Parlitz *et al.* [217], modelling the dynamics of a single spherical bubble of radius 10 μm in a sinusoidal sound field as the amplitude P_A and frequency ν of the field were varied, noted that the bubble oscillations would undergo such qualitative changes. These depended on P_A and ν, and the oscillation would either settle down to periodic solutions of different period, or remain in flux, ending with chaotic solutions. The parameter values where the oscillations undergo such changes are closely connected with the resonances of the bubble [217]. Ilyichev *et al.* [218] examined the motion of a single bubble in a compressible liquid as predicted by the Gilmore equation, and deduced the acoustic emissions, as the bubble progressed from periodic to nonperiodic solutions through a series of period-doubling bifurcations. Figure 4.51 shows the bubble motion $R(t)/R_0$ (on the left) and noise radiation (in the middle), as a function of time;

[63] The weather is a good example of such a system – it therefore is wrong to think that as our computers and knowledge improve over the years, we will one day be able to predict the weather precisely.

and the energy spectrum of the radiation (on the right), for a bubble of equilibrium radius 20 μm in a sound field of frequency $\nu = 31$ kHz for various ratios of P_A/p_0. The measurement point would be a distance $10R_0$ from the bubble centre. Evidence of frequencies at $n\nu/m$ occur. In Figure 4.51(a), the radial oscillation appears to be approximately sinusoidal. The bubble completes $n = 1$ oscillations in $m = 1$ cycles of the sound field. However, evidence of higher frequencies appears in the radiated signal. It should be noted that the bubble resonance frequency is roughly four times the insonation frequency. As P_A increases, the amplitude of pulsation increases, specific higher frequencies are evident as multiples of the driving frequency, and the noise spectrum broadens to ever higher frequencies. In 4.51(b) the bubble completes $n = 5$ oscillations in $m = 1$ cycles of the sound field, whilst in 4.51(c), the ratio is $n = 4$ to $m = 1$. After the first bifurcation, the bubble completes $n = 8$ oscillations in $m = 2$ cycles of the sound field, the period of the radiated pressure is doubled and the $\nu/2$ subharmonic appears in the spectrum. Following the next bifurcation, the subharmonics $\nu/4$ and $\nu/8$ are evident in the spectrum (4.51(e), where the bubble completes $n = 16$ oscillations in $m = 4$ cycles of the sound field; and 4.51(f), where the ratio is $n = 32$ to $m = 8$). Ilyichev *et al.* point out that the strongest ultraharmonics are close to ω_0 of the bubble, so that in 4.51(e) the $11\nu/2$ and $13\nu/2$ ultraharmonics are most intense. In 4.51(g) and 4.51(h), the motion becomes chaotic: the period nature disappears, and the harmonics, subharmonics and ultraharmonics disappear from the spectrum.

Through acoustic and high-speed holographic observations of cavitation in a cylindrical transducer system in a bubble system similar to those discussed in section 4.3.4, Lauterborn *et al.* were able to show that the entire population exhibits period doubling to chaos [195, 219]. Analysis of such phenomena enabled the characteristics of the chaos to be determined [220, 221]. One would expect there would be a significant amount of behaviour in the cloud contained within the cylindrical focused field that deviated from the spherical pulsations of isolated bubbles assumed in the theory given in section 4.2. Lauterborn and Holzfuss [195] observed in their cloud a range of behaviour, from violent collapses that emit shock waves, to relatively stable, almost pure shape oscillations at half the driving frequency. As Lauterborn and Holzfuss [195] comment, in the light of the theory by Longuet-Higgins that shape oscillations may give rise to monopole emissions at second order, the role of the surface oscillations may be significant.

The qualitative change in dynamics are often conveniently represented on a so-called bifurcation diagram. Figure 4.52 shows such a plot taken from Kamath and Prosperetti [76], who were comparing the suitability of the various models for the dynamics of a spherical bubble, particularly with respect to heat flow.[64] The figure shows *point measurements* of the instantaneous bubble radius as the forcing pressure is increased, for a 50 μm bubble driven at $\omega = 0.8\omega_0$ (i.e. $\nu \approx 49.76$ kHz). To interpret such figures it is important to understand that bifurcation diagrams are made up by taking point or spot measurements at fixed intervals, say at the driving frequency. One way of visualising this is to imagine illuminating the oscillator with a stroboscope which flashes at the driving frequency of the oscillator. If the motion were periodic at the same frequency as the driver, each point measurement (i.e. flash illumination by the stroboscope) would reveal the oscillator always in the same place. If, however, the oscillator were responding at half the driving frequency, between alternate measurements (flashes) it would complete only half its oscillation. Therefore the stroboscope image would show the oscillator in not one, but *two* different positions. The result of Kamath and Prosperetti [76] was calculated numerically, not visualised by stroboscope, but the principle is the same. The single position of the bubble oscillator is found for forcing pressures below $P_A \approx 1.72p_0$. However,

[64]See section 4.4.2(c).

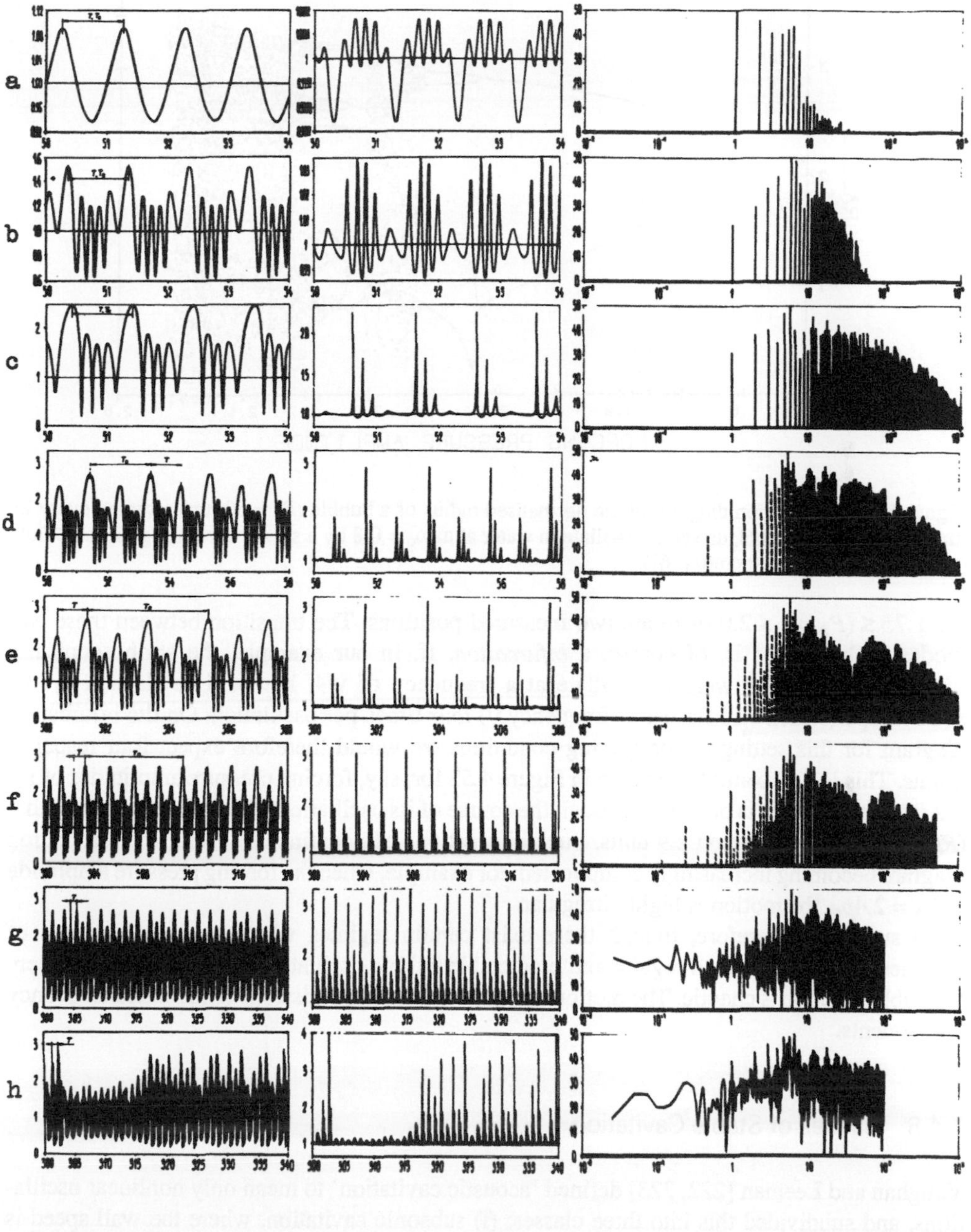

Figure 4.51 Bubble pulsations (on the left); noise radiation, i.e. the ordinate corresponds to acoustic pressure radiated by the bubble × 10^{-5}/Pa (in the centre); the energy spectrum of the acoustic radiation (on the right), the abscissa being the ratio of radiated frequency to insonation frequency. The bubble in question has $R_0 = 20$ µm, and is excited in a bifurcation zone at $\nu = 31$ kHz. The measurement point would be a distance $10R_0$ from the bubble centre. (a) $P_A/p_0 = 0.28$, $n/m = 1/1$; (b) $P_A/p_0 = 0.64$, $n/m = 5/1$; (c) $P_A/p_0 = 0.91$, $n/m = 4/1$; (d) $P_A/p_0 = 0.94$, $n/m = 8/2$; (e) $P_A/p_0 = 0.957$, $n/m = 16/4$; (f) $P_A/p_0 = 0.9617$, $n/m = 32/8$; (g) $P_A/p_0 = 0.9645$, chaos; (h) $P_A/p_0 = 0.9929$, chaos. (After Ilyichev *et al.* [218].)

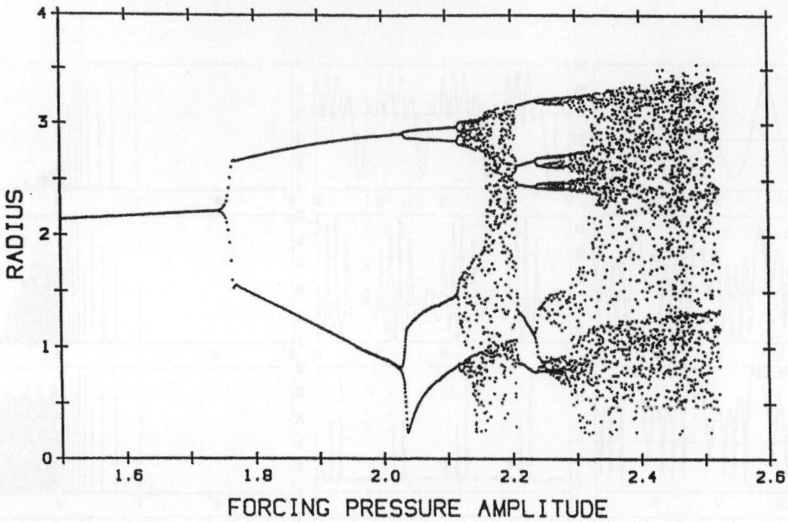

Figure 4.52 Bifurcation diagram of the normalised radius of a bubble ($R_0 = 50$ μm) sampled every 2π dimensionless time units, driven to oscillate in water at $\omega/\omega_0 = 0.8$ by a slowly increasing pressure field (after Kamath and Prosperetti [76]).

for $1.75 \le (P_A/p_0) \le 2.0$ there are two measured positions. The transition between these two modes of behaviour is, of course, a *bifurcation*. If, in our example, the stroboscopically illuminated oscillator were to oscillate at a frequency of $\nu/4$, it would be flashed by the stroboscope (which still flashes at frequency ν) *four* times per oscillation. On the bifurcation diagram for this setting of the driving conditions we would therefore expect four response points. This is, for example, the case in Figure 4.52 for, say, forcing pressure amplitudes of $P_A = 2.07p_0$: the bubble is observed through the course of its oscillation when it has radii such that $(R/R_0) \approx$ 0.6, 1.4, 2.8 and 2.9 units. Such bifurcations can continue to occur, the bifurcation diagram becoming increasingly complicated: for example, when the forcing pressure amplitude is $P_A = 2.4p_0$, the motion is highly irregular.

In summary, therefore, though there exist certain regimes where the combinations of parameters can give rise to regular and predictable bubble oscillations, there are others where the bubble motion is chaotic. The motion will induce acoustic emissions with related frequency components.

4.4.8 Classes of Stable Cavitation

Vaughan and Leeman [222, 223] defined 'acoustic cavitation' to mean only nonlinear oscillations, and subdivided this into three classes: (i) subsonic cavitation, where the wall speed is always less than the speed of sound in the gas and in the liquid (i.e. $\dot{R} < c_g < c$); 'gas phase cavitation' ($c_g \leqslant \dot{R} < c$); and (iii) 'liquid phase cavitation' ($\dot{R} > c > c_g$). This classification departs from the sometimes inappropriate division of cavitation into stable and transient, and Leeman and Vaughan [222] suggest that an advantage of this model is that each regime is associated with an observable: subsonic cavitation with the emission of harmonics and subharmonics, gas-phase cavitation with sonoluminescence, and liquid phase cavitation with both sonoluminescence and shock waves in the liquid. This relies, however, upon the proposition that shocks will be generated in a medium when the bubble wall velocity exceeds the speed of sound in that medium, and that sonoluminescence is generated only by the passage of shocks through the gas. These propositions are not to date conclusively supported by the evidence: shocks may be generated at wall speeds less than either speed of sound, and the current evidence does not support attributing sonoluminescence exclusively to shocks within the gas. This is further discussed in Chapter 5, section 5.2.1. Leeman and Vaughan were concerned primarily with the effects of the bubble-induced concentration of acoustic energy from an acoustic field, in order, for example, to study the potential for biophysical effects in an ultrasound field, and their definition reflects this. In particular, the restriction to nonlinear bubble oscillations, though justifiable in the context of the studies of Leeman and Vaughan, is not appropriate for the wider range of interactions between bubbles and sound fields discussed in this book, for two reasons. Firstly, the pulsating bubble is a nonlinear system, and so all bubble oscillations are essentially nonlinear: It is only in the limit of small amplitudes that the motion is approximately linear. Secondly, even in that case the linear model is extremely useful. For example, it has been used in this chapter to formulate the propagation of sound through bubbly media[65] and radiation forces on bubbles.[66] In particular, by far the most common acoustic bubble interactions are the class of phenomena discussed in Chapter 3, where bubbles generate sound through approximately linear free oscillations, and affect the propagation of sound through the medium in such a way that the bubbly liquid can be viewed as a continuum with uniform bulk properties of sound speed and attenuation which differ from those of the bubble-free liquid. Following the terminology of Chapter 2, section 2.1.2, these two processes may be called *passive acoustic cavitation*.

It would be wrong to equate passive acoustic cavitation to the assumption of linear oscillations. For example, to predict monopole passive acoustic emissions from shape oscillations requires nonlinear theory (Chapter 3, section 3.6). One might practically state that linear oscillations are a subclass of passive acoustic cavitation. An argument against this might cite the example of Bjerknes forces: the common descriptions of radiation forces outlined in section 4.4.1 assume linear oscillations, and these forces are of major importance in intense continuous-wave sound fields, where physical intuition would correctly rail against any suggestion that the cavitation is 'passive'. However, this argument is not relevant since the cavitation in such a sound field is prohibited from being classed as 'passive acoustic cavitation' on two counts: firstly, if radiation forces are significant, bubble aggregation will occur, and the medium cannot be viewed as a continuum with uniform bulk properties; and secondly, linear models are employed to describe the forces in order to make the formulations tractable, and in no way dictate to the bubbles the degree of nonlinearity of their oscillation! In an acoustic field where the radiation forces are practically important, the bubble oscillations are almost certainly deviating significantly from linearity.

As outlined above, passive acoustic cavitation comprises two features, those associated with entrainment emissions and those with propagation. Some phenomena, such as the collective oscillations[67] of bubble clouds, can be viewed in either sense, and it is interesting to note that the two methods of explaining the phenomenon, namely of coupled modes between bubbles, and of modes of oscillation of a body with sound speed differing from the surrounding water, fall respectively under the two classes of phenomena.

---

[65] See sections 4.1.2(d) and 4.1.2(e).
[66] See section 4.4.1.
[67] See Chapter 3, section 3.8.2(c)(i).

It is convenient to divorce passive acoustic cavitation from the behaviour of bubbles in stronger sound fields, where significant effects associated with individual bubbles (microstreaming, sonoluminescence) are observable. Despite the limitations of the distinction, we will proceed along the lines of the stable/transient division. Clearly, the propagation phenomena of passive acoustic cavitation are a form of stable cavitation, whereas the entrainment phenomena (where there is no driving sound field) in general are not. However, there comes a point when classification is inappropriate: the observable effects must take priority, and in order to emphasise these, conciseness in the following discussion of stable cavitation is sacrificed in an attempt to describe to those unfamiliar with the field the range of phenomena that can be encountered.

Stable cavitation is probably best characterised by the repetitive nature of the oscillations, which are therefore relatively long-term phenomena. It can be subdivided into three broad types. The first, high-energy stable cavitation, exhibits the repetitive pulsations associated with stable cavitation, but also exhibits energetic attributes (e.g. $R_{max}/R_o \gtrsim 2$ or 2.3, T_{max} and the maximum wall speed are high, the inertial function controls the start of the collapse etc.) and is therefore continuous with the class 'repetitive transient cavitation' discussed in section 4.3.2.

The second is characterised by the low energy of its oscillations (e.g. $R_{max}/R_o \lesssim 2$ or 2.3, T_{max} and the maximum wall speed are low, the inertial function does not control the start of the collapse etc.).

The third class is irregular oscillations of the type discussed in section 4.4.7. This class can exhibit a wide range of behaviour.

Low-energy stable cavitation can be subdivided into two classes. The first is 'repetitive stable cavitation' where the bubble pulsates in the sound field, either growing by rectified diffusion ('growing repetitive stable cavitation') or dissolving ('dissolving repetitive stable cavitation'). The second class is 'fragmentary stable cavitation', where the oscillation results in smaller bubble fragments. These might arise through a low-energy fragmentary collapse, such as is illustrated in Figure 4.53 ('collapse fragmentary stable cavitation'), or through the emission of microbubbles, which are sheared off from surface waves ('shear fragmentary stable cavitation'). In addition, the fragmentary classes can be further subdivided in 'cyclic' (where at least some of the bubble fragments are capable of growth by rectified diffusion) and 'non-cyclic' (where all the bubble fragments dissolve away). Clearly, the type of cavitation is greatly influenced by the environment surrounding the bubble (e.g. neighbouring bodies), and so it is not difficult for bubbles in their lifetime to transfer from one class to another.

4.4.9 Summary of the Theory for a Single Bubble

Much of this chapter, and the previous one, has dealt with the theory of the single bubble. It was necessarily mathematical in places, and so this chapter will end with a summary of the important qualitative features of the theory.

A spherical gas bubble within a liquid may perform volume pulsations, which can be either free or forced. The free oscillations were first studied by Minnaert [224]: given a mechanical exciting impulse, the amplitude of oscillation of the bubble wall is small, and so the bubble approximates to a linear oscillator. The pulsations occur at a well-defined resonance frequency ν_o, given approximately by equation (3.36), which reduces to $\nu_o R_o \approx 3$ Hz.m for air bubbles in water under one atmosphere. The acoustic output of such a freely-oscillating bubble is typically that of a lightly damped oscillator, an exponentially decaying sinusoid. This has been recorded in the sound of many phenomena, including the babbling of brooks and the impact of raindrops

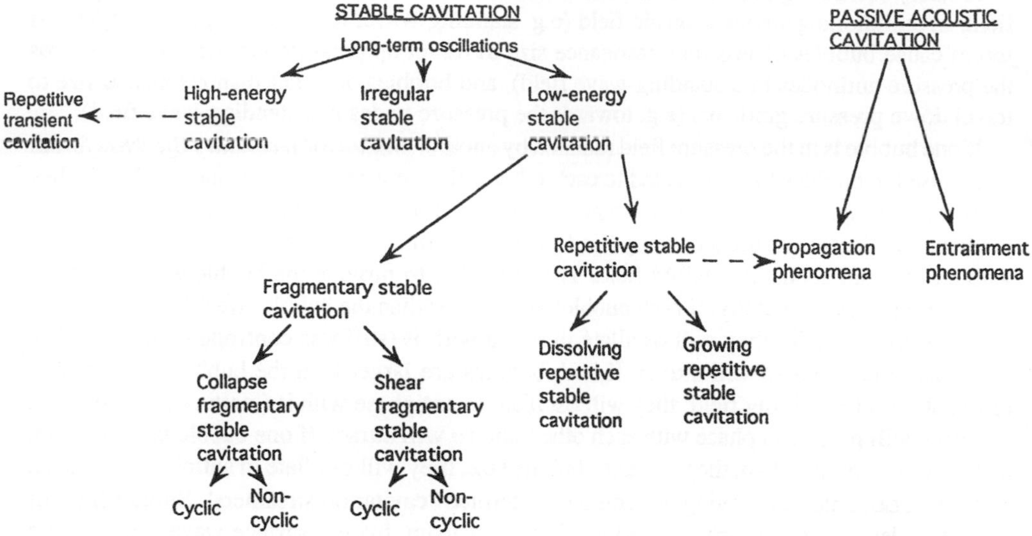

Figure 4.53 Classes of stable cavitation.

on a body of liquid. The damping arises through thermal, viscous and radiation losses. Shape oscillations with no initial volume change were thought to generate only short-range emissions, but nonlinear calculations predict monopole emissions at second order. The excitation impulse for bubbles in the natural world might arise through surface tension, hydrostatic pressure or effects associated with shape distortions.

As with any oscillating system, the bubble may also be driven into oscillation, for example, by being subjected to a periodic driving pressure in the form of an acoustic wave. At low acoustic pressure amplitudes, the radial oscillation is small and the bubble response approximates to the familiar form of the lightly damped simple harmonic oscillator. For such oscillators, the displacement is in phase with the driving force when the frequency of the driving force is much less than the bubble resonance ('stiffness controlled') and in antiphase when the driving frequency is much greater than the bubble resonance ('inertia controlled'). The phase change, which occurs as the driving force passes through resonance, becomes less rapid as the damping is increased. The presence of damping reduces the resonance frequency slightly. The damping in bubbles is dependent on the driving frequency.

Higher acoustic driving pressures can give rise to nonlinear oscillations of higher amplitude in the bubble. Approximate equations of motion of the bubble exist. In very simple terms, the nonlinearity can be seen in the fundamental asymmetry, where though there are no limits to the radius in expansion, the amplitude of the radial decrease in contraction is limited.

During stable cavitation, the bubble may undergo an increase in equilibrium size as gas dissolved in the liquid is pumped into the bubble through *rectified diffusion*. For given conditions of acoustic field and dissolved gas concentrations (below a supersaturation threshold), there is a radius above which the bubble may grow through rectified diffusion, and below which it will dissolve. This is the 'rectified diffusion threshold'. For a given bubble radius, a similar threshold minimum exists for the driving pressure.

A stably oscillating bubble will experience translational radiation forces in a travelling-wave field. In an inhomogeneous acoustic field (e.g. standing wave, focused) the *primary Bjerknes forces* cause bubbles of less than resonance size to travel up pressure gradients (e.g. towards the pressure antinodes in a standing-wave field), and bubbles of larger than resonance size to travel down pressure gradients (e.g. towards the pressure nodes in a standing-wave field).

If one bubble is in the pressure field radiated by another, *mutual* or *secondary Bjerknes forces* will cause the bubbles to be attracted to each other if they pulsate in phase, and repelled if they oscillate in antiphase. There are two main ways to lock bubbles in phase and so generate the predictable effects from the secondary Bjerknes interaction. The first is to drive two real bubbles with an external acoustic pressure field. The second is to provide the bubble with an image, using a boundary. Generally, if both bubbles are smaller than the bubble size which is resonant with the driving field, they will oscillate in phase with it (stiffness controlled) and therefore with each other, and so will attract. If both bubbles are larger than the bubble size which is resonant with the driving field, they will oscillate in antiphase with it (inertia controlled) and therefore will pulsate in phase with each other, and so will attract. If one bubble is larger than, and the other smaller than, the resonance bubble size, they will oscillate in antiphase and repel.

Lines of bubbles so moving are commonly termed 'cavitation streamers'. Similar lines of microbubbles generated through, and leading away from, intense surface wave activity on a source bubble are known as 'cavitation microstreamers'. These terms should not be confused with 'streaming'[68] and 'microstreaming',[69] which are fluid flows.

There are thresholds in bubble radius and acoustic parameters (e.g. ω, P_A) above which the oscillation of a single bubble is termed *transient* or *unstable*. In transient cavitation, the bubble may exist for several cycles, and then undergo a violent collapse. It might rebound intact, or fragment into smaller bubbles, which may act as seeds for growth by rectified diffusion in cyclic cavitation processes.

The various thresholds bound regimes of characteristic behaviour. Processes which change the bubble (e.g. rectified diffusion, fragmentation etc.) can cause a bubble to cross from one regime to another.

Bubbles may undergo surface oscillations, generate microstreaming and emit a variety of frequencies. These effects may sometimes be associated with resonances.

4.4.10 Population Phenomena

In several sections earlier in this book, bubbles in cloud have been treated as a population of entities which do not interact with each other, but which impart to the medium homogeneous bulk properties. Thus in section 4.1.2(e) we have considered the effect that such populations have on the effective acoustic impedance, and on the speed and attenuation of acoustic waves. In later sections we have seen that such bubbles can in fact interact with one another when intense sound is passed through the medium, for example, through the secondary Bjerknes forces. Other multi-bubble interactions, more relevant to erosion studies, can be found in Chapter 5, section 5.4.1.

During stable cavitation, the surface waves examined in section 4.4.6 may grow to such large amplitudes that microbubbles are ejected from the crests. This can happen so rapidly that the bubble appears to explode [21, 225]. Microbubbles may, of course, act as nuclei for further

---

[68] See Chapter 1, section 1.2.3(b).

[69] See section 4.4.5.

cavitating bubbles. Microbubbles ejected from the surface of a parent may be repulsed by the formation of ballo-electrical charges (proposed by Harvey [226], these charges occur whenever a closed meniscus is formed in a polar liquid). If the mother is larger than, and the daughters smaller than, resonance size, they will be repulsed by the mutual Bjerknes force. The translating microbubbles may follow a single pathway or stream, which is sometimes referred to as a 'cavitation microstreamer'. They will, of course, be responsible for gas loss from the parent bubble. However, in the situation shown in Figure 4.31 they are subsequently attracted by secondary Bjerknes forces to coalesce with a second bubble, so increasing its gas content.

Willard [60] made a fascinating observation in a focused 2.5-MHz acoustic field ($P_A \approx 70$ bar at the focus). Bubble nuclei were driven in cavitation streamers (see Figure 4.27 where a similar event is occurring). Following this so-called 'pre-initiation' stage, a suitable weak nucleus of larger than resonance size enters the focus ('initiation'), and with a loud snapping noise fragments into a cloud of microbubbles (the 'catastrophic' stage). This fragmentation is due to the emission of microbubbles from surface waves at an ever-increasing rate. The cloud takes a plume-shape, and passes through the focal region (at speeds of up to ~10 m/s). These 'Willard events' last for only a few milliseconds. Willard estimates the bubble fragments to be of near-resonance size. In aerated water, larger bubbles can be generated in the fragmentation process.

Microbubbles produced by larger bubbles, on entering the bulk medium as free-floating entities, may dissolve or grow by rectified diffusion, depending on their relation to the rectified diffusion threshold. If the latter occurs, they may themselves grow large enough to produce daughters.

Neppiras [21] outlined a variety of such simple cyclic bubble life-histories that could feasibly occur during acoustic cavitation. In any given situation, thresholds for stable and transient cavitation may or may not exist, depending on the acoustic and medium parameters. The cycles described below are dependent on the relationship between the initial bubble size and any attainable thresholds. The first two examples assume that the bubble initially lies above the rectified diffusion threshold, but that the transient cavitation threshold is for one reason or another unattainable.

The Degassing Cycle

A bubble may grow by rectified diffusion to a size where buoyancy forces cause it to separate out from the liquid. This process is actively pumping dissolved gas out of the liquid and into a bubble, which is then removed from the system.

Standing-wave Cycles

Walton and Reynolds [12] consider pulsating bubbles in a standing-wave field. Those smaller than resonance congregate at the pressure antinode, and grow by rectified diffusion to greater than resonance size. It is assumed that no fragmentation or shedding of microbubbles occurs. Strictly, the pressure gradient is zero at the pressure antinode, so there can be no force on the bubble. However, its position is one of unstable equilibrium, and any slight displacement from the antinode will cause the primary Bjerknes force to act. This drives these large bubbles to the pressure node. There being no time-varying pressure fields at that location, the bubbles simply dissolve. Once they are smaller than resonance, their position is unstable, and any slight perturbation from the nodal position will cause them to move under the primary Bjerknes force back to the pressure antinode. The process would then repeat.

The Gaseous Cavitation Cycle

A bubble of a size that places it between the stable and transient cavitation threshold will grow by rectified diffusion until it reaches the transient threshold. There it will undergo collapse into fragments. If any of these fragments lie above the rectified diffusion threshold, they too will grow, repeating the process.

In all likelihood these descriptions of cyclic processes greatly simplify the situation. It is clear that the behaviour is dependent on the threshold conditions: if, for example, the transient threshold lies below the resonance radius, the standing-wave cycle cannot occur. In the extreme case, violent microbubble emission can cause the bubble to explode into a mass of microbubbles [21, 225]. Such a process could prevent a bubble growing above resonance size, and so again stop the standing-wave cycle. If microbubbles, or the fragments of transient collapse, lie below the rectified diffusion threshold, the medium will not become re-nucleated by their generation. Despite the fact that acoustic radiation and Bjerknes forces are commonly observed to 'levitate' a bubble, preventing it from rising against buoyancy, it is nevertheless a fact that the degassing cycle can be made to occur, and is employed in several industrial processes (for example, in the production of photographic emulsions). None of these considerations have approached the question of the stabilisation of nuclei, discussed in Chapter 2. Therefore, though it may seem to be an obvious fact, acoustic cavitation depends critically on the bubble population and the medium, and on the acoustic parameters (such as the stable and transient thresholds, and to what extent surface waves will occur). What is not so obvious is, given a set of experimental conditions, how will a real bubble population behave?

The final chapter of this book illustrates the way some real acoustic bubble systems behave, through examination of their effects. As has been discussed, an examination of one such effect, the subharmonic emission, led to a considerable number of inspired theories and eventually to the discovery of chaos in ultrasonic cavitation. As with all observables, our entire interaction with bubbles comes about through their effects, whether they be the emissions or scattering of sound or light, or the way they change their environment, through chemical, biological or erosive action.

References

[1] French AP. Vibrations and Waves. Nelson, London, 1979, p 89

[2] Fanelli M, Prosperetti A and Reali M. Shape oscillations of gas–vapour bubbles in liquids. Part 1: Mathematical formulation. Acustica 1984; 55: 213–223

[3] Lu NQ, Prosperetti A and Yoon SW. Underwater noise emissions from bubble clouds. IEEE J Ocean Eng 1990; 15: 275–285

[4] Clay CS and Medwin H. Acoustical Oceanography: Principles and Applications. Wiley, New York, 1977

[5] Poritsky H. The collapse or growth of a spherical bubble or cavity in a viscous fluid. In: Proceedings of the First U.S. National Congress on Applied Mechanics, New York, 1952 (Sternberg E, ed.). pp 813–821

[6] Batchelor GK. An introduction to fluid dynamics. Cambridge University Press, London, 1967; p 147

[7] Lauterborn W. Numerical investigation of nonlinear oscillations of gas bubbles in liquids. J Acoust Soc Am 1976; 59: 283–293

[8] Rayleigh Lord. On the pressure developed in a liquid during the collapse of a spherical cavity. Phil Mag 1917; 34: 94–98

[9] Plesset MS. The dynamics of cavitation bubbles. J Appl Mech 1949; 16: 277–282
[10] Noltingk BE and Neppiras EA. Cavitation produced by ultrasonics. Proc Phys Soc 1950; B63: 674–685
[11] Neppiras EA and Noltingk BE. Cavitation produced by ultrasonics: theoretical conditions for the onset of cavitation. Proc Phys Soc 1951; B64: 1032–1038
[12] Walton AJ and Reynolds GT. Sonoluminescence. Adv Phys 1984; 33: 595–660
[13] Flynn HG. Physics of acoustic cavitation in liquids. In: Physical Acoustics, Vol. 1, Part B (Mason WP, ed.). Academic Press, New York, 1964; pp 57–172
[14] Kirkwood JG and Bethe HA. Office of Science Research and Development Report 558, USA, 1942
[15] Gilmore FR. Hydrodynamics Laboratory Report 26-4, California Institute of Technology, 1952
[16] Keller JB and Miksis M. Bubble oscillations of large amplitude. J Acoust Soc Am 1980; 68: 628
[17] Prosperetti A, Crum LA and Commander KW. Nonlinear bubble dynamics. J Acoust Soc Am 1986; 83: 502–514
[18] Flynn HG. Cavitation dynamics. I. A mathematical formulation. J Acoust Soc Am 1975; 57: 1379–1396
[19] Gaitan DF, Crum LA, Church CC and Roy RA. An experimental investigation of acoustic cavitation and sonoluminescence from a single bubble. J Acoust Soc Am 1992; 91: 3166–3183
[20] Church CC. The use of real vs ideal gas in predictions of cavitation dynamics. J Acoust Soc Am (submitted)
[21] Neppiras EA. Acoustic cavitation. Phys Rep 1980; 61: 159–251
[22] Flynn HG and Church CC. Transient pulsations of small gas bubbles in water. J Acoust Soc Am 1988; 84: 985–998
[23] Apfel RE and Holland CK. Gauging the likelihood of cavitation from short-pulse, low-duty cycle diagnostic ultrasound. Ultrasound Med Biol 1991; 17: 179–185
[24] Holland CK and Apfel RE. An improved theory for the prediction of microcavitation thresholds. IEEE Trans Ultrasonics Ferroelectrics Freq Control 1989; 36: 204–208
[25] Flynn HG. Cavitation dynamics. II. Free pulsations and models for cavitation bubbles. J Acoust Soc Am 1975; 58: 1160–1170
[26] Apfel RE. Methods in Experimental Physics, Vol. 19 (Edmonds PD, ed.). Academic Press, New York, 1981; pp 355–413.
[27] Apfel RE. Acoustic cavitation prediction. J Acoust Soc Am 1981; 69: 1624–1633
[28] Flynn HG and Church CC. A mechanism for the generation of cavitation maxima by pulsed ultrasound. J Acoust Soc Am 1984; 76: 505–512
[29] Flynn HG. Tech Memo 50, Acoustics Research Lab, Harvard University, Cambridge, Massachusetts, 1963
[30] Apfel RE. Possibility of microcavitation from diagnostic ultrasound. IEEE Trans Ultrasonics Ferroelectrics Freq Control 1986; UFFC-33: 139–142
[31] Holland CK and Apfel RE. An improved theory for the prediction of microcavitation thresholds. IEEE Trans Ultrasonics Ferroelectrics Freq Control 1989; 36: 204–208
[32] Suslick KS, Hammerton DA and Cline RE Jr. The sonochemical hot-spot. J Am Chem Soc 1986; 108: 5641–5642
[33] Šponer J. Theoretical estimation of the cavitation threshold for very short pulses of ultrasound. Ultrasonics 1991; 29: 376–380
[34] Šponer J. Dependence of ultrasonic cavitation threshold on the ultrasonic frequency. Czech J Phys 1990; B40: 1123–1132
[35] Šponer J, Davadorzh C and Mornstein V. The influence of viscosity on ultrasonic cavitation threshold for sonoluminescence at low megahertz region. Studia Biophys 1990; 137: 81–89
[36] Field JE, Gorham DA and Rickerby DG. High-speed liquid jet and drop impact on brittle targets. Erosion: Prevention and Useful Applications, Special Technical Publication 664, 1979, pp 298–319
[37] Steverding B and Lehnigk SH. Ceramic Bulletin, 1975; 49: 1057
[38] Carson PL, Rubin JM and Chiang EH. Fetal depth and ultrasound path lengths through overlying tissues. Ultrasound Med Biol 1989; 15: 629–639
[39] Duck FA and Bacon DR. A fundamental criticism of hydrophone-in-water exposure measurement. Ultrasound Med Biol 1988; 14: 305–307

[40] Aymé-Bellegarda EJ and Church CC. Nonmonotonic behavior of the maximum collapse pressure in a cavitation bubble. IEEE Trans Ultrasonics Ferroelectrics Freq Control 1989; 36: 561–564
[41] Aymé EJ and Carstensen EL. Cavitation induced by asymmetric, distorted pulses of ultrasound: a biological test. Ultrasound Med Biol 1989; 15: 61–66
[42] Aymé EJ and Carstensen EL. Cavitation induced by asymmetric, distorted pulses of ultrasound: theoretical predictions. IEEE Trans Ultrasonics Ferroelectrics Freq Control, 1989; 36: 32
[43] Aymé-Bellegarda EJ. Collapse and rebound of a gas-filled spherical bubble immersed in a diagnostic ultrasonic field. J Acoust Soc Am 1990; 88: 1054
[44] Lauterborn W. Zu einer Theorie der Kavitationsschwellen. Acustica 1969/70; 22: 48–54 (in German)
[45] Apfel RE. Some new results on cavitation threshold prediction and bubble dynamics. In: Cavitation and Inhomogeneities in Underwater Acoustics (Lauterborn W, ed.). Springer, Berlin, 1980, pp 79–83
[46] Rozenberg LD. High Intensity Ultrasonic Fields. Plenum, New York, 1971, Parts IV–VI, pp 203–419
[47] Akulichev VA. Pulsations of cavitation bubbles in the field of an ultrasonic wave. Sov Phys Acoust 1967; 13: 149–154
[48] Matsumoto Y and Aoki M. Growth and collapse of cavitation bubbles. Bull JSME 1984; 27(229): 1352–1357
[49] Taylor GI. The instability of liquid surfaces when accelerated in a direction perpendicular to their planes. I. Proc Roy Soc 1950; A201: 192–196
[50] Plesset MS. On the stability of fluid flows with spherical symmetry. J Appl Phys 1954; 25: 96–98
[51] Lewis DJ. The instability of liquid surfaces when accelerated in a direction perpendicular to their planes. II. Proc Roy Soc 1950; A202: 81–96
[52] Birkhoff G. Note on Taylor instability. Quart Appl Math 1954; 12: 306–309
[53] Birkhoff G. Stability of spherical bubbles. Quart Appl Math 1956; 13: 451–453
[54] Plesset MS and Prosperetti A. Bubble dynamics and cavitation. Ann Rev Fluid Mech 1977; 9: 145–185
[55] Plesset MS and Mitchell TP. On the stability of the spherical shape of a vapour cavity in a liquid. Quart Appl Math 1956; 13: 419–430
[56] Chapman RB and Plesset MS. Thermal effects in the free oscillations of gas bubbles. Trans ASME D J Basic Eng 1972; 94: 142–145
[57] Benjamin TB. PhD thesis, University of Cambridge, 1954
[58] Leighton TG, Walton AJ and Field JE. The high-speed photography of transient excitation. Ultrasonics 1989; 27: 370–373
[59] Blake FG Jr. Technical Memo 12, Acoustics Research Laboratory, Harvard University, Cambridge, Massachusetts, USA, September 1949
[60] Willard GW. Ultrasonically induced cavitation in water: a step-by-step process. J Acoust Soc Am 1953; 25: 669.
[61] Coakley WT and Nyborg WL. Chapter II: Cavitation; dynamics of gas bubbles; applications In: Ultrasound: its Applications on Medicine and Biology (Fry F, ed.). Elsevier, Amsterdam, 1978, Part 1, pp 77–159
[62] Yosioka K and Kawasima Y. Acoustic radiation pressure on a compressible sphere. Acustica 1955; 5: 167–173
[63] Leighton TG, Walton AJ and Pickworth MJW. Primary Bjerknes forces. Eur J Phys 1990; 11: 47–50
[64] Young FR. Cavitation. McGraw-Hill, London, 1989, pp 156–159
[65] Blake FG Jr. Bjerknes forces in stationary sound fields. J Acoust Soc Am 1949; 21: 551
[66] Miller DL. Stable arrays of resonant bubbles in a 1-MHz standing-wave acoustic field. J Acoust Soc Am 1977; 62: 12
[67] Crum LA and Eller AI. Motion of bubbles in a stationary sound field. J Acoust Soc Am 1969; 48: 181–189
[68] Prandtl L. Essentials of Fluid Dynamics. Blackie, London, 1954, pp 180, 342–345
[69] Bjerknes VJF. Zeit Phys Chem Unterricht 1930, 43: 1
[70] Bjerknes VFJ. Fields of Force. Columbia University Press, NewYork, 1906

[71] Bjerknes VFJ. Die Kraftfelder, No. 28 of the series Die Wissenschaft (see also Vol. 2 of the Vorlesungen). Brunswick, 1909
[72] Batchelor GK. An Introduction to Fluid Dynamics. Cambridge University Press, London, 1967; pp 452–455
[73] Crum LA. Bjerknes forces on bubbles in a stationary sound field. J Acoust Soc Am 1975; 57: 1363–1370
[74] Young FR. Cavitation. McGraw-Hill, London, 1989, p 160
[75] Kobelev YuA and Ostrovsky LA. Nonlinear acoustic phenomena due to bubble drift in a gas–liquid mixture. J Acoust Soc Am 1989; 85: 621–629
[76] Kamath V and Prosperetti A. Numerical integration methods in gas-bubble dynamics. J Acout Soc Am 1989; 85: 1538–1548
[77] Prosperetti A. The thermal behaviour of oscillating gas bubbles. J Fluid Mech 1991; 222: 587–616
[78] Devin C Jr. Survey of thermal, radiation, and viscous damping of pulsating air bubbles in water. J Acoust Soc Am 1959; 31: 1654–1667
[79] Eller AI. Damping constants of pulsating bubbles. J Acoust Soc Am 1970; 47: 1469–1470
[80] Prosperetti A. Thermal effects and damping mechanisms in the forced radial oscillations of gas bubbles in liquids. J Acoust Soc Am 1977; 61: 17–27
[81] Landau L and Lifschitz E. Fluid Mechanics. Addison-Wesley, Reading, Massachusetts, 1959, Section 39
[82] Plesset MS. In: Cavitation in Real Liquids (Davies R, ed.). Elsevier, Amsterdam, 1964, pp 1–18
[83] Crum LA. The polytropic exponent of gas contained within air bubbles pulsating within a liquid. J Acoust Soc Am 1983; 73: 116–120
[84] Pfriem H. Akust Zh 1940; 5: 202–207
[85] Chapman RB and Plesset MS. Thermal effects in free oscillation of gas bubbles. J Basic Eng 1971; 93: 373–376
[86] Prosperetti A. Bubble phenomena in sound fields: part 1. Ultrasonics 1984; 22: 69–77
[87] Prosperetti A. Bubble phenomena in sound fields: part 2. Ultrasonics 1984; 22: 115–124
[88] Devin C Jr. Survey of thermal, radiation, and viscous damping of pulsating air bubbes in water. J Acoust Soc Am 1959; 31: 1654–1667
[89] Plesset MS and Hsieh DY. Theory of gas bubble dynamics in oscillating pressure fields. Phys Fluids 1960; 3: 882–892
[90] Fanelli M, Prosperetti A and Reali M. Radial oscillations of gas vapour bubbles in liquids. Part I: Mathematical formulation. Acustica 1981; 47: 253–265
[91] Fanelli M, Prosperetti A and Reali M. Radial oscillations of gas vapour bubbles in liquids. Part II: Numerical examples. Acustica 1981; 49: 98–109
[92] Miksis MJ and Ting L. Nonlinear radial oscillations of a gas bubble including thermal effects. J Acoust Soc Am 1984; 76; 897–905
[93] Crum LA and Prosperetti A. Erratum and comments on 'Nonlinear oscillations of gas bubbles in liquids: an interpretation of some experimental results' (J Acoust Soc Am 1983; 73: 121–127). J Acoust Soc Am 1984; 75: 1910–1912
[94] Nigmatulin RI and Khabeev NS. Heat exchange between a gas bubble and a liquid. Fluid Dyn 1974; 9: 759–764
[95] Nigmatulin RI and Khabeev NS. Dynamics of vapour–gas bubbles. Fluid Dyn 1977; 12: 867–871
[96] Nagiev FB and Khabeev NS. Heat-transfer and phase transition effects associated with oscillations of vapour–gas bubbles. Sov Phys Acoust 1979; 25: 148–152
[97] Nigmatulin RI, Khabeev NS and Nagiev FB. Dynamics, heat and mass transfer of vapour–gas bubbles in a liquid. Int J Heat Mass Trans 1981; 24: 1033–1044
[98] Flynn HG. Cavitation dynamics. I. A Mathematical formulation. J Acoust Soc Am 1975; 57: 1379–1396
[99] Prosperetti A. Bubble dynamics: a review and some results. Appl Sci Res 1982; 38: 145–164
[100] Gupta R S and Kumar D. Variable time step methods for the dissolution of a gas bubble in a liquid. Computers and Fluids 1983; 11: 341–349
[101] Harvey EN, Barnes DK, McElroy WD, Whiteley AH, Pease DC and Cooper KW. Bubble formation in animals. J Cell Comp Physiol 1944; 24: 1–22
[102] Crum LA. Rectified diffusion. Ultrasonics 1984; 22: 215–223

[103] Wang T. Rectified heat transfer. J Acoust Soc Am 1974; 56: 1131–1143
[104] Epstein PS and Plesset MS. On the stability of gas bubbles in liquid–gas solutions. J Chem Phys 1950; 18: 1505–1509
[105] Blake FG Jr. Technical Memo 12, Acoustics Research Laboratory, Harvard University, Cambridge, Massachusetts, USA, September 1949
[106] Strasberg M. Onset of ultrasonic cavitation in tap water. J Acoust Soc Am 1959; 31: 163–176
[107] Strasberg M. Rectified diffusion: comments on a paper of Hsieh and Plesset. J Acoust Soc Am 1961; 33: 359–360
[108] Rosenberg MD. Technical Memo 25, Acoustic Research Laboratory, Harvard University, Cambridge, Massachusetts, USA, 1951
[109] Pode L. The deaeration of water by a sound beam. Report 854, David Taylor Model Basin, Washington DC, USA, 1953
[110] Jost W. Diffusion in Solids, Liquids and Gases. Academic Press, New York, 3rd printing, 1960, pp 16–28
[111] Hsieh D-Y and Plesset MS. Theory of rectified diffusion of mass into gas bubbles. J Acoust Soc Am 1961; 33: 206–215
[112] Kapustina OA. Gas bubble in a small-amplitude sound field. Sov Phys Acoust 1970; 15: 427–438
[113] Safar MH. Comments on papers concerning rectified diffusion of cavitation bubbles. J Acoust Soc Am 1968; 43: 1188–1189
[114] Skinner LA. Pressure threshold for acoustic cavitation. J Acoust Soc Am 1970; 47: 327–331
[115] Skinner LA. Acoustically induced gas bubble growth. J Acoust Soc Am 1972; 51: 378–382
[116] Eller AI. Effects of diffusion on gaseous cavitation bubbles. J Acoust Soc Am 1975; 57: 1374–1378
[117] Eller AI and Flynn HG. Rectified diffusion through nonlinear pulsations of cavitation bubbles. J Acoust Soc Am 1965; 37: 493–503
[118] Plesset MS and Zwick SA. A nonsteady heat diffusion problem with spherical symmetry. J Appl Phys 1952; 23: 95–98
[119] Eller AI. Bubble growth by diffusion in an 11-kHz sound field. J Acoust Soc Am 1972: 52; 1447–1449
[120] Lastman G and Wentzell RA. Comparison of five models of spherical bubble response in an inviscid compressible liquid. J Acoust Soc Am 1981; 69: 638–642
[121] Eller AI. Growth of bubbles by rectified diffusion. J Acoust Soc Am 1969; 46: 1246–1250
[122] Church CC. Prediction of rectified diffusion during nonlinear bubble pulsations at biomedical frequencies. J Acoust Soc Am 1988; 83: 2210–2217
[123] Nyborg WL. Mechanisms for nonthermal effects of sound. J Acoust Soc Am 1968; 44: 1302–1309
[124] Langmuir I and Blodgett KB. A mathematical investigation of water droplet trajectories. Army Air Forces Tech Rep 5418 (1946)
[125] Langmuir I. A mathematical investigation of water droplet trajectories. In: The Collected Works of Irving Langmuir, Vol. 10, Atmospheric Phenomena. Pergamon, New York, 1961, pp 348–393
[126] Gould RK. Rectified diffusion in the presence of, and absence of, acoustic streaming. J Acoust Soc Am 1974; 56: 1740–1746
[127] Cramer E. In: Cavitation and Inhomogeneities in Underwater Acoustics (Lauterborn W, ed.). Springer, Berlin, 1980, pp 54–63
[128] Church CC. A method to account for acoustic microstreaming when predicting bubble growth rates produced by rectified diffusion. J Acoust Soc Am 1988; 84: 1758–1764
[129] Crum LA and Hansen GM. Generalised equations for rectified diffusion. J Acoust Soc Am 1982; 72: 1586–1592
[130] Kapustina OA and Statnikov YuG. Influence of acoustic microstreaming on the mass transfer in a gas bubble–liquid system. Sov Phys Acoust 1968; 13: 327–329 (Akust Zh 1968; 13: 383–386)
[131] Davidson BJ. Mass transfer due to cavitation microstreaming. J Sound Vib 1971; 17: 261–270.
[132] Lewin PA and Bjørnø L. Acoustic pressure amplitude thresholds for diffusion in gaseous microbubbles in biological tissue. J Acoust Soc Am 1981; 69: 846–853

[133] ter Haar GR and Daniels S. Evidence for ultrasonically induced cavitation *in vivo*. Phys Med Biol 1981; 26: 1145–1149

[134] ter Haar GR, Daniels S, Eastaugh KC and Hill CR. Ultrasonically induced cavitation *in vivo*. Brit J Cancer 1982; 45: 151–155

[135] Crum LA and Hansen GM. Growth of air bubbles in tissue by rectified diffusion. Phys Med Biol 1982; 413–417

[136] Daniels S, Blondel D, Crum LA, ter Haar GR and Dyson M. Ultrasonically induced gas bubble production in agar based gels: Part I, Experimental investigation. Ultrasound Med Biol 1987; 13: 527–539

[137] Crum LA, Daniels S, ter Haar GR and Dyson M. Ultrasonically-induced gas bubble production in agar based gels: Part II, Theoretical analysis. Ultrasound Med Biol 1987; 13: 541–544

[138] Nyborg WL. Acoustic streaming near a boundary. J Acoust Soc Am 1958; 30: 329–339

[139] Lee CP and Wang TG. Near-boundary streaming around a small sphere due to two orthogonal standing waves. J Acoust Soc Am 1989; 85: 1081–1088

[140] Lee CP and Wang TG. Outer acoustic streaming. J Acoust Soc Am 1990; 88: 2367–2375

[141] Rayleigh Lord. The Theory of Sound. Dover, New York, 1945, pp 333–342

[142] Elder SA, Kolb J and Nyborg W. Physical factors involved in sonic irradiation of liquids. Phys Rev 1954; 93: 364(A) (An abstract from 1953 autumn meeting of the New England section of the American Physical Society at Storrs, Connecticut, November 7, 1953)

[143] Rozenberg LD. Physical Principles of Ultrasonic Technology, Vol. 1. Plenum, New York, 1973, p 399

[144] Kolb J and Nyborg W. Small-scale acoustic streaming in liquids. J Acoust Soc Am 1956; 28: 1237–1242

[145] Elder SA. Cavitation microstreaming. J Acoust Soc Am 1958; 31: 54–64

[146] Nyborg WL. Acoustic streaming near a boundary. J Acoust Soc Am 1958; 30: 329–339

[147] Rooney JA. Hemolysis near an ultrasonically pulsating gas bubble. Science 1970; 169: 869

[148] Miller DL. Microsteaming as a mechanism of cell death in *Elodea* leaves exposed to ultrasound. Ultrasound Med Biol 1985; 11: 285–292

[149] Miller DL, Nyborg WL and Whitcomb CC. Platelet aggregation induced by ultrasound under specialized conditions *in vitro*. Science 1979; 205: 505–507

[150] Williams AR. Ultrasound: Biological Effects and Potential Hazards. Academic Press, New York, 1983

[151] Hill CR (ed.). Physical Principles of Medical Ultrasonics. Ellis Horwood, Chichester (for Wiley, New York), 1986

[152] Lewin PA and Bjørnø L. Acoustically induced shear stresses in the vicinity of microbubbles in tissue. J Acoust Soc Am 1982; 71: 728

[153] Chang LS and Berg JC. Electroconvective enhancement of mass or heat exchange between a drop or bubble and surroundings in the presence of an interfacial tension gradient. AIChE J 1985; 31(1): 149

[154] Chang LS and Berg JC. The effect of interfacial tension gradients on the flow structure of single drops or bubbles translating in an electric field. AIChE J 1985; 31(4): 551

[155] Vivino AA, Boraker DK, Miller D and Nyborg W. Stable cavitation at low ultrasonic intensities induces cell death and inhibits ^{3}H-TdR incorporation by con-a-stimulated murine lymphocytes *in vitro*. Ultrasound Med Biol 1985; 11: 751–759

[156] Yeager E and Hovorka F. Ultrasonic waves and electro chemistry. I. A survey of the electrochemical applications of ultrasonic waves. J Acoust Soc Am 1953; 25: 443–455

[157] Elder SA, Kolb J and Nyborg W. Small-scale acoustic streaming effects in liquids. J Acoust Soc Am 1954; 26: 933(A) (Abstract of the 47th meeting of the Acoustical Society of America)

[158] Davidson BJ and Riley N. Cavitation microstreaming. J Sound Vib 1971; 15: 217–233

[159] Benjamin TB and Ursell F. The stability of the plane free surface of a liquid in vertical periodic motion. Proc Roy Soc, 1954; A225: 505–515

[160] Faraday M. On the forms and states assumed by fluids in contact with vibrating elastic surfaces. Phil Trans Roy Soc 1831; 121: 319–340

[161] Kornfeld M and Suvorov L. On the destructive action of cavitation. J Appl Phys 1944; 15: 495–506

[162] Hsieh DY. Dynamics and oscillation of nonspherical bubbles. J Acoust Soc Am 1972; 52: 151(A) (An abstract from the 83rd meeting of the Acoustical Society of America)

[163] Hsieh DY. On thresholds for surface waves and subharmonics of an oscillating bubble. J Acoust Soc Am 1974; 56: 392–393
[164] Benjamin TB and Strasberg M. Cavitation in real liquids (Davies R, ed.). Elsevier, Amsterdam, 1964, pp 164–180
[165] Eller A and Crum LA. Instability of the motion of a pulsating bubble in a sound field. J Acoust Soc Am 1970; 47: 762–767
[166] Sorokin VI. The effect of fountain formation at the surface of a vertically-oscillating liquid. Sov Phys Acoust 1957; 3: 281–291
[167] Eisenmenger W. Dynamic properties of the surface tension of water and aqueous solutions of surface active agents with standing capillary waves in the frequency range 10 kc/s to 1.5 Mc/s. Acustica 1959; 9: 327–340
[168] Strasberg M and Benjamin TB. Excitation of oscillations in the shape of pulsating gas bubbles; experimental work. J Acoust Soc Am 1958; 30: 697(A) (Abstract from the 55th meeting of the Acoustical Society of America)
[169] Gould RK. Heat transfer across a solid–liquid interface in the presence of acoustic streaming. J Acoust Soc Am 1966; 40: 219–225
[170] Esche R. Untersuchung der Schwingungskavitation in Flüssigkeiten. Acustica 1952; 2: (Akust Beih) AB208–AB218
[171] Bohn L. Acoustic pressure variation and the spectrum in oscillatory cavitation. Acustica 1957; 7: 201–216
[172] Akulichev VA. The structure of solutions of equations describing pulsations of cavitation bubbles. Akusticheskii J (Russian) 1967; 13: 533–537
[173] Niemczewski B. A comparison of ultrasonic cavitation intensity in liquids. Ultrasonics 1980: 107–110
[174] Vaughan PW and Leeman S. Some comments on mechanisms of sonoluminescence. Acustica 1986; 59: 279–281
[175] Negishi K. Experimental studies on sonoluminescence and ultrasonic cavitation. J Phys Soc Japan 1961; 16: 1450–1465
[176] Kuttruff H. Relation between sonoluminescence and cavitation-oscillations in liquids. Akustica 1962; 12: 230–254 (in German)
[177] Iernetti G. Pulsed ultrasonic cavitation. Part I: Cavitation noise, luminescence thresholds, nuclei distribution. Acustica 1970; 23: 189–207
[178] Ceschia M and Iernetti G. Cavitation threshold model verified through pulsed sound cavitation in water. Acustica 1973; 29: 127–137
[179] Margulis MA and Grundel LM. The ultrasonic luminescence of a liquid near the cavitation threshold. 1. The development of the pre-threshold luminescence of a liquid in an ultrasonic field. Russ J Phys Chem 1981; 55: 386–389
[180] Margulis MA and Grundel LM. Irradiation of a liquid with ultrasound near the cavitation threshold. 2. Fundamental irregularities of sonoluminescence at low ultrasound intensities. Russ J Phys Chem 1981; 55: 989–991
[181] Miller DL. Ultrasonic detection of resonant cavitation bubbles in a flow tube by their second harmonic emissions. Ultrasonics 1981; 19; 217–224
[182] Jordan DW and Smith P. Nonlinear Ordinary Differential Equations. Oxford University Press, 2nd edn, 1987, Chapter 7
[183] Du G and Wu J. Comparison between two approaches for solving nonlinear radiations from a bubble in a liquid. J Acoust Soc Am 1990; 87: 1965–1967
[184] Kumar S and Brennen CE. Nonlinear effects in the dynamics of clouds of bubbles. J Acoust Soc Am 1991; 89: 707
[185] Neppiras EA and Fill EE. A cyclic cavitation process. J Acoust Soc Am 1969; 46: 1264–1271
[186] Neppiras EA. Subharmonic and other low-frequency emission from bubbles in sound-irradiated liquids. J Acoust Soc Am 1969; 46: 587–601
[187] Neppiras EA. Subharmonic and other low-frequency signals from sound-irradiated liquids. J Sound Vib 1969; 10: 176
[188] Rayleigh Lord. On the crispations of fluid resting upon a vibrating support. Phil Mag Ser 5 1883; 16: 50–58
[189] Strasberg M. Gas bubbles as sources of sound in liquids. J Acoust Soc Am 1956; 28: 20–26
[190] Güth W. Nichtlineare Schwingungen von Luftblasen in Wasser. Acustica 1956; 6: 532–538

[191] Eller A and Flynn HG. Generation of subharmonics of order one-half by bubbles in a sound field. J Acoust Soc Am 1969; 46: 722–727
[192] Safar MH. Exploitation of the subharmonic pressure waves from pulsating gas bubbles in an acoustic field in liquids. J Phys D Appl Phys1970; 3: 635–636
[193] Nayfeh AS and Saric WS. Finite amplitude wave effects in fluids. IPC Science and Technology Press, Guildford, 1974, pp 272–276
[194] Leighton TG, Lingard RJ, Walton AJ and Field JE. Bubble sizing by the nonlinear scattering of two acoustic frequencies. In: Natural Physical Sources of Underwater Sound (Kerman BR, ed.). Kluwer, Dordrecht, The Netherlands, 1992
[195] Lauterborn W and Holzfuss H. Acoustic chaos. Int J Bifurcation Chaos 1991; 1: 3–26
[196] Lauterborn W. Subharmonische Schwingungen von Gasblasen in Wasser. Acustica 1969/70; 22: 238–239
[197] Lauterborn W and Cramer E. Subharmonic route to chaos observed in acoustics. Phys Rev Lett 1981; 47: 1445–1448
[198] Cramer E and Lauterborn W. On the dynamics and acoustic emission of spherical cavitation bubbles in a sound field. Acustica 1981; 49: 226–238 (in German)
[199] Lauterborn W. Cavitation bubble dynamics – new tools for an intricate problem. Appl Sci Res 1982; 38: 165–178
[200] Prosperetti A. Nonlinear oscillations of gas bubbles in liquids: steady-state solutions. J Acoust Soc Am 1974; 56: 878–885
[201] Samek L. Non-linear oscillations of gas bubbles in liquids. Czech J Phys 1980; B30: 1210–1226
[202] Samek L. A multiscale analysis of nonlinear oscillations of gas bubbles in compressible liquids II. First and second subharmonics and harmonics. Czech J Phys 1989; B39; 1366–1371
[203] Francescutto A and Nabergoj R. Steady-state oscillations of gas bubbles in liquids: explicit formulas for frequency response curves. J Acoust Soc Am 1983; 73: 457–460
[204] Francescutto A and Nabergoj R. Analytical predictions for the maximum amplitudes of oscillating bubbles. Acoust Lett 1983; 7: 43–46
[205] Francescutto A, Iernetti G and Nabergoj R. Subharmonic radial resonance of gas bubbles and related shape stability. 10th International Symposium on Nonlinear Acoustics, 1984, pp 189–191
[206] Plesset MS. Bubble dynamics and cavitation erosion. From Finite-amplitude Wave Effects in Fluids (Bjørnø L, ed.). IPC Science and Technology Press, Guildford, 1974, pp 203–209
[207] Leighton TG, Adlam M, Walton AJ and Field JE. The nonlinear oscillations of multiple bubbles in a simulated acoustic field. Proc IOA 1991; 13: 39–47
[208] Pedersen PO. Subharmonics in forced oscillations in dissipative systems. Part I. J Acoust Soc Am 1935; 6: 227–238; Part II. J Acoust Soc Am 1935; 7: 64–70
[209] Keolian R, Turkevich LA, Putterman SJ, Rudnick I and Rudnick JA. Subharmonic sequences in the Faraday experiment: departures from period doubling. Phys Rev Lett 1981; 47: 1133–1136
[210] Lauterborn W and Suchla E. Bifurcation superstructure in a model of acoustic turbulence. Phys Rev Lett 1984; 53: 2304–2307
[211] Parlitz U and Lauterborn W. Periodic and chaotic bubble oscillations. In: Proceedings of the 12th International Congress on Acoustics (ICA 12), Toronto, 1986, Vol III (Embleton TFW, Daigle GA, Stinson MR and Warnock ACC, eds.). Beauregard, Canada, 1986, paper I4-7
[212] Lauterborn W and Parlitz U. On the bifurcation structure of bubble oscillators. In: Proceedings of the XIIth International Symposium on Nonlinear Acoustics (Kedrinskii, ed.). Institute of Hydrodynamics, Novosibirsk, 1987, pp 71–80
[213] Smereka P, Birnir B and Banerjee S. Regular and chaotic bubble oscillations in periodically driven pressure fields. Phys Fluids 1987; 30: 3342–3350
[214] Lauterborn W and Parlitz U. Methods of chaos physics and their application to acoustics. J Acoust Soc Am 1988; 84: 1975–1993
[215] Lauterborn W. Acoustic turbulence. In: Frontiers in Physical Acoustics (Sette D, ed). North-Holland, Amsterdam, 1986, pp 123–144
[216] Smith CW, Tejwani MJ and Farris DA. Bifurcation universality for first-sound subharmonic generation in superfluid helium-4. Phys Rev Lett 1982; 48: 492–494
[217] Parlitz U, Englisch V, Scheffczyk C and Lauterborn W. Bifurcation structure of bubble oscillators. J Acoust Soc Am 1990; 88: 1061–1077

[218] Ilyichev VI, Koretz VL and Melnikov NP. Spectral characteristics of acoustic cavitation. Ultrasonics 1989; 27: 357–361
[219] Lauterborn W and Koch A. Holographic observation of period-doubled and chaotic bubble oscillations in acoustic cavitation. Phys Rev 1987; A35: 1974–1976
[220] Lauterborn W and Holzfuss J. Evidence of a low-dimensional strange attractor in acoustic turbulence. Phys Lett 1986; 115A: 369–372
[221] Holzfuss J and Lauterborn W. Liapunov exponents from time series of acoustic chaos. Phys Rev 1989; A39: 2146–2152
[222] Vaughan PW and Leeman S. Acoustic cavitation revisited. Acustica 1989; 69: 109–119
[223] Vaughan PW and Leeman S. Transient cavitation: fact or fallacy? Ultrasonics International 89 Conference Proceedings, 1989, pp 1259–1264
[224] Minnaert M. On musical air-bubbles and sounds of running water. Phil Mag 1933; 16: 235–248
[225] Nyborg W and Hughes DE. Bubble annihilation in cavitation streamers. J Acoust Soc Am 1967; 42: 891
[226] Harvey EN. Sonoluminescence and sonic chemiluminescence. J Am Chem Soc 1939; 61: 2392–2398.
[227] Leighton TG, Wilkinson M, Walton AJ and Field JE. The forced oscillations of bubbles in a simulated acoustic field. Eur J Phys 1990; 11: 352–358
[228] Leighton TG. Image intensifier studies of sonoluminescence, with application to the safe use of medical ultrasound. PhD thesis, Cambridge University, 1988
[229] American Institute of Ultrasound in Medicine/National Electrical Manufacturers Association (AIUM/NEMA). Standards for real-time display of thermal and mechanical acoustic indices on diagnostic ultrasound equipment. 1992

5

Effects and Mechanisms

For the fish keep disappearing,
And the Law's perturbed to hear,
When at last the shark's arrested,
That the shark has no idea.
For there's nothing he remembers,
And there's nothing to be done.
For the shark is not a shark if,
Nobody can prove he's one.

B. Brecht, *The Threepenny Opera*

In contrast to the legal system, the scientist cannot prove his theories to be true, since underpinning all of them are the natural laws of the universe, at which we can only make our best guess. Progress is made by interpreting the world in the light of known observations and so constructing models of the world which concur with those observations. When observations are made which conflict with a model, that model is modified or discarded. In practice the process is not so simple. Firstly, measurement is not always reliable. Secondly, what might be considered by a given worker to be a complete model may in fact be one facet of a larger process: the result is that a given observable might result from several mechanisms, contributing in proportions that vary as the system parameters are changed. One must therefore be wary of extrapolating to general statements from an observation which proves a certain mechanism to be either dominant or ineffective under the conditions of the experiment.

Examples of these issues can be found in the observables associated with acoustic cavitation. The phenomena discussed in this chapter are chosen purely from personal interest, and the list is by no means meant to represent either the complete or the most momentous effects of ultrasonic cavitation.[1] However, they are all interesting and informative, illustrative and important.

5.1 Bubble Detection

Several acoustic techniques for examining bubble populations were introduced in Chapter 3. Bubbles may be detected and sized upon entrainment from their passive acoustic emissions.[2] Since the freely oscillating bubble must first be excited in order to emit in this manner, it should

[1] Several texts are available in the literature where these and other effects associated with high-intensity underwater sound are discussed [1–3]. See also the reviews of bioeffects recommended in section 5.4.2.
[2] See Chapter 3, section 3.7.1.

be remembered that when passive emissions are used for bubble detection, the signal samples not the whole population, but only those that are suitably excited during the period of observation. All bubble detection techniques in some way fail to sample the entire population of bubbles that could *potentially* exist, for example, as a result of limits in sensitivity or resolution. Examples of techniques are shown in Figure 5.0(a). The technology is, however, currently progressing rapidly, and pushing back these limits: for example, recent advances in optical holography has enabled three-dimensional information to be obtained on objects (bubbles and particles) of radii down to about 10 μm [4, 5], and in high-speed holocinematography up to 300 000 holograms/s have been produced [6–8]. This is a far cry from the 1976 experiments of Kolovayev [9], who photographed near-surface oceanic bubbles after they had risen onto a glass plate, but illustrates that the hostility of the environment in which the measurements are made can be a very potent limiting factor. Other techniques, both acoustic and optical, that are appropriate for oceanic or laboratory deployment have been described,[3] and these generally measure the bubble size distribution of the population some time after entrainment. In addition, bubble clouds could be studied through sonar or microwave attenuation or scattering.[4] Young [10] cites a variety of techniques.

In this section some acoustic techniques for bubble detection are discussed, techniques specific to the nature of the cavities in question and their environment (Figure 5.0(b)). These methods fall into two broad classes. The first consists of those which detect pre-existing 'stable' bubbles. This is achieved in a range of ways, ranging from simple acoustic shadowgraphy, where the sound has minimal effect on the bubble itself, to ones where the sound reacts strongly with the bubble, exciting high-amplitude nonlinear oscillations and emissions. The second class involves the detection of what might be termed transient cavitation, where the bubbles are short-lived and undergo rapid and extreme changes in volume.

The reflection and scattering of ultrasound from a medium is the basis of clinical diagnosing with sound. The interaction of bubbles with incident sound has already been discussed: possible interactions include scattering and parametric effects [11–13]. Very closely related to the study of the scattering of sound from a bubble is the emission of sound by an oscillating bubble. Indeed, in some cases the two problems are indivisible. The wide variety of investigations to correlate effects range from investigations on the relationship between acoustic emission and cavitation erosion [14], to attempts to monitor the effect of ultrasound on biological materials. Examples of the latter include the attempts in 1982 by Morton *et al.* [15], who used subharmonic acoustic emissions, and in 1986 by Edmonds and Ross [16], who measured the $\nu/2$ subharmonic and noise levels combined, to relate the acoustic emission to the biological effects and damage

[3]See Chapter 2, section 2.1.2(b) and Chapter 3, sections 3.5.1(a)(ii), 3.8.1(b) and 3.8.1(c).
[4]See Chapter 3, sections 3.8.1 and 3.8.2(c)(ii).

Figure 5.0 (a) Example techniques for bubble detection. Though optical techniques are commonly employed, one may also use the effects of bubble excitation to obtain information about the sound field (see, for example, section 5.3). Such techniques may employ erosion, bioeffect, sonochemistry, or sonoluminescence (also an optical phenomenon) etc. However, the ability of such techniques to yield information on the sound field will be limited by our knowledge of the cavitational mechanism through which the effect comes about, and of the noncavitational mechanisms which will contribute. (b) Some idealised bubble histories, with examples. Acoustic emission, which may vary from the generation of the resonance frequency over many cycles to the formation of shock waves during a violent rebound, tends to be associated with histories where both the growth and collapse phases occur over a relatively short period of time. Since these histories relate to detection, only a few bubble cycles are shown. Therefore, effects of mass flux between gas and liquid that give significant effect only over many cycles (e.g. dissolution of the 'static' bubble; rectified diffusion during 'high-amplitude pulsation') are not illustrated.

(a)

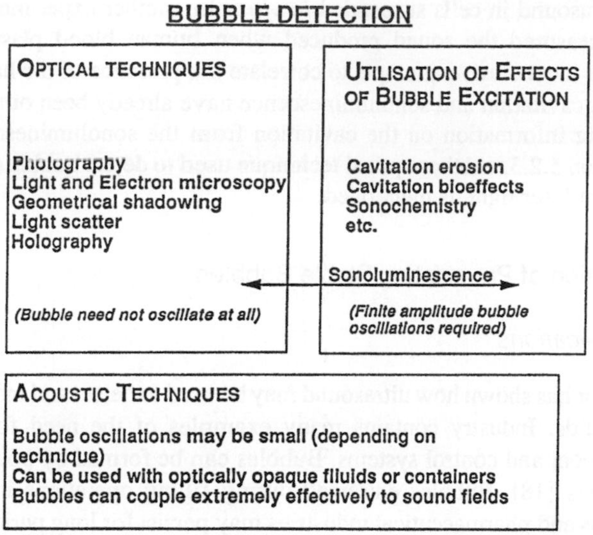
BUBBLE DETECTION
OPTICAL TECHNIQUES
Photography
Light and Electron microscopy
Geometrical shadowing
Light scatter
Holography
(Bubble need not oscillate at all)
UTILISATION OF EFFECTS OF BUBBLE EXCITATION
Cavitation erosion
Cavitation bioeffects
Sonochemistry
etc.
(Finite amplitude bubble oscillations required)
Sonoluminescence
ACOUSTIC TECHNIQUES
Bubble oscillations may be small (depending on technique)
Can be used with optically opaque fliuids or containers
Bubbles can couple extremely effectively to sound fields

(b)

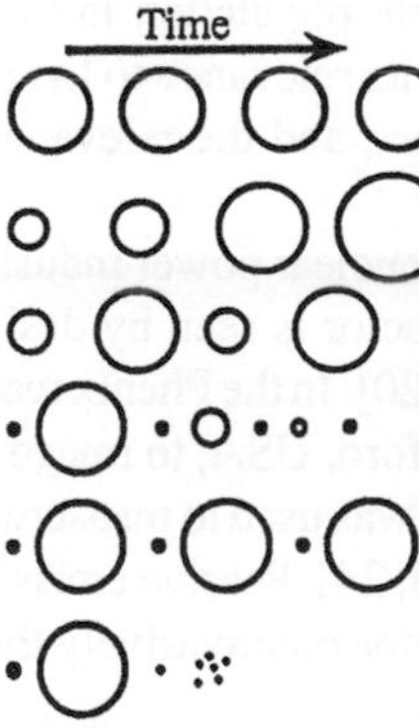
Bubble history
Time

Mode	Examples	Acoustic emission
Static (will dissolve unless stabilised)	"Old age" oceanic	
Steady growth	Decompression	
Low amp. pulsation	Brooks, sea surface, though actually low amp. versions of...	↑
Decaying oscillation	Hydrodynamic, short US pulse, explosions	
High amp. pulsation	Long US pulse	
Breakup on rebound	Short or long US pulse	↓

All may be modified by departures from sphericity:

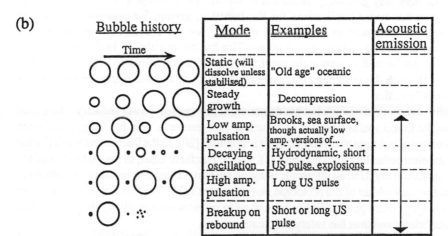

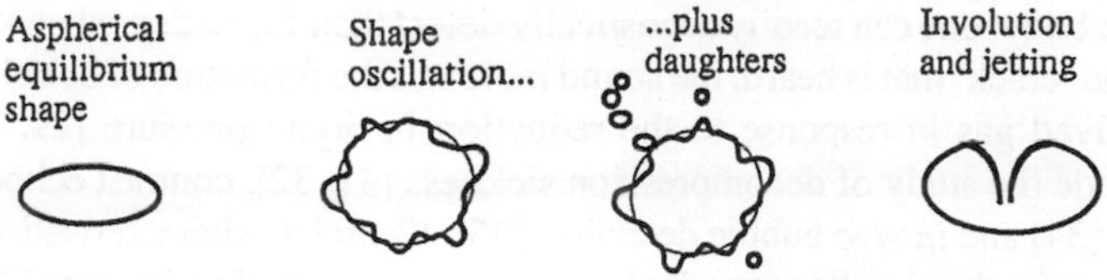
Aspherical equilibrium shape
Shape oscillation...
...plus daughters
Involution and jetting

All will geometrically scatter/shadow short wavelength sound and light (if medium transmits). If lower frequency sound is used (i.e. closer to the bubble resonance), the sound may modify the activity of the bubble (e.g. change 'static' to 'pulsation').

induced by the ultrasound in cells suspended *in vitro*. In another experiment, Eastwood and Watmough [17] measured the sound produced when human blood plasma was made to sonoluminesce. The inconclusive attempts to correlate the presence of the half-harmonic with the onset of violent cavitation and sonoluminescence have already been outlined.[5] The applicability of obtaining information on the cavitation from the sonoluminesce is discussed in section 5.2. In section 5.2.3, another optical technique used to detect bubble oscillation, that of the Mie scattering of laser light, is mentioned.

5.1.1 The Detection of Pre-existing Stable Bubbles

Introduction: Applications

The previous chapter has shown how ultrasound may be used to detect and influence gas bubble populations in liquids. Industry contains many examples of the need for reliable bubble detection, management and control systems. Bubbles can be formed by pouring molten glass or polymer solutions [18]. Bubbles entrained during filling operations in the paint, food, detergent, cosmetics and pharmaceutical industries may persist for long periods, degrading the product [19].

In the petrochemical industry alone, for example, bubbles may be nucleated through the exsolution of gas which has dissolved into the crude in the high pressures at the well base, and which comes out of solution as the crude is brought up to surface pressures. Knowledge of the bubble population is required to optimise harvesting and transportation resources. Bubble detection in the bore may give early indication of the presence of high-pressure gas pockets. Bubble detection has wider implications for the industry, giving, for example, warning of cavitation inception and erosion in pumping devices. Several decompression studies had attempted to assess bubble formation in humans. The bubble population in the near-surface layers of the sea has been discussed in the previous chapter: its relevance to the petrochemical industry arises through the environmental aspects of gas flux, and the relevance to undersea ultrasonic communication.

Specifically regarding ultrasonic bubble detection, in the nuclear power industry "the use of ultrasound in the liquid sodium coolant of a nuclear fast reactor is seen by designers of such systems as a potential solution to several difficult problems" [20]. In the Phenix reactor in France it is used to monitor refuelling manoeuvres [21], and in Hanford, USA, to image the top of the reactor [22]. At Dounreay, UK, an ultrasonic imaging system was used to measure the distortion of individual fuel assemblies of the prototype fast reactor [23, 24]. Passive emissions also have a use in the industry: noise generation has been used to monitor noninvasively the temperature reactivity coefficient, and thus has safety implications [25].

There are applications of ultrasonic techniques to fluid processing [26], the measurement of pressure changes [27] and detection of steam bubbles in pressure vessels [28]. Bubbles are also present within the body: one can readily acoustically detect their formation when knuckle joints are stressed, by the 'crack' that is heard; the sound indicates the formation of bubbles containing vapour and exsolved gas in response to the reduction in liquid pressure [29, 30]. Medical applications include the study of decompression sickness [31, 32], contrast echocardiography [33], and *in vitro* [34] and *in vivo* bubble detection [35]. Complete characterisation of a bubble population may require the simultaneous deployment of a number of techniques [34]. Particular examples of these techniques are examined below.

---

[5]See Chapter 4, section 4.4.7.

(a) The Detection of Stable Bubbles Through Bulk Acoustic Properties of the Medium, Imaging and Ultrasonic Tomography

In 1977 Medwin [36] published a review of acoustic techniques that had been used for bubble counting, emphasising the exploitation of the acoustic cross-sections of the bubble.[6] Medwin noted two facts which make the technique very suitable for bubble counting, provided that much larger scatterers are not present. Firstly, the scattering cross-section of a bubble is about 1000 times its geometrical cross-section at resonance, and about 10^{10} times that of a rigid sphere; and secondly the resonance frequency varies inversely with the radius if surface tension effects are negligible. Medwin also commented that a sound wave passing through a bubbly liquid will be attenuated more than in a bubble-free liquid, not just by the bubble-induced scattering, but also by the absorption of energy by the bubbles, and that this excess attenuation may be used to investigate the bubble population. Medwin also suggested that the bubble-induced change in compressibility might be suitable for population investigations. As shown in Chapter 4, section 4.1.2(e), the compressibility becomes complex in the presence of bubbles, and leads to dispersion. Medwin describes *in situ* experiments for measuring oceanic bubble populations, using pulse echo and continuous-wave techniques. The backscattering and extinction cross-sections per unit volume are simultaneously determined in the pulse-echo method by emitting a pulse from a source/receiver, the pulse being reflected from a plane steel surface a known distance away. Comparison of oceanic measurements with those in clean water reveals the excess absorption and scattering, whilst the relative reverberation between echoes, when compared with the preceding and succeeding echoes, gives the scatter along the path. Timing between pulses can give the local group velocity. Medwin found the actual dispersion too small to measure accurately with this system. Measurement of the continuous-wave signal received by two hydrophones spaced a known distance apart along the acoustic axis of the beam allows the extinction cross-section per unit volume to be found. The phase speed and attenuation could be found by this method. Other oceanic techniques are discussed in Chapter 3, section 3.8.1.

One year later, Mackay and Rubisson [37] published details of an experiment where 7.5-MHz pulses were used to image the bubbles resulting from decompression, *in vivo* in humans, fish and guinea pigs. The researchers took the returning echo strength to be a measure of the bubble size, calibrating their device through comparison of microscopic and ultrasonic measurements on bubbles produced by electrolysis in transparent media such as water and gelatin, or by other processes in transparent fish. However, small bubbles close together scattered incoherently such that they could not be distinguished from single, larger bubbles.

At about the same time, Daniels *et al.* [38] developed an 8-MHz mechanical ultrasonic sector scanner to study the formation of bubbles following decompression from exposure to increased ambient pressures. Bubble detection was optimised by adjusting the gain of the system to minimise the amount of fine tissue detail that was visualised. By the early 1980s there were indications of ultrasonically generated cavitation effects *in vivo* [39, 40]. ter Haar *et al.* [41–43] also employed the pulse-echo ultrasonic B-scan system to detect bubbles generated *in vivo* in the hind limbs of guinea pigs in response to 0.75-MHz insonation by continuous-wave (at I_{SAPA} = 110 mW cm^{-2}) or 1:1 duty cycle (repetition frequency v_{rep} = 250 Hz; pulse length τ_p = 2 ms; I_{SAPA} = 240 mW cm^{-2}) ultrasound from a commercial therapy device (Rank Sonacel Multiphon, Mk III). Figure 5.1 compares control images, showing the skin line and bones within the leg when the therapy device was not insonating, with frames taken during therapeutic

[6] See Chapter 4, section 4.1.2(d).

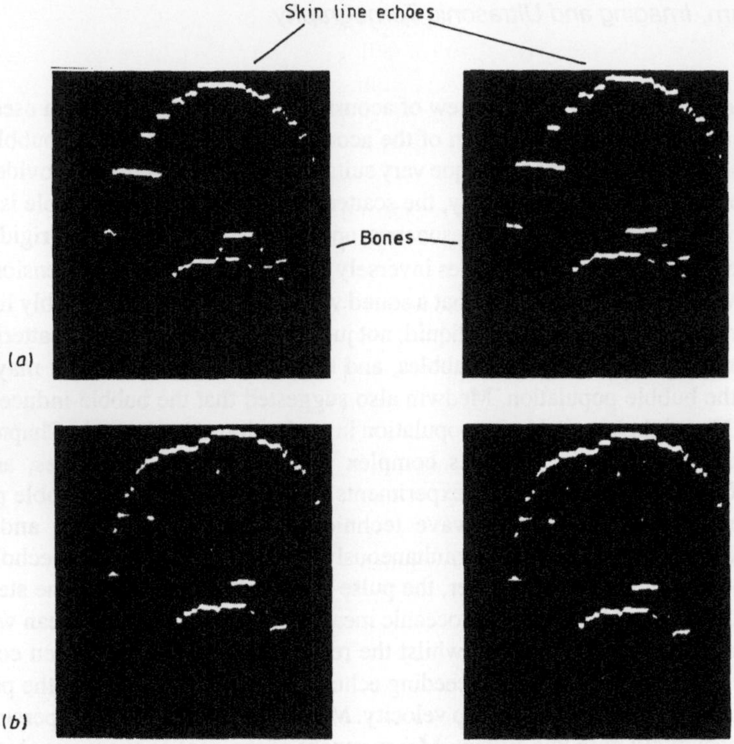

Figure 5.1 (a) Control images of the resting guinea-pig leg showing the skin line and bones within the leg. (b) Representative images of the guinea-pig leg following insonation with 0.75-MHz ultrasound ($I_{SA} = 300$ mW cm^{-2}). Additional echoes are seen from the interior of the leg (after ter Haar and Daniels [41]). Reprinted by permission from *Physics in Medicine and Biology*, vol. 26, pp. 1145–1149; Copyright © 1981 IOP Publishing Ltd.

insonation. The additional echoes from the inside of the leg are taken to be bubble events. Frames were recorded every two seconds on 35 mm film, and frame-by-frame analysis enabled the spatial distribution of bubbles to be followed in time [44]. Bubble-associated events were taken as being any record of a new echo at a single site. Such events lasted for intervals ranging from a single frame (taken to represent intra-vascular bubbles passing through the plane of the scan) to several successive frames (intra- or extra-vascular bubbles). The minimum detectable bubble radius using this technique was about 5 μm [45], which is larger than the size of bubble that would be resonant with the therapy beam (0.75 MHz) in water ($R_r \sim 4$ μm). Most of the detected bubbles were in the size range $5 < R_0 < 50$ μm, with less than 1% in the range $50 < R_0 < 250$ μm. It was not possible to distinguish between individual large bubbles and tight clusters of smaller bubbles separated by less than the resolution of the system [41], and no accurate measurements of bubble size could be made [34, 45]. The mechanism through which the bubbles were generated is debatable. ter Haar and Daniels [41] suggest that the bubbles were probably stable entities generated by rectified diffusion, or perhaps coalescence mechanisms from pre-exisiting nuclei. However, Watmough *et al.* [46] suggest that the effect is probably due to vapour and gas pressure changes resulting from heating. They observed temperature rises of up to 15° C W^{-1} cm^2 in guinea pig legs (*post mortem*), which would alone account for the observed bubble growth.

Fowlkes *et al.* [47] generated bubbles in excised canine urinary bladders which were sealed within a bag of degassed saline solution and then placed in a bath of degassed water at the common focus of a 555-kHz transducer (of radius 3.5 cm) and a brass reflector (of radius 5 cm). The bubbles were visualised on a diagnostic ultrasound scanner with a 5-MHz in-line mechanical scanhead. Pressure amplitudes as great as 10–20 bar were used to generate the largest bubbles detected which, from their rise times, had estimated radii of 50–70 μm. These bubbles, being very much larger than resonance, were probably generated through coalescence in the standing-wave field in the bladder. As regards detection of the smaller bubbles, the inability to distinguish bubble echoes from artefacts caused by the reverberant field within the bladder set resolution limits.

Bleeker *et al.* [48] used the attenuation coefficient, sound speed and backscatter coefficients at 5 and 7.5 MHz to examine a controlled population of Albunex™ spheres, a commercial echo-contrast agent for clinical applications, which consists of microbubbles in the size range $0.5 \leq R_0 \leq 5$ μm stabilised against dissolution by a shell approximately 28 nm thick made of coagulated human serum albumin.

In a technique similar to the optical ones, ultrasonic shadowgraphs can be obtained of bubbles which intercept an acoustic beam, the wavelength of which was very much less than the bubble radius. In 1988, Wolf [49] used the tomographic results obtained from several different directions of projection to obtain a two-dimensional cross-section of the bubble field. Wolf deployed a few dozen transducers circumferentially about a water pipe containing three simultaneous ascending bubble streams, each with a gas flow rate of 100–2000 litres/hour. By sequentially switching between the transducers to obtain the different emitter–receiver combinations required directions of interrogation, Wolf obtained a measure of the total surface area of bubble interface in the population as a function of position within the cross-section (Figure 5.2).

(b) The Detection of Pre-existing Bubbles Through Doppler Techniques and Resonance Excitation

In the 1970s two interesting examples of active ultrasonic bubble detection were published. In 1972, Nishi [50] employed the Doppler effect for bubble detection in blood. The signals from the moving bubbles exhibited a Doppler shift, which can distinguish them from reflections from static objects. Doppler techniques may be unable to distinguish between a single large bubble and a cluster of smaller ones [27].

In 1977, Fairbank and Scully [51] examined the emitted scattered signal from bubbles, assumed to be at resonance, subjected to broadband ultrasound (from 100 kHz to 1 MHz). They proposed using this technique in order to measure blood pressure changes in inaccessible regions of the heart through the resulting changes in the equilibrium volumes of injected bubbles. However, resonance techniques are not ideal. Simulations of common methods of bubble sizing through resonance techniques for hypothetical bubble size distributions by Commander and Moritz [52] suggest that the number of bubbles having $R_0 < 50$ μm can be significantly overestimated. More thorough analyses of data are required to compensate (see, for example, Commander *et al.* [53]).

The analysis of Chapter 4, section 4.1 illustrates one of the major problems with such resonance techniques. One might, for example, insonate a sample of bubbly liquid at a particular frequency ν, monitor variations in the scattered signal at ν, and assume these variations are caused by bubbles resonant with the frequency ν (i.e. those bubbles for which $\omega_0 = 2\pi\nu$). The deduction would be that the scattering of ν is yielding information about resonant bubbles only.

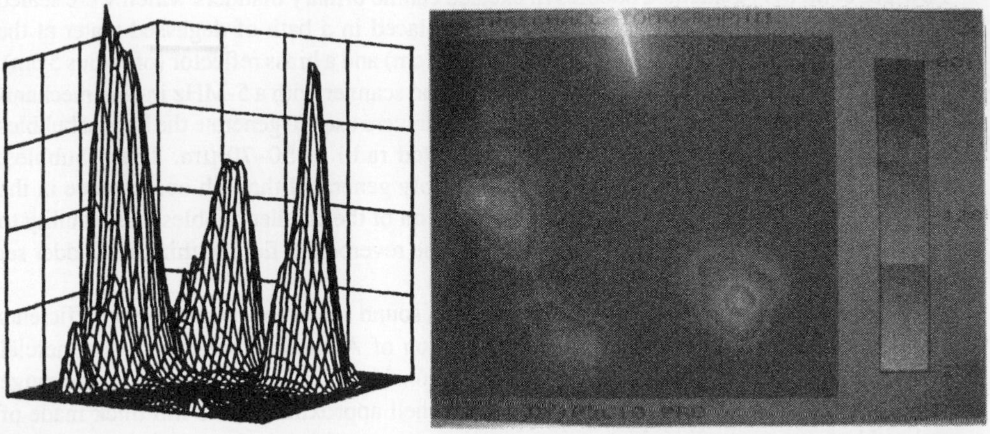

(a) (b)

Figure 5.2 Measurements made in a cross-sectional plane within a water pipe, about 30 cm above the point of injection of three simultaneous ascending bubble streams, each with a gas flow of 100 litres/hour. (a) Contour-plot representation of the total surface area of bubble interface in the population as a function of position within the cross section of a water pipe. (b) On-line output on TV monitor. (After Wolf [49].)

As can be seen from Figure 4.5, this might seem at first to be a reasonable assumption, since the acoustic scattering cross-section does peak sharply at resonance. However, this peak is only a *local*, and not a global, maximum, as can be seen from Figure 4.5: the rising tail of the cross-section plot may, with increasing insonation frequency, attain values comparable or greater than the resonance value. Therefore scattering of the signal at ν may occur not only through the response of resonant bubbles, but also contain contributions from bubbles which are much larger than resonance size. Indeed, a large bubble far from resonance size can generate a stronger signal than a smaller bubble at resonance, and so to interpret the scattering of the insonation frequency in terms of the resonant bubbles only may be a misinterpretation. This problem will not occur if the scattering of the monitored signal is a global maximum at resonance. This is the case for second harmonic emissions at low insonation powers.

(c) The Detection of Pre-existing Bubbles Through the Second Harmonic Emission

In Chapter 4, section 4.4.7(b) it was shown how, if the response of an oscillator has a component determined by the square of the driving force, of frequency ν, that response can contain elements at the second harmonic 2ν (convention labelling the fundamental ν the first harmonic). At low insonation powers, the emission of the second harmonic 2ν is a *global* maximum when the insonation frequency coincides with the bubble pulsation resonance (i.e. when $\nu = \omega_0/2\pi$). This result makes it a better indicator of bubble size in a population containing a wide range of bubble sizes than the emitted frequency ν [54]: the latter is only a local maximum at resonance, as discussed above. In 1981, Miller [55] incorporated this effect in a so-called 'resonant bubble detector' (RBD). Two transducers, a 1.64-MHz emitter and a 3.28-MHz receiver, were mounted with suitable acoustic absorbers, their axes perpendicular to each other and to the 4-mm diameter tube containing the flowing liquid that was to be investigated. Insonation was at low intensity (0.12 W cm^{-2}). The detector, tuned to twice the emitter frequency, was sensitive to bubble sizes resonant with the emitter ($R_0 \approx 2.1$ μm for 1.64 MHz). Miller *et al.* [56] used a larger, modified detector such that the interrogated region was 7.4 mm from each transducer, with the emitter operating at 0.89 or 1.7 MHz. Miller [55] tested the device for two bubble sizes in water: the second harmonic emitted by resonant bubbles (produced by electrolysis) was 43 times stronger than that produced by injected bubbles of 250 μm radius. By comparison, the ν emission from the resonant bubbles was 0.02 times that from the larger bubbles. Miller *et al.* [56] were able to semi-quantitatively detect bubbles produced by upstream ultrasonic cavitation, and by hydrodynamic cavitation at a detector tip, though they speculate that coalescence and radiation forces may have affected the population. In 1985, Gross *et al.* [35] were unable to detect bubbles downstream from the aorta when canine hearts were and were not insonated. In 1984, Vacher *et al.* [57] produced a similar bubble detector, though the frequency of interrogation (and therefore the size of bubble that was resonant) could be swept, the detector frequency being constantly twice that of the emitter. It should be noted that in the 1984 paper Miller *et al.* [56] estimate that their detector responded to some bubbles which were up to 25% larger or smaller than resonance. Spatial resolution is poor with the second harmonic technique [27]. Another drawback is that the second harmonic may arise through nonlinear effects when sound propagates through even bubble-free water, as noted in Chapter 1, section 1.2.3(a), which may generate spurious signals.

(d) The Detection of Pre-existing Bubbles Through Combination Frequencies

The simple description of the nonlinear oscillator given in Chapter 4, section 4.4.7(b), where the general response ¥ might be related to the driving force $\mathcal{F}$ through $¥(t) = s_0 + s_1.\mathcal{F}(t) + s_2.\mathcal{F}^2(t) + s_3.\mathcal{F}^3(t) + s_4.\mathcal{F}^4(t) + \ldots$ (equation (4.55)), was capable of indicating the source of the harmonics of frequencies that were integer multiples of the insonation frequency. These are the frequencies which were employed in the preceding section for bubble detection. This model can be used to illustrate how the use of *two* insonating frequencies can detect and size bubbles. The driving force $\mathcal{F}$ consists of the sum of two coherent forces of different frequency, i.e.

$$\mathcal{F} \equiv P(t) = P_1 \cos\omega_1 t + P_2 \cos\omega_2 t \qquad (5.1)$$

where $\omega_2 > \omega_1$ (the presence of phase constants would not alter the general result). After substituting equation (5.1) into equation (4.55), expansion of the quadratic component contains a term which can be expanded to generate the sum and difference frequencies:

$$2P_2P_1\cos\omega_1 t.\cos\omega_2 t = P_2P_1\{\cos(\omega_1 + \omega_2)t + \cos(\omega_2 - \omega_1)t\} \qquad (5.2)$$

Newhouse and Shankar [58] describe the sizing process using the scattered signals generated when a bubble is insonated with a pump frequency ω_p and an imaging frequency ω_i. Experimentally, Shankar *et al.* [27] insonated the bubble with the 'imaging' beam (fixed at a frequency of 2.25 MHz, far from the bubble resonance) and a second, 'pumping' beam, the acoustic axes being perpendicular. The pump frequency ω_p was scanned across the region where the bubble resonance could reasonably be expected to lie. As shown in Chapter 4, when the pump frequency is far from the bubble resonance, the bubble response is of small amplitude and it approximates to a linear oscillator.[7] In that situation, therefore, no sum and difference frequencies are detected. When ω_p is near the bubble resonance, the amplitude of oscillation of the bubble wall is large.

[7] This can be seen qualitatively: at small amplitude, the fundamental asymmetry between expansion and contraction is negligible.

The bubble oscillations are nonlinear, and they scatter sound nonlinearly. As with the second harmonic, these combination frequencies exhibit a global maximum at resonance, and Shankar *et al.* [27] therefore took the frequency of the pump signal when the sum and difference frequencies ($\omega_i \pm \omega_p$) are detected to be the bubble resonance, and so had a measure of the bubble size.

Chapelon *et al.* [59] found that when two bubble sizes were present, geometrical screening of smaller bubbles by larger ones was detectable. However, the size distribution of an electrolytically generated bubble cloud agreed well with photographic measurements. Schmitt *et al.* [60] and Siegert *et al.* [61] measured the smaller bubbles (of μm size) in clinical echocontrast agents using an imaging frequency of 10 MHz.

Quain *et al.* [62] use the difference frequency method on cylindrical gas pockets trapped in hydrophobic pores.[8] They were able to distinguish between two pore sizes, the larger of which had a diameter (≈ 5 μm) 66% greater than that of the smaller (≈ 3 μm), using the resonance properties of gas-filled micropores, as expounded in a series of theoretical and experimental discussions by Miller [63–65]. The mode of oscillation favoured by Quain *et al.* [62] is reminiscent of the motion of a clamped membrane. In this mode, the ratio of the amplitude of the second harmonic to that of the fundamental is lower than in the second mode of oscillation, which resembles a piston-like motion.

In 1988, Chapelon *et al.* [66] examined bubbles using the fractional Doppler shift, f_{Dop}, in the received signal, the detected frequencies including ω_i, $\omega_i \pm \omega_p$, $(1+f_{\mathrm{Dop}})\omega_i$ and $(1+f_{\mathrm{Dop}})\omega_i \pm \omega_p$. A certain amount of processing was required, though with eventual use of a pulsed imaging signal, Chapelon *et al.* [66] were able to obtain size, direction, number density and speed information from the bubbles. Cathignol *et al.* [67] describe a Doppler technique in which the imaging signal is pulsed, giving the range of the bubble on the axis of the imaging beam. They obtain lateral and longitudinal resolution of better than 1 mm, and discuss the contribution of signals at $\omega_i \pm 2\omega_p$ radiated by the bubble.

Phelps and Leighton [68] give a simple explanation of the production of signals at ($\omega_i \pm \omega_p$). As the bubble pulsates, the acoustic scattering cross-section it presents to the imaging beam varies periodically. Therefore the scattered imaging signal is modulated by the bubble pulsation at the driving frequency ω_p, and signals at $\omega_i \pm \omega_p$ are detected in the received spectrum. The closer ω_p is to the bubble resonance, the greater the amplitude of pulsation, and so the stronger the spectral components at $\omega_i \pm \omega_p$.

Leighton *et al.* [69, 70] observed a combination frequency involving the imaging signal and the subharmonic of the bubble resonance (at $\omega_i \pm \omega_p/2$). Phelps and Leighton [68] identified the source as Faraday waves on the bubble wall (Figure 4.47), which oscillate at half the driving frequency. Whilst subharmonic emissions from them do not propagate to distance,[9] such waves will modulate the target strength for the imaging beam at $\omega_p/2$, so that the received spectrum will contain additional components at $\omega_i \pm \omega_p/2$. The time-series data from a single tethered bubble insonated at resonance by pump and imaging signals reflects these descriptions (Figure 5.3). The amplitude-modulated 1.1 MHz scattered signal plots so densely as to appear continuously black. In Figure 5.3a the bubble is insonated at a pump signal amplitude of 25 Pa, which is below the threshold required to stimulate Faraday waves. The amplitude modulation reflects the bubble pulsation frequency, ω_p. However at a greater amplitude (40 Pa) the threshold is exceeded, and the subharmonic component in the amplitude modulation is clear (Figure 5.3b).

---

[8] See sections 5.4.3(b)(ii) and 5.4.3(b)(iii).
[9] See Chapter 4, section 4.4.7(b)(i).

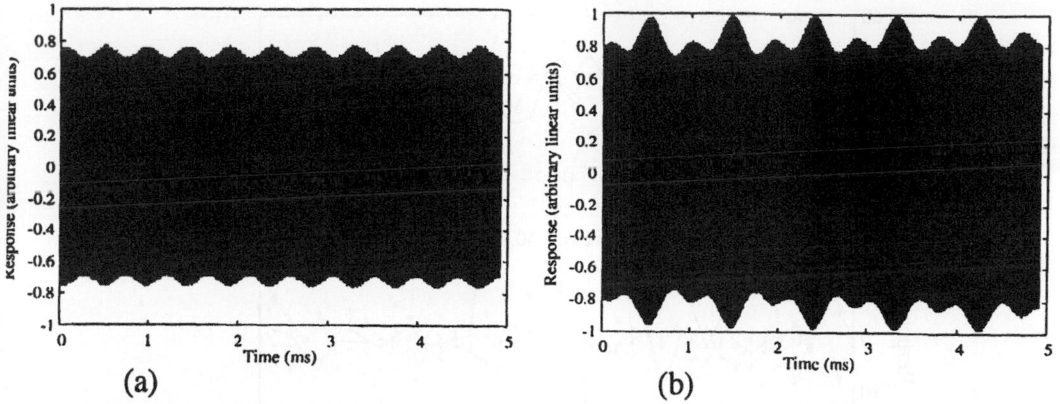

Figure 5.3 The signal scattered from a single, tethered bubble which is insonated by an $\omega_i/2\pi = 1.1$ MHz imaging signal, and simultaneously being driven at resonance ($\omega_p/2\pi$ = 2160 Hz) at a pump signal amplitude of (a) 25 Pa and (b) 40 Pa (zero-to-peak). The receiver is resonant at 1 MHz and has a half-power bandwidth of 450 kHz. The data was sampled at 10 MHz. The carrier frequency plots so densely as to appear black [68].

Figure 5.4 shows, in mesh plot, the scattered spectra from a single bubble insonated with fixed ω_i and a pump signal which is incremented in 25 Hz steps. To the right of the constant imaging signal is a broken ridge, corresponding to the $\omega_i + \omega_p$ signal (present for all 40 pumping tones). The difference in frequency between the two ridges equals the pumping frequency. The $\omega_i + \omega_p$ signal peaks towards the bubble resonance, but that critical frequency is more clearly indicated by the $\omega_i + \omega_p/2$ and $\omega_i + 3\omega_p/2$ signals, which peak sharply when the pumping frequency equals 1850 Hz. In certain circumstances therefore the $\omega_i + \omega_p/2$ signal may be a better sizing tool than $\omega_i + \omega_p$, as it is sharper and cannot readily be excited through non-bubble mechanisms, such as direct coupling of the transducers [68] and turbulence [71]. However, the generation of Faraday waves requires a threshold meniscus displacement amplitude to be

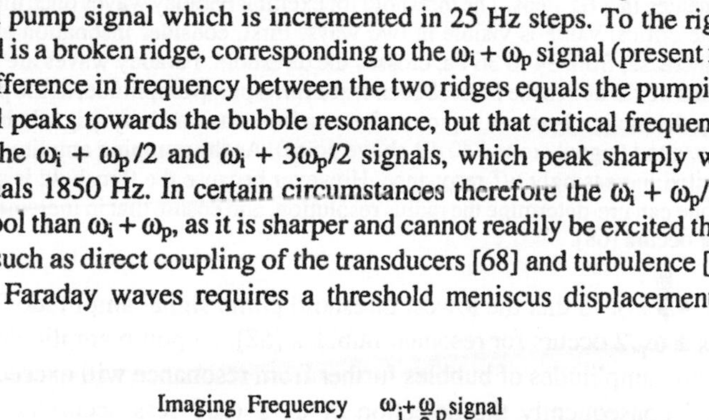

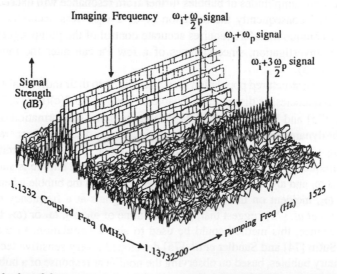

Figure 5.4 Mesh plot of the scattered signal from a single tethered bubble, using the same receiver as for Figure 5.3. The imaging frequency is 1.1 MHz. The pump frequency is incremented in 25 Hz steps from 1525 to 2500 Hz. For each setting of the pump signal, the high-frequency emissions from the bubble are shown in the frequency window from 1.1332 to 1.1373 MHz. The pump signal has amplitude 190 Pa (zero-to-peak) (A. D. Phelps and T. G. Leighton).

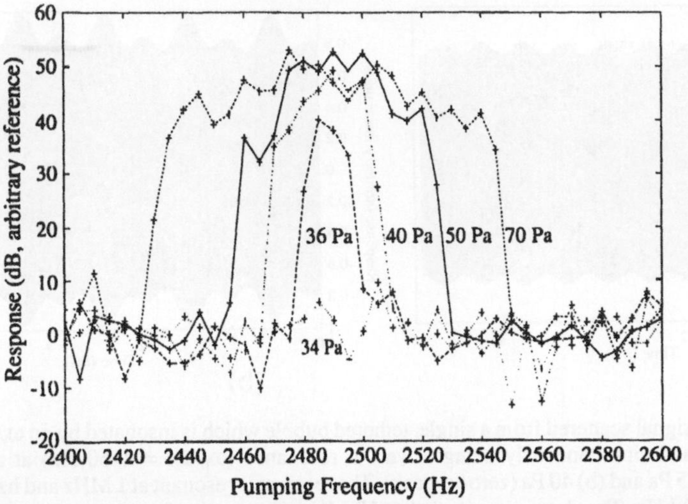

Figure 5.5 Plot showing the height of the signals at the frequency corresponding to $\omega_i \pm \omega_p/2$ as a function of the pump frequency setting, for insonation of a single tethered bubble at various pump signal amplitudes: 34 Pa (dotted), 36 Pa (large dashes), 40 Pa (dash-dot), 50 Pa (unbroken) and 70 Pa (small dashes). The pump signal is incremented in 5 Hz steps. The threshold for exciting Faraday waves once the wall pulsation amplitude exceeds the critical value is visible in two ways. First, consider insonation at the resonance (~2485 Hz). A 2 Pa increase, from 34 to 36 Pa, crosses the threshold: Faraday waves are excited and the $\omega_i \pm \omega_p/2$ signal appears 40 dB above the noise. Second, at a fixed pump amplitude, as the pump frequency varies there is a sharp distinction between regions of no surface wave activity (signal absent) and those where the waves are excited (signal around 40 dB above noise). As the pumping amplitude increases, the $\omega_i \pm \omega_p/2$ can be excited increasingly off-resonance. However because the threshold is sharp, by fixing the pump amplitude one can pre-determine the radius resolution, and ensure that in incrementing the pump frequency no aliasing occurs [68].

exceeded (equation 4.335), so that the lowest threshold pump signal amplitude for the generation of signals at $\omega_i \pm \omega_p/2$ occurs for resonant bubbles [68]. At pump amplitudes in excess of this, the wall pulsation amplitudes of bubbles further from resonance will exceed the displacement threshold, and consequently the detection of $\omega_i \pm \omega_p/2$ less accurately indicates the resonance. The technique therefore requires accurate control of the pump signal amplitude at the bubble under investigation, since changes of a few Pa can alter the range of bubbles stimulated (Figure 5.5).

Bubbles acting in a structured population may, in addition to their individual effect, give rise to combination frequencies, as discussed in Chapter 3, section 3.8.2(c)(ii) [12, 72].

Bunkin *et al.* [73] and Butkovsky *et al.* [13] relate in detail the formation of combination frequencies to the dynamics of the gas bubble. In particular they discuss the generation of signals at $\omega_1 - \omega_o$ and $\omega_1 + \omega_o$ [13] when a bubble of breathing-mode resonance is excited by a single pump wave of frequency ω_1. They also consider the case when a bubble is subjected to two incident signals at ω_1 and ω_2, which excite resonant oscillations in the bubble when $\omega_1 - \omega_2 = \omega_o$. A third signal, ω_3, incident on the bubble will cause signals at a frequency $\omega_3 \pm \omega_o$ to be generated. Bunkin *et al.* [73] suggest that, provided none of ω_1, ω_2, ω_3 or $(\omega_3 \pm \omega_o)$ are close to a bubble resonance, this method could be used to size a population. In the early 1980s, Ostrovskii and Sutin [74] and Sandler *et al.* [75] developed a very sensitive technique for the detection of solitary bubbles, based on observing the nonlinear response of a bubble excited by two sound waves, where the difference between the two wave frequencies is equal to the bubble resonance frequency. Naugol'nykh and Rybak [76] monitored resonance scattering of bubbles, using a secondary, low-frequency signal (that is, lower than the bubble resonance) to shift the bubble resonance and induce a low-frequency modulation on the scattered signal. The percentage modulation being proportional to the derivative of the bubble-radius distribution function, Naugol'nykh and Rybak predict that the distribution could be reconstructed from the scattered signal.

5.1.2 The Ultrasonic Detection of Short-lived Transient Cavitation

The acoustic emissions from transient cavitation were discussed in Chapter 4, section 4.4.7(a), as were the attempts to relate them to the violence of the cavitation.

This section deals with the acoustic detection of short-lived transient cavitation which is taken to be characterised by the initially explosive growth of a cavity from a small stabilised nucleus, bubble or gas pocket, followed by an appropriately energetic collapse. As seen in Chapter 4, section 4.1.2(d), the acoustic cross-section of bubbles near resonance are very much larger than their geometric cross-section, and so not surprisingly acoustic methods are popular in detecting bubbles having sizes of the order of a micrometre, where visual observation can be difficult. Where the bubbles are far from resonance, such acoustic techniques are similar to visual ones, in that only clouds of bubbles can be detected. However, the concept of a steady-state resonant response is inappropriate for the dynamics of transient cavitation, and the detection of cavitation in response to short, microsecond pulses of ultrasound is of particular importance. This is because of the clinical relevance of such pulses, and the need to assess the likelihood of cavitation-induced effects during exposure. By the mid 1980s Apfel [77–79] and Flynn and Carstensen [80, 81] had predicted that transient cavitation could occur in liquids in response to ultrasound pulses of only a few cycles. Free radical production in response to the high temperatures generated within the collapsing bubble had also been indicated for microsecond pulses in experiments published in 1985 to 1988 by Crum and Fowlkes [82–84] and by Carmichael *et al.* [85] and Christman *et al.* [86].

One technique for detecting such cavitation is the passive detector introduced in 1988 by Atchley *et al.* [87], which relies upon the scattering of the insonating sound field by bubbles associated with transient cavitation events. The object was to find acoustic pressure thresholds for the cavitation as a function of pulse duration (τ_p) and pulse repetition frequency (ν_{rep}) at insonation frequencies of 0.98 and 2.30 MHz. Their results for distilled, degassed, deionised water, filtered to 0.2 μm and seeded with hydrophobic carboxyl latex particles (1 μm diameter) indicated that the pressure threshold is independent of pulse duration and acoustic frequency for pulses longer than approximately ten acoustic cycles.

Holland and Apfel [88] used short tone bursts of ultrasound at 0.76, 0.99 and 2.30 MHz to find the acoustic pressure transient cavitation threshold. The host fluid preparation system is based on designs by Roy *et al.* [89] and Atchley *et al.* [87]. It had been noted [90] that experimentally determined transient cavitation thresholds in water had in the past exceeded theoretical predictions by a factor of two or more. Roy *et al.* [90] attributed this discrepancy to the *a priori* assumption in the models that a free-floating microbubble seed existed, from which the transient event was nucleated. Since such conditions are difficult to attain in the laboratory, it was decided to measure thresholds in clean water that had been deliberately seeded with a known nucleus population. Roy *et al.* [90] used Albunex™ spheres, tested at $\nu = 1$ and 2.25 MHz for $\tau_p = 10$ μs and $\nu_{rep} = 1$ kHz. The seeds were injected at a concentration of around 10^5 bubbles/ml into the streaming flow generated by a 15-MHz transducer, which convected the nuclei to the focal region of the second transducer, the one that was to excite them into transient collapse. Any cavitation events were recorded by a third transducer, which was the 1-MHz broadband passive detector, and the results tested against a model for the dynamical response of a bubble encapsulated by a layer of viscous, surface-active material.

Holland and Apfel [88] used in addition two other seeding species, making three in all. A monodisperse suspension of hydrophobic polystyrene spheres of nominal radius 0.5 μm were tested with insonating frequencies ν of 0.76, 0.99 and 2.30 MHz (ν_{rep} = 1 kHz), in distilled, filtered water and in a viscous aqueous solution of ethylene glycol. Holland and Apfel [88] tested two specific populations of Albunex™, with mean radii of 0.75 and 1.25 μm (for ν = 0.76 MHz; ν_{rep} = 1 kHz). They also tested the pre-exisiting nuclei in dilutions of whole blood and saline (ν = 0.76 MHz; ν_{rep} = 1 kHz). Pulse lengths τ_p of either 10 μs or 5 cycles (corresponding to 16.50, 14.02 and 8.28 μs at 0.757, 0.989 and 2.300 MHz respectively) were employed.

The cavitation cell used by Holland and Apfel is shown in Figure 5.6. It is cylindrical, 40 cm long, with 10 cm internal diameter. Potential nuclei were convected through the focal region of the insonating transducer (a focused PZT8 transducer) by a fluid jet injected into the test

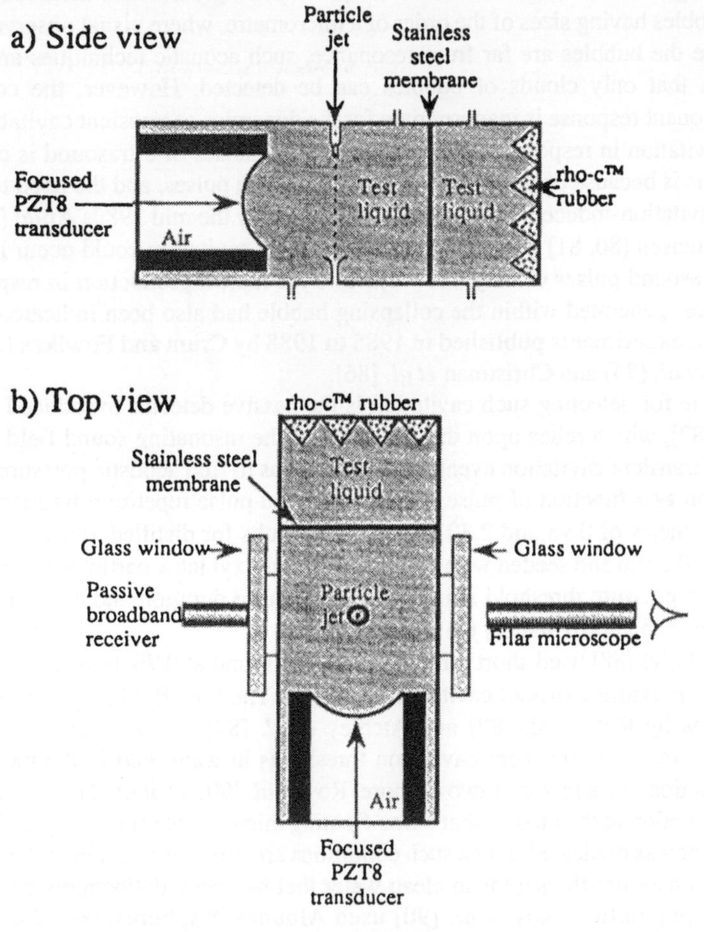

Figure 5.6 Schematic of the cavitation cell for passive detection of cavitation (after Holland and Apfel [88]).

chamber. A second transducer, a passive broadband receiver, was coupled to the test cell with coupling gel, and received signals generated near the focus of the insonating transducer. Once cavitation bubbles have been formed as a result of the insonating beam, they scatter any subsequent acoustic energy, which is detected by the passive transducer. The filar microscope views through a window and is used for alignment. Separated from the test chamber by a 12.7-μm thick stainless steel acoustic window is a second chamber containing an identical fluid and an absorber of rho-c™ rubber (so-called as it is impedance matched to water). This is done to minimise spurious reflections and inhibit the introduction of standing-waves (the steel window prevents contamination of the test liquid by dirt or small particles from the rubber). In addition, the detected signal is gated to eliminate spurious noise. As a detection technique this method relies on the strong interaction between the sound and the bubbles.[10] Holland and Apfel [88] found an average threshold of 0.59 MPa at 0.75 MHz, corresponding to a mechanical index[11] of 0.5.

A system which is more sensitive to bubbles in the micron size range is the active detector described by Roy *et al.* [91]. Subsequent to their production by the pulse from the first transducer ($\nu = 757$ kHz; $\nu_{rep} = 1$ kHz; $\tau_p = 10$ μs), the cavitation bubbles backscatter high-frequency pulses ($\nu = 30$ MHz; $\tau_p = 10$ μs) from a second transducer. Roy *et al.* [91] in fact deployed both active and passive acoustic detectors in their system, which is illustrated in Figure 5.7. The cavitation cell and the electronics associated with the transducers are shown in 5.7(a), whilst in 5.7(b) the fluid management system is illustrated. The fluid management occurs in a closed-flow system. From the reservoir, where the fluid is degassed using a vacuum pump, it is then sequentially deionised, cleared of organics, and filtered down to 0.2 μm. The fluid is then passed into the cell, which can be flushed of impurities using flow rates up to 5 litres/minute. Streaming speeds[12] of up to 3 cm/s could be used to convect nuclei into the focal region. The cell itself is similar to that shown in Figure 5.6, with the absorbing section separated from the main test section by a stainless steel membrane (9 μm thick in this particular case). The 757-kHz transducer, which generates the cavitation and is therefore labelled the 'primary', is mounted on a two-dimensional translating stage to enable its focus to coincide with that of the 30-MHz pulse-echo detector. The 1-MHz unfocused transducer is coupled with gel to the cell, opposite the 30-MHz active detector.

Figure 5.8 shows the output of the passive detector recorded by Roy *et al.* [91] with and without cavitation. The top trace (a) illustrates the primary pulse from the 757-kHz transducer, followed by a stable low-amplitude background resulting from multiple-path scattering and reverberation in the chamber. Its stability is testament to the stationary nature of the scattering surfaces. In the lower trace, 5.8(b), this scattered background contains a perturbation indicative of a time-varying scatterer, which Roy *et al.* showed to be at the focal region of the primary transducer. This signal is the result of transient cavitation. Figure 5.9 shows at the top the electrical signal which drives the 757-kHz primary transducer, which in turn generates the cavitation. Below this, the middle trace shows the signal from the active detector system in the absence of cavitation, the main pulse representing the interrogating 30-MHz signal. The bottom trace shows the signal from the active detector in the presence of the cavitation generated by the 757-kHz transducer. The reflected signal from the bubbles is clearly evident.

Holland and Apfel [88] comment that, when the thresholds measured by the active and passive systems for very similar fluid systems (e.g. having the same polystyrene sphere and gas

[10] As outlined in Chapter 4, section 4.1.2(d) in discussion of the acoustic cross-sections of bubbles.
[11] See Chapter 4, section 4.3.1(d).
[12] See Chapter 1, section 1.2.3(b).

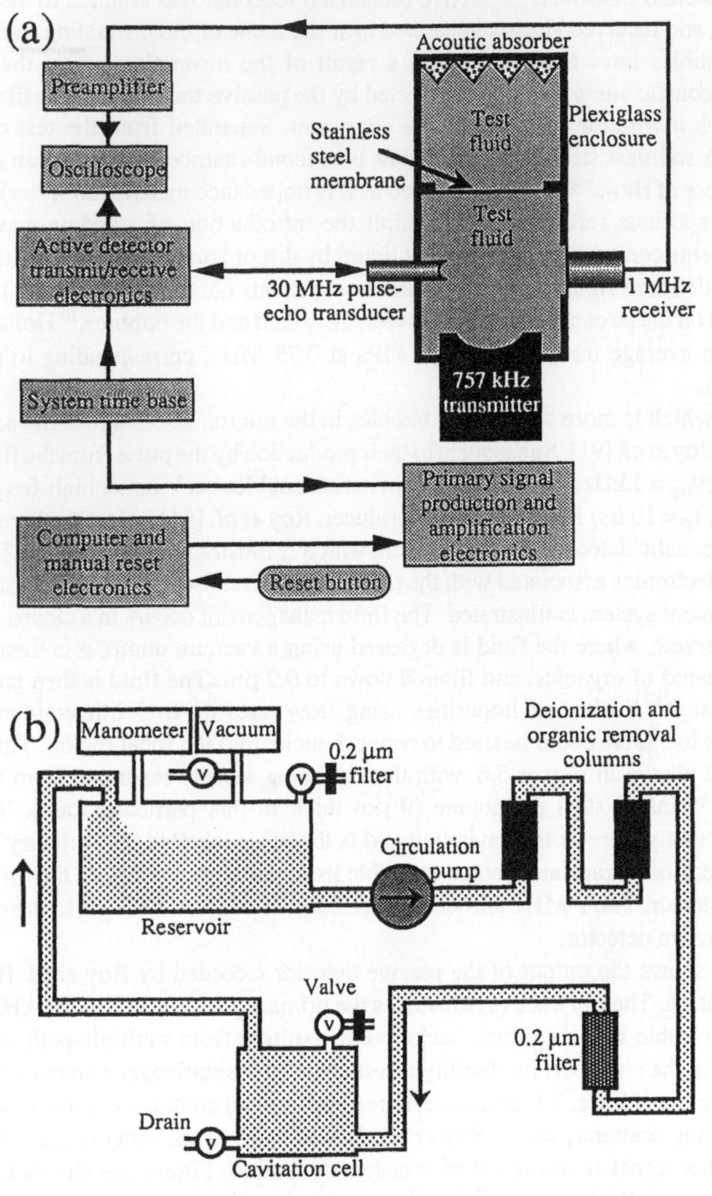

Figure 5.7 A system employing both active and passive acoustic detection. (a) The cavitation cell and associated electronics. (b) The closed-flow circulation system for cleansing and degassing the sample liquid (after Roy *et al.* [91]).

concentrations) were compared, they were found to be very close, suggesting that the acoustic pressure threshold to cause one bubble to go transient in this system is the same as that required to make a cloud of bubbles go transient.

Holland *et al.* [92] used the active detection system alone to investigate *in vitro* transient cavitation from short-pulse diagnostic ultrasound, both imaging M-mode and Doppler, under

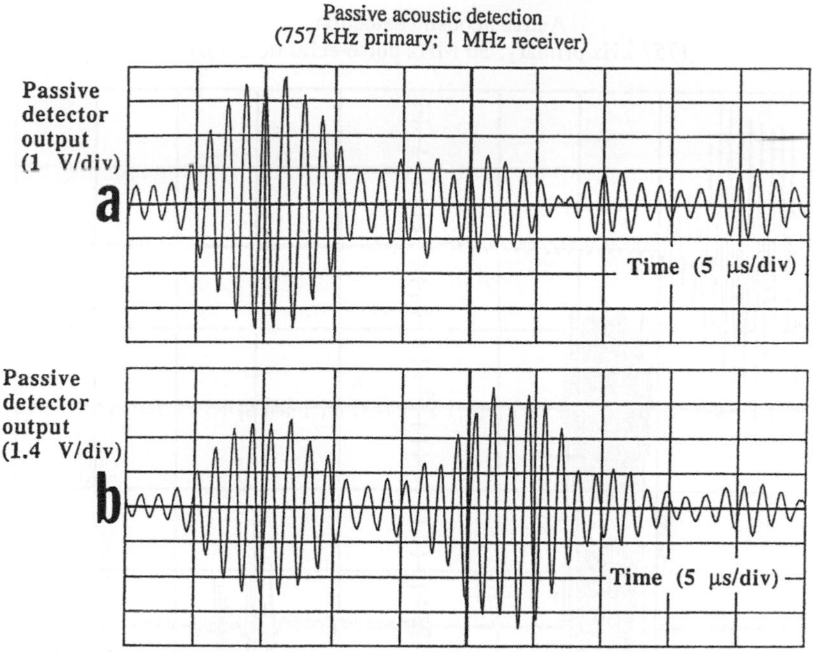

Figure 5.8 The output of the passive detector with and without cavitation. (a) The primary pulse from the 757-kHz transducer, followed by a stable low-amplitude background resulting from multiple-path scattering and reverberation in the chamber (scale: 1 V/div). (b) The scattered background contains a perturbation indicative of a time-varying scatterer (scale: 1.4 V/div). Note the difference in vertical scales between the two traces (after Roy *et al.* [91]).

conditions comparable with the clinical situation, with ν = 2.5 or 5 MHz. Cavitation was detected in water seeded with 0.125-μm mean radius hydrophobic polystyrene spheres at 2.5 MHz, with a threshold peak negative pressure of 1.1 MPa, in both M-mode and Doppler insonations. No cavitation was detected at 5 MHz, even at peak negative pressures as high as 1.1 MPa. No cavitation was detected in water seeded with Albunex™ at either frequency.

One interesting observation that was recorded by Madanshetty *et al.* [93] concerned the effect of the sensor system on the cavitation threshold for water containing a microparticle suspension. With the active system turned off, the passive system detected a cavitation threshold at around 15 bar; the active system detected a threshold at only 7 bar. This implies that either the active system is more sensitive, or that it encourages cavitation itself, or both. The fact that the passive system also detects a reduction in threshold (to around 8 bar) when the active system is on suggests that both might be true, and that the cavitational effect of the active system might be considerable. To investigate this, Madanshetty *et al.* [93] reduced the peak negative pressure from the active transducer to only 0.5 bar, so that its influence on transient cavitation should be negligible, and found that the active detector alone at this intensity could not generate cavitation events. They found that the active detector could effect cavitation through the streaming it develops (which is promoted by its high-frequency and focused nature). This flow convects nuclei into the cavitation zone. A second mechanism by which the active detector can affect cavitation is through the accelerations it imparts to particles. The combination of these accelerations with the density contrast generate kinetic buoyancy forces. These cause tiny gas

Active acoustic detection
(757 kHz primary; 30 MHz pulse-echo detector)

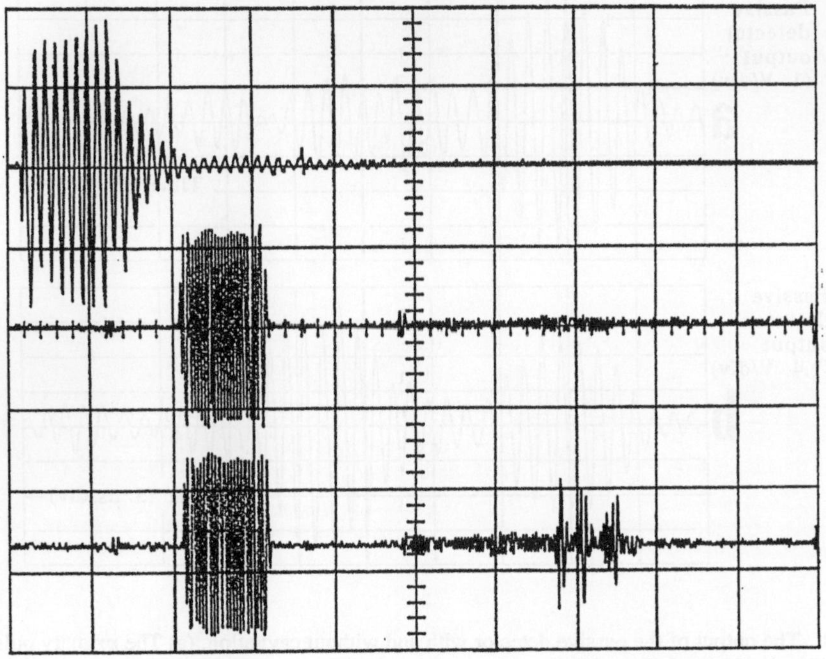

Time Base: 10 μsec/div
Top Trace: 50 V/div
Middle Trace: 100 mV/div
Bottom Trace: 100 mV/div

Figure 5.9 The top trace shows the electrical signal which drives the 757-kHz primary transducer (scale: 50 V/div). The middle trace shows the signal from the active detector system in the absence of cavitation, the main pulse representing the interrogating 30-MHz signal (scale: 100 mV/div). The bottom trace shows the signal from the active detector in the presence of the cavitation generated by the 757-kHz transducer (scale: 100 mV/div). The reflected signal from the bubbles is clearly evident. The oscilloscope digitising rate was 100 Msamples/s. (After Roy *et al.* [91].)

pockets on the surface of solid particles to aggregate into gas patches of a size suitable to act as nuclei for cavitation. Another possibility suggested by Madanshetty *et al.* [93] is that potential nuclei, which would normally be driven by the 0.75-MHz transducer, would be detained in the region where the fields are strongest by cross-streaming with the flow generated by the active transducer, so enhancing the likelihood of a cavitation event at lower insonating pressures. Citing simulation results of Church[13] [94], Madanshetty *et al.* [93] rule out the possibility of the active detector influencing cavitation through rectified diffusion.

In comparison with the passive detector, the active detector seems to be more sensitive, may affect cavitation itself, and is sensitive predominantly in its focal region, so may more readily be used to give a degree of spatial information. On the other hand, the passive detector is sensitive to a larger region of space, and remains continually alert for cavitation: for the active

[13]See Chapter 4, section 4.3.3(d)(iii).

detector to operate, its interrogating pulse must arrive at its focus at the instant when the transient bubbles are present. There are implications regarding the strain placed on the user of these systems inherent in the nature and display of the detected signals [93].

An interesting comparison can be made between the passive detection of prompt transient cavitation resulting from the sudden growth of a pre-existing nucleus in response to short-lived acoustic pulses, and the experiments of West and Howlett [95], made in 1968. The authors set up a 20.25-kHz (continuous-wave) standing-wave condition in a cylindrical transducer similar to the one described in Chapter 1, section 1.2.2(b)(i), which was filled with degassed tetrachloroethylene. At certain times, related to the phase of the continuous-wave field, they nucleated the medium using a pulsed neutron source. The shock waves emitted by these bubbles were used to count the number of cavitation events (up to 25 bubbles per second, compared to about 1 per minute when no neutron source was used).

Another detector system was developed by Crum and Fowlkes [83, 96], which demonstrated that ultrasonic pulses containing essentially one cycle at $\nu = 1$ MHz and a 1:10 duty cycle could generate cavitation with a 20-atm threshold acoustic pressure amplitude. Detection relied on photomultiplication of a chemiluminescent phenomenon, whereby free radical species generated within the collapsing bubble can react radiatively with a chemical dopant (called luminol, also tri-aminophthalic hydrazine) which is dissolved in the liquid. Indeed, the free radicals may themselves be used as an indicator [97]. Roy and Fowlkes [98] have demonstrated that the sensitivity of the chemiluminescent technique is lower than that of the passive detector.

The other disadvantage of the chemiluminescent system is that it necessitates considerable interference with the liquid, making it less suitable for use with, say, biological systems. In addition to the luminol, Fowlkes and Crum ensured that the host fluid was saturated with argon. The reason for this is that, assuming[14] that free radicals are generated in response to the high temperature that can be produced within the collapsing bubble, then as can be seen from equation (1.f1), the higher the value of γ, the greater the temperatures that are produced for a given compression, when a bubble collapses adiabatically. Therefore, all other things being equal, the greater the likelihood of significant free radical generation. Monoatomic gases such as argon have a very high γ, equal to 5/3.

In fact, luminol is generally used simply to enhance the luminescence. Bubble collapse alone in many media can generate detectable light emission: this sonoluminescence is discussed more fully in the next section. However, to close this section a mention will be made of two studies which combine sonoluminescence and passive detection.

The first employs a focused passive detector, the results from which have been temporally correlated with sonoluminescent emission. Coleman *et al.* [99] were attempting to use a passive remote hydrophone to detect cavitation produced by an electrohydraulic lithotripter. Extracorporeal shock wave lithotripsy (ESWL) is a technique by which short pulses of high pressure are focused into the body in order to break kidney or gall stones [100]. For the shock source, the energy can be applied through microexplosions, piezoceramic or electromagnetic elements [101]. Alternatively, the electrohydraulic lithotripter uses the pressure pulses emitted by a rebounding bubble, generated by an underwater spark (in 1956, Mellen [102] produced radius/time curves for just such a process, and measured bubble wall speeds in excess of about 0.9*c*). Clearly, the efficiency and reproducibility of the process will depend on the dissolved gas content and the conductivity of the water [103].

Figure 5.10 shows the pulses generated by a Lithostar (Siemens), in water and human bicep muscle, from hydrophone measurements taken in 1991 by Finney *et al.* [104]. The initial sharp

---

[14] See section 5.2.1.

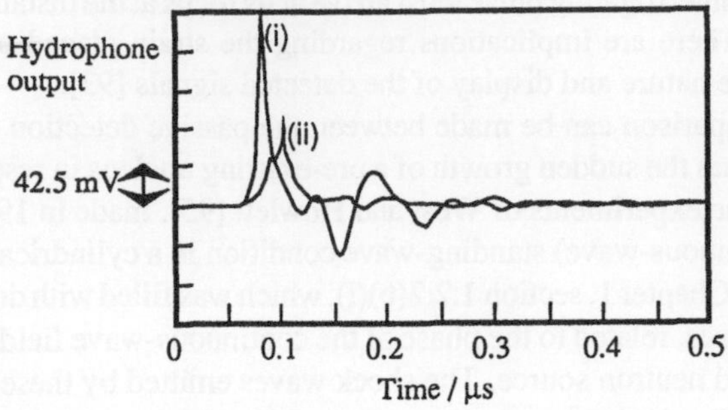

Figure 5.10 Extracorporeal lithotripter pressure waveforms obtained in (i) distilled water and (ii) human bicep muscle, obtained by an Imotec™ needle hydrophone (type 80-0.5-4.0), using 14.8-kV Lithostar discharges. (After Finney *et al.* [104]. Reprinted by permission from *Physics in Medicine and Biology*, vol. 36, pp. 1485–1493; Copyright © 1991 IOP Publishing Ltd.)

rise to high pressures, followed by a negative tail with oscillations, is typical. The peak negative pressures in lithotripsy can reach 10 MPa [105]. The mechanisms by which such pulses interact with tissue and stones is not well-understood. Acoustic cavitation in water near the focus of a lithotripter has been reported [106, 107], presumably brought about through the negative tail of the pulse. It is generally assumed in ESWL that cavitation plays some part in stone fragmentation [108] and bioeffects [106, 109–117].[15]

Williams *et al.* [118] attempted to use the resonant bubble detector described in section 5.1.1(c) to detect shock-induced cavitation bubbles downstream from the exposure site of a Dornier System lithotripter. Though they were able to detect bubbles *in vitro* in water and blood, and in blood pumped by the heart through a plastic arterio-venous shunt, they failed to detect cavitation *in vivo* in the canine abdominal aorta. Williams *et al.* attribute the absence of detectable bubbles to the high tensile strength of blood *in vivo*. The 1.65-MHz continuous-wave detector was sensitive to bubbles of radius $R_0 = 2 \pm 0.5$ μm at the detection site: bubbles would be expected to dissolve somewhat between the lithotripter exposure site and the detector [118]. The RBD effectively detects stable oscillations it excites in relatively long-lasting bubbles. Since lithotripter cavitation is characterised by transient cavitation and rapid changes in bubble size, the stable bubbles being the remnants from collapse, a more direct method for detection would employ the energetic emissions associated with rapid changes in bubble size.

Coleman *et al.* [99] attempted to do this through the introduction of a focused bowl lead zirconate titanate (PZT) piezoceramic transducer, of 100 mm diameter and 120 mm focal length in water, the focus measuring 5 mm (on axis) by 3 mm wide. This transducer had a 1-MHz resonance: the detection process would therefore correspond to its response to the white noise continuum generated by the violent bubble collapse.[16] There are three common ways to produce the shock wave in lithotripters: electromagnetic, piezoelectric and through the use of a spark gap, which was the type investigated by Coleman *et al.* [99]. The spark generates a bubble at the primary focus of the lithotripter, the focusing geometry of which is in this case similar to half of an ellipse. The shock waves radiated as the bubble rebounds successively at the primary focus converge at the secondary focus. In the laboratory lithotripter used by Coleman *et al.* (EEV Ltd.), which was designed to model the Dornier HM3 lithotripter for nonclinical

[15]See sections 5.4.1 and 5.4.2, respectively.
[16]See Chapter 4, section 4.4.7(a).

experimentation, the focus is approximately cylindrical, 12 mm long and 7.5 radius. It is aligned with the 1.38-cm semi-major axis of the half-ellipsoidal focusing bowl. Since the bowl is half of an ellipse, the 7.7-cm semi-minor axis is the internal radius of the aperture. The initial rebound gives the strongest shock, the second shock following some 3 ms later and reduced in amplitude by about 60%.

Figure 5.11(a) shows the signal detected by the passive hydrophone in response to the primary, secondary and tertiary (labelled 1°, 2° and 3° respectively). That there is emission resulting from violent cavitation is indicated by the form of light emission seen with a similar experimental system (Figure 5.11(b)): sonoluminescence is detected, indicating violent cavitation associated with the first and second shocks; there is no cavitation sufficient to generate

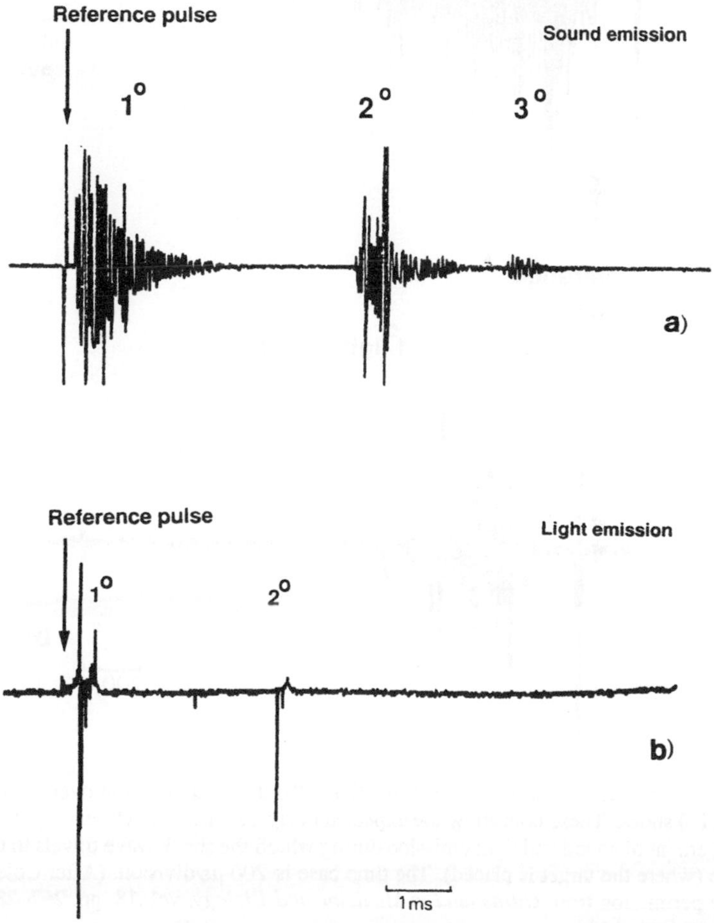

Figure 5.11 Signal traces representing (a) the 1-MHz acoustic signal detected by the passive hydrophone and (b) the sonoluminescence, emitted from the focal region in response to the primary, secondary and tertiary shocks (labelled 1°, 2° and 3° respectively). The acoustic and luminescence signals were generated by single discharges from two different (nominally identical) 20-kV electrohydraulic shock sources propagating in a water tank. The time delay between the emission of the primary (1°) and secondary (2°) shocks by each of the two sources is very sensitive to the discharge strength, and the difference in the time interval between the primary and secondary shocks in (a) and (b) is within the expected variability (±900 μs). (After Coleman *et al.* [99]. Reprinted by permission from *Ultrasound in Medicine and Biology*, vol. 18, pp. 267–281; Copyright © 1992 Pergamon Press Ltd.)

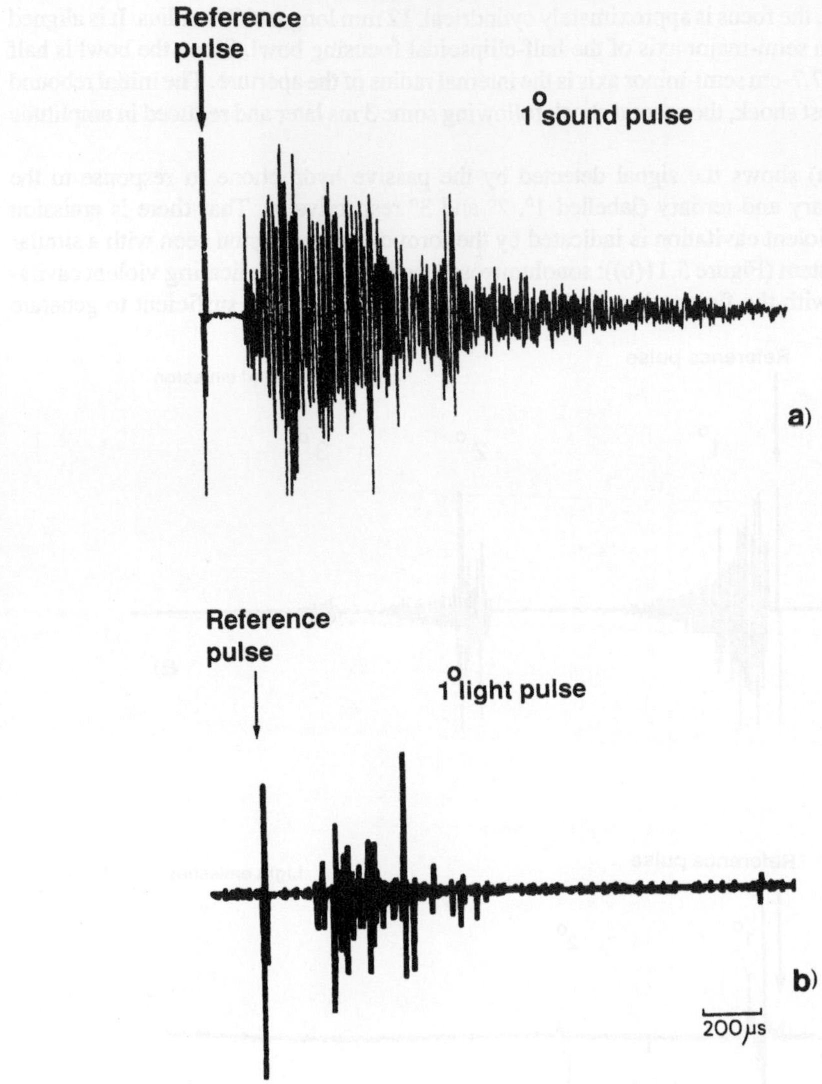

Figure 5.12 Signal traces of (a) the 1-MHz sound and (b) the light emission over a 2-ms period due to the primary (1 °) shock. These both show the expected delay between the reference (electrical) pulse and the commencement of sound and light emission during which the shock wave travels to the second focus of the ellipse (where the target is placed). The time base is 200 μs/division. (After Coleman *et al.* [99]. Reprinted by permission from *Ultrasound in Medicine and Biology*, vol. 18, pp. 267–281; Copyright © 1992 Pergamon Press Ltd.)

sonoluminescence associated with the third shock. Details of these emissions are shown for the first (Figure 5.12) and second (Figure 5.13) pulses. Though in general there was no correlation in the structure of the first pulse, there tended to be repetitive structure in both the acoustic and luminescence signals for the second pulse. The subdivision of the detected signals into two strong emission groups was always observed in the acoustic signal. Though common in the sonoluminescence signal, the two-peak structure was less reproducible: one trace in three

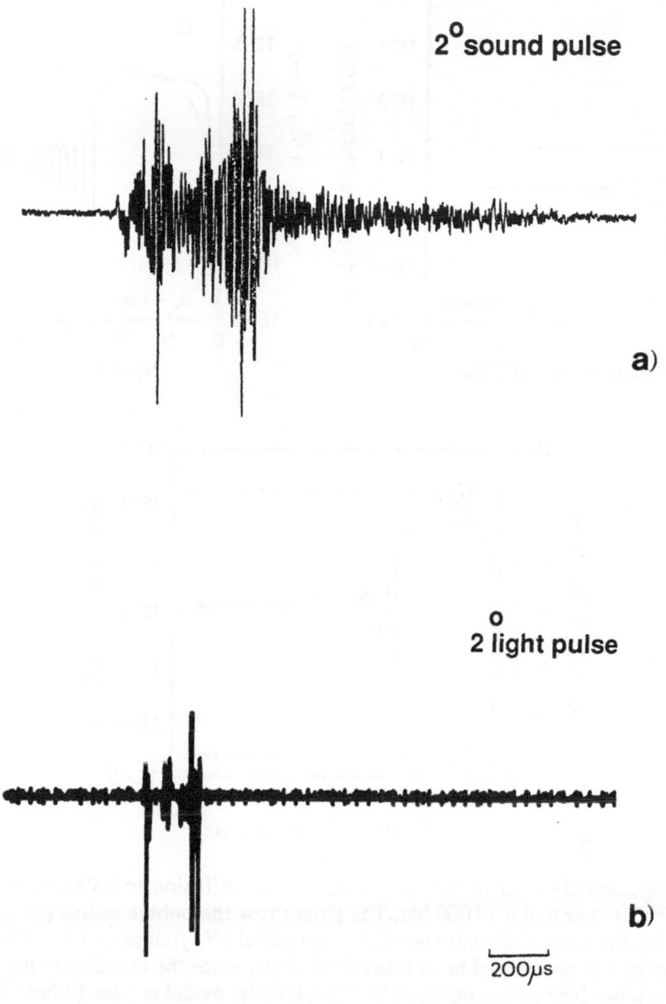

Figure 5.13 Signal traces of (a) the 1-MHz sound and (b) the light emission over a 2-ms period due to the secondary (2°) shock. Both the sound and light emission have a similar structure showing two reproducible peaks separated by times of the order of 100 μs. The time base is 200 μs/division. (After Coleman *et al.* [99]. Reprinted by permission from *Ultrasound in Medicine and Biology*, vol. 18, pp. 267–281; Copyright © 1992 Pergamon Press Ltd.)

showed a three-peak structure, and occasionally only one peak was detected. The two-peak structure can be explained using numerical models developed by Church [119] and based on the Gilmore equation.

Figure 5.14 shows the predicted response of a 3 μm bubble to an idealised lithotripter pulse of amplitude (a) 100, (b) 500 and (c) 1000 bar. From the plots of the bubble radius (normalised to the initial radius), it can be seen that in each case the bubble undergoes an initial compression in response to the positive pressure spike. It then grows to a maximum size, where it remains for a relatively long time, before rapidly collapsing. In the model, it then oscillates, though clearly in some circumstances fragmentation is a possibility. The figure also indicates the gas

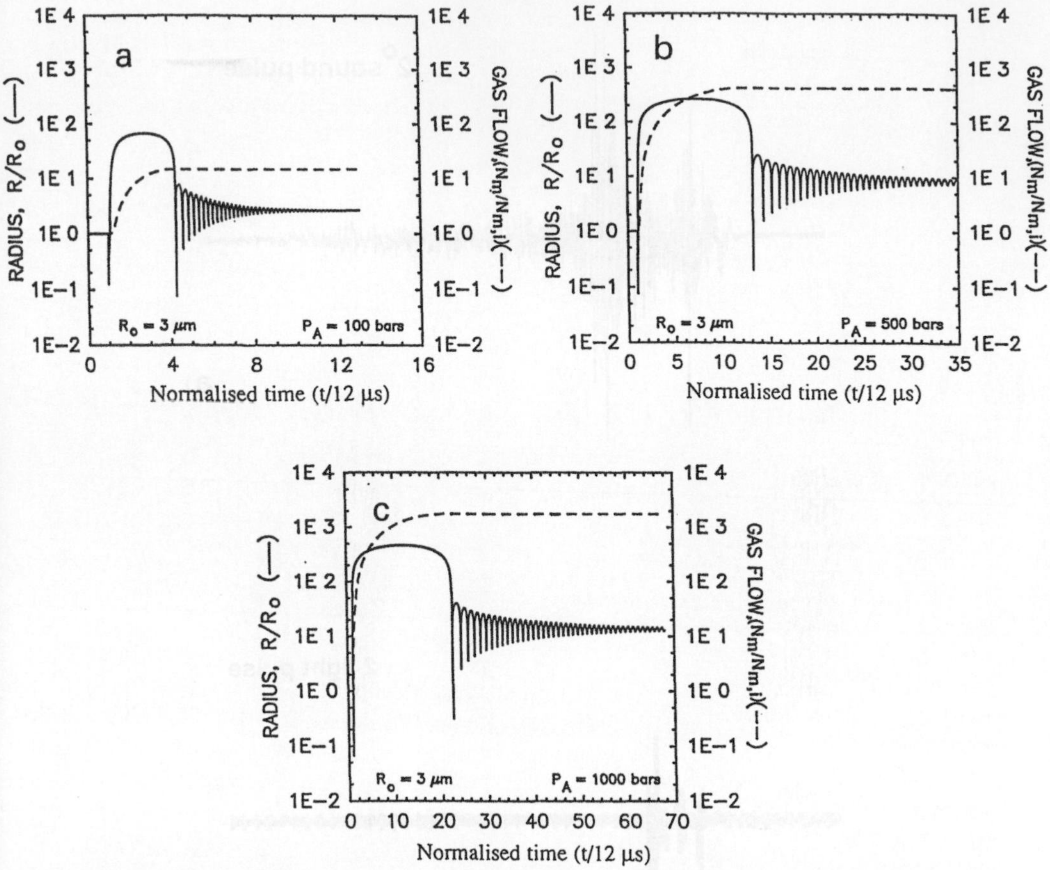

Figure 5.14 Responses of a 3-μm bubble, calculated with gas diffusion, to ESWL shocks with amplitudes of (a) 100 bar, (b) 500 bar and (c) 1000 bar. The plots show the bubble radius (—) and the gas flow (- - -), as given by the ratio of the current (N_m) to the initial ($N_{m,i}$) number of moles of gas within the bubble. The time axis is normalised to an interval of 12 μs, since the duration of the negative pressure component of the lithotripsy pulse employed by Church in the model is 6 μs. (After Church [119].)

content of the bubble, normalised to the initial amount of gas. It demonstrates the flux of dissolved gas from the liquid into the bubble, which is most rapid during the initial growth phase. As a result, in each case the final bubble radius is larger than the initial radius. Coleman *et al.* used this model to show that the duration of the extended growth phase in the bubble response was dependent on the initial bubble size, and that as a result of gas exsolution the bubbles at the end of the first shock all tended to be of a similar final size. Therefore the signals resulting from the secondary shock should display an emission resulting from the first compression, and then a 'quiet time' where no cavitation emissions were detected, because following rapid growth, all the bubbles would simultaneously be undergoing the 'pause' at maximum size before undergoing a violent collapse. The bubbles in the cavitation field would be expected to pause together since the nuclei from which they grow is a standardised population, produced by transient cavitation in response to the primary shock wave. The signals resulting from cavitation caused by the first pulse itself would be expected to show no such structure since the nuclei available for that pulse to act upon are not standardised in size, and so would be collapsing at different times (Figure 5.14).

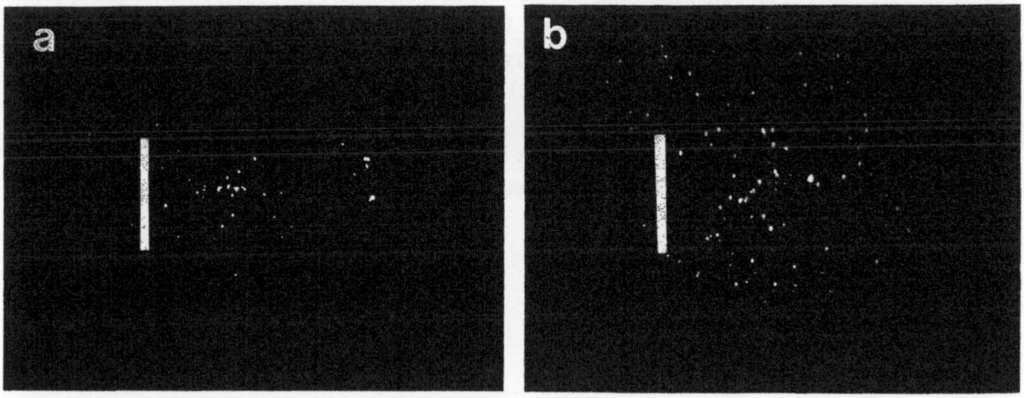

Figure 5.15 Image-intensified pictures of the luminescence from the laboratory lithotripter, set to 20 kV, (a) for a single shock and (b) for the total light emission from 10 shocks at a nominal pulse repetition frequency of 1 Hz. The white bar on the left represents the 15.4-cm internal aperture diameter of the ellipse, the right edge of this bar representing the front of the focusing bowl. The level of photon and ion-spot noise in the image intensifier can be estimated from the light-spots that occur to the left of the white bar. (Photograph: TG Leighton.)

Figure 5.15 shows image-intensified pictures of the sonoluminescence generated in a water-bath by this laboratory lithotripter at the 20-kV setting. Figure 5.15(a) shows the light emission resulting from a single shock, and Figure 5.15(b) the total emission from 10 shocks set with a nominal pulse repetition frequency of 1 Hz. In both cases the sonoluminescence tends to originate from the focus, though there is significant emission from other regions (this may be due to the translational motion of bubbles away from the focus through streaming).

It is interesting to note that Olsson *et al.* [120] record how a skilled operator can, by listening to the secondary acoustic emissions during clinical lithotripsy, determine whether the shock has hit the stone or not, though there is no quantification of the role of cavitation in this.

It is perhaps fitting to close this section with the work of Roy *et al.* [89], who in 1985 introduced the basic fluid management system (forms of which were used subsequently in the passive and active detector studies mentioned above), and who also employed a rudimentary form of passive acoustic detector. In contrast to the above pulsed systems, they subjected the liquid to a continuous-wave spherically symmetric stationary acoustic field generated at 61.725 kHz within a spherical resonator. The acoustic pressure amplitude was automatically ramped until cavitation occurred, as indicated by the detection of sonoluminescence through photomultiplication. In addition, the operator could listen in on the liquid contained within the spherical cavitation cell via headphones connected to a microphone. Audible 'clicks' or 'pops' were taken to indicate transient cavitation. They found that the pressure threshold for audible sound emission was always less than or equal to that for sonoluminescence. They deduce that "sound and light emission indicate thresholds for two different types of phenomena associated with transient cavitation." They suggest that sound emission would relate to the definition of transient cavitation of Apfel [77], which is that R_{max}/R_0 should exceed 2.3. In this situation, the wall velocity attains supersonic speeds on collapse which, as discussed in Chapter 4, section 4.4.7(a), will lead to broadband acoustic emission, including at audio frequencies. In comparison, sonoluminescence can be associated with the generation of high temperatures within the bubble.

Thus, in 1985, Roy *et al.* concluded that "if one desires a threshold for 'violent' cavitation, then sonoluminescence is a fitting criterion since the violence of a transient collapse is linked primarily to R_{max}, which corresponds to the energy stored in the liquid," and that "light emission may serve as an ideal indicator of what Apfel [77] calls the 'threshold for transient-violent cavitation'."

However, the nature of sonoluminescence and its relation to bubble dynamics are still controversial, with recent extraordinary developments. This subject will be explored in the next section.

5.2 Sonoluminescence

5.2.1 Mechanisms

As we have seen in the previous section, there is a light emission associated with a particular form of collapse, termed *sonoluminescence*. Over the years there have been a number of postulated mechanisms for the source of this emission. Though in recent years the great majority of workers interpret their observations of sonoluminescence and associated effects in terms of sonochemical processes involving reactive species (free radicals and electronically excited molecules) which may occur as a result of high temperatures generated within the collapsing bubble, the subject is still controversial.

Quite apart from the complexity of the problem, the controversy has over the years been fuelled by some unfortunate characteristics. Some key observations have been the subject of inaccurate measurement. A notable example is found in the historical determination of the point of light emission in the phase of the sound cycle. There have been several studies that attempt to relate the phase of the luminescent flashes with the phase of the sound field (summarised by Walton and Reynolds [121]), some of which were subject to the introduction of arbitrary phase shifts by the instrumentation. To differentiate mechanisms based upon the dynamics of individual bubbles, it is better to correlate the flashes directly to the phase of the bubble volume oscillation rather than the sound field.[17] The flashes were related to the volume oscillation by Meyer and Kuttruff [122], using flash photography and photomultiplication; by Negishi [123, 124], using light scatter and photomultiplication; and most recently in the elegant experiments of Gaitan *et al.* [125, 126] described in section 5.2.3. These results conclusively show that the sonoluminescent flash is associated with the collapse (see Figure 5.25).

Some workers have, in explaining the merits of their own theories, not fully examined the merits of others: a single observation can support more than one theory, and yet it is unlikely to be able conclusively to prove one theory right and all others wrong. An observation supports a theory to the limit of sophistication to which the model is expounded: particularly in an interdisciplinary field such as this, it would be uncommon to find a team of investigators that can explore their model to the full extent of its physical and chemical implications. The mechanisms proposed in the past have in general fallen into three classes: thermal, mechano-chemical and electrical.

[17] As seen even for the linear oscillations discussed in Chapter 4, sections 4.1.2(a) and 4.1.2(c), the relative phase of the volume oscillation and the acoustic field depends on several factors, such as the ratio of the insonation frequency to the bubble resonance, and on the damping of the bubble.

(a) Thermal Mechanisms

As discussed in Chapter 4, section 4.3.1, if there is sufficient energy concentration within the collapsing bubble from the spherical convergence of the surrounding liquid to outweigh dissipative losses, the compressed gas within the bubble can attain temperatures of the order of 1000 K or more. This has lead to the suggestion of two thermal mechanisms for the origin of sonoluminescence. In 1950, Noltingk and Neppiras [127] proposed that the emission was due to thermal incandescence, the radiation being an approximately black-body emission from the compressed gas. This hypothesis was initially supported by the first sonoluminescent spectra, which were taken with low resolution and revealed a broad emission which was correlated with black-body curves to estimate an effective source temperature of several thousand kelvin [128, 129]. Higher-resolution spectra, however, revealed emission bands from molecular species in the ultra-violet. Such observations correlate well with the so-called thermochemical theories, as proposed by Griffing [130, 131], where oxidising agents such as H_2O_2, formed by the high temperatures with the compressed bubble, give rise to chemiluminescent reactions in the local liquid.

(i) The Spectrum of Sonoluminescence. Spectral results have been extensively studied [132–135]. Walton and Reynolds [121] distinguish between the spectra from pure liquids saturated with gas, and from gas-saturated salt solutions.

The first series of spectra of sonoluminescence from pure liquids containing different dissolved gases was taken with a plate spectrophotometer by Paounoff [136] in 1939. From water saturated with air, nitrogen or oxygen, he obtained a continuum from 445 to 558 nm in the light wavelength. Following the exposition of the cavitational 'hot spot' by Noltingk and Neppiras [127] in 1950, the logical step was to fit black-body emission curves to the spectral data. Gunther *et al.* [137] found the spectrum of water saturated with noble gases to be a continuum between 300 and 700 nm, and a black-body curve at 6000 K was shown to fit these data [129], if the absorption of light is assumed to contribute to the fall-off below 400 nm. Gunther *et al.* [129, 137] found that the spectrum did not depend on acoustic frequency or intensity; also its form, though not its absolute intensity, was constant for all the noble gases. Srinivasan and Holroyd [128] fitted their spectral data to a 11 000 K black-body curve (though over a different part of the spectrum from that used by Gunther *et al.*) for Ar and He in water. For O_2 and N_2 dissolved in water, they fitted a black-body curve at 8800 K.

The ultra-violet emissions were studied in 1957 by Prudhomme and Guilmart [138], who found that water samples saturated with Xe, Kr, Ar, Ne and He emitted UV in order of decreasing intensity. Further studies on the sonoluminescence from water containing dissolved noble gases, however, revealed structure in the spectrum. Work carried out by Taylor and Jarman [134], Valladas-Dubois *et al.* [139] and Sehgal *et al.* [140] showed a spectral peak at 310 nm, superimposed on a continuum extending from 240 nm to the far infra-red. These features are dependent on the insonation frequency [141]. At 459 kHz, only the single peak at 310 nm is visible; at 333 kHz this peak is more intense, and three subsidiary peaks are also identifiable. As one moves up to the lighter noble gases, the peaks do not shift, but they do become stronger. Sehgal *et al.* [135] therefore concluded that the noble gases do not take part in the emission process, but instead the emission band at 310 nm results from the transitions of excited hydroxyl radicals (quenched by NO_3^- radical scavengers [142]); transitions of the triplet $^3B_1$ state of water give rise to the UV emissions; and the cut-off at 240 nm gives the heat of reaction of $H_2O \rightarrow H + OH$. Saksena and Nyborg [143] and Sehgal *et al.* [135, 140] deduced that the broad continuum is due to free radical combination of the type $H + OH + M \rightarrow H_2O + M + \textit{photon}$, where M is a gas or water molecule. Verrall and Sehgal [141] suggest that a transient collapse will favour electronic de-excitation and stable collapse radiative recombination as the source of the luminescence. As the insonation frequency changes, so the relative number of stable and transient events change [77]. Sehgal *et al.* [135] suggest that this is the reason for the frequency dependence of the spectrum. Margulis [133] criticises some of the more detailed aspects of the arguments pertaining to the origin of the continuum used by Sehgal *et al.* [140] regarding the role of radical quenching agents.

In their authoritative 1990 review, Suslick *et al.* [144] write of the spectrum of the sonoluminescence from water under an argon atmosphere. The spectrum "showed a well defined emission peak from OH at 306 nm. The continuum in the visible region was conclusively shown to be from molecular species because different regions of the spectrum could be selectively quenched by the addition of nitric acid [142]. It was proposed that radical and atomic reactions caused by the high temperature of the cavitation event were the source of the excited state molecules." Studying the spectrum of sonoluminescence from hydrocarbon and halocarbon liquids, Suslick and Flint [145] state that the spectra "originate unambiguously from excited-state molecules created during acoustic cavitation. These high-energy species probably result from the recombination of radical and atomic species generated during the high temperatures and pressures of cavitation."

The first systematic survey of the spectra of sonoluminescence from gas-saturated salt solutions was performed by Gunther *et al.* [137, 146] who took the spectra of 2-molar aqueous solutions of the chlorides of Li, Na, Rb, Co, Mg, Ca, Sr and Ba; each solution was saturated with Kr or Xe. Superimposed on the continuum they expected from a simple noble gas solution, they found a line spectra for the metal transitions $nP \rightarrow nS$. These were absent at low concentrations. Margulis and Dmitrieva [147] show that this is the result of there being too few atoms in the gas phase because of the low concentration of salt molecules at the bubble wall. Gunther *et al.* stated that the presence of the electrolytes did not affect the continuum; the results of Taylor and Jarman [134] contradict this (for example, the OH bands were often quenched). Heim [148] obtained sodium *D*-lines, showing pressure broadening consistent with internal bubble conditions of 5×10^8 Pa and 10 000 K [149]. Sehgal *et al.* [150] tried to repeat this work for sodium and potassium; though they obtained broadened and shifted peaks, their calculations on the conditions within the bubble are inconclusive [121]. Sehgal and his group have produced spectroscopic results on the photochemical reactions producing sonoluminescence in salt solutions. For example, they took the spectrum of a bromine-saturated solution containing argon, and suggested that the spectrum is formed from a combination of $A^2\pi \rightarrow X^2\pi$ transitions of BrO, and of transitions of Br_2 [151]. Similar investigations were made for saturated aqueous solutions of NO and NO_2 [152]; for nitrate-saturated water [142]; and for aqueous solutions of $CuSO_4$ containing argon [153].

An important feature of these discussions is that from the centre of the bubble (where the compressed gas attains a high temperature) to the bulk liquid far from the bubble (which is at ambient temperature), there will be a considerable temperature gradient. Since the type and rate of chemical reaction is strongly dependent on the temperature, spectral results from a specific reaction will reflect the local conditions where that reaction took place. The importance of the spatial variation in temperature, and the resulting reaction sites, arose primarily through sonochemical investigations [154, 155].

(ii) Sonochemistry. Currently, most workers are interpreting their observations in the light of a model which supposes the thermally induced formation of free radicals and molecules in excited states, sonoluminescence resulting from the subsequent development of these species

(radiative recombination of radicals, and relaxation/reaction of excited species etc.). It should be borne in mind, however, that there may be other processes occurring which could generate highly reactive species, for example, shock-wave propagation.[18] The use of sonochemistry to indicate the degree of cavitation, and some of the implications of these reactions, are discussed later.[19]

An acoustic field may accelerate chemical reactions that would occur in the absence of sound, or initial reactions that would, to all intents and purposes, not. The latter class are termed 'sonochemical' [156]. These effects were first reported in 1927 by Richards and Loomis [157], who had observed both the acceleration of conventional reactions (such as the hydrolysis of dimethyl sulphate) and the results of redox reactions in aqueous solution, including what we now know to be effects arising from the oxidation of sulphite [158]. Other classes of reactions which have been observed include the degradation of macromolecules in solution, first reported by Brohult [159] in 1937, and the decomposition of pure organic liquids, demonstrated in 1953 by Schulz and Henglein [160] despite three decades of dogma to the contrary [158]. Sonochemistry does not arise from a direct interaction of sound with molecular species [144]. Though bulk heating through the absorption of acoustic energy may affect the rates of conventional reactions, a more dramatic effect would be expected from the high temperatures attained within a collapsing bubble. In 1964, Flynn [161] concluded that the correlation between sonochemical yields and the intensity of sonoluminescence justified the assumption that they have a common source. Reviewing the data to that date, he decided that a thermal mechanism was most likely, though it should be noted that by interpreting some of the observations in a different way Prudhomme [162] favoured an electrical discharge mechanism as the common source. Indeed, electrical discharge theories were those favoured by the first investigators [144, 158, 163–165]. Reviews of sonochemistry are indicated in the references [2, 166–170] and there is no intention of surveying the field here. A thermochemical mechanism has been adopted by the majority of sonochemists, and to illustrate the success of this interpretation, and to outline the chemical implications of the spatial temperature gradients associated with the collapsing bubble, certain specific reactions are considered below.

Nitrous oxide in aqueous solution thermally decomposes far more rapidly than water, the yield (in common with many sonolytic reactions) increasing with increasing gas pressure[20] [171]. During insonation at 300 kHz under nitrous oxide/argon mixture atmospheres, Hart and Henglein [171] found that maximal yield is obtained for a volume ratio of 85% Ar/15% N_2O. Monoatomic gases such as argon have a high ratio of specific heats, and so exhibit considerable heating when adiabatically compressed, though consideration of the temperature attained within the bubble should include the thermal conductivity [172, 173]. Nitrous oxide has a lower γ, but is reactive, and all else being equal the higher the concentration of the reactive product, the greater the expected yield. Therefore whilst the inclusion of some noble gas would in general increase the sonochemical effect by achieving higher temperatures, above a certain point an increase in the noble gas component serves to dilute the reagent. Despite the greater thermal instability of N_2O, in argon bubbles containing a few percent N_2O similar decomposition rates are observed for H_2O and N_2O. This suggests that all water and nitrous oxide molecules are converted to radicals during a very short time, postulated to be the compression phase of the bubble, when the gas temperatures are high.

---

[18] See section 5.2.1(b).
[19] See sections 5.3 and 5.4, respectively.
[20] This is in accord with the increase in collapse energy seen for small increases in static pressure – see Chapter 4, section 4.3.1(b), and also section 5.2.2(c)(i).

Suslick and Schubert [174] note that ligand dissociation is induced by the transient high temperatures in species inert to photochemical or low-energy (< 200° C) thermal processes. They deduced reaction schemes for $Mn_2(CO)_{10}$ and $Re_2(CO)_{10}$ based upon hot-spot mechanisms where transient conditions exceeding 300 bar and 3000 K have been estimated. In consideration of the sonochemistry and sonocatalysis of metal carbonyls, Suslick *et al.* [175] assume a dissociative mechanism induced by the hot-spot. Further evidence of this comes about through an observed correlation between the vapour pressure of the solvent and the logarithm of the sonochemical reaction rate, since a high vapour content will tend to reduce the temperatures attained on collapse. This is because, during collapse, vapour will tend to recondense rather than be adiabatically heated, in contrast to the permanent gas phase.[21] In addition, a high vapour content will tend to encourage thermal transport during collapse, so reducing the temperature of the hot spot [144, 175]. The vapour pressure is affected by the ambient temperature.[22] Similar conclusions were reached when examining the vapour-pressure dependence of the sonochemistry on nonaqueous organic liquids by Suslick *et al.* [176]. They used radical trapping and decomposition as markers, and supported the hot-spot mechanism. Over a limited range of vapour pressure, they find evidence to support the relation between vapour and static pressures and the total free energy of cavity formation suggested by Sehgal *et al.* [177], but not over an extended range. Other models, which attempt to correlate sonoluminescent intensities with σ^2/p_v [178] and 'free molecular interaction energy' [179] are not well supported by the chemical dosimetry technique. Following the studies of Niemczewski [180], which were unable to demonstrate a correlation between broadband acoustic emission and physical or chemical properties, Suslick *et al.* [176] concluded that white noise[23] is not a good measure of the chemical effects of cavitation.

Evidence of thermal influence has been afforded through hydrogen/deuterium isotope exchange in the D_2–H_2O system in a 300-kHz acoustic field [181]. Maximal yields of H_2 and HD are found at 35% D_2 (H_2) and 60% D_2 (HD). The decrease in yields at higher D_2 concentrations is attributed to the reduction in maximum temperature attained by the collapsing cavity. This study was extended to study the formation of H_2 and D_2 under atmospheres of HD and HD–Ar mixtures [182]. The observed maximum in the yield as the concentration of atmospheric HD varied was indicative of a high temperature reaction. Exchange studies involving isotopes of nitrogen [183] and of oxygen [155] in aqueous solution have been interpreted from the assumption of thermally induced sonochemistry, where free radicals act as intermediates, and where specific chemical sequences are attributed solely to regions of different temperature during the collapse, namely the central gaseous hot spot and the cooler interface. Some radicals escape to the bulk solution.

Examining the scavenging of OH radicals produced through the sonolysis of water, Henglein and Kormann [184] commented on the relative state of ignorance concerning sonolysis radicals, as compared with radicals produced by ionising radiation. In the latter case, the kinetic effects resulting from the inhomogeneous initial distribution of radicals created by the absorption of ionising radiation is described by the so-called 'spur model'. In the case of sonolysis, the initial distribution of radicals is governed by the bubble population, and reaction may occur with solutes or with other radicals in different phases (e.g. gaseous bubble interior, the surrounding liquid etc.), so that the conditions are more complex than with ionising radiation. From the yield of hydrogen peroxide (H_2O_2) they concur with Sehgal *et al.* [140, 185] that in argon-saturated

[21] See Chapter 2, section 2.1.3(b)(iii).
[22] See section 5.2.2(b)(iii).
[23] See Chapter 4, section 4.4.7(a).

water sonoluminescence is caused by hydroxide (OH) radicals. These radicals migrate[24] into the liquid after their formation in the gas, and hydrogen peroxide is formed in the interfacial area, rather than in the gas phase. As a result, the peroxide is accessible to ions in solution. Henglein and Kormann [184] discuss the reaction kinetic relevant to the production of hydroxide within argon bubbles (the presence of, for example, oxygen within the bubbles would alter the scheme, leading to the production of HO_2 radicals, O atoms and ozone). Hart and Henglein [187] insonated aqueous iodide and formate solutions at 300 kHz, under atmospheres of argon, oxygen, or a mixture. Sonochemistry of iodide produced iodine and H_2O_2, and in the absence of oxygen generated hydrogen. A 70% Ar/30% O_2 mixture produced maximal yields. Insonation of formate solution produced CO_2 and H_2O_2 in the presence of O_2, and in the absence of O_2 produced those products and additionally hydrogen and oxalate. Stable ozone was not produced, and water previously containing ozone was found to lose it rapidly. They proposed a chemical mechanism based on the formation of O and OH radicals within the gas bubbles. If no oxygen is present these form the H_2 and H_2O_2 or react with nonvolatile substrates (such as iodide) at the liquid interface. If oxygen is available, uncharged HO_2, OH and O species are formed. The HO_2 radicals produce hydrogen peroxide, whilst the oxygen atoms react with iodide or formate ions.

Such reactions are concordant with a thermochemical mechanism. Hart and Henglein [188] compared gas reactions to those found in flames, and noted that the yields attained in this phase are higher than for liquid-phase reactions (such as the formation of hydrogen peroxide and the oxidation of solutes). They insonated water at 300 kHz under an atmosphere containing oxygen and ozone (O_3), and found an extremely rapid decomposition of O_3 which increased with the concentration of ozone in the liquid. Ozone decomposition was studied in oxygen–argon mixtures, and whilst argon acted merely as a dilutant in the oxygen–ozone system, the accompanying formation of H_2O_2 was maximal at 80% volume argon. These results are explained in terms of a thermochemical mechanism, whereby because of the thermal instability of ozone, its complete decomposition occurs within the bubble regardless of the composition of the bubble contents; in contrast, the efficiency of peroxide formation in ozone–argon bubbles increases with increasing collapse temperatures. They conclude that the decomposition of ozone is so rapid when oxygenated water is insonated that it is unlikely to be formed in significant amounts, though it may occur as a short-lived intermediate. In the 300-kHz sonolysis of carbon dioxide, nitrous oxide and methane, Henglein [189] observed the yield of products measured as a function of the gas atmosphere, and concluded that the gas mixture within the bubbles where chemistry occurs is not in Henry's Law equilibrium with the aqueous gas solution. The observations are concordant with thermally induced decomposition, free radicals often being the intermediate of the chemical reactions. The reaction mechanism is one of multi-radical attacks (alternatives are considered, but have never been observed in free radical chemistry). The postulated free radical reactions are typical of flames and so differ from photolithic and radiolithic processes in liquids. Radical concentrations are so high that the rate of radical–radical reaction is comparable with that of radical–molecule reaction.

The geometry of the sonochemical reaction can dramatically influence the reaction. As a result of the high surface-area to volume ratio in the small compressed bubbles, the sonochemical cavitation hydrogen–oxygen combustion reaction within bubbles differs markedly from that which occurs within flames [190]. As observed in a 300-kHz sound field, the chain length was only about ten: intermediate radicals rapidly reach the cooler interface where they no longer

[24]The diffusion of products from the inside of the cavity into the liquid would be aided by the generation of a shock wave at the bubble wall [150].

propagate as chains. A distinction is made between the reactions that can occur at the high temperatures of the compressed gas, and those which can occur at the cooler regions (the bubble wall). The yields of radicals which escape into the bulk solution, and also the yield of H_2O_2 which is initially formed by the radical–radical reaction in the interfacial region, are found through use of chemical indicators (Fe^{2+}, formate). Hydrogen peroxide is formed both at the hot-spot, and by HO_2 radicals in solution. Hydrogen atoms do not move to a substantial extent from the gas into the solution. The number of chemically active cavitation events was found to be 10^9 bubbles per litre per minute.

Distinguishing between the gaseous hot spot, the cooler interface and the bulk solution which is at ambient temperature, Fischer *et al.* [155] comment that the "interfacial region probably is not a thin interface consisting of one or two molecular layers of solvent molecules but is more voluminous." In 1987, Henglein [158] stated that "We do not know what kind of material state is represented by this 'interfacial region'. At temperatures of many hundreds or even some thousands of degrees and high pressures in this region, does it represent a very dense gas or is it still a liquid? Is the interfacial region a medium of relatively low polarity in which hydrophobic solute molecules are readily accumulated?" Suslick *et al.* [191] have managed to determine the size and lifetime of the various reaction sites, and also their 'effective temperature'. Since the high temperatures inside the bubble are too transient to be measured by conventional techniques, an effective temperature can be determined through the use of competing unimolecular reactions whose rate dependencies have already been measured [144]. This *comparative rate thermometry* was used to investigate sonochemical reactions of a range of volatile metal carbonyls which were employed as dosimeters in alkane solvents, revealing two reaction sites, namely, a gas-phase hot-spot (effective temperature 5200 ± 650 K [192]) and a heated liquid shell adjacent to, and around, the bubble (effective temperature ~1900 K). They predict the width and the lifetime of the liquid shell to be ~200 nm and < 2 μs respectively. Suslick *et al.* [191] note that earlier attempts to determine the temperature reached during cavitation included the employment of saturated aqueous solutions of alkali metal salts [193]: since Na^+ ions have no volatility, use of the sonoluminescence resulting from the reduction of these ions to determine the temperature will not reflect the conditions within the gas phase. The hypothesis of a two-site mechanism is reinforced by Suslick *et al.* [192] who note that the sonochemical rate coefficient with metal carbonyls increases linearly with the vapour pressure, indicating the gas-phase contribution to the reaction, as outlined earlier. However, there is a nonzero intercept, which indicates a contribution to the reaction rate which is independent of the vapour pressure, and therefore indicative of reactions in the liquid-phase component.

Comparing the thermal and electrical hypotheses in a 1990 review article, Suslick *et al.* [144] summarise the available experimental data and conclude that 'most observed effects originate from thermal processes associated with a localized hot-spot created by acoustic cavitation. Sonoluminescence is definitively due to chemiluminescence from species produced thermally during cavitational collapse and is not attributable to electrical microdischarge. Homogeneous sonochemistry follows the behaviour expected for high temperature thermal reactions' and "Homogeneous sonochemistry and sonoluminescence arise from hot molecules and radicals formed in the cavitational hot-spot by bond cleavage or rearrangement, followed by atomic and radical recombination, and thermal and chemical quenching." Suslick *et al.* go on to describe how the chemistry in the rapidly heated bubble is dominated by homolytic bond cleavage, which produces high-energy atomic and radical species: radicals recombine to form small molecules, which "emit because they are formed in the excited state, or because they are heated to temperatures where excited states are thermally populated." Suslick *et al.*, noting that "the radical and atomic nature of sonochemical reactions have been deduced from studies of sonoluminescence spectra, product analyses, and detection or trapping of intermediates," review the evidence supporting this theory, which, they state, includes the accumulated knowledge and experience with bubble dynamics, the nature of the observed chemistry and the effect of vapour pressure. They judge the theory to be consistent with the occurrence of the sonoluminescent flash during the collapse phase of the bubble, and with the effect of dissolved gases. It should be noted, however, that similar statements of confidence have recently been made by other workers in support of alternative models.[25] Contemporary workers should have roughly the same accumulated observational reports on which to base their judgements: the difference lies in interpretation of, and confidence ascribed to, individual results, which will be influenced by the worker's current research. All such statements should be interpreted in the light of: (i) our limited abilities to test the full extent of our models; (ii) observations may be wrong, misleading, or open to misinterpretation; (iii) there will always be observations that have yet to be made; and (iv) that observations are made for a finite range of parameters. The final point is particularly relevant where it is feasible that more than one mechanism may contribute to an effect, the contribution from some being dominant or insignificant for a certain set of parameters.

It is therefore interesting to compare statements made in recent years about an electrical and a mechanochemical theory with those quoted above from Suslick *et al.* [144] in support of a thermal mechanism. In 1985, Margulis [195] stated that "The model proposed ... of an electric charge localized on a cavitation bubble surface and the bubble's breakdown permits us to account for the above experimental facts which are difficult to explain within the bounds of the thermal theories." The nature of the experimental facts to which Margulis refers is discussed in section 5.2.1(c).

In the next section we will examine mechanochemical mechanisms, in particular discussing the 1986 publication of Vaughan and Leeman [196] who, following investigations of the effect of dissolved gas on sonoluminescence, favour a mechanism for sonoluminescence based upon the propagation of a shock through the gas within the bubble. Vaughan and Leeman [196] indicate an observation at odds with the predictions of a simple model of thermal collapse, which will now be outlined to conclude this discussion of thermal mechanisms. Young [173], incorporating thermal diffusivity into the collapse model of Neppiras and Noltingk [197], predicted the maximum temperatures attained by bubbles containing purely helium or xenon which collapse from a radius of 1 μm to 0.333 μm in a particular acoustic field ($\nu = 20$ kHz, $P_A = 6$ bar in his experiments). Vaughan and Leeman [196] used this formulation to calculate the following sequence of maximum temperatures which they predict will be attained by bubbles containing only the indicated gases: 350 K (hydrogen), 815 K (helium), 800 K (nitrogen), 1420 K (neon), 1600 K (argon), 1890 K (krypton) and 2000 K (xenon). Noting that Young observed only weak luminescence for hydrogen, they assume 350 K to be the threshold temperature for sonoluminescence,[26] and recast Young's formulation to give the acoustic pressure amplitude required to generate that temperature. With this model they predict pressure amplitude thresholds for sonoluminescence in the order H_2>He>N_2>Ne>Ar>Kr>Xe. They state

[25] As a historical note, in 1974 Degrois and Baldo [186], introducing an electrical mechanism that has since been shown by Sehgal and Verral [194] to be "at variance with the facts" [121], stated: "Chemical and sonoluminescent effects occur in gaseous (pseudo) cavitation but never in vaporous cavitation ... In order to understand these effects and some experimental results connected with them, the hypotheses put forward so far, i.e. thermal (adiabatic heating of the gas); thermochemical (decomposition and recombination of ions); mechano-thermal (shock waves) etc., have proved to be inadequate ... a thorough analysis of all the theoretical and experimental results known has enabled us to conclude that the various experimental facts observed: spectrum; ozone formation; duration of the flashes etc., may only be interpreted using a new electrical model."

[26] This is considerably lower than the threshold temperatures assumed by the workers discussed in Chapter 4, section 4.3.1(c)(iii) to indicate a collapse energetic enough to be termed 'transient cavitation'.

that the specific value of the threshold temperature for sonoluminescence affects only the absolute value of the threshold pressure, and not the ordering. However, that is still assuming that the different gases all have the same threshold temperature. Vaughan and Leeman [196], comparing the measured threshold sequence for a liquid saturated with argon, nitrogen and oxygen with the sequence predicted by Young's calculation, found disagreement. They therefore propose an alternative mechanism for sonoluminescence, which will be discussed below. It should be noted, however, that considering the success of the thermochemical interpretation outlined earlier, this evidence on its own is not strong enough to warrant discarding it. The model employed by Young is greatly simplified: the formulation is valid up to wall speeds of one-fifth of the velocity of sound in the liquid. The temperature gradient is assumed to occur over only about one mean free path in the gas adjacent to the bubble wall. Young neglected thermochemistry and assumed black-body emission. Similarly the chemical aspects of the thermochemical mechanism are ignored in the assumption that the attaining of 350 K, or any specific temperature which is the same for all gases, is the only criterion for generating sonoluminescence. Several key facts, such as the markedly different thermal instabilities and reactivities which different gases display, need to be incorporated. In addition there is the effect of the liquid vapour, which can react or alter the thermal properties of the collapse, as outlined above. It should also be noted that the threshold for cavitation is not simply a collapse phenomenon, but is also an inception phenomenon, the relative importance of these being dependent on the range of initial bubble nuclei present,[27] if the bubble nucleus is assumed to be a free-floating sphere. If, however, inception is dependent on skin- or crevice-stabilised nuclei,[28] and on the previous cavitation history of the sample, then the threshold can be complicated by many factors, including solubility of the gas (it should be noted that Leeman and Vaughan refer to gas-saturated solutions). Finally, unless great efforts are made to the contrary, sonoluminescence is a population phenomenon, and full discussion of a model should extend beyond single bubbles if populations are indeed involved. A discussion of the remarkable findings through the observation of single-bubble sonoluminescence is reserved for the final comment in this exploration of proposed mechanisms for sonoluminescence.

(b) Mechanochemical Mechanisms

Taking the analogy of the light emitted when many crystals are crushed, in 1936 Chambers [198] proposed the so-called 'triboluminescent theory', in which sonoluminescence arose from the breakdown of the quasi-crystalline structure which liquids were once thought to have. In 1949, Weyl and Marboe [199] suggested that the luminescence resulted from the radiative recombination of ions, generated through the mechanical fracture of molecules when this quasi-crystalline structure of the liquid was destroyed at the nascent surface of the growing bubble. However, both these theories predict that the emission should correspond to the growth phase of the bubble. In 1960, Jarman [200] proposed that sonoluminescence is caused by spherical micro-shock waves propagating within the bubble. A similar mechanism has recently been supported by Vaughan and Leeman [196], who criticised the thermal model on the grounds outlined in the previous section. Referring to Bradley [201], they cite the phenomenon of light emission associated with the propagation of shock waves through gases. Bradley states that: "The passage of a shock wave through a gas is always accompanied by a burst of visible radiation." Data on the radiation from shock-heated gases relate mainly to shock-tube experiments, where the emission is found to have two sources: the radiation is characteristic either of

[27] See Chapter 4, section 4.3.1(c)(iii).
[28] See Chapter 2, section 2.1.2(b).

chemicals and reactions occurring in the gas, or of impurities which have been scraped from tube walls,[29] and then heated by thermal collisions of the gas face.

Spectral measurements from shock-tube experiments reveal lines, bands and often a continuum. Interpretation of such results can be complicated. Bradley notes how electrons may be 'thermally insulated' from molecular species and hence possess a different temperature from that of the bulk gas, so that assessing the physical conditions pertinent to the emission is not simple [201]. The time-dependence of individual lines and bands can be related to chemical processes in the shock-heated gas, and the continua interpreted in terms of kinetic processes that occur in the gas. In general, the front of the shock is marked by discrete line emission and a rise in electrical conductivity. This decays, and after a short induction period is followed by a prolonged region displaying all types of emission, corresponding to equilibrium in the heated gas. The initial burst of light is usually attributed to the excitation in impurities with lower ionisation potentials.

Gas shocks may occur within the collapsing bubble, and Vaughan and Leeman [196] support the idea that this might be the cause of sonoluminescence (though they take as their criterion for gas shock formation that the bubble wall speed must exceed the speed of sound in the gas, though in fact shocks may form through lower wall-speeds). Vaughan and Leeman [196] note that their measured relative pressure thresholds for sonoluminescence in argon-, nitrogen- and oxygen-saturated water are in qualitative, but not quantitative, agreement with the ordering of gaseous sound speed. Departing from the stable/transient description of cavitation, which as discussed in Chapter 4 can cause confusion as a result of the multiplicity of energetic, temporal and shape stability descriptions that can become involved, Vaughan and Leeman propose a three-tier classification of acoustic cavitation. This system is based upon the maximum speed attained by the bubble wall ($\dot{R}_{max}$), and starts with the assumption that only nonlinear oscillations will be involved in the classification. The limitations of this definition when extended to cover the range of acoustic interactions of sound with bubbles is discussed in Chapter 4, section 4.4.8. Normally the speed of sound in the liquid, c, exceeds that in the gas, c_g. 'Subsonic cavitation' occurs when the oscillations are nonlinear, but the wall speed is always less than both sound speeds ($\dot{R}_{max} < c_g < c$). Since the oscillations are nonlinear, harmonics and possibly subharmonics may occur. If the maximum wall speed exceeds the sound speed in the gas but not in the liquid ($c_g < \dot{R}_{max} < c$), the regime is termed 'gas-phase cavitation'. *If* sonoluminescence were to occur solely as a result of gas shocks, and *if* shocks generated by wall speeds exceeding the speed of sound in the gas generate luminescence, but those generated by lower wall speeds do not, then the transition to this regime from subsonic cavitation would correlate with the onset of sonoluminescence. 'Liquid-phase cavitation' would occur when the wall speed is able to exceed the speeds of sound in both gas and liquid ($c_g < c < \dot{R}_{max}$). Vaughan and Leeman [196] suggest that the transition to this regime would be characterised by the emission of shock waves into the liquid,[30] though this does not allow for the generation of liquid shocks at lower wall speeds. Vaughan and Leeman [204] speculate that then sonochemistry could occur within the liquid.

Though shocks may indeed contribute to luminescence within the bubble, the evidence does not support the proposition that, in the general case, sonoluminescence is dominated by the emission of light behind gaseous shocks, with negligible contribution from the heating of

[29] Johansson [202] proposed that the presence of cavities might enhance explosive initiation through the spalling of small drops off the bubble wall and into the hot compressed gas, where they would react (see section 5.4.1).

[30] It is interesting to note that Margulis and Dmitrieva [203] propose an alternative interpretation, suggesting that observed shock waves in the liquid are due to phase explosion of superheated liquid.

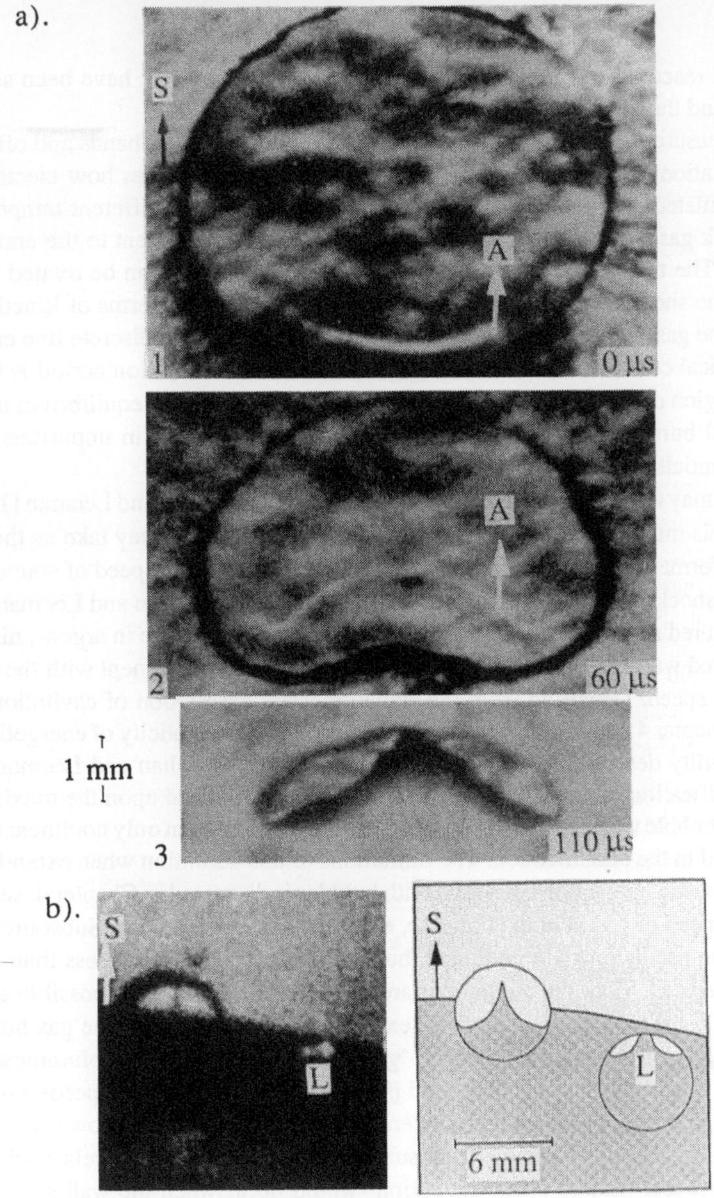

Figure 5.16 (a) Three frames from a sequence (frame exposure ≈ 0.5 µs) showing the asymmetric collapse of a two-dimensional disc-shaped (diameter 12 mm) air-filled cavity cut in gelatin, in response to the passage of a 0.26-GPa planar shock which travels up from the base of the picture. The schlieren was adjusted to reveal density variations in the air contained within the cavity. Frame 1: the incident shock (labelled 'S') is travelling up the frame. An air shock (labelled 'A') is revealed through schlieren techniques as a lighter region, moving up away from the upstream wall at the acoustic speed of air. Frame 2: taken 60 µs after frame 1, shows the air shock, having reflected from the downstream wall, after re-crossing the cavity and reflecting from the involuted upstream wall, now re-crossing the cavity. Involution of the upstream wall has produced a liquid jet which traverses the cavity at constant velocity. Frame 3: taken 110 µs after frame 1, shows the gas is trapped and compressed in two lobes as a consequence of the liquid jet impacting the downstream wall. This impact sent a shock wave into the liquid. The penetration of the downstream wall by the jet causes the formation of two linear vortices, which travel downstream in the following flow. (b) A pair of two-dimensional cavities (initially of diameter 6 mm) collapse as a result of the passage of a 1.88-GPa shock. The exposure of the frame is ≈20 ns. A schematic illustrates the geometry: whilst the cavity on the left is jetting, the collapse of the cavity on the right has progressed further (the shock encountered it first), and luminescence can be seen from the two lobes of compressed gas (after Bourne and Field [205]).

compressed gas. This can be seen, for example, through examination of photographs taken by Bourne and Field [205].

Figure 5.16 illustrates the collapse of a two-dimensional air-filled cavity in gelatine,[31] in response to an approximately planar 0.26-GPa shock wave which is incident from the base of the pictures. Figure 5.16(a) shows the collapse of the cavity, which was initially circular with a 6-mm radius: frames 2 and 3 are taken 60 μs and 110 μs respectively after the first frame. In Figure 5.16(a), schlieren photography is used to image the shock wave in air (labelled A), which moves into the cavity, and eventually bounces around inside it. The cavity collapses asymmetrically, in a manner described in section 5.4.1, the bottom wall involuting and passing through the cavity as a liquid jet, impacting the far wall and producing two lobes of compressed gas, which can be seen in frame 3. In Figure 5.16(b), the sonoluminescence resulting from such a collapse can be seen. The geometry is illustrated to the right of the photograph. A 1.88-GPa shock (labelled S) has traversed half the field of view, causing darkness in the photograph. The sample contains two cavities of initial radius 3 mm. The cavity on the right, which the shock encountered first, has collapsed into two lobes, which are luminescing (labelled L). The lower wall is traversing the cavity on the right as a liquid jet. Over the past few years that such experiments have been done by Field and co-workers, no luminescence has been observed from the known position of the initial gas shock [Field, private communication]: luminescence has only been observed from the compressed lobes and the point of impact of the liquid jet. The initial strength of the air shock is of the order of a few bars [Field, private communication], though as it bounces around the cavity it may intensify. Even so, the timescales indicate that the shock has bounced around the cavity several times before there is any luminescence, which occurs from the regions of gas compression (the lobes), and indeed the gas shock strengths required for luminescence are higher than those attained in most cavity collapse situations [Field, private communication]. It should however be noted that the increased propagation distance afforded by the bounces may strengthen the shock, and as the cavity collapses the sound speed of compressed gas will increase. Focusing may also occur within the lobe geometry.

The ignition of azide crystal, induced by bubble collapse, provides a good threshold thermometer for cavity collapse temperatures [206]. Ignition of the azide requires a hot-spot minimum temperature of ~700 K. Figure 5.17 shows a thin crystal of silver azide, mounted in a water tank on a tungsten rod. An air bubble (labelled B) and a solid tungsten particle (labelled P) are attached to the crystal. A shock wave, which is imaged through schlieren techniques, propagates to the right. The shock wave hits the bubble in frame 2 and in frame 3 initiation of fast decomposition occurs at the bubble hot-spot. There is no reaction through direct interaction of the incident shock with the bubble, nor as a result of the presence of the particle. The particle was placed on the far side of the wire since Mader [207, 208] had predicted that in this position it would be most effective at perturbing the shock and so possibly generate ignition through shock focusing. Tungsten was chosen since it thus ensured a high impedance mismatch between the particle and the water.

Though emission from gas shocks in a collapsing bubble is a feasible mechanism for the luminescence, the evidence to date is not strong enough to justify replacement of the thermochemical mechanism in most cavity collapse situations. The experimental observation relies on the sequence of thresholds obtained by Vaughan and Leeman, and they themselves comment on the marked change in luminescence that a mild contamination of the sample by air can make [204], and it is to this that they credit the discrepancy between their own results and those of Young, who took no precautions to prevent contamination. Observations of subharmonic emission at lower driving intensities than required for sonoluminescence would not be inconsistent with almost all models for cavitation,[32] and does not specifically support the mechanochemical

---

[31] This technique is further discussed in section 5.4.1.

[32] See, for example, Chapter 4, section 4.4.8.

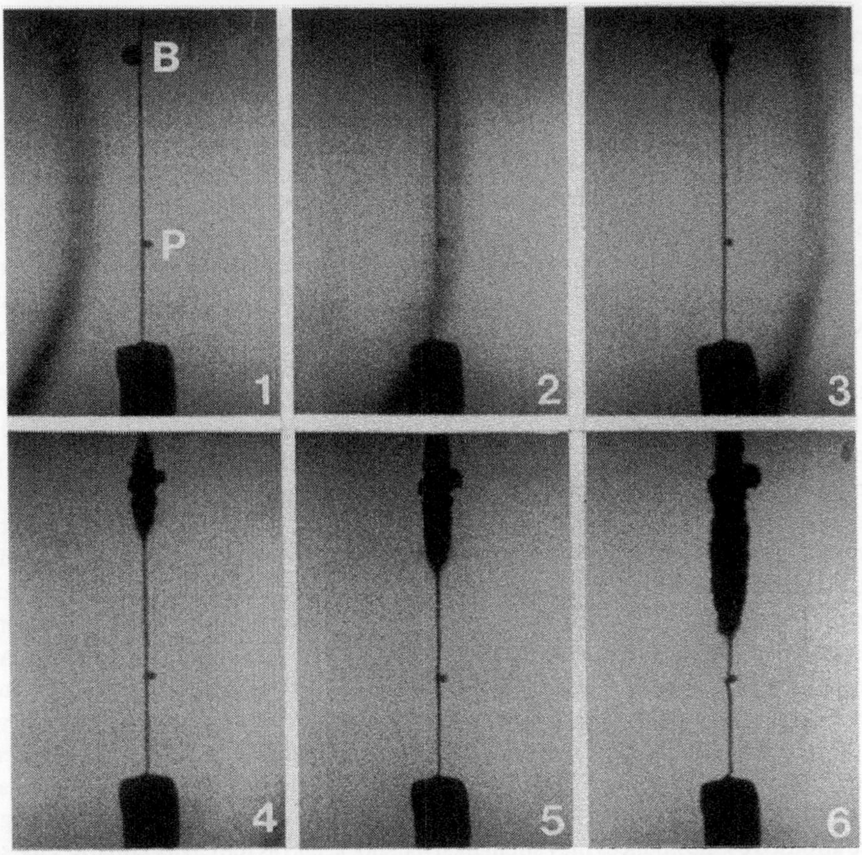

Figure 5.17 A gas bubble (B) and solid particle (P) are attached to a silver azide crystal, which is mounted on a thicker tungsten rod (visible at the base of each frame). Interaction with the incident shock wave causes ignition only through bubble collapse. Inter-frame time = 1.0 μs, horizontal extent of the final frame = 7.0 mm (after Chaudhri and Field [206]).

hypothesis. Though their observations of the relative onsets of sonoluminescence and liquid shocks by Vaughan and Leeman [209] support their scheme, there are conflicting observations. As described in section 5.1.2, Roy *et al.* [89] observed a higher threshold for sonoluminescence than for simple audio microphone detection of cavitation, though Vaughan and Leeman [210] argued that this did not represent a comprehensive detection scheme. Giminez [211], in a simultaneous study of the sonoluminescent flashes and emitted shock waves, found that not every shock-inducing collapse yielded light emission. He found that pressure pulses in the very low or very high range are more frequently accompanied by light pulses than pressure pulses in the middle range.

A new technique which would allow investigation of these processes is that reported by Gaitan *et al.* [125, 126], which allows measurement of wall motion and sonoluminescence of a single bubble.[33] It is hoped that such experiments will allow a clearer understanding of the relative importance of shock waves and adiabatic heating in the generation of sonoluminescence

[33]For example, a preliminary study of Figure 5.25, taken from Gaitan *et al.* [126], which shows experimentally measured values of the bubble radius simultaneously with the sonoluminescent output, suggests that the wall speeds associated with the sonoluminescence in that experiment are significantly less than the speed of sound in either gas or liquid. In the light of such observations the validity of the rigid definitions of gas-phase and liquid-phase cavitation must be re-examined.

(though care must be taken in generalising from this particular case to the mechanism for sonoluminescence in conditions involving bubble populations, shape distortions early in the collapse, fragmentations etc.). Recent studies with this system have produced remarkable observations of the timing of the sonoluminescent flash. This point will be expanded at the end of the next section, following a discussion of the proposed electrical mechanisms, which have historically been closely associated with speculation on the timing of the sonoluminescent flash.

(c) Electrical Mechanisms

There are a number of related theories based on the light emitted through electrical microdischarge in the interior of the cavity. In 1940, Frenkel [164] proposed that such discharge occurred between regions of opposite charge created by statistical fluctuations of charge distribution on the wall of a nonspherical, lens-shaped cavity. However, this would predict luminescence during the growth phase of the bubble.

In 1939, Harvey [212] proposed the creation of fields due to the generation of balloelectric charges on the bubble surface, generating electric fields in the surrounding liquid. The proposed field intensity at the bubble wall would increase in intensity as the bubble size decreased, until discharge occurred between the gas and the charged interface. Degrois and Baldo [186], speculate that ions, adsorbed from the liquid onto the inner surface of the bubble, are neutralised by the adsorption of gas molecules from the bubble interior, these molecules being "deformed by the induced polarization brought on by the free energy of the liquid." Sehgal and Verral [194] concluded that experimental observations are incompatible with such mechanisms. A recent interesting observation was made by Watmough *et al.* [213], who demonstrated using a chemical dye-paper that acoustically excited gas bubbles may carry a negative charge.

In 1987, Golubnichii *et al.* [214], though favouring in concept an electrical mechanism, comment on the dearth of reliable experimental evidence which might support it. They estimate the likelihood of being able to detect radiofrequency emission resulting from such breakdown if it occurred, and conclude that it might indeed be detectable in favourable conditions. They note, however, that if considerable thermal effects were present, electrically charged regions of the bubble might be neutralised by ionised gas, and suggest that a search for radiofrequency emissions should be undertaken in regimes of ultrasonic cavitation where the compression of the bubble contents is not attended by significant thermal effects.

Margulis [215], disagreeing with the previous electrical theories of Frenkel [164], Nathanson [216], Harvey [212] and Degrois and Baldo [186], proposed a new electrical theory [217] following a discussion of cavitation in low-frequency fields [195], where he distinguishes two classes of bubbles: 'small spherical' bubbles, which are generated from 'large deformed' bubbles. The latter are subject to significant departures from sphericity: those small in comparison with the bubble radius (which Margulis describes as "cracks, cavities and projections") impart to the bubble the appearance of a "spherical hedgehog" when imaged visually or through long-time exposure (0.1 s) photography. Following observations of sonochemistry and sonoluminescence from bubbles in low-frequency (7–200 Hz) acoustic fields [218–221], Margulis states that "the splitting of bubbles in the formation of small spherical bubbles leads to luminescence and chemical reactions" [195]. He came to this conclusion following the observation of only large bubble asphericity and repetitive pulsation of large and small bubbles at the acoustic frequency: he states that "in no case was the effect of a real collapse (disappearance) of the bubble observed." It should be noted that it is inappropriate to equate only such collapses to the thermochemical mechanism, since sonoluminescence has been observed from a repetitively pulsating cavity,[34] equivalently labelled either repetitive transient cavitation[35] or energetic stable cavitation.[36] Margulis's failure to observe fragmentary transient cavitation, where the bubble grows from a nucleus to many times its original size and then 'disappears', does not of course prove that it was not present: for example, Figure 4.19 illustrates how in that sound field such cavitation is far more difficult to observe than the repetitive pulsation. Margulis's observation is interesting, but not conclusive. Margulis proposes a *developed cavitation threshold*, where "the number of cavitation bubbles increases suddenly and many physiochemical effects are intensified." This definition does not, however, explicitly relate to the energy or 'violence' of individual collapses, and may be more dependent on the details of cavitation inception and nucleation.

In the new electrical theory [217], Margulis considers an electrical double layer adjacent to the neck that is formed between the large deformed bubble and the small spherical bubble which is separating from it. Water molecules orientate in the surface layer so that on average roughly 1/30th have their negative poles turned towards the gas phase [222]. The rate of formation of the smaller bubble is assumed to be faster than the rate of charge diffusion away from the neck, so that a large negative charge develops on the smaller bubble. A discharge through the smaller bubble occurs after separation to equalise the charges. Though Margulis prefers this to thermal theories, Suslick *et al.* [144] present evidence that suggests it would be difficult to adopt as a general theory for sonoluminescence. Margulis [195] states that "From the viewpoint of the thermal theories of cavitation, it is difficult to interpret some experimental facts: the luminescent, non-equilibrium character of the light emission, which requires the absence of high temperatures in the bubble during irradiation; the absence of the collapse of cavitation bubbles; the sharp increase in sonoluminescent flux in a constant electrical field; the emission of sonoflashes during the phases of contraction or expansion of the sound wave, etc." This last item follows from the 1969 observations of Golubnichii *et al.* [223]; it is therefore appropriate to refer to the observations in recent years of Gaitan *et al.* [125, 126], who related the timing of the flash directly to observations of the bubble wall motion, rather than to the sound field and from there trying to infer the bubble motion (which even in the linear regime is not necessarily in phase with the sound field[37]). Gaitan *et al.* consistently and repeatably observed the sonoflash upon bubble collapse (see Figure 5.25). Discussion of the discrepancies found among workers who have measured the phase of the flash, and of the misinterpretation, is given by Walton and Reynolds [121].

Suslick *et al.* [144] note that the sonoluminescence which has been observed in nonpolar liquids and liquid metals could not be electrical in origin. These authors also outline spectral evidence which disagrees with Margulis's electrical interpretation. Given recent developments, one item of evidence proposed by Margulis in support of the electrical theory, and contested by Suslick, is the temporal shape of the sonoluminescent pulse. Since simple hydrodynamic models suggest that the dynamics of the bubble wall are symmetrical about the minimum bubble size, Margulis suggests that so too should be the luminescence if it were thermal in origin: Suslick, however, appeals to the details of the thermochemistry, where a minimum temperature is required to initiate a reaction, which can then proceed even when the temperature falls to below the threshold value. Of particular interest are measurements of the duration of the pulse. The very short rise-time of the flash (< 2 ns), and longer fall (10–15 ns) reported by Kurochkin *et*

[34] See section 5.2.3.
[35] See Chapter 4, section 4.3.2.
[36] See Chapter 4, section 4.4.8.
[37] This was shown to be so even for small-amplitude linear oscillations of isolated bubbles in Chapter 4, sections 4.1.2(a) and 4.1.2(c).

al. [224] was cited by Margulis as evidence for the electrical theory, in that it resembled the pulse from an actual microdischarge. Suslick dismisses this as coincidence, noting that a common form does not necessarily imply a common origin. Margulis [195, 225] asserts that a thermal mechanism would predict a sonoluminescent rise-time of around 100 ns, though Suslick cites other studies which suggest 2 ns is consistent with thermal interpretations in a 20-kHz field [26, 226]. The luminescence itself will take energy away from the bubble, which may reduce the duration of the conditions which can excite species.

Recently, however, measurements have been made which point to the remarkably short duration of the sonoluminescent flash, and these raise a completely new set of questions regarding our understanding of the generation of sonoluminescence. Employing the discovery of Gaitan and Crum [125] that in a well-defined window in parameter space a single bubble could be made to emit a sonoflash each acoustic cycle,[38] Barber and Putterman [228] used a sufficiently intense 10.736-kHz sound field to generate precise, clock-like regular repetitive luminescent bursts of power 0.2 mW and containing 10^6 luminescence photons emitted uniformly in all directions, and measured the burst length to be less than 2.2 ns. The conversion of energy, incident sound to emitted light, is a factor of 10^{11}, with initial conversion efficiencies of 10^{-5}. These bursts are ten times faster than the visible 3–2 hydrogen atom transition. Further investigations, using microchannel plate photomultipliers with very fast response times and a 30-kHz sound field, suggested that the flash widths may be less than 50 ps, with the 'jitter' in time between flashes being much less than 50 ps [229]. The flashes displayed no ringing. Barber *et al.* [229] speculate on the nature of the cooperative/coherent optical (or fluid) phenomena involved in the mechanism. As Crum [230] states: "As the phenomenon may be too fast for the establishment of local thermal equilibrium, we may be facing a situation where focusing acoustic stress fields are transduced directly into quantum excitations."

5.2.2 Characteristics of Sonoluminescence

(a) Generation and Detection

(i) Producing Sonoluminescence by Acoustic Cavitation. A great variety of methods have been used to cavitate water to produce sonoluminescence. These include sparking [231], the use of lasers [232] and Venturi tubes [233], impacting water jets [234, 235] and collapsing glass spheres containing lower-pressure gas under water [236]. However, the most widely used method is to pass acoustic waves through a liquid, the technique which concerns us here. These waves are generated by some sort of transducer, the methods having been reviewed by Finch [237] and Walton and Reynolds [121]. Sonoluminescence has been produced from a variety of transducers: the commercial ultrasonic 'sonicator' probe which is used for biological cell disruption [238]; the magnetostrictive transducer [239, 240]; and the piezoceramic transducer used, for example, by Alfredsson [241]. These transducers may provide a plane or focused sound beam (see Bohn [242] and Rosenberg [243] for magnetostrictive and piezoceramic focusing respectively).

Referring back to the transducer discussed in Chapter 1, section 1.2.2(b)(i), and noting the form of cavitation generated by it discussed earlier,[39] it is interesting to examine the resulting luminescence. This is shown in Figure 5.18 for continuous-wave insonation in tap water at 10 kHz, with $P_A = 2.4$ bar along the central axis. The orientation is the same as for Figure 4.24,

[38] See section 5.2.3.
[39] See Chapter 4, sections 4.3.4 and 4.4.1(b).

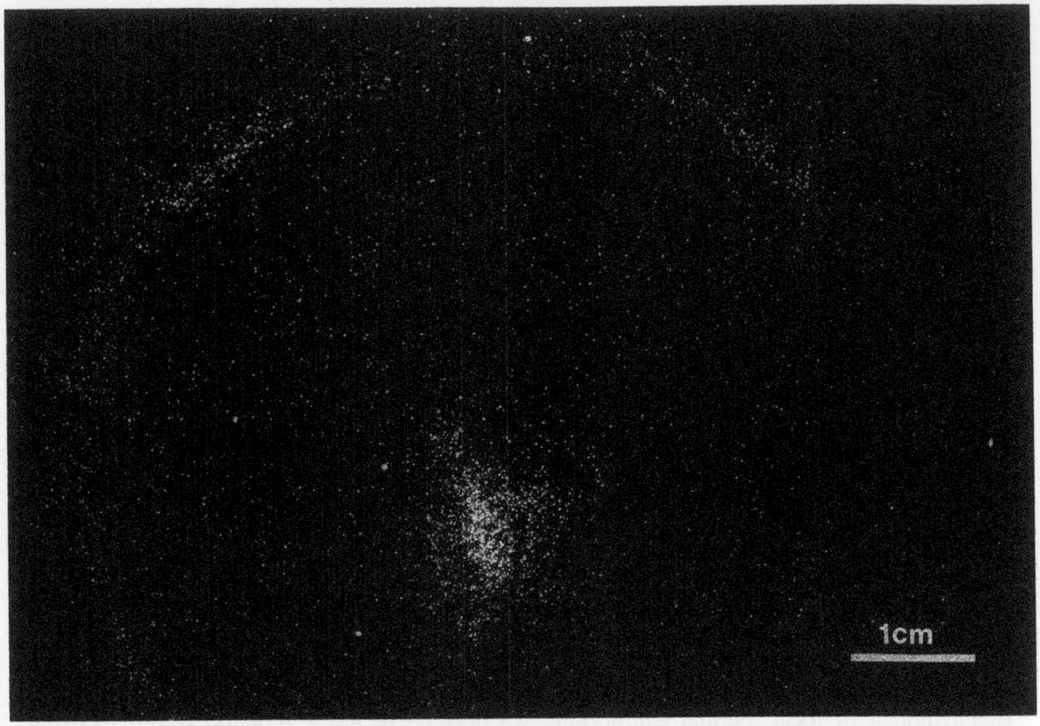

Figure 5.18 Sonoluminescence in aerated water occurring within the piezoceramic cylinder, as photographed through the image intensifier. The outer ring of light is due to the collapse of bubbles attached to the inner surface of the cylinder. The central light source is due to sonoluminescence at the focal pressure antinode. (This picture should be compared with Figure 4.24, which shows similar cavitation zones, taken in reflected illumination.) (After Leighton [294].)

the cylinder being viewed from above. Luminescence is clearly visible emanating from the focal cloud, where bubbles of smaller than resonance size have been driven by the primary Bjerknes forces. It is these bubbles that luminesce, as suggested by Figure 4.6, which illustrates that bubbles of larger than resonance tend not to attain as violent a collapse.[40] Some luminescence is also visible as an outer ring, which comes from small bubbles adhering to the inner wall of the piezoceramic. There is an element of stray light noise associated with such pictures, which will be discussed in the next section, and this contributes to the points of light seen in the body of the liquid, between the focal cloud and the cylinder wall. The outlying cavitation streamers are not luminescing significantly. Figure 5.19 shows the luminescence when the cylinder is filled, not with water, but with a mixture of 90% glycerol/10% water. There is a focal cloud in which, from the change of scale, the luminescing bubbles are more tightly packed, as one would expect because of the increased viscosity (around three orders of magnitude greater than that of water). Forces such as buoyancy, which would tend to disperse the focal cloud, will be less effective in the glycerol/water mixture. Though there appears to be evidence of luminescence from the streamers, no conclusions can be drawn as light originating from the focal cloud might be scattered off the bubbles in the streamers. However, one can say that there must be streamers

[40]See Figure 5.26.

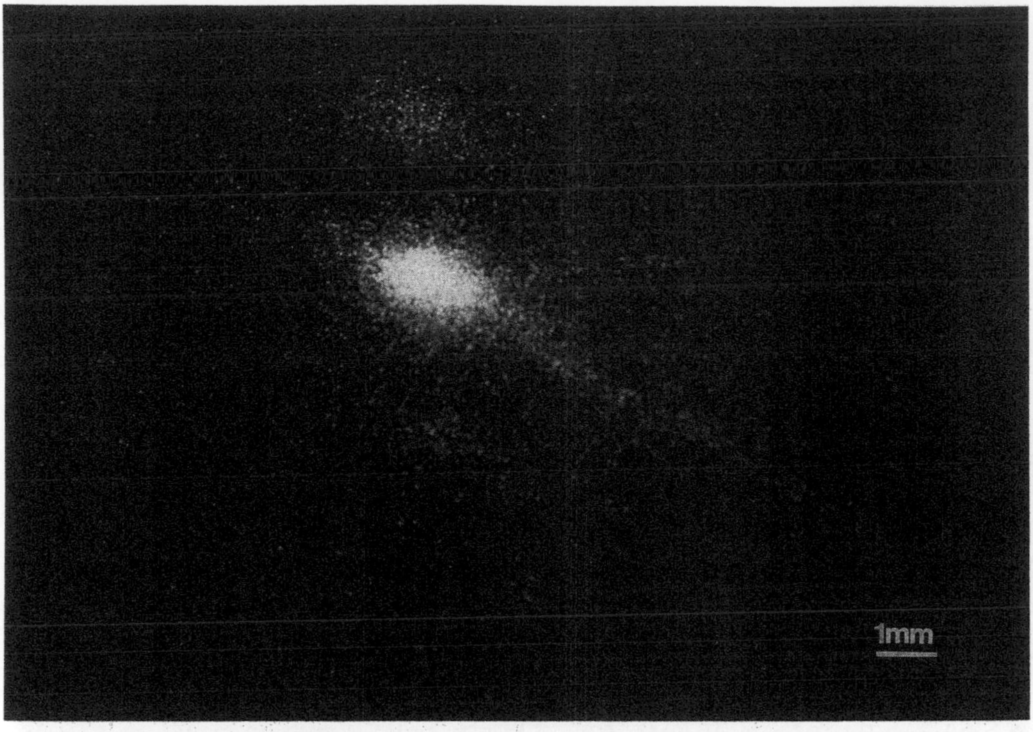

Figure 5.19 A photograph, taken through the image intensifier, of sonoluminescence in a 90% glycerine/10% water mixture, occurring within the piezoceramic cylinder. The bright sources of light arise from sonoluminescence from bubbles in the tight focal clump. This light is reflected off bubbles in streamers to the right of the central luminescence. The 'ghost' images above and below the focal clump are artefacts of the optics (explained in the text). (This picture should be compared with Figure 4.28, which shows similar cavitation zones, taken in reflected illumination.) (After Leighton [294].)

associated with this system. These conclusions are borne out in Figure 4.28, which shows a photograph of an illuminated focal cloud and associated streamer formation in the glycerol/water mix.

In contrast, Figure 5.20 shows the luminescence that results when the cylinder is filled with degassed water. The luminescence now does not occur predominantly at the focal cloud, but at random points within the liquid. The cavitation is therefore fundamentally different from that seen in aerated tap water and glycerine, where there are many free-floating nucleated bubbles, which oscillate stably within the body of the liquid for many oscillations and so are subjected to primary Bjerknes forces and driven to the focus, where the high acoustic pressures cause them to luminesce. Instead, the luminescence arises from short-lived cavitation events, nucleated from random points in the liquid. It should be noted that Figure 5.20 was taken with a longer exposure (1 s) than Figure 5.18: the luminescence from degassed water in this experiment was about 1/10th that from aerated water. In the next section, the method by which these photographs were taken is discussed.

(ii) Recording Sonoluminescence. In 1933, Marinesco and Trillat [244], investigating emulsification by ultrasonic cavitation, discovered latent images and fogging of silver halide emulsions on photographic films and plates that had been immersed in cavitated water. They

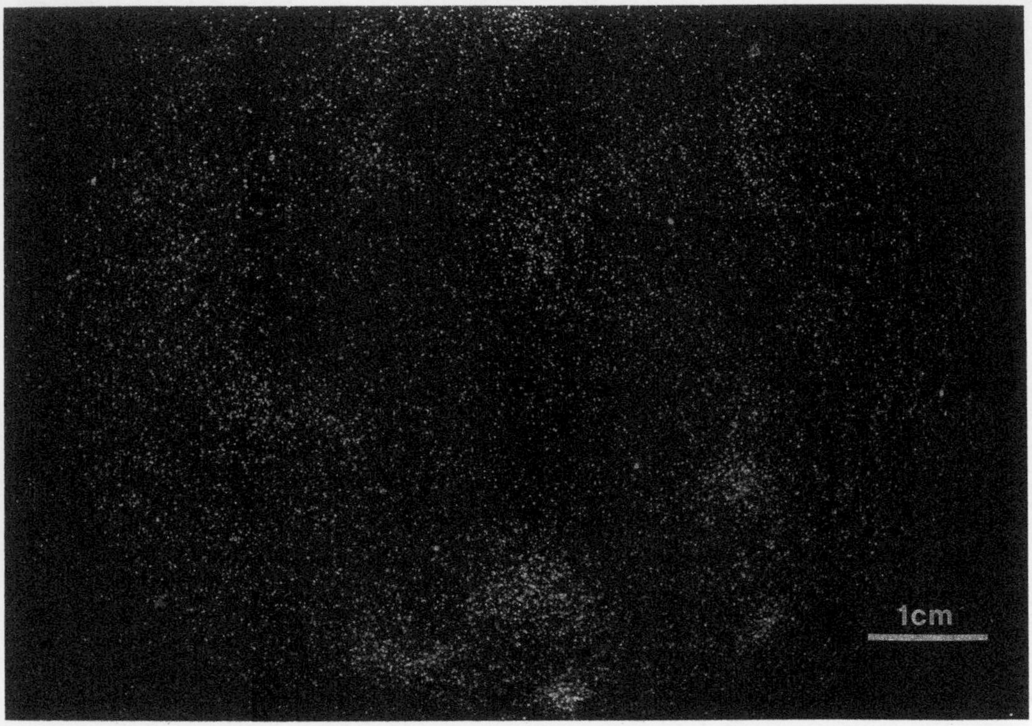

Figure 5.20 A photograph, taken through the image intensifier of sonoluminescence in distilled, degassed water occurring within the piezoceramic cylinder. The sonoluminescence occurs, not at the focal antinode, but at discrete sites within the body of the liquid. Exposure = 1 s (after Leighton [294]).

thought this was due to an acceleration of the oxy-reduction chemical process, though in the following year Frenzel and Schultes [163] suggested the effect was due to light exposure. The first quantitative application of the photographic technique soon followed, made in 1935 by Marinesco and Reggiani [245]. Chambers [198], who is reported to have waited 30 minutes to adapt his eyes to darkness in order to see sonoluminescence [246], called the emission sonic-luminescence, a term which was later contracted to its present form. It is interesting to note that Finch [237] considered the name inappropriate, preferring nomenclature that reflected a hot-spot origin, making it an 'incandescence' rather than a 'luminescence'. Walton and Reynolds [121] preferred 'cavitation luminescence', since it is not the sound but the cavitation which is the source of the emission, as evidenced by the fact that sonoluminescence has been generated through the various nonacoustic techniques listed in the previous section.

The use of luminol (tri-aminophthalic hydrazine) was discussed in section 5.1.2. Negishi [124] demonstrated its effectiveness, and it is generally employed in a solution of 0.2 g luminol and 5 g sodium carbonate per litre of water. This is not strictly a process that enhances sonoluminescence, but instead it introduces chemiluminescent reactions excited by the oxidation products of cavitation. After cavitation has ceased, the emission persists for around 1/20th of a second [237]. There are therefore two considerations which may in certain circumstances make its use inappropriate: (i) it is invasive, in that the cavitating medium must be changed, making it often unsuitable for biological studies of cavitation; (ii) the emission is one step further removed from the event, and so may not reflect the cavitation *per se*. The above example illustrates a temporal inconsistency. Indeed there could conceivably be a medium, the chemistry of which could generate spurious emission from luminol. This caution having been stated, many careful workers have used luminol to good advantage.

There are currently two main techniques for detecting sonoluminescence. In photomultiplication, photons entering the device stimulate the emission of electrons from the input photocathode. These are accelerated in evacuated conditions by an electric field so that they strike an electrode (known as the 'first dynode') with high kinetic energy. As a result, on average several 'secondary' electrons are emitted from the surface of the first dynode for each incident electron. In a series of such dynode interactions (typically around eleven), the number of accelerated electrons is increased through a cascade effect, and this group of electrons is finally collected at the tube anode. Thus the reception of a light pulse can generate a detectable electric pulse. In basic terms, the result given from a photomultiplier is a number, for example, an electrical current which increases as the light input increases. Another example signal is that which may be obtained from photomultipliers especially designed for quick response to an input signal: in such systems the electrical pulses are counted to give a measure of the number of photons detected. It should be noted that not every photon that enters the photomultiplier will cause an electrical signal at the output, nor is every electrical signal the response to a photon (it may, for example, be due to thermally stimulated electron emission from the input photocathode). However, the parameters of the photomultiplier can be suitably quantified and calibrated for photon counting. Photomultiplication is most useful in that it gives a quantitative measure of the light input, and can have very fast response times.[41] Griffing and Sette [247] were the first workers to use photomultipliers to observe sonoluminescence, providing researchers with accurate temporal resolution. However, the only spatial information available with the photomultiplier, that is the only way to tell where the light came from, comes through the restricted 'view' of the input. To examine spatial differences in light output using a photomultiplier, one must restrict the input and scan the device which, if the source changes significantly over the time of the scan, gives information of only limited value. A better way to obtain spatial information on the luminescence is through image intensification.

In image intensifiers, the electrons emitted from the input photocathode are not only accelerated by an electric field, but are also focused by a magnetic field. As a result, the electrons follow helical paths when they collide with a phosphor screen. The reason for this is that the photon to be detected impacts a specific small region on the input photocathode, and the magnetic field ensures that all the electrons released from that region are focused by the magnetic field to collide with a corresponding small region on a phosphor screen. This process ensures that the spatial information of where the original photon was is retained.

The phosphor screen releases photons in response to impact by the electrons. If a photocathode is placed against this screen, these photons will stimulate electron emission. Placing four photocathode/phosphor stages back-to-back can generate a photon gain of 10^7. The final image appears on a phosphor screen much like a small TV screen, and the high gain allows single photons at the input to generate an image on the screen. As with the photomultiplier, the device has finite efficiency, and there is also finite noise. The advantage of the image intensifier is that it readily provides spatial information, but the phosphors have slow response times, and so are unable to time-resolve the light emission to the same degree as appropriate photomultipliers.

---

[41]See concluding remarks to section 5.2.1.

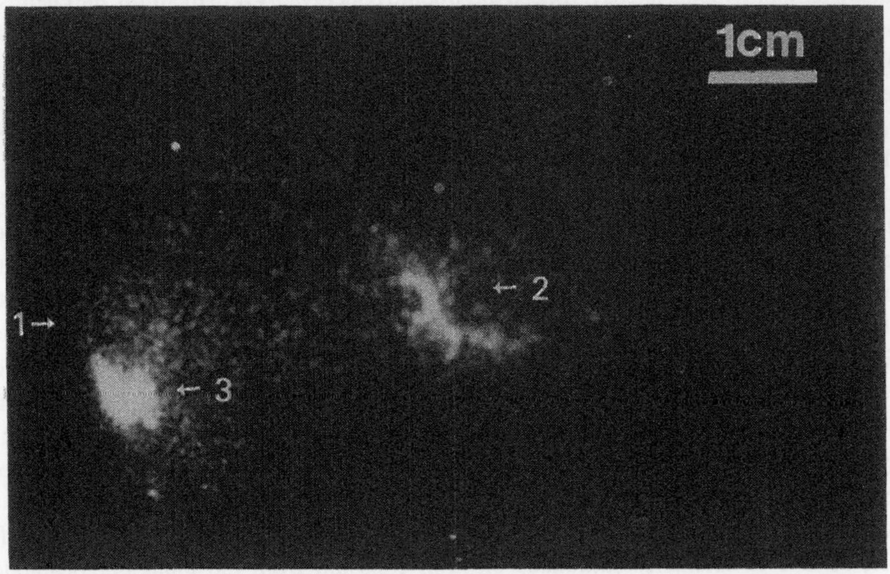

Figure 5.21 Sparks (3) are observed through the image intensifier to occur on the rim (1) of the piezoceramic cylinder. Sonoluminescence (2) from the aerated water contained within the cylinder can be seen at the axial pressure focus. (After Leighton [294].)

Image intensification was first applied to sonoluminesence in 1964 for a simple qualitative demonstration by Flynn [161]. The technique was re-introduced in the early 1980s by Reynolds *et al.* [248].

Figure 5.21 shows one way in which image intensification can be valuable. It shows sonoluminescence in aerated tap water from the cylindrical transducer described in Chapter 1, section 1.2.2(b)(i), the wall of which is labelled '1' in the figure. In comparison with Figure 5.18, it is clear that in addition to the focal cloud (2), there is light being produced by electrical sparking near the wall (3). If image intensification had not been used, photomultiplication alone may have erroneously included this electrically induced emission in the measure of sonoluminescence.

(b) Dependence of Sonoluminescence on the Medium

The sonochemical aspects of the effect of the medium on sonoluminescence have already been discussed. The recent history is reviewed by Walton and Reynolds [121].

(i) The Effect of the Dissolved Gas. The intensity of the luminescence varies greatly with the liquid medium [179, 249]. The dissolved gas content is important: it was thought in 1934 that degassed liquids could not luminesce [163] though later work showed that it was possible if greater acoustic pressure amplitudes were used (Rosenberg [243] required twice that needed to obtain sonoluminescence from aerated water).

Gas dissolved in the liquid will comprise a proportion of the gas content of a bubble and thus it influences the intensity of sonoluminescence (work in this field is reviewed by Finch [237] and Degrois [250]). This influence can occur in a number of ways. For example, the luminescence will depend on the thermal conductivity of the gas [173] since, all else being equal, the higher this conductivity, the lower the maximum temperature attainable within the bubble [251]. Similarly, the higher the value of κ (or γ) of a gas, the greater the temperature reached when the bubble volume is a minimum. These latter two demonstrate why, if a thermochemical origin is assumed, the monoatomic noble gases produce very intense sonoluminescence. However, the nature of the gas also affects the free radicals produced, and therefore the sonoluminescence, through the resulting chemical reactions (the noble gases are generally chemically inactive). The sonoluminescence from a variety of gas-saturated liquids has been investigated experimentally, workers including Weissler [252], Parke and Taylor [132], Gunther *et al.* [137], Prudhomme and Guilmart [138], Srinivasan and Holroyd [128], Young [173] and Vaughan *et al.* [253].

Several theoretical studies have been done on the influence that the gas content of the bubble has on sonoluminescence, including the effect of thermal diffusivity, though the authors restricted themselves to a discussion of the temperature reached inside the bubble, and not of the resulting radical reactions[42] [172, 173, 196, 203, 254].

(ii) Dependence of Sonoluminescence on the Liquid. The sonoluminescence intensity varies greatly with the liquid medium employed [179, 249], and one would certainly expect this given the effect of surface tension on cavitation inception,[43] and the role of nucleation.[44] Once the bubble is nucleated, its dynamics are dependent upon several physical constants of the liquid, as even an examination of the basic Rayleigh–Plesset equation (equation (4.81)) will suggest: one would logically expect the sonoluminescence to be similarly dependent. Given that a bubble is undergoing an energetic collapse, the earlier discussion of sonochemistry illustrates several ways in which the liquid might affect the resulting luminescence, through the roles of vapour pressure, the liquid shell etc. (assuming a thermal mechanism for the emission). Given the range of liquids which will sonoluminesce (in 1937, Levsin and Rzevkin [255], for example, observed sonoluminescence from twenty liquids, including concentrated sulphuric acid), one would expect these dependencies to be observable. Transparent biological fluids have been cavitated to produce sonoluminescence, including blood plasma [17] and human amniotic fluid [256]. Potato tubers [257] and liquid metals [258] have also sonoluminesced.

Chambers [246] first correlated luminescent intensity to certain of these parameters (viscosity, electrical dipole moment of liquid molecules and bulk liquid temperature). Jarman [178] found the output correlated closely to the parameter σ^2/p_v. Golubnichii *et al.* [179] correlated sonoluminescent intensity with the energy of a molecular dissociation, σ^2/p_v, viscosity and the reciprocal adiabatic compressibility of the liquid. As discussed earlier, Suslick *et al.* [176] were unable to support either the σ^2/p_v [178] or free molecular interaction energy [179] correlations using chemical dosimetry, suggesting that they would be difficult to justify as more than empirical fits to data. Their results do qualitatively support a correlation made by Sehgal *et al.* [177] between the vapour and static pressures and the total free energy of cavity formation, though over only a limited range of vapour pressures. Harvey [212] showed that, though suspensions increase the number of nucleation sites, the luminescent intensity is independent of the suspended substance. The concentration and nature of the salt used in an electrolytic solution can enhance or quench sonoluminescence [178]. Chemiluminescence can be used to detect radicals through the use of luminol, as outlined earlier.

---

[42] See the conclusion to section 5.2.1(a).
[43] See Chapter 2, sections 2.1.1, 2.1.2 and 2.1.3(a), and Chapter 4, sections 4.3.1(a) and 4.3.1(c).
[44] See Chapter 2, section 2.1.2(b), and section 5.1.2.

(iii) The Dependence on Ambient Temperature. Though the general statement is sometimes made that the light output decreases with increasing liquid temperature, this is too simplistic. This is based mainly on the work of Jarman [178], Iernetti [259], Sehgal *et al.* [260] and Chendke and Fogler [193]. However, even allowing for the variety of experimental conditions that were used (e.g. 16.5 kHz $\leqslant \nu \leqslant$ 700 kHz), an indisputable correlation of sonoluminescence with temperature has not been forthcoming. Sehgal *et al.* reported a sharp decrease around 10–20° C, with a gradual flattening off towards 70° C. Chendke and Fogler measured the initial and final temperatures of air-saturated water and measured the time-dependency of the sonoluminescence. Insonation warmed samples that were initially cold, and sonoluminescence was less at the end of the experiment than at the start. Samples that were initially hot, cooled during the course of the experiment, and the sonoluminescence increased. Jarman and Iernetti both reported maxima in the light output as the temperature varied. Having studied fifteen different liquids, Jarman considered secondary butyl alcohol, for which the maximum occurred at 25° C, to be typical. Iernetti found a maximum at 12° C for aerated water. Pickworth *et al.* [261] found that the sonoluminescence from areated water in a 1-MHz standing-wave field increased at both 1 and 3 W cm^{-2} (I_{SATA}) in the range 22–45° C. An experimental study of cavitation, rather than sonoluminescence, by Blake [262] showed that cavitation thresholds decrease with temperature in the range ~10–50° C.

The most immediate suggestion for the effect of ambient temperature on cavitation is through the solubility of dissolved gas, which increases with decreasing temperature.[45] It is clearly too simplistic to say that as the temperature rises more gas enters a given bubble, cushioning the collapse and making it less violent, so reducing sonoluminescence. Other mechanisms associated with solubility might occur. For example, as more gas comes out of solution, there will be an increased number of bubbles, causing increased luminescence. Rectified diffusion and bubble shielding effects might also have to be considered and quantified.[46] The effect of vapour pressure, as discussed in section 5.2.1(a)(ii), is also dependent on the ambient temperature.

(iv) The Dependence on the History of the Liquid. The tensile strength of a liquid is reduced just after it has been first cavitated, and continues to be so for several hours[47] [263]. However, the tensile strength in turn affects the cavitation, the latter being more likely as the strength is reduced [264]. The nuclei population, the opportunities for stabilisation and the gas content are all affected by the history, and will thus affect cavitational effects. For those liquids with a 'memory effect' of a time period likely to affect the results sufficiently, a standardised working method must be adopted by experimentalists.

One example of a history dependence is shown in Figure 5.22 from Pickworth *et al.* [265], showing the presence or absence of sonoluminescence in water as a function of the pulse length (measured as the number of cycles per pulse) for (a) a therapeutic transducer operating at 1.09 MHz, and (b) a diagnostic transducer operating at 0.97 MHz. The former was designed to operate with a Therasonic 1030 (Electro-Medical Supplies) unit, and the latter with a focused NE 4161 (Nuclear Enterprises), but for this experiment both were driven by a nonstandard pulsing unit, the driving voltage of which is given on the vertical axis.[48] In both cases there is a tendency to promote sonoluminescence as the driving intensity and pulse length are altered

[45]This is the reason for the appearance of bubbles attached to the wall of your bedtime glass of water by the morning. During the night, as the temperature decreases, more gas can dissolve into the liquid. When morning comes, the water temperature increases and the gas comes out of solution. Attachment to nucleation sites in the wall of the glass stabilises the bubbles against buoyancy. See Chapter 2, section 2.1.2(b)(ii).

[46]See section 5.3.

[47]See Chapter 2, section 2.1.2.

[48]See original source to convert this to acoustic intensity.

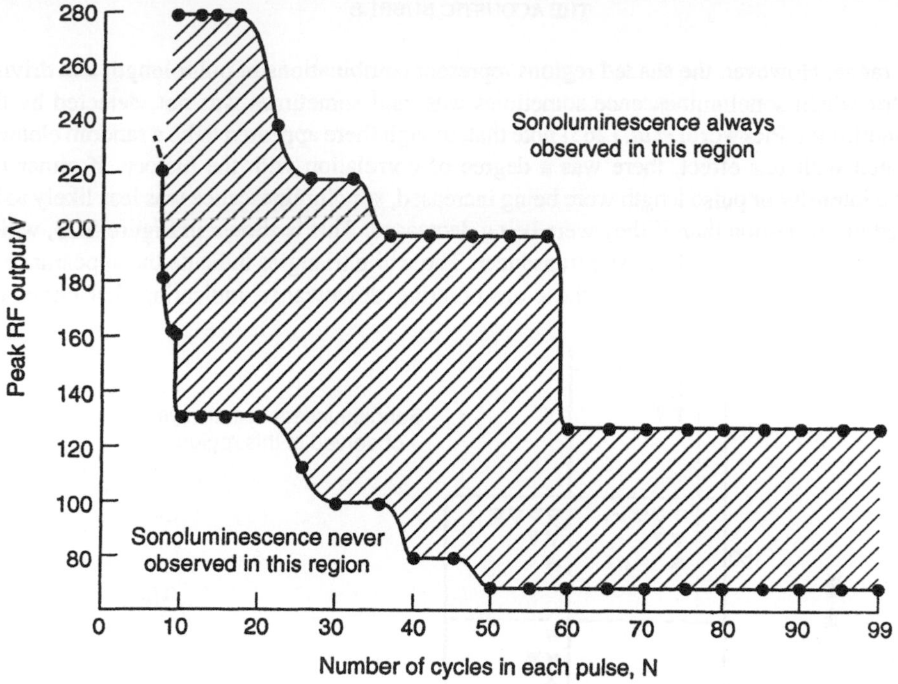

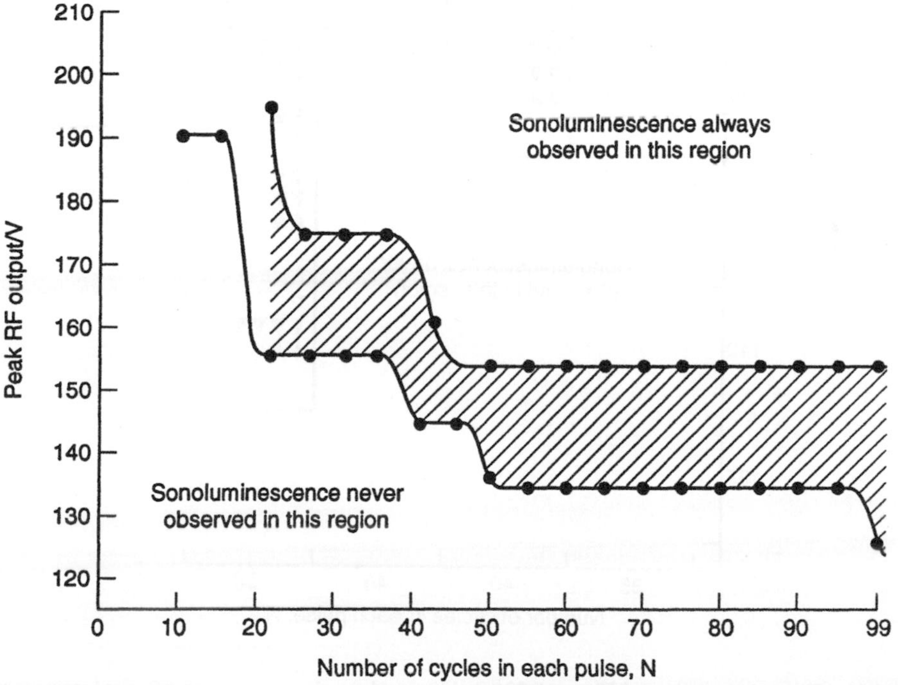

Figure 5.22 (a) Peak RF output against the number of cycles per pulse for the therapeutic transducer, showing three distinct regions. The shaded area represents settings where sonoluminescence sometimes occurs. (After Pickworth *et al.* [265].) (b) Peak RF output against the number of cycles per pulse for the diagnostic transducer, the shaded area again representing the region of uncertainty (after Pickworth *et al.* [265]). Reprinted by permission from *Physics in Medicine and Biology*, vol. 34, pp. 1139–1151; Copyright © 1988 IOP Publishing Ltd.

in this range. However, the shaded regions represent combinations of pulse length and driving force for which sonoluminescence sometimes was, and sometimes was not, detected by the photomultiplier. Pickworth *et al.* [265] note that, though there appeared to be a random element associated with this effect, there was a degree of correlation with the history. If either the acoustic intensity or pulse length were being increased, sonoluminescence was less likely to be detected in this region than if they were being decreased. This is shown in Figure 5.23, which is a detail of data for the diagnostic transducer. The white triangles indicate the appearance of sonoluminescence: twenty-two of these are pointing downwards, indicating that either the

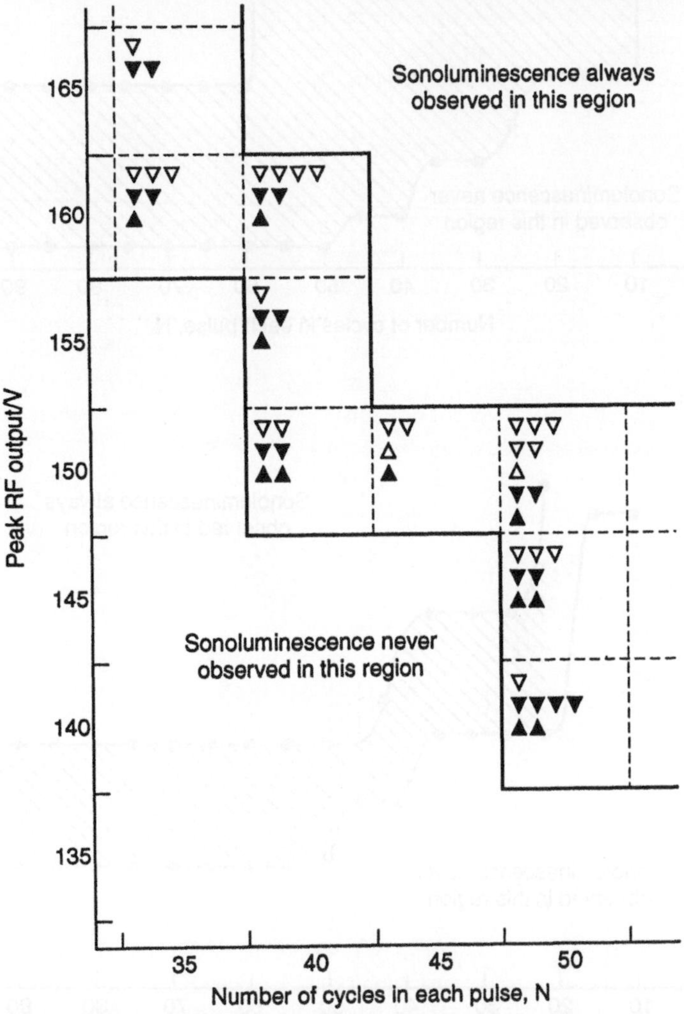

Figure 5.23 Expanded section of Figure 5.22(b) showing how the appearance of sonoluminescence depends on whether I (acoustic intensity) or N (number of cycles per pulse) are being increased or decreased: △, sonoluminescence present, I or N are being increased; ▲, sonoluminescence absent, I or N are being increased; ▽, sonoluminescence present, I or N are being decreased; ▼, sonoluminescence absent, I or N are being decreased. Note that nearly all the open triangles are pointing downwards. (After Pickworth *et al.* [265]. Reprinted by permission from *Physics in Medicine and Biology*, vol. 34, pp. 1139–1151; Copyright © 1988 IOP Publishing Ltd.)

acoustic intensity or pulse length were being decreased. Only two are pointing upwards, indicating that sonoluminescence was observed as the acoustic intensity or pulse length was increasing. The reason is probably that if one is reducing either acoustic intensity or pulse length in this transitional region, one has previously obtained sonoluminescence, and is going to a regime where there is little sonoluminescence. Since there has been a great deal of cavitational activity, particularly of an energy sufficient for sonoluminescence, one would expect the medium to contain a considerable population of nuclei, as a result of rectified diffusion or the fragmentation of bubbles on collapse. If one is increasing acoustic intensity or pulse length, from a regime where sonoluminescence was not detected, these extra nuclei will not be present in the transition regime. In an example of the history dependence of another cavitational effect, Henglein and Gutierrez [266] have shown 'memory effects' in the formation of NO_3^- and NO_2^- using pulsed ultrasound in aerated water. This will be discussed in section 5.3.1(a).

(c) Dependence of Sonoluminescence on the Pressure Field

(i) Dependence of Sonoluminescence on the Hydrostatic Pressure. Increasing the static pressure will tend to affect cavitation by (i) making it more difficult for the bubble to expand, and (ii) increasing the compressive forces that drive it to collapse. The first would tend to decrease, and the second to increase, the violent cavitation, and the two phases are discussed in Chapter 4, section 4.3.1. Equation (4.106), derived through consideration of both growth and collapse phases, relates the maximum temperature attained during the collapse (T_{max}) to the static pressure (p_0), but the trend is not obvious.

Walton and Reynolds [121] consider each phase individually as it relates to the transient collapse. For the effect of p_0 on the energetics of the collapse phase, substituting for R_{max} from equation (4.95) into equation (4.110) gives an approximation to T_{max} in terms of $(P_A - p_0)$:

$$T_{max,t} = \frac{T_0}{R_0^3}\left(p_0 + \frac{2\sigma}{R_0}\right)^{-1}$$

$$\times (\kappa - 1)(P_A + p_0)\left(\frac{4}{3\omega}\right)^3 (P_A - p_0)^3 \left(\frac{2}{\rho P_A}\right)^{3/2}\left(1 + \frac{2}{3p_0}(P_A - p_0)\right) \qquad (5.3)$$

This equation relates to the collapse of a single bubble, and if one assumes a thermal origin for sonoluminescence, then since T_{max} is dependent on p_0 so too will be the sonoluminescence. Clearly, a full discussion must also consider the thermal diffusivity and chemical nature of the gas.[49]

There is a second factor to consider. In addition to altering the energy of each individual collapse, changes in p_0 will alter the range of bubble nuclei that may collapse transiently, and so affect the total number of collapses. Having therefore considered the collapse, one must also consider the growth phase. Taking the simplest approximation, Walton and Reynolds [121] consider the Blake threshold to find how p_0 will affect the range of initial bubble radii which can give rise to transient cavitation through the lower radius threshold. As p_0 increases so does the Blake threshold pressure (equations (2.22) and (2.23)), and the larger the minimum bubble size that can undergo explosive growth (equation (2.21)). This would suggest that increasing the static pressure reduces the number of bubbles in the population that can go transient.

[49] See section 5.2.1(a).

Experimental observation began with Harvey [212] who noted that the intensity of luminescence increased up to $p_0 = 2.3 \times 10^5$ Pa, and then decreased to zero. This peak of activity was confirmed by Finch [267], though his maximum occurred between $p_0 = 1.5 \times 10^5$ to 1.7×10^5 Pa (depending on P_A). In addition to changing p_0 hydraulically, Finch used an increased gas pressure, a technique which differs in its effects on the concentration of dissolved gas. Finch related his results to the tendency of p_0 to increase the driving force on collapse, but decrease the maximum size attained by the bubble on expansion. Figure 5.24 shows the sonoluminescent measurements of Chendke and Fogler [268] at $\nu = 20$ kHz for (a) nitrogen-saturated water, and (b) a saturated solution of water in carbon tetrachloride, containing dissolved nitrogen, for hydraulically-increased pressures. The general trend is that initially increasing the static pressure increases the violence of each collapse, which counterbalances the reduction in their number: as the pressure is increased further, the latter effect dominates, until the cavitation is suppressed. They also fitted spectral measurements to black-body curves (though, as discussed in section 5.2.1(a), the spectra is not truly black body), and from the results Chendke and Fogler [268] suggested that the bubble temperatures were independent of p_0 in the pressure range considered. They inferred that changing p_0 did not change the violence of each individual cavitational event, but instead altered the number of events by changing the number of available nuclei. The dependence of cavitation thresholds relating to sonochemistry and sonoluminescence on the ambient pressure have been modelled by Šponer [269] whose results suggest a correlation between these thresholds and the conditions to generate collapse temperatures of 1550–1600 K in steady-state motion.

In a slight digression to a similar problem, flow-induced pressure fluctuations[50] can generate bubble growth and collapse, similar to transient cavitation.[51] If this happens with submarine propellers, the sound emission can give away the position of the submarine to the acoustic sensors of the opposition. It has long been known by submariners that submerging will tend to reduce this by suppressing the cavitation. However, when the cavitation is strong and the vessel is at high speed, increasing the depth of the vessel will first cause an increase in the cavitation noise, before suppression occurs [270]. This so-called 'anomalous depth effect' is due to the fact that, before the increasing pressure suppresses the growth phase, it first increases the violence of each collapse.

(ii) Dependence of Sonoluminescent Intensity on Acoustic Pressure Amplitude. As with the static pressure, the acoustic pressure amplitude will affect, firstly, the violence of the collapse, and so the temperature reached (see equation (5.3)); and secondly the range of bubble sizes that may go transient (to a first approximation – as regards the initial explosive growth, it must exceed the Blake threshold pressure, and will affect the upper limit of radius by determining R_{max} in equation (4.76)). These two mechanisms can affect the sonoluminescent output. Experimentally, Griffing and Sette [247] and Parke and Taylor [132] showed that sonoluminescent intensity varies more or less linearly with acoustic power. Jarman [178] supported these observations. Negishi [124] found that in certain fluids, sonoluminescence was suddenly quenched as the intensity rose above about 2 W cm^{-2}. This is due to an important phenomenon: intense cavitation around the transducer causes an acoustic impedance mismatch, so that the sound cannot reach the bulk liquid, and cavitation therefore ceases[52] [121]. Margulis and Akopyan [271] confirmed the linear dependence of sonoluminescence on acoustic intensity

[50] See Chapter 2, section 2.2.2.
[51] See Figure 5.58.
[52] Such shielding is discussed further in section 5.3.1(e).

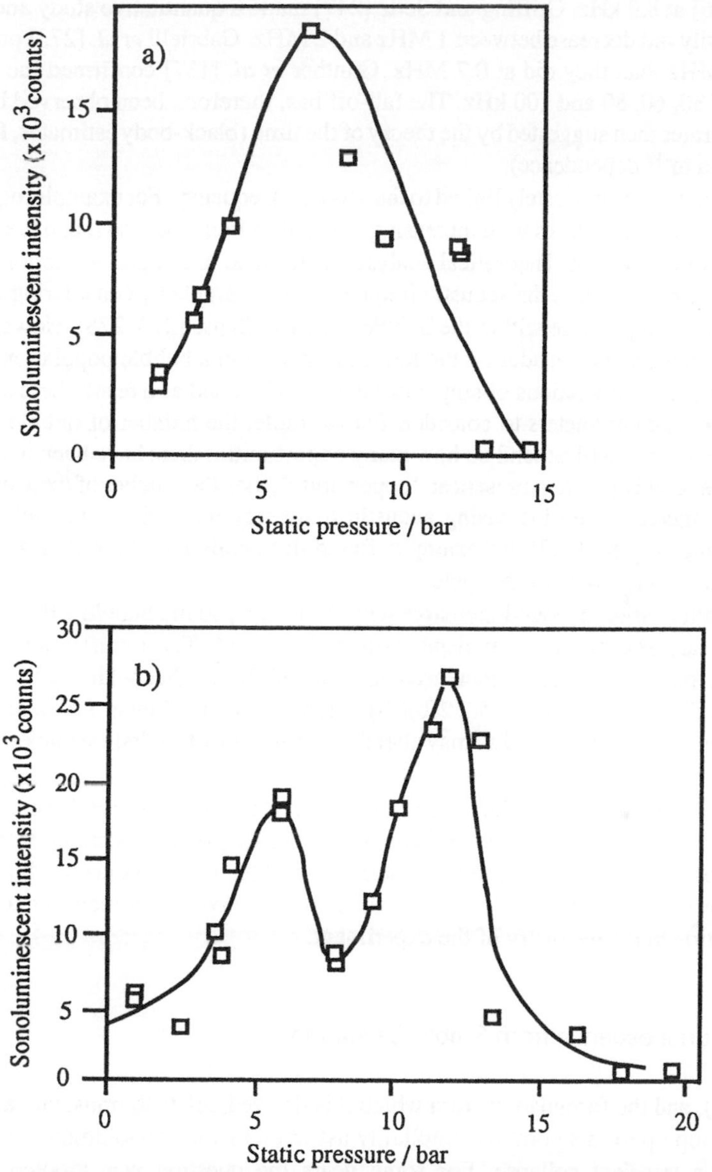

Figure 5.24 Variation of sonoluminescence with static pressure (hydraulically increased) at $\nu = 20$ kHz for (a) nitrogen-saturated water and (b) a saturated solution of water in carbon tetrachloride, containing dissolved nitrogen (after Chendke and Fogler [268]).

for levels greater than 0.7 W cm^{-2}; however, below this level the luminescent intensity varied linearly with the square of the acoustic power.

(iii) Dependence of Sonoluminescence on Acoustic Frequency. The first observations were made using dark-adapted eyes, where Kling [272] noted that the light output he obtained using sound at 103 kHz was of an estimated intensity less than that obtained by

Chambers [246] at 8.9 kHz. Griffing and Sette [247] made a quantitative study and found that the light intensity did decrease between 1 MHz and 2 MHz. Gabrielli *et al.* [273] produced half the flux at 2 MHz than they did at 0.7 MHz. Gunther *et al.* [137] confirmed the trend at the frequencies of 30, 60, 80 and 100 kHz. The fall-off has, therefore, been observed but it occurs at much lower rates then suggested by the theory of the time (black-body estimates, for example, would predict a ω^{-12} dependence).

Bubble dynamics are intricately linked to the acoustic frequency. For example, equation (5.3) suggests that the T_{max} decreases with increasing insonation frequency, so that one might expect light intensity to follow suit. Theoretical analyses mentioned in Chapter 4, section 4.3.1 relate the energetics of cavitation to the acoustic frequency, for example by considering the criterion to attain a certain temperature within the bubble during collapse [274, 275]. However, in most practical situations one is considering the luminescence from a bubble population, in contrast to these analytical considerations of single isolated bubbles, and as a result there are a number of physical acoustic parameters to consider. For example, the number of bubble collapses in any given time interval will depend on how many opportunities there have been in that interval: if one takes one sound cycle to represent an 'opportunity', then the number of these opportunities would tend to increase with increasing acoustic frequency for a given interval. Jarman and Taylor [149] and Negishi [123], for example, found light emissions occurring within roughly one-tenth of the sound cycle, every cycle.

Similarly, attenuation of sound increases with frequency, so the bubble will tend to experience reduced acoustic pressures at depth within the liquid. The ramifications of this are, however, that one might expect increased heating of the media, which will bring in the complications discussed in section 5.2.2(b)(iii); alternatively, the higher frequencies will tend to promote acoustic streaming, and so may alter the distribution of cavitation nuclei, as discussed in section 5.1.2.

As will be shown in section 5.2.4, if the field is standing wave, the sonoluminescence tends to originate from the pressure antinodes. The higher the frequency, the more of these sources are in a field of view of fixed size, though their size will tend to be reduced. Thus the amount of sonoluminescence detected as frequency changes may depend on many factors, from the nuclei population to the geometry of the experiment.

5.2.3 Sonoluminescence from Stable Cavitation

Equation (5.3), and the formulation from which it is derived, relate to transient cavitation, and so the discussion up to this point has implicitly assumed that the sonoluminescence has been the result of a transient collapse. For some years the question was mooted as to when sonoluminescence can occur through stable cavitation. However, as is clear from the discussions in Chapter 4, the definitions of these two terms are by no means unambiguous, and the discussion of sonoluminescence from stable cavitation illustrates a case when the distinction is particularly problematical. Sonoluminescence necessarily stems from energetic cavitation. In section 5.2.1, observations of a bubble undergoing periodic, repetitive pulsations and emitting sonoluminescence were discussed. This form of cavitation would of course correspond to the class 'repetitive transient cavitation' which, as proposed earlier,[53] is also a form of stable cavitation. Certainly, if one takes as a criterion for sonoluminescence that the gas content should achieve high

[53] See Chapter 4, sections 4.3.2 and 4.4.8 – alternative criteria are discussed in Chapter 4, section 4.4.8, and also section 5.2.1(b).

temperatures, and a sonoluminescing bubble would therefore satisfy a T_{max} criterion,[54] then as Walton and Reynolds [121] pointed out in 1984, studies of the Rayleigh–Plesset and related equations that model a pulsating spherical cavity show that there exist periodic solutions characteristic of stable cavitation, where the gas inside the bubble can attain temperatures similar to those associated with sonoluminescence from transient collapse. Theoretical predictions suggesting the possibility of sonoluminescence from a stable bubble were done by Flynn [276], who in 1975 modelled a bubble's response to a Gaussian pressure pulse, and in 1982 [80] analysed its response to a sinusoidal pressure pulse.

Saksena and Nyborg [143] provided in 1970 the first experimental evidence when, with a photomultiplier, they detected light emission from a column of rising bubbles. These had been introduced with varying diameters into a glycerine–water mixture insonated at 30 kHz. The acoustic intensity required for sonoluminescence was 1.25×10^5 Pa, less than half that required when no bubbles were injected. (This latter pressure amplitude, 2.8×10^5 Pa coincided with that necessary for streamer activity.) They detected no sonoluminescence from stable cavitation in water. Margulis and Grundel [277, 278] recorded the sonoluminescence from air-saturated water as a function of increasing acoustic power. At a critical level, the amount of sonoluminescence increased dramatically. The authors suggest that below this level the light emission is from stable cavitation; above it, from transient cavitation.

In 1985, Crum and Reynolds [279] insonated aerated water at 20 kHz and viewed the sonoluminescence using an image intensifier. They were prevented from making accurate measurements of acoustic pressure levels, but estimated that for an intermediate stage of light emission it was around 0.17 ± 0.03 MPa. The authors saw some light patterns which persisted for several seconds. These they took to be "From what most of us would call 'stable' cavitation" [279]. Visual observations of the illuminated bubble field revealed some cavitation streamers. Only occasionally was the sharp 'snap' associated with transient collapse observed,[55] and that sound, the authors estimated, did not emanate from the region of the sound field where they observed their form of stable cavitation. The inability to determine the mode of oscillation and any associated activity (e.g. surface waves, cavitation microstreamers etc.) of this luminescing air bubble in water was not the case with a single bubble in a glycerine/water mixture observed in 1990 by Gaitan and Crum [125]. In an elegant and important experiment[56] these workers used acoustic pressure amplitudes up to 0.15 MPa to observe the stable cavitation of a bubble that exhibited no detectable surface wave or streamer activity, and over measurement intervals of thousands of acoustic cycles the bubble did not break up. The bubble was levitated in a cylindrical transducer of the type described in Chapter 1, section 1.2.2(b)(i), driven at 21–25 kHz. The radius–time behaviour of the bubble, which pulsated at the frequency of insonation and with amplitudes up to $(R_{max}/R_0) \approx 7$, was measured from the Mie scattering of a 3-watt argon-ion laser.[57] Time-resolved measurements of the sonoluminescence, which was bright enough to be seen in a dark room by the unaided eye, and which microscope observations showed to originate from the geometric centre of the bubble [125, 126], were compared with radius–time measurements of the nonlinear pulsation, made a few seconds later. Through comparison with a common phase reference, the sonoluminescent flashes were shown to be simultaneous with the bubble collapse. Figure 5.25 shows simultaneous plots of the sound field, the bubble radius and sonoluminescence for $P_A = 1.2$ atm and $\nu = 22.3$ kHz.

---

[54] T_{max} criterion of Apfel and Holland, and of Šponer, are discussed in Chapter 4, section 4.3.1(c)(iii).
[55] See Chapter 4, sections 4.3.4 and 4.4.7(a).
[56] See section 5.2.1.
[57] Optical scattering from bubbles and Mie theory are discussed by Marston, Langley and Kingsbury [280–284].

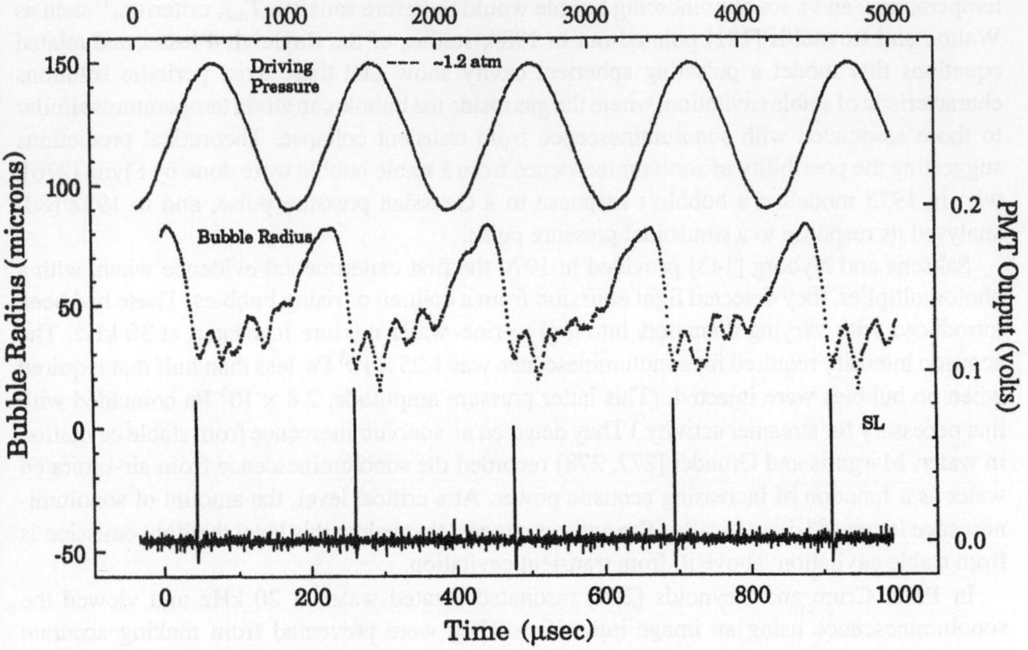

Figure 5.25 For insonation at $P_A = 1.2$ atm and $\nu = 22.3$ kHz, the acoustic field (top), bubble radius (middle) and photomultiplier output (bottom) are plotted against a common time-axis. The sonoluminescent flash is coincident with bubble collapse. (After Gaitan *et al.* [126].)

Gaitan *et al.* [126] used the results from a similar experiment to compare with theoretical predictions of Keller and Miksis [285] (which use a linear polytropic exponent approximation), Prosperetti *et al.* [286] (which introduces a more exact description of the internal pressure) and Flynn [276] (which includes thermal effects within the bubble). The compressibility of the liquid is incorporated, and the model used to tabulate temperature and pressures attained within the bubble for R_0 equal to 15–20 μm. Comparison with observed threshold lead Gaitan *et al.* to suggest that the minimum temperature required for observable luminescence was between 2000 and 3000 K. The models predicted temperatures up to 10 000 K for the experimental conditions (e.g. bubble sizes employed). This can be compared with the measured value of 5200 ± 650 K measured by Suslick *et al.* [192] using comparative rate thermometry in aqueous solutions in 20-kHz sound fields at 24 W cm^{-2} (corresponding to $P_A \sim 8$ atm). Such temperatures are predicted by Gaitan *et al.* for $P_A \approx 1.3$ atm. As mentioned in section 5.3.1(c)(iii), Šponer *et al.* [274, 287, 288] employed $T_{max} = 1550$ K as the minimun requirement for sonoluminescence in models of the adiabatic collapse of an isolated gas bubble in a viscous compressible liquid.

One attractive feature of the stable oscillations obtained is that, because the sonoluminescent source is a single spherical isolated bubble, the absence of surface waves and disturbing sources in the sound field means that the bubble does not undergo the erratic 'dancing' motion commonly seen when bubbles are driven at high amplitude, nor is fragmentation renucleation an issue. Gaitan *et al.* [126] give a clear and informative description of how the qualitative form of the experimentally observed cavitation depends on the water/glycerine mixture and the driving pressure. If the concentration of glycerine was greater than 60%, at $P_A = 0.6$ atm the injected bubble would dance, indicating surface wave activity and asymmetric collapses [289].

As P_A increased, the dancing became more vigorous, and the bubble fragmented to form a cluster of daughters, moving around the parent. The cloud often took on a form which the authors refer to as 'shuttlecock', an object which, in contrast to the single bubble, was not on the axis of the cell, but 5–10 mm off-axis. The shuttlecock consisted of a larger bubble roughly 50 μm in radius, surrounded by a cloud of microbubbles, which would move from one side of the bubble to the other under radiation forces. Such forces also caused the microbubbles which were occasionally ejected from the cloud, away from the pressure antinode, to move back around the cloud to the antinode, and re-enter the cloud at the position opposite that from which they were ejected, so that a rotational orbital motion was established. The shuttlecock emitted a low rate of sonoluminescent flashes.

Gaitan *et al.* infer from observations of single bubbles and clouds of bubbles that, as discussed earlier, surface oscillations and such asymmetries do reduce the temperatures attained by the collapsing bubble, and consequently the sonoluminescence. Thus sonoluminescence from stable cavitation is more likely in bubbles smaller than resonance than in bubbles at resonance, where surface oscillations can be readily excited.

If there was less than 60% glycerine in the mix, the bubble cluster became tighter as P_A increased. It would collapse in upon itself, so that at a stability threshold a single bubble appeared, and the luminescence increased to one flash per acoustic cycle. The authors state that at this point "the pulsation amplitude was estimated to be between 4 and 5 times the equilibrium radius," presumably referring to the expansion phase. Gaitan *et al.* [126] also note that once stability had been obtained in this way, P_A could be reduced to below the threshold, and stability would be maintained down to acoustic pressure amplitudes 0.1 atm below the stability threshold. There was also an upper stability threshold, such that at high acoustic pressures the bubble became unstable once more. The authors concluded from threshold evidence that rectified diffusion plays an important part in the stability process.

5.2.4 Cavitation in Standing-wave Fields

To close this section on sonoluminescence, an examination will be made of sonoluminescence and other cavitation effects in standing-wave systems. The aim of this is to form introductory links to following sections, so that the examples of research mentioned are chosen with this in mind, rather than to provide a review of cavitation in standing-wave fields. It is debatable how relevant standing-wave studies are to clinical situations: though in the body there do exist plane-like interfaces between media of substantially different acoustic impedance, attenuation and the use of short acoustic pulses[58] reduce the likelihood of a significant standing-wave component. However, the standing-wave system is important in that it can often form a component (known or unknown) of experimental systems, can be exploited in certain laboratory and industrial arrangements, and above all is interesting.

The first recorded production of sonoluminescence from a standing-wave field was by Marinesco and Reggiani [245] in 1935. They directed plane continuous-wave 717-kHz ultrasound at a photographic plate which was immersed at an angle θ to the beam in a bath of

[58]Roughly speaking, a pulse of length τ_p can only set up standing waves a distance $c\tau_p/2$ in front of the reflecting boundary, since the standing-wave field is set up by the interference of the reflected half of the pulse with the incoming half. Since $c = \nu\lambda = \lambda/\tau_\nu$, then the standing waves form a distance equal to $\lambda\tau_p/(2\tau_\nu) = \lambda n_p/2$, where n_p is the number of cycles in the pulse, an obvious result! Averaging over the decay of this distance as the pulse progresses through reflection, the standing waves exist over a distance equal to $\lambda n_p/4$, and last for roughly only $\lambda n_p/(4c) = \tau_p/4$.

developer. After about 15 minutes, bands appeared on the plate, aligned perpendicular to the acoustic axis and at a spacing of $\lambda/(2\cos\theta)$. In 1947, Pinoir and Pouradier [290] obtained a similar result at 300 kHz for film immersed parallel to the beam. This therefore indicated light sources spaced $\lambda/2$ along the beam, which is the node–node or antinode–antinode spacing in a standing-wave field.[59] From the discussions in Chapter 4 and this chapter we would expect the sonoluminescing bubbles to be those of less than resonance size, which would be found at the pressure antinodes.[60] Henglein [291] demonstrated sonochemical effects to be located at the acoustic pressure antinodes. It is also clear that the standing-wave systems need not be confined to planar ones: such systems as the cylinder of Chapter 1, section 1.2.2(b)(i) and the spherical resonator of Roy *et al.* [89], which have specific modal pressure distributions, will exhibit the same effect. Wagner [292] in 1958 first demonstrated that sonoluminescing bubbles tended to cluster around the pressure antinode through photographing at ten-minute exposures a known planar field, set up between a planar 260-kHz transducer and a $\lambda/4$ reflector.

Using photomultiplication, Gunther *et al.* [137, 293] found that in a standing-wave field, flashes of sonoluminescence were occurring at twice the acoustic frequency. This is simply because the adjacent pressure antinodes oscillate π out-of-phase in time (see Figure 1.11); therefore during the first half of each acoustic cycle, one set of alternate antinodes are in compression, whilst in the other half the other set are in compression. The time of emission of flashes varies within about 10% of the acoustic cycle, owing to the range of bubble sizes present [149].

Sonoluminescence from other standing-wave systems has been recorded by several workers. Eastwood and Watmough [17] recorded sonoluminescence from human blood plasma using therapeutic ultrasound of frequencies 0.75 and 1.5 MHz and intensity 2 W cm^{-2} (spatial average). First applied by Flynn [161] in 1964, image intensification was re-introduced for sonoluminescence studies in 1982 by Reynolds *et al.* [248]. Crum *et al.* [256] obtained image-intensified pictures of sonoluminescence from standing waves in human amniotic fluid and pig blood plasma at 37° C, using a video camera to record the output of the image intensifier. Leighton *et al.* [294, 295] recorded similar results from water, shown in Figure 5.26. The apparatus, which is substantially similar to that used by Crum *et al.* [256], is shown in Figure 5.27.

The liquid sample is contained within a cell, placed within a water bath for temperature control. Insonation from a Therasonic 1030 physiotherapeutic unit occurs through an acetate acoustic window, opposite to which is placed a brass acoustic reflector. The cell is painted black to prevent reflected luminescence confusing the image, which is formed through a low-power microscope and projected onto the input photocathode of the image intensifier. The output phosphor was photographed onto 35 mm film which when subsequently scanned by a microdensitometer (Joyce, Loebl and Co. Ltd, Mk IIIb) generated an aligned measure of the light intensity along the acoustic axis (Figure 5.26). Hydrophone measurements gave a measure of the variation in the acoustic pressure amplitude along the axis. The sonoluminesence is clearly aligned with the pressure antinodes. Having characterised such a system, these workers [296, 297] used it in an investigation of the effect of such a field on the growth of mouse tumour cells *in vitro*.

[59] See Chapter 1, section 1.1.6.
[60] See Chapter 4, section 4.4.1(b).

Figure 5.26 Results obtained with the brass block reflector at 22 °C, showing: (a) peak-to-peak acoustic pressure as measured with the needle hydrophone; (b) photograph of the appearance of the corresponding sonoluminescence as seen through the intensifier; (c) light intensity (as measured in relative units). (a), (b) and (c) have a common abscissa. Nominal intensity setting on Therasonic = 3.3 W cm^{-2}. (After Leighton *et al.* [295]. Reprinted by permission from *Physics in Medicine and Biology*, vol 33, pp. 1239–1248; Copyright © 1988 IOP Publishing Ltd.)

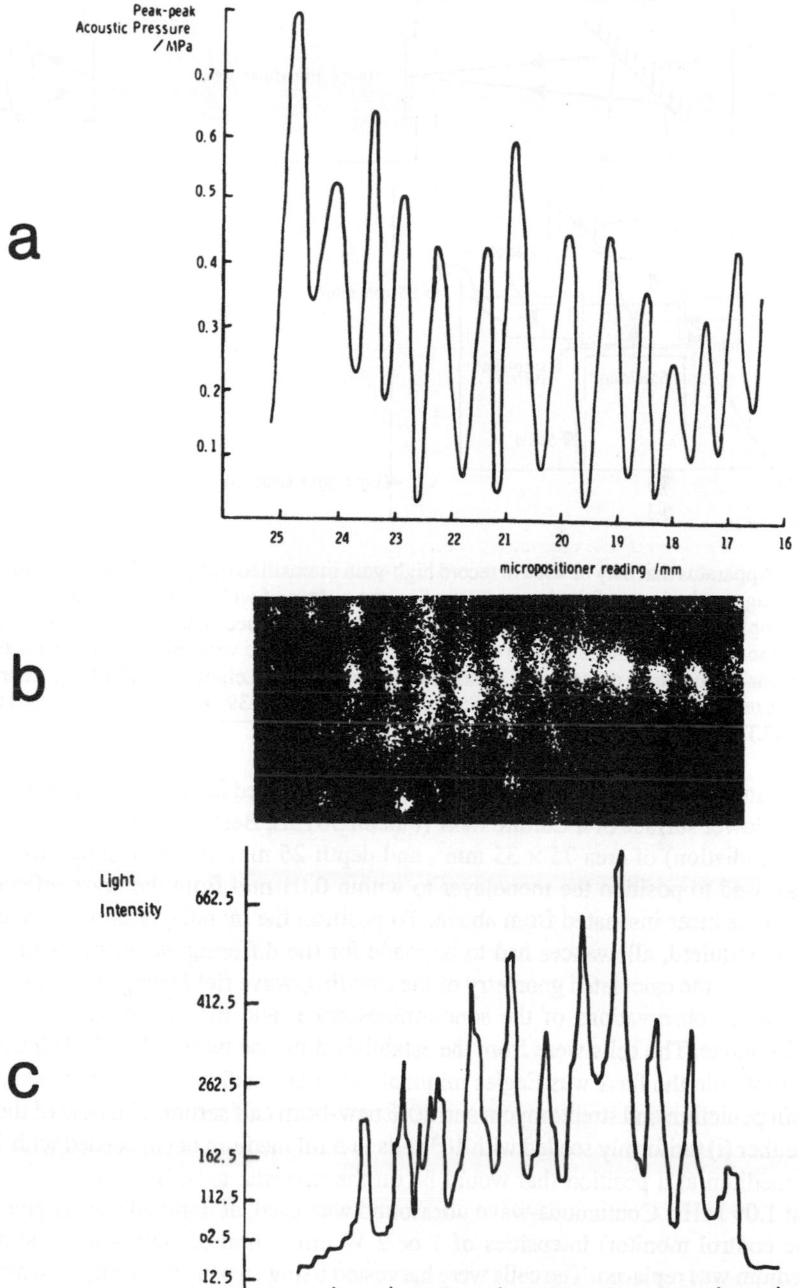
a
Peak-peak
Acoustic Pressure
/ MPa
0.7
0.6
0.5
0.4
0.3
0.2
0.1
25
24
23
22
21
20
19
18
17
16
micropositioner reading / mm
b
c
Light
Intensity
662.5
412.5
262.5
162.5
112.5
o2.5
12.5
24
23
22
21
20
19
18
17
16
micropositioner reading / mm

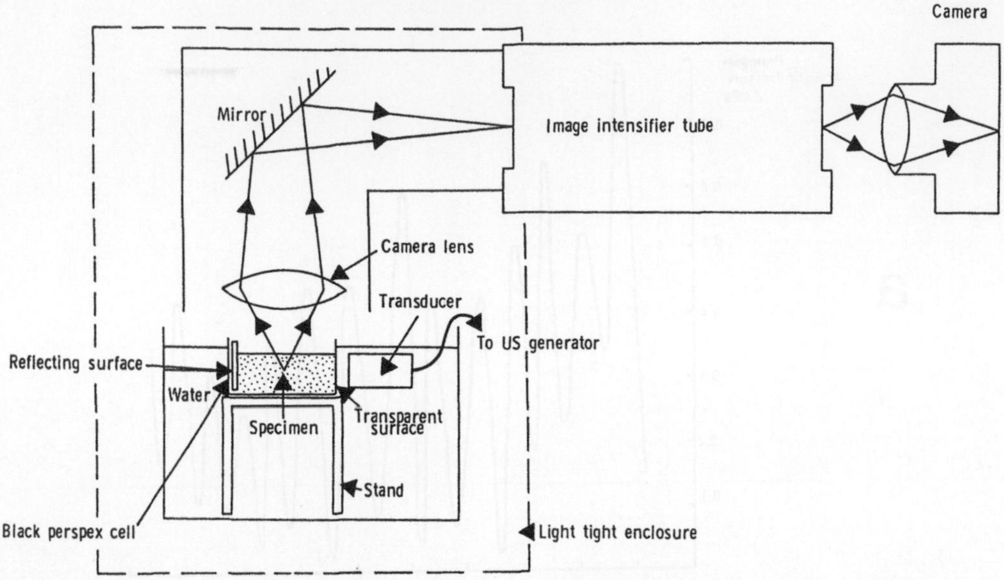

Figure 5.27 Apparatus that may be used to record high-gain intensified images. A low-power microscope focuses the image to the input of the image intensifier, the output of which may be photographed. In the diagram, a sample cell is equipped with an acoustic reflector to produce standing waves: the ultrasound, generated by the transducer, enters the cell through an acetate acoustic window. The cell is placed within a water bath for temperature control and acoustic coupling. (After Leighton *et al.* [295]. Reprinted by permission from *Physics in Medicine and Biology*, vol 33, pp. 1239–1248; Copyright © 1988 IOP Publishing Ltd.)

The apparatus is shown in Figure 5.28. The cells were attached in monolayer on tissue culture plastic to the lower surface of a Culture flask (Falcon 3013E, Becton-Dickinson Ltd, sterilised by gamma irradiation) of area 75 × 35 mm^2, and depth 26 mm. A stand, supported by three screws, was used to position the monolayer to within 0.01 mm from the brass reflector. The Therasonic transducer insonated from above. To position the monolayer at pressure nodes or antinodes, as required, allowances had to be made for the differing wavelengths of sound in water and plastic, the calculated geometry of the standing-wave field being confirmed through image-intensified observations of the sonoluminescence, and through measurement with a needle hydrophone. The cells were from the established mouse tumour line EMT6/Ca/VJAC. The medium within the flask was Eagles minimal essential medium with Earles salts, supplemented with penicillin and streptomycin and 20% new-born calf serum. The base of the culture flask was either (i) uniformly seeded with 10^5 cells in 5 ml medium or (ii) seeded with 10^4 cells in 0.1 ml medium at a position that would be on the acoustic axis. Insonation lasted for 10 minutes, at 1.09 MHz. Continuous-wave ultrasound was used, at nominal (i.e. as given by the Therasonic control monitor) intensities of 1 or 2 W cm^{-2}. Immediately after treatment, the culture medium was replaced. The cells were harvested using trypsin at seventy-two hours after insonation, resuspended and counted Preliminary checks were made to check that the cells were not being dislodged from the plastic simply by mechanical action (streaming, microstreaming, or bubble involution etc.).

Figure 5.29 shows the results, averaged and normalised to the controls, for insonation at (a) the pressure antinodes and (b) the pressure nodes. The results show that there is no statistical reduction in the number of cells when insonated at the pressure node, as compared with the

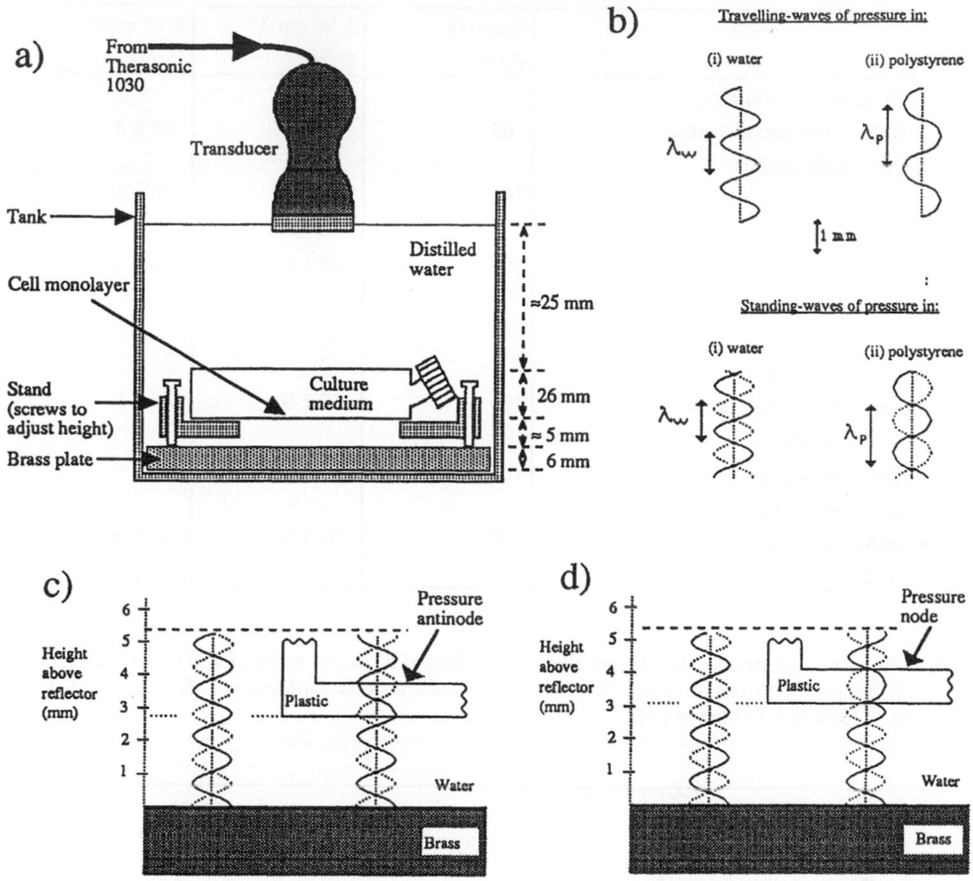

Figure 5.28 (a) Apparatus used to insonate cell culture. (b) Schematic of relative acoustic wavelengths in water (λ_w) and polystyrene (λ_p). These wavelength scales are then used to show position of cell for insonation of cells at (c) pressure antinodes and (d) pressure nodes.

control. However, cells from the pressure antinode are reduced to around 80% at 1 W cm^{-2}, and about 70% at 2 W cm^{-2}. There is no statistical difference between cells seeded on the acoustic axis and those seeded uniformly in the beam.

To put these reductions in cell numbers in perhaps a more familiar perspective: the fourteen flasks were seeded with 10^5 cells, and ten of them subjected to various doses of 250-kVp X-rays twenty-four hours later (four were controls). The dose rate was 0.6 Gy min^{-1}. The remaining four flasks acted as controls. The cells were harvested and counted 72 hours later. The number of cells, normalised to the control, is shown in Figure 5.30. Doses of 1.4 Gy and 2.0 Gy produced reductions in cell numbers comparable to the ten-minute insonations at 1 W cm^{-2} and 2 W cm^{-2} respectively. These experiments indicated a reduction in the number of cells harvested 72 hours after insonation, the reduction increasing with increasing ultrasonic intensity. The effect was seen only for cells positioned at the pressure antinodes, and not for those at the pressure nodes (separated by a distance of 0.34 mm). Sonoluminescence is produced by such fields, and is confined to the region around the pressure antinode. However, one cannot simply assign the

	Control (%)	1 W cm ^{2} (%)	2 W cm ^{2} (%)
Pressure antinodes			
Cells uniformly seeded (4 experiments)	100	75 ± 6	63 ± 8
Cells locally seeded (3 experiments)	100	82 ± 7	70 ± 5
Grand average (7 experiments)	100	78 ± 4	68 ± 4
Pressure nodes			
Cells uniformly seeded (3 experiments)	100	100 ± 8	104 ± 5
Cells locally seeded (3 experiments)	100	95 ± 7	99 ± 6
Grand average (6 experiments)	100	97 ± 5	101 ± 2

Figure 5.29 Average number of cells per flask normalised to a control value of 100 (after Pickworth *et al.* [297]). Reprinted by permission from *Physics in Medicine and Biology*, vol. 34, pp. 1553–1560; Copyright © 1989 IOP Publishing Ltd.

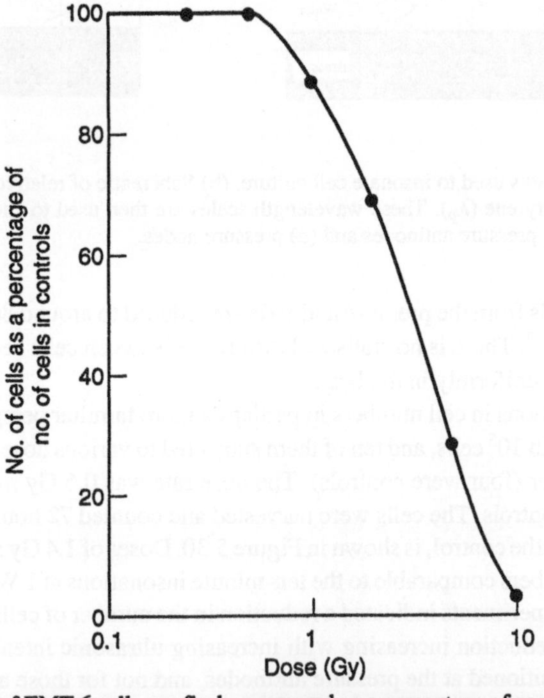

Figure 5.30 Number of EMT 6 cells per flask, expressed as a percentage of controls, 72 h after various doses of 250 kVp X-rays (after Pickworth *et al.* [297]). Reprinted by permission from *Physics in Medicine and Biology*, vol. 34, pp. 1553–1560; Copyright © 1989 IOP Publishing Ltd.

reduction in cell numbers to the free radicals associated with ultrasonic cavitation generated by this clinical physiotherapeutic instrument. Mechanical forces are exerted on the cells during insonation, as evidenced by their detachment under 1 W cm^{-2}. Such forces might produce damage sufficient to inhibit subsequent cell division, without causing detachment, at the 2 W cm^{-2} or 1 W cm^{-2} levels. Microscopic examination of the cells after ten-minute insonation at 2 W cm^{-2} did not suggest significant mechanical damage. Other mechanisms might conceivably be responsible for the inhibition, such as the release of toxins from the insonated plastic, though to minimise the chance of this the culture medium was changed immediately after insonation, and no such toxins were in evidence at the pressure nodes.

These researchers went on to investigate the possibility of setting up a similar standing-wave field *in vivo* in humans, and searching for sonoluminescence from it. The region of insonation was the human cheek. The transducer was held against the outer surface, and the light detection system examined the inside surface (i.e. the interior of the mouth) for sonoluminescence. The human cheek was chosen for insonation for four reasons. Firstly, it attenuates light passing through it to a lesser extent than other regions, because it is only about 6 mm thick. At 700 nm wavelength, 8% intensity is transmitted, and at 600 nm, 4% [298]. Secondly, because it is thin, acoustic attenuation as 1 MHz sound passes through the cheek is only about 3% in pressure amplitude. Thirdly, because the attenuation is low, and because there is a considerable acoustic impedance mismatch at the air–flesh (or air–light guide) interface at the end remote from the transducer, a partial standing-wave system will be set up. Fourthly, if the light sensor (an optic fibre, leading to a photomultiplier) is held against the inner surface of the cheek, the chances of detecting sonoluminescence are improved. This is because this moist, living tissue would be a better site for cavitation than the dry, dead skin that covers most regions of the body. Despite using a variety of arrangements to insonate the cheek from outside, including the insertion of water bags between the transducer and the skin, no statistically significant sonoluminescent was detectable with the photomultiplication system. No sonoluminescence was detected. Statistical analysis enabled the following conclusions to be drawn. If sonoluminescence was to have occurred in the cheek, it would be at a level five times the noise, that is about 300 c.p.s. as an upper limit. Since the volume of tissue insonated here was about 3 cm^3, this corresponds to 100 photons per second per cm^3 tissue. Since the same sound field in water produces about 2×10^4 photons per second per cm^3 [299], at most the photon production rate from sonoluminescence occurring in cheek tissue must be 1/40th the level found in aerated tap water.

When living tissue is exposed to ultrasound, radiation forces can arrest the motion of cells flowing in the blood vessels. The cells have been observed to cluster into bands at half-interval spacings despite the pressure of the blood pulse, by Dyson *et al.* [300] in 3.5-day-old chick embryos, and by ter Haar [301] in the blood vessels of mouse uteri. Unless the ultrasonic intensity is high enough, the bands break up and reform each cardiac cycle as a result of blood pulse pressure, but sufficiently high intensities will arrest the cells (Figure 5.31). ter Haar and Wyard [302] analysed the forces involved in the formation of these bands, including viscous drag, standing-wave radiation pressure, inter-particle forces, forces associated with gas bubbles, forces due to the temperature-dependence of viscosity, and Oseen-type forces. The latter arise because, as discussed in Chapter 1, section 1.2.3(a), sound waves propagating through a medium may be distorted from the sinusoid, and in the extreme case will form a 'saw-tooth'. Therefore the rate of change of momentum of a particle in the sound beam will be greater at the leading edge than the trailing edge, and so there will be a net force imbalance on the particle [302, 303]. ter Haar and Wyard [302] produced models for the cell banding in standing-wave fields.

Dinno *et al.* [304] found reversible changes to occur in the transport of sodium across the epidermis of abdominal frog skin, insonated with continuous-wave ultrasound at 1 MHz and

(a)

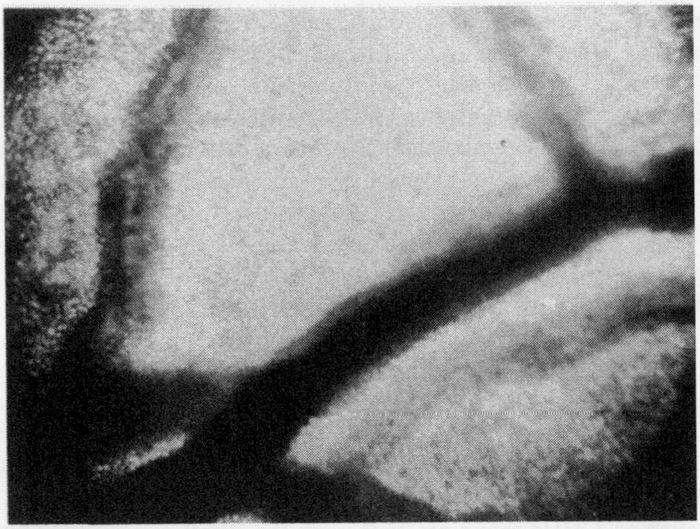

(b)

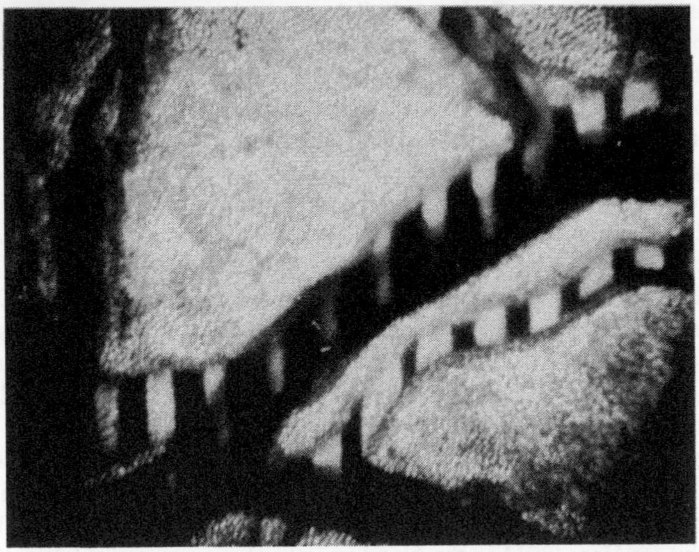

Figure 5.31 (a) Blood vessels of the area vasculosa of 3.5-day-old chick embryo in the absence of ultrasonic irradiation (after ter Haar and Wyard [302]). (b) Red blood cell banding in the blood vessels of a 3.5-day-old chick embryo in the presence of an ultrasonic standing wave field. Frequency = 3 MHz; intensity = 1 W cm^{-2}; separation of the bands = 0.25 mm. (After ter Haar and Wyard [302]. Reprinted with permission from *Ultrasound in Medicine and Biology*, vol. 4, pp. 111–123; Copyright © 1978 Pergamon Press Ltd.)

60–480 mW cm^{-2} from a commercial therapeutic source, in a field that contained a standing-wave component. Child *et al.* [305] found that continuous-wave 1-MHz travelling waves (3 W cm^{-2}) and nominally 1-W cm^{-2} standing waves killed about one-third of exposed fruitfly eggs.

Kerr *et al.* [306] insonated *in vitro* HeLa cells in suspension and surface culture at 0.75 MHz and high intensities (up to I_{SP} = 6.5 W cm^{-2}), and found no effects. However, in a standing-wave field in the same intensity range, cell membrane damage occurred in the suspension, but not in the surface culture. The cells had been placed in a chamber at the last axial maximum of the progressive-wave field. In an attempt to explain why cells exposed to such high intensities exhibited no membrane damage, Watmough *et al.* [307] in 1990 discussed the effect of streaming on the cavitation threshold in a chamber placed in distilled water at the last axial maximum of a 0.75-MHz standing-wave field. This study was in responso to their observation of bubble formation outside the chamber, both above and below it, at intensities lower than those required to form bubbles within the chamber. A theoretical study of streaming in travelling- and standing-wave fields lead to the discovery that, because of the lack of an energy gradient there, no bulk streaming would be observed if the liquid at the last axial maximum were enclosed. The authors suggest that this reduces the cavitation threshold, as streaming might (i) convect nuclei to the pressure maximum where they may grow by rectified diffusion, (ii) flush out partially degassed water from the pressure antinodes, and so promote further growth by rectified diffusion,[61] (iii) allow Bernoulli forces to bring about agglomeration. The results also suggest that the threshold may depend on the interval between filling the chamber with the cell suspension and the onset of insonation. These observations bear relation to the discussion of cavitation enhancement in the next section, and Watmough *et al.* [307] themselves include an interesting discussion on the results of rotation enhancement experiments.

As a final note to this section, Figure 5.32 demonstrated how fields can arise when one tries to set up a particular MHz acoustic field in a tank of a size commonly used in laboratories, as a result of finite beam sizes, multiple reflections (including from the free upper surface of the liquid) and near-field effects. A needle hydrophone was used to take measurements of the near-field generated in a 15 cm × 15 cm × 30 cm long water tank by a commercial physiotherapeutic unit at $I_{SATA} = 0.5$ W cm^{-2} continuous wave, the transducer of which had an effective

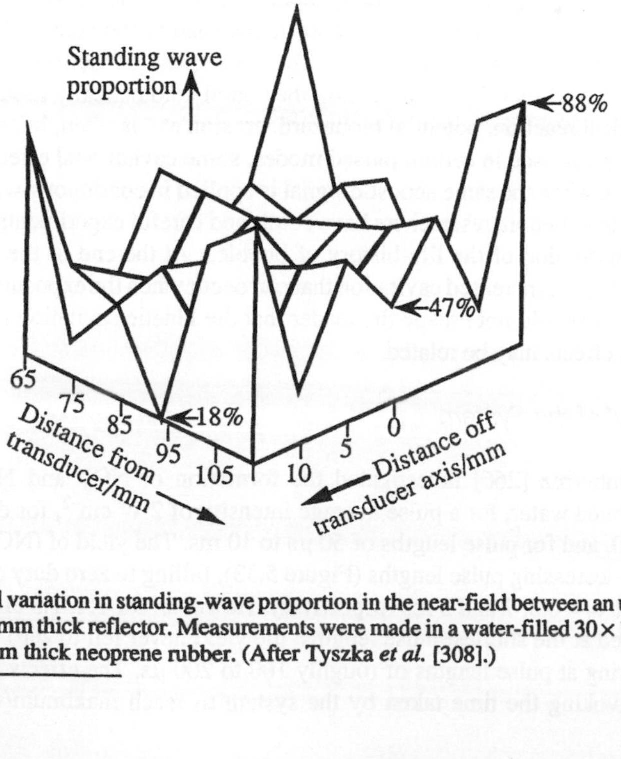

Figure 5.32 Spatial variation in standing-wave proportion in the near-field between an unfocused 1-MHz transducer and a 10-mm thick reflector. Measurements were made in a water-filled 30 × 15 × 15 cm^3 water tank, lined with 3-mm thick neoprene rubber. (After Tyszka *et al.* [308].)

[61]See section 5.3.1(e).

radiating area of 4 cm^2. The sides of the tank were coated with neoprene rubber, which is often used to minimise acoustic reflections, except for the side opposite the transducer faceplate, where various reflectors were located. Figure 5.32 shows the variation in the proportion of standing wave (as calculated from equation (1.63)) in the horizontal plane that passes through the acoustic axis. If cavitational effects are dependent on the local standing-wave component, clearly they will vary throughout the field.

Curiously, there have been several cases in the literature where experiments on bubbles in standing-wave fields have been associated with another phenomenon, that of the enhancement of cavitational effects through acoustic pulsing, and through the institution of a relative rotation between the beam and the sample. This enhancement of effects of the type discussed in this section will be the topic of the next section.

5.3 The Enhancement of Cavitation by Acoustic Pulsing and Sample Rotation

5.3.1 Pulse Enhancement

Introduction

At first glance, it may seem obvious that if identical intensities of sound are transmitted into a liquid for different durations, the amount of cavitational effect (e.g. sonoluminescence) observed will be reduced if the period of insonation is reduced. Ultrasound is often applied not in a continuous-wave mode, but instead with the sound divided up into discrete pulses, separated by an off-time. In any, say, thirty-second application at the same I_{SAPA}, less acoustic energy will be transmitted into the liquid in the pulsed mode than in the continuous-wave (because during the off-times, no sound is generated). The consequent deduction that the cavitational effects will be less, and therefore that pulsing the sound automatically reduces sonoluminescence, sonochemical reaction, potential biohazard, erosion etc. is often, but not always, true.

When ultrasound is used in certain pulsed modes, some cavitational effects may occur to a greater degree than when the same acoustic signal is applied in continuous-wave. In examining this interesting phenomenon, researchers have combined careful experimentation with innovative theoretical discussion of the life-history of bubbles. At the end of the section, a second phenomenon, that of the increased cavitation that can occur when the exposure vessel is rotated, is also discussed. It should not escape the reader that the kinetic cavitation mechanisms at the root of these two effects may be related.

(a) The 'Activity' of the System

Henglein and Gutierrez [266] investigated the formation of NO_3^- and NO_2^- using pulsed ultrasound in aerated water, for a pulse average intensity of 2 W cm^{-2}, for duty cycles of 1 to 1, 3, 8, 20 and 40, and for pulse lengths of 50 μs to 10 ms. The yield of ($NO_2^- + NO_3^-$) tended to decrease with decreasing pulse lengths (Figure 5.33), falling to zero duty cycles of 1:8, 1:20 and 1:40 at about $\tau_p = 0.5$, 2 and 3 ms respectively. However, for 1:1, 1:3 and 1:8 duty cycles, the yield increased at the shortest pulse lengths: the yield never fell to zero for 1:1 or 1:8, the minimum occurring at pulse lengths of roughly 100 to 200 μs. The effects were explained in simple terms, invoking the time taken by the system to reach maximum/saturation activity

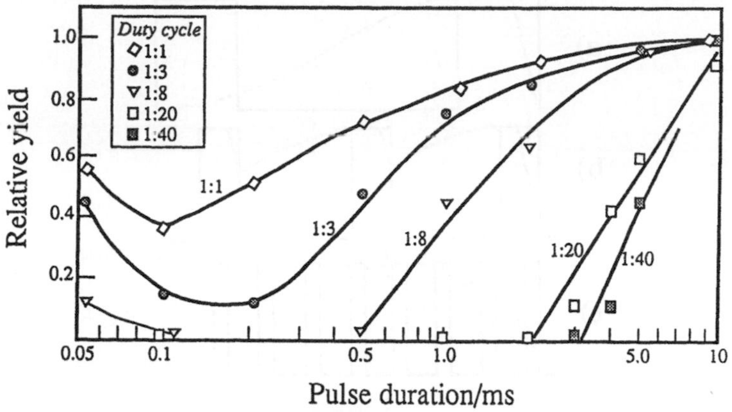

Figure 5.33 Relative yield of (NO_2^- and NO_3^-) as a function of pulse duration at various (on:off) pulsing ratios. The results are normalised such that the yield under continuous-wave insonation equals unity. (After Henglein and Gutierrez [266].)

following the onset of insonation, and of persistence of activity beyond the end of the pulse. The finite rate of increase in activity following insonation was attributed to the time taken for the growth of bubbles to resonance size, and their migration to the pressure antinodes in a standing-wave field. The persistence in activity beyond the end of insonation, which gradually decays as the off-time progresses, Henglein and Gutierrez [266] attributed to the continued presence of ultrasound in the medium as a result of multiple reflections, though later Henglein [158] invoked the far more likely reason, which is the presence, survival and gradual loss of suitable cavitation nuclei. This would, however, represent a persistence in the potential for an enhancement of the cavitational activity that will occur at the next insonation.

Figure 5.34 illustrates the proposed activity of the system in three situations. In 5.34(a) the pulse and off-time are long, so that the activity is at saturation for much of the pulse, and the decay occurs to such an extent that one pulse has negligible influence on the next. In 5.34(b) the pulse is sufficiently long for the activity to reach saturation, though the off-time is short enough for the proceeding pulse to influence its successor. Since the first pulse ceases with activity at saturation, a steady-state variation in activity is established by the second pulse. This is not the case in 5.34(c), where at first the pulse is too short for activity to saturate. However, the off-time is also short, and the persistence in activity allows the next pulse to build up to a finite level: after several pulses, a steady state is reached, where saturation is reached during the on-time. It is important to note that the insonating transducer was specifically damped to reduce ringing at the end of each pulse. Depending on the amount of ringing, there will always be an off-time so short that ringing from one pulse to the next will produce effectively continuous-wave insonation. Henglein and Guttierez use the illustrated scheme to explain their results. Figure 5.34(a) illustrates why the yield is zero, except for the longest on-times, when the off-time is long (duty cycles of 1:20, 1:40): only for the longer pulses does the finite rise-time allow sufficient activity to occur. Figure 5.34(b) is used to explain why the yield with 1:8 duty cycle falls off more slowly with decreasing pulse length than for 1:20 and 1:40 duty cycles (where the off-time is longer for the same pulse length): the shorter off-time at 1:8 allows more persistence of activity from one pulse to the next. Figure 5.34(c) explains why short pulses may produce significant yields if the off-time is short (as seen for 1:2, 1:4 and 1:8 duty cycles).

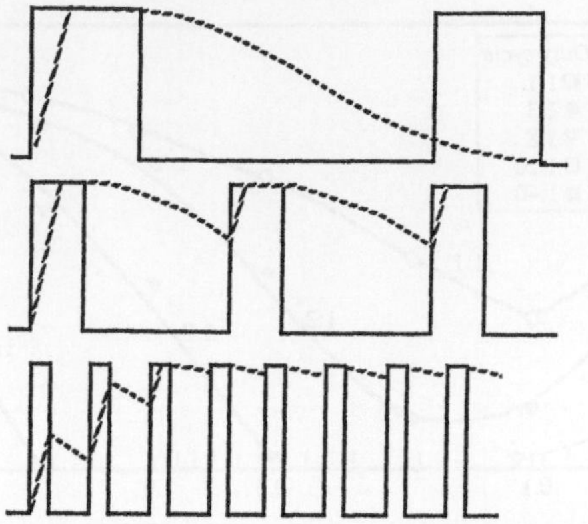

Figure 5.34 Postulated activity of the system (- - -) as a function of the insonation (solid line: high = on; low = off) (after Henglein and Gutierrez [266]).

Though adequate as far as it goes, this falls short of giving a full insight into the mechanisms, in that it relates the sound field directly to the 'activation' of the system, a quantity which is to all intents and purposes the chemical effect (since variations in the activity are used to explain the observed yields). However, the intermediate step is one of acoustic cavitation, which is too complicated to encapsulate in a simple finite rise to saturation following the onset of insonation, and by a gradual decay after its cessation. The bubble activity itself will involve nucleation and inception, threshold and population phenomena which, as will be shown in the remainder of this section, can produce a variety of effects when ultrasound is pulsed.

(b) Pulse Enhancement by Recycling Periodic Cavities Through Resonance

Interestingly enough, the first study on the pulse enhancement of cavitation also incorporated the rotation effect. In 1969, Hill *et al.* [309] published an investigation into the effect of changing the pulse length on three ultrasonically stimulated processes:

(a) the degradation of deoxyribonucleic acid macromolecules (commercially prepared polydisperse calf thymus and salmon sperm DNA, with effective molecular weights of 10^7 and 5.5×10^6 respectively);

(b) the sonochemical release of free iodine from a KI solution, which provided a more sensitive indicator for the presence of free radicals; and

(c) the generation of the first subharmonic of the driving frequency (which was investigated as a possible measure of cavitational activity).

Working at 1 MHz, in far-field travelling-wave conditions (checked by hydrophone) and with duty cycles of 1:1, 1:2 and 1:10, Hill *et al.* rotated the sample cell about a vertical axis within the sound field and water bath. This rotation technique was originally devised for a different experiment, in which the need to stir cell suspensions had been anticipated. The fortuitous incorporation of rotation generated a remarkable observation (which will be discussed more fully at the end of this chapter). This is that a rotation of 0.3–3 rev s^{-1} induced an 'all or nothing' response in all three effects investigated. The positive results discussed below are for rotated samples.

The initial experiments studied the decrease in molecular weight as the DNA was degraded, as a function of insonation time ($\leqslant$ 20 minutes) and acoustic intensity ($\leqslant$ 8 W cm^{-2} spatial average for 3 minutes) for a continuous wave. In the latter, a threshold of $I_{SA} \approx 0.4$ W cm^{-2} was observed, below which no effect was observed. The effect increased rapidly to a maximum at around 3 W cm^{-2} with a gentle decrease up to intensities of 8 W cm^{-2}. In similar conditions the release of free iodine was found to threshold $I_{SA} \approx 0.5$ W cm^{-2} ($I_{SP} \approx 1.5$ W cm^{-2}), with a maximum effect at $I_{SA} \approx 2.25$ W cm^{-2}. No sonoluminescence could be detected with the equipment. Subharmonic emission increased sharply to a maximum at $I_{SA} \approx 2$ W cm^{-2}. The similarities in these continuous-wave observations are clear, as expected if the effects have, directly or indirectly, a common source (e.g. cavitation).

The use of pulsed ultrasound on DNA and KI solution dramatically affected the results. Hill *et al.* found that decreasing the pulse length increased the effect down to pulse lengths of about 30 ms. The increase manifested itself, for example, in the rate of degradation and terminal weight of the DNA. Beyond the maximum at 30 ms, that is for smaller pulse lengths, a decrease in effect was observed. Below 10 ms, the effectiveness decreases rapidly to become negligible at pulse lengths of 1 ms.

The authors suggested that the absence of sonoluminescence, and the fact that the mechanism required an insonation period of the order of milliseconds to become established, indicated that stable rather than transient cavitation was occurring.

The explanation postulated to explain the pulse enhancement suggests that bubbles were maximally effective at resonance size [309]. During insonation, some bubbles would grow by rectified diffusion to greater than resonance size. Such bubbles would be ineffective at producing the sonochemical changes, and in a continuous-wave sound field many bubbles would be nullified in this manner. However, if the sound were pulsed, the dissolution of bubbles that would occur during the off-time would bring the radius of many of these bubbles back to below resonance size, so that when insonation restarted, they would again grow to, and beyond, resonance. In this manner a given bubble could be continually re-cycled through resonance (Figure 5.35). If the pulses are too short, not enough bubbles grow to resonance.

In its simplest terms, this theory discusses bubbles which remain as single, whole noninteracting entities. Application of later knowledge to this theory shows some interesting facets. For example, violent surface oscillations, which can set in as the bubbles approach resonance size in intense sound fields, could prevent growth to larger than resonance by gas loss through microbubble emission. However, there are enough dependent parameters to produce scenarios whereby the presence of such activities as surface oscillations, fragmentation and coalescence could in fact enhance the theory. A larger bubble is in general more prone to fragmentation. The emission of microbubbles from a large bubble undergoing surface oscillations during insonation would contribute to gas loss from the bubble, and so a decrease in bubble size, which could assist the dissolution that occurs during the off-time. Similarly the attraction of such bubbles by the mutual Bjerknes force to a bubble of less than resonance size could in turn assist with its growth. This illustrates an important point: with bubble populations in intense sound fields there are many interacting factors, which are often difficult to quantify, that will influence the effects produced by the cavitation to a greater or lesser extent.

It is also interesting to consider in detail the process of rectified diffusion. For example, the theories outlined in Chapter 4, section 4.4.3 suggest that bubbles generally cannot grow to much larger than resonance by rectified diffusion. This would suggest the possibility of a population of bubbles of just larger than resonance which accumulates as time goes by which, provided conditions and the relevant thresholds were appropriate for bubbles to grow to this point, would make this mechanism more effective. At the the end of each pulse one might expect to find a considerable number of bubbles at just greater than resonance size, ready to dissolve to below resonance.

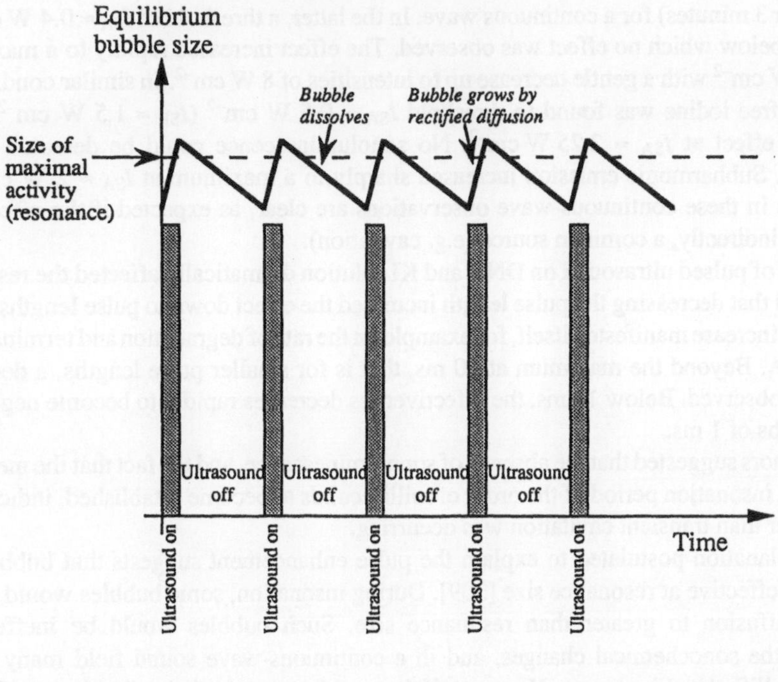

Figure 5.35 A bubble which grows to larger than resonance size through rectified diffusion during insonation may dissolve to less than resonance size during the off-time and so be continually recycled through resonance.

One must also, however, consider the later work of Church [310], outlined in Chapter 4, section 4.4.3(d)(iii), which suggests that the classic paradigm of bubbles growing to resonance size and then collapsing violently is true for only a small range of bubble radii and acoustic pressures at frequencies above 1 MHz. To relate Hill's experimental intensities to Church's model, which employs the acoustic pressure amplitude, in a plane-wave approximation, equation (1.39) can be used to show that $I_{SA} = 2.25$ W cm^{-2} (where maximum effect was obtained) is roughly equivalent to 1.8×10^5 Pa. Examination of Church's predictions at 1 MHz (Figure 4.43(a)) suggest that the actual behaviour of the bubble depends on the initial bubble size, and is by no means the simple growth to resonance that had previously been thought to be the case. This does not, however, necessarily invalidate Hill's theory, which requires the existence of some specific size (not necessarily the resonance), to which the bubble may grow through rectified diffusion, where the activity is such as to produce a maximum effect without the bubble breaking up significantly. Thus, for example, if by transient cavitation Church's threshold refers to repetitive transient cavitation,[62] where the collapse is highly energetic but the bubbles survive the rebound intact, the bubbles of initial radius 2.17–2.38 μm, which initially lie above the rectified diffusion threshold but below the transient cavitation threshold for $P_A = 1$ bar and $\nu = 1$ MHz, could be those to which the theory of Hill *et al.* refer. It should also be noted that Hill *et al.* themselves suggested that the effect might arise through a form of stable cavitation.

This illustrates a second point. Since all such theories include assumptions, it is important to understand which are critical and which are not for a given theory, since this will determine

[62] See Chapter 4, section 4.3.2.

the specific cavitation conditions for which it is valid. The theory of Hill *et al.* requires that the bubbles which bring about the effect are undergoing violent but repetitive cavitation, and that there exists at least one specific radius pertinent to each bubble (which may be the resonance, or some threshold) where the activity is a maximum. The other assumptions are not critical. It is interesting to note that strict adherence to a classical distinction between transient and stable cavitation would tend to make the theories of Hill and Church in this particular case incompatible: however, with real bubble populations, rigid divisions can be inappropriate as anything more than guidelines.

In a similar experiment, Clarke and Hill [311] in 1970 examined the ultrasonically induced release of free iodine from KI solution, DNA degradation and the disruption of mouse lymphoma cells in a rotating sample. The apparatus was similar to that employed by Hill *et al.* [309]. Pulse length was increased from 1 ms in factors of 10 for duty cycles of 1:1 and 1:10. Both cell death and iodine release showed enhancement for pulses of length 10 ms and 100 ms (Figure 5.36). Longer pulses and continuous-wave were less effective, and pulse lengths of 1 ms virtually ineffective.

From a biological viewpoint, Clarke and Hill concluded from their results that cell death and iodine release both resulted from cavitation effects. In an analogous manner to Hill *et al.* [309], they proposed that stable, rather than transient, cavitation was the cause. The mechanism of damage was the result of microstreaming, which occurs maximally at one particular bubble size (e.g. the bubble pulsation frequency resonant with the acoustic field). The high amplitude of radial pulsation generating maximum microstreaming results in high hydrodynamic shear and tensile stresses in the region of the bubble, which could disrupt cells. The large amplitude of

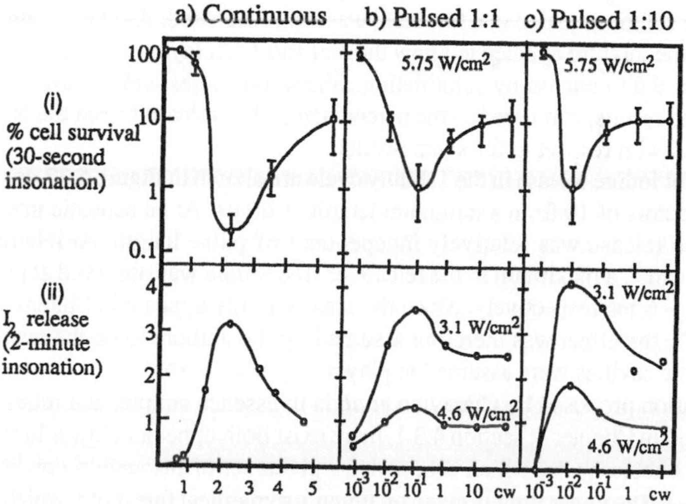

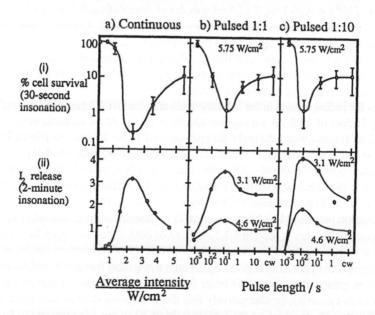

Figure 5.36 The variation in (i) cell survival and (ii) iodine release for variation in (a) continuous-wave ultrasound intensity, for (b) 1:1 duty cycle pulse length and (c) 1:10 duty cycle pulse length. All intensities are spatially averaged over the sample region, with $\nu = 1$ MHz. Cell survival in (b) and (c) was for $I_{PA} = 5.75$ W cm^{-2}. Iodine release in (b) and (c) was measured at $I_{PA} = 3.1$ and 4.6 W cm^{-2}. (After Clarke and Hill [311].)

pulsation might well cause any free radical production to be maximal at this point. Following Hill *et al.* [309], they proposed that the continual recycling of bubbles through resonance as a result of rectified diffusion during insonation, and dissolution during the off-time, causes enhanced effects with pulsed ultrasound over continuous-wave. They suggest that longer pulses are less effective than those of length 10–100 ms because bubbles grow well beyond resonance size during each pulse. This is inconsistent with later work which showed that the threshold increases rapidly, and the growth rate decreases, as bubbles grow past resonant size, though as shown above, the process should be re-examined in the light of Church's analyses. It should be remembered when considering such an effect that with a fixed duty cycle such as was used by Clarke and Hill, a change in pulse length means a change in the off-time, and so any observed effect might be due to one or the other, or a combination.

(c) Pulse Enhancement by Survival of Unstabilised Nuclei

In 1981, Ciaravino *et al.* [312] published the results of an experiment that measured the release of an iodine isotope from NaI, where the pulse length was varied, and the duty cycle fixed at 1:1. They used a fixed acoustic frequency of 1.06 MHz and a pulse duration of between 60 μs and 60 s (varied in powers of ten). Spatial peak intensities of 10 W cm^{-2}, 20 W cm^{-2} and 30 W cm^{-2} were employed for total exposure times of 60 s, so that each experiment lasted 2 minutes (the sum of the total off-time and the total on-time, which at a 1:1 duty cycle are equal). As an indicator of cavitational activity, Ciaravino *et al.* chose the release of free iodine, modified, however, to incorporate radioactive iodine tracers for quantification of the effect. ^{131}I-labelled NaI was added to the stock solution of 0.005 N NaI, adjusted to a pH of about 8, to give a final concentration of 10^4 c.p.m./0.1 ml. To 2.8 ml samples of the solution, 0.1 ml carbon tetrachloride was added. The solution, sealed in a test tube, was then insonated. After exposure, 0.9 ml CCl_4 was added and the mixture agitated to allow the NaI and CCl_4 layers to separate. Aliquots from each layer were then counted by scintillation. Sham exposures and controls were taken for comparison. The results, expressed as the percentage of ^{131}I released from the NaI to the CCl_4, were normalised with respect to the sham results.

The results for iodine release in the 1:1 duty cycle are shown in Figure 5.37, for pulse lengths increasing by factors of 10 from a minimum length of 60 μs. At an acoustic intensity of 10 W cm^{-2}, the iodine release was relatively independent of pulse length. At intensities of 20 W cm^{-2} and 30 W cm^{-2}, a maximum in the release of free iodine was observed at pulse lengths of ~ 6–60 ms and ~ 6 ms respectively. Since the maxima only appeared at intensities of greater than 10 W cm^{-2}, the effect was therefore assumed by the authors to be the result of transient cavitation. Stable cavities were assumed to play no significant role.

The explanation proposed by Ciaravino *et al.* is in essence simple, and relies upon the fact that, as discussed in Chapter 4, section 4.3.1, there exist both upper and lower limits to the range of equilibrium bubble nuclei which can, under a given set of insonation conditions, grow to become transient. The upper limit is given by Flynn's dynamical threshold, which demands that R_{max}/R_0 must be greater than about 2 in order for there to be significant energy concentration within the collapsing cavity, or alternatively that the collapsing bubble wall reach supersonic speeds. Ciaravino *et al.*, considering the 1-MHz field of 30 W cm^{-2} (equivalent to $P_A = 9.5$ bar) where pulse enhancement was observed, calculated that a nucleus of initial radius $R_0 = 1$ μm expands to a maximum $R_{max} \approx 14R_0$, whilst one with $R_0 = 5$ μm expands to a maximum radius $R_{max} \approx 3R_0$ (i.e. both reach similar maximum radii, in keeping with the observations of Chapter 4, section 4.3.1). Provided a bubble is larger than the lower limit and so will undergo initial explosive growth, then the smaller the initial radius R_0 of the nucleus, the greater the ratio

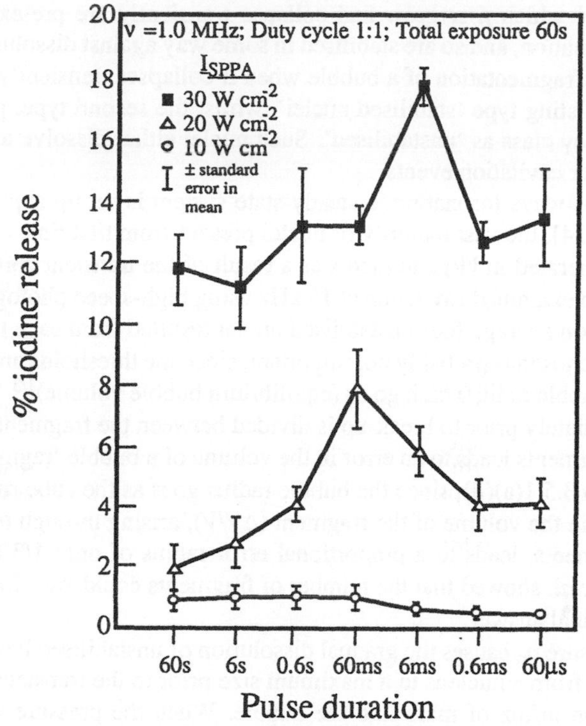

Figure 5.37 Iodine release as a function of ultrasound exposure duration (60 s to 60 μs) at intensities of 10, 20 and 30 W cm^{-2}, 1:1 duty cycle (after Ciaravino *et al.* [312]). Reprinted by permission from *Ultrasound in Medicine and Biology*, vol. 7, pp. 159–166; Copyright © 1981 Pergamon Press Ltd.

R_{max}/R_0, and consequently the more violent the subsequent collapse. Since to effect iodine release the bubble activity must exceed some threshold energy, there is a range of initial equilibrium bubble radius that will bring about iodine release. Referring back to Figure 4.16(a), produced by Flynn and Church [313] to explain the results of Ciaravino *et al*, the upper and lower limits to the range of initial bubble sizes that can go transient is clear. Flynn and Church show the bubble response to a pressure amplitude of 10 bar, and to 7.5 bar for comparison. This they relate to the insonation conditions of Ciaravino *et al.*, equating the 10, 20 and 30 W cm^{-2} of the latter to acoustic pressure amplitudes of 5.44, 7.69 and 9.42 bar (equation (1.39)), noting that pulse enhancement was only observed at 20 and 30 W cm^{-2}. Flynn and Church took the upper limit from Apfel's criterion [26] that an appropriate upper threshold for cavitation may be taken at the equilibrium radius for which $R_{max} = 2.3R_0$ (see Chapter 4, section 4.3.1). From Figure 4.16 it is clear that the upper limit is around 7 μm. The lower limit, set by the condition that the nucleus must first undergo initial explosive growth as discussed in Chapter 4, was estimated by Ciaravino *et al.* to be $R_0 = 0.055$ μm. Using Figure 4.16(a), Flynn and Church found that for such growth R_0 should exceed 0.06 μm. As is clear from Figure 4.16, it is the bubbles of initial radius R_0 lying between these limits which undergo energetic collapse, and so generate the chemical effect.

The authors note that for the bubble sizes in question in this experiment ($\leqslant$1 μm), the threshold for rectified diffusion is almost identical to that for transient cavitation, and so they infer that rectified diffusion does not play an important role in this pulse enhancement mechanism.

The bubble nuclei which *first* grow and collapse transiently are pre-existing within the medium prior to insonation, and so are stabilised in some way against dissolution. Other nuclei are generated by the fragmentation of a bubble when it collapses transiently. Ciaravino *et al.* term the first, pre-existing type 'stabilised nuclei', whilst the second type, produced through transient collapse, they class as 'unstabilised'. Such nuclei either dissolve away, coalesce, or survive to seed further cavitation events.

During continuous-wave insonation, a steady-state system is set up approximately within the first 10 cycles [314], the vast majority of nuclei present from that time onwards being the unstabilised type generated in large numbers as a result of the transient collapses. Following Akulichev [314] who examined cavitation at 15 kHz using high-speed photography, Ciaravino *et al.* suggested that on average four unstabilised nuclei resulted from each transient collapse. The precise number chosen is probably not important, since the thresholds are defined in terms of the equilibrium bubble radii, which go as (equilibrium bubble volume)$^{1/3}$. The mass content of the bubble immediately prior to break-up is divided between the fragments, and so an error in the number of fragments leads to an error in the volume of a bubble fragment. As explained in Chapter 3, section 3.5.1(a)(ii), since the bubble radius goes as the cube root of the volume, a proportional error in the volume of the fragment ($\Delta V/V$), arising through incorrect choice in the number of fragments, leads to a proportional error radius of only 1/3 as much ($\Delta R/R \approx \Delta V/3V$). Ciaravino *et al.* showed that the number of fragments could vary from 2 to 8 without much effect on the final answer.

The Laplace pressure p_σ causes the gradual dissolution of unstabilised bubbles.[63] However, as a bubble expands from a nucleus to a maximum size prior to the transient collapse, the fall in pressure causes an influx of mass into the bubble. When the pressure within the bubble increases to above equilibrium during the collapse phase, mass is lost from the bubble. In general, the expansion phase is slower than the collapse, in which case there could be a net flow of gas into the bubble. Thus the bubble immediately prior to fragmentation could have a greater or lesser mass of gas contained than had the original nucleus.

The combined effects of mass flow and bubble division into fragments means that after a nucleus of equilibrium radius R_{01} has undergone a transient collapse, the resulting unstabilised nuclei (its daughter fragments) may have equilibrium radii *immediately* following the collapse of R_0^{dhtr}, which may be larger or smaller than R_{01}. Ciaravino *et al.* calculate that in their sound field:

$$R_0^{\text{dhtr}} > R_{01} \quad \text{if} \quad R_{01} < 5\ \mu\text{m}, \quad \text{but}$$

$$R_0^{\text{dhtr}} < R_{01} \quad \text{if} \quad 0.1\ \mu\text{m} < R_{01} < 0.5\ \mu\text{m}$$

If the original nucleii population has $0.055\ \mu\text{m} < R_{01} < 1\ \mu\text{m}$ (i.e. the range which can produce transient collapse, as described above), then the tendency is for the generation of nuclei with the narrow distribution $0.2\ \mu\text{m} < R_0^{\text{dhtr}} < 0.7\ \mu\text{m}$. Comparing this with the range of nuclei that can undergo transient collapse ($0.055\ \mu\text{m} \geq R_0 \geq 7\ \mu\text{m}$), it is clear that *continuous-wave* insonation will remove from the population the very smallest nuclei.

The steady-state population of unstabilised nuclei (those generated by bubble fragmentation) is established within the first ten acoustic cycles. If the pulse length is longer than this 'start-up' time, cavitation at times after this interval during the on-time will be based upon the same population distribution as for continuous-wave insonation. In Ciaravino's 1-MHz experiment, this 'start-up' time is 10 μs, so that if the maximum in cavitational effects observed at pulse

[63] See Chapter 2, sections 2.1.1 and 2.1.2 and Chapter 4, section 4.4.3(b)(ii).

lengths of ~ 6–60 ms is due to an additional element in the nuclei population (as seems reasonable), then it must result from the survival of unstabilised nuclei from the previous pulse.

The underlying basis of the theory is in essence simple. Ciaravino *et al.* [312] postulate that the maximum is due to the persistence of unstabilised nuclei capable of undergoing transient collapse, generated by the previous pulse. As stated above, if the pulse is longer than the time taken to establish a steady-state population, the population will be the same as that which results from the use of continuous-wave ultrasound. However, since the duty cycle was *fixed* in Ciaravino *et al.*'s experiments, then a longer pulse means that the off-time is longer. During this off-time, nuclei will be dissolving away. If the pulses are long, significant dissolution will occur, in the limit removing all the unstabilised nuclei generated by the last pulse from the medium. In contrast, if short pulses are used, then the off-time will be short, and little dissolution will occur between pulses. Therefore when either very long or very short pulses are used in this 1:1 duty cycle, the nuclei distribution will be very similar to that generated in the continuous-wave mode. In some way an optimum is reached for pulses and off-times of intermediate length, when the bubble fragments generated during the previous pulse will be present at the start of the next pulse and provide a population that will increase the amount of violent cavitation to a level above that found with continuous-wave.

Ciaravino *et al.* propose the following mechanism by which this optimum may arise. Since rectified diffusion plays no part in this process (see above), then if the pulse length is τ_p, the bubble fragment has a time interval in which it will dissolve that is (i) a minimum of τ_p, corresponding to a fragment created at the very end of one pulse, and which goes transient at the start of the next; and (ii) a maximum of $3\tau_p$, corresponding to a fragment created at the very start of one pulse, and which goes transient at the end of the next. Ciaravino *et al.* then consider a nuclei of size R_{01}, the possible initial sizes of the daughter fragments generated by the collapse of the mother, and the final sizes these fragments can reach by dissolving for a time interval which is greater than τ_p but less than $3\tau_p$, using the work[64] of Epstein and Plesset [315].

They calculated that:

(i) For pulse lengths (and off-times) of 60 ms, only nuclei with initial radii greater than about 0.75 μm survive to the next pulse, and 58% of these will have radius less than 0.3 μm.

(ii) For pulse lengths (and off-times) of 6 ms, only nuclei with initial radii greater than about 0.2 μm survive to the next pulse, and 47% of these will have radii less than 0.3 μm.

(iii) As expected, when τ_p is very long or very short in the 1:1 duty cycle, the nucleus population during insonation is similar to the steady state during continuous-wave insonation. When $\tau_p = 0.6$ ms, nuclei generated by one pulse survive relatively unchanged to the next pulse. When $\tau_p = 130$ ms, all the unstabilised nuclei generated by one pulse with initial radii less than 1 μm dissolve away before the start of the next pulse.

Thus a pulse of length 6–60 ms will cause, in the population encountered by a second pulse, the presence of unstabilised nuclei of a size smaller than that being generated by the second pulse (or by a continuous-wave). The smaller nuclei would generate more violent collapses, and so enhance the cavitational effects observed. That the collapses of these unstabilised nuclei are of a sufficient number to produce an observable enhancement is supported by the estimates of Sirotyuk [316], which suggest that during insonation the number of unstabilised nuclei generated by collapse greatly exceeds the population of pre-existing stabilised nuclei.

Thus the theory of Ciaravino *et al.*, substantiated by quantitative calculation based upon a simple model, predicts enhancement for pulses of length 6–60 ms in their 1-MHz field, compared with other pulse lengths. In addition, the removal from the population of the smallest

[64]See Chapter 4, section 4.4.3(b)(ii).

nuclei (≲ 0.3 μm) which occurs during continuous-wave insonation would depress cavitation below that found for the 6–60 ms pulses (where the population is the steady-state continuous-wave one *plus* smaller nuclei surviving from the previous pulse).

The mechanism described above [312] was supported in 1984 by more thorough calculations by Flynn and Church [313]. By considering the threshold pressure for transient cavitation as a function of the initial bubble radius and noting that the enhancement was observed at 20 and 30 W cm^{-2}, but not at 10 W cm^{-2}, Flynn and Church were able to say that the nuclei which were making a difference, and were responsible for the observed maximum in activity, were in the size range 0.063 μm ≲ R_0 ≲ 0.13 μm (see Figure 4.16(b)). The lower limit is set by the threshold radius at 30 W cm^{-2}, and the upper limit by that at 10 W cm^{-2}. It should be noted that the accuracy of these limits is imposed by the discrete nature of the experiments performed by Ciaravino *et al.*: since, for example, no measurements were made between 10 and 20 W cm^{-2}, one cannot say with any greater accuracy what was the minimum power to generate the enhancement. They also found that the model of a cavity fragmenting into 50 daughters on collapse was more consistent with the proposed mechanism than one where only 4 daughters are formed.

(d) Pulse Enhancement by Transient Excitation

Another mechanism by which pulse enhancement can occur was suggested by Leighton *et al.* [294, 317], and is indicated in the photographs included in Figure 4.19. The figure shows the response of a bubble field to the start of insonation, and in fact is taken at the start of a sound pulse. The 10-kHz field, produced on the axis of the cylindrical transducer described in Chapter 1, section 1.2.2(b)(i), had an amplitude of 2.4 bar. As described in Chapter 4, section 4.3, the larger bubbles undergo stable cavitation, whilst the smaller ones undergo explosive growth, followed by violent collapse. All this begins soon after the start of insonation, the bubble behaving in a manner predicted in Figure 4.8. Key to the mechanism is the fact that the explosive growth and rapid collapse occur only immediately after every insonation: despite the fact that insonation continues, no more such transient events were observed,[65] the necessary conditions for the event being both the sound pulse and preceding quiet period. This suggests that transients in the response are a key element.

Evidence for such pulse enhancement, associated with increased effects and emissions very soon after the start of insonation, was provided by examining the sonoluminescence produced by this system using photomultiplicative techniques to time-resolve the light emission [318]. The results of the photomultiplier output, and the simultaneous hydrophone signal, can be seen in Figure 5.38 for (a) the start of 10 kHz continuous-wave insonation and (b) pulsing conditions (16 cycles on, 16 cycles off). Vertical deflection in the photomultiplier trace corresponds to the detection of light, which in the figure is clearly associated with the onset of insonation. During pulsing, sonoluminescence associated with the first pulse is the most intense since the steady-state motion will not be completely damped out during these off-times, and so the transients will have a less violent effect for the subsequent pulses.

This mechanism was suggested following the real-time observation of the sonoluminescence produced by the *continuous-wave* field from a 1-MHz physiotherapeutic generator by Leighton

[65] Owing to the Nyquist criteria, this experiment was performed repeatedly to verify that violent cavitation from initially small bubbles occurred only at the start of insonation. In the sample of aerated water used, it was assumed that, since the nuclei population proved to be sufficient for transient cavitation whenever the sound field was first activated, it was similarly sufficient throughout the insonation. The whole story probably incorporates elements of this and of the previous theory.

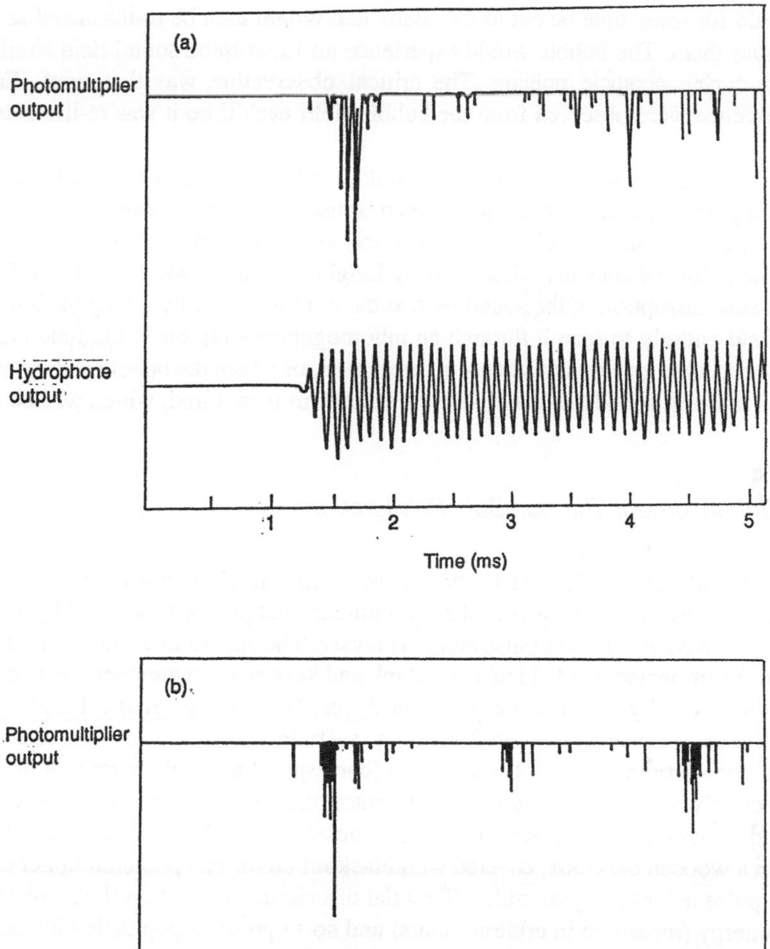

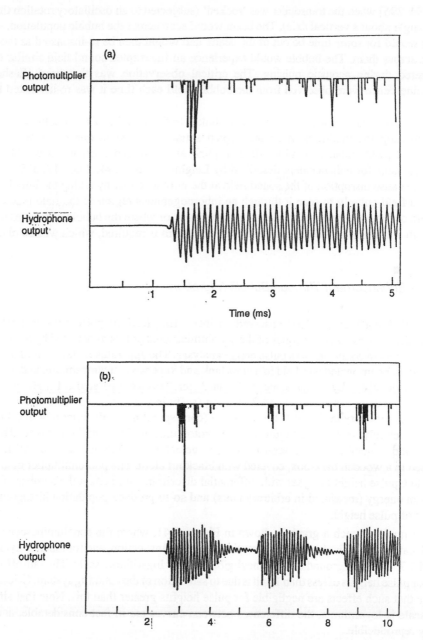

Figure 5.38 (a) Simultaneous photomultiplier and hydrophone outputs are shown on a common time-axis for the insonation of water at 10 kHz, with 2.4 ± 0.1-atm pressure amplitude, at the start of continuous-wave insonation. (b) Simultaneous photomultiplier and hydrophone outputs are shown on a common time-axis for the insonation of water at 10 kHz, with 2.4 ± 0.1-atm pressure amplitude. Three pulses of sound are used: both the pulse length and the separation are equal to 16 acoustic cycles. (After Leighton [318].)

et al. [294, 295] when the transducer was 'rocked' (subjected to an oscillatory rotation through a small angle about a vertical axis). The beam would scan across the bubble population, so that bubbles would for some time be out of the beam, and would then be re-insonated as the beam scanned across them. The bubble would experience an interrupted sound field similar to that encountered during acoustic pulsing. The critical observation was that brief 'flashes' of sonoluminescence were observed from the bubble field each time it was re-insonated by the beam.

It is interesting how pulsing and rotation are linked in this way for this particular mechanism, both providing the bubble with an interrupted sound field, when one considers how the discovery of pulse enhancement by Hill *et al.* [309] also involved a rotation effect. However, the mechanisms for enhancement described by Leighton *et al.* [294, 295, 317, 318] requires rotation to cause disruption of the sound field at the bubble, either by taking the bubble out of the sound field entirely to scan it through an inhomogeneous region of the field (such as the near-field or focus) to generate transients in the motion, or where the bubble undergoes relative rotation in a uniform sound field, a different mechanism is required, which will be discussed later.

(e) Pulse Enhancement Through Bubble Migration

In 1988, Pickworth *et al.* [261] observed an interesting form of pulse enhancement when studying the *distribution* of energies of the sonoluminescent pulses as detected by a photomultiplication system connected to a pulse height analyser. The apparatus is shown in Figure 5.39. The liquid to be insonated was held in a glass tank and kept at room temperature, and insonated vertically from above by a Therasonic 1030 transducer. This was operated at 1 MHz, producing continuous-wave or a pulse of sound 2 ms long (sufficient to set up 45–62% standing-wave conditions), with duty cycles of 1:2, 1:4 and 1:7 (corresponding to off-times of 4 ms, 8 ms and 14 ms respectively, as shown in Figure 5.40). A photomultiplier (RCA 8575) was placed beneath the tank and aligned along the acoustic axis, to detect any sonoluminescence. All this was contained in a wooden dark-box, covered with blackout cloth. The photomultiplier output was passed to a pulse height analyser with differential discriminator to count the number of pulses at a given energy (measured in arbitrary units) and so to produce population histograms as a function of pulse height.

An example of such a graph is shown in Figure 5.41, where the sonoluminescence from aerated water at 22° C is shown for the continuous-wave insonation, and for the duty cycles 1:2, 1:4 and 1:7. The background count is negligible on this logarithmic scale. The fall-off in each curve for pulse heights of less than 1 unit is due to saturation of the counting system. Calculations indicate that such effects are negligible for pulse heights greater than this. Note that since the count scale is logarithmic, the differences between curves are in fact considerable, and were entirely reproducible.

Without pulse enhancement, one would expect the magnitude of the sonoluminescence to increase in ascending order for 1:7, 1:4 and 1:2 and continuous-wave, and the count rate ratios to be respectively 1/8:1/5:1/3:1 (i.e. in the ratios of the total insonation times). This can be seen from Figure 5.40. If the light output were to depend only on the total insonation time, then the magnitude of the sonoluminescence expected in each case would be in the ratios of the shaded areas in Figure 5.40 (taken over a sufficiently long time). These expected results are seen for the smaller pulse heights, but not for the larger pulses when the light output in increasing order is for continuous-wave and then for the duty cycles 1:2, 1:4 and finally 1:7.

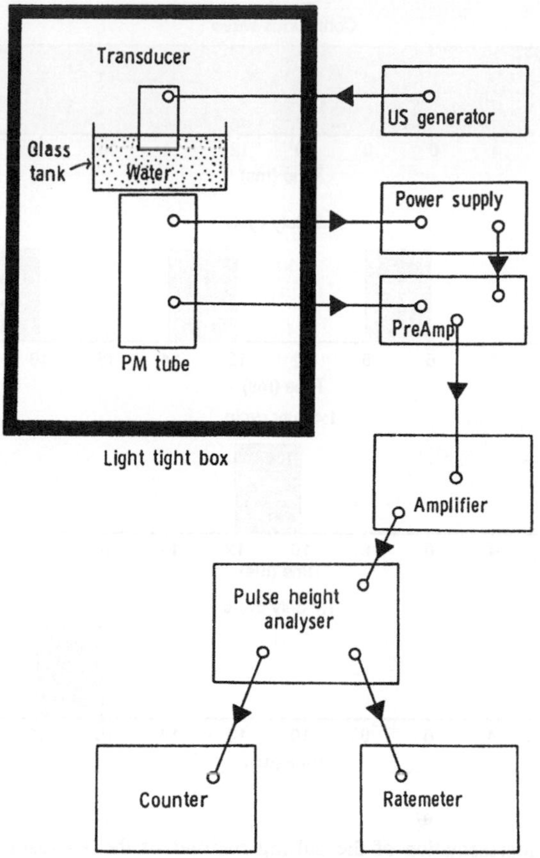

Figure 5.39 Apparatus: sonoluminescence, generated from the standing-wave field within the water tank, is detected by the photomultiplier and graphs of the form of Figure 5.41 produced through pulse height analysis (after Pickworth *et al.* [261]). Reprinted by permission from *Physics in Medicine and Biology*, vol. 33, pp. 1249–1260; Copyright © 1988 IOP Publishing Ltd.

If transient excitation were the sole cause of this pulse enhancement, one would expect the count rate to increase with the number of starts of pulses contained within a given interval. From Figure 5.40 it can readily be seen that this would lead to the magnitude of the luminescence occurring in increasing order for continuous-wave, 1:7, 1:4 and then 1:2. This is indeed the case for pulse heights between 3.2 and 4.2 units. The fact that this is not the order for the larger pulses suggests that, while transient excitation may be a potent mechanism in some acoustic regimens, its effects in this 1-MHz system are observable only in a limited range of pulse heights. Outside this region, other factors must dominate. The work of Ciaravino *et al.* [312] and Flynn and Church [313] is not applicable, since they observed no pulse enhancement at intensities of less than 10 W cm^{-2}, and Hill *et al.* [309] detected no sonoluminescence whatsoever, and negligible pulse enhancement effect on iodine release at pulse lengths of 2 ms.

In 1988, Leighton *et al.* [261, 294, 319] proposed two mechanisms for the pulse enhancement that had been observed, both based on the migration of bubbles away from bubble aggregates during the off-time of the insonation. Such migrations would: (a) reduce local degassing of the liquid at these aggregates, so promoting bubble growth by rectified diffusion there; (b) remove

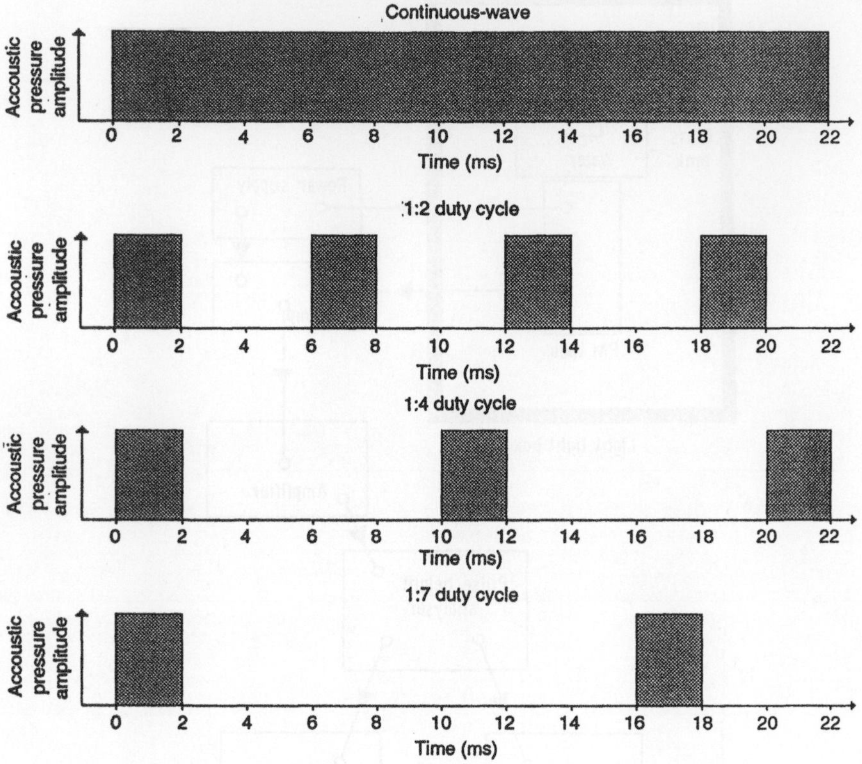

Figure 5.40 Schematic representation of the pulsing regimes of the Therasonic 1030 (shaded area represents on-time for 1-MHz insonation) (after Leighton *et al.* [319]).

regions which have an acoustic impedance very different from that of the pure liquid (and would thus otherwise attenuate the passage of sound to the bubbles). Both mechanisms rely on the assumption that the brightest light flashes are the product of the most intense cavitational collapses.

The Degassing Mechanism. The first mechanism refers to the cyclic process of a bubble growing by rectified diffusion from a nucleus to a certain size where activity is a maximum (e.g. the transient cavitation threshold), where it collapses unstably to produce bubble fragments. These fragments then grow to the threshold, to collapse themselves, and so on. Within each oscillatory cycle, there is a finite chance of the bubble undergoing an unstable collapse which causes it to break up. This, for example, could be due to shape instabilities induced by inhomogeneities in the medium, such as the approach of another bubble, in the manner shown in Figure 4.29. Broadly speaking, this chance increases as the bubble grows towards threshold size, since the surface instabilities are easier to induce on larger bubbles. Therefore, the more cycles it takes for a bubble to grow to threshold size by rectified diffusion, the more likely it will collapse before reaching threshold size. Such a collapse would by definition be less energetic than one that occurs at the radius of maximum activity (e.g. the transient threshold), and so would produce a lower-energy light pulse.

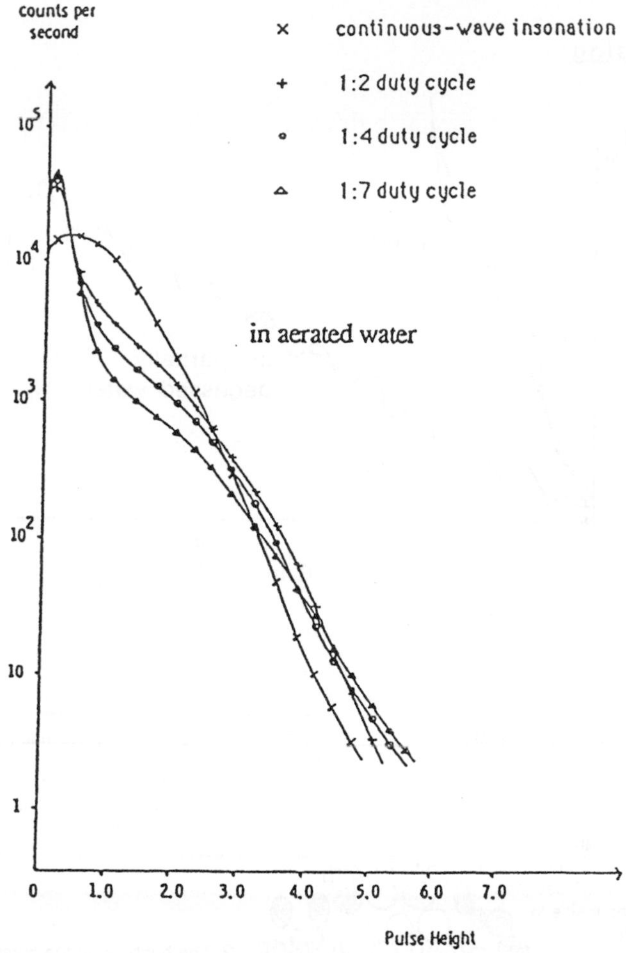

Figure 5.41 The luminescence count rate from aerated water as a function of the pulse height, for continuous-wave insonation (×), and for 1:2 (+), 1:4 (∘) and 1:7 (△) duty cycles (pulse length = 2 ms). For clarity, only every third data point taken is shown. (After Pickworth *et al.* [261]. Reprinted with permission from *Physics in Medicine and Biology*, vol. 33, pp. 1249–1260; Copyright © 1988 IOP Publishing Ltd.)

During a prolonged period of insonation, bubbles of less than resonant size, growing by rectified diffusion, cluster at the pressure antinodes under the influence of Bjerknes forces,[66] as shown for this particular sound field in Figure 5.26. As they grow, they will deplete the water in the region of dissolved gases. This can be seen from the fact that the diffusion length of air in water after a time t is given by $\sqrt{4Dt}$, where D is the diffusivity of dissolved air in water, and equal to about 10^{-5} cm^2 s^{-1} [4]. Thus in 2 ms (the length of each sound pulse) the gas diffusion length in water is 2.8 μm, which is very much less than the node–antinode spacing in a 1-MHz field (≈ 0.75 mm). Therefore during insonation the pressure antinodes will become partially

[66] See Chapter 4, section 4.4.1(b).

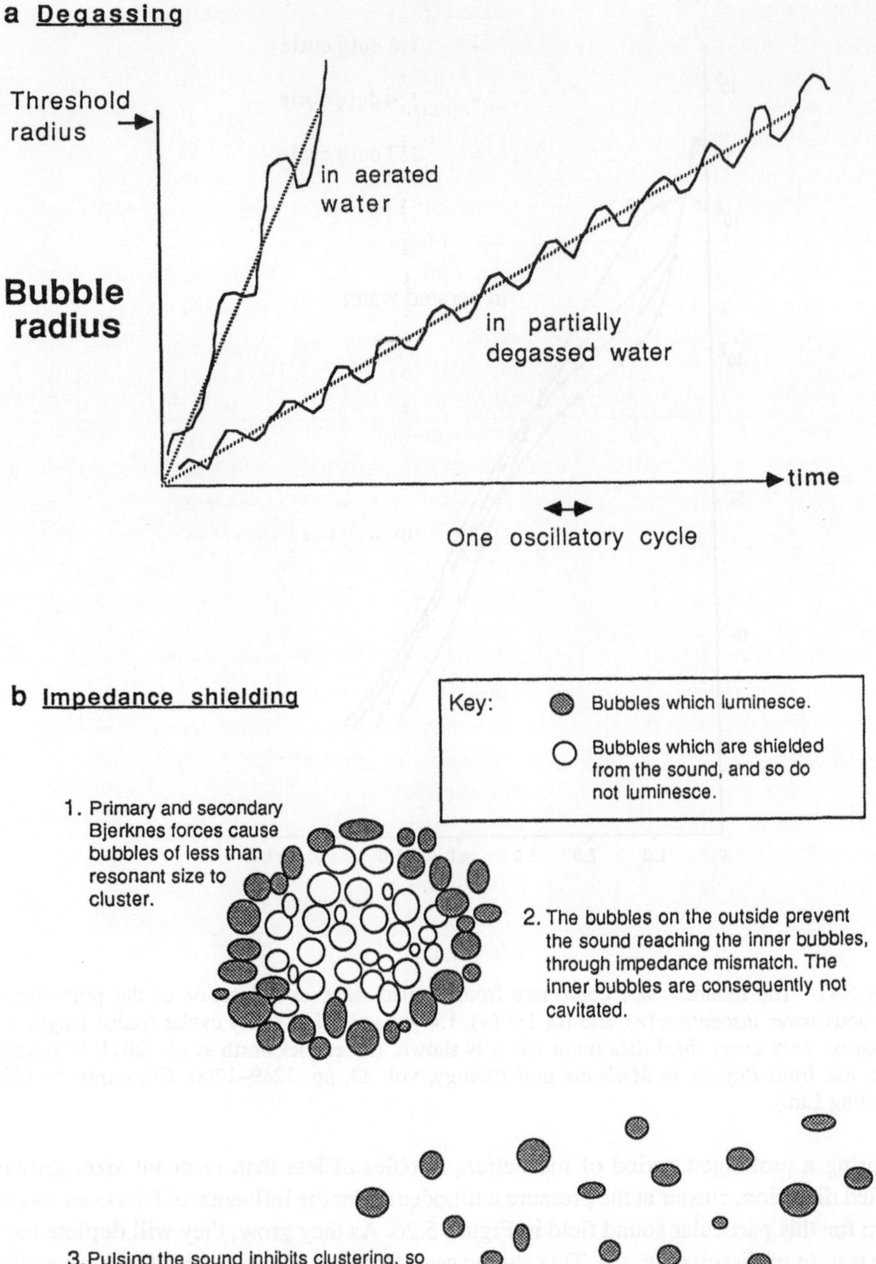

a Degassing
Threshold radius
Bubble radius
in aerated water
in partially degassed water
time
One oscillatory cycle
b Impedance shielding
Key:
Bubbles which luminesce.
Bubbles which are shielded from the sound, and so do not luminesce.
1. Primary and secondary Bjerknes forces cause bubbles of less than resonant size to cluster.
2. The bubbles on the outside prevent the sound reaching the inner bubbles, through impedance mismatch. The inner bubbles are consequently not cavitated.
3. Pulsing the sound inhibits clustering, so the medium has a uniform acoustic impedance. No bubbles are shielded.

Figure 5.42 (a) The 'degassing' mechanism, and (b) the 'impedance' mechanism, for pulse enhancement through bubble migration (after Leighton *et al.* [294, 319]).

degassed. This will make each bubble oscillation less efficient in drawing dissolved gas into itself from the liquid, and so it will take more oscillations to grow to threshold size than would a bubble in water that remained fully aerated (Figure 5.42(a)). Therefore, on average, this type of growth would result in more of the lower-energy light pulses.

If, when the sound was pulsed, the bubbles were to migrate during the off-time in such a way as to break up these antinodal clusters, and to disturb the liquid distribution at the antinodes, then at the start of the next pulse when the Bjerknes forces drove them back to the antinode, this region would be flushed with fresh aerated water. Rectified diffusion would consequently be more efficient, and the bubbles would be more likely to reach threshold size before collapsing unstably. This would result in more higher-energy flashes of sonoluminescence.

Thus the form of Figure 5.41 would be explained: continuous-wave insonation causes bubble clustering which leads to antinodal degassing, resulting in more low-energy sonoluminescent pulses. On the other hand, when the sound is pulsed, degassing does not occur to so great an extent, bubble growth is thus more rapid, and so more of the higher-energy sonoluminescent pulses are detected. The longer the off-time, the greater the effect.

This theory is substantiated by observations by Blake [320] who describes a phenomenon called 'ultrasonic degassing'. Here liquid in the locality of cavitating bubbles becomes degassed. Blake noticed that once strong cavitation was obtained in water which was continually stirred, the intensity of the cavitation would show a marked decrease a few seconds after the stirrer was switched off. The speculation by Watmough *et al.* [307] that acoustic streaming might flush gassy water to pressure antinodes and so aid rectified diffusion has already been outlined.[67]

The Shielding Theory. The second theory is schematically illustrated in Figure 5.42(b). During insonation, the bubbles of less than resonant size (i.e. the source of sonoluminescence) cluster at the pressure antinodes. As has already been discussed,[68] bubbly liquids have an effective sound speed and acoustic impedance different from that of bubble-free liquid. An acoustic impedance mismatch can occur at the interface between regions of high and sparse bubble concentrations (i.e. at the outer layers of the antinodal clusters), hindering the transfer of sound from one region to the next. Sound from the body of the liquid is attenuated before it reaches the bubbles at the heart of the cluster. The reduced acoustic pressure amplitudes will result in reduced bubble activity, and so less sonoluminescence (particularly of the most intense type). However, if bubbles migrate during the off-times when the sound is pulsed, these clusters will break up. In the limit, the migration will tend to produce a homogeneous distribution of bubbles, eliminating the macroscopic spatial variations in bulk properties that were encountered when the bubbles formed aggregates. At the start of the next pulse, the sound will reach all the bubbles at full amplitude, and the most intense sonoluminescence pulses will be detected (particularly when the effect of transient excitation is considered). Thus, the form of Figure 5.41 is produced: antinodal aggregates form during continuous-wave insonation, and the bubbles in the interior are shielded by an impedance mismatch, and so fewer high-energy collapses occur; when the sound is pulsed, these aggregates are broken up, and more of the high-energy sonoluminescent pulses are recorded. The longer the off-time, the greater the effect.

Both theories rely on the flushing of bubbles from the antinodal regions during the off-times. The further the migration, the more the enhancement, so both theories work to increase the

[67] See section 5.2.4.
[68] See Chapter 4, section 4.1.2(e).

luminescence as the off-time increases, though in diminishing increments as the bubbles are decelerated through drag. If the experiment were to be performed in a more viscous medium, the migrations would be greatly reduced. One might expect the cross-over seen in Figure 5.41 to disappear, and the count rates to be increased in the order of duty cycle 1:7, 1:4 and 1:2, with the most luminescence from continuous-wave insonation, the count rate now being simply a function of the total time of insonation. Pickworth *et al.* [261, 296] obtained the pulse-energy distribution spectrum in fluid 1.875 g/l agar at 22° C. This substance had a viscosity of 170 ± 10 cp, and so was still fluid; since viscosity is a second-order effect when describing the actual bubble oscillations (equation (4.81)), then the actual cavitational behaviour will be similar to that in water. What will be different is the scale of the migrations, which will be greatly reduced in agar, because of the increased viscosity [319]. The results are shown in Figure 5.43; the cross-over has disappeared, the count rates increasing in the order of duty cycle of 1:7, 1:4 and 1:2, with continuous-wave insonation the most luminescence. Therefore the hypothesis that cross-over is associated with bubble migrations is strengthened. Experiments with agar gels of varying viscosity suggest that both mechanisms may work together.

The driving force for these migrations could arise through buoyancy, acoustic streaming forces, the inertia of bubbles moving in streamers, or a coupling of the acoustic pressure field with the bubble oscillation at the end of each sound pulse. The particular driving force dominating in a given regimen depends on the size of the active bubbles, and so is frequency dependent. For physiotherapeutic ultrasound (operating at 1 MHz), Leighton *et al.* [319] calculated that the last process is probably responsible, causing the motion of large bubbles at the pressure nodes, which in turn generates secondary motions of small bubbles away from the pressure antinodes. At 10 kHz, where this phenomenon was also observed [294], the migration would more likely be driven by the inertia of bubbles in moving streamers, which continue to move into the antinodal aggregate after the sound has ceased, and break it up.

In 1989, Miller *et al.* [321] came up with a similar shielding theory to explain the observation, noted by several workers [311, 322–325] that, in the insonation of a rotating sample, as the ultrasonic intensity was increased the effect (cell lysis) firstly increased, then decreased as the intensity was further increased.

For many years such experimentalists working with MHz ultrasound in liquids had often found it convenient to contain the liquid sample they wished to cavitate within some thin-walled vessel, placed in a larger liquid bath. Once the sample is physically isolated from the bath, the latter can provide acoustic coupling to a transducer and also temperature stability. For example, water within the bath can be circulated through thermostatically controlled heating elements. To do this to the liquid sample directly would in many cases be inappropriate: the sample might become contaminated with surfactants etc. or seeded with nuclei and motes, or the sample itself might be a chemical in short supply. Most often a thermal water bath is used when the isolated sample is biological, to avoid contamination.

Miller *et al.* [321] proposed that the observed effect of reduced cell lysis at the highest intensities was due to the development of a cloud of bubbles in the medium outside the exposure vessel, between the vessel and the transducer. In their standing-wave experimental system they demonstrated a maximum in cell lysis at 5 W cm^{-2}, the amount of lysis decreasing when bubbles were artificially introduced between transducer and vessel using an aquarium aerator. Photo-multiplier observations of sonoluminescence (with no aerator) emitted from the region between vessel and transducer indicated a dramatic increase above 10 W cm^{-2}: the sonoluminescence from the exposure vessel itself showed a maxima at around 10 W cm^{-2}, a minimum at around 25 W cm^{-2}, followed by a rise towards 35 W cm^{-2}. They concluded that a cloud of bubbles

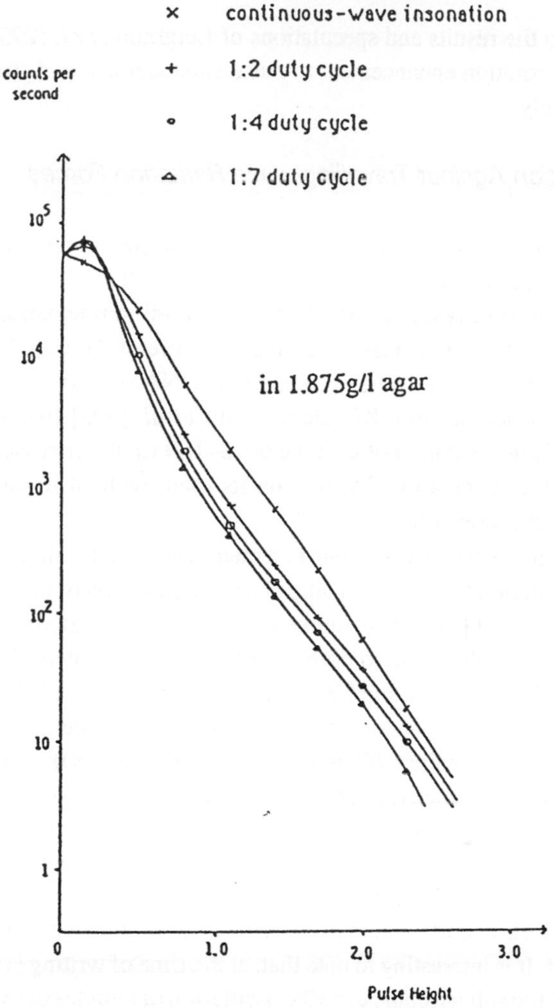

Figure 5.43 The luminescence count rate from 1.875% agar as a function of the pulse height, for continuous-wave insonation (×), and for 1:2 (+), 1:4 (∘) and 1:7 (△) duty cycles (pulse length = 2 ms). For clarity, only every third data point taken is shown. (After Pickworth *et al.* [261]. Reprinted by permission from *Physics in Medicine and Biology*, vol. 33, pp. 1249–1260; Copyright © 1988 IOP Publishing Ltd.)

generated at the higher intensities between the transducer and the exposure vessel attenuated the ultrasound that would reach the exposure vessel, and so reduced the lysis that occurred there.

The above experimenters rotated their samples. The connection between pulse enhancement and the related rotation enhancement has already been noted, as has the potential of certain pulse enhancement mechanisms (e.g. transient excitation) to bring about enhancement during a rotation-like motion. The effect of that rotation will now be discussed.

5.3.2 Rotation Enhancement of Acoustic Cavitation

Certain observations of rotation enhancement, and possible common mechanisms, were introduced in section 5.3.1(d). The phenomenon will now be discussed in more detail. Attention has already been drawn to the results and speculations of Leighton *et al.* [295] and Watmough *et al.* [307] which relate rotation enhancement to pulse enhancement and streaming in standing-wave fields, respectively.

(a) Bubble Stabilisation Against Travelling-wave Radiation Forces

As mentioned in section 5.3.1(b), in an experiment to measure the release of free iodine from KI solution, DNA degradation and subharmonic emission, Hill *et al.* [309] found that a simple slow rotation of the exposure vessel in travelling-wave conditions enhanced biological effects. Such rotations, designed to homogenise ultrasonic exposure of the vessel and to maintain cell suspensions, could enhance such diverse processes as DNA degradation, subharmonic emissions and the release of iodine from KI solution. Hill *et al.* [309] found that at 1 MHz and intensities of a few W cm^{-2}, rotation of the tube at 18–180 r.p.m. increased both cell lysis and subharmonic emission considerably. Rotation on its own, without insonation, had no effect whatsoever on any of the three tests.

Hill *et al.* [309] suggested that the enhancement was the result of stabilisation of the cavitation bubble population by the continually changing direction of insonation. This was also proposed by Clarke and Hill [311] who, unlike Hill *et al.*, used relatively thick-walled culture tubes which would perturb the beam to observe rotation enhancement. In a stationary vessel, travelling-wave radiation forces on bubbles would force them out of the body of the liquid, where they are required to bring about the sonochemical effects etc., and, say, hold them up against the back wall of the container. If the container is rotated, then the direction of insonation, and of the radiation forces, effectively change continuously, and so the bubbles are retained within the body of the sample. This of course assumed predominantly travelling-wave conditions. As Williams and Miller [326] point out, Hill *et al.* [309] recorded an increase in subharmonic generation, showing that in fact cavitational activity is itself increased, so whilst the mechanism may well operate, the explanation could not be a complete exposition of the mechanisms involved. It is interesting to note that, at the time of writing (1969) Hill *et al.* [309] believed the source of the subharmonic to be the oscillation of bubbles of twice resonance size.[69]

(b) Bubble Motion and Standing-wave Radiation Forces

For standing-wave conditions, Ciaravino *et al.* [312] proposed, and Church *et al.* [327] gave theoretical backing for, the idea that rotation enhancement was the result of cell and bubble behaviour in a standing-wave field set up within the exposure vessel. The action of bubbles within such a field is well known: bubbles smaller than resonance size would aggregate at the pressure antinodes, and it is these which collapse violently.[70] Cells, they postulated, would aggregate at the pressure nodes where the only bubbles to be found would be of larger than resonance size, and not subject to violent collapse. Thus in the stationary tube, the cells would be kept well-separated from the violent cavitation. Rotation of the tube would sweep the cells through the bubble aggregates (the acoustic forces restraining the cells are weaker than those on the bubbles). In this way, rotation would bring about increased cell lysis.

---

[69] See Chapter 4, section 4.4.7(b)(ii).
[70] See Chapter 4, section 4.3, and also section 5.2.4.

(c) High-speed Bubbles from Travelling-wave Radiation Forces

The theory of Ciaravino *et al.* [328] and Church *et al.* [327] relies upon a standing-wave situation. Therefore (as Williams and Miller [326] point out) this could not be the mechanism for the rotation enhancement in the travelling-wave conditions found within the thin-walled exposure vessels employed by Hill *et al.* [309], Morton *et al.* [15] and Miller and Williams [329]. Church and Miller [330, 331] gave a detailed formulation which involved micron-sized bubbles travelling at speeds of 1.4 m s^{-1} as a result of random thermal motion, but mistakenly neglected the liquid viscosity. Miller [332] showed that in fact the bubbles will undergo a 'random walk' which leaves them only a few microns from their starting position after about a second. Thus, although the theory still maintains elements of use [333], it does not represent a satisfactory explanation for this rotation enhancement.

Miller and Williams [329] suggested a novel explanation for the rotation enhancement observed in travelling-wave conditions. Their experimental arrangement is illustrated in Figure 5.44. The exposure vessel was a segment of 15-mm diameter standard dialysis tubing, having 20 μm thick walls, when dry, and an intensity reflection coefficient of 0.5%. The ends were plugged with plastic. The tube was angled at 30° from the vertical to help maintain the cell suspension, and remained straight as a result of magnetically induced tension (magnets were placed in the lower plug and the stationary stub). The ultrasound was 1.45 MHz (focused, at up to 512 W cm^{-2}) or 1.61 MHz (unfocused, at up to 16 W cm^{-2}). An absorber was placed behind the vessel to try to attain travelling-wave conditions. Ultrasonic pulses could be triggered from

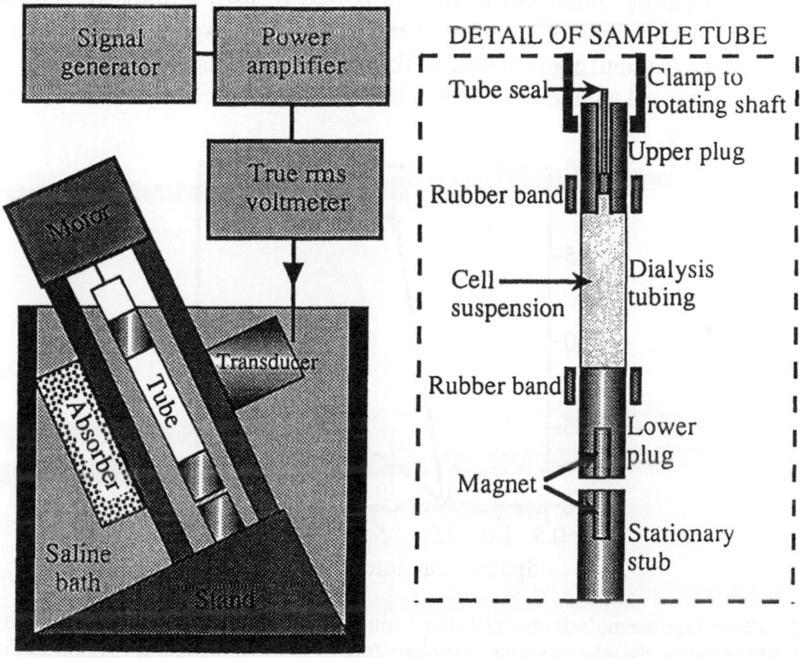

Figure 5.44 An illustration of the exposure apparatus. The 17 L saline-filled exposure bath was maintained at 37° C. (After Miller and Williams [329]. Reprinted with permission from *Ultrasound in Medicine and Biology*, vol. 15, pp. 641–648; Copyright © 1989 Pergamon Press Ltd.)

the rotation drive of the vessel, so that the ultrasound could turn on and off twice each rotation (180° apart) or only once. The duration of each burst was half a period of rotation (0.42 s) or a quarter (0.21 s), with a 1:1 duty cycle. Being synchronised to the rotation, the two burst lengths caused the vessel to receive either one unidirectional exposure each rotation, or two exposures of a region, the second in the direction opposite to the first, each rotation. With the focused beam, the first situation corresponds to a normal, non-rotating exposure, with a halved on-time, since each sample always receives exposure from the same direction. The second situation relates more to a rotating experiment, since the sample experiences a beam from two different directions.

The absorber could be replaced by a styrofoam reflector to generate fields with a higher proportion of standing-wave within the vessel. The vessel was filled with heparinised canine blood samples. The erythrocytes were washed twice with isotonic buffered saline and resuspended to a haematocrit of 0.5%. Following exposure, the samples were centrifuged, and the supernatant analysed for free haemoglobin, which would indicate haemolysis. Sham exposures were used for control.

The results of 5 minute exposures are shown in Figure 5.45. The results for focused ultrasound in a rotating tube tended to be 'all-or-nothing', in that exposure either gave tens of percent haemolysis, or was practically indistinguishable from the control. Therefore, taking haemolysis as an indicator of cavitation, a number indicating the probability that cavitation bioeffects occurred has been associated with each point. For spatial peak intensities of greater than a threshold (which is greater than 2 W cm^{-2}), exposure to unfocused ultrasound in a tube rotating at 72 r.p.m. always produced cavitation bioeffects (100% probability), whilst unfocused exposures in stationary tubes consistently produced no cavitation (0%).

The results comparing continuous-wave with pulsed focused ultrasound in rotating and nonrotating tubes are shown in Figure 5.46. Continuous-wave exposure of a stationary tube produced no haemolysis until high intensities. The authors note that exposures higher than 256 W cm^{-2}, which correlated with levels at which haemolysis occurred regardless of whether the

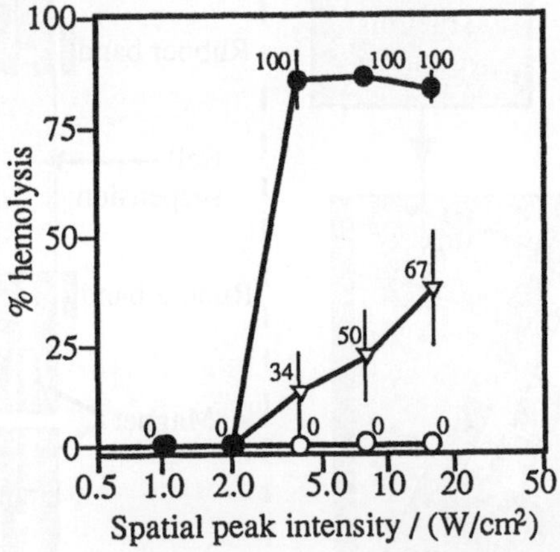

Figure 5.45 Percentage haemolysis obtained after 5-min exposure in a dialysis tubing exposure chamber. For unfocused exposures, the tube was either stationary (open circles) or rotated at 72 r.p.m. (solid circles). Focused exposures (crosses) were in a rotating tube. Error bars extend one standard error above and below the mean of six measurements. The numbers adjacent to each datum specify the probability of occurrence of cavitation bioeffects. (After Miller and Williams [329]. Reprinted with permission from *Ultrasound in Medicine and Biology*, vol. 15, pp. 641–648; Copyright © 1989 Pergamon Press Ltd.)

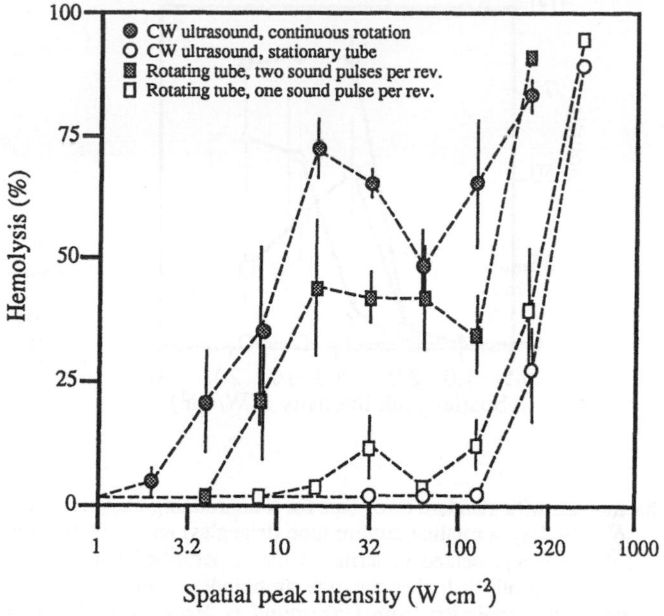

Figure 5.46 Haemolysis with standard error bars for focused exposures in a dialysis tube. Continuous exposures were made with the tube rotating (solid circles) or not rotating (open circles). Burst mode exposures were made in the rotating tube for the ultrasound switched on and off once (open squares) or twice (solid squares) per revolution. (After Miller and Williams [329]. Reprinted with permission from *Ultrasound in Medicine and Biology*, vol. 15, pp. 641–648; Copyright © 1989 Pergamon Press Ltd.)

tube was stationary or rotating, was also the approximate threshold for the onset of audible cavitation noise.[71] In the rotating tube, pulsed exposure twice per rotation consistently generated more haemolysis than pulsed exposure once per revolution. Continuous-wave exposure in the stationary tube gave the highest haemolysis. However, the on-time was twice that of the mode that pulsed twice per revolution, and four times the mode that pulsed once per revolution.

There is a threshold for haemolysis of around 4 W cm^{-2}. High intensities (256 W cm^{-2} and 512 W cm^{-2}) of ultrasound caused haemolysis regardless of whether or not the tube was rotated or the ultrasound pulsed. Miller and Williams [329] attribute this to Willard events.[72] In general, rotation of the tube has a dramatic effect on haemolysis, and therefore (by inference) on the acoustic cavitation which occurs.

Given the low reflectivity of the tube walls, the authors consider the conditions to be travelling wave (free field). The theories of Ciaravino *et al.* [328] and Church *et al.* [327] are inappropriate when the standing-wave component of the field is not significant. Miller and Williams [329] in fact demonstrated that strong standing-wave conditions, when attained, actually reduced the rotation effect, in that the enhancement was less at the higher spatial peak intensities (Figure 5.47).

Miller and Williams [329] propose their own hypothesis, based on the finding that pulsing the sound twice per rotation gives significant rotation enhancement, whilst pulsing once per

[71] See Chapter 4, section 4.4.7, and also sections 5.1.1(c) and 5.1.2.
[72] See Chapter 4, section 4.3.4.

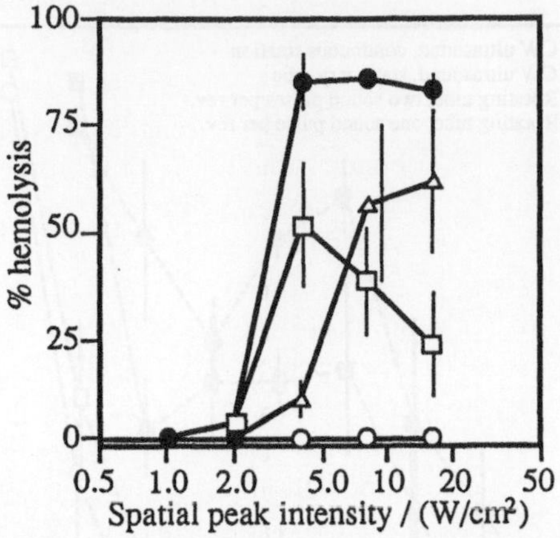

Figure 5.47 Mean haemolysis with standard error bars for a nonrotating dialysis tube (open circles), a rotating dialysis tube (solid circles), a rotating culture tube (triangles) and a rotating dialysis tube with reflector (squares). All data are represented in terms of the unperturbed intensity, even though the thick-walled tube and reflector result in higher pressure amplitudes in the sample. (After Miller and Williams [329]. Reprinted with permission from *Ultrasound in Medicine and Biology*, vol. 15, pp. 641–648; Copyright © 1989 Pergamon Press Ltd.)

rotation does not. As explained above, the latter mode only receives sound from a single direction. The changing of direction of reception of sound by the sample (two opposing directions in the case of the beam pulsed twice per revolution) is therefore requisite for rotation enhancement.

As discussed in Chapter 4, section 4.4.1(a), the radiation force in a travelling-wave field translates bubbles in the direction of propagation of the sound, and is greatest for bubbles of resonance size.

Miller and Williams [329] suggest the following explanation for their result. Owing to the method of preparation, their samples contained scant nuclei. Cavitation bioeffects only occur if nuclei are present in the beam (which accounts for the 'all or nothing' effect described above, the smaller insonation volume of the focused beam giving less opportunity of finding a nucleus and so fewer events than the unfocused beam). Bubble populations generated from these nuclei travel through the cell population. For stationary tubes, or those pulsed singly per revolution, the direction of bubble motion is always the same, relative to the internal geometry of the tube and cell population. Therefore the bubbles are driven to the tube wall remote from the transducer when insonation occurs, and are then removed from the region of the cells. The bioeffect is therefore small. However, in the rotating tube where insonation occurs continuously or twice per revolution, the bubbles translate back and forth through the cell population, bringing about lysis.

Williams and Miller [326] demonstrated that a partial standing-wave field within the rotating tube reduced the lysis effect. This is in accord with their theory, since the net unidirectional radiation force would be reduced if there were a reflected component present, which would reduce bubble speeds. They also demonstrated an effect of chamber length, virtually no lysis being observed if the chamber were less than 4 mm long. The authors proposed that this could be due to scarcity of bubble nuclei in such a small sample volume. Alternatively, they suggested that to cause an effect the bubbles must necessarily be translating freely, and at the intensities employed this required distances in excess of 4 mm. Subdivision of a longer chamber into 4-mm lengths proved inconclusive, but the authors did tentatively make the observation that short pathlengths did tend to reduce cavitational activity. This makes interesting comparison with the results of Leighton *et al.* [299], who found[73] sonoluminescence only at distances greater than 3 mm or so from the transducer faceplate. They experimented at 2 W cm^{-2}, the lower range of intensities investigated by Williams and Miller.

Subsequent to this study, Miller *et al.* [334] further examined the effect of short chambers. The authors used three tube lengths (5 mm, 14 mm and 40 mm). They found that haemolysis was only reduced in the 5-mm chamber, which exhibited the 'all or nothing' response described above, reinforcing the idea that suppression of rotation enhancement in short chambers is due to scarcity of bubble nuclei. Application of a 10-MPa over-pressure to the chamber using a nitrogen gas head, in order to reduce the population of cavitation nuclei, reduced the haemolysis. From their results, Miller *et al.* [334] suggest that haemolysis occurs primarily in the bulk of the medium, rather than on the surfaces.

Miller *et al.* [334] examined the possible mechanisms by which this continual traversing of bubbles through a medium might increase cell lysis. The passage of bubbles might re-nucleate the suspension, and mix bubbles and cells into closer spatial contact. The authors rule out cell damage resulting from microstreaming from bubbles attached to tube surfaces [335], since the effect occurs in the body of the suspension, though Miller *et al.* suggest that the microstreaming flow resulting from the back-and-forth oscillation of free bubbles might cause damage [336]. They cite Kondo *et al.* [337] to suggest that lysis is in these circumstances not a result of transient cavitation. The large volume change in a pulsating bubble could tear a cell membrane [334]. Inhomogeneities in the sound field, caused by other bodies or surfaces, can cause asymmetric bubble collapse.[74] Involution of the bubble can give rise to a liquid jet, impact of which can create high 'water hammer' pressures.[75] Such a mechanism could perforate cell membrane close to collapsing bubbles. Church and Miller [330] conceived of the cell as a target for 'bubble bullets', and a theoretical treatment followed [333]. Miller *et al.* [334] explain how the translations of single high-speed bubbles, or bubble clouds, could generate hydrodynamic stresses within the liquid capable of damaging cells.

Consider the bubble pulsating at low amplitude described in Chapter 4, section 4.1. The velocity of the bubble wall, $\dot{R}$, is in phase with the driving force when the bubble is resonant.[76] Therefore, in a similar manner to the calculation shown in Chapter 4, section 4.4.1, it is clear that maximum radiation force is transmitted to the resonant bubble. The radiation force in a travelling-wave field for a bubble at resonance, substituting for ω_0 into equation (4.142), is

$$\overline{F}_{\text{trav}} = \frac{V_0 k P_A^2}{2\kappa\delta p_0} \qquad \text{(at resonance)} \qquad (5.4)$$

Coakley and Nyborg [156] equate this radiation force to the Stokes drag force $6\pi\eta R_0 u_b$ to give u_b, the terminal translational bubble speed of the bubble. Obviously the derivation has limited itself to small-amplitude, linear bubble pulsations, and to small translational velocities. In fact, the value for u_b obtained by this first approximation to the speed is large enough to make the use of the Stokes formula inappropriate. Following Neppiras and Coakley [338], Miller *et*

[73] See section 5.2.4.
[74] See section 5.4.1.
[75] See section 5.4.1.
[76] See Chapter 4, sections 4.1.1(b) and 4.1.2(a).

al. multiply the drag force by ($1+0.2\mathcal{Re}$) to yield the translational velocity of the bubble to be approximately

$$u_b \approx \frac{kR_oP_A^2}{9\kappa\eta\delta p_o(1+0.2\mathcal{Re})} \qquad (5.5)$$

Iterative solution of this gives, for $P_A = p_o = 10^5$ Pa (1 bar[77]) and $\mathcal{Re} = 2.8$, that $u_b = 1.35$ m s^{-1}. Though only a rough estimation, this calculation demonstrates the high speeds that a bubble can achieve. Neppiras and Coakley [338] and Miller *et al.* [334] both obtained evidence (photographic and acoustic) suggesting that such speeds might have been attained in their experiments. Hydrodynamic forces resulting from such speeds could generate shear stresses capable of damaging cells. Miller *et al.* [334] consider a bubble, resonant in a MHz sound field, travelling in a suspension of 10 μm cells at 10 m s^{-1} through their apparatus. Though the bubble would pass a given cell in 1 μs, the stresses would persist for longer as the cell experienced wake effects. Consideration of the cylindrical target volume swept out by a single bubble demonstrates that a population of only a few thousand travelling bubbles might generate the observed cell lysis.

5.3.3 Summary

Several theories have arisen for both pulse and rotation enhancement following experimental observation. It is clear that on a mechanistic level there is some overlap between the two phenomena. For example, if rotation of the tube removes regions of the liquid sample containing cells and bubble nuclei from the ultrasonic beam, then effectively that sample experiences an off-time in its ultrasonic exposure, so that the various theories of pulse enhancement (particularly transient excitation) might come into effect. However, if the exposure beam were broader than the cell, and uniform over the sample volume, no pulse enhancement mechanisms would operate. Hill *et al.* [309] did note that, from beam profile scans performed both inside and outside their sample container, the acoustic intensity in the irradiation volume varied from 0.2 to 3.0 times the spatial average measured over the container cross-section. Therefore transients may have been excited in the bubbles as they were scanned through the field. The bubble motions proposed by Ciaravino *et al.* [328] and Church *et al.* [327], which concern the migration of bubbles from aggregates formed in standing-wave conditions, might not just bring cells and bubbles into closer proximity, but in addition cause enhancement through the migration mechanisms introduced in section 5.3.1(e).

Because of the nature of the observations, the theories have tended to be applicable to specific sets of conditions such as travelling or standing wave etc. Since in the experimental data it is possible to find enhancement occurring in conditions where any given theory is inapplicable, it seems likely that most of the theories contain an element of truth, testament to the rich complexity of interaction possible with bubble populations in high intensity sound fields, and the importance of knowing the details of the sound field and the medium.

As an illustration of this, Miller and Williams [329] discuss some other mechanical system adjustments which can affect the activity of the cavitation population. Kaufman and Miller [324] found that exposure of cells in a thin-walled bag was similar in effect to exposure in a rotating culture tube. Graham *et al.* [339] observed that cavitational activity could be initiated in a system

[77] Such high acoustic pressure amplitudes preclude the possibility of small-amplitude, linear pulsations for bubbles near resonance.

that had previously shown none, by altering the frequency slightly, or by a rotation of a few degrees. Bioeffects were enhanced and activity thresholds lowered when Williams [340] stirred a sample with a mechanical stir-bar, which Saad and Williams [341] suggested was due to tribonucleation (i.e. nucleation through rubbing).

Pulse enhancement of the effects of cavitaton is a complex phenomenon with a variety of mechanisms, the dominance of which vary with the experimental parameters (e.g. acoustic frequency, intensity, pulse length, nature of the liquid etc). The enhancement effects are not generally linear in nature. For example, the dispersion of bubbles from aggregates decays exponentially with time, so that continually increasing the pulse length gives ever decreasing returns: beyond a certain duration of off-time, the bubbles do not migrate significantly further from the clusters. Transient excitation operates only within the first few cycles of insonation, so that provided the pulse length is greater than this there will be no effect of increasing pulse length. Several mechanisms may act together, and potentially interact. Because of the strong dependence on population phenomena and radiation forces, several mechanisms are critically dependent on whether conditions are standing- or travelling-wave. If the pulse length is shorter than approximately the ratio of the tank dimension to the speed of sound in the liquid, standing-wave systems will not persist. There may be a finite $\vec{\nabla}P$ resulting from near-field acoustic pattern, acoustic reflections, or focusing, which could give rise to primary Bjerknes forces. In a discussion of any of these effects, it must be remembered that any increase in activity resulting from an increased offtime by an appropriate mechanism must be offset by a decrease in the time of insonation.

In consideration of this discussion, it is not surprising that work on the potential bioeffects of ultrasound has not always yielded obvious and repeatable results.

5.4 Acoustic Cavitation: Effects on the Local Environment

5.4.1 Erosion and Damage

It was the erosion of ship propellers which first introduced Rayleigh to the subject of cavitation.[78] Since then there have been extensive studies of cavitation erosion, covering a wide range of its aspects, from materials testing to bubble dynamics. Two examples of cavitation erosion are shown in Figure 5.48. The microscopic damage produced by a single cavitation event can be seen in Figure 5.48(a): the stellite sample was damaged in the cavitation vortex generator at the Institut de Machines Hydrauliques et de Mechanique des Fluids, EPFL, Lausanne. The time-integrated erosive effects of many such events can readily be observed in the focused cylindrical transducer which has been used to illustrate various aspects of cavitation throughout this book. Figure 5.48(b) shows a sample of aluminised mylar sheet[79] which has been subjected for around 1 minute to a sound field which, at the axial focus, had an acoustic pressure amplitude of 2.4 bar. Roughly circular regions, where the aluminium has been eroded away to leave only the transparent mylar, are testimony to the presence of eroding cavities.

Given here is a brief outline of some of the work in the field. Two main sources of physical damage are now recognised. The first arises from the pressure pulse emitted into the surrounding

[78] See Chapter 2, section 2.1.3(b).

[79] Aluminised mylar sheet is commercially available as 'survival blanket': it consists of a thin transparent sheet, one side of which is coated with a thin reflective layer of aluminium.

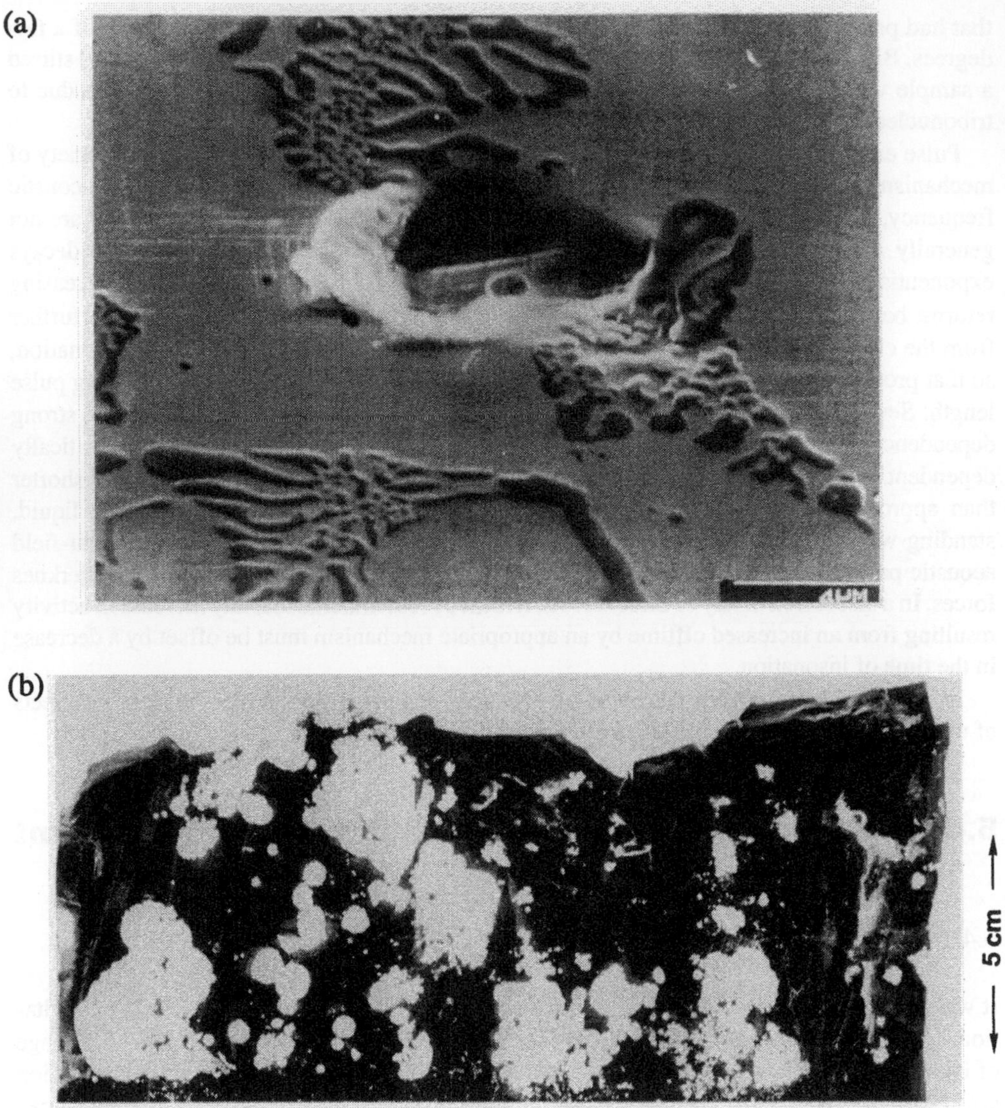

Figure 5.48 (a) Cavitation erosion damage produced in stellite sample which was damaged in the cavitation vortex generator at the Institut de Machines Hydrauliques et de Mechanique des Fluids, EPFL, Lausanne (photograph courtesy of M Farhat and F Avellan). The scale bar (bottom right) measures 4 μm. (b) A back-lighted sample of the aluminised mylar sheet. In the dark regions, the sheet has remained intact. However, large circular regions of light can be seen where cavitational erosion has removed the aluminium and left only the transparent mylar. (After Leighton [294].)

liquid when a cavity collapses [161, 342, 343]. The case of the spherical collapsing bubble has been considered at length earlier in this book. Much research on this has been done since the early work of Rayleigh [342], and is reviewed by Plesset and Prosperetti [344] and Mørch [345].The emitted spherical shock wave may have an amplitude of up to 1 GPa (10 kbar). Cavitation events and shock-wave propagation can be recorded using holography and schlieren photography [346]. However, when an isolated bubble collapses, this shock is so rapidly attenuated that only surfaces within about the initial bubble radius of the centre of collapse may be damaged [343, 347]. On the other hand, in a concentrated mass of bubbles, the collapse of one may initiate the collapse of a neighbour, and under certain circumstances the combined shocks from this 'cloud cavitation' can cause damage at much greater distances [348, 349]. This was observed experimentally by Brunton [350] when cavitation occurred near a solid surface. The model proposed for this effect is of a large hemisphere of cavities collapsing inwards as consecutive shells; the collapse of each shell releases the hydrostatic pressure onto the adjacent inner neighbour shell. The energy of each collapsing shell is passed on to the next shell, so that cavitation in the centre of the cloud is an order of magnitude more energetic than the collapses in the outermost shell. This model was formulated by Mørch [345], Hansson and Mørch [351] and Hansson *et al.* [352], and supported by the experimental observations of Ellis [353]. Noting the reported variability in the pressures generated even by laser- and spark-induced single-bubble collapses, where the bubbles all initially attained the same R_{max}, Zhang *et al.* [354] employed a statistical model for erosion by multibubble collapses.

The second mechanism for cavitation erosion is through liquid jets formed by bubble involution.[80] The first suggestion that the asymmetric collapse of cavities might generate such jets came in 1944 from Kornfield and Suvorov [355], though it was not until the 1960s that Naudé and Ellis [356], Walters and Davidson [357, 358], and later Benjamin and Ellis [359], provided the first real experimental and theoretical evidence for this. Figure 5.49 shows the historic comparison by Lauterborn and Bolle [360] of their high-speed photographic observations of the initial stages of the collapse of a laser-generated bubble near a solid boundary with the predictions of Plesset and Chapman [361]. This jet may traverse the bubble interior, to penetrate the remote bubble surface.

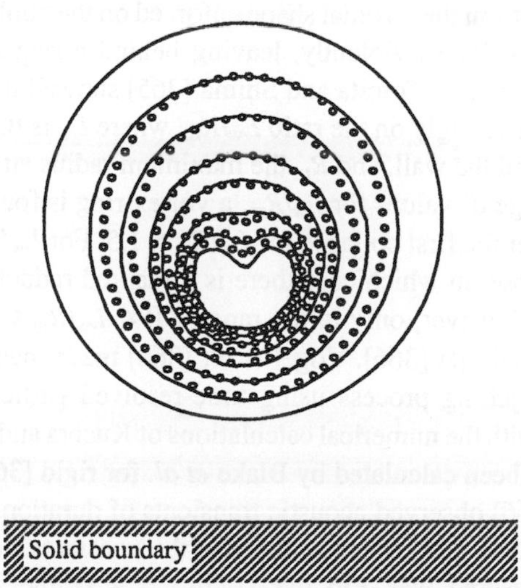

Figure 5.49 Comparison of measured (open circle) [360] with predicted (line) [361] bubble shape for asymmetric collapse of bubble near wall. (After Lauterborn and Bolle [360].) Framing rate: 300 000 f.p.s.; maximum bubble radius: 2.6 mm; distance from bubble centre to wall: 3.9 mm.

[80] Though in the asymmetric collapse shown in Figure 4.35(b), the bubble did not produce a jet and break up, but instead rebounded intact; nevertheless the general bubble shape is suggested of the form of involution which can lead to jetting.

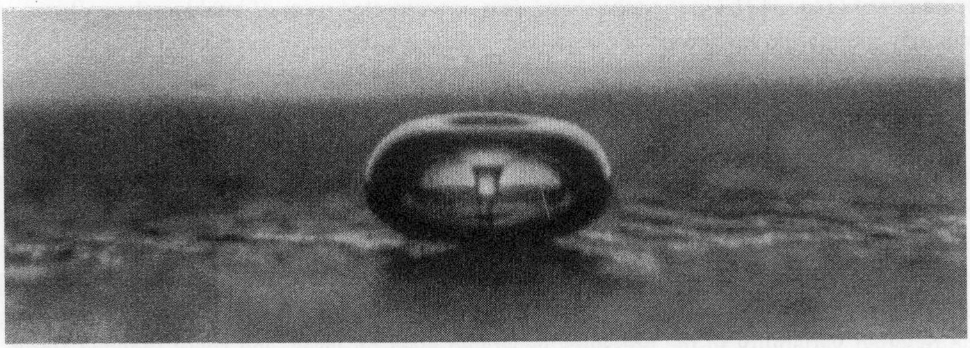

Figure 5.50 Jet formation during the collapse of an oscillating gas–vapour bubble at low pressure (0.04–0.05 bar) in a 60-Hz sound field. The bubble size is approximately 0.2 cm. (Photograph courtesy of Professor LA Crum, University of Mississippi, previously published in reference [362].)

A beautiful picture of this effect, taken by Crum, is shown in Figure 5.50. Figure 5.51 shows the collapse of an initially spherical bubble generated by laser action a distance of 2.3 mm from a boundary, filmed at 305 000 f.p.s. In the figure, jetting first occurs towards the solid surface, and on rebound a counter-jet is generated. After that, the bubble invariably disintegrates [363]. Benjamin and Ellis [359] predicted that a ring vortex would emerge from the jet flow, and Lauterborn [289] and Olson and Hammit [364] observed how jetting may lead to the formation of a 'bubble cavitation ring' from the toroidal shape enforced on the bubble by the jet. The torus itself can then expand and collapse violently, leaving behind a ring-shaped cloud of small bubbles. Naudé and Ellis [356] and Tomita and Shima [365] showed that the dynamics of the asymmetric collapse depend strongly on the ratio L_w/R_m, where L_w is the distance between the point of bubble formation and the wall, and R_m the maximum radius attained by the bubble on expansion. Over a wide range of values for L_w/R_m, a vortex ring is found to emerge from the jet flow. The ring forms after the first collapse for $L_w/R_m > 1.5$. For $L_w/R_m \approx 0.9$, a vortex ring forms before the first collapse, in which case there is a marked reduction in sound emission during the collapse [366]. However, only in the range $2.0 \leqslant L_w/R_m \leqslant 1.0$ does a counter-jet appear as regularly as the main jet [366]. Vogel *et al.* [366] made measurements of the fluid flow velocities during the jetting process using time-resolved particle image velocimetry, obtaining good agreement with the numerical calculations of Kucera and Blake [367]. Pathlines and pressure contours have been calculated by Blake *et al.* for rigid [368] and free [670, 671] boundaries. Vogel *et al.* [366] observed acoustic transients of duration 20–30 ns, comparable with earlier measurements [369–371]. The rise times of the transients were less than 10 ns. The highest jet speed they measured was 156 m/s (L_w/R_m = 2.3), the speed decreasing with decreasing L_w/R_m (as predicted by Plesset and Chapman [361] and Blake *et al.* [368]). The highest pressure amplitudes measured at the solid boundary were found for bubbles attached to the boundary, when there is no cushioning liquid film between bubble and boundary, the pressure within the bubble and at the boundary being calculated to reach about 0.25 GPa for a bubble which had L_w/R_m = 0.2 and a maximum radius R_m of 3.5 mm, collapsing down to a volume equivalent to a radius of 0.6 mm. These high pressures are in agreement with the findings

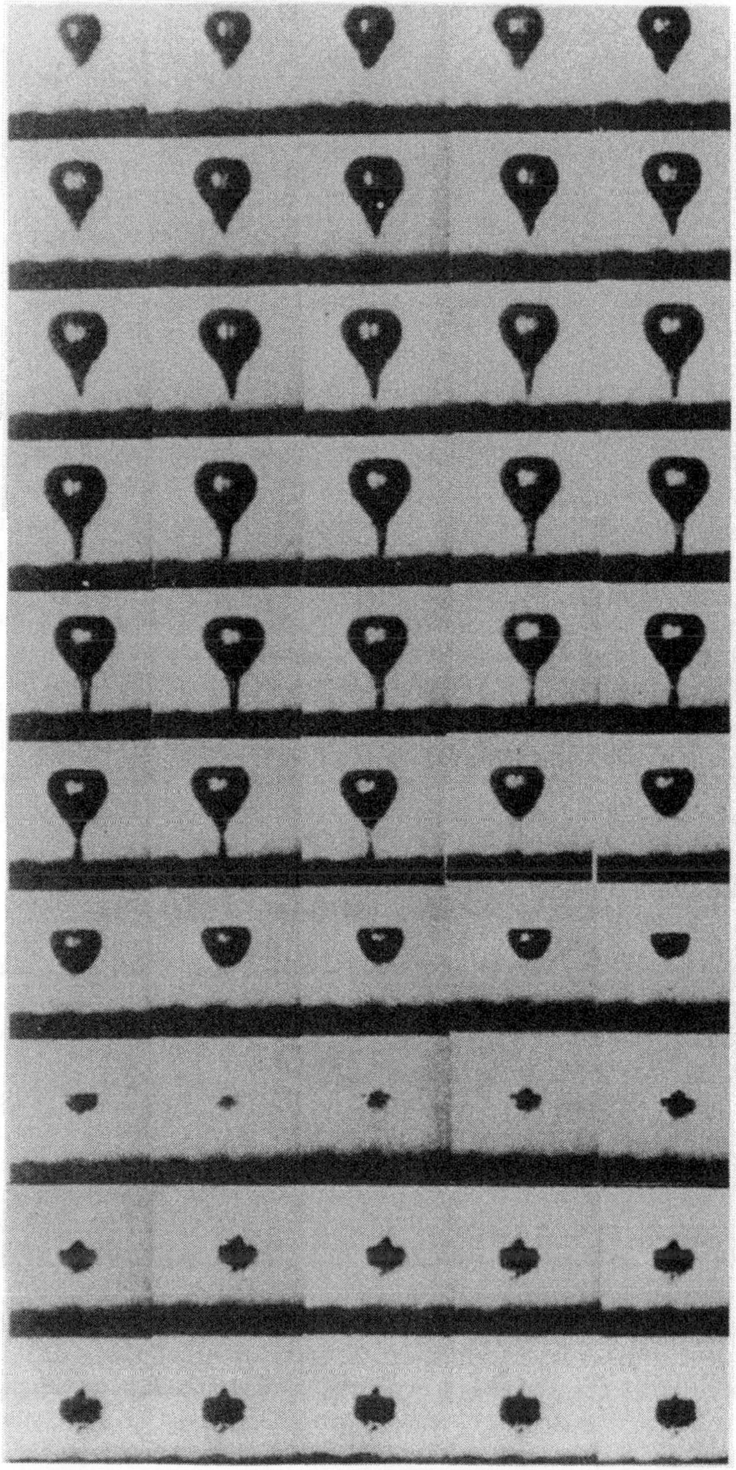

Figure 5.51 Dynamics of an initially-spherical bubble generated by laser action a distance of 2.3 mm from a plane solid boundary, after the first collapse (filmed at 305 000 f.p.s.). Consecutive frames read from left to right, proceeding line by line down the page. (After Neppiras [363].)

of Jones and Edwards [372], though for the same values of L_w/R_m and R_m, Tomita and Shima obtained only 15 MPa (0.15 kbar) (though their transducer may have had significant rise time limitations [366]). Only when L_w/R_m is less than about 0.9 does the jet impact the boundary without being cushioned by a water film, or being transformed into a ring vortex before the end of the collapse. For estimating the erosive damage capability of the jet, Vogel *et al.* [366] had therefore to presume a jet speed of 100 m/s or less. Tomita and Shima observed jet speeds of 200–370 m/s after a gas bubble was subjected to a 5-MPa (50 bar) pressure pulse, and Dear and Field [373] observed jet speeds of 400 m/s when shocks of 0.26 GPa (2.6 kbar) passed over two-dimensional cavities (see also below).

The production of a jet through asymmetric collapse can arise through a constant pressure gradient across the cavity, or through the transient pressure pulse engendered by a shock wave [205]. In the simplest terms, one needs to break down the spherical symmetry and through the asymmetry define some direction for the jet to follow. There are two related but different scenarios for jet formation. The first, as analysed, for example, by Plesset and Chapman [361], occurs when a cavity collapses near a rigid surface: the closer the cavity was to the surface, the more pronounced the jetting [361, 374]. The jetting here results from the asymmetry of the flow caused by the surface. The second situation arises when a shock passes over a cavity, as illustrated in Figure 5.16. In cavitation, particularly when there are several bubbles collapsing in close proximity near a boundary, both effects are likely to be present since the shock induced by the collapse of one cavity can interact with an adjacent cavity [373, 375].

Erosion and damage can occur when the high-speed liquid jet impacts a surface, leaving the characteristic 'pits' that are approximately one-tenth the radius of the original cavity [365, 376–379]. Lush *et al.* [380] positioned single air bubbles in water, near the surface of various target materials. These bubbles were then collapsed by the shock wave produced on the detonation of small silver or lead azide charges. The resulting pit dimensions were compared with those predicted by a model based on liquid impact on an elastic–plastic solid that has a bilinear stress–strain relationship (the plastic and elastic wave-speeds are known). The pit radius appears to be roughly independent of the hardness of the material. Hansson *et al.* [381] suggest that jetting would prevail during low-intensity cavitation, and when only a few cavities are involved, whereas the effect of the cloud would dominate during high intensities of cavitation. The concerted collapse of a cavity cloud can, of course, cause rather larger damage marks than those due to the jetting of single bubbles. In fact, it was the presence of relatively large pits which lead to studies of the collapse of cavity clouds.

Erosion through jetting arises as a result of the high pressures which can be generated when a liquid impacts a solid, and the two are moving at high speed relative to each other. The basic process is symmetrical with respect to role-reversal: it is as relevant for cases where the stationary liquid drops contact fast moving bodies, as for those where high-speed jets impact a relatively stationary material, so that applications include the rain erosion of aircraft, the erosion of steam turbine blades, and the cutting and cleaning of surfaces. Field [382] provides a clear and succinct account of the history and physical processes involved.

The pressures that can arise through liquid–solid impact can be appreciated through simple analysis. Many people will be familiar with the knocking that can be produced when a water tap is closed rapidly. This is the result of a shock wave, generated by the sudden arrest of the water column which was moving through the pipe. Consider the liquid shown in Figure 5.52. A column of liquid of density ρ was moving to the right at speed v in the pipe of cross-sectional area A, but the far right end of the pipe has just been suddenly terminated by a rigid wall. A shock moves back through the liquid at speed c_s (to the left). The liquid to the right of the shock, through which the shock has passed, is motionless; the liquid to the left of the shock is still

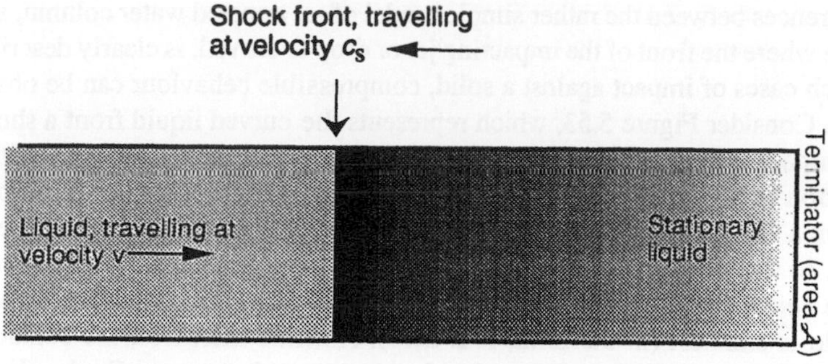

Figure 5.52 A column of liquid of density ρ, moving to the right at speed v in a pipe of cross-sectional area A, is arrested by the passage of a shock (speed c_s) to the left, which is the product of the sudden termination of the pipe by a rigid wall.

moving towards the terminating wall at speed v. In time Δt, the shock moves a distance $c_s\Delta t$ to the left, and a mass of liquid $\rho A c_s \Delta t$ to a standstill. That liquid was moving at speed v, its momentum has changed by an amount $\rho A c_s v \Delta t$. Thus the rate of change of momentum in the fluid is $\rho A c_s v$. From Newton's Second Law, this must equal the force applied to the liquid. Thus the wall exerts on the liquid a force to the left of $\rho A c_s v$. From Newton's Third Law, upon arrest the liquid must exert an equal and opposite force on the wall. Since the wall has area A, the pressure exerted on the wall by the liquid, the so-called 'water-hammer' pressure, is to the right, and has magnitude

$$p_{w\text{-}h} = \rho c_s v \tag{5.6}$$

If the pipe is not terminated by a perfectly rigid wall, but instead by a solid of density ρ_d and sound speed c_d, the pressure is [383]

$$p_{w\text{-}h} = \frac{(\rho c_s \rho_d c_d)v}{\rho c_s + \rho_d c_d} \tag{5.7}$$

If instead the column of liquid that is arrested is an impacting liquid drop or jet, the situation is complicated by the free surfaces at the sides of the liquid. These surfaces take a time to communicate their presence to any part of the liquid, and until they do, the pressure at that point in the liquid retains its initial value. That communication comes in the form of release waves, so assuming the jet is a flat-ended cylinder of liquid of radius r_c, the pressure at the centre of a jet remains at the initial value for a time

$$t_{col} = r_c/c_s \tag{5.8}$$

following the initial impact. Once flow is set up, the pressure is given simply by the Bernoulli term, $\rho v^2/2$.

It is important to realise that the shock speed c_s in equations (5.6) and (5.7) can be much higher than the acoustic wave speed c for water (1470 m/s at room temperature and under one atmosphere). As pointed out by Heymann [384], $c_s \sim c + \varphi_s v$ where φ_s is a constant of value ~2. Example values for the ratio of the water-hammer to the Bernoulli pressure are given by Field [382]: for the impact of a 2-mm diameter water-drop at impact velocities of 50, 100 and 500 m/s, the ratio is 64, 34 and 10 respectively, emphasising the importance of the initial stages of impact to the damage.

The differences between the rather simple model of an arrested water column, and the more realistic case where the front of the impacting jet or drop is curved, is clearly described by Field [382]. In such cases of impact against a solid, compressible behaviour can be observed in the liquid [385]. Consider Figure 5.53, which represents the curved liquid front a short time after the initial impact against a solid. The contact edge moves supersonically (with respect to communication *in the liquid*) and, following Lesser [386], the position of the shock envelope is found from a Huygens-type construction.[81] The liquid beneath this shock-wave envelope is compressed, so that high pressures of the order of $\rho c v$ are exerted on the solid beneath it. The speed of the contact edge is highest when the liquid first touches the solid, because then the contact angle is smallest (in the limit, it is zero). As time increases, the contact angle also increases, and the speed of the contact edge decreases, so that eventually the shock envelope will overtake the contact edge, and release waves will move into the liquid.

If r_{cur} is the local radius of curvature of the front of the original liquid surface, then it takes a time $(vr_{cur})/(2c^2)$ after impact for the shock envelope to overtake the contact edge, at which point the latter has a radius of about $(vr_{cur})/c$. As it takes the release waves a further $(vr_{cur})/c^2$ to reach the central axis, the total time over which the liquid behaves compressibly is [382]

$$t_{cur} = \frac{3vr_{cur}}{2c^2} \tag{5.9}$$

and more than this for a compliant target.

The geometrical acoustics model by Lesser and Field [386, 387] provides the most complete treatment to date of the impact of a liquid drop. This demonstrates that, although the centre of the impact is the water-hammer pressure, the solid is subjected to even greater pressures at the expanding contact edge, which reach a maximum of about $3\rho c$ just before the shock envelope overtakes the contact edge, physically because "the wavelets ... bunch up in this region as the

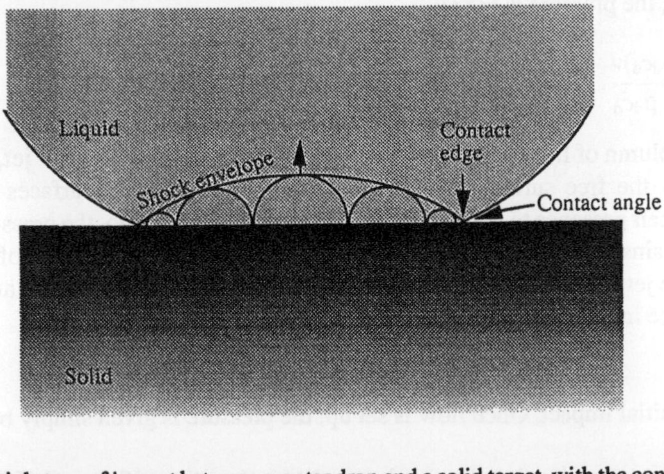

Figure 5.53 Initial stage of impact between a water drop and a solid target, with the contact edge moving faster than the shock velocity in the liquid (i.e. supersonically). The shock envelope is made up of many wavelets and can be found from a Huygens-type construction. The liquid behind the envelope is compressed and the target beneath this area subjected to high pressure. (After Field [382]. Reprinted with permission from *Physics in Medicine and Biology*, vol. 36, pp. 1475–1484; Copyright © 1991 IOP Publishing Ltd.)

[81] Similar to those used in Chapter 1, section 1.1.1(a) and Chapter 3, section 3.8.2(b)(ii).

edge velocity decreases" [382]. Though these pressures are higher than the water-hammer pressure, they are of much shorter duration, and have a much smaller physical damage effect. However, in liquid-drop impact considerable damage can occur after the shock has detached from the solid and moved into the liquid, because jetting occurs at the contact edge, liquid spurting outwards in a flow that can have speeds several times the impact velocity [382, 388]. This principle is shown in Figure 5.54 for the impact of a high-speed metal slider against a

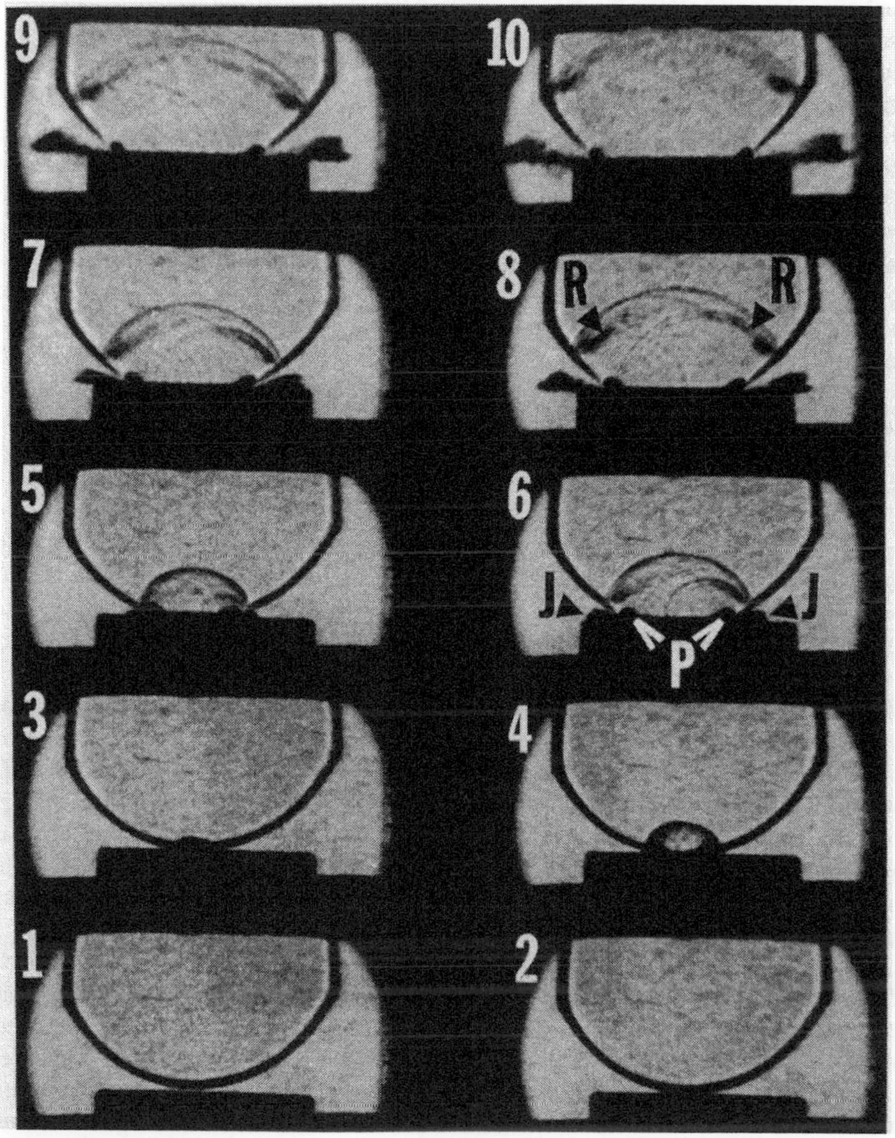

Figure 5.54 Impact at 150 m/s of a metal slider with a disc-shaped gel layer having an 11-mm radius of curvature. Initially the shock envelope expands supersonically, and there is no jetting. The shock envelope overtakes the contact edge between frames 4 and 5, and the high-speed jets develop. J = high-speed jet (≈1800 m/s); P = high 'edge' pressure lobes; R = release waves. Inter-frame time = 1 μs. (After Dear and Field [389].)

stationary disc-shaped gel. High-speed jetting at the contact edge, high pressure edge lobes and release waves can clearly be seen.

In studies of the collapse of initially spherical bubbles, a variety of methods have been employed to ensure that the bubble is positioned and collapsed in a controllable manner. The degree of control is critical, as the event is high-speed: it must be made to occur at a predictable time to ensure that the instrumentation (e.g. high-speed cameras etc.) will record the event. The bubble must also be localised in space, for example, to be within sensor range of cameras and hydrophones etc. and because the spacing between the cavity and any boundary can critically affect the collapse [390]. Techniques for ensuring predictable events include the use of lasers, as illustrated earlier, to produce a bubble in liquid [232, 391, 392]. As discussed in section 5.1.2, electrical sparking can also be used to produced cavitation bubbles predictably, and this has been used in several jetting studies (see, for example, Chahine [393]). Shima *et al.* [390], for example, use a spark to produce a single bubble, which subsequently collapses. This creates a spherical shock wave, which then collapses a second bubble asymmetrically. This bubble, which approaches the case of a free-floating, spherical bubble, is in fact attached to a thin iron wire. Interactions between such collapsing bubbles have since been the subject of much work. Lauterborn [374] produced bubble pairs of varying sizes, which then collapsed under hydrostatic pressure. The relative sizes of the bubbles, and the presence of solid surfaces, greatly influenced the direction of the jets. Chaudhri *et al.* [415] observed similar effects when bubble pairs were collapsed with planar shock waves. Jetting has been modelled [394] for a variety of geometries, including the collapse of a spherical bubble placed midway between two solid walls [395, 396], the collapse of a bubble between narrow parallel solid walls, but with the bubble attached to one of these walls [397], and the hemispherical bubble attached to a single wall [398].

Brunton [350] introduced the technique of collapsing disc-shaped cavities, a method which has proved useful in studying the mutual interactions between collapsing cavities. It facilitates optical studies since the complications arising from spherical refraction of light are avoided. Circular air bubbles were trapped in water between two flat parallel transparent plates. Impact was used to generate the shock. Coupling this technique with high-speed photography and metallographic analysis of the eroded surfaces, Brunton [399] calculated the jet velocity from the collapse of single bubbles to be ≈ 500 m s^{-1}. This would generate a shock pressure of 5×10^8 Pa in a free-floating bubble, and 10^9 Pa if the bubble were attached to a solid surface. Brunton also found that once the jet had impacted, the cavity would begin to expand again, creating a net flow in the direction of the jet. The hydrodynamic pressure from this flow is an order of magnitude less than the shock pressure, but lasts longer, and so could contribute significantly to the damage (particularly if the collapse were to occur near a notch or crevice in the specimen). This is discussed later.

An improved version of Brunton's method of using disc-shaped cavities was introduced by Field *et al.* [400]. The bubbles were trapped between transparent plates, but in gelatin instead of water. This allowed greater accuracy in the size and position of the bubbles, which could simply be cut out of a sheet of gelatin. A striker or flyer plate, fired from a rectangular-bore gas gun, passes between the transparent plates which contain the gelatin, impacts the gelatin and produces the required shock wave. Alternatively the shock wave can be generated using explosives. High-speed photography records the cavity collapse, and schlieren optics image the shock. Figure 5.55 shows the collapse of a cavity of initial radius 1.5 mm in response to a 0.26-GPa shock wave, the inter-frame time being 0.96 μs. The liquid jet (labelled J) can clearly be seen. Variations on this method have been used to study liquid wedge impact [388, 402]; liquid drop impact [403]; and shaped-charge configurations [404].

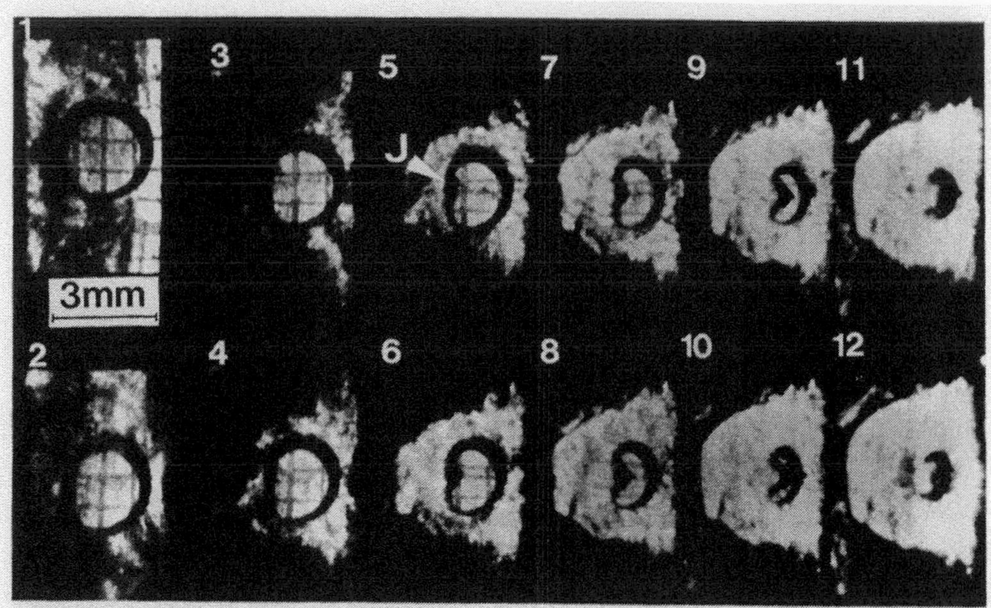

Figure 5.55 Two-dimensional cavity of diameter 3 mm collapsed by a shock wave which travels from left to right. The rear surface involutes to form a jet, J, labelled in frame 5. The jet travels at a speed of 400 m/s. Inter-frame time = 0.96 μs. (After Dear *et al.* [401]. Reprinted by permission from *Nature* vol. 332, pp. 505–508; Copyright © 1988 Macmillan Magazines Limited.)

As well as the collapse and jet formation from single bubbles, this technique has been used to study the behaviour of lines of bubbles (both parallel and perpendicular to the shock front), and of rectangular and triangular bubble arrays [373]. An analytical treatment of cavity collapse by Lesser and Finnström [405] agrees well with the findings of these studies. The influence that one jetting cavity has on another is important, as the cluster collapse model outlined at the beginning of this section might not only concentrate pressure pulses but also bring about intensified multiple jet impacts on the solid surface [382]. The 2-D gel technique is ideal for studying this, as specific required cavity geometries can be tailored and cut out of the gel. Figure 5.56 shows the collapse of a 3 × 3 array of cavities of initially 1.5 mm radius, as a 0.26-GPa shock wave passes through gelatin, with an inter-frame time of 5 μs. The cavities collapse one row at a time, the later rows being shielded from the incident shock by the previous cavities. When shielded cavities do eventually collapse, it is in response to the shock waves emitted by the rebounding cavities in the previous row. The collapse is, therefore, a step-by-step process, and the collapse velocity can be very different from that of the initial shock. For weak shocks, the collapse wave can be much slower than that of the initial shock. However, as Bourne and Field [205] have recently shown, it is possible with very intense shocks to generate new shocks ahead of the main shock. This is because the collapse can produce jets which can travel faster than the collapsing shock (see also below).

Tomita *et al.* [406] used the shock waves produced by the rebound of a spark-induced gas globe to collapse two-dimensional arrays of three-dimensional air bubbles in water held by buoyancy beneath a solid wall. In all these studies it is clear that inter-bubble interactions will affect the direction of the jet, perturbing it from the normal to the wall.

Extracorporeal shock wave lithotripsy was introduced in section 5.1.2. Field [382] outlines how, in addition to the conventional mechanisms for stone breaking which involve tensile

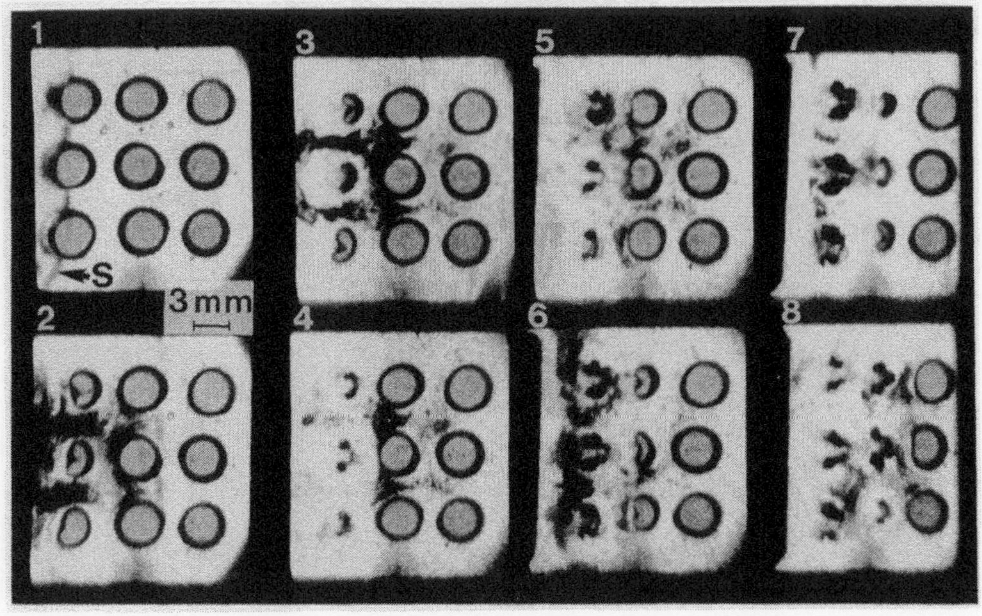

Figure 5.56 Rectangular array of nine cavities, diameter 3 mm, collapsed by shock wave S. Note the layer-by-layer collapse. Inter-frame time = 4.25 μs. (After Dear and Field [373].)

stresses generated within the stone by direct shock interaction, cavitation damage may also occur. When the initial positive pulse from the lithotripter (Figure 5.10) interacts with the solid it may set up surface waves, which can generate cavities at the interface between the stone surface and the surrounding liquid medium [407]. Also, immediately following the pressure pulse is a tensile tail, and pressure oscillations.[82] The tension could cause cavitation inception in its own right, or excite the cavities generated at the interface. Noting that lithotripter treatment incorporates many successive pulses, any cavities remaining from one pulse would undoubtedly be excited by the high pressures in the subsequent pulse. Cavities close to the stone surface would damage through involution and jetting, and cluster effects could occur. Field [382] also draws attention to the case introduced above, where a cavity might collapse into a crevice on a stone surface, which can result in the greatest damage as a result of the large tensile stresses generated at the root of the crevice by the expanding cavity which, on rebound, is confined by the crevice walls. This is illustrated by Brunton and Camus [408] in Figure 5.57, where a gas bubble under the influence of a pressure wave collapses into a notch in a plate of PMMA (polymethylmethacrylate). Shock waves radiated from the bubble can be seen in both liquid and PMMA. Stresses set up by the expansion of the bubble causes fracture of the plate, visible cracks beings formed at the base of the notch.

Much interest in the shock-induced asymmetric collapse of cavities has arisen through investigations into the initiation of explosives. In the 1950s, Bowden and Yoffe [409, 410] and co-workers demonstrated how ignition can be the result of the degradation of the mechanical energy of a shock wave propagating through a medium into heat. Even if the bulk average temperature rise was insufficient to cause general ignition, thermal explosion could arise through the concentration of energy into hot-spots. Amongst the various mechanisms for

[82] The presence and form of these features is dependent on the type of lithotripter – see section 5.4.2.

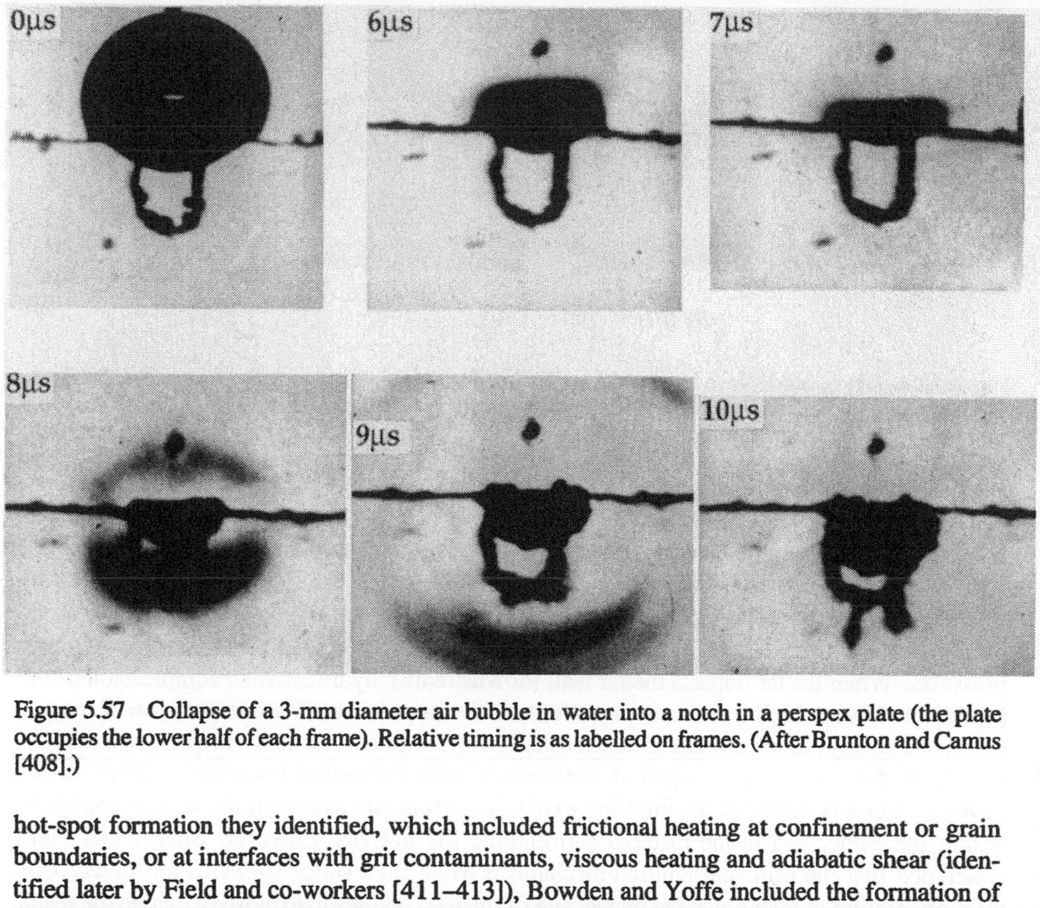

Figure 5.57 Collapse of a 3-mm diameter air bubble in water into a notch in a perspex plate (the plate occupies the lower half of each frame). Relative timing is as labelled on frames. (After Brunton and Camus [408].)

hot-spot formation they identified, which included frictional heating at confinement or grain boundaries, or at interfaces with grit contaminants, viscous heating and adiabatic shear (identified later by Field and co-workers [411–413]), Bowden and Yoffe included the formation of hot-spots through the adiabatic compression of entrapped gas pockets when the shock wave passed. Studies by Coley and Field [407], Chaudhri, Field and co-workers [414–416], Starkenberg [417] and others demonstrated the efficacy of this process.[83]

The mechanism by which cavitation-induced ignition comes about depends on several factors, including conductive and convective heat losses from the collapsing cavity [414, 417], the spalling of liquid droplets from the cavity walls [202], heating through hydrodynamic flow and compression [418–421], inviscid plastic work and visco-plastic work in the material [421]. In turn, these processes will be dependent on the initial size and rate of collapse of the cavity, and the thermodynamic and physical properties of the liquid, gas and vapour.

Bourne and Field [205] photographed at microsecond framing rates the shock-induced asymmetric collapse of cylindrical cavities (the radius being greater than the length), formed out of thin sheets of gelatine or of an ammonium nitrate/sodium nitrate emulsion confined between transparent blocks. Figure 5.58 shows the emission of sonoluminescence from a collapsing 2-D cavity in gel, as evidence of the high temperatures that can be achieved. The inter-frame time is 0.2 μs. In the first frame the 1.88-GPa shock (labelled 'S'), moving upwards, can be seen, as can luminescence resulting from the compression of gas trapped between the impacting jet tip and the downstream cavity wall. By frame 3 this site has ceased to luminesce,

[83]See Figure 5.17.

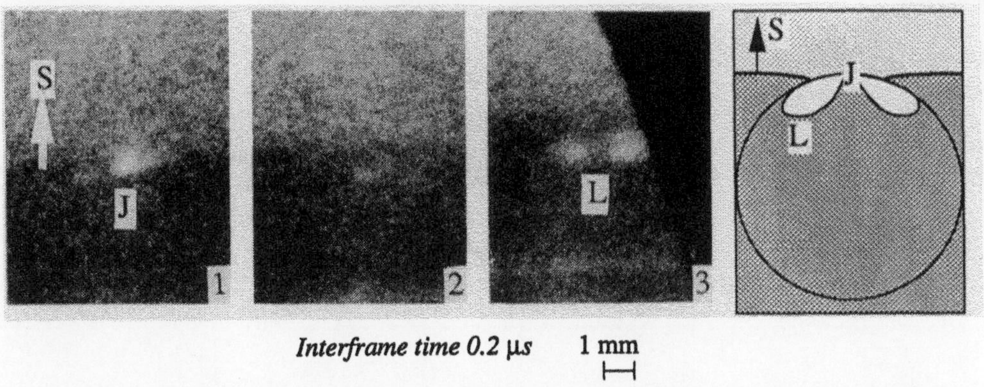

Figure 5.58 Three consecutive frames of the collapse of a 6-mm cavity in gelatine by a 1.88-GPa shock, S. Frame 1 shows the luminescence, J, from gas trapped between the jet tip and the downstream cavity wall just before the jet strikes the downstream wall. The luminescence is not apparent in frame 2. By frame 3, two lobes of entrapped gas, L, have been isolated and begin to luminesce. The schematic indicates the relative positions of the flashes J and L. (After Bourne and Field [205].)

but light emission can be detected from the lobes of trapped gas. The typical collapse began with the formation of a high-speed jet from the wall which the shock first meets. This jet traversed the cavity and impacted the far wall, causing the emission of another shock wave. This shock could travel ahead of the initial shock wave if the latter was at sufficiently high pressures. When the jet impacts the far wall (downstream), hydrodynamic compression of the liquid could cause high transient temperatures at the impact site. The compression of the gas within the cavity causes sonoluminescence at high rates of compression. The penetration of the jet into the far wall generates a pair of vortices which are convected downstream. When such a cavity collapse occurred in an emulsion explosive, a reaction started in the vapour contained within the cavity and in the material surrounding the hot gas. However, Bourne and Field concluded that the principal ignition mechanism when high shock strengths are involved is through ignition of material at the impact site of the jet. Cavities in explosives can be the result of accident or design [205, 422]. Emulsion explosives can be sensitised by the addition of gas-filled glass spheres of radius 50–100 μm.

Hydrodynamically induced cavitation can cause erosion, the phenomenon which prompted Rayleigh's pioneering study [342] outlined in Chapter 2, section 2.1.3(b). The pressure reductions caused by the flow around the blades can cause bubbles to expand, and the blade may be damaged through the subsequent cavity collapse. A beautiful example of this effect is shown in Figure 5.59, which shows the effects that can occur when water flows at 40 m/s over a fixed NACA009 blade in the high speed cavitation tunnel at Institut de Machines Hydrauliques et de Mechanique des Fluids, EPFL, Lausanne. The angle of incidence of the blade relative to the water is +4°, and the cavitation number is $\sigma_c = 0.9$. The geometry is shown schematically in Figure 5.59(a). As can be seen from the photographs, cavitation is generated on a cloud at the pressure drop following the leading edge of the blade (on the right). The cavities are then convected with the flow along the blade. Any subsequent cavity collapses which occur sufficiently close to the blade will damage it.

In section 5.2.2(c)(i), propeller noise was related to bubble dynamics, and the measurement of the noise generated by these processes has proved valuable in determining the exact nature of the mechanisms (see, for example, Leggat and Sponagle [423] and Bark [424]). Such a system, where large numbers of bubbles oscillate in close proximity, is a form of cloud cavitation [425]. In such cases, bubbles no longer behave in the way predicted by formulations involving single bubbles only. Chahine [426] showed numerically that in a cloud, the driving pressure on a bubble is firstly reduced, so delaying the implosion. Then, in the later parts of the sound cycle it is dramatically increased, so that when the implosion does occur its violence is greatly increased. Arakeri and Shanmuganathan [427] used electrolysis to introduce bubbles into the hydraulic equivalent of a wind tunnel; there, they were acoustically cavitated. The authors found that the smaller the spacing between adjacent bubbles, the lower the maximum sound frequency up to which single-bubble dynamics hold true for the collapse. Haenscheid and Rouvé [428] produced a 'hybrid' model of hydraulic erosion, dividing the process into two parts: the hydrodynamic component, and the bubble dynamic one (which includes the interaction between the two parts).

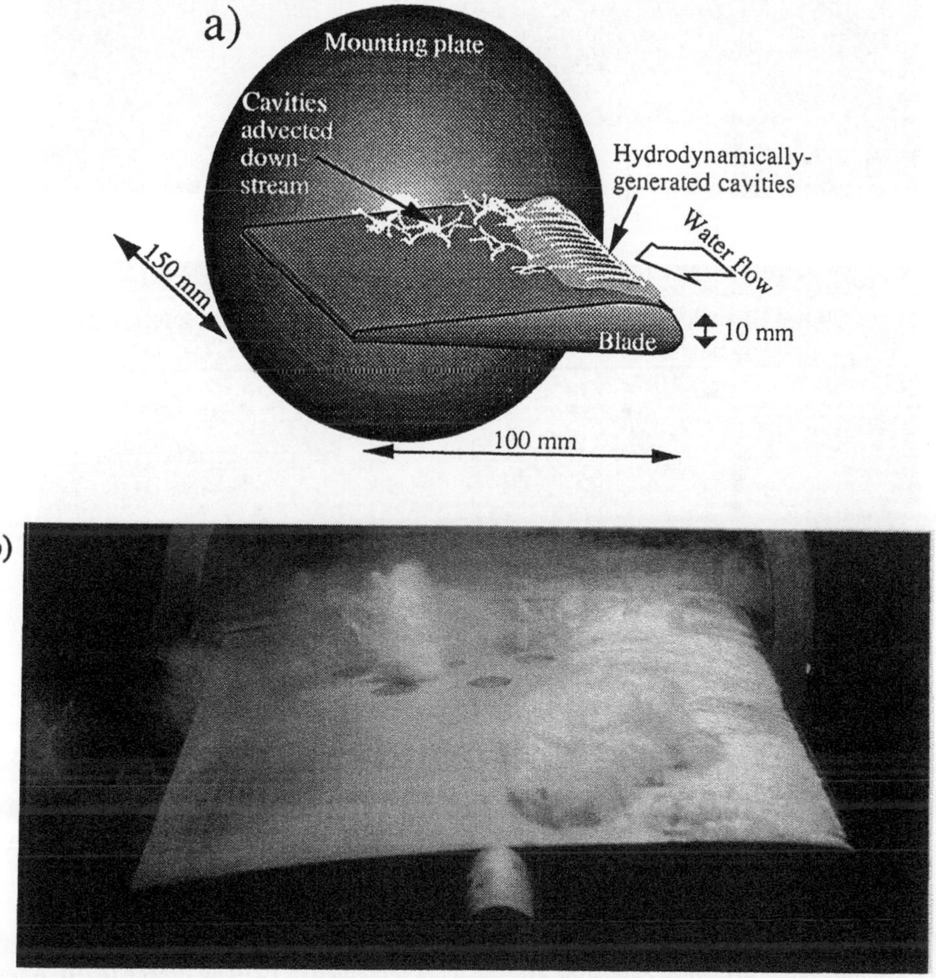

Figure 5.59 Hydrodynamic cavitation in the high-speed cavitation tunnel at Institut de Machines Hydrauliques et de Mechanique des Fluids, EPFL, Lausanne, when water flows (from right to left) at 40 m/s over a fixed NACA 009 blade (truncated at 90%): angle of incidence is + 4 °, hydrodynamic cavitation number is $\sigma_c = 0.9$. (a) Schematic of the geometry. The cavitation is shown in side view (b) and from above ((c) and (d)). (Photograph courtesy of M Farhat and F Avellan.)

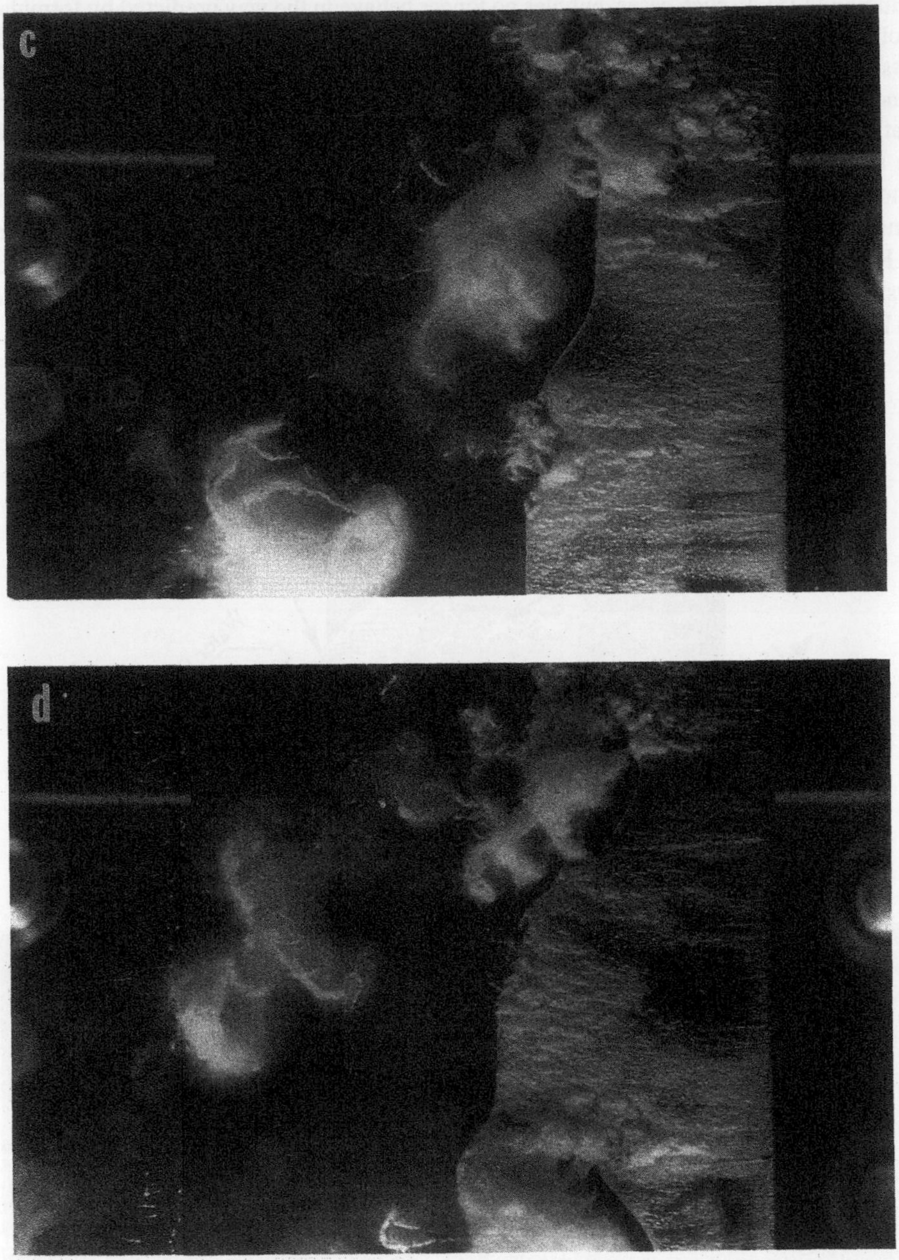

Figure 5.59 *continued*

The macroscopic effects of cavitation erosion were reviewed in 1979 by Preece [429]. A few examples of the considerable body of research undertaken since then has included attempts to develop a size-scaling relation to describe cavitation erosion which would, for example, facilitate the creation of any model systems for engineering purposes [430, 431], and attempts to correlate acoustic emission of jet impact with damage when protective coatings are applied to solids [14]. Acoustic cavitation may also be exploited to improve material properties. For example, Namgoong and Chun [432] found that the application of ultrasound whilst chromium is being electroplated onto a surface improved the protective properties of the chromium surface, in that microhardness, resistance to microcracking and stain-proofing were enhanced.

The erosive nature of acoustic cavitation has been exploited for many years in industrial and laboratory cleaning baths, cell disrupters etc. The effect is also utilised in dentistry [433–436]. Richman [437] first suggested the exploitation of ultrasound in endodontic therapy in 1957, but it was not until 1976 that Martin [438] developed a commercial ultrasonic device for the preparation and cleaning of the root canal. So-called 'endosonic' techniques are now well-established [439, 440]. Both magnetostrictive and piezoelectric devices have been devised, operating at 20–42 kHz, with a few tools (powered by pressurised air) operating at either 3–6 kHz or 16–20 kHz [441]. Lumley *et al.* [441] describe the ultrasonic endosonic file, the tip of which is mounted at 60°–90° to the primary direction of motion of the driver, so that the abrasive tip undergoes primarily a translational motion, the greatest amplitude (of order μm) being at the very tip when used in an unconstrained mode (though in a tight-fitting canal, constraint may occur [442]). The sonic files operate in an elliptical oscillation when free, becoming translational upon constraint in the canal [442]. The dental scaler has also been adapted to operate ultrasonically [227, 443].

The erosive mechanism is thought to be through direct contact. However, dental ultrasonic devices are generally continuously bathed in an irrigant/coolant liquid flow, and in addition to direct contact, cavitation and microstreaming can occur [438, 441, 444]. Microstreaming, the circulation being driven directly by the oscillating tool rather than by bubble oscillations, has been visualised around the file through the motion of 15-μm diameter polystyrene spheres dispersed as a layer on the surface of a methylene blue solution [445, 446], and is thought to be significant. Some cavitation has also been detected by using spectrophotometry with chemical dosimetry [447], though Ahmad *et al.* [445] detected no observable sonoluminescence, and cavitation has been deemed to have negligible contribution to the efficacy of ultrasonic files [434, 447].

Cavitation is, however, readily demonstrated with the dental scaler [448] where its erosive effects can be utilised to assist in the removal of solid deposits from the surface of the tooth [449, 450]. Detection systems include chemical dosimetry, erosion of model systems of vacuum-deposited aluminium on glass [449] and sonoluminescence [445, 451].

Walmsley *et al.* [452] demonstrated indentation caused by the scaling tip, and cavitation erosion resulting in a characteristic pitted surface, on polished gold. They also observed that cavitation and microstreaming in the cooling water caused the superficial removal of root surface constituents. Figure 5.60(a) shows a silhouette[84] of the scaler (backlit using a beta-light, which consists of a tritium radioactive source coated with phosphor), as seen through the image intensifier. The luminescence is shown for gradually increasing power supply to the scaler in Figures 5.60(b) to 5.60(f). The scaler is immersed in water. Luminescence was also observed *in vitro* when the scaler was held against a tooth (Figure 5.61), though occasionally a luminescence that was very much brighter could be observed (Figure 5.62). In this case an eroded groove, rather than pitting, resulted on the tooth: the light was probably triboluminescence.

---

[84] Use of the silhouette precludes the ambiguities in the perimeter of a highly reflective object that can arise if the object is illuminated directly by low-level light sources for viewing through the image intensifier.

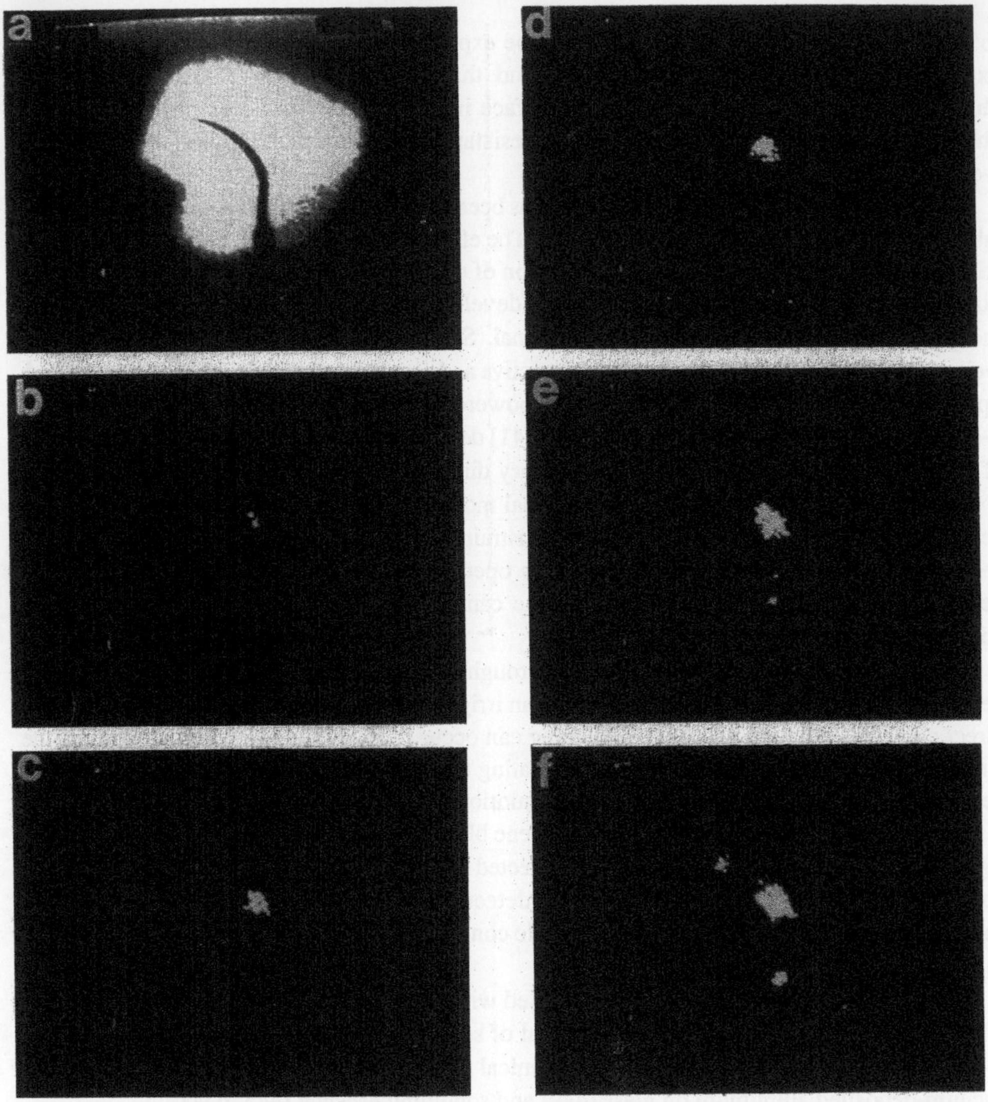

Figure 5.60 (a) The silhouette of the scaler (tip type TFi10 in a Dentsply Cavitron 2002™, backlit using a beta-light, which consists of a tritium radioactive source coated with phosphor), as seen through the image intensifier. The length of the scaler tip is approximately 2 cm. Frames (b)–(f) show the luminescence from the scalar, immersed in water with no backlighting but the same positioning as in (a), operating at 25 kHz and at power settings on the Cavitron (arbitrary units) of (b) 0.75, (c) 1.50, (d) 2.00, (e) 2.75, (f) 3.25. (Photograph: PJ Byrne and TG Leighton.)

Therefore both cavitation and acoustic microstreaming can contribute to the direct mechanical action of ultrasonic dental instruments. One other possibility is that the chemical products[85] that can result from energetic cavitation could have a sterilising action in the mouth. That the destructive effects of ultrasonic cavitation can be so readily utilised in clinical practice raises the question of the bioeffects of ultrasonic cavitation, which will be the topic of the final section of this book.

---

[85]See section 5.2.1(a)(ii).

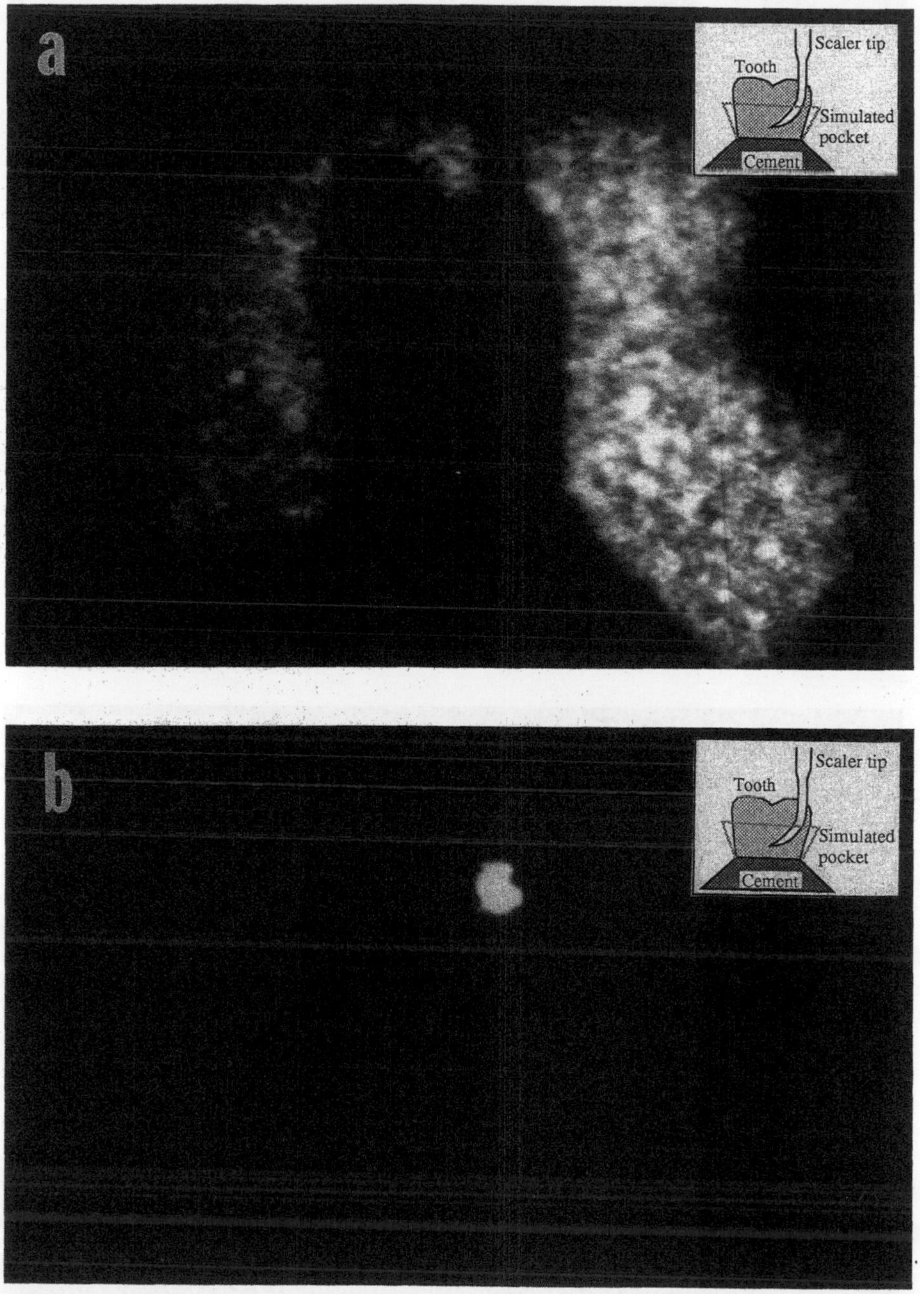

Figure 5.61 Image-intensified views when the scaler (type as in Figure 5.60) is held against the tooth, immersed in a water tank, and the cooling spray is activated. The insert shows the geometry, which is the same for both images. The tooth is mounted in dental cement. A short length of PVC tubing simulates the periodontal pocket, and fills with water from the cooling spray. (a) Silhouette of the scaler tip, the end of which is not visible owing to the silhouette of the tooth. (b) With no backlighting and with the scalar and cooling spray activated, the sonoluminescence is visible. The geometry is identical to (a). (Photograph: PJ Byrne and TG Leighton.)

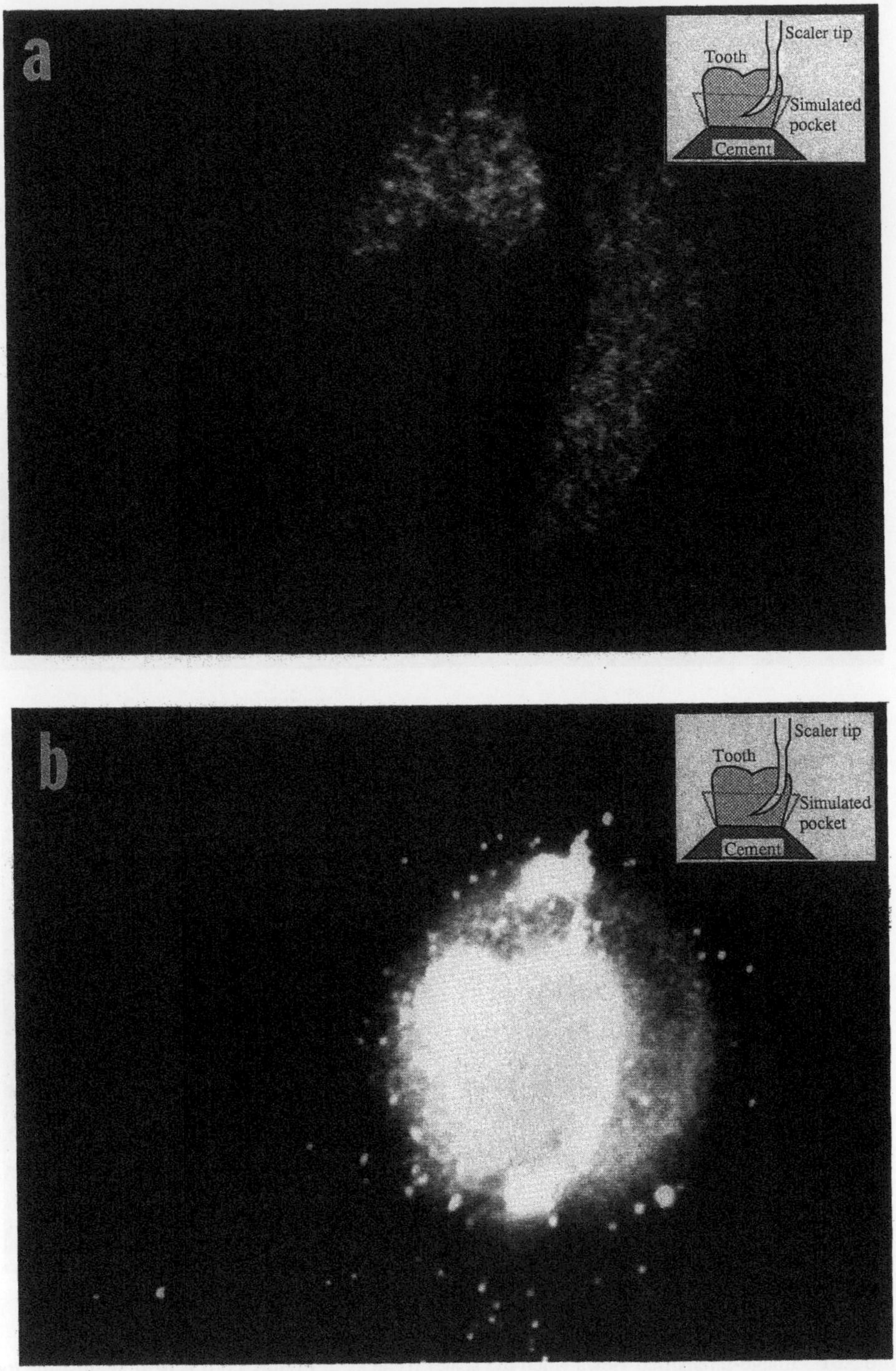

Figure 5.62 Image-intensified views when the scalar (type as in Figure 5.60) is held against the tooth, and not immersed, and the cooling spray is not activated. (a) Silhouette of scalar and tooth (mounted from above). (b) The luminescence is reflecting from the entire tooth surface, illuminating it. This very bright light emission (probably triboluminescence) was occasionally detected under these conditions. (Photograph: PJ Byrne and TG Leighton.)

5.4.2 Bioeffects of Ultrasound: the Role of Cavitation

Having discussed the chemical and physically erosive effects that can result from acoustic cavitation,[86] it is only natural to question the potential bioeffects of ultrasound. A great deal of work has been done in this field, and there is no intention of producing here a review of the huge body of literature on the subject. This book has been concerned with the physics of acoustic cavitation, and the biological consequences of cavitation are beyond its terms of reference. However, this final chapter has explored certain effects and mechanisms associated with acoustic cavitation, and given the wide use of clinical ultrasound, it would be inappropriate not to discuss the potential for bioeffect within the framework of physical mechanisms. In doing so, the discourse extends outside the physics of acoustic cavitation in two areas. Firstly, it is necessary to introduce noncavitational mechanisms for bioeffect (particularly hyperthermia), since to interpret the role of acoustic bubble interactions in the observed effects one must appreciate other, concurrent mechanisms which may dominate. Secondly, some historical and social aspects of the subject are mentioned to give perspective: however, specific studies and nonreview publications are cited only as examples to assist in the examination of the possible roles of acoustic cavitation, and are in no way meant to represent the current, complete, or most important studies with respect to the bioeffects of ultrasound as a whole. It must be emphasised that this section is simply an exploration of the physical mechanisms through which cavitation may cause bioeffect, presented in the light of the preceding chapters on the behaviour of bubbles in acoustic fields, and that readers must look elsewhere for historical and current reviews of ultrasonic bioeffects, and for information on non-cavitational mechanisms. Suggested sources include books by Hill [1], Williams [453], Nyborg and Ziskin [454], Repacholi *et al.* [455], Suslick [2]; and review articles by Hill [456], Coakley [457], Stewart *et al.* [458], Miller [335], Sikov [459, 460], Carstensen [461], and Miller [462]. Valuable reports have been published by the National Council on Radiation Protection (NCRP) [463], the American Institute of Ultrasound in Medicine (AIUM) Bioeffects Committee [464] and the World Federation for Ultrasound in Medicine and Biology (WFUMB) Symposia [465–468]. Bibliographies are periodically published [469–486].

(a) History

Though high levels of ultrasound can undoubtedly induce bioeffect, which led the National Council on Radiation Protection and Measurements (NCRP) 1983 report to recommend that "research should be carried out to investigate the possibility that biologically significant cavitation or bubble activity occurs in human tissue under conditions of diagnostic and therapeutic ultrasound" [463], nevertheless there is no conclusive evidence of bioeffects in humans at clinically relevant diagnostic exposures. As a result, the same NCRP report quoted above goes on to say "there is no firm evidence that any physiological change ... is produced" [463].

As regards therapeutic ultrasound, in 1976 Eastwood and Watmough [17] wrote "therapy safety precautions are generally poor, even the power delivered to the patient rarely being known [487]. It is usually assumed that the main hazard is one of overheating, and that safety is ensured by keeping the ultrasound output below that at which the patient feels pain. Nonthermal effects of ultrasound have, however, been noted in tissues [488], and these may cause damage". In

[86] See sections 5.2.1(a)(ii) (sonochemistry) and 5.4.1 (erosion); section 5.3 contains examples of mechanical, chemical and biological actions.

1987, ter Haar *et al.* [489] published the results of a survey on the practices of physiotherapists in England and Wales in 1985. They concluded that the number of treatments was very considerable (about a million per year in the English and Welsh National Health Service departments, and 150 000 in the private practices that replied to the questionnaire), and commented "It seems poor that for such a widely used modality, the training in its use is so limited. The replies highlighted the need for more clinical trials into the efficacy of ultrasound in physiotherapy, and also the need for a more basic understanding of its potential as a therapeutic agent. Some of the knowledge held by physiotherapists using ultrasound is not based on scientific fact as found in the literature, but has been passed from user to user. Such information may in some cases be inaccurate. The whole field needs to be put on a more scientific footing. There is an outstanding need for physiotherapists to be more vigilant about the need for calibration if ultrasound is to be used safely and effectively and to receive appropriate pre- and post-registrational training in its effects and the mechanisms underlying them."

Public attention has, however, focused on diagnostic ultrasound. By 1984, there were probably well over a quarter of a million foetuses exposed to ultrasound each year in Britain alone [490], and public concerns prompted Mr John Patten, Junior Health Minister of the United Kingdom, to warn against the routine use of ultrasound in pregnancy in a letter to the Association for Improvements in Maternity Services. Quoted in the *Daily Mail* (October 22nd, 1984) and five days later in *Lancet*, he said, "Given the publicity there has recently been about the possible risk of ultrasound scanning we would not expect any health authority to be advocating screening for all mothers as a routine procedure." However, in the same year Davies [491] stated in the *British Medical Journal* that "the accumulated clinical experience of the past quarter century should be reassuring enough." In response to Davies, however, Chalmers [490] warned against complacency in the light of a lack of evidence to the contrary. He cited two interesting historical cases. X-rays were first used in obstetrics in 1899, and by 1935 clinicians were recommending their routine use. Then in 1956 it was suggested that such a practice might predispose to the development of leukaemia in children [492]. Subsequent research supported this hypothesis [493]. It may be noted that both X-rays and ultrasound have at times in the past been perceived to be noninvasive. Citing in addition the history of the use of the drug diethylstilboestrol, which was first used in obstetrics in the early 1940s [494], and in 1969 was proposed to predispose young women to vaginal adenocarcinoma [495], Chalmers noted that none of the adverse effects in either case were 'clinically obvious' to radiologists and obstetricians, and in fact were identified through the research of non-clinicians. In 1987, Ziskin [496] commented that "There is nothing that I'm aware of that has a safer record than that of diagnostic ultrasound."

In the 1970s there was a broad consensus on safety, surveys of diagnostic devices indicating outputs much less than $I_{SPTA} = 0.1$ W cm^{-2} [497], the level below which in 1984 the American Institute of Ultrasound in Medicine (AIUM) stated that mammalian bioeffects had not been observed [498]. However, in the 1980s many instruments in the new generation of devices were capable of exceeding that intensity [499–501]. The underlying reason for the trend to increase powers lies in the tendency for manufacturers to produce diagnostic systems which utilise ever increasing frequencies, so that the shorter wavelengths will lead to an improvement in resolution.[87] However, since these higher frequencies are more strongly attenuated, there is an associated tendency to utilise higher acoustic pressures [502]. Figure 5.63 shows the median and overall maximum outputs from clinical devices surveyed. Figure 5.63(a) shows the change in peak negative (labelled 'P–') and peak positive (labelled 'P+') pressures measured in water

[87] See Chapter 1, section 1.1.1(a).

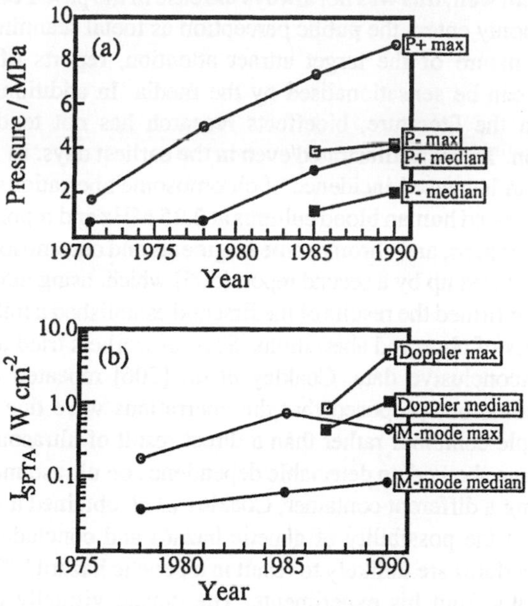

Figure 5.63 The median and overall maximum outputs from each survey of clinical devices up to 1990. (a) The change in peak negative (labelled 'P–') and peak positive (labelled 'P+') pressures measured in water for pulse-echo diagnostic equipment. (b) The change in I_{SATA} output, measured in water, for M-mode operation and for pulsed Doppler modes. (After Duck and Martin [105]. Reprinted by permission from *Physics in Medicine and Biology*, vol. 36, pp. 1423–1432; Copyright © 1991 IOP Publishing Ltd.)

for pulse-echo diagnostic equipment. Figure 5.63(b) shows the change in I_{SATA} output, measured in water, for M-mode operation and for pulsed Doppler modes (note that the scale is logarithmic, showing that the maximum from pulsed Doppler has increased roughly tenfold).

Commenting on diagnostic devices in 1990, Williams [503] notes that "the time-averaged intensities emitted by many devices in certain operating modes can exceed the time-averaged intensities to which patients are being subjected during typical physiotherapeutic treatments. We are consequently in a state of confusion where the physiotherapists are convinced (albeit based on tenuous or anecdotal evidence) that their treatments change the functioning of tissues and are therefore efficacious, while the clinical users are equally convinced (based on even less evidence) that their diagnostic investigations are not changing the tissues at all. Clearly both groups of users cannot be correct in their beliefs."

In discussing the possible bioeffects of ultrasound, one must distinguish between those systems where bioeffects are desirable (such as in ultrasonic cell disrupters), and those where it is not. In clinical practice, diagnostic systems fall into the latter category: in therapy, certain effects are desirable, and others are not, so that the question is one of control, mechanisms and degree.

There are several unfortunate factors which make research into ultrasonic bioeffects difficult. The interdisciplinary nature of the subject means that some experiments do require a research team of broad but detailed conceptual and technical expertise: failure to provide this can mean that reported work may, for example, specify the biologically relevant conditions well but provide inadequate consideration of the acoustics, and vice versa. Though current research is approaching this problem well, this was not always the case in the past. Perhaps because clinical ultrasonics most commonly enters the public perception as foetal scanning, where its frequent use and the sensitive nature of the target attract attention, reports of research indicating detrimental bioeffects can be sensationalised by the media. In addition, though detrimental effects are reported in the literature, bioeffects research has not tended to lend itself to independent verification. This was illustrated even in the earliest days. In 1970, Macintosh and Davey [504] reported an increased incidence of chromosome aberrations in insonated human lymphocytes. They insonated human blood cultures at 2.25 MHz and reported chromatid breaks and achromatic lesions or gaps, and chromosome fragments and chromosome (or isochromatid) gaps. This work was followed up by a second report [505] which, using ultrasound at $\nu = 2$ MHz at 8.2 – 40 mW cm^{-2}, confirmed the results of the first and established a link between ultrasound intensity and the number of observed aberrations. Several workers tried to repeat the findings, and most presented inconclusive data. Coakley *et al.* [506] repeated the experiments and obtained the same results, but proposed that the aberrations were due to a chemical agent released from the sample container rather than a direct result of ultrasound interaction. They also reported that damage showed no detectable dependence on ultrasound intensity. When the work was repeated using a different container, Coakley *et al.* obtained a negative result [507]. Thacker [508] reviewed the possibility of genetic hazard and concluded that "it seems that current diagnostic procedures are unlikely to result in a genetic hazard." Thacker himself [509] obtained negative results from his experiments. The debate virtually ended when in 1975 Macintosh *et al.* [510], working in Coakley's laboratory, performed a double-blind experiment (this had not been the case with the earlier work) and found no increase in chromosomal abnormalities. Macintosh consequently professed to have no faith in the original results. Since then an abundance of reported ultrasonic bioeffects has been published, though there can still be difficulties in obtaining independent verification [511–513].

There are conceivably many mechanisms of acoustic interaction with matter through which ultrasound could generate a bioeffect [514], and these potentially include both cavitational and noncavitational ones. There are many possibilities for the former, bioeffects having been observed during stable, as well as transient, cavitation [515]. Mechanisms involving stable cavitation have been introduced in earlier sections of this book. There are firstly those effects associated simply with the formation of stable bubbles, which may bring about decompression-type symptoms. As discussed in section 5.1.1(a), there are several mechanisms through which these bubbles can form, such as rectified diffusion, and the effect of ultrasonically induced temperature rises on gas solubility and vapour pressure. Other potential mechanisms for bioeffect are associated with the *motions*, either oscillatory or rectilinear, of stable cavities. These include microstreaming, and effects related to radiation forces, which can generate destructive high-speed translation of bubbles, or cause accumulations of particles (platelet aggregation has, for example, been induced under specialised *in vitro* conditions in the field of a foetal heart monitor [516]). One mechanism that has not so far been discussed in detail in this book are the stresses that can be induced by the growth of bubbles or gas pockets. If the medium resists bubble growth (for example, it is a liquid constrained to a fixed volume by structure, such as a mammalian cell; or the gas pocket is itself part-bounded by rigid walls, for example the larval trachae), then stresses can be induced by the growth which may damage the restraining structure [517–519]. This latter mechanism clearly applies to both stable and transient cavitation. An example was shown in Figure 5.57. Other mechanisms, more usually associated with violent collapses, are: the pressure wave emitted on cavitation rebound; the impact pressures associated with bubble involution and jetting; the generation of free radicals and other, longer-lived, sonochemicals; and the hot-spots attained upon collapse. Though it is interesting to speculate on the very considerable temperature rises that can occur on a microscopic scale through cavitation hot-spots, particularly as macroscopic instrumentation would greatly reduce any measured temperature rise through spatial averaging, there can be no doubt that bioeffect concerns are frequently more suitably attributed to the bulk temperature rise associated with attenuation of the beam and the absorption of ultrasonic energy by the tissue. Some of the implications for and of tissues heating were introduced in Chapter 1, section 1.2.3(a) in discussion of the propagation of finite-amplitude waveforms. This illustrates the variety of mechanisms that may be involved in the conversion of acoustic energy to heat: the investing of energy to higher frequencies through nonlinear propagation, such frequencies being more strongly absorbed; hot-spot cavitation; self-focusing etc. The specific mechanism will dictate the spatial and temporal characteristics of the heating, and the appropriate measurement techniques and bioeffects to be considered. Other noncavitational mechanisms[88] include mechanical effects due to nonlinear propagation, acoustic streaming and microstreaming, and radiation forces, as discussed earlier. Clearly, the relative efficacy of these mechanisms will depend on the acoustic parameters (e.g. frequency, pressure, duty cycle etc.) involved. The medium will also be important, not just because, as has been noted in this book, heating, streaming, cavitation inception etc. all depend on the liquid, but also because the sensitivity of the material insonated will be important. As Crum *et al.* [502] point out, though the energy associated with a single cavitation event is very large on a microscopic scale (tens of MeV) [79], it is localised and likely to affect only a few cells. Though normally this might be inconsequential, in the first eight weeks of embryonic development it may be significant [502, 521].

Judgement of which mechanism acts is vital to the safety consideration. For example, prompt transient cavitation of the type used in the models of Chapter 4, section 4.3.1 is an almost instantaneous response to acoustic pressure changes, though it exhibits thresholds. In contrast, bulk heating can have a somewhat slower response time, and tends to be a fairly predictable if often unavoidable aspect of insonation.[89] Therefore one might expect cavitation to dominate in regimes where high acoustic pressures are employed, but where long off-times make the I_{TA} low. However, if continuous-wave sound is employed below a cavitation threshold, thermal effects will dominate. As regards the medium, one might expect in the general case for cavitation to be more common in cell suspension than it would be in tissue, where thermal effects might be expected to dominate. Clearly, these illustrate general trends only: to judge the likely importance of cavitation in any given situation one would have to consider all aspects of the acoustic field, nuclei population, fluid properties and target sensitivity.

The closing sections of this book includes some examples of bioeffects, chosen not as a review but to illustrate potential mechanisms. For convenience, four broad and inexact insonation regimes are chosen, starting from a class where one would expect macroscopic thermal effects to dominate, and ending with a class where one might expect them to be negligible

[88] Re-introducing the main theme of this book, and in no way undermining the importance of the direct noncavitational mechanisms, it is interesting to note that earlier sections have illustrated how, at least in liquid, these processes can in general also affect cavitation: for example, bubbles may generate microstreaming; streaming may convect bubble nuclei to focal regions; bulk temperature changes may affect nucleation, cavitation and chemistry through alteration of gas solubility and vapour pressure etc. One case of particular interest is the effect of the nonlinearly-distorted wave on a bubble. The formation of a shock might well cause concern, particularly in the light of the erosive processes outlined above. However, to every bubble there is a response time, related to its resonance, and the nonlinear distortion of sine waves is consistent with pumping energy into higher frequencies (Chapter 1, section 1.2.3(a)). Once the energy is at frequencies much higher than the bubble resonance, inertia will prevent a prompt response, and the harmonic content of such a nonlinearly distorted wave may be unimportant with respect to the bubble pressure [520].

[89] Although it may be moderated, for example by the use of water baths *in vitro*, and through the heat dissipative effects of blood flow *in vivo*.

compared with the cavitation, with this general trend reflected in the two intermediate classes. The four classes are: (i) ultrasonic surgery, where high intensity focused sound is used deliberately to damage tissue; (ii) continuous-wave or millisecond pulses of a few MHz or so, and at I_{SATA} of up to a few W cm^{-2}, a subset of which therefore comprises the output of physiotherapy units; (iii) systems employing μs pulses of up to a few MPa (a subset of which includes diagnostic systems); and (iv) extracorporeal shock wave lithotripsy.

(b) Clinical Acoustic Regimes

(i) Ultrasonic Surgery. The damaging potential of focused ultrasound having very high I_{TA}, for example very high intensity continuous-wave insonation at high pressure amplitudes, can be beneficially exploited in ultrasonic surgery to treat specific regions of tissue [522]. Developed in the 1950s and 1960s for use in neurosurgical research by Fry, Dunn and others [523–526], exposures have been used in studies including ophthalmology work [527–531], the nervous system [40, 532, 533] and the liver [534, 535]. Meniere's disease, an episodic debilitating disorder affecting balance, has been treated through the use of high-intensity focused ultrasound [536–542]. Barnett, using 20 minute exposures of 3.35-MHz ultrasound at average intensities of 5–10 W cm^{-2} in a beam of diameter 1–2 mm, has determined the mechanism of therapeutic action to be tissue ablation primarily due to heating [541, 542].

High-intensity sound has also been used on tumours [543, 544]. In the 1950s Nightingale [545] considered that attempts to treat malignant tumours with ultrasound had failed owing to the production of metastases, and Herrick [546] reported no inhibition of tumour growth unless they were cauterised by heat generated by the ultrasound. Nevertheless, in 1978, Fry and Johnson [547] obtained a 29.4% cure rate from hamster tumours exposed to 1.11-MHz ultrasound at spatial peak intensities of 907 W cm^{-2}. There is currently much interest in this field. For example, ter Haar *et al.* [548] used high-intensity 1.7-MHz ultrasound, focused to peak intensities of 1400–3500 W cm^{-2} in 10-s exposures, to arrest histologically detectable growth in rat tumours that had been exposed over the whole volume. Where the lesions were not contiguous, regrowth occurred. The controlled lesions that may be produced are evidence of local high tissue temperatures [529, 549]. Macroscopic hyperthermia is undoubtedly a potent effect (see, for example, in addition to the reviews listed at the start of section 5.4.2, also [550–554]). Though care must be taken when applying such methods to clinically relevant situations [467], the hyperthermia mechanism is amenable to modelling so that the effects of perfusion, tissue and thermal transfer characteristics can be examined [532, 555]. However, as Billard *et al.* [555] point out, at the higher intensities, thermal effects resulting from nonlinear propagation[90] [556, 557] and high-energy cavitation [558, 559] may significantly increase the temperature rise, and both of these effects are difficult to predict. Hynynen [560] (using 246 kHz $\leqslant \nu \leqslant$ 1.68 MHz; $\tau_p = 1$ s; and I_{SPPA} up to about 2500 W cm^{-2}) noted that the strong emission of wideband noise, taken to signify the presence of cavitation *in vivo* in canine thigh muscle, appeared to be related to an increase in acoustic energy absorption in tissue at the transducer focus. During the 1-s sound pulse, there was a significant increase in temperature, a loss of the smooth temperature rise and a reduction in the acoustic power transmitted through the thigh. Strong ultrasonic echoes appeared during these pulses, all of which indicated bubble activity. Temperatures were measured using an embedded thermocouple, the twisted wires of which each had a diameter of 50 μm. The probe is therefore not small on the scale of possible cavitation hot-spots.

---

[90] See Chapter 1, section 1.2.3(a).

(ii) Continuous-wave and Millisecond Pulses. (I_{SATA} of the order of 1 W cm^{-2}). At lower intensities there is a tradition of beneficial response of soft tissue to ultrasound, and it is often used to accelerate the repair of injured tissue, to reduce pain and to modify the formation of scar tissue [561–563]. Scar tissue is the end product of wound healing in mammalian skin, and can cause immense physical and psychological problems [564]. It is cosmetically and elastically inferior to normal tissue, can impair movement, and may contract abnormally, which can have very serious effects [564, 565]. However, many of the ultrasonic treatments are grounded in personal opinion and experience, rather than controlled research [566]. Controlled experiments have indicated, amongst other things, that ultrasound may affect tendon healing [567], wound contraction [568], the formation of new blood vessels [569] and macrophage action [570]. Quantitating wound healing through changes in the relative number of polymorphonuclear leucocytes, macrophages, fibroblasts and endothelial cells, Young and Dyson [564] investigated the healing of full-thickness excised lesions in rat flanks, which were either insonated ($\nu = 0.75$ or 3 MHz, $I_{SATA} = 0.1$ W cm^{-2}, $\tau_p = 2$ ms, duty cycle = 1:4) for five minutes daily, or were sham-insonated. The cell counts provided an objective, quantitative measure of the rate of healing. Though the cell count in the insonated wounds five days after injury showed marked differences from the sham, there were no significant differences after seven days, suggesting that ultrasonic therapy may be useful in accelerating the inflammatory and early proliferative stages of repair. The researchers suggest the mechanisms were not purely thermal. Thermocouples (size ~ 0.5 mm, which is considerably larger than possible cavitation hot-spots) inserted into the dermis of otherwise uninjured tissue during treatment indicated a temperature rise of 1°C over the 5 minutes. Young and Dyson estimate that cavitation may possibly have occurred.

Sonoluminescence from human blood plasma in a roughly travelling-wave field was recorded by Eastwood and Watmough [17] at intensity thresholds of $I_{SA} = 0.8 \pm 0.1$ W cm^{-2} for $\nu = 0.75$ MHz. Parallel investigations in distilled water with pulsed ultrasound at $\nu = 0.75$ or 1.5 MHz and $\tau_p = 2$ ms, duty cycle = 1:4 or $\tau_p = 5$ ms, duty cycle = 1:1 indicated that pulsing raised the threshold. Results in standing-wave fields[91] include sonoluminescence using a physiotherapeutic unit at $\nu = 1$ MHz continuous-wave and pulsed ($\tau_p = 2$ ms, duty cycle = 1:4) in pig blood plasma and amniotic fluid for 1–1.5 W cm^{-2} [256]. Blood cell stasis or banding was detected in standing-wave fields by Dyson *et al.* [300], who also reported endothelial damage to the blood vessels of chick embryos at $I_{SA} \geqslant 0.5$ W cm^{-2}. In 1953, Lehmann and Herrick [571] reported blood vessel rupture at therapeutic intensities ($I_{SA} = 1$–2 W cm^{-2}, $\nu = 1$ MHz). Several investigators have subjected blood *in vivo* to therapeutic intensities of continuous-wave ultrasound and have been unable to detect significant amounts of damage to any of the cellular elements of blood [572–574]. Sarvazyan *et al.* [575] reported damage to chick embryos caused by ultrasound, though Garrison *et al.* [576] and Barnett [577] saw no such effect.

Reductions in mean foetal weight in mice as a result of *in utero* exposure to ultrasound have been reported [578–583]. However, the effects are by no means established with any certainty, as maternal damage can occur. For example, a re-evaluation of the data of Stolzenberg *et al.* [582, 583] confirms that maternal damage gives smaller foetuses, whereas undamaged mothers produced normal weight foetuses [584]. Sikov *et al.* [585] detected foetal abnormalities and a prenatal death threshold with continuous-wave ultrasound at $I_{SATA} = 3$ W cm^{-2} for 5 minute exposures. However, when applied directly to exteriorised rat uteri, no effect was noted by these workers, suggesting that the deleterious effects resulted from uterine hyperthermia [586]. Noting therefore that all previous positive results were described when the maternal abdomen

[91] See section 5.2.4.

was exposed (usually ventrally across the spinal column), and that negative results (Sikov) from exteriorised uteri were obtained when ultrasound did not pass the maternal verterbrae and were probably not due to a difference in hyperthermia in these studies, Barnett and Williams [587] duplicated the maternal compromise reported by Stolzenberg and O'Brien and clearly showed a relationship between foetal weight reduction and maternal distended bladder syndrome. This was found to be the case even when the uterus was not exposed directly and was not heated. Exposure to 2.15-MHz continuous-wave ultrasound at $0.5 \leqslant I_{SA} \leqslant 5.5$ W cm^{-2} for 3 minutes occurred at day 8 of gestation. Direct insonation of the uterus caused resorptions, abortions and gross developmental abnormalities, and occasionally foetal weight reduction (in this case, apparently not associated with distended bladder syndrome).

From observations that survival curves for sham-insonated hyperthermic (43° C) V-79 cells differed significantly from those insonated *in vitro* at 3 MHz and 1 atm ambient pressure [588] and 1 MHz and 3 atm ambient pressure [15], ter Haar *et al.* [588] and Morton *et al.* [15] postulated a nonthermal, noncavitation mechanism. Though there was an increased rate of decrease in surviving fraction with heating time during insonation, they demonstrated no significant temperature rise (< 0.1° C) during insonation, and took failure to detect an acoustic subharmonic as a indication of an absence of cavitation. A similarity between cell survival fractions obtained for heat and shear stress, and for heat and insonation, has been demonstrated by Dunn [589]. Inoue *et al.* [590] insonated suspensions of V-79 cells in 10 or 20% serum at 1- or 3.4-MHz continuous-wave (0.7–1 W cm^{-2}) for 0–90 minutes, at 37° C or 43° C, and tested for cell lysis, plating efficiency and surviving fraction. A 0.5° C temperature rise was detected at 37° C with 1-MHz, 10% serum. As with the earlier results, they found an increase in the effectiveness of ultrasound on surviving fraction at the higher temperature, and with data from the full experiment deduced that it was due to a thermal effect of ultrasound exposures. They argue that cells exposed to ultrasound whilst under hyperthermic conditions experience an additional slight increase in temperature.

If one is, in addition to hyperthermia, discussing the potential for cavitational effects, one must consider how these studies relate to the bubbles that can occur in tissue. Of particular interest in the studies with relatively long (ms) exposures at relatively low intensities are resonance and stable cavitation effects. Miller [591–594] introduced the idea of modelling natural gas pockets in the water plant *Elodea* by the acoustic response of cylindrical gas pockets trapped and stabilised in µm pores in Nuclepore™ filters, the resonances of which were examined by Miller [592]. These were used to study the effect of stable cavitation of bubbles in suitable MHz fields. The study showed that, for example, microstreaming from stable cavitation could cause cell death in *Elodea* [594]. Vivino *et al.* [515] demonstrated similar results using Nuclepore™ to induce cell lysis in spleen cells *in vitro* at a threshold intensity of $I_{SP} \geqslant 75$ mW cm^{-2}, $\nu = 1.6$ MHz continuous-wave. These experiments, using Nuclepore™, are characterised by the very low intensities at which bioeffects can be detected: Williams and Miller [595] used photometry to detect ATP release from *in vitro* human erythrocytes, a technique which has probably produced effects at the lowest continuous-wave intensity recorded for a detectable bioeffect, 4 mW cm^{-2} [453]. The mechanism was probably rupture or a change in permeability of the cell membrane brought about through microstreaming stresses at the cell wall. Other effects reported with this technique include platelet aggregation [516] and haemolysis [596, 597]. In an examination of mechanisms in the latter process, Miller *et al.* [597] suggest that the increased haemolysis observed at elevated ambient temperatures is a result of the reduced viscosity, and consequently modified liquid flow near the cell, rather than increased cell fragility. Direct variation of viscosity tended to support this. It should be remembered that these effects relate to exposures at very low intensities, where hyperthermia and transient cavitation are negligible.

At higher intensities, more usually considered as 'therapeutic levels', both pulsed and continuous-wave ultrasound may affect placental processes in small mammals [598, 599]. Atkinson *et al.* [600] found no effect on placental transfer in rats whose body temperature was maintained at 37° C, but a significant effect if the body temperature was allowed to fall to 32° C during a 3-minute exposure to 1.1-MHz continuous-wave ultrasound at I_{SATA} = 1 W cm^{-2}. Mortimer *et al.* [601] have reported changes in the resting cardiac muscle tension in rats (2.3 MHz, 1.1, 2.2 and 3.3 W cm^{-2}, spatial and temporal average). Revell and Roberts [602] observed that 1.5-MHz continuous-wave ultrasound at 250 mW cm^{-2} could increase the discharge rate of the miniature end-plate potential of the frog neuromuscular junction. Stella *et al.* [603] induced sister–chromatid exchanges (SCE) in human lymphocytes exposed *in vitro* in culture, and *in vivo* in patients, to therapeutic ultrasound at ν = 0.86 MHz, 1 W cm^{-2}, 40–60 s duration. Each of ten patients, with a wide range of symptoms (e.g. elbow epicondylitis, chin healing, knee arthrosis), had a statistically significant increase in frequency of SCEs at mid and end of therapy, but none of the seven patients sampled 3 months after the end of therapy exhibited a significant increase. Ciaravino *et al.* [604] failed to confirm independently the *in vitro* results. Miller *et al.* [605] failed to observe an increase in SCEs in four patients recommended for therapeutic ultrasound (three with bursitis, one with bone spur). Neither the patients, who were exposed for 7 minutes per application at I_{SA} = 1 or 1.2 W cm^{-2}, nor the four sham-exposed humans, showed an increase in the frequency of SCEs.

Other effects include placebo [606] and phonophoresis, where at frequencies of a few MHz and I_{SATA} of the order of 1 W cm^{-2} intensities, ultrasound can increase the rate of penetration of a wide range of pharmacologically active agents through the skin [607–609]. Watmough *et al.* [213], have recently suggested that this may be related to a negative charge which may be associated with an acoustically excited bubble.

(iii) Microsecond Pulses. Writing in 1987 on the safety implications of diagnostic ultrasound, one authoritative experimenter [610] stated that "The temporal average intensities used in most of the laboratory studies have exceeded typical diagnostic levels by an order of magnitude or more. Many claims of effects have been made. However, the magnitudes of the claimed effects have been small, i.e., they have been comparable in magnitude to natural biological variability. Furthermore, there have been many studies at the same exposure levels which failed to find effects." For example, contention can be found in the reported bioeffects on the immunological system resulting from exposure to microsecond pulses of ultrasound. On insonating mouse spleens *in vivo* with 2-MHz pulsed ultrasound for 1.6, 3.3 and 5 minutes with I_{SATA} = 8.9 mW cm^{-2}, I_{SPTP} = 28 W cm^{-2}, ν_{rep} = 691 Hz, Anderson and Barrett [611] observed a slight decrease in the IgM compartment of immunoglobulin response, and a decrease in the number of direct antibody plaque-forming cells in the spleen, compared with the controls. Child *et al.* [612], who used τ_p = 1 μs and ν_{rep} = 690 Hz in an attempt to reproduce the findings of Anderson and Barrett [611], reported no immunosuppressive effect of ultrasound. Berthold *et al.* [613] similarly reported no effect. There are a great many other studies involving the response to microsecond pulses of ultrasound, some of which will be examined at the end of this book, when mechanisms are discussed.

Pinamonti *et al.* [614] and Rosenfeld *et al.* [615] demonstrated antigen changes in human erythrocytes exposed *in vitro* to diagnostic ultrasound. Romanowski and Rosenfeld [616] and Miller *et al.* [617] could not confirm Pinamonti *et al.*'s results under similar experimental conditions. An example of the experimental problems that can occur is found in the report by

Bause *et al.* [618], whose finding that erythrocytes of women who were exposed to Doppler (12 mW cm^{-2}, spatial average) for seven or more hours during labour were less osmotically fragile, is probably the result of the administration of fat-soluble analgesics which would make the cells more resistant to osmotic lysis, rather than an ultrasonic effect. Indeed, *in vitro* results showed increased cell fragility following exposure: however, the blood samples contained no antibiotics, so that the result could have been caused by contamination.

Epidemiological studies of the human response to diagnostic ultrasound are thoroughly discussed in the reports by the WFUMB [465] and by the AIUM [464], published in 1986 and 1988 respectively. Individual studies of obstetric ultrasound by Bernstine [619] in 1969, and by Bakketeig *et al.* [620] and Cartwright *et al.* [621] in 1984, suggested no difference between those mothers and children who were insonated, and those who were not. In 1984, Kinnier Wilson and Waterhouse [622] summarised the evidence by stating that ultrasound does not tend to cause cancer or leukemia in children up to the age of six years, though there is no evidence for this with older children. In the same year, Stark *et al.* [623] investigated short- and long-term risks after exposure to diagnostic ultrasound *in utero*. They used ultrasound with $I_{SATA} = 2.1–7.1$ W cm^{-2}, and $I_{SPTP} = 200–1000$ W cm^{-2}, and reported an increase in the incidence of dyslexia in young children, although they investigated so many possible manifestations, that statistically it would have been surprising if they had not been able to report at least one adverse effect. Indeed, the authors themselves warn against over-interpretation of their data. It should be remembered that epidemiological studies on the routine use of ultrasound are complicated by large numbers of interacting variables that can affect the study. Care must be taken in interpretation, as there is often an underlying bias, in that the medical need for an examination would suggest an effect is already present prior to examination: taking the example of Miller [462], if diagnostic ultrasonography is used to study problem pregnancies, a survey could erroneously indicate an association between the exposure and pregnancy problems.

Carstensen and Child [624] used a simple analysis to show that some diagnostic devices may generate significant tissue heating, particularly if there is bone within the path of propagation: rats were used to model contact exposure of the human skull experimentally, and mice were used to model foetal exposures. The speed (and possibly attenuation) of ultrasound propagating in bone depends on the mineral content [625]. As diagnostic outputs increase, the heating of foetal bone is likely to become an issue of particular concern [462]. Miller [462] re-examines the safety implications of diagnostic ultrasound in the light of current research and instrumentation. In 1987, Wells [626] assessed the prudent use of diagnostic ultrasound, providing a framework against which the advantages and disadvantages of a given treatment can be judged. In the same year, Carstensen [610] reviewed the safety of diagnostic ultrasound, emphasising the cavitation aspects. Writing in 1987, Carstensen states that "temporal average intensities in some pulse Doppler devices in use today are close to levels which produce measurable heating of tissues, and the temporal peak pressures in some modern instruments approach the levels of acoustic saturation, i.e., the condition in which nonlinear losses of the medium set upper limits on local sound source levels regardless of the output from the source."

Though it has been shown that transient cavitation may occur in a liquid medium containing adequate nuclei in response to microsecond pulses of ultrasound of the type used in diagnosis,[92] Carstensen writes "it is not clear at this time whether the appropriate conditions exist within the body." In the light of this, it is interesting to conclude this section with a discussion of the conditions and potential for nucleation. Consider the experiments described in section 5.1.2, where contrast agents were used to nucleate transient cavitation in response to such exposures.

[92] See section 5.1.2.

It is not uncommon to employ contrast agents to enhance a diagnostic ultrasound image at source. These include free and encapsulated gas bubbles, colloidal suspensions, emulsions (including dense spherical particles of iodipimide ethyl ester with diameters between 0.05 and 3.0 μm [627]) and aqueous solutions. They operate through backscatter, attenuation and sound speed contrast. These were reviewed in 1989 by Ophir and Parker [628], who concluded that "to date there are no completely satisfactory materials for clinical imaging." As discussed earlier, the contrast agent Albunex™ has been introduced to nucleate transient cavitation, an event which *in vivo* would have safety implications. The first reported use of *free* gas bubbles to enhance contrast was in 1968 [629]. Ophir and Parker [628] comment that whilst being very efficient acoustic scatterers, they are effectively removed by the lungs. Though it would therefore be impractical to use them to enhance contrast in soft tissue through venous injection, intra-arterial injection is a viable possibility for localised contrast enhancement [628]. However, the potential for nucleation of cavitation through added free-floating bubbles should be considered.

As regards nuclei within tissue, in the previous section the Nuclepore™ model of Miller and co-workers was described. There it was used particularly to model the bioeffects observed at low-intensity continuous-wave exposures as a result of the stable oscillations of bubbles. In an interesting study, Child *et al.* [630] suggested that larval gas pockets might be used to model the response of bubbles within mammalian cells, particularly in respect to μs diagnostic-type pulses. Insect larvae supply their tissues with oxygen and eliminate unwanted gaseous products of cellular respiration by taking up air through tubes (tracheae), which branch throughout the body. Whereas in the plant tissue studies it had been found that the stiffness of the walls contributed significantly to the dynamics, preliminary low-level continuous-wave studies had shown that the walls of the tracheae do not contribute significantly to the dynamics of larval gas pockets, and model the case of human tissues more closely. Child *et al.* [630] demonstrated the susceptibility of fruit fly larvae (*Drosophilia melanogaster*) to diagnostic-type insonation ($\tau_p = 1$ μs, low I_{TA}), with a sharp threshold (around 1 MPa in a 2.5-minute exposure). Evidence, including the coincidence of the killing and transient cavitation thresholds [80, 630], suggests a cavitation-type mechanism on the larvae [631–634], probably acting primarily through these micrometre-sized gas bodies within the organisms [634]. The effect was related to temporal peak, rather than the temporal average, intensity. These results are further interpreted in the light of subsequent lithotripsy results [635], outlined below. It should be pointed out that though these larval nuclei may model the mechanics of gas pockets in mammalian tissue, they do not reflect the distribution, which in test animals is generally unknown.

(iv) Extracorporeal Shock Wave Lithotripsy. In 1990, Hartman *et al.* [116] stated that "Overall, up to the present time, there is no clear evidence that ultrasound influences the development of the vertebrate embryo/fetus unless there is an associated, biologically significant temperature rise [636, 637]." However, they noted that in lithotripsy the patient is subjected to pressures an order or so greater than previously, and demonstrate that though the time-averaged intensities are so small as to have insignificant bulk heating effect, the exposures can drastically effect embryo development with as few as three double shocks of amplitude 10 MPa.

Discussions of the production, propagation and focusing of lithotripsy shocks can be found in the literature [100, 101, 638]. In a typical clinical lithotripsy procedure, more than 1000 shocks with pressure amplitudes in excess of 20 MPa are deployed [116]. Temporal peak pressures have been reported as high as 100–300 MPa [639], in contrast to diagnostic ultrasonics, where the largest peaks are no more than about 10 MPa. The pulse length in lithotripsy is of microsecond order (Figure 5.10); in clinical practice the pulse repetition frequency rarely exceeds 2 Hz [640], so that I_{TA} is small. Clearly, the interpretation of observed bioeffects must explore mechanisms other than hyperthermia and stable cavitation. This is, on the other hand, clearly a situation where short-lived violent cavitation events have a potential role. Certainly lithotripsy can cause violent short-lived cavitation in water. Some evidence of this was presented in section 5.1.2. Though bulk transient temperature rises during lithotripsy may be calculated [641], the role of cavitation hot-spot hyperthermia may be significant. The detection of sonoluminescence from single pulses also suggests that they are capable of generating free radicals [99]. The presence of these agents has been correlated, using chemiluminescence and scavengers, with cell viability in suspension [642]. Coleman, Crum and co-workers [106, 643] have shown that single lithotriptic pulses are capable of bringing about bubble involution and jetting. Lithotripsy fields in water can damage metal surface through cavitation [106]. Physical and erosive damage to solid material is not unexpected: lithotriptic pulses are designed to damage kidney stones and gall stones, and the role that cavitation may have in stone fragmentation was discussed in section 5.4.1. However, research has demonstrated that lithotripsy might also produce unwanted effects in soft tissue with as little as a single shock. Possible side effects of lithotripsy include, for example, canine lung [109] and kidney [112, 113, 644] haemorrhage. Workers have researched the effect of lithotriptic exposure on the development of chick embyros [116], effects in canine [113, 640] and piglet livers [114], and canine gall bladders [113, 640]. Lithotriptic effects on the electrophysiological parameters across abdominal frog skin [645], where exposure to a single shock caused a reversible increase in the total ionic conductance and a decrease in the short circuit current, were similar to those found in a 1-MHz therapeutic ultrasound field [304], as discussed in section 5.2.4. Crum *et al.* [502] present results which implicate acoustic cavitation as the source of ion transport changes in the frog skin experiments. In tissue, Kuwahara *et al.* [646] and Delius *et al.* [114] showed that lithotripsy can generate bubbles in tissues that are large enough to be detected by ultrasonic pulse-echo equipment. Hartman *et al.* [116] subjected chick embryos at 72-h incubation to three double shock waves generated by the 9 French probe of a Wolf electrohydraulic lithotripter (Model 2137.50), the fields of which have been characterised by Campbell *et al.* [647]. A significant increase in early deaths, delayed deaths and malformations was seen at pressures of 10 MPa, with possible effects at lower pressures. Hartman *et al.* also found that comparatively mild lithotriptic exposures were required to haemorrhage mouse lung (less than 2 MPa for only 10 pulses) [117]. Severe damage occurred at 5–6 MPa. In contrast, foetal lung did not show damage at 20 MPa. Following the conventional assumption that the sensitivity of the lung to damage results from the presence of air bodies, which represent nucleation sites, the reduced sensitivity of foetal lung suggests a cavitation-related mechanism for the damage. Since the pressures generated by this system are mainly positive, these workers propose that measurement of negative pressures may not represent the best indicator of the behaviour of gas bodies in tissues.

The same conclusion was reached by Carstensen *et al.* [635], who, following the suggestion of Child *et al.* [630] that larval gas pockets might be used to model the response of bubbles within mammalian cells, killed roughly one-half of a population of fruit fly larvae (*Drosophilia melanogaster*) subjected to 3–10 double shocks at 2–3 MPa from a Wolf electrohydraulic lithotripter (Model 2137.50). Addition of negative pressure to the exposure, which would increase cavitation from free-floating bubble nuclei, had little effect. This indicates that the mechanism of damage was not tensile-excited expansion of the small gas bodies stabilised within the respiratory system of the larvae. Carstensen *et al.* [635] suggest that this is because the containment inhibits the expansion of the bodies, that this system might provide a good model for the behaviour of gas bubbles within mammalian cells and that the bioeffects from lithotripsy may be related to the positive pressures. Earlier workers had suggested that the negative pressure, relating as it does to the energy of the cavitation response of a free-floating bubble nucleus in an infinite liquid, is the better predictor of bioeffects associated with the activation of gas bodies within tissues [116, 117, 635, 648–650]. Hypothesised killing mechanisms for the lithotripter through simple crushing (tested by application of a pseudostatic pressure) and indirectly through drowning (by external water entering the trachea, tested by the use of dye) have proven unlikely [635]. Bulk heating is, of course, negligible. It is interesting to note that experimental observations of lithotriptic tumour therapy suggest the effectiveness of the component of the wave reflected from the free surface of the liquid bath [651], which will cause inversion of the pressure pulse. It is important to understand the roles of the positive and negative pressures in lithotripsy, and the mechanisms through which they act, as the devices may be engineered to enhance either pressure cycle [652], and the choice of which one requires understanding.

(c) Mechanisms for Bioeffect: Final Comments on the Role of Cavitation

If a suspension of biological cells is exposed to intense ultrasound, the violent mechanical effect of cavitation bubbles can decimate a suspension of biological cells. These mechanical effects may arise through the erosion process mentioned earlier (emission of pressure pulses on rebound, bubble involution and jetting), or through the action of bubbles translating at high speed, as discussed in section 5.3.2(c), or through the growth of intra-cellular bubbles, or through acoustic microstreaming. Bubble activity may cause bioeffect through contribution to temperature rises which, as discussed in section 5.4.2(b)(i), can be difficult to predict. As discussed in section 5.4.2(a), any thermal effects (on both the medium and the measuring instrument) that come about through a hot-spot mechanism may be complicated. Bubbles may also bring about bioeffect through degassing the medium, for example during rectified diffusion. In many bioeffects studies the cell lysis is dominant, with other possible effects in the surviving fraction of cells masked by the burden of dead cell [653].

Pinamonti *et al.* [654] demonstrated DNA damage in human leukocytes exposed *in vitro* to pulsed ultrasound. Miller *et al.* [655] detected no significant single strand breaks in DNA of suspended human leukocytes exposed to relatively high-frequency or pulsed-mode ultrasound. Single strand breaks, cell disruption and loss of viability were detected in a rotating tube system (of the type discussed in section 5.3.2) where cavitation occurred. Such strand breaks were localised in cells that remained intact after exposure, but it was uncertain whether any of the breaks resided in the viable fraction of cells (strand breaks in dead cells have no genetic consequence). In order to elucidate this, Miller *et al.* [656] compared sham-exposed suspensions of Chinese hamster ovary cells to sample exposed for 10 minutes in rotating tube systems (72 r.p.m.) to 1.61-MHz pulsed ultrasound (τ_p = 10.5 μs; duty cycles of 1:2 or 1:4; I_{SPTA} = 2.0, 2.8, 4.0, 5.6 or 8 W cm^{-2}), or continuous-wave (where viability fell dramatically at $I_{SP} \geqslant 5.6$ W cm^{-2}). Subharmonic emissions were used to monitor cavitation. Cells could be tested for viability through a 30-minute post-exposure incubation period at 37° C in which viable cells could repair breaks. They concluded that the observed ultrasonically induced single strand breaks resided primarily in the nonviable fraction of cells, since no repairs were undertaken.

The question of whether sonochemical action induced by cavitation could generate a bioeffect was addressed by Miller *et al.* [657]. Free radicals are short-lived, and observation of their effect would require the cell to be present during insonation: in such conditions, the effect would probably be swamped by the mechanical lysis effect. However, longer-lived sonochemicals may also be produced, and residual species may remain after insonation has ceased and generate a bioeffect in cells not directly exposed to ultrasound. This was determined by Clarke and Hill [311], who found that prolonged insonation of a cell culture medium reduced its ability to support cells. Miller *et al.* [653] demonstrated in 1991 that single strand breaks could be induced in Chinese hamster ovary cells after continuous-wave insonation (ν = 1.61 MHz, I_{SA} = 8 W cm^{-2}) had ceased, presumably through the action of hydrogen peroxide. The breaks resided in viable cells. The effect was comparable with that of 1 Gy of ^{60}Co γ-rays. Carmichael *et al.* [85] and Christman *et al.* [86] used electron spin resonance and spin trapping to detect free radicals produced in aqueous solution by microsecond pulses.

In an investigation into the relevant mechanisms for bioeffect, particularly temperature elevation, Barnett *et al.* [658] subjected rat embryos, in a temperature-controlled bath containing 50 litres of water at 38.5° C, to pulsed ultrasound (ν = 3.14 MHz, τ_p = 3.2 s, ν_{rep} = 2 kHz, I_{SPTA} = 1.2 W cm^{-2}) for 5, 15 or 30 minutes, and observed no major morphological changes. Changes in protein synthesis and delayed development were indicative of embryonic stress, effects which were enhanced when the temperature during insonation was raised by 1.5° C. This temperature rise could, however, not be responsible for the effect, firstly since elevations of $\leqslant$ 1.5° C are universally agreed to be safe for embryonic exposures [467], and secondly because Barnett *et al.* demonstrated that foetal growth impairment was related to maternal compromise, the severity of which could not be caused by a rise of 1.5° C. Barnett [private communication] suggests that these results might be interpreted as (i) a synergism between ultrasonically induced shear stresses and mild elevations in bulk temperature and (ii) microscopic heating due to bubble activity coating the embryos. The complexities of hot-spot hyperthermia have already been introduced. Given the number of possible mechanisms in which cavitation may be involved, the prospect of synergistic interactions is not unexpected. Loverock and ter Haar [659] observed an ultrasonic synergy, where acoustically induced bulk temperature elevations could not have been responsible, when suspension cultures of V79 Chinese hamster lung fibroblasts were exposed to 43° C hyperthermia with or without simultaneous continuous-wave ultrasound (ν = 2.6 MHz, I_{SA} = 2.3 W cm^{-2} travelling wave) either before or after ^{60}Co γ-exposure (428 c Gy). Though the ultrasound alone only contributed a bulk temperature rise of less than 0.1° C, when the two types of exposure immediately followed one another (regardless of which went first), the ultrasound increased the amount of cell killing produced by the combination of heat and ionising radiation. The enhancement did not occur if an interval of 72 minutes occurred between the heat and the γ-irradiation. Ultrasonic exposure had no synergistic effect on γ-irradiation if the exposure occurred at 37° C.

Watmough *et al.* [660] insonated HeLa cells with therapeutic ultrasound (ν = 0.75 MHz, $I_{SP} \gtrsim$ 0.7 W cm^{-2}), and suggested that at intensities below the threshold for collapse cavitation, ultrasonically induced degassing of the culture medium and/or cells might contribute to the observed damage. They also observed evidence of microstreaming. Rooney [661, 662] reported microstreaming-induced haemolysis caused by near-resonant bubbles at 20 kHz.

It is clear from the discussion of the acoustic regimes, and of these particular examples, that a variety of mechanisms can potentially operate to cause a bioeffect. Under any given circumstances, one mechanism may dominate to such an extent as to mask the effect of all others. Two that clearly can dominate are:

(i) Bulk heating (e.g. when I_{TA} is high). This is dramatically illustrated in ultrasonic surgery, though it has probably been the effective mechanism in many ultrasonic bioeffect studies that employed more modest exposures. The biological consequences of macroscopic hyperthermia are reviewed by Miller and Ziskin [663]. The role of microscopic cavitation hot-spot heating is not certain.

(ii) Bubble-induced mechanical damage (when I_{TP} is high). This may occur through erosive jetting, high-speed translations etc. Other cavitation mechanisms may be more subtle (sonochemical changes) or may be associated with high-amplitude stable cavitation (e.g. microstreaming).

Taking just the hyperthermic and cavitation-related mechanisms, it is interesting to speculate in a general way as to which may be significant in the four groups. With lithotripsy, which can generate short, high-pressure pulses but have a low I_{TA}, one might expect cavitation to be far more significant than hyperthermia. Though sonoluminescence has been detected, indicating the potential for a sonochemical effect, the cavitation damage is likely to be mechanical and erosive (the brevity of individual pulses, which have been shown to cause damage, limits the time in which to induce high-speed translations or set up significant streaming or microstreaming). Cavitational effects probably play a supportive role to the direct action of the shock when it comes to the stone (see Figure 5.57), but may have a very significant influence where gas pockets are concerned [664]. Following high-speed photographic observations (10 000 f.p.s.) of the response of a stone in a degassed water bath to shocks from a commercial electromagnetic shock source, where the focal peak pressure was 640 bar, Sass *et al.* [665] suggested that the interaction of the shock with the stone firstly produces fissures, which liquid penetrates. Disintegration occurs later, and is the result of imploding cavitation bubbles within the small split lines. If this mechanism is important in clinical practice, it has implications for the timing between shocks, which should be sufficient to allow fluid penetration of the fissures [665]. The relative importance of the compressional and tensile parts of the pulse, as regards the excitation of gas pockets in tissue, is a matter of debate (see section 5.4.2). As discussed by Holmer *et al.* [666] (who in addition to high-energy cavitation discuss the possible roles of streaming and microstreaming, the latter being induced both by cavitation and the presence of the stone), an understanding of the clinically relevant mechanisms is required so that account can be taken of the local environment of the stone (e.g. fluid, tissue) when lithotripsy is applied.

With microsecond pulses, both hyperthermia and cavitation can in principle occur, but the action will be very dependent on the precise conditions of the medium and sound field. Cavitation effects are likely to be those associated with transient collapse, such as erosion, jetting, rebound pressures, hot-spots and sonochemistry. This is because for energetic stable cavitation, one is relying on the bubble to exhibit the characteristics of a driven nonlinear oscillator, and in general one would expect the pulses to be too short to do this.[93] However, there are some pulsed Doppler devices where increasing pressure amplitudes and pulse lengths, and high repetition frequencies, may enable this behaviour to occur. Those bubbles with radius such that they do not undergo transient collapse may be driven into high-amplitude stable oscillations, and the associated bioeffects mechanisms may be possible: hot-spots, sonochemistry, high-speed translations, microstreaming etc.

This situation will be more likely with the use of high-amplitude continuous-wave or millisecond pulses. A high-amplitude sound field could generate both energetic stable and transient cavitation (such as observed in Figure 4.19), the behaviour depending on the initial bubble size. Hyperthermia of course could be significant, probably becoming dominant at the

[93] Cartensen *et al.* observed the lysis of *Elodea* cells subjected to continuous-wave and μs pulsed ultrasound, and found higher thresholds with the pulsed exposure, illustrating the qualitative difference in the response of bubbles to continuous-wave, and to μs pulses, of ultrasound.

power levels associated with ultrasonic surgery. There is evidence that at low-amplitude continuous-wave insonations, microstreaming and radiation forces are dominant.

The mechanical index was discussed in Chapter 4, section 4.3.1(d). This describes the likelihood of prompt transient cavitation. As such, it eliminates such effects as rectified diffusion from consideration and, assuming an optimum nuclei population, presents an assessment of an immediate cavitational response. Quite naturally, therefore, it depends on the pulse-peak pressures. Though there are other measures [467], in a summary of the potential heating effect of diagnostic ultrasound Miller [462] describes the 'thermal index' [668, 669] which, because thermal bulk heating is a continuous and immutable companion to insonation, will represent a temporal average parameter. The use of these two indices to describe clinical equipment represents an improvement over the old equivalent parameters, I_{SPPA} and I_{SPTA}, though primarily they are indicative of the implications for use with respect to prompt transient cavitation and bulk hyperthermia mechanisms only. Key to the design and implementation of these indices are the mechanisms through which cavitation and heating come about. Key to the design and interpretation of experiments of the type described in this section, and throughout this chapter, is an understanding of the mechanisms. The potential for application of any technique should not cause the question of effects and mechanisms to be neglected.

References

[1] Hill CR (ed.). Physical Principles of Medical Ultrasonics. Ellis Horwood, Chichester (for Wiley, New York), 1986

[2] Suslick KS (ed.). Ultrasound: its Chemical, Physical, and Biological Effects. VCH Publishers, New York, 1988

[3] Young FR. Cavitation. McGraw-Hill, London, 1989

[4] Hüttmann G, Lauterborn W, Schmitz E and Tanger H. Holography with a frequency doubled Nd:YAG laser. SPIE (Holography techniques and applications) 1988; 1026: 14–21

[5] Nyga R, Schmitz E and Lauterborn W. In-line holography with a frequency doubled Nd:YAG laser for particle size analysis. Appl Opt 1990; 29: 3365–3368

[6] Hentschel W and Lauterborn W. High speed holographic movie camera. Opt Eng 1985; 24: 687–691

[7] Lauterborn W and Koch A. Holographic observation of period-doubled and chaotic bubble oscillations in acoustic cavitation. Phys Rev 1987; A35: 1974–1976

[8] Hentschel W, Merboldt K-D, Ebeling K-J and Lauterborn W. High speed holocinematography with the multiply cavity-dumped argon-ion laser. J Photogr Sci 1982; 30: 75–78

[9] Kolovayev PA. Investigation of the concentration and statistical size distribution of wind produced bubbles in the near-surface ocean layer. Oceanology 1976; 15: 659–661

[10] Young FR. Cavitation. McGraw-Hill, London, 1989, pp 373–388

[11] Fanelli M, Prosperetti, A and Reali M. Shape oscillations of gas–vapour bubbles in liquids. Part 1: Mathematical formulation. Acustica 1984; 55: 213–223

[12] Donskoi DM, Zamolin SV, Kustov LM and Sutin AM. Nonlinear backscattering of acoustic waves in a bubble layer. Acoust Lett 1984; 7: 131

[13] Butkovsky OYa, Zabolotskaya EA, Kravtsov YuA, Petnikov VG and Ryabikin VV. Possibilities of active nonlinear spectroscopy of inhomogeneous condensed media. Acta Physica Slovaca 1986; 36: 58

[14] Basu S. Accelerated cavitation screening of organic coatings using acoustic emission technique. Trans ASME, J Vib Stress Reliability Design 1984; 106: 560–564

[15] Morton KI, ter Haar GR, Stratford IJ and Hill CR. Subharmonic emission as an indicator of ultrasonically-induced biological damage. Ultrasound Med Biol 1983; 9: 629–633

[16] Edmonds PD and Ross P. Acoustic emission as a measure of exposure of suspended cells *in vitro*. Ultrasound Med Biol 1986; 12: 297–305

[17] Eastwood LM and Watmough DJ. Sonoluminescence, in water and in human blood plasma, generated using ultrasonic therapy equipment. Ultrasound Med Biol 1976; 2: 319–323

[18] Detsch RM and Sharma RN. The critical angle for gas bubble entrainment by plunging liquid jets. Chem Eng J 1990; 44: 157–166

[19] Lin TJ and Donnelly HG. Gas bubble entrainment by plunging laminar liquid jets. AIChE J 1966; 12: 563–571

[20] Watkins RD, Barrett LM and McKnight JA. Ultrasonic waveguide for use in the sodium coolant of fast reactors. Nucl Energy 1988; 27: 85–89

[21] Lions N *et al.* Special Instrumentation for Phenix. Fast Reactor Power Stations. Telford, London, 1984, pp 525–535

[22] Day CK and Smith RW. Undersodium viewing. IEEE Trans 1974; SU-21(3, July)

[23] McKnight JA *et al.* The use of ultrasonics for visualising components of the prototype fast reactor whilst immersed in sodium. Ultrasonics International 83. Butterworth Scientific, 1983, pp 135–140

[24] McKnight JA *et al.* Recent advances in the technology of undersodium inspection in LMFBRs. Liquid Metal Engineering and Technology, Vol. 1. BNES, London, 1984, pp 423–430

[25] Aguilar A and Por G. Monitoring temperature reactivity coefficient by noise methods in a NPP at full power. Ann Nucl Energy 1987; 14: 521–526

[26] Apfel RE. Methods in Experimental Physics, Vol. 19 (Edmonds PD, ed.). Academic Press, New York, 1981, pp 355–413

[27] Shankar PM, Chapelon JY and Newhouse VL. Fluid pressure measurement using bubbles insonified by two frequencies. Ultrasonics 1986; 24: 333–336

[28] Hulshof HJM and Schurink F. Continuous ultrasonic waves to detect steam bubbles in water under high pressure. Kema Scientific and Technical Reports 1985; 3: 61–69

[29] Dowson D, Unsworth A and Wright V. Ann Rheumatic Dis 1971; 30: 348–358

[30] Dowson D, Unsworth A and Wright V. Cracking of human joints – cavitation in metacarpio-phalangeal joint. Ind Lub Tri 1971; 26: 212

[31] Belcher EO. Quantification of bubbles formed in animals and man during decompression. IEEE Trans Biomed Eng 1980; 27: 330–338

[32] Kisman H. Spectral analysis of Doppler ultrasonic decompression data. Ultrasonics 1977; 15: 105–110

[33] Tickner EG. Precision microbubbles for right side intercardiac pressure and flow measurements. In: Contrast Echocardiography (Meltzer RS and Roeland J, eds). Nijhoff, London, 1982

[34] Leighton TG, Ramble DG and Phelps AD. Comparison of the abilities of multiple acoustic techniques for bubble detection. Proc Inst Acoustics 1995; 17(8): 149–160

[35] Gross DR, Miller DL and Williams AR. A search for ultrasonic cavitation within the cardio-vascular system. Ultrasound Med Biol 1985; 11: 85–97

[36] Medwin H. Counting bubbles acoustically: a review. Ultrasonics 1977; 15: 7–14

[37] Mackay RS and Rubisson JR. Decompression studies using ultrasonic imaging of bubbles. IEEE Trans Biomed Eng 1978; BME-25: 537–544

[38] Daniels S, Paton WDM and Smith EB. An ultrasonic imaging system for the study of decompression induced gas bubbles. Undersea Biomed Res 1979; 6: 197

[39] Fry FJ and Goss SA. Further studies of the transkull transmission of an intense focused ultrasonic beam: lesion production at 500 kHz. Ultrasound Med Biol 1980; 6: 33–38

[40] Frizzel LA, Lee CS, Aschenbach PD, Borrelli MJ, Morimoto RS and Dunn F. Involvement of ultrasonically induced cavitation in hind limb paralysis of the mouse neonate. J Acoust Soc Am 1983; 74: 1062–1065

[41] ter Haar GR and Daniels S. Evidence for ultrasonically induced cavitation *in vivo*. Phys Med Biol 1981; 26: 1145–1149

[42] Morton KI, ter Haar GR, Stratford IJ and Hill CR. The role of cavitation in the interaction of ultrasound with V79 Chinese hamster cells *in vitro*. Brit J Cancer 1982; 45(Suppl V): 147

[43] ter Haar GR, Daniels S and Morton K. Evidence for acoustic cavitation *in vivo*: thresholds for bubble formation with 0.75-MHz continuous-wave and pulsed beams. IEE Trans Ultrasonics Ferroelectrics Freq Control 1986; 33: 162–164

[44] Daniels D, Davies JM, Paton WDM and Smith EB. The detection of gas bubbles in guinea-pigs after decompression from air saturation dives using ultrasonic imaging. J Physiol 1980; 308: 369

[45] Beck TW, Daniels D, Paton WDM and Smith EB. The detection of bubbles in decompression sickness. Nature 1978; 276: 173
[46] Watmough DJ, Davies HM, Quan KM, Wytch R and Williams AR. Imaging microbubbles and tissues using a linear focussed scanner operating at 20 MHz: possible implications for the detection of cavitation thresholds. Ultrasonics 1991; 29: 312
[47] Fowlkes JB, Carson PL, Chiang EH and Rubin JM. Acoustic generation of bubbles in excised canine urinary bladders. J Acoust Soc Am 1991; 89: 2740–2744
[48] Bleeker HJ, Shung KK and Barnhart JL. Ultrasonic characterization of Albunex™, a new contrast agent. J Acoust Soc Am 1990; 87: 1792–1797
[49] Wolf J. Investigation of bubbly flow by ultrasonic tomography. Part Syst Charact 1988; 5: 170–173
[50] Nishi RY. Ultrasonic detection of bubbles with Doppler flow transducers. Ultrasonics 1972; 10: 173–179
[51] Fairbank WM and Scully MO. A new non-invasive technique for cardiac pressure measurement: resonant scattering of ultrasound from bubble. IEEE Trans Biomed Eng 1977; BME-24: 107–110
[52] Commander KW and Moritz E. Off-resonance contribution to acoustical bubble spectra. J Acoust Soc Am 1989; 85: 2665–2669
[53] Commander KW and McDonald RJ. Finite-element solution of the inverse problem in bubble swarm acoustics. J Acoust Soc Am 1991; 89: 592–597
[54] Tucker DG and Wesby VG. Ultrasonic monitoring of decompression. Lancet 1968; 1: 1253
[55] Miller DL. Ultrasonic detection of resonant cavitation bubbles in a flow tube by their second harmonic emissions. Ultrasonics 1981; 19; 217–224
[56] Miller DL, Williams AR and Gross DR. Characterisation of cavitation in a flow-through exposure chamber by means of a resonant bubble detector. Ultrasonics 1984; 22: 224–230
[57] Vacher M, Giminez G and Gouette R. Nonlinear behaviour of microbubbles: application to their ultrasonic detection. Acustica 1984; 54: 274–283
[58] Newhouse VL and Shankar PM. Bubble size measurement using the nonlinear mixing of two frequencies. J Acoust Soc Am 1984; 75: 1473–1477
[59] Chapelon JY, Shankar PM and Newhouse VL. Ultrasonic measurement of bubble cloud size profile. J Acoust Soc Am 1985; 78: 196–201
[60] Schmitt RM, Schmitt HJ and Siegert J. *In vitro* estimation of bubble diameter distribution with ultrasound. IEEE Eng Med Biol Soc 1987; 9th Ann Conf: 13-6
[61] Siegert J, Schmitt RM, Schmitt HJ and Fritzsch T. Application of a resonance method for measuring the size of bubbles in an echocontrast agent. IEEE Eng Med Biol Soc 1987; 9th Ann Conf: 13-6
[62] Quain RM, Waag RC and Miller MW. The use of frequency mixing to distinguish size distributions of gas-filled micropores. Ultrasound Med Biol 1991; 17: 71–79
[63] Miller DL. Experimental investigation of the response of gas-filled micropores to ultrasound. J Acoust Soc Am 1982; 71: 471–476
[64] Miller DL. Theoretical investigation of the response of gas-filled micropores and cavitation nuclei to ultrasound. J Acoust Soc Am 1983; 73: 1537–1544
[65] Miller DL. On the oscillation mode of gas-filled micropores. J Acoust Soc Am 1985; 77(3): 946–953
[66] Chapelon JY, Newhouse VL, Cathignol D and Shankar PM. Bubble detection and sizing with a double frequency Doppler system. Ultrasonics 1988; 26: 148–154
[67] Cathignol D, Chapelon JY, Newhouse VL and Shankar PM. Bubble sizing with high spatial resolution. IEEE Trans Ultrasonics Ferroelectrics Freq Control 1990; 37: 31
[68] Phelps AD and Leighton TG. High-resolution bubble sizing through detection of the subharmonic response with a two-frequency excitation technique. J Acoust Soc Am 1996; 99: 1985–1992
[69] Leighton TG, Lingard RJ, Walton AJ and Field JE. Bubble sizing by the nonlinear scattering of two acoustic frequencies. In: Natural Physical Sources of Underwater Sound (Kerman BR, ed.). Kluwer, Dordrecht, The Netherlands, 1992
[70] Leighton TG, Lingard RJ, Walton AJ and Field JE. Acoustic bubble sizing by the combination of subharmonic emissions with an imaging frequency. Ultrasonics 1991; 29: 319–323

[71] Korman MF and Beyer RT. Nonlinear scattering of crossed ultrasonic beams in the presence of turbulence in water: II Theory. J Acoust Soc Am 1989; 85: 611–620

[72] Ostrovsky LA and Sutin AM. Nonlinear sound scattering from subsurface bubble layers. In: Natural Physical Sources of Underwater Sound (Kerman BR, ed.). Kluwer, Dordrecht, The Netherlands, 1992

[73] Bunkin FV, Vlasov DV, Zabolotskaya EA and Kravstov YuA. Active acoustic spectroscopy of bubbles. Sov Phys Acoust 1983; 29: 99–100

[74] Ostrovskii L and Sutin A. Nonlinear acoustic methods in diagnostics. In: Ultrasound Diagnostics (Grechova, ed.). Gorky, pp 139–150 (in Russian)

[75] Sandler B, Selivanovskii D and Sokolov A. Measurement of gas bubble concentration on the sea surface. Dokl Akad Nauk SSSR 1981; 260(6): 1474–1476 (in Russian)

[76] Naugol'nykh KA and Rybak SA. Interaction of sound waves in scattering by bubbles. Sov Phys Acoust 1987; 33: 94

[77] Apfel RE. Acoustic cavitation prediction. J Acoust Soc Am 1981; 69: 1624–1633

[78] Apfel RE. Acoustic cavitation: a possible consequence of biomedical uses of ultrasound. Brit J Cancer 1982; 45(Suppl V): 140–146

[79] Apfel RE. Possibility of microcavitation from diagnostic ultrasound. IEEE Trans Ultrasonics Ferroelectrics Freq Control 1986; UFFC-33: 139–142

[80] Flynn HG. Generation of transient cavities in liquids by microsecond pulses of ultrasound. J Acoust Soc Am 1982; 72: 1926–2932

[81] Carstensen EL and Flynn HG. The potential for transient cavitation with microsecond pulses of ultrasound. Ultrasound Med Biol 1982; 8: L720–L724

[82] Crum LA and Fowlkes JB. Cavitation produced by short acoustic pulses. Ultrasonics Intl Proc 1985; 1: 237–242

[83] Crum LA and Fowlkes JB. Acoustic cavitation generated by microsecond pulses of ultrasound. Nature 1986; 319: 52–54

[84] Fowlkes JB and Crum LA. Cavitation threshold measurements for microsecond pulses of ultrasound. J Acoust Soc Am 1988; 83: 2190–2201

[85] Carmichael AJ, Massoba MM, Reisz P and Christman CL. Free radical production in aqueous solutions exposed to simulated ultrasonic diagnostic conditions. IEEE Trans Ultrasonics Ferroelectrics Freq Control 1986; UFFC-33: 148–155

[86] Christman CL, Carmichael AJ, Mossoba MM and Riesz P. Evidence for free radicals produced in aqueous solutions by diagnostic ultrasound. Ultrasonics 1987; 25: 31–34

[87] Atchley AA, Frizzell LA, Apfel RE, Holland CK, Madanshetty S and Roy RA. Thresholds for cavitation produced in water by pulsed ultrasound. Ultrasonics 1988; 26: 280–285

[88] Holland CK and Apfel RE. Thresholds for transient cavitation produced by pulsed ultrasound in a controlled nuclei environment. J Acoust Soc Am 1990; 88: 2059–2069

[89] Roy RA, Atchley AA, Crum LA, Fowlkes JB and Reidy JJ. A precise technique for the measurement of acoustic cavitation thresholds and some preliminary results. J Acoust Soc Am 1985; 78: 1799–1805

[90] Roy RA, Church CC and Calabreses A. Cavitation produced by short pulses of ultrasound. Frontiers of Nonlinear Acoustics, 12 ISNA (Hamilton MF and Blackstock DT, eds). Elsevier, Amsterdam, 1990, pp 476–491

[91] Roy RA, Madanshetty S and Apfel RE. An acoustic backscattering technique for the detection of transient cavitation produced by microsecond pulses of ultrasound. J Acoust Soc Am 1990; 87: 2451–2455

[92] Holland CK, Roy RA, Apfel RE and Crum LA. *In-vitro* detection of cavitation induced by a diagnostic ultrasound system. IEEE Trans Ultrasonics Ferroelectrics Freq Control 1992; 39: 95–101

[93] Madanshetty SI, Roy RA and Apfel RE. Acoustic microcavitation: its active and passive detection. J Acoust Soc Am 1991; 90: 1515–1526

[94] Church CC. A method to account for acoustic microstreaming when predicting bubble growth rates produced by rectified diffusion. J Acoust Soc Am 1988; 84: 1758–1764

[95] West C and Howlett R. Some experiments on ultrasonic cavitation using a pulsed neutron source. Brit J Appl Phys (J Phys D) Ser 2 1968; 1: 247

[96] Fowlkes JB and Crum LA. An examination of cavitation due to short pulses of megahertz ultrasound. Tech Rep NCPA LC.01 for NIH, University of Mississippi, 1988

[97] Suslick KS, Hammerton DA and Cline RE Jr. The sonochemical hot spot. J Am Chem Soc 1986; 108: 5641–5642

[98] Roy RA and Fowlkes JB. A comparison of the thresholds for free-radical generation and transient cavitation activity induced by short pulses of ultrasound. J Acoust Soc Am 1988; 84(Suppl 1): S36

[99] Coleman AJ, Choi MJ, Saunders JE and Leighton TG. Acoustic emission and sonoluminescence due to cavitation at the beam focus of an electrohydraulic shock wave lithotripter. Ultrasound Med Biol 1992; 18: 267–281

[100] Sturtevant B. The physics of shock wave focusing in the context of extracorporeal shock wave lithotripsy. Intl Workshop Shock Wave Focusing, March 1989, Sendai, Japan, pp 39–64

[101] Kuwahara M. Extracorporeal shock wave lithotripsy: recent advanced strategy using piezo-ceramic shock wave generator. Intl Workshop Shock Wave Focusing, March 1989, Sendai, Japan, pp 65–89

[102] Mellen RH. An experimental study of the collapse of a spherical cavity in water. J Acoust Soc Am 1956; 28: 447–454

[103] Cathignol D, Mestas JL, Gomez F and Lenz P. Influence of water conductivity on the efficiency and the reproducibility of electrohydraulic shock wave generation. Ultrasound Med Biol 1991; 17: 819–828

[104] Finney R, Halliwell M, Mishriki SF and Baker AC. Measurement of lithotripsy pulses through biological media. Phys Med Biol 1991; 36: 1485–1493

[105] Duck FA and Martin K. Trends in diagnostic exposure. Phys Med Biol 1991; 36: 1423–1432

[106] Coleman AJ, Saunders JE, Crum L and Dyson M. Acoustic cavitation generated by an extracorporeal shockwave lithotripter. Ultrasound Med Biol 1987; 13: 69–76

[107] Riedlinger R. Cavitation in the field of focused pulsed high-power-sources: photo documents and their relationship to biological effects. Proc Workshop Biological Effects and Physical Characterization of Shock Waves, Stuttgart, 1990

[108] Koch H and Grunewald M. Disintegration mechanisms of weak acoustical shock waves. Ultrasound International 89 Conference Proceedings, 1989

[109] Delius M, Enders G, Heine G, Stark J, Remberger K and Brendel W. Biological effects of shock waves: lung hemorrhage by shock waves in dogs – pressure dependence. Ultrasound Med Biol 1987; 13: 61–67

[110] Brummer F, Staudenraus J, Nesper M, Suhr D, Eisenmenger W and Hulser DF. Biological effects and physical characterization of shock waves generated by an XL-1 experimental lithotripter. Ultrasound International 89 Conference Proceedings, 1989, pp 1130–1135

[111] Delius M, Heine G and Brendel W. A mechanism of gall stone destruction by extracorporeal shock waves. Gastroenterology 1988; 94: A93

[112] Delius M, Enders G, Xuan Z, Liebich HG and Brendel W. Biological effects of shock waves: kidney damage by shock waves in dogs – dose dependence. Ultrasound Med Biol 1988; 14: 117–122

[113] Delius M, Jordan M, Eizenhoefer H, Marlinghaus E, Heine G, Liebich HG and Brendel W. Biological effects of shock waves: kidney hemorrhage by shock waves in dogs – administration rate dependence. Ultrasound Med Biol 1988; 14: 689–694

[114] Delius M, Denk R, Berding C, Liebich H, Jordan M and Brendel W. Biological effects of shock waves: cavitation by shock wave in piglet liver. Ultrasound Med Biol 1990; 16: 467–472

[115] Gambilher S, Delius M and Brendel W. Biological effects of shock waves: cell disruption viability, and proliferation of L1210 cells exposed to shock waves *in vitro*. Ultrasound Med Biol 1990; 16: 587–594

[116] Hartman C, Cox CA, Brewer L, Child SZ, Cox CF and Carstensen EL. Effects of lithotripter fields on development of chick embryos. Ultrasound Med Biol 1990; 16: 581–585

[117] Hartman C, Child SZ, Mayer R, Schenk E and Carstensen EL. Lung damage from exposure to the fields of an electrohydraulic lithotripter. Ultrasound Med Biol 1990; 16: 675–679

[118] Williams AR, Delius M, Miller DL and Schwarze W. Investigation of cavitation in flowing media by lithotripter shock waves both *in vitro* and *in vivo*. Ultrasound Med Biol 1989; 15: 53–60

[119] Church CC. A theoretical study of cavitation generated by an extracorporeal shock wave lithotripter. J Acoust Soc Am 1989; 86: 215–227

[120] Olsson L, Almquist L-O, Grennberg A and Holmer N-G. Analysis and classification of secondary sounds from the disintegration of kidney stones with acoustic shock waves. Ultrasound Med Biol 1991; 17: 491–495

[121] Walton AJ and Reynolds GT. Sonoluminescence. Advan Phys 1984; 33: 595–660

[122] Meyer E and Kuttruff H. On the phase relation between sonoluminescence and the cavitation process with periodic excitation. Z Angew Phys 1959; 11: 325–333 (in German)

[123] Negishi K. Phase relation between sonoluminescence and cavitating bubbles. Letters to the Editors, Acoustica 1960; 10: 124

[124] Negishi K. Experimental studies on sonoluminescence and ultrasonic cavitation. J Phys Soc Japan 1961; 16: 1450–1465

[125] Gaitan DF and Crum LA. Observation of sonoluminescence from a single cavitation bubble in a water/glycerine mixture. In: Frontiers of Nonlinear Acoustics, 12th ISNA (Hamilton MF and Blackstock DT, eds). Elsevier, New York, 1990, p 459

[126] Gaitan DF, Crum LA, Church CC and Roy RA. An experimental investigation of acoustic cavitation and sonoluminescence from a single bubble. J Acoust Soc Am 1992; 91: 3166–3183

[127] Noltingk BE and Neppiras EA. Cavitation produced by ultrasonics. Proc Phys Soc 1950; B63: 674–685

[128] Srinivasan D and Holroyd LV. Optical spectrum of sonoluminescence emitted by cavitated water. J Appl Phys 1961; 32: 446–449

[129] Gunther P, Heim E and Borgstedt HU. Über die kontinuierlichen Sonolumineszenzspektren wäßriger Lösungen. Z Electrochem 1959; 63: 43–47 (in German)

[130] Griffing V. Theoretical explanation of the chemical effects of ultrasonics. J Chem Phys 1950; 18: 997–998

[131] Griffing V. The chemical effects of ultrasonics. J Chem Phys 1952; 20, 939–942

[132] Parke AVM and Taylor D. The chemical action of ultrasonic waves. J Chem Soc 1956; 4: 4442–4450

[133] Margulis NA. The mechanism of the formation of a continuum in sonoluminescence spectra. Russ J Phys Chem 1980; 54: 859–861

[134] Taylor KJ and Jarman PD. The spectra of sonoluminescence. Acust J Phys 1970; 23: 319–34

[135] Sehgal C, Sutherland RG and Verrall RE. Optical spectra of sonoluminescence from transient and stable cavitation in water saturated with various gases. J Phys Chem 1980; 84: 388–395

[136] Paounoff P. Luminescence – La luminescence de l'eau sous l'action des ultrasons. C R Hebd Séanc Acad Sci Paris 1939; 209: 33 (in French)

[137] Gunther P, Heim E and Eichkorn G. Phasenkorrelation von Schallwechseldruck und Sonolumineszenz. Z Angew Phys 1959; 11: 274–277 (in German)

[138] Prudhomme RO and Guilmart Th. Photogènese ultraviolette par irradiation ultrasonore de l'eau en présence des gas rares. J Chim Phys 1957; 54: 336–340 (in French)

[139] Valladas-Dubois S, Haug R and Pruhomme RO. Spectra of sonoluminescence from water in visible range. J Chim Phys 1978; 75: 855–858

[140] Sehgal C, Steer RP, Sutherland RG and Verrall RE. Sonoluminescence of aqueous solutions. J Phys Chem 1977; 81: 2618

[141] Verrall RE and Sehgal CM. Sonoluminescence. Ultrasonics 1987; 25: 29–30

[142] Sehgal C, Sutherland RG and Verrall RE. The selective quenching of species that produce sonoluminescence. J Phys Chem 1980; 84: 529–531

[143] Saksena TK and Nyborg WL. Sonoluminescence from stable cavitation. J Chem Phys 1970; 53: 1722

[144] Suslick KS, Doktycz SJ and Flint EB. On the origin of sonoluminescence and sonochemistry. Ultrasonics 1990; 28: 280–290

[145] Suslick KS and Flint EB. Sonoluminescence from non-aqueous liquids. Nature 1987; 330: 553

[146] Gunther P, Zeil W, Grisar U, Langmann W and Heim E. Über Sonolumineszenz. Z Naturf 1956; 11A: 882–883 (in German)

[147] Margulis MA and Dmitrieva AF. Dynamics of the cavitation bubble. 3. Mechanism of line emission of metals on sonoluminescence spectra of solutions of salts. Russ J Phys Chem 1982; 56: 531

[148] Heim E. Asymmetrisch verbreiterte Emissionslinien in den Sonolumineszenzspektren wäßriger Salzlösungen. Z Angew Phys 1960; 12: 423–424 (in German)

[149] Jarman P and Taylor KJ. The timing of the main and secondary flashes of sonoluminescence from acoustically cavitated water. Acustica 1970; 23: 243
[150] Sehgal C, Steer RP, Sutherland RG and Verrall RE. Sonoluminescence of argon saturated alkali metal salt solutions as a probe of acoustic cavitation. J Chem Phys 1979; 70: 2242
[151] Sehgal C, Sutherland RG and Verrall RE. Sonoluminescence from aqueous solutions of Br_2 and I_2. J Phys Chem 1981; 85: 315–317
[152] Sehgal C, Sutherland RG and Verrall RE. Sonoluminescence of NO- and NO_2-saturated water as a probe of acoustic cavitation. J Phys Chem 1980; 84: 396–401
[153] Sehgal C, Sutherland RG and Verrall RE. Sonoluminescence of argon-saturated copper-sulphate solution. J Phys Chem 1980; 84: 227–228
[154] Suslick KS and Hammerton DA. The site of sonochemical reactions. IEEE Trans Ultrasonics Ferroelectrics Freq Control 1986; 33: 143–147
[155] Fischer Ch-H, Hart EJ and Henglein A. Ultrasonic irradiation of water in the presence of $^{18,18}O_2$: isotope exchange and isotopic distribution of H_2O_2. J Phys Chem 1986; 90: 1954–1956
[156] Coakley WT and Nyborg WL. Chapter II: Cavitation; dynamics of gas bubbles; applications. In: Ultrasound: its Applications in Medicine and Biology (Fry F, ed.). Elsevier, New York, 1978, Part 1, pp 77–159
[157] Richards WT and Loomis AL. The chemical effects of high frequency sound waves I. A preliminary survey. J Am Chem Soc 1927; 49: 3086–3100
[158] Henglein A. Sonochemistry: historical developments and modern aspects. Ultrasonics 1987; 25: 6–16
[159] Brohult S. Splitting of the Hæmocyanin molecule by ultra-sonic waves. Nature 1937; 140: 805
[160] Schulz R and Henglein A. Über den Nachweis von freien Radikalen, die unter dem Einfluβ von Ultraschallwellen gebildet werden, mit Hilfe von Radikal-Kettenpolymerisation und Diphenyl-pikryl-hydrazyl. Z Naturforsch 1953; 8b: 160–161 (in German)
[161] Flynn HG. Physics of acoustic cavitation in liquids. In: Physical Acoustics, Vol. 1 (Mason WP ed.). Academic Press, New York, 1964, Part B, pp 57–172
[162] Prudhomme RD. Vijnana Parishad Anusandhan Patrika 1972; 15: 3
[163] Frenzel J and Schultes H. Luminescenz im ultraschall beschickten Wasser. Kurze Mitteilung. Zeit Phys Chem 1934; B27: 421–424 (in German)
[164] Frenkel YaI. Acta Phisiochemica URSS 1940; 12: 317–323 (in Russian)
[165] Bresler S. Acta Physicochim (USSR) 1940; 12: 323 (in Russian)
[166] Bergmann L. Der Ultraschall S. Hirzel, Stuttgart, 1954
[167] El' Piner I. Ultrasound: Physical, Chemical and Biological Effects. Consultants Bureau, New York, 1964
[168] Weissler A. Encyclopedia of Chemical Technology, Vol. 15 (Kirk RE and Othmer DF, eds). Wiley Interscience, New York, 1981, p 773
[169] Boudjouk P. Nachr Chem Techn Lab 1983; 31: 797
[170] Young FR. Cavitation. McGraw-Hill, London, 1989, pp 353–372
[171] Hart EJ and Henglein A. Sonolytic decomposition of nitrous oxide in aqueous solution. J Phys Chem 1986; 90: 5992–5995
[172] Hickling R. Effects of thermal conduction in sonoluminescence. J Acoust Soc Am 1963; 35: 967–974
[173] Young FR. Sonoluminescence from water containing dissolved gases. J Acoust Soc Am 1976; 60: 100–104
[174] Suslick KS and Schubert PF. Sonochemistry of $Mn_2(CO)_{10}$ and $Re_2(CO)_{10}$. J Am Chem Soc 1983; 105: 6042–6044
[175] Suslick KS, Goodale JW, Schubert PF and Wang HH. Sonochemistry and sonocatalysis of metal carbonyls. J Am Chem Soc 1983; 105: 5781–5785
[176] Suslick KS, Gawienowski JJ, Schubert PF and Wang HH. Sonochemistry in non-aqueous liquids. Ultrasonics 1984; 22: 33–36
[177] Sehgal C, Yu TJ, Sutherland RG and Verrall RE. Use of 2,2-diphenyl-1-picrylhydrazyl to investigate the chemical behaviour of free-radicals induced by ultrasonic cavitation. J Phys Chem 1982; 86: 2982–2986
[178] Jarman P. Measurements of sonoluminescence from pure liquids and some aqueous solutions. Proc Phys Soc 1959; B72: 628

[179] Golubnichii PI, Goncharov VD and Protopopor KhV. Sonoluminescence in various liquids. Sov Phys Acoust 1971; 16: 323–326
[180] Niemczewski B. A comparison of ultrasonic cavitation intensity in liquids. Ultrasonics 1980; 18: 107–110
[181] Fischer ChH, Hart EJ and Henglein A. H/D isotope exchange in the D_2–H_2O system under the influence of ultrasound. J Phys Chem 1986; 90: 222–224
[182] Fischer ChH, Hart EJ and Henglein A. H/D isotope exchange in the HD–H_2O system under the influence of ultrasound. J Phys Chem 1986; 90: 3059–3060
[183] Hart EJ, Fischer ChH and Henglein A. Isotopic exchange in the sonolysis of aqueous solutions containing $^{14,14}N_2$ and $^{15,15}N_2$. J Phys Chem 1986; 90: 5989–5991
[184] Henglein A and Kormann C. Scavenging of OH radicals produced in the sonolysis of water. Int J Radiat Biol 1985; 48: 251–258
[185] Sehgal C, Sutherland RG and Verrall RE. Cavitation-induced oxidation of aerated aqueous Fe^{2+} solutions in the presence of aliphatic alcohols. J Phys Chem 1980; 84: 2920–2922
[186] Degrois M and Baldo P. A new electrical hypothesis explaining sonoluminescence, chemical actions and other effects produced in gaseous cavitation. Ultrasonics 1974; 12: 25–28
[187] Hart EJ and Henglein A. Free radical and free atom reactions in the sonolysis of aqueous iodide and formate solutions. J Phys Chem 1985; 89: 4342–4347
[188] Hart EJ and Henglein A. Sonolysis of ozone in aqueous solution. J Phys Chem 1986; 90: 3061–3062
[189] Henglein A. Sonolysis of carbon dioxide, nitrous oxide and methane in aqueous solution. Z Naturforsch 1985; 40b: 100–107
[190] Hart EJ and Henglein A. Sonochemistry of aqueous solutions: H_2–O_2 combustion in cavitation bubbles. J Phys Chem 1987; 91: 3654–3656
[191] Suslick KS, Cline RE Jr and Hammerton DA. Determination of local temperatures caused by acoustic cavitation. IEEE Ultrasonics Symposium Proceedings 1985; 4: 1116
[192] Suslick KS, Hammerton DA and Cline RE Jr. The sonochemical hot-spot. J Am Chem Soc 1986; 108: 5641–5642
[193] Chendke PK and Fogler HS. Variation of sonoluminescence intensity of water with the liquid temperature. J Phys Chem 1985; 89: 1673–1677
[194] Sehgal CM and Verrall RE. Ultrasonics 1982; 20: 37–39
[195] Margulis MA. Sonoluminescence and sonochemical reactions in cavitation fields. A review. Ultrasonics 1985; 23: 157–169
[196] Vaughan PW and Leeman S. Some comments on mechanisms of sonoluminescence. Acustica 1986; 59: 279–281
[197] Neppiras EA and Noltingk BE. Cavitation produced by ultrasonics: theoretical conditions for the onset of cavitation. Proc Phys Soc 1951; B64: 1032–1038
[198] Chambers LA. Phys Rev 1936; 49: 881
[199] Weyl WA and Marboe EC. Research, 1949; 2: 19
[200] Jarman PD. J Acoust Soc Am 1960; 32: 1459–1463
[201] Bradley JN. Shock Waves in Chemistry and Physics. Methuen, London, 1968, pp 246–263
[202] Johansson CH. The initiation of liquid explosives by shock and the importance of liquid break up. Proc Roy Soc 1958; A246: 160–167
[203] Margulis MA and Dmitrieva AF. Dynamics of the cavitation bubble. 2. Results of a quantitative integration of equations for dynamics of bubble taking into account heat exchange. Russ J Phys Chem 1982; 56: 198
[204] Vaughan PW and Leeman S. Acoustic cavitation revisited. Acustica 1989; 69: 109–119
[205] Bourne NK and Field JE. Bubble collapse and the initiation of explosion. Proc Roy Soc 1991; A435: 423–435
[206] Chaudhri MM and Field JE. The role of rapidly compressed gas pockets in the initiation of condensed explosives. Proc Roy Soc 1974; A340: 113–128
[207] Mader C. Shocks and hot spot initiation of homogeneous explosives. Phys Fluids 1963; 6: 375–381
[208] Mader C. Initiation of detonation by the interaction of shocks with density discontinuities. Phys Fluids 1965; 8: 1811–1816
[209] Vaughan PW and Leeman S. Transient cavitation: fact or fallacy? Ultrasonics International 89 Conference Proceedings, 1989, pp 1259–1264

[210] Vaughan PW and Leeman S. Sonoluminescence: violent light or gentle glow? IEEE 1986 Ultrasonics Symposium, pp 989–992

[211] Giminez G. The simultaneous study of light emissions and shock waves produced by cavitation bubbles. J Acoust Soc Am 1982; 71: 839–846

[212] Harvey EN. Sonoluminescence and sonic chemiluminescence. J Am Chem Soc 1939; 61: 2392–2398.

[213] Watmough DJ, Shiran MB and Quan KM. Circumstantial evidence showing that acoustically excited gas bubbles carry a negative charge. Acoust Bull 1992; April: 5–8

[214] Golubnichii PI, Sytnikov AM and Filonenko AD. Electrical nature of ultrasonic luminescence and possibility of testing the hypothesis experimentally. Sov Phys Acoust 1987; 33(3)

[215] Margulis MA. Investigation of electrical phenomena connected with cavitation. I. On electrical theories of chemical and physiochemical actions of ultrasonics. Russ J Phys Chem 1981; 55: 154–158

[216] Nathanson GL. Value of electric field in cavities produced by ultrasonic cavitation in a liquid. Dokl Akad Nauk SSSR 1948; 59: 83–85

[217] Margulis MA. Investigation of electrical phenomena connected with cavitation. II. On theory of appearance of sonoluminescence and sonochemical reactions. Russ J Phys Chem 1984; 58: 1450

[218] Margulis MA and Grundel LM. Chemical action of low-frequency acoustic vibrations. Dokl Acad Nauk SSSR 1982; 265: 914–917

[219] Margulis MA and Grundel LM. Appearance of light in liquid in low-frequency acoustic fields. Dokl Akad Nauk SSSR 1983; 269: 405–407

[220] Margulis MA and Grundel LM. Investigation of ultrasonic light in liquid near the cavitation threshold. I. Appearance of prethreshold glow of liquid in ultrasonic field. Russ J Phys Chem 1981; 55: 687–691

[221] Margulis MA and Grundel LM. Physiochemical processes in liquids under the influence of low-frequency acoustic vibrations. I. Growth and pulsation of gas bubbles in liquid. Russ J Phys Chem 1982; 56: 1445–1449

[222] Loeb LB. Static Electrification. Springer, Berlin, 1958

[223] Golubnichii PI, Gonsharov VD and Prothopopov HV. Sonoluminescence in liquids – effect of dissolved gases and recoiling from heat theory. Russ J Acoust 1969; 15: 534–541

[224] Kurochkin AK, Smorodov EA, Valitov RB and Margulis MA. Investigation of sonoluminescence mechanism. 2. Study of the form of the light sonoluminescence pulse. Zh Fiz Khim 1986; 60: 1234–1238

[225] Margulis MA. The investigation of electrical phenomena related to cavitation. 2. Theory of the sonoluminescence and sound-chemical reaction origin. Zh Fiz Khim 1985; 59: 1497–1503

[226] Atchley AA and Crum LA. Acoustic cavitation and bubble dynamics. In Ultrasound: its Chemical, Physical and Biological Effects (Suslick KS, ed.). VCH, New York, 1988, p 1

[227] Walmsley AD, Laird WRE and Williams AR. Dental plaque removal by cavitational activity during ultrasonic scaling. J Clin Periodont 1988; 15: 539–543

[228] Barber BP and Putterman SJ. Observation of synchronous nanosecond sonoluminescence. Nature 1991; 352: 318

[229] Barber BP, Hiller R, Arisaka K, Fetterman H and Putterman S. Resolving the picosecond characteristics of synchronous sonoluminescence. J Acoust Soc Am 1992; 91: 3061–3063

[230] Crum LA. Sonoluminescence. J Acoust Soc Am 1992; 91: 517

[231] Benkovskii VG, Golubnichii PI, Maslennikov SI and Olzoec KF. J Appl Spectrosc 1974; 24: 964

[232] Buzukov AA and Teslenko VS. Sonoluminescence following focussing of laser radiation into a liquid. Sov Phys JETP Lett 1971; 14: 189

[233] Jarman PD and Taylor KJ. Light emission from cavitating water. Brit J Appl Phys 1964; 15: 321–322

[234] Schmid J. Kinematographische Untersuchung der Einzelblasen – Kavitation. Acustica 1959; 9: 321–326

[235] Camus JJ. PhD thesis, University of Cambridge, 1971

[236] Schmid J. Gasgehalt und Lumineszenz einer Kavitationsblase (Modellversuche an Glaskugeln). Acustica 1962; 12: 70–83 (in German)

[237] Finch RD. Sonoluminescence. Ultrasonics 1963; 1: 87–98

[238] Jarman PD and Taylor KJ. Light flashes and shocks from a cavitating flow. Brit J Appl Phys 1965; 16: 675–682

[239] Golubnichii PI, Goncharov VD and Protopopov KhV. Sonoluminescence in liquids: influence of dissolved gases, departures from thermodynamic theory. Sov Phys – Acoust 1970; 15: 464

[240] Samek L and Taraba O. Measurements of the dependence of sonoluminescence on acoustic pressure of ultrasound field and measurements of the acoustic spectrum of cavitation noise. Czech J Phys 1973; B23: 287

[241] Alfredsson B. A study of soluminescence in ultrasonic induced cavitation. Acustica 1965; 16: 127–133

[242] Bohn L. Acoustic pressure variation and the spectrum in oscillatory cavitation. Acustica 1957; 7: 201–216

[243] Rosenberg MD. La génération et l'étude des vibrations ultra-sonores de très grande intensité. Acustica 1962; 12: 40–49 (in French)

[244] Marinesco N and Trillat JJ. Chimie Physique – Action des ultrasons sur les plaques photographiques. C R Hebd Séanc Acad Sci Paris 1933; 196: 858 (in French)

[245] Marinesco N and Reggiani M. Chimie Physique – Impression des plaques photographiques par les ultrasons. C R Hebd Séanc Acad Sci Paris 1935; 200: 548 (in French)

[246] Chambers LA. The emission of visible light from cavitated liquids. J Chem Phys 1937; 5: 290–292

[247] Griffing V and Sette D. Luminescence produced as a result of intense ultrasonic waves. J Chem Phys 1955; 23: 503–509

[248] Reynolds GT, Walton AJ and Gruner S. Observations of sonoluminescence using image intensification. Rev Sci Instrum 1982; 53: 1673–1676

[249] Singal SP and Panchaloy M. Luminescence in liquids irradiated by sonic waves. J Sci Ind Res 1967; 26: 101–109

[250] Degrois M. Études sur la sonoluminescence. L'Onde électrique 1968; 48: 3–13 (in French)

[251] Fitzgerald ME, Griffing V and Sullivan J. Chemical effects of ultrasonics – 'hot spot' chemistry. J Chem Phys 1956; 25: 926–933

[252] Weissler A. Sonochemistry: the production of chemical changes with sound waves. J Acoust Soc Am 1953; 25: 651–657

[253] Vaughan PW, Graham E and Leeman S. The effect of dissolved gases on the dynamics of acoustic emission and sonoluminescence from cavitating liquids. Appl Sci Res 1982; 38: 45–52

[254] Margulis MA and Dmitrieva AF. Dynamics of the collapse of cavitation bubbles. I. Derivation of the equations of motion of the bubble with allowance for heat exchange. Russ J Phys Chem 1981; 55: 83

[255] Levsin VL and Rzevkin SN. C R (Dokl) Acad Sci URSS 1937; 16: 399

[256] Crum LA, Walton AJ, Mortimer A, Dyson M, Crawford DC and Gaitan DF. Free radical production in amniotic fluid and blood plasma by medical ultrasound. J Ultrasound Med 1987; 6: 643–647

[257] Akopyan VB. Determination of the cavitation threshold in biological tissue from the luminescence appearing on exposure to ultrasound. Biophysics 1980; 25: 891–894

[258] Smith RT, Webber GMB, Young FR and Stephens RWB. Sound propagation in liquid metals. Adv Phys 1967; 16: 515–522

[259] Iernetti G. Temperature dependence of sonoluminescence and cavitation erosion in water. Acustica 1972; 26: 168–169

[260] Sehgal C, Sutherland RG and Verrall RE. Sonoluminescence intensity as a function of bulk solution temperature. J Phys Chem 1980; 84: 525–528

[261] Pickworth MJW, Dendy PP, Leighton TG and Walton AJ. Studies of the cavitational effects of clinical ultrasound by sonoluminescence: 2. Thresholds for sonoluminescence from a therapeutic ultrasound beam and the effect of temperature and duty cycle. Phys Med Biol 1988; 33: 1249–1260

[262] Blake FG Jr. Technical Memo 12, Acoustics Research Laboratory, Harvard University, Cambridge, Massachusetts, September 1949

[263] Overton GDN, Williams PR and Trevena DH. The influence of cavitation history and entrained gas on liquid tensile strength. J Phys D 1984; 17: 979–987

[264] Trevena DH. Cavitation and the generation of tension in liquids. J Phys D 1984; 17: 2139–2164

[265] Pickworth MJW, Dendy PP, Leighton TG, Worpe E and Chivers RC. Studies of the cavitational effects of clinical ultrasound by sonoluminescence: 3. Cavitation from pulses of a few microseconds in length. Phys Med Biol 1989; 34: 1139–1151
[266] Henglein A and Gutierrez M. Chemical reactions by pulsed ultrasound: memory effects in the formation of NO_2^- and NO_3^- in aerated water. Int J Radiat Biol 1986; 50: 527–533
[267] Finch RD. The dependence of sonoluminescence on static pressure. Brit J Appl Phys 1965; 16: 1543–1553
[268] Chendke PK and Fogler HS. Sonoluminescence and sonochemical reactions of aqueous carbon-tetrachloride solutions. J Phys Chem 1983; 87: 1362
[269] Šponer J. Dependence of ultrasonic cavitation threshold on raised ambient pressure. Studia Biophys 1990; 137: 91–97
[270] Urick RJ. Principles of Underwater Sound. McGraw-Hill, New York, 3rd edn, 1983, p 337
[271] Margulis MA and Akopyan VB. Experiments of the dependence of acoustic–chemical reaction-rate and sonoluminescence flow on ultrasonic wave intensity. Russ J Phys Chem 1978; 52: 339 (Zh Fiz Khim 1978; 52: 601–604)
[272] Kling R. Revue Scient Paris 1947; 85: 364
[273] Gabrielli I, Iernetti G and Lavenia A. Sonoluminescence and cavitation in some liquids. Acustica, 1967; 18: 173–179
[274] Šponer J. Dependence of ultrasonic cavitation threshold on the ultrasonic frequency. Czech J Phys 1990; B40: 1123–1132
[275] Holland CK and Apfel RE. An improved theory for the prediction of microcavitation thresholds. IEEE Trans Ultrasonics Ferroelectrics Freq Control 1989; 36: 204–208
[276] Flynn HG. Cavitation dynamics. I. A mathematical formulation. J Acoust Soc Am 1975; 57: 1379–1396
[277] Margulis MA and Grundel LM. The ultrasonic luminescence of a liquid near the cavitation threshold. 1. The development of the pre-threshold luminescence of a liquid in an ultrasonic field. Russ J Phys Chem 1981; 55: 386–389
[278] Margulis MA and Grundel LM. Russ J Phys Chem 1981; 55: 989
[279] Crum LA and Reynolds GT. Sonoluminescence produced by 'stable' cavitation. J Acoust Soc Am 1985; 78: 137–139
[280] Marston PL. Critical angle scattering by a bubble: physical-optics approximation and observations. J Opt Soc Am 1979; 69(9): 1205
[281] Marston PL and Kingsbury DL. Scattering by a bubble in water near the critical angle: interference effects. J Opt Soc Am 1981; 71(2): 358
[282] Marston PL, Langley DS and Kingsbury DL. Light scattering by bubbles in liquids: Mie theory, physical-optics approximations, and experiments. Appl Sci Res 1982; 38: 373–383
[283] Langley DS and Marston PL. Scattering of laser light from bubbles in water at angles from 68 to 85 degrees. SPIE 1984; 489(Ocean Optics VII): 142
[284] Langley DS and Marston PL. Critical-angle scattering of laser light from bubbles in water: measurements, models, and application to sizing of bubbles. Appl Opt 1984; 23(7): 1044
[285] Keller JB and Miksis M. Bubble oscillations of large amplitude. J Acoust Soc Am 1980; 68: 628
[286] Prosperetti A, Crum LA and Commander KW. Nonlinear bubble dynamics. J Acoust Soc Am 1986; 83: 502–514
[287] Šponer J. Theoretical estimation of the cavitation threshold for very short pulses of ultrasound. Ultrasonics 1991; 29: 376–380
[288] Šponer J, Davadorzh C and Mornstein V. The influence of viscosity on ultrasonic cavitation threshold for sonoluminescence at low megahertz region. Studia Biophys 1990; 137: 81–89
[289] Lauterborn W. Cavitation bubble dynamics – new tools for an intricate problem. Appl Sci Res 1982; 38: 165–178
[290] Pinoir R and Pouradier J. Action des ultrasons sur les couches sensibles. J Chim Phys 1947; 44: 261 (in French)
[291] Henglein A. Chemische Wirkungen von kontinuierlichen und impulsmodulierten horbaren Schallwellen. Z Naturforsch 1955; 10b: 20–26 (in German)
[292] Wagner W. Phasenkorrelation von Schalldruck und Sonolumineszenz. Z Angew Phys 1958; 10: 445–452 (in German)
[293] Gunther P, Heim E, Schmitt A and Zeil W. Versuche über Sonolumeszenz. Z Naturf 1957; 12A: 521–522 (in German)

[294] Leighton TG. Image intensifier studies of sonoluminescence, with application to the safe use of medical ultrasound. PhD thesis, Cambridge University, 1988

[295] Leighton TG, Pickworth MJW, Walton AJ and Dendy PP. Studies of the cavitational effects of clinical ultrasound by sonoluminescence: 1. Correlation of sonoluminescence with the standing-wave pattern in an acoustic field produced by a therapeutic unit. Phys Med Biol 1988; 33: 1239–1248

[296] Pickworth MJW. Studes of the cavitational effects of clinical ultrasound by sonoluminescence. PhD thesis, University of Surrey, 1988

[297] Pickworth MJW, Dendy PP, Twentyman PR and Leighton TG. Studies of the cavitational effects of clinical ultrasound by sonoluminescence: 4. The effect of therapeutic ultrasound on cells in monolayer culture in a standing wave field. Phys Med Biol 1989; 34: 1553–1560

[298] Cartwright CH. Infra-red transmission of the flesh. J Opt Soc Am 1930; 29: 81–84

[299] Leighton TG, Pickworth MJW, Tudor J and Dendy PP. Studies of the cavitational effects of clinical ultrasound by sonoluminescence: 5. Search for sonoluminescence *in vivo* in the human cheek. Ultrasonics 1990; 28: 181–184

[300] Dyson M, Pond JB and Woodward B *et al.* The production of blood cell stasis and endothelial damage in the blood vessels of chick embryos treated with ultrasound in a stationary wave field. Ultrasound Med Biol 1974; 1: 133–148

[301] ter Haar GR. The effect of ultrasonic standing wave fields on the flow of particles. PhD thesis, University of London, 1977

[302] ter Haar GR and Wyard SJ. Blood cell banding in ultrasonic standing wave fields: a physical analysis. Ultrasound Med Biol 1978; 4: 111–123

[303] Westervelt PJ. Theory of steady forces caused by sound waves. J Acoust Soc Am 1951; 23: 312–315

[304] Dinno MA, Crum LA and Wu J. The effect of therapeutic ultrasound on electrophysiological parameters of frog skin. Ultrasound Med Biol 1989; 15: 461–470

[305] Child SZ, Carstensen EL and Smachlo K. Effects of ultrasound on *Drosophila* – I. Killing of eggs exposed to traveling and standing wave fields. Ultrasound Med Biol 1980; 6: 127–130

[306] Kerr CL, Gregory DW, Shammiri M, Watmough DJ and Wheatley DN. Differing effects of ultrasound irradiation on suspension and monolayer cultured HeLa cells, investigated by scanning electron microscopy. Ultrasound Med Biol 1989; 15: 397–401

[307] Watmough DJ, Quan KM and Shiran MB. Possible explanation for the unexpected absence of gross biological damage to membranes of cells insonated in suspension and in surface culture in chambers exposed to standing and progressive wave fields. Ultrasonics 1990; 28: 142–148

[308] Tyszka M, Dendy PP, Leighton TG and Pickworth MJW. Standing-wave fields in a vessel of finite dimensions and the implications for sonoluminescence. Presented at the 1989 Bioeffects Group Meeting, Torquay, UK (4 December 1989)

[309] Hill CR, Clarke PR, Crowe MR and Hannick JW. Ultrasonics for Industry Conference Papers 26–30, 1969

[310] Church CC. Prediction of rectified diffusion during nonlinear bubble pulsations at biomedical frequencies. J Acoust Soc Am 1988; 83: 2210–2217

[311] Clarke PR and Hill CR. Physical and chemical aspects of ultrasonic disruption of cells. J Acoust Soc Am 1970; 47: 649–653

[312] Ciaravino V, Flynn HG, Miller MW and Carstensen EL. Pulsed enhancement of acoustic cavitation: a postulated model. Ultrasound Med Biol 1981; 7: 159–166

[313] Flynn HG and Church CC. A mechanism for the generation of cavitation maxima by pulsed ultrasound. J Acoust Soc Am 1984; 76: 505–512

[314] Akulichev VA. Investigation of the onset and development of acoustic cavitation. Candidate's dissertation (in Russian), Akust Inst Acad Nauk SSSR Moscow, 1966. Cited by Margulis MA, Kinetics of the number of cavitation bubbles in an ultrasonic field. Sov Phys Acoust 1976; 22: 145–147

[315] Epstein PS and Plesset MS. On the stability of gas bubbles in liquid–gas solutions. J Chem Phys 1950; 18: 1505–1509

[316] Sirotyuk MG. Energetics and dynamics of the cavitation zone. Sov Phys Acoust 1967; 13: 226–229

[317] Leighton TG, Walton AJ and Field JE. High-speed photography of transient excitation. Ultrasonics 1989; 27: 370–373

[318] Leighton TG. Transient excitation of insonated bubbles. Ultrasonics 1989; 27: 50–53

[319] Leighton TG, Pickworth MJW, Walton AJ and Dendy PP. The pulse enhancement of unstable cavitation by mechanisms of bubble migration. Proc Inst Acoust 1989; 11: 461–469
[320] Blake FG Jr. Technical Memo 12, Acoustics Research Laboratory, Harvard University, Cambridge, Massachusetts, September 1949
[321] Miller MW, Church CC, Brayman AA, Malcuit MS and Boyd RW. An explanation for the decrease in cell lysis in a rotating tube with increasing ultrasound intensity. Ultrasound Med Biol 1989; 15: 67–72
[322] Fu Y-K, Miller MW, Lange CS and Griffiths TD. Ultrasound lethality to synderonous and asynderonous Chinese hamster V-79 cells. Ultrasound Med Biol 1980; 6: 39–46
[323] Kaufman GE, Miller MW, Griffiths TD and Ciaravino V. Lysis and viability of cultured mammalian cells exposed to 1-MHz ultrasound. Ultrasound Med Biol 1977; 3: 21–25
[324] Kaufman GE and Miller MW. Growth retardation in Chinese Hamster V-79 cells exposed to 1-MHz ultrasound. Ultrasound Med Biol 1978; 4: 139–144
[325] Kaufman GE. Mutagenicity of ultrasound in cultured mammalian cells. Ultrasound Med Biol 1985; 3: 497–501
[326] Williams AR and Miller DL. The role of non-acoustic factors in the induction and proliferation of cavitational activity *in vivo*. Phys Med Biol 1989; 34: 1561–1569
[327] Church CC, Flynn HG, Miller MW and Sacks PG. The exposure vessel as a factor in ultrasonically-induced mammalian cell lysis – 2. An explanation of the need to rotate exposure tubes. Ultrasound Med Biol 1982; 8: 299–309
[328] Ciaravino V, Miller MW and Carstensen EL. Pressure-mediated reduction of ultrasonically-induced cell lysis. Radiat Res 1981; 88: 209–213
[329] Miller DL and Williams AR. Bubble cycling as the explanation of the promotion of ultrasonic cavitation in a rotating tube exposure system. Ultrasound Med Biol 1989; 15: 641–648
[330] Church CC and Miller MW. The kinetics and mechanics of ultrasonically-induced cell lysis produced by non-trapped bubbles in a rotating culture tube. Ultrasound Med Biol 1983; 9: 385–393
[331] Church CC and Miller MW. The kinetics of ultrasonically-induced cell lysis produced by non-trapped bubbles in a rotating culture tube. In: Proceedings of the Ultrasonics Symposium, Institute of Electrical and Electronic Engineers, New York, 1983, pp 744–748
[332] Miller DL. On the thermal motions of small bubbles. Ultrasound Med Biol 1984; 10: L377–L378
[333] Church CC and Miller MW. On the thermal motions of small bubbles – authors reply. Ultrasound Med Biol 1984; 10: L378–L379
[334] Miller DL, Thomas RM and Williams AR. Mechanisms for haemolysis by ultrasonic cavitation in the rotating exposure system. Ultrasound Med Biol 1991; 17: 171–178
[335] Miller DL. A review of the ultrasonic bioeffects of microsonation, gas-body activation, and related cavitation-like phenomena. Ultrasound Med Biol 1987; 13: 443–470
[336] Davidson BJ and Riley N. Cavitation microstreaming. J Sound Vib 1971; 15: 217–233
[337] Kondo T, Gamson J, Mitchell JB and Riesz P. Free radical formation and cell lysis induced by ultrasound in the presence of different rare gases. Int J Radiat Biol 1988; 54: 955–962
[338] Neppiras EA and Coakley WT. Acoustic cavitation in a focused field in water at 1 MHz. J Sound Vib 1976; 45: 341–373
[339] Graham E, Hedges M, Leeman S and Vaughan P. Cavitational bioeffects at 1.5 MHz. Ultrasonics 1980; 18: 224–228
[340] Williams AR. Absence of meaningful thresholds for bioeffect studies on cell suspensions *in vitro*. Brit J Cancer 1982; 45: 192–195
[341] Saad AH and Williams AR. Possible mechanisms for the agitation-induced enhancement of acoustic cavitation *in vitro*. J Acoust Soc Am 1985; 78: 429–434
[342] Rayleigh Lord. On the pressure developed in a liquid during the collapse of a spherical cavity. Phil Mag 1917; 34: 94–98
[343] Hickling R and Plesset MS. Collapse and rebound of a spherical bubble in water. Phys Fluids 1964; 7: 7–14
[344] Plesset MS and Prosperetti A. Bubble dynamics and cavitation. Ann Rev Fluid Mech, 1977; 9: 145–185
[345] Mørch KA. Erosion (Preece CM, ed.). Academic Press, London, 1979, pp 309–353
[346] Dmitriev AP, Dreiden GV, Ostrovskii YuI and Étinberg MI. Profile of a shock wave produced by collapse of a spherical bubble in a liquid. Sov Phys Tech Phys 1985; 30: 224

[347] Ivany RD. Tech Rep 15, University of Michigan, Dept of Nuclear Engineering, 1965
[348] Vyas B and Preece CM. Stress produced in a solid by cavitation. J Appl Phys 1976; 47: 5133–5138
[349] Mørch KA. Euromech Colloq 98, Eindhoven, The Netherlands, 1977
[350] Brunton JH. Proc 2nd Int Conf Rain Eros (Fyall AA and King RB, eds). Royal Aircraft Establishment, Farnborough, United Kingdom, 1967, p 291
[351] Hansson I and Mørch KA. The dynamics of cavity clusters in ultrasonic (vibratory) cavitation erosion. J Appl Phys 1980; 51: 4651–4658
[352] Hansson I, Kendrinskii V and Mørch KA. On the dynamics of cavity clusters. J Phys D 1982; 15: 1726–1734
[353] Ellis AT. Proc 6th Symp Naval Hydrodyn, Washington DC, 1966, pp 137–161
[354] Zhang YJ, Li SC and Hammitt FG. Statistical investigation of bubble collapse and cavitation erosion effect. Wear 1989; 133: 257–265
[355] Kornfeld M and Suvorov L. On the destructive action of cavitation. J Appl Phys 1944; 15: 495–506
[356] Naudé CF and Ellis AT. On the mechanism of cavitation damage by nonhemispherical cavities collapsing in contact with a solid boundary. Trans ASME D: J Basic Eng 1961; 83: 648–656
[357] Walters JK and Davidson JF. The initial motion of a gas bubble formed in an inviscid liquid. Part 1. The two-dimensional bubble. J Fluid Mech 1962; 12: 408–416
[358] Walters JK and Davidson JF. The initial motion of a gas bubble formed in an inviscid liquid. Part 2. The three-dimensional bubble and the toroidal bubble. J Fluid Mech 1963; 17: 321–336
[359] Benjamin TB and Ellis AT. The collapse of cavitation bubbles and the pressures thereby produced against solid boundaries. Phil Trans Roy Soc 1966; A260: 221–240
[360] Lauterborn W and Bolle H. Experimental investigations of cavitation-bubble collapse in the neighbourhood of a solid boundary. J Fluid Mech 1975; 72: 391–399
[361] Plesset MS and Chapman RB. Collapse of an initially spherical vapour cavity in the neighbourhood of a solid boundary. J Fluid Mech 1971; 47: 283–290
[362] Prosperetti A. Bubble phenomena in sound fields: part 2. Ultrasonics 1984; 22: 115–124
[363] Neppiras EA. Acoustic cavitation. Phys Rep 1980; 61: 159–251
[364] Olson HG and Hammitt FG. High-speed photographic studies of ultrasonically induced cavitation. J Acoust Soc Am 1969; 46: 1272–1283
[365] Tomita Y and Shima A. Mechanisms of impulsive pressure generation and damage pit formation by bubble collapse. J Fluid Mech 1986; 169: 535–564
[366] Vogel A, Lauterborn W and Timm R. Optical and acoustic investigations of the dynamics of laser-produced cavitation bubbles near a solid boundary. J Fluid Mech 1989; 206: 299–338
[367] Kucera A and Blake JR. Computational modelling of cavitation bubbles near boundaries. In: Computational Techniques and Applications CTAC-87 (Noyce J and Fletcher C, eds). North-Holland, Amsterdam, 1988, pp 391–400
[368] Blake JR, Taib BB and Doherty G. Transient cavities near boundaries. Part 1. Rigid boundary. J Fluid Mech 1986; 170: 479–497
[369] Radek U. Kavitationserzeugte Druckpulse und Materialzerstörung. Acustica 1972; 26: 270–283 (in German)
[370] Hinsch K and Brinkmeyer E. Investigation of very short cavitation shock waves by coherent optical methods. High Speed Photography, SPIE Vol. 97, Toronto, 1976, pp 166–171
[371] Ebeling KJ. Zum Verhalten kugelförmiger, lasererzeugter Kavitationsblasen in Wasser. Acustica 1978; 40: 229–239 (in German)
[372] Jones IR and Edwards DH. An experimental study on the forces generated by the collapse of transient cavities in water. J Fluid Mech 1960; 7: 596–609
[373] Dear JP and Field JE. A study of the collapse of arrays of cavities. J Fluid Mech 1988; 190: 409–425
[374] Lauterborn W. Proc 5th Int Conf Erosion by Liquid and Solid Impact (Field JE, ed.). Paper 58, Cavendish Laboratory, Cambridge, 1979
[375] Tulin MP. On the creation of ultra-jets. In: L. I. Sedov 60th Anniversary Volume: Problems of hydrodynamics and continuum mechanics. Soc Ind Appl Maths Philadelphia, pp 725–747
[376] Preece CM (ed.). Treatise on materials science and technology. In: Erosion, Vol. 16, Academic Press, New York, 1979

[377] Preece CM and Hansson I. Advan Mech Phys Surfaces, Harwood Academic Publishing 1981; 1: 199–254

[378] Okada T and Iwai Y. A study of cavitation bubble collapse pressures and erosion. Part 1: A method for measurement of collapse pressures. Wear 1989; 133: 219–232

[379] Iwai Y and Okada T. A study of cavitation bubble collapse pressures and erosion. Part 2: Estimation of erosion from the distribution of bubble collapse pressures. Wear 1989; 169: 535–564

[380] Lush PA, Sanada N and Takayama K. Proc 7th Int Conf Erosion by Liquid and Solid Impact (Field JE and Dear JP, eds). Paper 24, Cavendish Laboratory, Cambridge, 1987

[381] Hansson I, Mørch KA and Preece CM. Ultrasonics. IPC Science and Technology Press, Guildford, 1978, p 267

[382] Field JE. The physics of liquid impact, shock wave interactions with cavities, and the implications to shock wave lithotripsy. Phys Med Biol 1991; 36: 1475–1484

[383] de Haller P. Untersuchungen über die durch Kavitation hervorgerufenen Korrosionen. Schweizerische Bauzeitung 1933; 101: 243–260 (in German)

[384] Heymann FJ. High speed impact between a liquid drop and a solid surface. J Appl Phys 1969; 40: 5113–5122

[385] Bowden FP and Field JE. The brittle fracture of solids by liquid impact, by solid impact and by shock. Proc Roy Soc 1964; A282: 331–352

[386] Lesser MB. Analytical solutions of liquid drop impact problems. Proc Roy Soc 1981; A337: 289–308

[387] Lesser MB and Field JE. The impact of compressible liquids. Ann Rev Fluid Mech 1983; 15: 97–122

[388] Field JE, Lesser MB and Dear JP. Studies of two-dimensional liquid-wedge impact and their relevance to liquid-drop impact problems. Proc Roy Soc 1985; A401; 225–249

[389] Dear JP and Field JE. High speed photography of surface geometry effects in liquid/solid impact. J Appl Phys 1988; 63: 1015–1021

[390] Shima A, Tomita Y and Takahashi K. The collapse of a gas bubble near a solid wall by a shock wave and the induced impulsive pressure. Proc Instn Mech Engrs 1984; 198C: 81–86

[391] Felix MP and Ellis AT. Laser-induced liquid breakdown – a step-by-step account. Appl Phys Lett 1971; 19: 484–486

[392] Giovanneschi P and Dufresne D. Experimental study of laser-induced cavitation bubbles. J Appl Phys 1985; 58: 651–652

[393] Chahine GL. Experimental and asymptotic study of non-spherical bubble collapse. Appl Sci Res 1982; 38: 187

[394] Chahine GL and Bovis AG. Pressure field generated by nonspherical bubble collapse. Trans ASME I: J Fluids Eng 1983; 105: 356–364

[395] Kuvshinov GI, Prokhorenko PP and Dezhkunov NV. Collapse of a cavitation bubble between two solid walls. Int J Heat Mass Transfer 1982; 25: 381–387

[396] Ueki H, Kimoto H and Momose K. Behaviour of spark-induced bubble between parallel walls. Bull JSME 1984; 27: 1358–1365

[397] Shima A and Sato Y. The collapse of a bubble between narrow parallel plates. Rep Inst High Speed Mech Tohoku Univ Ser B (Japan) 1984; 49: 1–21

[398] Kimoto H, Momose K and Ueki H. A study of a cavitation bubble on a solid boundary. Bull JSME 1985; 28: 601–609

[399] Brunton JH. Proc 3rd Int Conf Rain Eros (Fyall AA, ed.). Royal Aircraft Establishment, Farnborough, 1970, p 821

[400] Field JE, Lesser MB and Davies PNH. Theoretical and experimental studies of two-dimensional liquid impact. Proc 5th Int Conf Erosion by Liquid and Solid Impact (Field JE, ed.). Paper 2, Cavendish Laboratory, Cambridge, 1979

[401] Dear JP, Field JE and Walton AJ. Gas compression and jet formation in cavities collapsed by a shock wave. Nature 1988; 332(6164): 505–508

[402] Field JE, Dear JP, Davies PNH and Finnström M. An investigation of shock structures and the conditions for jetting during liquid impact. Proc 6th Int Conf Erosion by Liquid and Solid Impact (Field JE and Corney NS, eds). Paper 19, Cavendish Laboratory, Cambridge, UK, 1983

[403] Dear JP, Field JE and Swallowe GM. High speed photographic studies of liquid/solid impact and cavity collapse using two-dimensional gelatine configurations. Proc 16th Int Conf High-speed Photography and Photonics, Strasbourg, France, 1984 (SPIE Vol 491).
[404] Dear JP. PhD thesis, University of Cambridge, 1985
[405] Lesser M and Finnström M. Proc 7th Int Conf Erosion by Liquid and Solid Impact (Field JE and Dear JP, eds). Paper 23, Cavendish Laboratory, Cambridge, UK, 1987
[406] Tomita Y, Shima A and Takashi O. Collapse of multiple gas bubbles by a shock wave and induced impulsive pressure. J Appl Phys 1984; 56: 125
[407] Coley GD and Field JE. The role of cavities in the initiation and growth of explosion in liquids. Proc Roy Soc 1973; A335: 67–86
[408] Brunton JH and Camus JJ. The application of high speed photography to the analysis of flow in cavitation and drop impact studies. Proc 9th Int Congr High-speed Photography, Denver, Colorado, 2–7 August, 1970
[409] Bowden FP and Yoffe AD. Initiation and Growth of Explosion in Liquids and Solids, 1952. Reprinted Cambridge University Press, London, 1985
[410] Bowden FP and Yoffe AD. Fast Reaction in Solids. Butterworths, London, 1958
[411] Winter RE and Field JE. The role of localised plastic flow in the impact initiation of explosives. Proc Roy Soc 1975; A343: 399–413
[412] Heavens SN and Field JE. The ignition of a thin layer of explosive by impact. Proc Roy Soc 1974; A338: 77–93
[413] Field JE, Swallowe GM and Heavens SN. Ignition mechanisms of explosives during mechanical deformation. Proc Roy Soc 1982; A383: 231–244
[414] Chaudhri MM and Field JE. The role of rapidly compressed gas pockets in the initiation of condensed explosives. Proc Roy Soc 1974; A340: 113–128
[415] Chaudhri MM, Almgren L-A and Persson A. High-speed photography of the interaction of shocks with voids in condensed media. In: 15th Int Congr High-speed Photography and Photonics, San Diego 1982. SPIE Vol 348, pp 388–394
[416] Chaudhri MM. The initiation of fast decomposition in solid explosives by fracture, plastic flow, friction and collapsing voids. In: Proc 9th Symp (Int) Detonation, Portland, Oregon, 1989, pp 857–868
[417] Starkenberg J. Ignition of solid high explosive by the rapid compression of an adjacent gas layer. In: Proc 7th Symp (Int) Detonation, 1981, pp 3–16
[418] Mader CL, Taylor RW, Venable D and Travis JR. Theoretical and experimental two-dimensional interactions of shocks with density discontinuities. Los Alamos National Laboratory Rep La-3614, 1967
[419] Mader CL and Kershner JD. The three-dimensional hydrodynamic hot-spot model. In: Proc 8th Symp (Int) Detonation, Albuquerque, 1985, pp 42–52
[420] Mader CL and Kershner JD. The heterogeneous explosive reaction zone. In: Proc 9th Symp (Int) Detonation, Portland, Oregon, 1989, pp 693–700
[421] Frey RB. Cavity collapse in energetic materials. In: Proc 8th Symp (Int) Detonation, Albuquerque, 1985, pp 68–80
[422] Field JE, Palmer SJP, Pope PH, Sundararajan R and Swallowe GM. Mechanical properties of PBXs and their deformation during drop weight impact. In: Proc 8th Symp (Int) Detonation, Albuquerque, 1985, pp 635–644
[423] Leggat LJ and Sponagle NC. The study of propeller cavitation noise using cross-correlation methods. J Fluids Eng 1985; 107: 127–133
[424] Bark G. Prediction of propeller cavitation noise from model tests and its comparison with full scale data. Trans ASME 1985; 107: 112–120
[425] Arai C. An acoustic detection method of cloud cavitation. Trans ASME J Fluids Eng 1984; 106: 466–476
[426] Chahine GL. Cloud cavitation: theory. 14th Symposium on Naval Hydrodynamics, 23–27 August, 1982, pp 1–21
[427] Arakeri VH and Shanmuganathan V. On the evidence for the effect of bubble interference on cavitation noise. J Fluid Mech 1985; 159: 131–150
[428] Haenscheid P and Rouvé G. Hydrodynamic studies of cavitating flow interfacing with a mathematical model of bubble growth. 2nd Intl Conf Computational Methods and Experimental Measurements, pp 67–78

[429] Preece CM. Erosion (Preece CM, ed.). Academic Press, New York, 1979, pp 249–308
[430] Rao P, Buckley DH and Matsumura M. A unified relation for cavitation erosion. Int J Mech Sci 1984; 26: 325–335
[431] Rao P and Buckley DH. Cavitation erosion size scale effects. Wear 1984; 96: 239–253
[432] Namgoong E and Chun JS. The effect of ultrasonic vibration on hard chromium plating in a modified self-regulating high speed bath. Thin Solid Films 1984; 120: 153–159
[433] Brown FH, Lubow RM and Cooley RL. A review of applied ultrasonics in periodontal therapy. Journal of the Western Society of Periodontology Periodontal Abstracts 1987; 35(2): 53
[434] Walmsley AD. Ultrasound and root canal treatment: the need for scientific evaluation. Int Endodont J 1987; 20: 105–111
[435] Laird WRE and Walmsley AD. Ultrasound in dentistry. Part 1 – Biophysical interactions. J Dent 1991; 19: 14–17
[436] Walmsley AD, Laird WRE and Lumley PJ. Ultrasound in dentistry. Part 2 – Periodontology and endodontics. J Dent 1992; 20: 11–17
[437] Richman MJ. The use of ultrasonics in root canal therapy and root resection. J Dent Med 1957; 12: 12–18
[438] Martin H. Ultrasonic disinfection of the root canal. Oral Surg 1976; 42: 92–99
[439] Nehammer CF and Stock CJR. Preparation and filling of the root canal. Brit Dent J 1985; 158: 285–291
[440] Stamos DG, Haash GC, Chenail B and Gerstein H. Endosonics: clinical impressions. J Endodont 1985; 11: 181–187
[441] Lumley PJ, Walmsley AD and Laird WRE. Ultrasonic instruments in dentistry: 2. Endosonics. Dent Update 1988; 15: 362–369
[442] Walmsley AD and Williams AR. The effect of constraint on the oscillatory pattern of endosonic files. J Endodont (in press)
[443] Suppipat N. Ultrasonics in periodontics. J Clin Periodont 1974; 1: 206–213
[444] Cunningham WT and Martin H. A scanning electron microscope evaluation of root canal debridement with the endosonic ultrasonic synergistic system. Oral Surg 1982; 53: 527–531
[445] Ahmad M, Pitt Ford TR and Crum LA. Ultrasonic debridement of root canals: an insight into the mechanisms involved. J Endodont 1987; 13: 93–101
[446] Ahmad M, Pitt Ford TR and Crum LA. Ultrasonic debridgement of root canals: acoustic streaming and its possible role. J Endodont 1987; 13: 490–499
[447] Lumley PJ, Walmsley AD and Laird WRE. An investigation into the occurrence of cavitational activity during endosonic instrumentation. J Dent 1988; 16: 120–122
[448] Walmsley AD, Laird WRE and Williams AR. Displacement amplitude as a measure of the acoustic output of ultrasonic scalers. Dental Materials 1986; 2: 97–100
[449] Walmsley AD, Laird WRE and Williams AR. A model system to demonstrate the role of cavitational activity in ultrasonic scaling. J Dent Res 1984; 63: 1162–1165
[450] Walmsley AD, Laird WRE and Williams AR. Dental plaque removal by cavitational activity during ultrasonic scaling. J Clin Periodont 1988; 15: 539–543
[451] Byrne PJ. Acoustic cavitation associated with ultrasonic scalers. MSc thesis, Institute of Dental Surgery, University of London, 1990
[452] Walmsley AD, Walsh TF, Laird WRE *et al.* Effects of cavitational activity on the root surface of the teeth during ultrasonic scaling. J Clin Periodont 1990; 17: 306–312
[453] Williams AR. Ultrasound: Biological Effects and Potential Hazards. Academic Press, New York, 1983
[454] Nyborg WL and Ziskin MC (eds). Biological Effects of Ultrasound. Churchill Livingstone, New York, 1985
[455] Repacholi MH, Grandolfo M and Rindi A (eds). Ultrasound: Medical Applications, Biological Effects, and Hazard Potential. Plenum, New York, 1986 (papers from the 6th International School of Radiation Damage and Protection)
[456] Hill CR. The possibility of hazard in medical and industrial applications of ultrasound. Brit J Radiol 1968; 41: 561–569
[457] Coakley WT. Biophysical effects of ultrasound at therapeutic intensities. Physiotherapy 1978; 64: 6
[458] Stewart HD, Stewart HF, Moore RM Jr and Garry J. Compilation of reported biological effects data and ultrasound exposure levels. J Clin Ultrasound 1985; 13: 167–186

[459] Sikov MR. Effects of ultrasound on development. Part I. Introduction and studies in inframammalian species. J Ultrasound Med 1986; 5: 577–583

[460] Sikov MR. Effects of ultrasound on development. Part II. Studies in mammalian species and overview. J Ultrasound Med 1986; 5: 651–661

[461] Carstensen EL. Acoustic cavitation and the safety of diagnostic ultrasound. Ultrasound Med Biol 1987; 13: 597–606

[462] Miller DL. Update on safety of diagnostic ultrasonography. J Clin Ultrasound 1991; 19: 531–540

[463] National Council on Radiation Protection (NCRP). Biological Effects of Ultrasound: Mechanisms and Implications. NCRP Report 74, 1983, pp 94–115

[464] American Institute of Ultrasound in Medicine (AIUM), Bioeffects Committee. Bioeffects considerations for the safety of diagnostic ultrasound. J Ultrasound Med 1988; 7(Suppl 9): S1–S38

[465] WFUMB First Symposium on Safety and Standardisation of Ultrasound in Obstetrics (Kossoff G and Barnett SB, eds). Ultrasound Med Biol 1986; 12: 673–721

[466] WFUMB Second Symposium on Safety and Standardisation of Ultrasound. (Kossoff G and Nyborg WL eds). Ultrasound Med Biol 1989; 15(Suppl 1)

[467] WFUMB Working Group Geneva Report on Safety and Standardisation in Medical Ultrasound; Issues and Recommendations Regarding Thermal Mechanisms for Biological Effects of Ultrasound (1991 draft, replaced by reference [468])

[468] WFUMB Symposium on Safety and Standardisation in Medical Ultrasound: Issues and Recommendations Regarding Thermal Mechanisms for Biological Effects of Ultrasound. (Barnett SB and Kossoff G, eds). Ultrasound Med Biol 1992; Suppl (in press)

[469] White E and White D. Bibliography of biomedical ultrasound .72. Ultrasound Med Biol 1988; 14(2): pp 147–162

[470] White E and White D. Bibliography of biomedical ultrasound .73. Ultrasound Med Biol 1988; 14(3): pp 229–244

[471] White E and White D. Bibliography of biomedical ultrasound .74. Ultrasound Med Biol 1988; 14(4): pp 311–330

[472] White E and White D. Bibliography of biomedical ultrasound .75. Ultrasound Med Biol 1988; 14(5): pp 431–450

[473] White E and White D. Bibliography of biomedical ultrasound .78. Ultrasound Med Biol 1988; 14(8): pp 747–762

[474] White D and White E. Bibliography of biomedical ultrasound 1985. Ultrasound Med Biol 1988; 14(9): pp 769–947

[475] White E and White D. Bibliography of biomedical ultrasound. Ultrasound Med Biol 1988; 15(1): pp 77–82

[476] White E and White D. Bibliography of biomedical ultrasound .80. Ultrasound Med Biol 1989; 15(2): pp 149–153

[477] White E and White D. Bibliography of biomedical ultrasound .82. Ultrasound Med Biol 1989; 15(4): pp 403–409

[478] White E and White D. Bibliography of biomedical ultrasound .83. Ultrasound Med Biol 1989; 15(7): pp 695–697

[479] White E and White D. Bibliography of biomedical ultrasound .84. Ultrasound Med Biol 1989; 15(8): pp 775–788

[480] White D and White E. Bibliography of biomedical ultrasound 1986. Ultrasound Med Biol 1989; 15(9): pp 795–958

[481] White D and White E. Bibliography of biomedical ultrasound .85. Ultrasound Med Biol 1990; 16(1): pp 99–114

[482] White E and White D. Bibliography of biomedical ultrasound .86. Ultrasound Med Biol 1990; 16(2): pp 199–207

[483] White E and White D. Bibliography of biomedical ultrasound .87. Ultrasound Med Biol 1990; 16(3): pp 321–326

[484] White E and White D. Bibliography of biomedical ultrasound .89. Ultrasound Med Biol 1990; 16(5): pp 527–535

[485] White E and White D. Bibliography of biomedical ultrasound .90. Ultrasound Med Biol 1990; 16(6): pp 629–635

[486] White D and White E. Bibliography of biomedical ultrasound 1987. Ultrasound Med Biol 1990; 16(9): pp 851–1018
[487] Stewart HF, Harris G, Robinson R and Garry J. Survey of use and performance of ultrasonic therapy units in the Washington D.C. area. USDHEW Publication (FDA) 73-8029, Health Physics in the Healing Arts, 1973, pp 467–472
[488] Lehmann JF and Guy AW. Ultrasound Therapy. Interaction of Ultrasound and Biological Tissues/Workshop Proceedings. USDHEW Publication (FDA) 73-8008, 1972, pp 141–152
[489] ter Haar G, Dyson M and Oakley EM. The use of ultrasound by physiotherapists in Britain, 1985. Ultrasound Med Biol 1987; 13(10): 659–663
[490] Chalmers I. Hazards of ultrasound. Brit Med J 1984; 289: 184–185
[491] Davies P. Hazards of ultrasound. Brit Med J 1984; 288: 2001–2002
[492] Stewart A, Webb J, Giles D and Hewitt D. Malignant disease in childhood and diagnostic irradiation *in utero*. Lancet 1956; ii: 447
[493] Bithell JF and Stewart AM. Prenatal irradiation and childhood malignancy: a review of British data from the Oxford Survey. Brit J Cancer 1975; 31: 271–287
[494] Smith OW, Smith GV and Hurwitz D. Increased excretion of pregnanediol in pregnancy from diethylstilboestrol with special reference to prevention of late pregnancy accidents. Am J Obstet Gynecol 1946; 51: 411–415
[495] Herbst AL and Scully RE. Adenocarcinoma of the vagina in adolescence. Cancer 1970; 25: 745–757
[496] Ziskin MC. The prudent use of diagnostic ultrasound. J Ultrasound Med 1987; 6: 415
[497] Nyborg WL. Physical Mechanisms for Biological Effects of Ultrasound. HEW Publication (FDA) 78-8062, US Department of Health, Education and Welfare, US Government Printing Office, Washington DC, 1977
[498] American Institute of Ultrasound in Medicine. Safety Considerations for Diagnostic Ultrasound. Bethesda, Maryland, 1984
[499] American Institute of Ultrasound in Medicine. Acoustical Data for Diagnostic Ultrasound Equipment. Bethesda, Maryland, 1985
[500] Zagzebski JA. Acoustic ouput of ultrasound equipment – summary of data reported to the AIUM. Ultrasound Med Biol 1989; 15(Suppl 1): 55
[501] Duck FA. Output data from European studies. Ultrasound Med Biol 1989; 15(Suppl 1): 61
[502] Crum LA, Roy RA, Dinno MA and Church CC. Acoustic cavitation by microsecond pulses of ultrasound: a discussion of some selected results. J Acoust Soc Am 1992; 91: 1113–1119
[503] Williams AR. Ultrasonics 1990; 28: 131 (Editorial to Bioeffects special issue)
[504] Macintosh IJC and Davey DA. Chromosome-aberrations induced by an ultrasonic fetal pulse detector. Brit Med J 1970; 4: 92–93
[505] Macintosh IJC and Davey DA. Relationship between intensity of ultrasound and induction of chromosome aberrations. Brit J Radiol 1972; 45: 320–327
[506] Coakley WT, Hughes DE, Slade JS and Laurence KM. Chromosome abberations after exposure to ultrasound. Brit Med J 1971; 1: 109–110
[507] Coakley WT, Slade JS, Braemen JM and Moore JL. Examination of lymphocytes for chromosome damage. Brit J Radiol 1972; 45: 328–332
[508] Thacker J. The possibility of genetic hazard from ultrasonic radiation. Current Topics Rad Res Quarterly 1973; 8: 235–258
[509] Thacker J. An assessment of ultrasonic radiation hazard using yeast genetic systems. Brit J Radiol 1974; 47: 130–138
[510] Macintosh IJC, Brown RC and Coakley WT. Brit J Radiol 1975; 48: 230–232
[511] Child SZ, Carstensen EL and Davis H. A test for the effects of low-temporal-average-intensity pulsed ultrasound on the rat fetus. Exp Cell Biol 1984; 52: 207–210
[512] Child SZ, Carstensen EL, Gates AH and Hall WJ. Testing for the teratogenicity of pulsed ultrasound in mice. Ultrasound Med Biol 1988; 14: 493–498
[513] Miller MW, Azadniv M, Pettit SE, Church CC, Carstensen EL and Hoffman D. Sister chromatid exchanges in Chinese hamster ovary cells exposed to high intensity pulsed ultrasound: inability to confirm previous positive results. Ultrasound Med Biol 1989; 15: 255–262
[514] ter Haar GR. Therapeutic and surgical applications. In: Physical Principles of Medical Ultrasonics (Hill CR, ed.). Ellis Horwood, Chichester (for Wiley, New York), 1986, Part III: Biophysical implications and applications

[515] Vivino AA, Boraker DK, Miller D and Nyborg W. Stable cavitation at low ultrasonic intensities induces cell death and inhibits ^{3}H-TdR incorporation by con-a-stimulated murine lymphocytes *in vitro*. Ultrasound Med Biol 1985; 11: 751–759

[516] Miller DL, Nyborg WL and Whitcomb CC. Platelet aggregation induced by ultrasound under specialized condition *in vitro*. Science 1979; 205: 505

[517] Ward CA, Johnson WR, Venter RD, Ho S, Forest TW and Fraser WD. Heterogeneous bubble nucleation and conditions for growth in a liquid–gas system of constant mass and volume. J Appl Phys 1983; 54: 1833–1843

[518] Venter RD, Ward CA, Ho S, Johnson WR, Fraser WD and Landolt JP. Fracture studies on mammalian semicircular canal. Undersea Biomed Res 1983; 10: 225–240

[519] Aymé EJ and Carstensen EL. Cavitation induced by asymmetric distorted pulses of ultrasound: theoretical predictions. IEEE Trans Ultrasonics, Ferroelectrics Freq Control 1989; 36: 32–40

[520] Aymé EJ, Carstensen EL, Parker KJ and Flynn HG. Microbubble response to finite amplitude waveforms. Proc IEEE Symp Ultrasound, Williamsburg, Virginia, 1986

[521] Dinno MA, Dyson M, Young SR, Mortimer AJ, Hail J and Crum LA. The significance of membrane changes in the safe and effective use of therapeutic and diagnostic ultrasound. J Phys Biol Med 1989; 34: 1543–1552

[522] ter Haar GR. Therapeutic and surgical applications. In: Physical Principles of Medical Ultrasonics (Hill CR, ed.). Ellis Horwood, Chichester (for Wiley, New York), 1986, Chapter 13

[523] Fry WJ, Mosberg WH, Barnard JW and Fry FJ. Production of focal destructive lesions in the central nervous system with ultrasound. J Neurosurg 1954; 11: 471–478.

[524] Fry WJ and Dunn F. Ultrasonic irradiation of the central nervous system at high sound levels. J Acoust Soc Am 1956; 28: 129–131.

[525] Basauri L and Lele PP. Simple method for the production of trackless focal lesions with focused ultrasound – statistical evaluation of effects of irradiation on central nervous system of a cat. J Physiol 1962; 160: 513–534

[526] Warwick R and Pond JB. Trackless lesions produced by high intensity focused ultrasound (high-frequency mechanical waves). J Anat 1968; 102: 387–405

[527] Coleman DJ, Lizzi FL, Silverman RH, Dennis PH, Driller J, Rosada A and Iwamoto T. Therapeutic ultrasound. Ultrasound Med Biol 1986; 12: 633–638

[528] Lizzi FL, Coleman DJ, Driller J, Franzen LA and Jacobiec FA. Experimental, ultrasonically-induced lesions in retina, choroid and sclera. Invest Ophthalmol Vis Sci 1978; 17: 350–360

[529] Lizzi CA, Coleman DJ, Driller J, Ostromogilsky M, Chang S and Greenall P. Ultrasonic hyperthermia for ophthalmic therapy. IEEE Trans Sonics Ultrason 1984; SU-31: 473–481

[530] Polack PJ, Iwamoto T, Silverman RH, Driller J, Lizzi FL and Coleman DJ. Histologic effects of contact ultrasound for the treatment of glaucoma. Investigative Ophthalmology Visual Science 1991; 32(7): 2136–2142

[531] Lizzi FL, Driller J, Lunzer B, Kalisz A and Coleman DJ. Computer-model of ultrasonic hyperthermia and ablation for ocular tumors using B-mode data. Ultrasound Med Biol 1992; 18(1): 59–73

[532] Lizzi FL and Ostromogilsky M. Analytical modelling of ultrasonically induced tissue heating. Ultrasound Med Biol 1987; 13: 607–618

[533] Fry FJ, Kossoff G, Eggleton RC and Dunn F. Threshold ultrasonic dosages for structural changes in the mammalian brain. J Acoust Soc Am 1970; 48: 1413–1417.

[534] Chan SK and Frizzell LA. Ultrasonic thresholds for structural changes in mammalian liver. IEEE Trans Sonics Ultrasonics 1978; 25: 240

[535] Frizell LA, Linke CA, Carstensen EL and Fridd CW. Thresholds for focal ultrasonic lesions in rabbit kidney, liver and testicle. IEEE Trans Biomed Eng 1977; BME-24: 393–396

[536] Sjöberg A, Stahle J, Johnson S and Sahl R. The treatment of Meniere's disease by ultrasonic radiation. Acta Otolaryngol 1963; Suppl 178

[537] James JA, Dalton GA, Freundlich HF, Bullen MA, Wells PNT, Hughes JA and Chow JTY. Histological, thermal and biochemical effects of ultrasound on the labyrinth and temporal bone. Acta Otolaryngol 1964; 57: 306

[538] Arslan M. Ultrasonic destruction of the vestibular receptors in Meniere's disease. Laryngoscope 1964; 74: 1962

[539] Kossoff G, Wadsworth JR and Dudley PF. The round window ultrasonic technique for the treatment of Meniere's disease. Arch Otolaryngol 1967; 86: 534

[540] Barnett SB and Kossoff G. Round window ultrasonic treatment of Meniere's disease. Arch Otolaryngol 1977; 103:124–127
[541] Barnett SB. The effect of ultrasonic irradiation on the structural integrity of the inner ear labyrinth. Acta Otolaryngol 1980; 89: 424–432
[542] Barnett SB. The influence of ultrasound and temperature on the cochlear microphonic response following a round window irradiation. Acta Otolaryngol 1980; 90: 32–39
[543] Burov AK and Andreevskaya GD. Effect of high intensity supersonic oscillations on malignant tumours in animals and in man. Dokl Akad Nauk 1956; 106: 445–448
[544] Wagai T and Kaketa K. Annual Report (1970) of Med Ultrasonics Research Center (Juntendo University School of Medicine, Hongo, Japan), 1971, pp 35–39
[545] Nightingale A. Physics and Electronics in Physical Medicine. Bell, London, 1959, p 275
[546] Herrick JF. Temperatures produced in tissues by ultrasound. J Acoust Soc Am 1953; 25: 12
[547] Fry FJ and Johnson LK. Tumour irradiation with intense ultrasound. Ultrasound Med Biol 1978; 4: 337–341.
[548] ter Haar G, Rivens I, Chen L and Riddler S. High intensity focused ultrasound for the treatment of rat tumours. Phys Med Biol 1991; 36(11): 1495–1501
[549] Goss SA and Fry FJ. The effect of high intensity ultrasonic irradiation on tumour growth. IEEE Trans Sonics and Ultrason 1984; SU-31: 491–496
[550] Gautherie M, Frederiksen F, Frey P, Hand JW, Hill CR, Hynynen K, Lagendijk J, Marchal C and Watmough D. Ultrasound Hyperthermia – Summary Proceedings of a Comac-Bme Workshop – 12–14 September 1991, Strasbourg, France. Ultrasonics 1992; 30(2): 113
[551] Fan XB and Hynynen K. The effect of wave reflection and refraction at soft-tissue interfaces during ultrasound hyperthermia treatments. J Acoust Soc Am 1992; 91(3): 1727–1736
[552] Lin WL, Roemer RB, Moros EG and Hynynen K. Optimization of temperature distributions in scanned, focused ultrasound hyperthermia. Int J Hyperthermia 1992; 8(1) 61–78
[553] Dorr LN and Hynynen K. The effects of tissue heterogeneities and large blood-vessels on the thermal exposure induced by short high-power ultrasound pulses. Int J Hyperthermia 1992; 8(1): 45–59
[554] Diederich CJ and Hynynen K. The feasibility of using electrically focused ultrasound arrays to induce deep hyperthermia via body cavities. IEEE Trans Ultrasonics Ferroelectrics Freq Control 1991; 38(3): 207–219
[555] Billard BE, Hynynen K and Roemer RB. Effects of physical parameters on high temperature ultrasound hyperthermia. Ultrasound Med Biol 1990; 16: 409–420
[556] Hynynen K. Demonstration of enhanced temperature elevation due to nonlinear propagation of focussed ultrasound in dog's thigh *in vivo*. Ultrasound Med Biol. 1987; 13: 85–91
[557] Hynynen K. The role of nonlinear ultrasound propagation during hyperthermia treatments. Med Phys 1991; 18(6): 1156–1163
[558] Frizzell LA, Lee CS, Aschenbach PD, Borrelli MJ, Morimoto RS and Dunn F. Involvement of ultrasonically induced cavitation in the production of hind limb paralysis of the mouse neonate. J Acoust Soc Am 1983; 74: 1062–1065
[559] Lee CS and Frizzell LA. Exposure levels for ultrasonic cavitation in the mouse neonate. Ultrasound Med Biol 1988; 14: 735–742
[560] Hynynen K. The threshold for thermally significant cavitation in dog thigh muscle *in vivo*. Ultrasound Med Biol 1991; 17(2): 157–169
[561] Dyson M. Mechanisms involved in therapeutic ultrasound. Physiotherapy 1987; 73: 116–120
[562] ter Haar G, Dyson M and Oakley S. Ultrasound in physiotherapy in the United Kingdom; results of a questionnaire. Physiotherapy Practice 1988; 4: 69–72
[563] Dyson M. The use of ultrasound in sports physiotherapy. In: Sports Injuries (Grisogono V, ed.). Churchill-Livingstone, Edinburgh, 1989, 213–232
[564] Young SR and Dyson M. Effect of ultrasound on the healing of full-thickness excised skin lesions. Ultrasonics 1990; 28: 175–180
[565] Hunt TK and Dunphy JE. Fundamentals of Wound Management. Appleton Century Crofts, New York, 1979, pp 142–144
[566] Partridge CJ. Evaluation of the efficacy of ultrasound. Physiotherapy 1981; 73: 22–24
[567] Enwemeka CS, Rodriguez O and Mendosa S. The biomechanical effects of low-intensity ultrasound on healing tendons. Ultrasound Med Biol 1990; 16: 801–817

[568] Dyson M and Smalley DS. Effects of ultrasound on wound contraction. In: Ultrasound Interactions in Biology and Medicine (Millner F, Rosenfeld E and Cobet U, eds). Plenum, New York, 1983; pp 151–158
[569] Young SR and Dyson M. The effect of therapeutic ultrasound on angiogenesis. Ultrasound Med Biol 1990; 16: 261–269
[570] Young SR and Dyson M. Macrophage responsiveness to therapeutic ultrasound. Ultrasound Med Biol 1990; 16: 809–816
[571] Lehmann JF and Herrick JF. Biological reactions to cavitation, a consideration for ultrasonic therapy. Arch Phys Med Rehabil 1953; 34: 86–98
[572] Howkins SD and Weinstock A. The effect of focused ultrasound on human blood. Ultrasonics 1970; 8: 174–176
[573] Williams AR, Chater BV, Allen KA and Sanderson JH. The use of beta-thromboglobulin to detect platelet damage by therapeutic ultrasound *in vivo*. J Clin Ultrasound 1981; 9: 145–151
[574] Chater BV and Williams AR. Absence of platelet damage *in vivo* following the exposure of non-turbulent blood to therapeutic ultrasound. Ultrasound Med Biol 1982; 8: 85–87
[575] Sarvazyan AP, Beloussov LV, Petropavlovskaya MN and Ostroumova TV. The action of low-intensity pulsed ultrasound on amphibian embryonic tissues. Ultrasound Med Biol 1982; 8: 639–654
[576] Garrison BM, Bo WJ, Kreuger WA, Kremkau FW and McKinney WM. J Clin Ultrasound 1973; 1: 316–319
[577] Barnett SB. The influence of ultrasound on embryonic development. Ultrasound Med Biol 1983; 9: 19–24
[578] O'Brien WD. Dose-dependent effect of ultrasound on fetal weight in mice. J Ultrasound Med 1983; 2: 1–8
[579] Muranaka A, Tachibana M and Suzuki F. Effects of ultrasound on embryonic development and fetal growth in mice. Teratology 1974; 10: 91–96
[580] O'Brien WD Jr. Ultrasonically induced fetal weight reduction in mice. In: Ultrasound in Medicine, Vol. 2 (White DN and Barnes R, eds). Plenum, New York, 1976, pp 531–532
[581] Stratmeyer ME, Simmons LR, Pinkavitch FZ, Jessup GL and O'Brien WD. Growth and development of mice exposed *in utero* to ultrasound. Symp Biological Effects and Characterizations of Ultrasound Sources, 1977. US Department of Health, Education and Welfare Publication 78-8048
[582] Stolzenberg SJ, Torbit CA, Edmonds PD and Taenzer JC. Effects of ultrasound on the mouse exposed at different stages of gestation: acute study. Radiat Environ Biophys 1980; 17: 245–270
[583] Stolzenberg SJ, Torbit CA, Pryor GT and Edmonds PD. Toxicity of ultrasound in mice: neonatal studies. Radiat Environ Biophys 1980; 18: 37–44
[584] Williams AR, Edmonds P and Barnett S. Private communication from Williams
[585] Sikov MR, Hildebrand BP and Stearns JD. Postnatal sequelae of ultrasound exposure at 15 days of gestation in the rat. In: Ultrasound in Medicine, Vol. 3B (White D and Brown R, eds). Plenum, New York, 1977, pp 2017–2023
[586] Sikov MR, Collins DH and Carr DB. Measurement of temperature rise in prenatal rats during exposure of the exteriorized uterus to ultrasound. IEEE Trans Sonics Ultrason 1984; SU-31: 497–503
[587] Barnett SB and Williams AR. Identification of mechanisms responsible for fetal weight reduction in mice following ultrasound exposure. Ultrasonics 1990; 28: 159–165
[588] ter Haar G, Stratford IJ and Hill CR. Ultrasonic irradiation of mammalian cells *in vitro* at hyperthermic temperatures. Brit J Radiol 1980; 53: 784–789
[589] Dunn F. Cellular inactivation by heat and shear. Radiat Environ Biophys 1985; 24: 131–139
[590] Inoue M, Miller MW and Church CC. An alternative explanation for a postulated non-thermal, non-cavitational ultrasound mechanism of action on *in vitro* cells at hyperthermic temperature. Ultrasonics 1990; 28: 185–189
[591] Miller DL. The effects of ultrasonic activation of gas bodies in *Elodea* leaves during continuous and pulsed irradiation at 1 MHz. Ultrasound Med Biol 1977; 3: 221–240
[592] Miller DL. A cylindrical bubble model for the response of plant-tissue gas-bodies to ultrasound. J Acoust Soc Am 1979; 65: 1313–1321
[593] Miller DL. Cell death thresholds in *Elodea* for 0.45–10 MHz ultrasound compared to gas-body resonance theory. Ultrasound Med Biol 1979; 5: 351–357

[594] Miller DL. Microsteaming as a mechanism of cell death in *Elodea* leaves exposed to ultrasound. Ultrasound Med Biol 1985; 11: 285–292

[595] Williams AR and Miller DL. Photometric detection of ATP release from human erythrocytes exposed to ultrasonically activated gas-filled pores. Ultrasound Med Biol 1980; 6: 251–256

[596] Miller DL. The influence of hematocrit on hemolysis by ultrasonically activated gas-filled micropores. Ultrasound Med Biol 1988; 14: 293–297

[597] Miller DL and Thomas RM. The influence of variations in biophysical conditions on hemolysis near ultrasonically activated gas-filled micropores. J Acoust Soc Am 1990; 87(5): 2225–2230

[598] Englehardt S, Grunewald R and Wynsberghe D. The effects of continuous and pulsed ultrasound on fetal rats in hypothyroid and euthyroid maternal rats. Health Phys 1977; 33: 661

[599] Kelman BJ, Pappas RA and Sikov MR. Effects of ultrasound on placental function. IEEE Trans Ultrasonics Ferroelectrics Freq Control 1986; UFFC 33: 218–224

[600] Atkinson DE, Sibley CP and Williams AR. Effects of ultrasound on placental transfer during the last third of gestation in the rat. Ultrasonics 1990; 28: 171–174

[601] Mortimer AJ, Roy OZ, Trollope BJ, McEwen JI *et al.* A relationship betwen ultrasonic intensity and changes in myocardial mechanics. Can J Physiol Pharmacol 1980; 58: 67–73

[602] Revell WJ and Roberts MG. Ultrasound effects on miniature end plate potential discharge frequency are contingent upon acoustic environment. Ultrasonics 1990; 28: 149–154

[603] Stella M, Trevison L, Montaldi A, Zaccaria G, Rossi G, Bianchi V and Levis AG. Induction of sister-chromatid exchanges in human lymphocytes exposed to *in vitro* and *in vivo* therapeutic ultrasound. Mutation Res 1984; 138: 75–85

[604] Ciaravino V, Miller MW and Carstensen EL. Sister-chromatid exchanges in human lymphocytes exposed to *in vitro* therapeutic ultrasound. Mutation Res 1986; 172: 185–188

[605] Miller MW, Azadniv M, Cox C and Miller WM. Lack of induced increase in sister chromatid exchanges in human lymphocytes exposed to *in vivo* therapeutic ultrasound. Ultrasound Med Biol 1991; 17: 81–83

[606] Hashish I and Harvey W. Anti-inflammatory effects of ultrasound therapy – evidence for a major placebo effect. Brit J Radiology 1987; 60: 613

[607] Skauen DM and Zenter GM. Phonophoresis. Int J Pharmaceut 1984; 20: 235–245

[608] Williams AR, Rosenfeld EH and Williams KA. Gel-sectioning technique to evaluate phonophoresis *in vivo*. Ultrasonics 1990; 28: 132–136

[609] Williams AR. Phonophoresis: an *in vivo* evaluation using three topical anaesthetic preparations. Ultrasonics 1990; 28: 137–141

[610] Carstensen EL. Acoustic cavitation and the safety of diagnostic ultrasound. Ultrasound Med Biol 1987; 13: 597–606

[611] Anderson DW and Barrett JT. Ultrasound – new immunosuppressant. Clin Immunol Immunopathol 1979; 14: 18–29

[612] Child SZ, Hare JD, Carstensen EL, Vives B, Davis J, Adler A and Davis HT. Clin Immunol Immunopathol 1981; 18: 299–302

[613] Berthold F, Berthold R, Matter I *et al.* Effect of spleen exposure to ultrasound on cellular and antibody-mediated immune reactions in man. Immunobiology 1982; 162: 46–55

[614] Pinamonti S, Gallenga PE and Mazzeo V. Effect of pulsed ultrasound on human erythrocytes *in vitro*. Ultrasound Med Biol 1982; 8: 631–638

[615] Rosenfeld E, Romanowski U and Williams AR. Positive and negative effects of diagnostic intensities of ultrasound on erythrocyte blood group markers. Ultrasonics 1990; 28: 155–158

[616] Romanowski U and Rosenfeld E. The effect of diagnostic ultrasound on the detectability of the blood group antigen N. In: Proc UBIOMED VII Eisenach GDR (1986) (Hein HJ, Millner R and Pahl L, eds). Wiss Beitr 1987; 29: 143–146 (in German)

[617] Miller DL, Lamore BJ and Boraker DK. Lack of effect of pulsed ultrasound on ABO antigens of human erythrocytes *in vitro*. Ultrasound Med Biol 1986; 12: 209–216

[618] Bause GS, Niebyl JR and Sanders RC. Doppler ultrasound and maternal fragility. Obstet Gynecol 1983; 62: 7–10

[619] Bernstine RL. Safety studies with ultrasonic Doppler technique – a clinical follow-up of patients and tissue culture study. Obstet Gynecol 1969; 34: 707–709

[620] Bakketeig L, Eik-Nes SH, Jacobsen G, Ulstein MK, Brodtkorb CJ, Balstad P, Eriksen BC and Jorgensen NP. Randomised controlled trial of ultrasonographic screening in pregnancy. Lancet 1984; 2: 207–211

[621] Cartwright RA, McKinney PA, Hopton PA *et al.* Lancet 1984; 2: 999–1000
[622] Kinnier Wilson LM and Waterhouse JAH. Lancet 1984; 2: 997–998
[623] Stark CR, Orleans M, Haverkamp AD and Murphy J. Short- and long-term risks after exposure to diagnostic ultrasound *in utero*. Obstet Gynecol 1984; 63: 194–200
[624] Carstensen EL and Child SZ. Ultrasonic heating of the skull. J Acoust Soc Am 1990; 87: 1310–1317
[625] Tavakoli MB and Evans JA. Dependence of the velocity and attenuation of ultrasound in bone on the mineral content. Phys Med Biol 1991; 36(11): 1529–1537
[626] Wells PNT. The prudent use of diagnostic ultrasound. Ultrasound Med Biol 1987; 13(7): 391–400
[627] Parker KJ, Baggs RB, Lerner RM, Tuthill TA and Violante MR. Ultrasound contrast for hepatic tumors using IDE particles. Investig Radiol 1990; 25: 1135–1139
[628] Ophir J and Parker KJ. Contrast agents in diagnostic ultrasound. Ultrasound Med Biol 1989; 15(4): 319–333
[629] Gramiak R and Shah PM. Echocardiography of the aortic root. Investig Radiol 1968; 3: 356–366
[630] Child SZ, Carstensen EL and Lam SK. Effects of ultrasound on *Drosophila* – III. Exposure of larvae to low-temporal-average-intensity, pulsed irradiation. Ultrasound Med Biol 1981; 7: 167–173
[631] Berg RB, Child SZ and Carstensen EL. The influence of carrier frequency on the killing of drosophilia larvae by microsecond pulses of ultrasound. Ultrasound Med Biol 1983; 8: L448–L451
[632] Carstensen EL, Child SZ, Lam S, Miller DL and Nyborg WL. Ultrasonic gas-body activation in *Drosophilia*. Ultrasound Med Biol 1983; 9: 473–477
[633] Carstensen EL, Berg RB and Child SZ. Pulse average vs maximum intensity. Ultrasound Med Biol 1983; 9: L451–L455
[634] Child SZ and Carstensen EL. Effects of ultrasound on *Drosophila* – IV. Pulsed exposures of eggs. Ultrasound Med Biol 1982; 8: 311–312
[635] Carstensen EL, Campbell DS, Hoffman D, Child SZ and Aymé-Bellegarda, E.J. Killing of *drosophila* larvae by the fields of an electrohydraulic lithotripter. Ultrasound Med Biol 1990; 16(7): 687–698
[636] Carstensen EL and Gates AH. The effects of pulsed ultrasound on the foetus. J Ultrasound Med 1984; 3: 145–147
[637] Carstensen EL and Gates AH. Ultrasound and the foetus. In: Biological Effects of Ultrasound (Nyborg WL and Ziskin MC, eds). Churchill Livingstone, New York, 1985, pp 85–95
[638] Pfeiler M, Matura E, Iffländer H and Seyler G. Lithotripsy of renal and biliary calculi: physics, technology and medical–technical application. Electromedica 1989; 57: 52–63
[639] Chaussy C, Schmiedt E, Jocham D, Fuchs G, Brendel W, Forssmann B and Hepp W. Extracorporeal Shock Wave Lithotripsy. Karger, Basel, 1986
[640] Delius M, Jordan M, Liebich H-G and Brendel W. Biological effects of shock waves: effect of shock waves on the liver and gallbladder wall of dogs – administration rate dependence. Ultrasound Med Biol 1990; 16: 459–466
[641] Filipczyński L and Wójcik J. Estimation of transient temperature elevation in lithotripsy and in ultrasonography. Ultrasound Med Biol 1991; 17: 715–721
[642] Suhr D, Brümmer F and Hülser DF. Cavitation-generated free radicals during shock wave exposure: investigations with cell-free solutions and suspended cells. Ultrasound Med Biol 1991; 17: 761–768
[643] Crum LA. Cavitation microjets as a contributory mechanism for renal calculi disintegration in ESWL. J Urol 1989; 85: 1518–1522
[644] Delius M, Mueller W, Goetz A, Liebich H-G and Brendel W. Biological effects of shock waves: kidney haemorrhage in dogs at fast shock-wave administration rate of 15 Hz. J Litho Stone Dis 1990; 2: 103–210
[645] Dinno MA, Kennedy W, Crum LA and Church CC. The effect of extracorporeal shock wave lithotripsy on electrophysiological parameters across abdominal frog skin: real time observations. J Acoust Soc Am 1990; 88: S166
[646] Kuwahara M-A, Ioritani N, Kambe K *et al.* Hyperechoic region induced by focused shock waves *in vitro* and *in vivo*: possibility of acoustic cavitation bubbles. J Litho Stone Dis 1989; 1: 282

[647] Campbell DS, Flynn HG, Blackstock DT, Linke C and Carstensen EL. The acoustic fields of the Wolf Electrohydraulic Lithotripter. J Litho Stone Dis 1990

[648] Aymé EJ and Carstensen EL. Cavitation induced by asymmetric, distorted pulses of ultrasound: theoretical predictions. IEEE Trans Ultrasonics Ferroelectrics Freq Control 1989; 36: 32

[649] Aymé EJ and Carstensen EL. Cavitation induced by asymmetric, distorted pulses of ultrasound: a biological test. Ultrasound Med Biol 1989; 15(1): 61–66

[650] Mayer R, Schenk E, Child S, Norton S, Cox CA, Hartman C, Cox CF and Carstensen E. Pressure threshold for shockwave induced renal hemorrhage. J Urol 1990; 144: 1505–1509

[651] Weiss N, Delius M, Gambihler S, Dirschedl P, Goetz A and Brendel W. Influence of the shock wave application mode on the growth of A-MEL 3 and SSK2 tumours *in vivo*. Ultrasound Med Biol 1990; 16: 595–605

[652] Lewin PA, Chapelon J-Y, Mestas J-L, Birer A and Cathignol D. A novel method to control P+/P– ratio of the shock wave pulses used in extracorporeal peizoelectric lithotripsy. Ultrasound Med Biol 1990; 16: 473–488

[653] Miller DL, Thomas RM and Frazier ME. Ultrasonic cavitation indirectly induces single strand breaks in DNA of viable cells *in vitro* by the action of residual hydrogen peroxide. Ultrasound Med Biol 1991; 17(7): 729–735

[654] Pinamonti S, Caruso A, Mazzeo V, Zebini E and Rossi A. DNA damage from pulsed sonication of human leukocytes *in vitro*. IEEE Trans Ultrasonics Ferroelectrics Freq Control 1986; UFFC 33: 179–184

[655] Miller DL, Reese JL, Frazier ME. Single strand DNA breaks in human leukocytes induced by ultrasound *in vitro*. Ultrasound Med Biol 1989; 15: 765–771

[656] Miller DL, Thomas RM and Frazier ME. Single strand breaks in CHO cell DNA induced by ultrasonic cavitation *in vitro*. Ultrasound Med Biol 1991; 17(4): 401–406

[657] Miller DL, Thomas RM and Frazier ME. Single strand DNA breaks in CHO cells after exposure to ultrasound *in vitro*. J Ultrasound Med 1990; 9: S21

[658] Barnett SB, Walsh DA and Angles JA. Novel approach to evaluate the interaction of pulsed ultrasound with embryonic development. Ultrasonics 1990; 28: 166–170

[659] Loverock P and ter Haar G. Synergism between hyperthermia, ultrasound and γ irradiation. Ultrasound Med Biol 1991; 17: 607–612

[660] Watmough DJ, Dendy PP, Eastwood LM, Gregory DW, Gordon FCA and Wheatley DN. The biophysical effects of therapeutic ultrasound on HeLa cells. Ultrasound Med Biol 1977; 3: 205–219

[661] Rooney JA. Hemolysis near an ultrasonically pulsating gas bubble. Science 1970; 169: 869

[662] Rooney JA. Shear as a mechanism for sonically induced biological effects. J Acoust Soc Am 1972; 52: 1718

[663] Miller MW and Ziskin MC. Biological consequences of hyperthermia. Ultrasound Med Biol 1989; 15(8): 707–722

[664] Pfeiler M, Matura E, Iffländer H and Seyler G. Lithotripsy of renal and biliary calculi: physics, technology and medical–technical application. Electromedica 1989; 57: 52–63

[665] Sass W, Bräunlich M, Dreyer H-P, Matura E, Folberth W, Priesmeyer H-G and Seifert J. The mechanisms of stone disintegration by shock waves. Ultrasound Med Biol 1991; 17: 239–243

[666] Holmer N-G, Almquist L-O, Hertz TG, Holm A, Lindstedt E, Persson HW and Hertz CH. On the mechanism of kidney stone disintegration by acoustic shock waves. Ultrasound Med Biol 1991; 17: 479–489

[667] Carstensen EL, Child SZ, Crane C, Miller MW and Parker KJ. Lysis of cells in *elodea* leaves by pulsed and continuous wave ultrasound. Ultrasound Med Biol 1990; 16(2): 167–173

[668] National Council on Radiation Protection. Exposure Criteria for Medical Diagnostic Ultrasound Part I: Criteria Based on Thermal Mechanisms. Bethesda, Maryland (in press)

[669] American Institute of Ultrasound in Medicine/National Electrical Manufacturers Association. Standard for Real-time Display of Thermal and Mechanical Indices of Diagnostic Ultrasound Equipment (in preparation)

[670] Blake JR and Gibson DC. Growth and collapse of a vapour cavity near a free-surface. J Fluid Mech 1981; 111: 123–140

[671] Blake JR, Taib BB and Doherty G. Transient cavities near boundaries. 2. Free-surface. J Fluid Mech 1987; 181: 197–212

INDEX

Note: Figures and Tables are indicated by *italic page numbers*; footnotes by suffix 'n'

Printed and bound by CPI Group (UK) Ltd, Croydon, CR0 4YY
03/10/2024
01040426-0020